# INTRODUCTION TO ANIMAL SCIENCE

## Global, Biological, Social, and Industry Perspectives

### FOURTH EDITION

**W. Stephen Damron**
**Oklahoma State University**

PEARSON
Prentice
Hall

Upper Saddle River, New Jersey
Columbus, Ohio

**Library of Congress Cataloging-in-Publication Data**

Damron, W. Stephen.

Introduction to animal science: global, biological, social, and industry perspectives / W. Stephen Damron. — 4th ed.

p. cm.

Includes bibliographical references and index.

ISBN-13: 978-0-13-513486-3 (casebound)

ISBN-10: 0-13-513486-2 (casebound)

1. Livestock. 2. Animal industry. I. Title.

SF61.D36 2009

636—dc22

2008023677

**Editor in Chief:** Vernon Anthony
**Acquisitions Editor:** William Lawrensen
**Editorial Assistant:** Nancy Kesterson
**Production Coordination:** Thistle Hill Publishing Services
**Project Manager:** Alexandrina Benedicto Wolf
**AV Project Manager:** Janet Portisch
**Operations Specialist:** Laura Weaver
**Art Director:** Candace Rowley
**Cover Designer:** Anne DeMarinis
**Cover Image:** Shutterstock
**Director of Marketing:** David Gesell
**Marketing Coordinator:** Alicia Dysert

This book was set in Times by S4Carlisle Publishing Services and was printed and bound by Edwards Brothers. The cover was printed by Phoenix Color Corp.

**Pearson Prentice Hall**™ is a trademark of Pearson Education, Inc.
**Pearson**® is a registered trademark of Pearson plc
**Prentice Hall**® is a registered trademark of Pearson Education, Inc.

Pearson Education Ltd., London
Pearson Education Singapore Pte. Ltd.
Pearson Education Canada, Inc.
Pearson Education—Japan

Pearson Education Australia Pty. Limited
Pearson Education North Asia Ltd., Hong Kong
Pearson Educacíon de Mexico, S.A. de C.V.
Pearson Education Malaysia Pte. Ltd.

10 9 8 7 6 5 4 3 2 1
ISBN-13: 978-0-13-513486-3
ISBN-10:    0-13-513486-2

*For my wife, Rebecca, and our children, Joshua and Aubryana.*
*I am truly blessed.*

# Preface

This book provides a text for introductory courses in Animal Science and reflects what Animal Science has come to be in the modern age. The major, traditional biological disciplines and how the science of each of these contributes to the whole of Animal Science are included. There is information on how to feed, manage, breed, and care for animals. Because the scope of Animal Science is so broad, this book examines how animals fit into all of society and how they contribute to the well-being of humans from a worldwide perspective. The text takes a brief tour to look at the various types of agriculture found around the world. It explores other uses that we have found for our domesticated animals. It individually considers the species of primary importance to humans. This text also discusses the industries that have arisen around those species and their effects on our society and our economy.

In the last 50 years or so, some profound changes have taken place in the animal industries. Traditional animal husbandry has been revolutionized, and has become less driven by tradition and more driven by business judgment and profound advances in science-based technology. Restructuring of various animal industries has occurred to accommodate changes in the tastes and habits of consumers, the economic upheaval in the agricultural sector, and changes in the relative costs of animal products. There is also an attitude shift that recognizes the animal industries for what they are—dynamic, integrated parts of a greater food-providing system that is increasingly reaching out to other countries and, conversely, being influenced by other countries.

This text also acknowledges changing viewpoints toward the animals in our care. Profound societal changes have affected the animal industries and the people who work in them. Concerns over animal welfare, animal rights, food safety, biotechnology, ethical resource allocation, sustainability of agriculture as it uses the earth's resources, and other issues now affect the usage of animals in a very real way. These issues and others are dealt with in the text.

There is another nontraditional aspect to this text. A more affluent population has reached for new animals and new uses for animals, and consequently has turned to animal scientists to demand information. This text provides information concerning llamas, companion animals, and other species that traditionally has not been provided in the animal sciences. These species are now a part of Animal Science as surely as are the cow and pig. Many of these subjects would not have been considered in Animal Science classes just a few years ago.

Written with sufficient flexibility, this text accommodates the three major approaches to Animal Science—the biological approach, the industry approach, and the species approach. The worldview information from Part One and the societal issues from Part Four round out all approaches.

This fourth edition accomplishes several goals. First, the statistics have all been updated to retain the currency I feel is vitally important. Second, many figures have been either added or improved. In addition, new information has been added to virtually all chapters. One important thing that did not change was the educational philosophy of the previous editions. Those who used the earlier editions will quickly find all the aspects previously contained, including the "flavor."

To access supplementary materials online, instructors need to request an instructor access code. Go to **www.pearsonhighered.com/irc**, where you can register for an instructor access code. Within 48 hours after registering, you will receive a confirming e-mail, including an instructor access code. Once you have received your code, go to the site and log on for full instructions on downloading the materials you wish to use.

## ACKNOWLEDGMENTS

Writing a textbook requires the assistance of many people. In particular, I would like to thank colleagues for their continued support and help. Special thanks to Don Wagner, Jerry Fitch, David Freeman, Clement Ward, Steven Cooper, Leon Spicer, Joe Berry, Bob Kropp, Stanley Gilliland, Udaya De Silva, Christina DeWitt, Melanie Brashears, Michael Davis and David Buchanan from Oklahoma State University. Jim Horne of the Noble Research Center deserves special thanks. Through the years, these people have graciously contributed ideas, figures, photos, and text for the book; reviewed chapters; helped find obscure materials I was seeking; and offered advice on everything. They did so with grace and enthusiasm for the book, and I will be forever in their debt for their contributions. A special thanks to my information assistant, Patricia Bane, for her assistance on the text and the many ways she helps me better serve my students and the academy.

Invaluable assistance was also provided by Jon Beckett of Cal Poly, San Luis Obispo; Robert McDowell of North Carolina State University; Temple Grandin of Colorado State University; Ronald Brown, Martin Brunson, Robert Martin, Louis R. D'Abramo, and William Daniels of Mississippi State University; LaDon Swann of Illinois-Indiana Sea Grant Program, Purdue University; An Peischel of the University of Tennessee; and Tony Seykora of the University of Minnesota. Dr. Seykora's advice and assistance was especially valuable.

Lesa Griffiths, Ph.D., University of Delaware; Doug Parrett, University of Illinois; Tonya S. Amen, University of Wisconsin-River Falls; Michael O. Smith, The University of Tennessee; W. L. Flowers, North Carolina State University; Brian J. Rude, Mississippi State University; and Joe C. Paschal, Texas A&M University provided feedback that was absolutely priceless. This text is better for their efforts and their diligence is appreciated.

Several individuals at Prentice Hall provided invaluable assistance in ushering this book through the development and publishing stages. I would like to thank specifically Nancy Kesterson and Alex Wolf. Many thanks also to Angela Williams Urquhart of Thistle Hill Publishing Services for her efforts.

Finally, a special thanks to my family. My wife, Rebecca L. Damron, Professor of English at Oklahoma State University, continues to provide assistance in response to "Hey hon, will you read this and tell me what you think?" and "Hon, how do you spell . . . ?"

And to my children, Joshua and Aubryana, who have become active participants in the development of this text, a big thanks for their contributions.

W. Stephen Damron, Ph.D.
Oklahoma State University

# Brief Contents

# Contents

# PART ONE

# The Place of Animals and Animal Science in the Lives of Humans

# Introduction to the Animal Sciences

## LEARNING OBJECTIVES

After you have studied this chapter, you should be able to:

1. Define *animal science* and all of its component parts.

2. Describe how, why, and when domestication occurred.

3. Give an overview of the distribution of agricultural animals worldwide.

4. Explain to a nonagriculturist the contributions of domestic animals to humankind and state why domestic animals are so important to life as we know it.

5. Describe the worldwide livestock revolution and its implications.

## KEY TERMS

| | | |
|---|---|---|
| Agriculture | Dairy product science | Hunter-gatherer |
| Animal behavior | Diet | Livestock revolution |
| Animal breeding | Domestic animals | Meat |
| Animal health | Draft animal | Meat science |
| Animal science | Essential amino acids | Nutrient density |
| Applied ethology | Ethology | Nutrition |
| Biofuel | Farmer | Omnivore |
| Biometry | Genetic code | Physiology |
| Biotechnology | Genetics | Renewable resources |
| Civilization | Green revolution | |
| Culture | Heredity | |

## INTRODUCTION

Animals. We live with them, worship them, consume them, admire them, fear them, love them, care for them, and depend on them. They are part of our sustenance, our sociology, and our day-to-day lives. Because they are so important to us, we also study

**Animal science**

The combination of disciplines that together comprise the study of domestic animals.

**Agriculture**

The combination of science and art used to cultivate and grow crops and livestock and process the products.

**Domestic animals**

Those species that have been brought under human control and that have adapted to life with humans.

**Culture**

In this context, culture refers to the set of occupational activities, economic structures, beliefs/values, social forms, and material traits that define our actions and activities.

**Hunter-gatherer**

Hunter-gatherer peoples support their needs by hunting game, fishing, and gathering edible and medicinal plants.

them and apply what we learn to improve their lives and their roles in our lives. The branch of science that deals with domestic animals is called *animal science,* which is the topic of this book.

Much of our use for animals revolves around their contributions to our food supply. Food comes from the land. To coax a more stable food supply from the land, humans have developed a complicated resource management system called *agriculture.* In agriculture, domestic plants and animals are kept to produce for humankind's needs. Humans have practiced agriculture for thousands of years and, either directly or indirectly, every person on the planet depends on agriculture for his or her daily food (Figure 1–1). Because that is true, it is also ultimately true that all of humankind's other occupations are tied to agriculture. This is especially true in the developed countries of the world. In fact, the entire urban industrial complex of the developed world is sustained only because of food surpluses generated by agriculturists. Humans have found many other uses for *domestic animals* in such areas as sports, recreation, manufacturing, religion, and as companions. Add these uses to food production and we discover that animals are at the core of virtually all of our lives, whether or not we are aware of it. Because agriculture and its animals are such an integral part of our existence, they have become a dominating part of our *culture,* our influence on the landscape, and, either directly or indirectly, our day-to-day activities.

Exactly when the individual animal species were domesticated is unknown. DNA sequencing technology suggests that the dog may have been domesticated from the wolf as long as 135,000 years ago, but archaeological evidence suggests that the dog was domesticated about 14,000 years ago (12000 B.C.). The earliest domestic food species (as most Westerners currently define it) was the sheep (somewhere around 8000 B.C.), followed closely by goats, hogs, and cattle (6500 B.C.), llama (5500 B.C.), horses (3500 B.C.), donkeys (4000 B.C.), and chickens (6000 B.C.).

Humans' dependence on the animals they tamed and then domesticated was not planned. *Hunter-gatherers* (who first domesticated animals) used the meat, bones, and skins just as they had done before domestication. The only difference after domestication was convenience. The additional uses (milk, clothing, power, war, sport, and prestige) came later. This happened after humans had lived in the company of animals for a long time in a more sedentary lifestyle. Humans had hunted and consumed animals

**FIGURE 1–1**    Bolivian farmers cultivating potatoes on old Inca terraces. They use the same tools as those used by their ancestors. (FAO photo 22399/Roberto Faidutti. Used with permission.)

for 2 million years before domesticating them. The behavioral change required for hunters and gatherers to become *farmers* was a major cultural revolution. In fact, animal domestication represented a major step toward what we call *civilization.* With the acquisition of domestic animals came the need to ultimately manage them, care for them, and learn to use them to our best advantage. Those needs caused the development of the discipline of study that we call animal science.

## ANIMAL SCIENCE SPECIALTIES

Animal science is simply the collective study of domestic animals. This includes every aspect from conception to death, behavior to management, physiology to nutrition, and reproduction to product distribution. Animal science represents an accumulation of knowledge that began with observations of those hunter-gatherers who began the process of domestication long ago. As animal scientists have learned more and more about animals, the accumulated wealth of information has become too large for anyone to comprehend completely. Out of necessity, its study is divided into disciplines, or specialties, as a means of creating manageable pieces. These specialties may be broken down several ways, but the following categories illustrate the point:

- *Genetics* is the science of *heredity* and the variation of inherited characteristics. *Animal breeding* is the use of *biometry* and genetics to improve farm animal production. Genetics is an expanding field due largely to steady progress in deciphering the *genetic code.*
- *Nutrition* is the study of how organisms take in and use food/feed for body needs. Whether or not animals develop their genetic potential depends on their environment. The most important environmental factor is feed. Nutrition is the science that combines feeds with feeding management to bring about the economical production of livestock and/or health and long life to animal companions.
- *Physiology* is the study of the mechanisms of life from the single biochemical reactions in cells to the coordinated total of specialized cells that constitute a living animal. Because physiology is complex, we usually break down the study to the workings of physiological systems. Examples include reproductive physiology, renal physiology, and exercise physiology.
- *Animal health* is the study of how diseases, parasites, and environmental factors affect productivity and animal welfare. Disease is defined as any state other than a state of health. Once animals were domesticated, diseases and parasites began taking their toll.
- *Ethology* is the study of the biology of animal behavior. The specific study of behavior in domestic animals is *applied ethology.* This discipline developed along with the livestock industry's increased dependence on confinement rearing systems, which provide greater control over animals, reduce labor and feed costs, and help maximize genetic potential. They also present problems associated with behavior. Applied ethology includes many aspects of animal behavior, including animal welfare assessment, optimizing production, behavioral control, behavioral disorders, and behavioral genetics.
- *Meat science* deals with the handling, distribution, and marketing of finished meat products. *Meat* is defined as the edible flesh of animals that is used for food. Meat

**Farmer**

Anyone who practices agriculture by managing and cultivating livestock and/or crops.

**Civilization**

In modern context this refers to what we consider a fairly high level of cultural and technological development.

**Genetics**

The science of heredity and the variation of inherited characteristics.

**Heredity**

The transmission of genetic characteristics from parent to offspring.

**Animal breeding**

The use of biometry and genetics to improve farm animal production.

**Biometry**

The application of statistics to topics in biology.

**Genetic code**

The set of rules by which information encoded in genetic material (DNA or RNA sequences) is translated into proteins (amino acid sequences) by living cells.

**Nutrition**

The study of nutrients and how the body uses them.

**Physiology**

The study of the physical and chemical processes of an animal or any of the body systems or cells of the animal.

### Animal health

The study and practice of maintaining animals as near to a constant state of health as is possible and feasible.

### Ethology

The study of animals in their natural surroundings.

### Applied ethology

The study of behavior in domestic animals.

### Meat science

The science of handling, distributing, and marketing meat and meat products.

### Meat

The flesh of animals used for food.

### Dairy product science

The science of providing milk and milk products as food.

### Biotechnology

A collective set of tools and applications of living organisms, or parts of organisms, to make or modify products, improve plants or animals, or develop microorganisms for specific uses.

### Renewable resources

Those resources that can be replaced or produced by natural ecological cycles or management systems.

### Omnivore

An animal that eats both animal- and plant-based feeds.

"by-products" are all of the products other than the carcass meat, some of which are edible and some of which are not.

- **Dairy product science** deals with the collection, handling, and marketing of milk in its many forms to the consuming public.
- **Biotechnology** involves technological applications of biology. This discipline has received new attention in animal science because of recombinant DNA technology and its many promises. Each of the other disciplines of animal science has benefited from biotechnology and will continue to do so at an ever-increasing rate.

Obviously tremendous overlap occurs in these areas, and separations are made for our convenience. However, this convenience can also be a hindrance. By breaking the discipline of animal science down into smaller units, we have made it easier to learn but harder to grasp—we know the pieces of the puzzle better, but it is harder to put the pieces together. Always remember that it is the combination of the specialties that constitutes the whole discipline we call animal science.

## ANIMAL DISTRIBUTION

There are approximately 4.4 billion large farm animals and 17.6 billion poultry distributed throughout the world (Table 1–1). The number of large farm animals has been increasing at a very modest rate for nearly three decades. However, there have been, and continue to be, shifts in the size of individual species populations. Poultry numbers have increased fairly consistently and at a more rapid rate. Until just very recently, greater than two thirds of the large farm animals were found in developing countries, but they produced only about a third each of the meat, milk, and wool produced in the world. Reasons for the low productivity include environmental stresses, disease challenges, lack of access to technology, and different objectives of livestock production. However, the world agricultural order is undergoing profound changes, which are causing a greater percentage of the world's livestock to be found in the developing world. In addition, the productivity of the livestock in the developing world is improving.

Agricultural animals have made a major contribution to the welfare of human societies for millennia by providing a variety of products and services, as shown in Table 1–2. They are a **renewable resource,** and they utilize another renewable resource—plants—to produce these products and services.

## CONTRIBUTIONS OF ANIMALS TO HUMANITY

A detailed look at animal use comes later in this book. This section briefly surveys some of the many contributions of livestock and other animals to humans.

### FOOD SOURCE

Humans are **omnivores,** consuming both plant- and animal-based foods. Figure 1–2 shows the contributions of different food sources to the world food supply. Although food is the most important contribution of agricultural animals to humans, plants supply a greater total quantity of food. Plants supply 82.9% of the total food energy consumed by the world's people, primarily because such a high percentage of the human

**TABLE 1 – 1** **Agricultural Animal Numbers in the World**

| | World Total | South America | North and Central America | Oceania | Africa | Europe | Asia |
|---|---|---|---|---|---|---|---|
| **LARGE FARM ANIMALS** | | | | | | | |
| Cattle (Head)[1] | 1,357,943,210 | 334,289,953 | 162,064,719 | 37,417,161 | 238,450,131 | 136,499,674 | 449,221,573 |
| Sheep (Head) | 1,049,279,277 | 69,662,736 | 17,266,488 | 139,726,740 | 252,620,097 | 137,387,703 | 432,615,514 |
| Pigs (Head) | 942,637,353 | 50,923,009 | 94,541,139 | 5,417,650 | 22,500,949 | 194,996,122 | 574,258,486 |
| Goats (Head) | 780,755,333 | 21,111,397 | 15,528,529 | 879,180 | 230,477,516 | 18,293,619 | 494,465,092 |
| Buffalos (Head) | 171,105,572 | 1,141,611 | 5,675 | 65 | 3,811,025 | 284,151 | 165,863,045 |
| Horses (Head) | 54,967,628 | 15,650,311 | 14,243,071 | 378,513 | 3,678,604 | 6,476,644 | 14,540,485 |
| Asses (Head) | 41,181,019 | 4,056,217 | 3,770,126 | 9,000 | 14,498,580 | 752,818 | 18,094,278 |
| Camels (Head) | 19,090,610 | | | | 15,604,644 | 12,000 | 3,473,966 |
| Mules (Head) | 12,748,369 | 2,783,554 | 3,767,630 | | 1,023,107 | 239,598 | 4,934,481 |
| Other Camelids (Head)[2] | 6,290,000 | 6,290,000 | | | | | |
| Total | 4,435,998,370 | 505,908,787 | 311,187,376 | 183,828,309 | 782,664,652 | 494,942,328 | 2,157,466,919 |
| **RODENTS** | | | | | | | |
| Rabbits (Head) | 519,144 | 3,498 | 1,405 | | 12,497 | 116,513 | 385,231 |
| Other Rodents (1,000 Head)[3] | 16,325 | 16,325 | | | | | |
| **POULTRY** | | | | | | | |
| Chickens (1,000 Head) | 16,094,406 | 1,741,090 | 2,767,610 | 113,636 | 1,342,868 | 1,817,573 | 8,311,631 |
| Ducks (1,000 Head) | 1,017,860 | 7,375 | 16,481 | 915 | 16,360 | 65,528 | 911,203 |
| Geese (1,000 Head) | 277,176 | 333 | 340 | 69 | 12,282 | 14,947 | 249,206 |
| Turkeys (1,000 Head) | 273,848 | 42,552 | 98,056 | 1,676 | 9,220 | 109,007 | 13,338 |
| Total (1,000 Head) | 17,663,291 | 1,791,349 | 2,882,486 | 116,295 | 1,380,730 | 2,007,054 | 9,485,377 |
| **INSECTS** | | | | | | | |
| Beehives (Number) | 61,499,656 | 5,089,400 | 5,627,756 | 706,756 | 15,473,305 | 15,924,848 | 18,677,592 |
| Silkworm Cocoons (MT) | 317,754 | 9,787 | | | 166 | 1,250 | 306,551 |

SOURCE: FAO, 2007.
[1]Includes yaks
[2]Includes both llamas and alpacas
[3]Primarily guinea pigs

**TABLE 1-2    Contributions of Animals to Human Societies**

### Food

| | |
|---|---|
| Eggs | Blood |
| Meat | Fat |
| Milk | Edible slaughter by-products |

### Body Coverings

Wool
Leather, pelts, hides
Hair, fur, feathers

### Work

Fieldwork and other labor
Transportation

### Body Wastes

| | |
|---|---|
| Fuel | Construction material |
| Fertilizer | Animal feed |

### Other Uses

| | |
|---|---|
| Income | Slaughter by-products |
| Storage of capital | Recreation and sport |
| Storage of food | Pest and weed control |
| Contributions to the economy | Companionship and service |
| Buffer for fluctuating grain supplies | Human health research |
| Soil fertility enhancement | Conservation |
| Prestige | |
| Religion and other cultural needs | |

*SOURCE:* McDowell, 1991, and Turman, 1986.

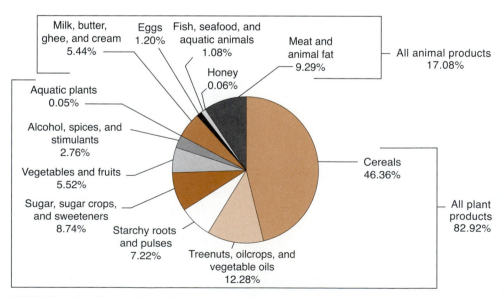

**FIGURE 1–2**   Contributions of food sources to human energy (calorie) consumption. (*SOURCE:* FAO, 2007.)

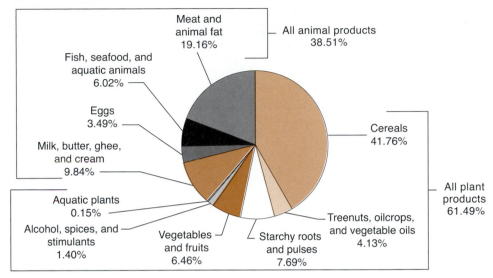

**FIGURE 1–3**   Contributions of food sources to human protein consumption. (*SOURCE:* FAO, 2007.)

diet in the developing countries is of plant origin. Animal products supply the remaining 17.1%. In developed countries, animals contribute a greater percentage of the total food energy. In the United States, for instance, they provide 24%. Animals are a more important source of protein than they are of calories (Figure 1–3), supplying 38.5% of the protein consumed in the world. Of the animal protein sources, meat provides approximately 54%, milk provides approximately 32%, fish supplies approximately 6%, eggs supply 7%. Developed countries obtain a greater percentage of their total protein from animal products. The United States, for example, gets approximately 64% of its protein from animal products. Table 1–3 shows a more complete picture of the contribution of various foods to the U.S. food supply. Meat, eggs, and dairy products are important food sources because they are **nutrient dense.** This means they have many nutrients compared to their calories, and the nutrients are digestible and readily available. High-quality protein and biologically available levels of vitamins and minerals are supplied to the **diet** by animal foods as well as a significant amount of energy.

Animal foods are generally preferred over plant foods by human populations, and the vast majority of the world's population routinely chooses food produced by animals in its diet. A country's living standards can be gauged by the proportion of its food supply that consists of animal foods. Over and over, people have demonstrated that increasing animal-produced foods in their diet is one of the first things they will do when their income increases. Not only are animal foods palatable and delicious, they are also the most nutritionally complete foods. They are an important source of vitamins and minerals, and the protein in animal foods is more likely than are plant proteins to include the **essential amino acids** in the correct proportions.

In some countries, the consumption of animal products is greater than in others. Nearly 30% of the calories in the average diet in the United States is from animal products compared to 7% for the average African. There is also a tremendous difference in food distribution throughout the world. The population of a "poor" (developing) continent like Africa eats only about 64% of the daily calories eaten by the "richer" (developed) population of a country like the United States. Because this is an average, it

**Nutrient density**

A measurement of the nutrients provided in a food compared to the calories it contains.

**Diet**

The total of the foods and water being consumed by an individual or group.

**Essential amino acids**

Those amino acids required by the body that must be consumed in the diet.

**TABLE 1-3**    Percentage Contribution of Food Grouping to Various Nutrients in the U.S. Food Supply, 2000

| Nutrient | Meat, Poultry, Fish | Dairy (excl. butter) | Eggs | Total Animal | Vegetables | Legumes, Nuts, Soy | Grains | Fruits | Fats and Oils | Sugars and Sweeteners | Misc[1] |
|---|---|---|---|---|---|---|---|---|---|---|---|
| Energy | 14 | 9 | 1 | 24 | 5 | 3 | 24 | 3 | 20 | 19 | 1 |
| Protein | 40 | 19 | 4 | 63 | 5 | 6 | 22 | 1 | —[2] | 0 | 2 |
| Total fat | 23 | 12 | 2 | 37 | — | 4 | 2 | 1 | 56 | 0 | 1 |
| SFAs[3] | 26 | 23 | 2 | 51 | — | 2 | 2 | — | 44 | 0 | 1 |
| MUFAs[3] | 25 | 8 | 2 | 35 | — | 4 | 1 | 1 | 59 | 0 | 1 |
| PUFAs[3] | 14 | 2 | 1 | 17 | 1 | 5 | 4 | 1 | 72 | 0 | 1 |
| Cholesterol | 44 | 16 | 35 | 95 | 0 | 0 | 0 | 0 | 5 | 0 | 0 |
| Vitamin A | 27 | 22 | 5 | 54 | 24 | — | 4 | 2 | 9 | 0 | 6 |
| Vitamin E | 4 | 2 | 2 | 8 | 7 | 5 | 4 | 3 | 72 | 0 | 1 |
| Vitamin C | 2 | 3 | 0 | 5 | 45 | — | 4 | 42 | 0 | 0 | 4 |
| Thiamin | 18 | 5 | 1 | 24 | 9 | 5 | 59 | 4 | 0 | — | 1 |
| Riboflavin | 17 | 26 | 6 | 49 | 5 | 2 | 39 | 2 | — | 1 | 1 |
| Niacin | 36 | 1 | — | 37 | 9 | 4 | 45 | 2 | 0 | 0 | 3 |
| Vitamin $B_6$ | 35 | 9 | 2 | 46 | 22 | 4 | 18 | 10 | 0 | — | 2 |
| Folate | 6 | 6 | 4 | 16 | 21 | 18 | 62 | 10 | 0 | 0 | 1 |
| Vitamin $B_{12}$ | 75 | 20 | 4 | 99 | 0 | 0 | — | 0 | 0 | 0 | 0 |
| Calcium | 3 | 72 | 2 | 77 | 7 | 5 | 5 | 3 | — | 1 | 3 |
| Phosphorus | 25 | 33 | 4 | 62 | 8 | 6 | 19 | 2 | — | — | 4 |
| Magnesium | 13 | 16 | 1 | 30 | 14 | 14 | 22 | 6 | — | 1 | 14 |
| Iron | 16 | 2 | 2 | 20 | 10 | 8 | 52 | 2 | — | 1 | 7 |
| Zinc | 38 | 17 | 3 | 58 | 6 | 6 | 26 | 1 | — | 1 | 4 |
| Copper | 14 | 3 | — | 17 | 15 | 21 | 22 | 6 | 0 | 4 | 15 |

*SOURCE:* USDA, 2007.

[1] Coffee, tea, chocolate-liquor equivalent of coca beans, spices, and fortification of foods not assigned to a specific group.

[2] = less than 0.5%.

[3] SFAs = saturated fatty acids; MUFAs = monounsaturated fatty acids; PUFAs = polyunsaturated fatty acids.

means many people in the developing countries consume far less food per day and are severely undernourished, if not on the verge of starvation. Approximately 865 million people are undernourished.

Most people include meat and dairy products in their diet whenever they can. Exceptions are almost always because of religious prohibitions (cattle in India, for example) or because of prohibitive costs. The world's meat (excluding fish) is predominantly supplied by pigs, cattle, and poultry with lesser amounts from sheep, goats, buffalos, and horses. Several other species provide a significant amount of meat to the people of various geographic regions. Most milk comes from cows, but buffalos, goats, and sheep provide significant amounts of milk, and most domestic mammals are milked somewhere in the world.

## OTHER USES

In addition to food, other animal products are also of great importance to humans, who have used wool, hair and other fibers, feathers, and hides for millennia for clothing and other uses (Figure 1–4). Manure from animals is a valuable by-product. Estimates place its value in the United States alone as high as $10 billion for fertilizer (Figure 1–5).

Slaughter by-products are the source of a large number of industrial and consumer products. Some examples include insecticides, crayons, cosmetics, plastics, cellophane, glass, water filters, plywood adhesive, soap, and animal feed (Figure 1–6).

In developing countries, 80% of the energy needs are provided by the muscles of humans and animals. ***Draft animals*** are vitally important to many Asian, African, and South American countries. Oxen plow fields, water buffalos work in rice paddies, yaks and camels still trudge over ancient trade routes, and dogs still pull sleds (Figure 1–7).

Animals are used as models for humans in biomedical research. Thirty years have been added to the average American life span since 1900. In addition, the quality of life for people afflicted with chronic diseases has been helped immeasurably. Medical research depends on the use of animals as models. It will continue to do so in the foreseeable future (Figure 1–8). In addition, animals are used in research to benefit animal health, resulting in healthier, longer-lived pets and healthier, more productive livestock.

**Draft animal**

An animal whose major purpose is to perform work that involves hauling or pulling. An ox or horse pulling a plow or wagon is a draft animal.

**FIGURE 1–4** Hides are a slaughter by-product that humans have used since long before domestication. (Photo courtesy of Adele M. Kupchik.)

**FIGURE 1–5** The fertilizer value of livestock wastes produced in the United States may be as high as $10 billion. (Photographer R. L. Kane. Courtesy of U.S. Department of Agriculture.)

**FIGURE 1–6** Slaughter by-products are used in the manufacture of a variety of industrial and consumer products. (Photo courtesy of Adele M. Kupchik.)

Companionship is provided to humans by several million, perhaps even billions, of animals around the world. Specially trained animals, including dogs and monkeys, assist people with visual and hearing disabilities and paralysis, helping them to live more independently (Figure 1–9).

In addition, many entertainment industries such as racing, rodeos, and bullfighting are based on animal use (Figure 1–10).

Agricultural animals convert inedible feeds to valuable products. About two thirds of the feed used in the U.S. livestock industry is not suitable for human consumption. Hay, pasture, coarse forages, by-products, garbage, and damaged food are examples. Animal use diversifies the food supply and the economy. Diversified agriculture is more stable and more sustainable.

**FIGURE 1–7**   Draft animals are still the most important nonhuman power source in the developing countries. (Photo courtesy of the Philippine American Chamber of Commerce, Inc.)

**FIGURE 1–8**   Biomedical research depends on the use of animals as research models. (Photo courtesy of Stock Boston.)

# THE FUTURE OF LIVESTOCK PRODUCTION

The first two decades of the 21st century appear to be a boom time for global livestock production with huge increases in livestock production occurring and further increases predicted. This increasing animal production is being referred to as the *livestock revolution,* likening it to the cereal grains boom of the *green revolution,* which began in the 1960s and is credited with saving millions of lives while building national economies.

Global demand for meat, milk, and eggs is increasing very rapidly. The forces driving this increase in demand are simple: people and money. Unprecedented economic development around the world is increasing family income. This trend contributes to increased per capita consumption of animal products, which, along with rapid population growth, results in large increases in total demand. Most of the new demand for

**Green revolution**

Dramatic improvements in grain production in developing countries during the 1960s to the 1980s because of technological innovation and application.

**FIGURE 1–9** Service dogs, such as these seeing-eye dogs shown with a trainer and a man with a vision disability, help people with disabilities. (Photographer C. Prescott-Allen.)

**FIGURE 1–10** Many entertainment industries are based on animal use. Jockey Gary Stevens is shown here, just after he won the 1995 Kentucky Derby atop Thunder Gulch. (Photographer Andy Lyons. Courtesy of Getty Images, Inc.)

meat, milk, and eggs is in developing countries. By 2020, developing countries may produce the majority of both the world's meat and milk. Along with the increased demand for animal products is an increased demand for other agricultural commodities. In addition, the world's developing demand for *biofuel* production will increasingly play a role in food availability and prices.

The challenges associated with these profound changes in agriculture are significant. Much will depend on research and technology development. Crop yield breakthroughs will be a key issue because we will not be cultivating more area. The prime agricultural lands are already in use, and, worldwide, the amount of acreage available

**Biofuel**

Gas or liquid fuel made from biological materials such as crops and animal waste.

for agriculture is decreasing yearly. Agriculture of all kinds has the potential to affect the environment. For the sake of future generations we must achieve these massive increases in yield at the same time protecting air, soil, and water quality. Combined, the opportunities and the challenges suggest an unprecedented dynamic period in the world agricultural order.

## SUMMARY AND CONCLUSION

Animal science has its roots in the challenges that the first domesticators of animals encountered many millennia ago when they permanently brought in animals from the wild. Today, animal science is a vital field with specialties in genetics and animal breeding, nutrition, physiology, animal health, animal behavior, meat and dairy product science, and biotechnology. Animals are used for a myriad of purposes, including food, fiber, work, research, companionship, and entertainment. Although agricultural animals have come under attack in recent years by those who feel they are a luxury, agricultural animals are not destined to disappear or even be reduced in numbers and importance. What is happening is that they are becoming even more important in helping to alleviate human hunger. For this reason, we should learn something about the factors that determine the kinds of agricultural animals found throughout the world. Chapter 2 explores in depth the contributions animals make to humankind.

## STUDY QUESTIONS

1. Define animal science. When did animal science begin?
2. Explain why all of the world's occupations are tied to agriculture.
3. When did animal domestication occur? When were each of the major species domesticated? Was domestication a conscious decision by humans?
4. Define the specialties of animal science.
5. Why is specialization of animal science disciplines both a help and a hindrance?
6. Study Table 1–1, which gives livestock numbers in the world. Notice the relative numbers of each species. Offer some reasons why animals are distributed as they are. Based on the numbers in this table, what are the world's major farm species?
7. Table 1–2 gives an overview of the goods and services provided by domestic animals to humans. (These are explored in detail later in the text.) Develop a list of uses from this table that are ranked from "most useful" to "least useful" from your current perspective. (At the end of the book, come back to your list and see if your perspective has changed.)
8. What proportion of human food energy and protein come from animal products?
9. What proportion of the U.S. calorie and protein supply comes from animal products? How do other countries compare?
10. If a person doesn't eat meat, what are generally the reasons?
11. Most of the world's meat supply is provided by seven species. Name them.
12. Meat is important as a food for the human population because it is nutrient dense. What does "nutrient density" mean?

13. List some of the important products made from by-products of the slaughter industry.

14. What proportion of the nonmachinery power for the world is provided by humans and work animals? Name six important draft animals.

15. What is the role of animals to medical research?

16. Briefly discuss some of the ways that animals provide companionship, recreation, and entertainment to humans.

17. What types of humanly unusable feeds do animals convert to valued products?

18. Why is a diversified agriculture so important, and what role do animals have in diversification?

19. What is the livestock revolution, and what are some of its challenges?

## REFERENCES

Bourden, R. M. 2000. *Understanding animal breeding*. Upper Saddle River, NJ: Prentice Hall.

Council for Agricultural Science and Technology. 1980. *Food from animals: Quantity, quality and safety.* Report No. 82. Ames, IA: Council for Agricultural Science and Technology.

Council for Agricultural Science and Technology. 1997. *Contribution of animal products to healthful diets.* Report No. 131. Ames, IA: Council for Agricultural Science and Technology.

Council for Agricultural Science and Technology. 1999. *Animal agriculture and global food supply.* Report No. 135. Ames, IA: Council for Agricultural Science and Technology.

Delgado, C., M. Rosegrant, H. Steinfeld, S. Ehui, and C. Courbois. 1999. *Livestock to 2020: The next food revolution.* Washington, DC: International Food Policy Research Institute.

FAO. 2006. *The state of food insecurity in the world 2006*. Accessed online July 9, 2007. http://www .fao.org/docrep/009/a0750e/a0750e00.htm.

FAO. 2007. *FAOSTAT statistics database: Agricultural production and production indices data.* http://apps.fao.org/.

Foundation for Biomedical Research. 2006. *Proud achievements of animal research*. 5th ed. Washington, DC: Foundation for Biomedical Research.

Gillespie, J. R. 1987. *Animal nutrition and feeding.* Albany, NY: Delmar.

Houpt, K. A. 2005. *Domestic animal behavior*. 4th ed. Ames, IA: Blackwell Publishing Professional.

Jensen, P. 2002. *The ethology of domestic animals.* Wallingford, UK: CAB International.

Jordan-Bychkov, T. G., M. Domosh, R. P. Neumann, and P. L. Price. 2008. *The human mosaic: A thematic introduction to cultural geography.* 10th ed. New York: W. H. Freeman.

Lawrence, A. B., and J. Rushen, eds. 1993. *Stereotypic animal behavior—Fundamentals and applications to welfare.* Wallingford, UK: CAB International.

McDowell, R. E. 1984. The need to know about animals. In *World food issues,* M. Droskoff, ed. Ithaca, NY: Center for Analyses of World Food Issues, Cornell University.

McDowell, R. E. 1991. *A partnership for humans and animals.* Raleigh, NC: Kinnic Publishers, Kinnickinnic Agri-Sultants, Inc.

National Research Council. 1989. *Recommended dietary allowances.* 10th ed. Washington, DC: National Academy Press.

Otten, J. J., J. P. Hellwig, and L. D. Meyers, eds. 2006. *Dietary reference intakes: The essential guide to nutrient requirements.* Washington, DC: The National Academies Press.

OECD-FAO. 2007. *OECD-FAO agricultural outlook 2007–2016.* Paris: Organization for Economic Co-operation and Development; Rome: Food and Agricultural Organization of the United Nations.

Reed, C. A. 1974. *The beginnings of animal domestication in animal agriculture in the biology, husbandry, and use of domestic animals.* 2nd ed. H. H. Cole and W. N. Garrett, eds. San Francisco: W. H. Freeman.

Taylor, R. E., and T. G. Field. 2004. *Scientific farm animal production.* 8th ed. Upper Saddle River, NJ: Prentice Hall.

Turman, E. J. 1986. *Agricultural animals of the world.* Stillwater: Oklahoma State University.

USDA. 2007. *Agricultural statistics.* Washington, DC: National Agricultural Statistics Service. Accessed online July 7, 2007. http://www.usda .gov/nass/pubs/agstats.htm.

# 2

# The Value of Animals to Humanity

## LEARNING OBJECTIVES

After you have studied this chapter, you should be able to:

1. Describe the value of animal products in providing for the world's food.

2. Explain the current rates of growth or decline of animal products on a worldwide basis.

3. Elaborate on the milk-producing species, state their importance to world milk production, and understand what is happening to world milk production.

4. Describe the value of eggs in feeding the world's people.

5. Develop a modest understanding of some miscellaneous food uses for the world's animals.

6. Explain the value of animal products in the human diet.

7. Give a good overview of all the many nonfood uses humans have for the world's animals.

## KEY TERMS

By-products
Civilization
Companion animal
Compost
Conservation
Draft
Essential fatty acids
Food and Agricultural
   Organization of the
   United Nations (FAO)

Health research
Hides
Meat
Metric ton
Milk
Nutrient density
Nutrients
Per capita
Pest control

Pesticides
Poultice
Power
Recombinant DNA
Spectator sport
Storage of capital
Wool
Xenotransplantation

# INTRODUCTION

What did humans derive from domestication and why do we need domestic animals in today's world? The answer to both questions is essentially the same, and the answer comes in several parts. Food is the first answer. When humans turned from hunter to farmer by domesticating animals and plants, we created a much more readily available food supply and set the stage for great advancement in culture. What we think of as "civilization" began to occur. Goods and services from the world's animal populations supply many social, religious, and economic functions in addition to food. How important each good or service is varies, depending on many factors, including ethnicity of the owners, country, and ecological condition. Nonfood uses for animals are generally more important in developing countries than they are in developed countries. However, some nonfood uses are also extremely important in richer countries. In this chapter, we examine the major contributions of agricultural animals to humanity.

# THE FOOD USES OF AGRICULTURAL ANIMALS

### RED MEAT AND POULTRY PRODUCTION

The **nutrients** provided by meat are important for human survival. Protein and energy are quantitatively and qualitatively important. However, a substantial share of the vitamins and minerals in our diet are also contributed by meat. Annual **per capita** meat supply ranges from over 270 lbs per year in affluent countries to very little in poor countries (Figure 2–1). World meat production has increased steadily for many years at a rate of 3–4% per year. Table 2–1 shows the meat production for most meat-producing species. The pig is the most important meat producer, producing 39% of the world's meat (Figure 2–2). Chicken is next at 27%, followed by beef with approximately 24% (Figure 2–3). Together, these three sources produce 90% of all meat. Poultry meat

**Nutrients**

Chemical substances that provide nourishment to the body. Essential nutrients are those necessary for normal maintenance, growth, and functioning.

**Per capita**

Per unit of population.

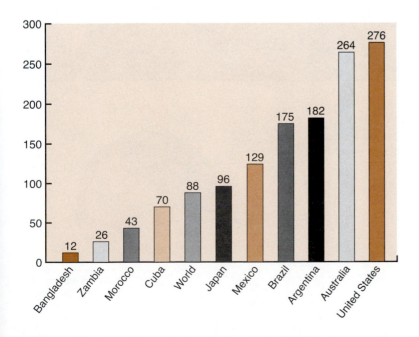

**FIGURE 2–1** Per capita meat supply in pounds, selected countries and the world. (*SOURCE:* FAO, 2007.)

**TABLE 2 – 1    World Meat Production (1,000 metric tons)**

|  | 1990 | 1995 | 2000 | 2005 |
|---|---|---|---|---|
| Meat, total | 155,778.59 | 206,843.36 | 236,955.16 | 270,697.45 |
| Pig meat | 60,879.90 | 79,470.66 | 90,738.36 | 104,923.66 |
| Chicken meat | 31,500.47 | 46,601.15 | 59,505.66 | 72,159.57 |
| Bovine meat | 46,366.14 | 58,005.62 | 60,956.18 | 64,099.01 |
| Sheep and goat meat | 8,329.89 | 10,308.65 | 11,117.64 | 12,852.56 |
| Duck, goose, and guinea fowl meat | 1,875.23 | 3,627.28 | 5,050.23 | 6,302.18 |
| Turkey meat | 3,768.42 | 4,670.69 | 5,358.22 | 5,360.28 |
| Rabbit meat | 608.53 | 922.28 | 1,026.01 | 1,411.91 |
| Equine meat | 638.33 | 957.97 | 1,065.75 | 1,040.29 |
| Meat, all other | 1,811.68 | 2,279.08 | 2,137.12 | 2,548.00 |

*SOURCE:* FAO, 2007.

**FIGURE 2–2** The pig is the most important meat-producing species. Many pigs are unimproved animals that either scavenge or are fed household wastes, like this fellow photographed in the Cook Islands.

**FIGURE 2–3** Relative contribution of the major meat species to world meat supplies. (*SOURCE:* FAO, 2007.)

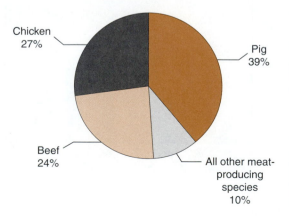

Chicken 27%

Pig 39%

Beef 24%

All other meat-producing species 10%

**FIGURE 2–4** World poultry-meat production is increasing by over 4% annually, the most rapid increase of all major meats. (Photo courtesy of UN/DPI.)

production is growing the most rapidly at almost 5% per year, more than doubling since 1990 (Figure 2–4). Some developing countries are undergoing phenomenal production increases. Several are currently producing three times as much poultry meat per year as they were producing 15 years ago. This increase is primarily in response to two things. First, the technology, genetics, and management know-how associated with the poultry industry have made its production much more cost effective. Second, economic development in many countries has increased incomes, thus providing a better diet to more people. Time and time again, people demonstrate their willingness to purchase more meat for their diet as soon as they are able to afford it. Pork production is increasing at the rate of 3–4% per year. Production from most other meat-producing species is increasing 1–2% yearly.

## EDIBLE SLAUGHTER BY-PRODUCTS

Once the products of greatest value are removed from a carcass, the substances and products remaining still have value. We commonly refer to these as *by-products*. Some of these by-products are edible and others are not. Table 2–2 provides a list of edible

| TABLE 2–2 | Edible By-Products from Animals | |
|---|---|---|
| Brain | | Lungs |
| Cheek meat, ears, and snouts | | Melts (spleen) |
| Feet, pig's feet, knuckles, and calves' feet | | Pancreas |
| | | Stomach |
| Head meat | | Sweetbreads (thymus) |
| Heart | | Tail or oxtails |
| Intestines | | Tallow and lard |
| Kidney | | Testicles |
| Lips | | Tongue |
| Liver | | Tripe |

**FIGURE 2–5**  Offered side by side in this Mexican meat market are hearts, lungs, other organ meats, feet, tripe, and carcass meat, fresh and still warm from the morning's premarket slaughter. Food tastes and traditions vary, and not all people agree on what is the product of a carcass and what is a by-product.

by-products of meat animals. In developed countries, these by-products are usually considered specialty foods and are called "variety meats." Most are organ meats such as liver, kidney, tongue, sweetbread, brain, heart, and tripe. An additional large share comes from edible fat. These products are often considered delicacies (Figure 2–5). However, in the United States, we are not as fond of them as is the rest of the world, so we export large quantities to Europe and other countries.

## MILK AND MILK PRODUCTS

Milk provides much-needed protein, energy, minerals, and vitamins to humankind's diet. Annual per capita whole milk supply ranges from over 385 lbs per year in some countries to little or none in others. The total production of milk in the world has increased slowly for the last two decades. Total world milk production is increasing by approximately 1% per year. World milk production for the major species is given in Table 2–3. On a worldwide basis, approximately 84% of milk is from cattle, 12% is from buffalos, and most of the remainder is from sheep, camels, and goats (Figure 2–6). Humans use almost 640 million **metric tons (MT)** of milk per year. This includes fluid milk and processed milk. Table 2–4 shows the world production of the major processed milk products. Cheese is obviously the most important, although to many people, milk fat represents the major source of fat in the diet. In

**Metric ton (MT)**
Approximately 1.1 U.S. tons. Equal to 1 million grams, or 1,000 kilograms.

**TABLE 2–3**    **Milk Produced in Metric Tons by the World's Major Species**

|  | 1980 | 1990 | 2000 | 2005 | Rank 2005 |
|---|---|---|---|---|---|
| Cow milk (whole, fresh) | 422,323,803 | 479,159,972 | 487,883,119 | 537,291,694 | 1 |
| Buffalo milk | 27,525,084 | 44,075,742 | 67,401,462 | 77,360,244 | 2 |
| Goat milk | 7,707,908 | 9,961,376 | 11,655,018 | 12,582,027 | 3 |
| Sheep milk | 6,812,862 | 7,992,220 | 8,037,704 | 8,582,173 | 4 |
| Camel milk | 1,195,772 | 1,337,039 | 1,251,691 | 1,276,634 | 5 |
| Total milk | 465,565,429 | 542,526,349 | 576,228,994 | 637,092,772 |  |

*SOURCE:* FAO, 2007.

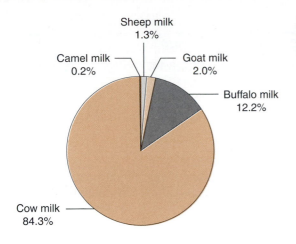

Sheep milk 1.3%

Camel milk 0.2%

Goat milk 2.0%

Buffalo milk 12.2%

Cow milk 84.3%

**FIGURE 2–6** Relative percentage of world milk supply by species. (*SOURCE:* FAO, 2007.)

**TABLE 2–4    World Dairy Product Production in Metric Tons**

|  | 1980 | 1990 | 2000 | 2005 | Rank in 2005 |
|---|---|---|---|---|---|
| Cheese | 11,505,050 | 14,828,628 | 16,467,678 | 18,627,141 | 1 |
| Butter and ghee | 6,955,880 | 7,822,057 | 7,413,584 | 8,424,725 | 2 |
| Evaporated and condensed milk | 4,211,870 | 4,085,099 | 4,008,361 | 4,146,633 | 3 |
| Skim milk and buttermilk, dry | 4,177,811 | 4,252,419 | 3,521,299 | 3,038,442 | 4 |
| Dry whole cow milk | 1,726,792 | 2,085,752 | 2,503,615 | 2,922,545 | 5 |

*SOURCE:* FAO, 2007.

addition to adding flavor to high-starch foods such as rice and root crops, fats also provide needed calories and ***essential fatty acids*** to the diet.

Dairying is being promoted in the developing countries through the efforts of several major international lending agencies. Dairy industry promotion assists development in several ways. In addition to improving nutrition, it also provides year-round employment (as opposed to crops that have seasonal labor needs). The year-round income bolsters the overall economy. Dairying is also a very efficient means of converting animal feed to people food.

**Essential fatty acids**
Fatty acids required in the diet.

### POULTRY AND EGGS

In addition to meat (Table 2–1), the various poultry species produce a second high-quality food for human consumption—eggs. Poultry offer great potential for improving the nutritional levels of all the world's peoples. As food producers, they have many advantages. Poultry require a low initial investment and, if necessary, can produce some food for their owners with only a minimal input in terms of feed, equipment, and housing. They are also generally fairly hardy and prolific.

Eggs are a very important global food source from both a volume and quality perspective. Chickens produce most of the eggs for human consumption (Table 2–5). Their production is increasing over 3% per year worldwide. Ducks, turkeys, geese, and guinea fowl also contribute but to a much lesser degree. Eggs are an excellent source of high-quality protein and fat. The protein is important to all economic classes. For poor societies, this fat content is a positive factor because it provides calories and

**TABLE 2–5**     **World Egg Production in Metric Tons**

|                     | 1980        | 1990        | 2000        | 2005        |
|---------------------|-------------|-------------|-------------|-------------|
| Hen eggs            | 26,215,606  | 35,231,843  | 51,728,147  | 58,998,595  |
| Eggs, excluding hen | 1,199,337   | 2,280,657   | 4,119,528   | 5,132,513   |
| Total               | 27,414,943  | 37,512,500  | 55,847,675  | 64,131,108  |

SOURCE: FAO, 2007.

essential fatty acids. Note that the number of metric tons of eggs produced in the world is slightly less than the amount of meat from chickens and exceeds meat production from many other species.

### MISCELLANEOUS FOOD USES

**Food and Agricultural Organization of the United Nations (FAO)**

The largest autonomous agency within the United Nations system. FAO works to alleviate poverty and hunger by promoting agricultural development.

Honey is another important animal food. World production is estimated by the *Food and Agricultural Organization of the United Nations (FAO)* at just over 1.4 million MT per year (Figure 2–7). A rather exotic way in which living animals are utilized for food in some primitive areas is as a source of blood. Veins are tapped by one means or another and the blood is collected. Tibetan nomads and African tribesmen (especially the Masai) do this. Blood is also collected at slanghter and used (Figure 2–8). The fat from the tails of fat-tailed sheep is also collected from live animals in Africa and Asia (Figure 2–9). Hides and skins (especially roasted pig) are used for food in many countries, and cattle hides are eaten in Africa and a few other places (Figure 2–10).

### THE VALUE OF ANIMAL PRODUCTS IN THE HUMAN DIET

**Nutrient density**

A measurement of the nutrients found in a food compared to the caloric content.

Animal products are a very important part of human diets because they are such high-quality food. They are excellent sources of protein, energy, minerals, and essential vitamins. The nutrient availability is good and the *nutrient density* is high. In wealthy

**FIGURE 2–7** Honey is favored around the world for its sweetness and unique flavor. When fermented with water and yeast it becomes mead or "honey wine." Beeswax also has many uses, including candles and craft-making. (FAO/ 19181/M. Marzot. Used with permission.)

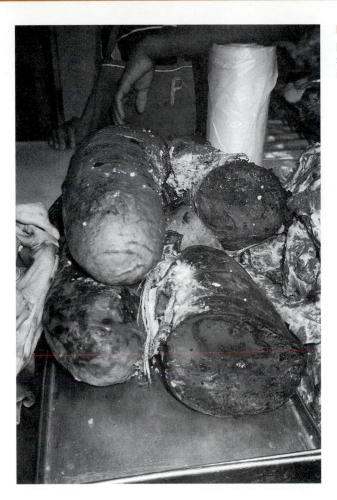

**FIGURE 2–8** Blood collected at slaughter is used for a variety of purposes, including soup stock, and a favorite in many places: blood sausage. This blood sausage was photographed at a meat market in Saltillo, Mexico.

**FIGURE 2–9** The large tails of the fat-tailed sheep breeds are fat storage organs. To acquire this fat, the tail is cut and a hollow reed used to suction out the fat for use in cooking.

**FIGURE 2–10** Roasted pigskins are a popular food in many countries. They are eaten as is or used in cooking. These were roasted and spiced and are on sale in bulk.

societies, meat is the food item around which meals are built—the meat is chosen first, and the rest of the meal is selected to complement the meat. In most developed countries, 30–40% of the total cost of food is for meat and fish. Affluent families in developing countries spend a similar amount; low-income groups spend less (only 5–15%). Cost prevents animal products from being the main dish for the world's low-income groups. However, these groups identify animal products as desirable or essential food items. In addition to the nutritive value, animal products supplement the taste of the bland, starchy foods that, of necessity, form the bulk of their diets.

## THE NONFOOD USES OF AGRICULTURAL ANIMALS

Table 2–6 lists the great variety of important contributions that various species of domestic animals provide for humans in addition to food. Some of these uses are outside the social context of people who live in developed countries, which makes it difficult for us to understand the great magnitude of these contributions to people's lives. This section explores some of these contributions in greater detail and gives us all the opportunity to have an enhanced appreciation for our dependence on domestic animals.

### BODY COVERINGS

The most important products from some animals are their body coverings. Examples are sheep (wool, Persian lambskins, Figure 2–11), goats (cashmere and mohair), and alpacas (Figure 2–12). For other species, the body covering is a by-product rather than the main product. A major advantage of the fiber products is that they are renewable—they can be harvested repeatedly from the same animal. Skins and hides are nonrenewable, at least from the same animal. In countries with a highly developed agriculture, the skins and hides obtained at the time of slaughter are a valuable by-product. However, their contribution to the total value of the slaughter is relatively small compared to the value of the carcass. Often, they are shipped to a developing country for processing. In poorer countries, hides or skins (rather than the meat) may be the most valuable animal export

**TABLE 2–6** **Animal Contributions of Services and Nonfood Products**

| Classification | Contribution | Main Sources |
|---|---|---|
| Draft power | Crop production | Cattle, buffalo, yaks, camel, horse, donkey, mule |
| | Cartage | Cattle, buffalo, yak, camel, horse, mule, donkey, reindeer |
| | Packing | Camel, yak, horse, mule, donkey, reindeer, llama, sheep, goat |
| | Herding | Horse, mule, camel, ass |
| | Irrigation pumping | Buffalo, cattle, camel, ass |
| | Threshing grains | Cattle, horse, ass |
| | Transportation | Horse, donkey, mule, buffalo, camel, reindeer |
| | Human leisure time | All working species |
| Storage in animals | Capital | All domestic species |
| | Grains | Buffalo, cattle, sheep, pigs, poultry |
| Conservation | Grazing | All domestic herbivores |
| | Seed distribution | All domestic herbivores |
| | Soil through crop rotation | Most grazing animals |
| Ecological | Maintenance | Most animals |
| | Restoration | Most animals |
| Pest control | Weeding crops | Domestic ruminants, ducks, geese |
| | Insects in crops | Poultry, ducks, geese |
| | Irrigation canals | Buffalo, ducks, geese |
| Cultural uses | Exhibitions and sports | Horse, cattle, goat, sheep, pig, buffalo, chickens, dog, cat |
| | Bride price | Cattle, goats, sheep, camels |
| | Religious sacrificial | Sheep, goats, cattle, poultry, buffalo, pigs |
| | Fighting | Chickens, cattle, buffalo, sheep, dogs, camels |
| | Pet | All species |
| | Racing | Horse, cattle, dogs, camel, buffalo |
| | Status symbol | Horse, cattle, pigs, buffalo, camel, fighting cocks, most species to some degree |
| | Blood and death restitution | Horse, cattle, buffalo, camel, pig |
| Income | Security | Cattle, buffalo, goat, sheep, pig, all poultry |
| | Liquidity | All species |
| | Reduce risks of cropping | All species |
| Nonarable lands | Income | Cattle, camels, goats, sheep |
| | Soil fertility in cropping | Cattle, goats, sheep, buffalo |
| Research | Biomedical and numerous other | All domestic species and fowl |
| Fiber | Wool, hair, feathers | Sheep, camels, llamas, alpacas, yaks, horses, goats and other mammals |
| | | All fowl |
| Skins | Hides, pelts | Most all species |
| Wastes | Fertilizer, fuel, methane gas, construction material, feed (recycled) | Most all species |
| Inedible products | Fat, horns, hooves, bones, tankage, endocrine extracts | All species |

*SOURCE:* McDowell, 1991, pp. 17, 45. Various other sources.

**FIGURE 2–11** The Karakul sheep produces Persian lambskins that are used in the manufacture of luxurious coats. (Photo courtesy of New Mexico Department of Tourism.)

**FIGURE 2–12** Alpaca and wool fibers are used to make many products, including wool blankets such as those held by these Ecuadorians. (Photographer Hodge. Courtesy of SuperStock, Inc.)

product, and they may provide valuable foreign currency. In addition, wool, hair, and feathers are often used in cottage industries to produce clothing, pillows, bedding, carpets, brushes, and handicrafts. World hide and fiber production is shown in Table 2–7. As one might expect, cattle hides, wool, and sheepskins are produced in greatest volume.

Goats yield two types of body coverings that are used for making fabrics—mohair and cashmere. Mohair is clipped from the Angora goat each year just as the fleece is shorn from sheep (Figure 2–13). The average yield is about 5 lbs. Angora goats are raised in several parts of the world. Cashmere, also called pashmina, is the very fine underfur of the Kashmir goat. Kashmir goats are really a type rather than a breed. Almost all goats produce some cashmere. It must be combed out of the coarse, outer guard hair by hand. The average yield is only about 3 oz. Kashmir goats are raised in the high plateaus of Tibet, China, and northern India.

**TABLE 2–7**  **World Hide and Fiber Production in Metric Tons**

|  | 1980 | 1990 | 2000 | 2005 | Rank 2005 |
|---|---|---|---|---|---|
| Cattle hides, fresh | 5,654,775 | 6,309,193 | 7,415,173 | 8,149,577 | 1 |
| Wool, greasy | 2,794,158 | 3,347,708 | 2,322,746 | 2,130,860 | 2 |
| Sheepskins, fresh | 1,105,811 | 1,337,416 | 1,644,503 | 1,778,241 | 3 |
| Goatskins, fresh | 390,157 | 579,714 | 839,876 | 1,017,988 | 4 |
| Buffalo hides, fresh | 488,176 | 622,492 | 796,062 | 827,588 | 5 |
| Cocoons, reelable | 341,660 | 363,587 | 342,454 | 434,500 | 6 |
| Hair of horses | 110 | 100 | 130 | 140 | 7 |

SOURCE: FAO, 2007.

**FIGURE 2–13** The Angora goat produces mohair, which is a valuable fiber in fine clothing manufacture.

Yaks produce a coarse fiber from which clothing and one of the best horse blankets are made. The long hair of the camel is woven into cloth or blankets primarily for use by the people who own the camels. Camel-owning nomads also use camel hair as tent fabric material. Camel hair paintbrushes are highly valued by some artists. The other domestic animals of the camel family that yield fiber are the llama and alpaca native to the Andes Mountains of South America. The alpaca produces wool that is of higher quality than most sheep wool.

The Karakul and Kuche are fur-bearing breeds of sheep. Lambs are slaughtered within a few days of birth and the pelts are removed. The Karakul is the best known of the two, yielding Persian lambskins used in the manufacture of very expensive coats. The Karakul thrives in very dry environments. Most are produced in Afghanistan and Namibia. The skins are exported to the developed countries. The pelt of the Kuche is used in making fur hats, collars, and coats. It is seldom exported out of northern Asia.

Feathers are used for pillows, bedding, and clothing (down coats), for jewelry and adornments on clothing (hats), for fishing lures, and as a protein supplement to feed animals. Silk from silkworms is important in production agriculture in China, India, Uzbekistan, and Brazil (Figure 2–14).

**FIGURE 2–14** Sri Lankan women workers display silkworm cocoons in a small silk factory. (FAO Photo/17014/G. Bizzarri. Used with permission.)

## POWER SOURCES

In a developed agriculture, almost all farm power is provided by machines rather than by animals. In developing countries, however, 80% of the total nonhuman power used in crop production and agriculture is still provided by animals. Animals are used for plowing, pulling carts or sleds, and packing. Many Asian, African, and Central and South American countries still use 95% or greater animal power. As a general rule, the poorer the country, the greater the dependence on animal power. Oxen (steers of cattle), donkeys, horses, buffalos, mules, and camels are the major animal species used for *draft* purposes. Several other species are used in specific geographic locations, such as yaks, llamas, goats, reindeer, and dogs. Oxen and buffalos are the most powerful in pulling, with oxen being faster on dry land. Horses and mules move the fastest, which is an advantage if high-quality implements are available. Donkeys and camels carry the heaviest loads in relation to body size and are preferred in drier climates. Yaks are best adapted to high altitudes. For most of the developing world, there is no alternative to draft power (Figure 2–15).

## BODY WASTES

### Manure

As many as half of the world's farmers depend on animal manure and *compost* from plant residues as their only source of fertilizer (Figure 2–16). In fact, manure is preferred over chemical fertilizers by small farmers who till the land by hand or use crude plows because manure improves soil texture. Manure is also used in the developed countries as a source of soil fertility but not very efficiently. In many of the poorer areas of the world, manure must be utilized as a fuel. This is particularly true in very densely populated areas where trees have been cut and used as fuel or for other purposes. In these areas, dried manure, particularly from cattle and buffalo, is an important source of fuel for cooking and heating (200+ million tons each year). At higher elevations, yak dung becomes most important. The sale of this product is an important source of income for many farm families who take dried dung cakes to the cities and sell them. A third use of manure is for construction. Many people in Africa and Asia use cow and/or buffalo

**Draft**

To move loads by drawing or pulling. A draft animal is one that is used to draw or pull loads.

**Compost**

Decayed organic matter used for fertilizing and conditioning land.

**FIGURE 2–15** Family coming to market in a donkey-powered cart. Oxen, donkeys, horses, water buffalos, mules, and camels are the major draft species, but yaks, llamas, goats, reindeer, and dogs are also used.

**FIGURE 2–16** Body wastes and compost from crops are preferred crop fertilizers for millions of farmers around the world. (FAO Photo/18658/G. Blank. Used with permission.)

manure as plaster in their huts (Figure 2–17). In the upper elevations of Peru and Uruguay, houses are built of blocks made from a mixture of approximately 50% manure plus soil and straw. In other cultures, cow manure is used as an important ingredient in *poultices* for wound healing.

## Urine

Urine is used to enhanced soil fertility and also has some specialized uses. In Asia, cow urine is sprinkled on dirt floors to control dust and pests. In Africa, urine and blood are used to extend the low volume of milk available for human food. It is even used as a rinse for hair and in religious ceremonies.

## PEST AND WEED CONTROL

Land that is being left to fallow (rest) can be grazed by animals to control weeds. Their manure rejuvenates the soil. Grazing animals are also frequently used along the edges

**Poultice**

A soft moist mass held between layers of cloth, usually warm, and applied to some area of the body.

**FIGURE 2–17** In many parts of the world, manure is used as a construction material. This Masai woman and child are sitting in front of a manure-plastered hut. (Photo courtesy of Dr. Thomas Thedford. Used with permission.)

of canals to control vegetation. Grazing the canals also helps control snailborne diseases in the human population because large numbers of snails are killed under the hooves of the grazing animals. Ducks are also used in irrigation canals because they eat snails and leeches. It is common to use ducks in rice fields after harvest to pick up missed grain and reduce insect populations (Figure 2–18). Animals provide a tremendous service globally by reducing weeds and crop residues on cropland, which must be removed if the land is to be prepared by hand tools or animal-drawn plows. One method used to accomplish this is to stock the land heavily with animals to remove the weeds, grasses, and residues just before planting. Farmers and pastoralists in Africa often have a centuries-old arrangement to provide mutual service to each other in this way. Consumption of these nuisances adds additional economic value to the cropland through the animal product gained. In Africa, close grazing of vegetation around cropped areas reduces insect numbers by creating a vegetation-free barrier around the crop. The use of animals to control insects and weeds has potential for the developed world as a means of reducing the use of ***pesticides*** in certain areas.

## Pesticides

Any agent or poison used to destroy pests including fungicides, insecticides, herbicides, and rodenticides.

**FIGURE 2–18** Animals serve important roles in pest and weed control. Ducks are invaluable in rice paddies. They pick up dropped grain after harvest, reduce snail and insect populations, and provide their manure for the next crop. (Photo courtesy Suzanne Tolleson. Used with permission.)

## STORAGE OF CAPITAL AND FOOD

In many of the poorer developing countries, banks are nonexistent. Even if they exist, the poorer people either do not have access to them or do not trust them. As an alternative to banks, people use livestock (particularly cattle) as a vehicle for storing their surplus capital. If money is needed, animals can be sold or often bartered for other needed items (Figure 2–19). Use of livestock for storage of capital has several undesirable effects. Obviously, large herds are desirable and are acquired and maintained, which often results in severe overgrazing. This is a major reason why desertification increases when the human population increases. A second very undesirable effect is the reluctance of people to sell animals unless money is needed. This prevents the development of an orderly program of marketing and slaughtering of animals such as that found in the developed countries. For maximum utilization of the livestock of a country, such a marketing program must be in place.

The use of animals as storage of food is very important in many countries of the world because animals represent one of the most reliable reserve food supplies. In the richer countries of the world, livestock populations can support the energy and protein requirements of the population for a year or more. Most poor countries have inadequate facilities for storing grain (especially for any long period) without suffering large losses from pests. This is not a problem with animals. Animals assume even greater importance as a storage of food in countries subjected to periodic and prolonged droughts. When the rains do not come, grains cannot be planted or, if planted, they die. Animals can continue to survive for several months, even years, and may be the only means of survival of the human population.

## CULTURAL USES

For some of the world's peoples, the roles of animals in food supplies and other services can be less important than the cultural roles the animals hold. The most important reason to a given group of people for keeping animals may be cultural.

Religion is a very important cultural role that is given high priority by the world's people. The following is a partial listing of the contributions animals make to cultures; a more detailed explanation is given in Chapter 3.

**FIGURE 2–19** Cattle are often used to store wealth. When cash is needed, they are sold or bartered, like these animals photographed at a trade day in a remote part of Luzon, Philippines.

## Exhibitions and Spectator Sports

The animal industries have long used shows as a means of promoting their breeding stock to potential buyers. However, the love of competition is also an important part of this activity. Some people believe the purebred livestock industry in the United States would disappear without this social dimension. The excitement of this activity is catching, as evidenced by 4-H and FFA club animal projects (Figure 2–20). Horses, dogs, cats, small mammals, llamas, and other species are also exhibited in many countries. Activities such as rodeos, fighting (bulls, buffalos, sheep, and chickens), and racing (camels, horses, buffalos, dogs, fowl, turtles, and frogs), together provide billions of person days of recreation annually. In developing countries, these combined animal recreation events represent the largest source of entertainment for people. In developed countries, they provide viable outlets for viewers and participants and contribute to the economy.

## Companionship and Service

The dog and the cat have long and venerable histories of companionship and service to humans (Figure 2–21). However, other species, such as ferrets and the small rodents (hamsters, gerbils, rabbits, and so on), and several new species, such as alpacas, llamas, small pigs, hedgehogs, and small primates, are also used. Domestic livestock are also used extensively as pet species in both developed and developing countries. Modern research has demonstrated a myriad of benefits to humans from animal companionship. In addition, seeing-eye dogs, hearing-ear dogs, monkeys to assist paraplegics, drug-search dogs, and a variety of others make valuable contributions to humanity.

## Social Structure

The less developed the agriculture, the more livestock are a part of the cultural fabric of the people. In pastoral societies, animals are the wealth left to the next generation. This

**FIGURE 2–20** Exhibition of livestock has a social dimension, in addition to helping promote breeding stock to potential buyers.

**FIGURE 2–21** Dogs and other animals have long and venerable histories of companionship and service to humans.

is especially true of nomads. Animals play an important role in the traditional way that formal contracts are sealed and social obligations met. Marriage contracts involve exchanges of livestock. In tribes where polygamy is practiced, there is pressure to accumulate animals to obtain additional wives. In some tribes, the traditional bride price is so high that young men must work for several years to obtain the necessary livestock for payment. This thus serves an important social function in birth control.

## NONFOOD OR INEDIBLE SLAUGHTER BY-PRODUCTS

Opinions vary about what is edible on a carcass and what is not. Many things that people in developed countries consider inedible are eaten in developing countries. However, a large number of nonfood products are derived from slaughter at all levels of development. A partial listing of sources and products is provided in Table 2–8. In developing countries, various by-products become jewelry, religious implements, tools, fuel, construction material, fly swatters, or musical instruments. These products are often considered indispensable for daily life. In developed countries, many pharmaceuticals and drugs are extracted from animal organs and other tissues. Intestines become sutures; pigskin is used to help burn victims; heart valves are harvested from

**TABLE 2–8    Nonfood Slaughter By-Products**

| Source | Some Products |
| --- | --- |
| Blood | Glue, feed, vaccines, serums |
| Bones | Buttons, jewelry, gelatin, glue |
| Digestive tract | Feed, fertilizer |
| Endocrine glands | Pharmaceuticals |
| Hides | Leather |
| Inedible fat | Fatty acids, fuel, soap, candles |
| Hooves | Gelatin, glue |
| Horns | Handicrafts, cultural symbols |

*SOURCE:* Adapted from McDowell, 1991, p. 42. Used with permission.

pigs and used to replace defective valves in human hearts. Many of the inedible meat products are used in pet foods and livestock feeds. Bones and hooves are converted to fertilizer and mineral supplements; inedible fat is used for a variety of purposes, such as lubricants, candles, detergents, crayons, plastics, insecticides, shaving cream, and livestock feeds. These represent multibillion-dollar industries in the United States and other developed countries.

## HUMAN HEALTH RESEARCH

The number of animal applications to health research are simply too numerous to mention. The average American life span increased by 30 years in the 20th century. Much of that was owing to health research conducted on animals. In addition, the quality of life for people afflicted with chronic diseases has been helped immeasurably. During World War II, biomedical research began making extraordinary strides in the pursuit of human health. Animals have been used extensively as models in that research. Rats, mice, guinea pigs, hamsters, and rabbits have been the most important species used. In recent decades, more dogs and cats have been used, and currently, pigs and other traditional livestock species are being used more and more. Medical research depends on the use of animals as models and will continue to do so for the foreseeable future. *Recombinant DNA* technology is making animals in research an even more promising aid in fighting disease; for example, small rodents have been genetically engineered to mimic human disease conditions. Such developments greatly reduce the amount of time needed by scientists to offer cures and treatments for diseases. In addition, *xenotransplantation,* the replacing of human organs with those from animals, is an active and promising area of research.

**Recombinant DNA**

DNA molecules that have had new genetic material inserted into them. A product and tool of genetic engineering.

**Xenotransplantation**

The transplanting of animal organs into humans.

## INCOME

In developed countries, the economic structure depends on specialization. This is true of agriculture as well as other sectors of the economy. Animal agriculture is a highly specialized means of using land and labor to generate earnings. The yearly earnings from animal products represent roughly half the total from agriculture in the United States. Annual cash receipts from animal agriculture exceed $100 billion, with a general upward trend. The average annual increase in cash receipts from animal agriculture was $1.8 billion for the decade of the 1990s.

Developing-country agriculture also depends on animals for income. This is evident in the most developed of the developing countries. Their economic systems are operating in much the same way as the developed countries they hope to join in economic status. However, it is also true of the least-developed countries as well. In subsistence systems, animals are frequently the major source of income generated from the farm. Extra income from the livestock is used to buy items such as fertilizer and improved seeds that help the farmers increase overall productivity and increase the standard of living for the family. The local economy is stimulated through increased availability and demand of jobs and products.

## CONSERVATION

Domestic animals play vital roles in many land- and water-conservation practices. These roles include controlling pests and adding manure to lands being left to fallow; using alley-cropping systems where leguminous trees are planted in strips with animals

**FIGURE 2–22** Sheep grazing under a stand of rubber trees eat weeds and other plants that would otherwise compete with the rubber trees. Valuable food and fiber products are then harvested from the sheep flock.

and crops alternated between them; and providing alternative use in tree-cropping systems and "weeding" stands of palms and other trees (Figure 2–22). The entire concept of sustainable agriculture often depends on having animals in the system.

## SUMMARY AND CONCLUSION

Agricultural animals help humans extend their use of the available resources by converting nonedible material to humanly edible food. This conversion is not always efficient or fast. However, allowing animals to eat wheat straw or corn stover, or to browse brush or even catch their own insects, reptiles, and rodents, is the only way these resources can be turned to food for humans. We, as humans, have historically consumed animal products and will go to great lengths to procure them. In so doing, we are balancing our diet, which might otherwise be of a very low quality. The high biological value of the protein, coupled with needed vitamins and minerals, makes animal products an essential part of the human diet.

Animals provide humans with a myriad of services in addition to providing food. Without the goods and services provided by animals in addition to food, peoples in developed and developing countries would find it difficult to maintain their living standard and perhaps even to survive. New products are constantly being developed that increase that dependence.

## STUDY QUESTIONS

1. Using numbers, compare and contrast the quantity of meat consumed by people in developed and developing countries.
2. What species provide most of the world's meat?
3. Describe the rates of growth in the production of the various meats used by human societies in the world.

4. Cattle produce approximately what percentage of the total milk produced in the world? Which species produces the second largest amount of total milk?

5. Describe the value that dairying can bring to economic development of a country.

6. What species is most important to world egg production?

7. Describe some miscellaneous food uses of animals.

8. Describe the differences in the ways in which meals are selected in wealthy versus poor societies.

9. What is the physiological and nutritional significance of animal products in the human diet?

10. List 10 nonfood uses for animals that are most important to you.

11. What are the various body coverings provided by animals? Which animal produces which product?

12. What is meant when we say that skins and hides are nonrenewable but wool is renewable?

13. What is different about the Karakul and the Kuche breeds of sheep as compared to other sheep?

14. What is the value of animal fibers, feathers, and so on, in the cottage industries of poor countries?

15. What are the various uses of draft power? What places in the world rely most on draft power? What species provide the bulk of draft power?

16. Describe the various uses humans have found for the body wastes of animals.

17. Explain how animals are used in pest and weed control. With the expanding number of ranchettes and hobby farms in the United States and other developed countries, can you see opportunities for a resurgence of interest in this low-tech approach to pest control?

18. Explain how animals can be used as storage for capital and food.

19. Describe the significance of animals as a provider of cultural dimensions to people's lives.

20. Compare and contrast the various uses the people of developed and developing countries have for nonfood slaughter by-products.

21. Human health research depends on the use of animals as models for humans. Most diseases for which we have a cure are under control because of the contributions of animals. Which animals are the most important to this research?

22. How do animal industries contribute to income generation in developed and developing countries?

23. How do animals aid in land conservation?

## REFERENCES

Bradford, G. E. 1999. Contributions of animal agriculture to meeting global human food demand. *Livestock Production Science* 59: 95–112.

CAST. 1977. *Energy use in agriculture—Now and for the future.* Report No. 68. Ames, IA: Council for Agricultural Science and Technology.

CAST. 1995. *Waste management and utilization in food production and processing.* Report No. 124. Ames, IA: Council for Agricultural Science and Technology.

FAO. 2000. Animal production service fact-file—draught power. http://www.fao.org/WAICENT/FAOINFO/AGRICULT/aga/agap/factfile.draught.htm.

FAO. 2007. *FAOSTAT statistics database: Agricultural production and production indices data.* http://apps.fao.org/.

Hirst, E. 1974. Food-related energy requirements. *Science* 184: 34.

Jasiorowski, H. A. 1998. Constraints and opportunities for the future use of animals in human society. In *Proceedings, Special Symposium and Plenary Sessions, The 8th World Conference on Animal Production.* Seoul: Seoul National University.

Kaushik, S. J. 1999. Animals for work, recreation, and sports. *Livestock Production Science* 59:145–154.

McDowell, R. E. 1977. Ruminant products: More than meat and milk. Winrock Report, Sept. Morrilton, AR: Winrock International Livestock Research and Training Center.

McDowell, R. E. 1991. *A partnership for humans and animals.* Raleigh, NC: Kinnic Publishers, Kinnickinnic Agri-Sultants, Inc.

*Author's note.* This chapter borrows very heavily from R. E. McDowell's 1991 work and is done so with his permission. The text cited here is an excellent book. However, Dr. McDowell has no intentions of revising and updating the text. Therefore, he has graciously granted consent for extensive use of the material.

Ramaswany, N. 1994. Draught animals and welfare. *Review of Science and Technology* 13:195–216.

Sansoucy, R., S. Ehui, and H. Fitzhugh. 1996. The contribution of livestock to food security and sustainable development. Keynote paper, Proceedings of the Joint FAO/ILRI Roundtable on Livestock Development Strategies for Low Income Countries. Rome: FAO.

Seré, C., and H. Steinfeld. 1996. World livestock production systems. FAO Animal production and health paper 127. Rome: FAO.

Taylor, R. E. 1995. *Scientific farm animal production.* 5th ed. Upper Saddle River, NJ: Prentice Hall.

Turk, K. L. 1975. Significance of animals in farming systems and in the provision of farm power and pleasure. Prepared for the Rockefeller Foundation Review of the Role of Animals in the Future World Food Supply, April 21–22, New York.

USDA. 1997. *Agricultural income and finance summary.* Washington, DC: Economic Research Service, U.S. Department of Agriculture.

# Factors Affecting World Agricultural Structure

## LEARNING OBJECTIVES

After you have studied this chapter, you should be able to:

1. Explain the process of adaptation.

2. List the five major categories of environmental stressors and tell how animals react to them.

3. Describe the climatic environments of the world.

4. Summarize how climate and natural vegetation are tied together.

5. Explain how social and cultural differences affect agriculture.

6. Integrate information on levels of economic and agricultural development to explain how they are linked.

## KEY TERMS

| | | |
|---|---|---|
| Adaptation | Economic development | Pathogen |
| Animal rights movement | Economic institutions | Primitive agriculture |
| Artificial environment | Grading up | Religion |
| Centrally planned economy | Kosher | Stress |
| Climate | Literacy | Subsistence agriculture |
| Desertification | Market economy | Symbiotic relationship |
| Developed agriculture | Nomadism | Taboo |
| | Parasite | Transhumance |

## INTRODUCTION

To understand how animals are useful to humanity, we must understand the factors that influence where and how animals live, and their uses to humans. An extraordinarily rich and complex set of cultural and environmental factors influence both humans and our animals and, consequently, animal usage. The various social and cultural norms of societies have a tremendous influence on what people value, tolerate, and eat. These norms also affect animal use and animal agriculture throughout the world. Religion, recreation, and social customs are very important in establishing an animal's value. A value established by one of these factors is often greater than the animal's food or work value. An animal species that has become a successful domesticated species has adapted to the conditions of its environment (natural and human-made) and has been of some utility to humans. This chapter discusses the factors to which animals have adapted to become successful domestic species and provides a rationale for why specific animals have value in various places.

# ADAPTATION

*Adaptation* is the sum of the adjustments occurring in an organism that promotes its welfare and favors its survival in a specific environment. These adjustments must be made to the unfavorable features, or stresses, of the environment. The environment is all of the combinations of conditions under which an organism must live. There are many conditions, and each is identifiable as part of either the natural, or physical, environment, or the artificial environment imposed by humans. The ability of an animal to adapt to the specific stresses of the environment in which it lives is very important. Agricultural animals must have the ability to adapt to the natural environment and the artificial environment imposed by humans if they are to be useful. This ability to adapt to a ***symbiotic relationship*** with humans and the stresses imposed by that association is a key element of domestication.

The natural environment refers to climate, geography, altitude, feed, and other such factors. The natural environment is very important because agriculture must always be practiced within its constraints. The artificial environment to which agricultural animals must adapt is a mixture of factors linked to the economic level and culture of a given society, as well as the steps taken to control the natural environment.

The stresses of the natural and/or artificial environment may be classified into five categories: (1) climatic stresses, (2) nutritional stresses, (3) internal stresses, (4) geographical stresses, and (5) social stresses. Each ***stress*** affects the distribution and utilization of the agricultural animals of the world. A brief discussion of each follows.

**Climatic stresses.** *Climate* is the long-time pattern of meteorological factors; *weather* is the immediate condition of these factors in a given area. The most important of these meteorological factors from the standpoint of animal comfort and performance are ambient temperature, precipitation, solar radiation, wind, and relative humidity. Each factor has both a direct and an indirect effect on animals (Figure 3–1). Indirect effects include those that affect the plants that the animal must use for food.

**Nutritional stresses.** These are the stresses related to the quantity and quality of available feedstuffs. In most cases, climate primarily determines these stresses, with temperature and rainfall the most important. However, soil type, fertility, topography, and geographic location are also important (Figure 3–2).

**Internal stresses.** These are the stresses that affect an individual animal by gaining entrance into its body. The two most common internal stresses are ***pathogens*** and toxins (Figure 3–3).

**Geographical stresses.** These are the nonclimatic stresses associated with a particular geographic location. The most important geographic stresses are the stresses of high altitudes or the stresses associated with the type of terrain.

**Adaptation**

The sum of the changes an animal makes in response to environmental stimuli.

**Symbiotic relationship**

When organisms live together in a mutually beneficial relationship.

**Stress**

A state of physical, emotional, or chemical strain or tension on the body.

**Pathogen**

A living organism that causes disease in another.

**FIGURE 3–1**   These cattle are in a dust storm on a sparse range in New Mexico. They are being stressed by the climate and its effects. (Photo courtesy of U.S. Department of Agriculture.)

**FIGURE 3–2**   This cow is suffering nutritional stress due to the lack of grass on this range. (Photo courtesy of U.S. Department of Agriculture.)

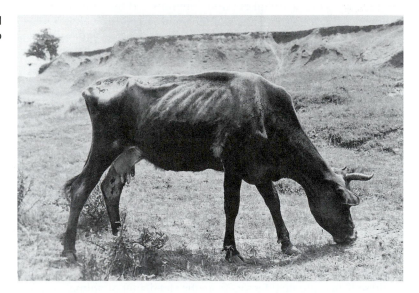

Examples of the latter include constant snow or ice cover; soft, shifting terrain such as sand; soft, wet terrain such as swampy areas; steep areas; and rocky areas.

**Social stresses.**   These are stresses associated with the interactions of an animal with other animals (Figure 3–4). The most important social stresses that affect domesticated animals are those associated with humans. Interactions with the same species include competition for available space and food. Social stresses associated with animals of other species include the predator–prey relationship.

### ARTIFICIAL ENVIRONMENTS

As humans, we create artificial environments to make animals best serve us (Figure 3–5). For example, we can irrigate very dry areas or provide shelter for our livestock in very

**FIGURE 3–3** Internal stresses affect animals by gaining entrance into the body. These cows are infected with dermatophilosis caused by the bites of the tropical bont tick and *Cowdria ruminantium,* which causes infectious heartwater disease. The cows suffer fever, poor appetite, nervous disorders, and eventually die. (FAO photo/17295/ Photographer unknown. Used with permission.)

**FIGURE 3–4** Animals suffer social stress due to their interactions with other animals. The sheep shown here are suffering stress.

cold areas. We can carry out disease prevention measures to reduce the detrimental effects of a disease environment. We can also impose management practices that increase the detrimental effects of the physical environment, for example, confining animals and not permitting them to escape from unfavorable conditions and increasing their exposure to diseases and *parasites* by overcrowding.

The specific nature of the artificial environment imposed by humans depends on the nature of the "service" demanded from the animals. In developed areas such as the United States and Europe, this service is usually the maximum production of some useful product, with the most common being food or some type of body covering. In developing countries, the nature of the service is usually quite different, such as work, transportation, or religion, with food sometimes being much less important. Because the major emphasis differs between developed and developing countries, it is not surprising

**Parasite**

An organism that lives at the expense of a host organism. Generally must live on or in the host.

**FIGURE 3–5**  Artificial environments like these poultry houses are created to help coax maximum production of some useful product. (Photographer Kevin Fleming. Courtesy of Corbis.)

(a)

(b)

**FIGURE 3–6**  (a) Holstein dairy cows have been selected for developed agriculture conditions to produce the optimum amount of milk per cow. To best do this, they require considerable alteration of their environment as evident in this photo. These cows were giving about 80 lbs of milk a day. (b) These Holsteins in conditions where they are not well adapted were giving about 6 lbs of milk per day. Developed and developing agricultures emphasize different roles for livestock and provide different conditions. Animals well adapted to artificial environments of developed agriculture are not equally adapted to harsher conditions.

that animals well adapted to the artificial environments of developed countries are not equally adapted to the artificial environments of poorer nations (Figure 3–6).

### ADAPTIVE CHANGES

The adjustments (changes) that occur in individual animals to make them adapt to specific environments (Figure 3–7) are of three types:

1. **Morphological** or **anatomical changes** are changes in form and structure. These changes are external in nature and are easily seen. They include changes in quantity, quality, and the nature of body coverings and body appendages such as differences in color and presence or absence of body parts such as horns or humps.

**FIGURE 3–7** Animals adapt to their environment in a variety of ways. The Brahma has adapted through anatomical changes such as pendulous ears and sheaths to dissipate heat, light coloring to reflect solar radiation, special muscles under the skin that help is shake off parasites, and horns for protection. Changes in its physiology help conserve water and nitrogen. Behavioral adaptations include modified grazing behaviors. (Photos courtesy of UN/DPI.)

2. **Physiological changes** are primarily changes in the biochemistry of the body. These are internal changes, such as changes in blood chemistry, and are not often readily apparent.

3. **Behavioral changes** can be either genetically induced behaviors or learned responses of the animal to the environment. Examples include self-protection behaviors and food- and water-seeking behaviors. Different species show different levels of instinctive behaviors versus learned behaviors. For instance, the poultry species rely more on instinct than the pig with its larger and more complex brain. A very important effect of domestication on behavior is reduced responsiveness to environmental change. This single behavioral change is found in virtually all populations of domestic animals (Price, 1998).

## CLIMATIC ENVIRONMENTS OF THE WORLD

Human societies live and produce agricultural animals in a wide variety of climates. The majority of these climates impose some degree of climatic stress. It is important to consider the geographic location of the world's climates and consider briefly some of their more significant effects on agricultural animals serving human societies.

### TROPICAL CLIMATES

Figure 3–8 presents the locations of the areas of the world that have climates classified as *tropical*, with humid and subhumid areas indicated. Dense rain forests cover most of the humid tropics. They are only slightly used for agriculture. The subhumid areas are the major portions of the tropics used for agriculture.

The major climatic variation from season to season in the tropics is not in ambient temperature but rather in the amount and distribution of rainfall. This means there are only two seasons—rainy and dry, with little difference between the two in ambient temperature. These names are very descriptive. It rains nearly every day during the rainy season, and during the dry season, it may not rain at all (Table 3–1).

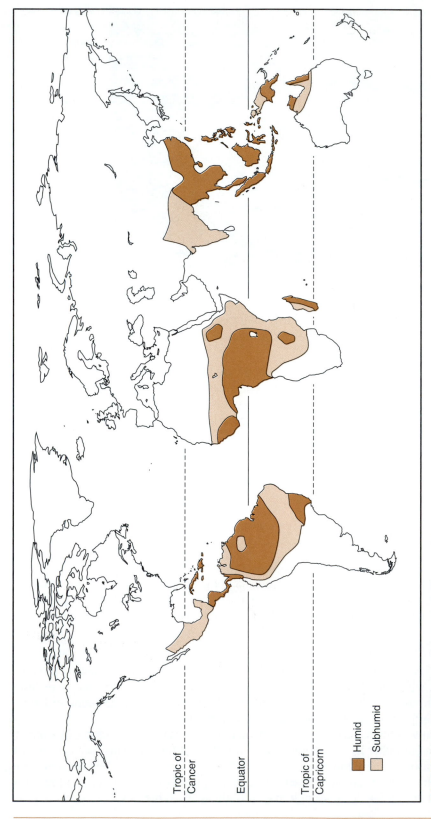

**FIGURE 3–8** Distribution of the tropical climates of the world.
(*SOURCE:* USDA, 1941 *Yearbook of Agriculture.*)

**TABLE 3–1**    **Classification of Regions Between the Two Tropics Based on the Length of the Rainy Season**

| Regions | Length of Rainy Season |
| --- | --- |
| Tropical rain forest | 11 months or more |
| Humid tropics | 7–11 months |
| Semihumid tropics | 4 1/2–7 months |
| Semiarid tropics | 2–4 1/2 months |
| Semidesert to desert | 0–2 months |

SOURCE: Turman, 1986.

The length of the dry portion of the cycle has a very significant effect on the type and amount of vegetation produced, which largely determines the system of livestock production. In many of the developing countries, this cycle is the major factor that forces pastoralists to practice *transhumance* or *nomadism*. If the dry cycle is too long, the people are forced to move the animals from area to area to find feed (Figure 3–9).

One of the two most important problems associated with any attempt to produce agricultural animals in the humid tropics is extreme heat stress. The second major problem limiting livestock production in the humid tropics is diseases and parasites, which thrive in the hot, humid environment and are a constant threat. Diseases and parasites are probably the major factor limiting the introduction of nonadapted animals for the purpose of *grading up* local stock.

## DESERTS

Figure 3–9 presents the areas of the world that are either desert or very dry.

Two major problems limit livestock production in the semiarid regions of the world. To a lesser extent, they are also problems of the semihumid tropics. These problems are (1) seasonal and limited rainfall, resulting in a shortage of drinking water and water to grow food for the animals; and (2) stress caused by the heat.

The semiarid tropics contain parasites and insects but not to the extent of the humid tropics. This problem is also of less magnitude in the subhumid tropics. This is one of the important reasons why a high percentage of the livestock in the tropics is in the subhumid and semiarid regions.

Another problem in the drier areas of the world is high winds because of the association between the wetness of an area and amount of wind. These winds are a major cause of erosion in very dry, overgrazed areas and are a contributing factor to *desertification* (Figure 3–10).

The problems of the semiarid regions are of even greater magnitude in the desert areas. However, there is little animal agriculture in deserts, so this is of little practical significance.

**Transhumance**

The practice of moving animals seasonally from a permanent base to more abundant feed and water and then returning to the permanent base as the season changes.

**Nomadism**

The practice of people without a permanent home base moving from place to place, generally in a pattern dictated by climate and/or season to find feed for their subsistence herds of livestock.

**Grading up**

In animal species, the process of improving a stock of animals for some productive function by consecutive matings with animals considered genetically superior.

**Desertification**

The degradation or destruction of the biological potential of land, leading to desert-like conditions.

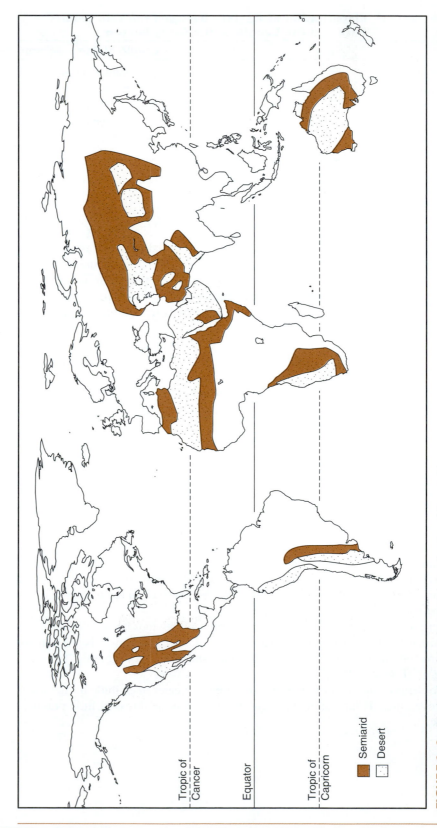

**FIGURE 3–9** Distribution of the desert and dry climates of the world. (*SOURCE:* USDA, 1941 *Yearbook of Agriculture.*)

**FIGURE 3–10**    Desertification in Mauritania. A Mauri writing in the sand the old name of Chinguetty: Arbweir. Considered the seventh holy city of Islam, the city is now completely buried by sand. (FAO photo/18828/I. Balderi. Used with permission.)

### COLD ENVIRONMENTS

The distribution of cold climates of the world is shown in Figure 3–11. The major problems associated with livestock production in these areas of the world are cold stress and availability of food. From a practical standpoint, the very cold areas are so devoid of feed that they are largely devoid of livestock except for reindeer (caribou) (Figure 3–12).

### TEMPERATE CLIMATES

Figure 3–13 shows the areas of the world with temperate climates. Generally, these are the most productive areas of the world. Because of this fact, we generally use the temperate regions as the standard to which all other climatic regions are compared (Figure 3–14). All developed countries have a temperate climate in at least part of their boundary.

## SOCIAL AND CULTURAL DIFFERENCES

Many social and cultural influences on agriculture are region and/or people specific, although certainly not all. They are also dynamic and evolving and are influenced by the economic situation. An example is the *animal rights movement*. This is more an issue in wealthy Western societies than in less-fortunate countries. However, regardless of the issue or the motivation, social and cultural structures have tremendous influence on what specific people eat. We rarely eat simply to satisfy hunger or fulfill nutritional needs, even though nourishment is one of our basic needs. The influences of culture and economics are often difficult to clearly separate because they are closely associated. It is also difficult to separate their effects on the number and kinds of agricultural animals of a given area and how they are produced and used by people of the area.

**Animal rights movement**

A political and social movement that concerns itself with philosophy, sociology, and public policy as each deals with the standing of animals in relation to human society.

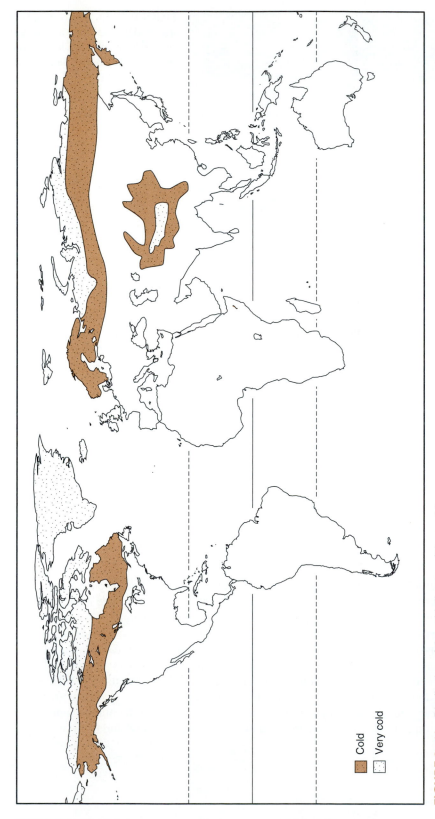

**FIGURE 3–11** Distribution of the cold climates of the world. (SOURCE: USDA, 1941 *Yearbook of Agriculture*.)

Cold

Very cold

**FIGURE 3–12**    Reindeer (caribou) are the only livestock found in any numbers in the cold climates of the world. Uniquely adapted creatures, they thrive in conditions that defeat all other domestic livestock.

## RELIGION

Religion is an important part of the culture of human societies that affects agriculture. Figure 3–15 presents the worldwide distribution of the major religions of the world.

## Effects of Religion on Agriculture

Three of the great religions of the world affect the numbers and utilization of livestock. Islam, whose followers are called Muslims, forbids all contact with swine. As a result, in Muslim areas (primarily North Africa and the Near East), pigs are completely absent. Judaism also considers pork unclean and forbids its consumption. Likewise, there are very few pigs in Israel. The Hindu religion is primarily practiced in India and is the state religion. About 90% of the Indian population is Hindu and in this religion the cow is sacred and can neither be slaughtered nor sold for slaughter. As a result, there is a vast accumulation of cattle in India (Figure 3–16).

## Food and Religion

We owe many modern-day taboos about food and the way it is served to ancient rituals and customs that are part of religious codes. In some religions (cultures), the ritual may simply be blessing food to give it special powers. In others, religious law dictates selection and judgment of the quality of food before the gods will approve. In the past, these religious dictates often had a practical basis. It seems much more than coincidence that religious food laws frequently reflect a commonsense approach to the prevention of food contamination and foodborne disease. For instance, *kosher* food inspection and certification were in place long before modern food inspection (Figure 3–17). Many other examples exist in which religious leaders have affected the relationship of humans to animals, some with profound, long-term influence. Let's use China and India as examples.

When a low-density population has access to grasslands, they historically have high beef consumption. These conditions fit the people of ancient northern India, and the ancestral Chinese of the Yellow River basin. In ancient times, both peoples raised cattle and ate beef. The populations grew in both areas, and farming intensified. Cattle became too valuable in both places for food use because the need for draft animals to produce crops was too great. In China, pork gradually became the preferred meat. This was probably because pigs used household wastes and agricultural by-products more efficiently than did cattle. Cattle were used for draft, if at all. Gradually their numbers diminished. Religion was not significantly involved in the process. In India, a different course was taken. The fierce summer heat, the monsoons, and periodic droughts were not good conditions for hogs and they did not flourish. As the cattle population density increased, cow's milk became the major source of animal protein to India's people.

**Kosher**

Kosher food is considered ritually fit for use as sanctioned by Jewish religious law.

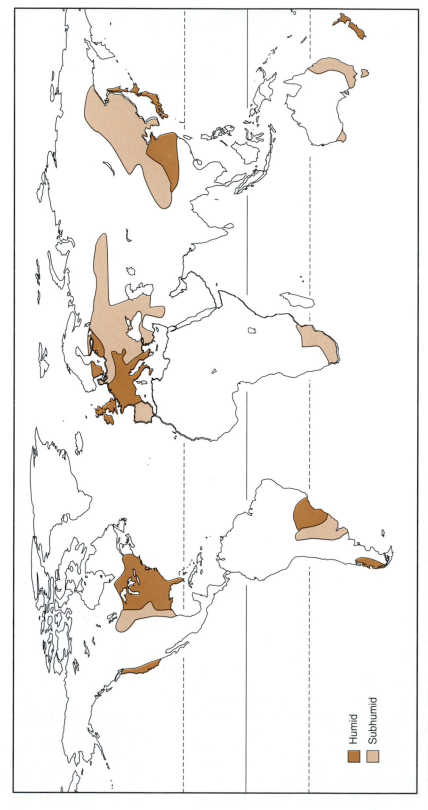

**FIGURE 3–13** Distribution of the temperate climates of the world.
(*SOURCE:* USDA, 1941 *Yearbook of Agriculture.*)

Humid
Subhumid

**FIGURE 3–14** Beef cows grazing a temperate zone pasture. Because temperate regions present the best overall set of conditions for agriculture, the conditions and production of all other zones are compared to temperate regions.

Prohibiting the slaughter of cattle soon became a religious principle of Hinduism, and subsequently, Buddhism (which sprang from Hinduism), to protect the supply of draft animals. The Hindus carried this to the greatest extreme and a civilization arose in which beef eating is *taboo* to the faithful.

Examples abound where mistakes were made (from a practical perspective, that is) in the name of religion. Although there is logic in Hinduism's taboo on beef slaughter, for protection of the milk and work animals, fish eating is also taboo to most Hindus. Even though fish-containing oceans surround India, protein and general food deficiencies are a way of life for its people because they do not consume fish. Seventh-Day Adventists also have a finless fish taboo and a taboo against pork. After their missionaries converted the Pitcairn islanders, the self-sufficient economy of the island disintegrated. Before their conversion they had depended on pork and finless fish in their diet.

It is a human tendency to compare values. Our own religious beliefs make it difficult to fully understand and appreciate the beliefs of others, particularly if the beliefs differ radically. The important thing to remember is that there is no chance that major segments of any religion will easily change their religious laws. Thus the situations that exist are the situations that are likely to continue. Whether the general population thinks it is right is beside the point. One of the great lessons of life is to learn to accept graciously when one's opinions are considered irrelevant.

## LEVELS OF ECONOMIC DEVELOPMENT

Two terms, *developed countries* and *developing countries*, appear repeatedly in this book. The terms refer to a country's level of economic development. In this book, the classification of countries into one or the other category is the same as that used by the Food and Agricultural Organization of the United Nations (FAO) and is summarized in Table 3–2.

Although not used by the FAO, three other classifications in common usage are *First World*, *Second World*, and *Third World*. First World refers only to the developed countries with a ***market economy***; Third World countries are developing countries with a market economy. All Second World countries have a ***centrally planned economy***, whether developed or developing. The term *Second World* has been fading from use since the fall of the Berlin Wall in 1989.

**Taboo**

A prohibition imposed by social custom against some action or object. Frequently found as part of religious codes.

**Market economy**

Economies in which prices are freely determined by the laws of supply and demand.

**Centrally planned economy**

An economy under government control. Prices, labor, and other economic inputs are controlled and not allowed to fluctuate in accordance with supply and demand.

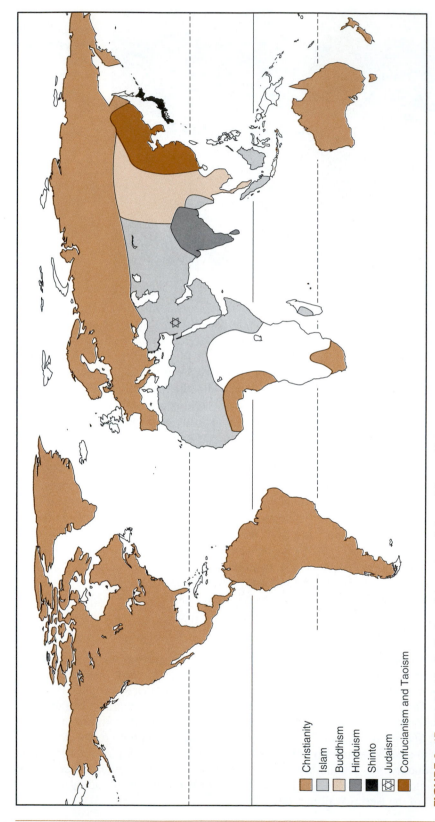

**FIGURE 3–15** Worldwide distribution of the major religions of the world.
(*SOURCE: USDA, 1941 Yearbook of Agriculture.*)

Christianity
Islam
Buddhism
Hinduism
Shinto
Judaism
Confucianism and Taoism

**FIGURE 3–16** The Hindu religion prohibits the slaughter of cattle, which results in a vast accumulation of cattle in India. (Photographer Eugene Gordon. Courtesy of Pearson Education/PH College.)

**FIGURE 3–17** Modern-day kosher slaughter has its roots in ancient Jewish law and custom. (Photo courtesy of Empire Kosher Poultry, Inc.)

As the name implies, the developed countries have reached a higher level of economic development than have the developing countries. The same is true of the countries with a market economy as compared to those with a centrally planned economy. Considerable variation is seen in the economic development of different nations within each category, particularly among the developing nations. Third World countries range from those that are poor but enjoy a fairly high standard of living to the extremely poor countries where per capita income is less than $100 per year.

**TABLE 3–2    FAO Classification of Developed and Developing Countries**

**Developed Countries**

North America: Canada and the United States
Europe
Oceania: Australia and New Zealand
Asia: Israel and Japan
Africa: South Africa
Russian Federation

**Developing Countries**

Africa: All countries except South Africa
Asia: Except Israel and Japan
Latin America: All countries of Central America and South America
North America: Bermuda, Greenland, St. Pierre et Miquelon, and Mexico
Oceania: All islands except Australia and New Zealand

*SOURCE:* FAO, 1997.

## LEVELS OF AGRICULTURAL DEVELOPMENT

The three levels of agricultural development are developed, subsistence, and primitive. The livestock systems in a developed agriculture are further subdivided into intensive and extensive. The intensive systems are usually small, labor-intensive farm units. The extensive systems are usually large range units. In extensive systems, the type and breed(s) of livestock that are most popular are determined by whether the range area is humid or dry.

Most developed countries also have a developed agriculture. The exceptions are some of the nations in eastern Europe and the Russian Federation. Several countries in these two groups are marginal between a developed and a subsistence agriculture. There are no countries that have a developed agriculture classified by FAO as being a developing country.

A *developed agriculture* is usually associated with the following characteristics:

- A very small proportion (usually less than 10%) of the total population on the farm actively engaged in farming
- A highly specialized agriculture with each unit producing only one or two products
- A highly mechanized agriculture with little or no animal or hand labor
- A high per capita income
- A high literacy rate for the total population

Subsistence agriculture is the level of agricultural development found in most of the developing countries outside of Africa.

*Subsistence agriculture* is usually associated with the following characteristics:

- Approximately half of the total population is engaged in farming.
- Each farm family produces roughly what it consumes with only a small surplus for sale or barter. This surplus is barely enough to supply the needs of the portion of

the population that is not on a farm producing its own food. Thus the country has little or no agricultural products for export.

- Little mechanization and much hand and animal labor
- A relatively low per capita income
- A relatively low literacy rate

Primitive agriculture is found in the most undeveloped of the developing countries. Most of the primitive agriculture is in Africa.

*Primitive agriculture* has the following characteristics:

- Almost the entire population involved in producing their own food because no one produces a surplus that can be purchased
- Generally a scarcity of food and a low nutritional level
- No mechanization and very little animal power is utilized in farming. Almost all of the labor is hand labor, which means that only a small amount of acreage can be farmed by one individual.
- Extremely low per capita income
- Very few literate individuals

Table 3–3 presents a comparison of the percentage of the population in various levels of economic development that is engaged in agriculture. Generally speaking, the developed countries have a developed agriculture and the developing countries have either subsistence or primitive agriculture. Several things stand out in Table 3–3 relative to the percentage of the population engaged in agriculture. Farmers are in a definite minority in the United States and other developed countries. Although most developed countries have less than 15% of their population engaged in agriculture, several countries have a considerably higher percentage. FAO sets 30% as the upper limit for classification as a developed country. Included in the group that has more than 15% of its population engaged in agriculture are a number of countries in eastern and southeastern Europe that practice some subsistence agriculture. It was stated earlier that almost all of a population farms in primitive agriculture. Table 3–3 does not appear to completely support this statement. The overall average for Africa is far from "almost all" of the population. However, many African countries have 80–90% or more of their population engaged in agriculture. Because it is unlikely that the agriculture is entirely primitive in any country, the 80–90% figure suggests that in the areas with primitive agriculture, the entire population is involved.

*Literacy* refers only to the ability to read and write and indicates nothing about native intelligence or the ability to learn. However, it is a definite handicap to a nation to have an illiterate population. Illiteracy is one of the major obstacles in efforts to introduce new technology to improve levels of agricultural development. If people cannot read, they learn new technology only by being told and shown. An extension specialist can write agricultural bulletins that can reach hundreds of thousands of people to tell them of improved agricultural methods. However, if it requires personal contact and verbal explanation, that same extension worker can reach only a few hundred people per month. Most people in developing countries realize that the first step to improving their level of agricultural and economic development is to improve their educational level. This is the main reason why developed-country universities have so many students from developing countries.

**Literacy**

The ability to read and write.

**T A B L E  3 – 3**    **Percentage of Total Population Engaged in Agriculture**

| Country/Continent/Classification | Agricultural Population as a Percentage of Total Population |
|---|---|
| **All Developed Countries** | 6.6 |
| **All Developing Countries** | 49.8 |
| **Africa, Developing Countries** | 55.9 |
| **Asia** | 50.7 |
| **World** | 38.4 |
| United States | 1.9 |
| Canada | 2.2 |
| Denmark | 3.2 |
| Australia | 4.3 |
| South Africa | 12.4 |
| Argentina | 9.2 |
| Cuba | 14.8 |
| Brazil | 14.3 |
| Greece | 11.7 |
| Poland | 17.4 |
| Azerbaijan | 25.1 |
| Turkmenistan | 31.8 |
| Bangladesh | 51.8 |
| China | 64.3 |
| Rwanda | 90.1 |
| Bhutan | 93.6 |

SOURCE: FAO, 2007.

## ECONOMIC INSTITUTIONS AND AGRICULTURAL DEVELOPMENT

In countries with a developed agriculture, a number of essential economic institutions are highly developed and functional. In contrast, these institutions are relatively poorly developed, if present at all, in countries with subsistence agriculture and are almost totally absent in countries with a high incidence of primitive agriculture. These economic institutions are (1) financial institutions, (2) marketing agencies, (3) industrial institutions, and (4) governmental agencies. The following is a very brief discussion of their importance.

### Financial Institutions

Included in this group are all the institutions that handle money. Financial institutions are a source of loans to finance agricultural enterprises. They provide a place for the investment of the profits from agriculture that are not reinvested in agriculture. They serve as a safe repository of money and provide a means for the safe and orderly transfer of money in financial transactions. The importance of these functions to the economic development of a country's agriculture should be obvious. One of the undesirable side effects of the lack of banks and other financial institutions in the developing countries is the use of livestock, particularly cattle, for "storage of capital."

## Marketing Agencies

Marketing agencies are essential to provide a means for people engaged in agriculture to sell their products at a satisfactory price and at a time when they want to sell. Included are agencies such as local auction sale barns or large terminal markets. In most of the poorer countries, farmers do not have access to such competitive markets and are at the mercy of local traders who offer very low prices (Figure 3–18). Marketing agencies also provide the means by which those engaged in agriculture can purchase the goods they need and want.

## Industrial Institutions

It is impossible for a nation to reach a high level of agricultural development without also having a high level of industrial development. Three important reasons explain why this is true. The first is that industrial institutions, such as factories, provide nonfarm employment for part of the working population not engaged in agriculture. Second, the wages paid to these workers provide a cash market for agricultural products. Third, these industrial workers produce the consumer goods needed in agriculture, either as essentials or as luxuries. The availability of these consumer goods is also a stimulus to high agricultural productivity to make more money to purchase them.

## Governmental Agencies

Governmental agencies include the agencies responsible for transportation (road building and maintenance, and so on), education, protection (both police for internal and an army for external protection), equitable land policies, and sound money. Unfortunately, too often the people who have the power to make changes in government to improve

**FIGURE 3–18** In many developing countries, marketing agencies are nonexistent. Farmers are at the mercy of local traders who offer very low prices.

these services are unwilling to do so. They profit from the fact that the masses are illiterate or lack good markets. They may be large landholders who do not favor land redistribution. In most of the developing countries, the task is formidable. Change will require a generation to grow up with greatly improved educational opportunities and, hence, be able to push for the needed changes.

## SUMMARY AND CONCLUSION

Agricultural animals must adapt to a wide variety of stresses if they are to be of use to humans. Our domestic animals and improved varieties of plants are influenced by a wide variety of factors, including cultural issues, economic development, and religion. Levels of economic development of various peoples influence the different levels of agricultural development. The world's economic institutions also affect development. In the next chapter, we assimilate this information to see exactly what types of agricultures have developed to accommodate all of these factors.

## STUDY QUESTIONS

1. What is adaptation?
2. What is the difference between the natural and the artificial environment?
3. List and briefly describe the five stresses of the physical environment under which organisms must live.
4. Study the figures that show the climatic environments of the world. Be able to identify the regions.
5. What are the three major religions of the world that have a direct and marked effect on the numbers and utilization of livestock in the countries where they predominate? What class of livestock does each religion primarily affect and what is the effect on numbers and utilization of each class of livestock?
6. What is the difference between a market economy and a centrally planned economy? What is meant by the terms *First World, Second World*, and *Third World?*
7. Which continents are primarily composed of developing countries? What are the only two countries of Asia and the one country of Africa that are classified as developed?
8. What are the three levels of agricultural development? Describe the characteristics of each from the standpoint of percentage of the population that is engaged in agriculture and type of farm power used.
9. Describe the three essential services provided by financial institutions and explain why each is important.
10. Describe the two essential services provided by marketing institutions and explain why they are important.
11. Discuss the three major reasons why industrial development is essential for the existence of a highly developed agriculture.
12. What are the functions of government as it relates to development?

## REFERENCES

Allen, T., and A. Warren. 1993. *Deserts: The encroaching wilderness*. New York: Oxford University Press.

Bosley, G. C., and M. G. Hardinge. 1992. Seventh-day Adventists: Dietary standards and concerns. In *Overview: Religious and philosophical bases of food choices*. Chicago: Institute of Food Technologists.

Chaudry, M. M. 1992. Islamic food laws: Philosophical basis and practical implications. In *Overview: Religious and philosophical bases of food choices*. Chicago: Institute of Food Technologists.

Cohen, Y. A. 1977. Dietary laws and food customs. In *Encyclopedia Britannica: Macropedia*. 15th ed. Chicago: University of Chicago.

FAO. 1997. *FAOSTAT statistics database: Agricultural production and production indices data*. http://apps.fao.org/.

FAO. 2007. *FAOSTAT statistics database: Agricultural production and production indices data*. http://apps.fao.org/.

Harris, M. 1990. Mother cow. In *Conformity and conflict, readings in cultural anthropology*, 7th ed., James P. Spradley and Dave W. McCurdy, eds. New York: HarperCollins.

Harris, M., and Eric B. Ross. 1978. How beef became king. *Psychology Today*, p. 53.

Huang, Y., and C. Y. W. Ang. 1992. Vegetarian foods for Chinese Buddhists. In *Overview: Religious and philosophical bases of food choices*. Chicago: Institute of Food Technologies.

Jordan, T. E., M. Domosh, and L. Rowntree. 1994. *The human mosaic: A thematic introduction to geography*. 6th ed. New York: HarperCollins.

Kibra, A., and K. K. Iye. 1992. Food and dietary habits of the Hindu. In *Overview: Religious and philosophical bases of food choices*. Chicago: Institute of Food Technologies.

Knutson, A. L. 1981. The meaning of food. In *People and food*, D. Kirk and E. Eliason, eds. San Francisco: Boyd and Fraser.

Mares, M. H., ed. 1999. *Encyclopedia of deserts*. Norman: University of Oklahoma Press.

McDowell, R. E. 1991. *A partnership for humans and animals*. Raleigh, NC: Kinnic Publishers, Kinnickinnic Agri-Sultants, Inc.

Page, J. 1984. *Arid lands*. (Planet Earth). Alexandria, VA: Time-Life Books.

Pike, O. A. 1992. The Church of Jesus Christ of Latter-day Saints: Dietary practices and health. In *Overview: Religious and philosophical bases of food choices*. Chicago: Institute of Food Technologies.

Price, E. O. 1998. Behavioral genetics and the process of animal domestication. In *Genetics and the behavior of domestic animals*, Temple Grandin, ed. San Diego: Academic Press.

Regenstein, J. M., and C. E. Regenstein. 1992. The kosher food market in the 1990's—A legal view. In *Overview: Religious and philosophical bases of food choices*. Chicago: Institute of Food Technologies.

Smith, H. 1994. *The illustrated world's religions: A guide to our wisdom and traditions*. San Francisco: Harper.

Turman, E. J. 1986. *Agricultural animals of the world*. Stillwater: Oklahoma State University.

USDA. 1941. Climate and man. In *1941 yearbook of agriculture*. Washington, DC: USDA.

# 4

# Worldwide Systems of Agricultural Production

## LEARNING OBJECTIVES

After you have studied this chapter, you should be able to:

1. Identify the broad types of agricultural production systems found worldwide.

2. Describe the use of animals within each agricultural type.

3. Compare and contrast the livestock industries of developed and developing countries.

4. Differentiate between commercial systems of agriculture and subsistence systems and their goals.

## KEY TERMS

| | | |
|---|---|---|
| Agricultural systems | Extensive agriculture | Self-sufficient |
| Barter | Intensive agriculture | Shifting cultivation |
| Cash crop | Market gardening | Slash-and-burn agriculture |
| Commercial livestock finishing | Pastoralism | Steppes |
| Commercial plantation | Primitive agriculture | Subsistence agriculture |
| Developed agriculture | Rudimentary sedentary tillage | Transhumance |
| Drylot | Savanna | Urban agriculture |

## INTRODUCTION

This chapter describes the systems of agricultural production that have developed in response to the complex set of environmental, cultural, and economic conditions of the world.

The types of livestock systems found in an area are primarily determined by the principal agricultural system of that area. In turn, the agricultural system is determined by a number of factors, including climate, topography, soil type, and socioeconomic

**TABLE 4-1** **Agricultural Systems Found in Areas with Primitive, Subsistence, and Developed Agricultures**

| Agriculture | Goal | Agricultural Systems |
|---|---|---|
| Primitive | Subsistence, but usually agriculture is supplemented with nonagricultural food sources (hunting, gathering, stealing) | 1. Nomadic herding, transhumance<br>2. Shifting cultivation<br>3. Rudimentary sedentary tillage |
| Subsistence | Subsistence | 1. Subsistence crop and livestock farming<br>2. Paddy rice farming (Intensive subsistence tillage with paddy rice)<br>3. Peasant grain, tuber and livestock farming (Intensive subsistence tillage without paddy rice)<br>4. Urban agriculture |
| Developed | To make money with the agriculture | 1. Ranching<br>2. Commercial grain farming<br>3. Commercial livestock and crop farming<br>4. Commercial livestock finishing<br>5. Commercial dairy farming<br>6. Commercial plantation<br>7. Specialized horticulture<br>8. Mediterranean agriculture<br>9. Market gardening |

NOTE: This listing of the systems proposed by Whittlesey has been modified by grouping the systems according to the level of agricultural development in which they are found. It has been further modified to reflect changes in agriculture since 1936.

factors such as level of economic development, demand for products, political systems, and religions. Although all of the factors are important, climate (particularly rainfall and temperature) and level of economic development are most critical. Climate determines to a large degree what can be grown and, therefore, the basic type of agriculture practiced. But the level of economic (and agricultural) development determines the specific system. Economic development also determines demand for products.

There are a number of classifications of agricultural systems, with one of the most widely accepted that of Whittlesey (Table 4–1). Their worldwide distribution is shown in Figure 4–1. Figure 4–2 summarizes the relationship of these systems to amount of rainfall and temperature.

The most common reason for devoting an area to grazing livestock rather than cultivation is that it is too dry, but three other reasons for this usage are that these areas are (1) too rough to cultivate, (2) located at high elevations where the growing season is too short, or (3) so far removed from good markets that the cost of shipping the products to market would be unprofitable (Figure 4–3). In the latter case, grazing is usually only a temporary utilization of the land and is discontinued once transportation facilities are built. For example, much of the Corn Belt region of the United States was devoted to livestock grazing for a number of years following the Civil War. Farmers knew they could grow corn, but there was no economical way to ship vast amounts of corn to the eastern seaboard where there was a good market. Once the railroads were extended into the Midwest, the land was plowed and planted to corn. Even today, much of Brazil is ranched because of lack of infrastructure to transport products.

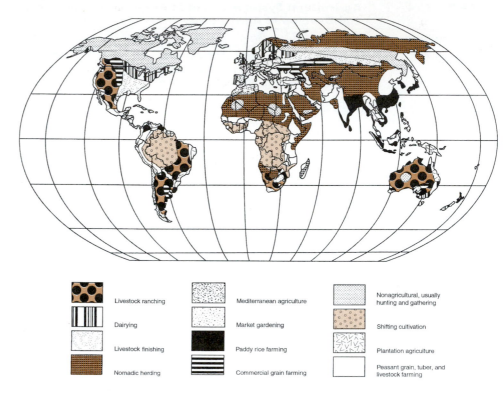

**FIGURE 4–1**    Distribution of world agricultural systems. (*SOURCE:* Adapted from Whittlesey, 1936; modifications according to Jordan-Bychkov et al., 2008.)

Commercial grazing of livestock can rarely compete with crop production as the most profitable way to utilize land if crop production is possible. Many of the shifts of agriculture that have occurred in Argentina in the last quarter century have been to crop production and have occurred for precisely this reason.

## NOMADISM AND TRANSHUMANCE

Nomadism is irregular, opportunistic movement in response to availability of feed, usually of people without a permanent base. Nomadic herding is primarily practiced in the deserts, **steppes**, and **savannas** of Africa, Arabia, and the interior of Eurasia and in the very cold areas north of the tree line in the countries of the former Soviet Union. Nomadic people move their families and belongings with their herds and flocks of cattle, sheep, goats, camels, yaks, horses, and/or reindeer to areas where feed and water are available. The herd may consist entirely of a single species or a mixture of species. Peoples of sub-Saharan Africa are the only nomadic group that depends almost exclusively on cattle. Nomadic people who inhabit the tundra of Eurasia are the only group who use the reindeer. The goal of these peoples is subsistence, rather than profit from the sale of their livestock. Nomads live in tents or snow huts that are easy to transport or abandon and replace. They own no land, although certain tribes may claim specific areas as their lands and resist trespassing. This lack of ownership sometimes creates serious problems when arbitrary political boundaries are constructed that divide traditional grazing areas

**Steppes**

Short grass vegetation zones.

**Savanna**

Tall grass vegetation belts in the hot areas of the world.

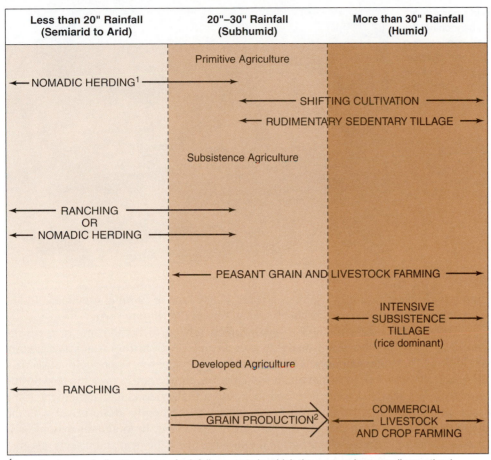

|  Less than 20" Rainfall (Semiarid to Arid) | 20"–30" Rainfall (Subhumid) | More than 30" Rainfall (Humid) |
| --- | --- | --- |

**Primitive Agriculture**

← NOMADIC HERDING[1] ──────→

←─── SHIFTING CULTIVATION ───→

←─ RUDIMENTARY SEDENTARY TILLAGE ─→

**Subsistence Agriculture**

←───── RANCHING OR NOMADIC HERDING ─────→

←──── PEASANT GRAIN AND LIVESTOCK FARMING ────→

INTENSIVE SUBSISTENCE TILLAGE (rice dominant)

**Developed Agriculture**

←───── RANCHING ─────→

GRAIN PRODUCTION[2]

COMMERCIAL LIVESTOCK AND CROP FARMING

[1] Length of arrow indicates range of rainfall amounts in which the system is generally practiced.

[2] Width of arrow indicates extent to which the system is practiced in an area with a given amount of rainfall.

**FIGURE 4–2**   Influence of amount of rainfall on agricultural systems in each of the levels of agricultural development. (*SOURCE:* Adapted from Turman, 1986.)

**FIGURE 4–3**   Livestock are used to graze land when it is too dry or too rough to cultivate, when high elevations cause a short growing season, or when the land is too far from good markets to make crop production profitable.

**FIGURE 4–4**   Many of the world's nomads are disappearing. Young people are abandoning the traditional way of life, and political conflicts and natural disasters take a huge toll. These destitute nomads came from drought-stricken Mali to northern Upper Volta in search of pasture but found a wasteland. (FAO photo 6715/F. Botts. Used with permission.)

into two different countries. Seldom, if ever, is any type of farming practiced. Today, nomadism is a declining, and probably dying, way of life (Figure 4–4).

**Pastoralism**

Herding grazing animals.

**Transhumance**

Seasonal moving of animals from a permanent base to more abundant feed and water and then returning to the permanent base.

Transhumance is a method of ***pastoralism*** related to nomadism. In ***transhumance***, pastoralists move seasonally to take advantage of shifting water and grazing conditions. Transhumants have a permanent base of residence where cultivation may be practiced. In some transhumanant systems, the entire family may move with the livestock to a seasonal settlement and then return to the permanent settlement at the appropriate time. In others, only a few members of the family group travel with the animals and later return them to the permanent settlement. In mountain regions, animals are moved to higher elevations in the summer to escape heat and insects and find grazing and then returned to lower elevations for the winter. In Africa, livestock may be moved south in the dry season but then back north in the wet season to avoid the tsetse fly. Transhumance is practiced in the Mediterranean region, the Alps and other mountain areas of Europe, the Himalayas, the western United States, dry portions of Asia, and in other geographic regions as well. In the past in the United States and Australia, cattle were often released into mountain pastures in the spring and rounded up in the fall, with little herding in between. Today there is more supervision and management of grazing.

## SHIFTING CULTIVATION AND RUDIMENTARY SEDENTARY TILLAGE

**Slash-and-burn agriculture**

The practice of clearing a plot of land from the forest by cutting the trees and shrubs and then burning them. The ash from the burning fertilizes the soil.

Shifting cultivation occurs when farmers clear small plots of land, farm them for a few years until the land's fertility declines, then abandon that plot and clear another. This practice is often called ***slash-and-burn agriculture*** because the forest is cut and then burned as a means of clearing (Figure 4–5). The ash provides nutrients for the crops. The farmers may move to a new spot or remain in the same place. Eventually, the abandoned plots grow up in brush, trees, and other plants that help to renew soil fertility; such renewed plots are later cleared, and the cycle is repeated. In its best form, this is a good system of land rotation. The land should have 10 to 20 years to rejuvenate. Shifting cultivation is practiced in the remote tropical areas of South America, Africa,

**FIGURE 4–5** Slash-and-burn method of clearing a field. Large trees were usually killed by ringing them. The smaller trees, underbrush, and cover plants were burned to clear the land for planting. (FAO photo/14185/R. Faidutti. Used with permission.)

Southeast Asia, and Indonesia. The goal of shifting cultivation is subsistence. However, nonagricultural sources usually supplement the diet.

Many areas previously devoted to shifting cultivation have been converted to rudimentary sedentary tillage.

Rudimentary sedentary tillage differs from shifting cultivation in that the farmers remain in one place. This system requires land that is basically more fertile. However, the farming methods and crops produced are usually similar to those of shifting cultivation and are very primitive, involving mainly hand labor and very crude tools. With increasing human population, the cycles of production become ever shorter and the recovery of fertility less complete, which can lead to breakdown of the system. In addition, lands have been cleared in areas where rejuvenation of the cover does not occur readily, and these plots often erode and become unusable.

Farmers in both of these types of primitive farming usually also own some animals. Generally, all animals owned by the people of a village graze together on common grazing grounds near the village. One of the jobs assigned to the young boys and old men of the village is herding the grazing animals. Usually, the animals are brought into the village compound at night for protection from predators—human as well as wild animals. In addition, milk and manure can be collected. The animals are not generally used for draft purposes and are infrequently used for meat. Most meat comes from fishing and hunting.

## PADDY RICE FARMING

Paddy rice production (Figure 4–6) is concentrated in South and East Asia. Large acreages are found in India, China, Japan, and the Philippines. Rice is grown if the necessary water is available because it will outyield any other grain. This system is very labor intensive, both for human and animal labor. Vast numbers of work animals, primarily cattle and buffalos, are needed. The animals may also be utilized for food, but their primary use is for draft purposes. Pigs and poultry are kept as scavengers and are used as food by the human population (Figure 4–7). A few cattle or goats may also be kept to graze dry land nearby and are generally a prized source of milk. Water buffalo

**FIGURE 4–6**   Rice is very high yielding and grown in paddies in most of the densely populated areas of Asia. This system is very labor intensive, requiring much hand labor and draft animal power. Here an Indonesian farmer and his work animals return home across the rice terraces after a day in the paddy. (FAO Photo/17343/R. Faidutti. Used with permission.)

**FIGURE 4–7**   In many primitive and subsistence agricultural systems, unimproved pigs are kept as scavengers.

cows also provide milk. Fish are taken from the irrigation reservoirs and canals and some aquaculture is practiced in some areas (Figure 4–8). It is common for paddy rice farmers to raise a cash crop such as tea, jute, sugarcane, or mulberry bushes for silkworm production. Paddy rice farming has been and continues to be one of the world's most successful and sustainable subsistence systems of agriculture.

## PEASANT GRAIN, TUBER, AND LIVESTOCK FARMING

Peasant grain, tuber, and livestock farming, also called subsistence crop and livestock farming, is also a highly successful subsistence agriculture and involves small farms,

**FIGURE 4–8** In intensive paddy rice farming, much needed protein comes from fish taken from the irrigation reservoirs and canals. Here, one man fishes while the other weeds the rice field. (FAO Photo/22575/J. Van Acker. Used with permission.)

each with a few head of livestock. It is practiced primarily in the poorer regions of Europe and in the colder, drier parts of Asia, the river valleys of the Middle East, throughout Africa, and in Latin America. Most of the produce of the land and animals is utilized by the farmer, with only small amounts available for sale or **barter. Self-sufficiency** rather than profit is the objective of farmers engaged in this system of agriculture. The dominant crops are wheat, barley, millet, oats, corn, sorghum, soybeans, and potatoes. Dual-purpose animals (for work and food) are used (Figure 4–9). Milk is a more important food than meat. Cattle, pigs, and sheep are the large animals usually raised. Chickens and ducks are often kept. In South America, llamas and alpacas are raised. A **cash crop** such as cotton, flax, hemp, or tobacco is raised to generate a meager income to purchase items the farmers cannot grow.

## RANCHING

*Ranching* is the commercial use of the dry areas of the world (Figure 4–10). In very general terms, ranching is the predominant system in areas with less than 20 inches of yearly rainfall, although it is practiced in areas that receive as much as 25 inches of yearly rainfall. Livestock is the primary enterprise in ranching. Generally, ranching is based on cattle or sheep, and only occasionally are goats or other animals involved. Management is very **extensive,** which means that the animals are spread out and do not receive much individual attention. Usually, only one species is involved in a given ranching enterprise. However, mixing species often produces better use of the land. Ranching is always a for-profit enterprise even if found in a developing country. Thus species decisions on ranches are made with a profit motive.

Unlike nomads, ranchers are permanently located on land they either own or lease. Ranchers very definitely seek to make a profit from their livestock. Although ranchers are primarily concerned with livestock production, some farming may be practiced, if possible, to produce forage to feed in winter. Most cattle ranches are found in the

**Barter**

Trading services or commodities for each other.

**Self-sufficient**

Providing for one's own needs.

**Cash crop**

A crop grown specifically with the intent of marketing its product.

**Extensive agriculture**

Agriculture systems practiced in a manner that spreads human time and attention across vast acreages and/or many animals.

**FIGURE 4–9** The Hariana cow of India (a) is kept for her milk production and the bullocks she produces (b), which are used as draft animals. (Photos courtesy of Dr. Eric Bradford, Animal Science Department, University of California, Davis.)

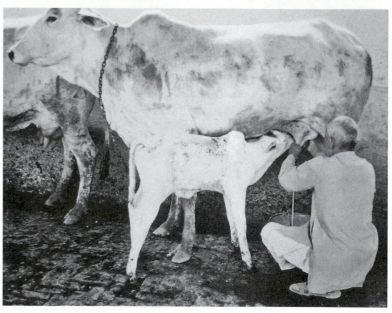

(a)

(b)

United States, Canada, tropical and subtropical Latin America, and the hotter parts of Australia. Most sheep ranches are located in Australia, New Zealand, South Africa, and Argentina. Ranching is practiced in some countries that are not developed. In many of these undeveloped areas, the income from ranching has helped the economic development of those nations tremendously.

**FIGURE 4–10** Ranching is a for-profit, extensive system of agriculture practiced in the dry areas of the world.

# COMMERCIAL CROP AND LIVESTOCK FARMING

Commercial crop and livestock farming predominates in areas with more than 30 inches of rainfall per year but is also practiced in areas that receive 25–30 inches of yearly rainfall. Crop production is the primary enterprise, with livestock secondary to crops. Beef cattle, sheep, dairy cattle, pigs, horses, or poultry are secondary enterprises on these farms. Often, animals are finished for slaughter in the winter when the farmer is not working the crops. Farmers may specialize in only one species or may be diversified, raising two or more species. The production is usually *intensive,* which generally means the animals are confined so that mechanization and the latest technology can be maximized and individual attention given to the animals without greatly increasing labor.

**Intensive agriculture**

Any agriculture system in which much human attention and focus is directed to a small plot of land or to each animal.

## COMMERCIAL LIVESTOCK FINISHING

Commercial livestock finishing is a modification of commercial crop and livestock farming that has only been in existence since the last half of the 20th century and represents a further specialization of agriculture. Generally, these producers do not raise any crops; they buy commodities and by-products to feed their livestock. The Corn Belt, the South, and the Plains area of the United States, parts of western and central Europe, and now some of Australia are the major areas where commercial livestock finishing is found. The United States is the largest practitioner of this agricultural system. The cattle feedlots found in the High Plains of the United States are the classic example of this system. Large numbers of cattle are finished for market in this area by systems that are highly specialized for the purpose (Figure 4–11). With the expanding influence of vertically integrated poultry and swine operations, these species are commonly raised in this type of specialized production as well. In the case of poultry and to a lesser degree pigs, this system is becoming increasingly important around the world, including many developing countries. Because this system is so closely interspersed with commercial crop and livestock farming, Figure 4–1 does not show this region separately.

**FIGURE 4–11** Commercial livestock finishing is a modification of commercial crop and livestock farming. The cattle feedlots in the High Plains of the United States are an example. (Photo courtesy of Brett A. Gardner, Department of Animal Science, Oklahoma State University.)

## COMMERCIAL DAIRY FARMING

Commercial dairy farming is also a modern modification of commercial crop and livestock farming and has much in common with commercial livestock finishing operations including their intensive nature. It is generally found intermingled with commercial crop and livestock farming. Proximity to markets dictates the primary dairy products produced in a specific region. For instance, fluid milk is perishable and expensive to transport so it tends to be produced near human population centers. Dairies far removed from dense populations specialize in processed dairy foods such as butter, cheese, and condensed milk. For example, geographically remote New Zealand produces butter and other processed products that can be kept fresh and shipped to population centers. Wisconsin, a major dairy state in the United States, produces much of the cheese shipped to and consumed in other regions of the United States.

As practiced traditionally in the United States, Europe, Australia, and New Zealand, dairying has depended on maximizing the use of pastures with some crop production to produce winter feed. Confinement *drylot* (Figure 4–12) systems have replaced pasture systems in many commercial dairies. Feed is either purchased or grown, harvested, and brought to the cows who are in confinement drylot conditions.

**Drylot**

A confined area generally equipped with feed troughs, automatic watering devices, shelter, and working facilities where animals are fed and managed.

**FIGURE 4–12** In a confinement dairy system, the feed is brought to the confined animals rather than relying on the animals to graze pasture.

This resembles livestock finishing in many ways, except for the products. Around the world these dairy farms are often situated close to human population centers. In addition to feed, farmers often buy replacement females for their herd or contract with other producers to grow them. Economies of scale dictate that these be large-scale operations with thousands of cows rather than the 200 to 300 cows found on small family dairies. Clearly, these operations are in the process of further specialization.

## COMMERCIAL GRAIN FARMING

Commercial grain production is practiced in areas receiving between 20 and 30 inches of yearly rainfall. Some livestock may be produced. The animals are usually stocker cattle and/or lambs brought in to graze small grain pastures in the fall and winter, or cows/calves used to clean up the crop residues. The major emphasis is on producing corn, small grains, usually wheat, or rice. Australia has very large wheat belts, as do the American plains, the steppes of Russia, and the pampas of Argentina. Commercial rice farms can be found in the United States, Argentina, Australia, Egypt, India, Pakistan, Spain, Vietnam, and other countries.

## MEDITERRANEAN AGRICULTURE

Mediterranean agriculture is very different today than it used to be. In ancient times, it was peasant subsistence agriculture based on wheat and barley for rainy winter season crops, drought-resistant trees and vines (grape, olive, and fig) for the summer (Figure 4–13), along with sheep, goats, and some pigs. Grain was often raised in alleys in the orchards, and animals were not really integrated into the land. Farmers did not raise feed, collect manure, or keep draft animals. The animals were kept in the mountains, and crops were planted in the hills and valleys below. However, irrigation has changed all of this. Today much of the Mediterranean region is referred to as *market gardening* and is a commercial area. Some animals are still grazed in the mountains, but citrus is now a dominant crop. Grapes are an important crop, making

**Market gardening**

Specialized production of fruits, vegetables, or vine crops for sale.

**FIGURE 4–13** Olives were once a mainstay of traditional Mediterranean agriculture. Livestock had marginal use, generally relegated to the mountainous areas. (FAO Photo/19836/R. Faidutti. Used with permission.)

this area famous for its wines. These areas are usually mountainous and often use mountain pastures through transhumance by slowly grazing livestock up the mountains during the summer. Once they reach the highest elevations (generally midsummer), they graze slowly back down the mountain to reach the valley before snow falls.

## MARKET GARDENING

Market gardening is a specialized agriculture of developed zones, which produces and sells products to large urban centers. These farms do not have any livestock but rather produce nontropical fruit, vegetables, vine crops, and flowers in intensive tillage systems. Common products include various flowers, potatoes, tomatoes, lettuce, melons, beets, broccoli, celery, radishes, onions, cabbage, strawberries, table grapes, raisins, wine, citrus fruits, peaches, and plums. These operations tend to be labor intensive and often depend on seasonal farm laborers who move from area to area depending on the work available. Most countries have significant market gardening areas. Often market gardening is practiced in an emerging form of agriculture called *urban agriculture*.

## COMMERCIAL PLANTATION

Plantation agriculture is a form of commercial agriculture that uses large land holdings and labor-intensive practices, most often in developing countries, to produce a single crop for the commercial market. Today, most of this agriculture is found in the tropics and subtropics and produces crops to be marketed in Europe, Japan, and the United States. Profit is the objective of this system of agriculture. Most plantations are owned by governments or large corporations. Crops include various fruits, coffee, tea, spices, sugar, fiber, and vegetable oil products. Plantation agriculture is found in South America, Central America, Indonesia, the Philippines, the Caribbean, India, Sri Lanka, and West Africa (Figure 4–14). Animals have little place in plantations other

**FIGURE 4–14** Sri Lankan women carry sacks of harvested tea on a plantation. Plantation agriculture is commercial agriculture that uses the cheap labor force of developing countries to produce crops for the commercial market. (FAO Photo/17022/G. Bizzarri. Used with permission.)

than as pack animals. Most of the labor has traditionally been provided by humans. In modern times, even the human labor requirement has been reduced substantially through mechanized farming practices. The term *neo-plantation* is used to describe these modern mechanized plantations.

## URBAN AGRICULTURE

The urbanization of the population of the world continues in both developed and developing countries. Many of these new city dwellers have brought farming practices with them to their urban/city homes. With this is a newly developed type of agriculture referred to as *urban agriculture*. Vegetables, fruit, herbs, ornamental plants, meat, eggs, and milk are commonly produced. Backyards are the most common location urban agriculture is practiced, but roadsides, along waterways, rooftops, and public areas are also used. Some urban farmers even practice hydroponics production.

Both intraurban and peri-urban spaces are used. A variety of animals are raised, including poultry, rabbits, goats, sheep, cattle, pigs, guinea pigs, and fish. Urban agriculture is distinguished from rural agriculture in that it is integrated into the urban economic and ecological system. The UNDP United Nations Development Program estimates that worldwide 800 million people rely on urban agriculture for food and income (Figure 4–15). Much is subsistence oriented but income is also generated by many, and full-fledged commercial ventures are also common. Urban agriculture is being promoted in many developing countries for its positive roles in alleviating urban poverty, promoting social inclusion, and providing for urban food security, urban waste management, and urban greening.

**FIGURE 4–15** Urban agriculture has become an important part of the equation in providing food and income for the world's population. (FAO photo/DSC-037/G. Bizzarri. Used with permission.)

## DEVELOPED VERSUS SUBSISTENCE AGRICULTURE

In comparison with the agriculture of the developed countries, subsistence agriculture appears to be very inefficient and unproductive. However, it is a serious error to judge all economic systems in terms of what is considered to be the best in a developed country such as the United States.

Primitive and subsistence systems have proven to be substantial systems for centuries. The animals have been integrated into the systems to achieve maximal returns for their upkeep. There is a very low capital investment in animals and housing for the animals and farm equipment. This is vital to the systems because the majority of the farmers have a low income, so little or no money is available to finance capital improvements of any kind in the farm enterprise. Of even greater importance, when repairs are needed, they can usually be done locally using relatively cheap, readily available materials and with a minimum of delay.

There are several reasons why animals, rather than tractors, are the predominant source of power in subsistence agriculture. Obviously, they cost less, both initially and to replace. Animals are a renewable resource that can be produced locally, usually by the farmer. Machines require fossil fuels that are both expensive and are usually in short supply. The fuel for animals, largely consisting of plants that cannot be consumed by humans, can be produced by the farmer.

The final reason is that labor has no alternate use. Hand labor is a major contribution of agriculture to poor human populations. It provides work for a vast number of people who would otherwise be unemployed. This labor pool will continue to be utilized in agriculture as long as human labor is more economical than machine labor. If the time ever comes when there are nonagricultural demands for the service of these people, they will no longer be a cheaper source of labor than machines and they will be replaced in agriculture by machines. Until such time, however, the utilization of vast amounts of hand labor is a very efficient use of human hands.

## COMPARISON OF THE LIVESTOCK INDUSTRY IN THE DEVELOPED AND THE DEVELOPING COUNTRIES

Table 4–2 compares the livestock industry in the developed and developing countries. The two are similar in that each has a very few units that are very large in size, but differ in that such units in the developed countries are usually much more productive. However, even though they are not managed as efficiently as are such operations in the developed countries, the very large units are usually the only livestock-producing units in the developing countries that utilize any modern technology or have improved breeds of livestock.

Developed and developing countries differ radically in the numbers of livestock units that are moderate to large in size. Such units make up the majority of the units in the developed countries, but they are almost totally absent in the developing countries. In the developed countries, these are the farms and ranches owned and operated by the middle class, which is almost nonexistent in the poor countries of the world.

By far, the most numerous type of livestock enterprise in the developing countries is the small unit. Small units are also fairly numerous in the developed countries but for a different reason. In the developed countries, small units are seldom the major source of income for the owner/operator. In sharp contrast, almost all of the small units

**TABLE 4–2**   **Comparison of the Livestock Industry in the Developed and Developing Countries**

| Characteristics of the Livestock Industry | | |
|---|---|---|
| **SIZE OF UNIT** | **DEVELOPED COUNTRIES** | **DEVELOPING COUNTRIES** |
| Very large | Very few | Very few |
| Moderate to large | The majority | Almost none |
| Small | Fairly numerous | By far the most numerous type of livestock enterprise |
| | Usually not the major source of income for the owner/operator | The major source of income for the owner/operator |
| | 1.  Utilize limited land area | |
| | 2.  Operated almost entirely by family labor | |
| | 3.  Cash returns to livestock enterprise are low as a result of: | |
| | a.  Small size | a.  Small size |
| | b.  Inefficient marketing | b.  Poor markets available |
| | | c.  Low level of animal productivity |
| | |     1.  Genetically poor animals |
| | |     2.  Low nutritional levels |
| | |     3.  Diseases and parasites |
| | Little or no desire to expand | Little opportunity to expand |
| | 1.  Livestock usually more of a hobby than a business | 1.  Total income needed for subsistence; none available for expansion |
| | | 2.  Little or no collateral for loans (if loans are available) |

SOURCE: Adapted from Turman, 1986.

in the developing countries are not only a major source of income, but they are usually the only source of income for their owners.

As shown in Table 4–2, the small units in both groups of countries are similar: (1) they utilize a limited land area, (2) they are operated almost entirely by the labor of the family, and (3) cash returns are low. The most important reason for low cash returns is the small size of the operation. However, returns per animal unit are usually less than those of larger units. One of the main reasons for this low return in the developed countries is that small producers do a poor job of marketing their animals even though good markets are available. In the developing countries, low returns are the result of the combined effect of the unavailability of good markets and a general low level of animal productivity. Usually the only cash market available to the producers in the developing country is from traders who offer a very low price; hence the producer has no choice because the traders are the only buyers.

The small units in both groups of countries are likely to remain small, but for different reasons. In the developed countries, small units are usually a sideline—many are a hobby rather than a business. The operator has a job in town but lives on a few acres in the country. He keeps a few animals because he has the space to do so and enjoys working with animals, but he has no desire to expand. In contrast, most small farmers in the developing countries would like to expand but have little, or no, opportunity to do so.

Two additional points should be made about the livestock industry in the developing counties: (1) livestock are kept for many reasons other than to produce food, and (2) animal performance is usually submaximal from a biological standpoint, so there is room for improvement while staying within the system that has developed.

## SUMMARY AND CONCLUSION

Agriculture as it is practiced in the United States is almost exclusively one of the developed systems. However, understanding the various agricultural systems throughout the world has value from several perspectives. Understanding the other systems gives us better knowledge of our own systems because it gives us a frame of reference. It also helps us to gain perspective on where we have been, and assists us in knowing how to help others achieve higher levels of development, presuming that they wish to. Perhaps most important, understanding how others live and practice agriculture given the constraints they are under offers us an opportunity to appreciate their way of doing things.

## STUDY QUESTIONS

1. What are the major factors that determine the type of livestock systems found in a given location?
2. Based on your knowledge of the world's climate regions, write a short summary of how climate influences each of the agricultural types.
3. What are the characteristics that differeniate primitive, subsistence, and developed agricultures?
4. Consider the use of animals in each of the agricultural types described in this chapter. Which systems could use more species of animals? Why do you think this is so? Which could substitute different species than are now used? Why would your suggested changes be effective?

## REFERENCES

Behnke, R. H., and I. Scoones. 1993. Rethinking range ecology: Implications for rangeland management in Africa. In *Range ecology at disequilibrium: New models of natural variability and pastoral adaptation in African savannas.* London: Overseas Development Institute.

Cole, J. 1999. *Global 2050: A basis for speculation.* Nottingham: Nottingham University Press.

Cooksley, D. G., J. H. Prinsen, C. J. Paton, and J. A. Pini. 1991. Performance of a beef cattle production system on native pasture using *Leucaena leucocephala* as a supplement. *Livestock Production Science* 28: 65–72.

De Boer, J. 1992. Technological and socioeconomic changes: Including urbanization, as they impact upon animal production in Asia. In *Animal production and rural development:* Proceedings of the Sixth AAAP Animal Science Congress, Vol. I, pp. 57–72. Bangkok.

de Haan, C., H. Steinfeld, and H. Blackburn. 1997. *Livestock and the environment.* Commission of the European Community.

El-Sarafy, A. M., A. M. Aboul-Naga, and E. S. E. Galal. 1992. Cereal grain input in the small ruminant production system in the coastal zone of western desert of Egypt. Proceedings of the Joint ANPA-EAAP-ICEMAS Symposium, Rabat, October 1990. No. 49.

FAO. 1982. *Report on the agro-ecological zones project—Methodology and results for Africa.* Rome: FAO.

FAO. 2007. *FAOSTAT statistics database: Agricultural production and production indices data.* http://apps.fao.org/.

FAO/IFAD. 1982. *Report of the Sudan Stock Route Development Project: Preparation Mission.* Investment Centre. Report No. 10/82 DDC/SUD 18.

Fernandez-Baca, S., G. R. De Lucia Silva, and L. C. Jara. 1986. Milk and beef production from tropical pastures: An experience in the humid tropics. *World Animal Review* (FAO). Rome. No. 58: 2–12.

Finzi, A. 2002. *Integrated backyard systems.* Rome: FAO. http:www.fao.org/ag/aga/AGAP/FRG/ibys/default.htm.

Hanks, L. M. 1972. *Rice and man.* Chicago: Aldine.

Hecht, S. 1993. The logics of livestock and deforestation in Amazonia: Directly unproductive, but profitable investments (DUPS) and development. In Proceedings VII *World Conference on Animal Production,* Vol. 1, pp. 41–62. Edmonton, Alberta, Canada.

Humphrey, J. 1980. *The classification of world livestock systems.* A study prepared for the Animal Production and Health Division. FAO. AGA/MISC/80/3.

International Monetary Fund. 1992. *International Financial Statistics Yearbook.*

Jahnke, H. E. 1982. *Livestock production systems and livestock development in tropical Africa.* Kiel, Germany: Kieler Wissenschaftsverlag Vauk.

Jasiorowski, H. A. 1998. *Constraints and opportunities for the future use of animals in human society.* In Proceedings of the 8th World Conference on Animal Production. Seoul, South Korea: Seoul National University.

Johnson, J. E. 1975. Conflict and complementary in Asian animal feeding and food production. Prepared for Rockefeller Foundation Review of the Role of Animals in the Future World Food Supply, April 21–22, New York.

Jordan-Bychkov, T. G., M. Domosh, R. P. Neumann, and P. L. Price. 2008. *The human mosaic: A thematic introduction to cultural geography.* 10th ed. New York: W. H. Freeman.

Jordan-Bychkov, T. G., and M. Domosh. 1999. *The human mosiac: A thematic introduction to cultural geography.* 8th ed. Reading, MA: Addison-Wesley-Longman.

McDowell, R. E. 1991. *A partnership for humans and animals.* Raleigh, NC: Kinnic Publishers, Kinnickinnic Agri-Sultants, Inc.

Otchere, E. O. 1984. Traditional cattle production in the sub-humid zone of Nigeria. ILCA/NAPRI *Symposium on Livestock Production in the Sub-humid Zone.* Kaduna, Nigeria.

Owen, E., T. Smith, M. A. Steele, S. Anderson, A. J. Duncan, M. Herrero, J. D. Leaver, C. K. Reynolds, J. I. Richards, and J. C. Ku-Vera. 2004. *Responding to the livestock revolution.* Nottingham: Nottingham University Press.

Pradham, S. L. 1987. Integrated crop and small ruminant systems in Nepal. In *Small Ruminant Production Systems in South and Southeast Asia,* pp. 144–147. IDRC Proceedings of a workshop held in Bogor, October 1986.

RUAF Foundation. 2007. What is urban agriculture? Accessed online July 24, 2007. http://www.ruaf.org/.

Ruthenberg, H. 1980. *Farming systems in the tropics.* Oxford: Clarendon Press.

Sere, C., and H. Steinfeld. 1996. *World livestock production systems.* Rome: FAO.

Turman, E. J. 1986. *Agricultural animals of the world.* Stillwater: Oklahoma State University.

United Nations. 1991. *World Urbanization Prospects 1990.* Department of International Economic and Social Affairs. ST/ESA/SER.A/121. New York.

Whittlesey, D. 1936. Major agricultural regions of the earth. *Annals of the Association of American Geographers* 26: 199.

Wilson, T. 1994. *Integrating livestock and crops for the sustainable use and development of tropical agricultural systems.* AGSP-FAO.

Winrock International. 1992. *Animal agriculture in developing countries: Technology dimensions.* August 1992. Morrilton, Arkansas, USA. 45 pp.

Wint, W., and D. Bourn. 1994. *Livestock distribution and the environment in sub-Saharan Africa.* ERGO, Oxford.

Yunlong, C., and B. Smith. 1994. Sustainability in agriculture: A general review. *Agriculture, Ecosystems & Environment* 49: 299–307.

# PART TWO

# The Biological Sciences of Animal Science

# Introduction to Nutrition

## LEARNING OBJECTIVES

After you have studied this chapter, you should be able to:

1. Define nutrition and understand the reasons for studying nutrition.

2. Explain what a nutrient is and know the difference between dietary essential and nonessential nutrients, classify the nutrients, and list the 50 dietary essential nutrients.

3. Describe the general uses of nutrients in the body and discuss the major factors that affect an animal's needs for nutrients.

4. Explain in detail the three major types of animal trials that nutritionists use.

5. Define and explain feedstuff analysis.

6. List sources of current information about nutrition.

7. Summarize the feeds evaluation procedures described in this chapter.

## KEY TERMS

Applied or production
  nutritionist
Ash
Balance trial
Basic nutritionist
Bomb calorimeter
Carbohydrates
Cellulose
Crude fiber
Crude protein
Diet
Digestibility
Digestion trial
Dry matter
Energy

Estrous cycle
Ether extract
Fats
Feed
Feed analysis
Feeding trial
Feedstuff
Finishing
Growth
Maintenance
Metabolism trial
Minerals
Monogastric
Nitrogen-free extract
  (NFE)

Nutrient
Nutrient requirement
Nutrition
Palatability
Peer review
Production
Protein
Ration
Reproduction
Ruminant
Van Soest fiber
  analysis
Vitamin
Work

# INTRODUCTION

**Nutrition**

The study of the body's need and mechanism of acquiring, digesting, transporting, and metabolizing nutrients.

*Nutrition* is the study of how the body uses the nutrients in feed to sustain life and for productive purposes. Nutrition is a very complicated science. To study nutrition adequately, we must study the nutrients themselves and also look at how animals consume, digest, absorb, transport, metabolize, and excrete them. We must further consider how the nutrients are used in the body for productive purposes, and then consider the economics of feeding and the potential effects to the human population of various feeds, additives, medicinals, and so on. Good nutrition is basic to good health and production.

In commercial livestock production, it is important to study nutrition to be able to feed livestock cost effectively. Depending on the type of enterprise, the cost of feed and feeding is 45–75% of the total cost of livestock production. You cannot make money in animal production agriculture without properly feeding the livestock. Of the total costs of various enterprises, the portions attributed to feed include ranges of 65–80% for swine, 55% for layers, 65% for broilers and turkeys, 50–60% for dairy cattle, 70% for feedlot finishing of beef cattle, and 50% for feeding lambs.

The study of nutrition is important for several other reasons. Nutrition affects general health and well-being, physical abilities, and susceptibility to and ability to recover from disease. Nutrition is also an intricate part of many body systems. If we understand nutrition, we can understand those systems. Studying nutrition provides an excellent opportunity to understand the basic "bio-logic" of life—you can't understand nutrition without understanding the basis of life, and you can't understand the basis of life without understanding nutrition.

**Monogastric**

Having only one stomach.

**Ruminant**

Hooved animals that have a rumen and chew their cud.

**Basic nutritionist**

A nutritionist interested in elucidating basic metabolism and nutrient action and interaction.

**Applied or production nutritionist**

The practical nutritionist. An applied nutritionist works on practical questions such as cost effectiveness, method of delivery, and carcass effects.

The field of nutrition is so broad and encompasses so much information that it has been necessary for nutritionists to specialize. This specialization allows the individual nutritionist to keep up with the rapidly changing industries and the rapidly advancing science. The areas of specialty for animal nutritionists are divided into two categories based on whether the animals studied are monogastric or ruminant. *Monogastric* nutritionists focus on one-stomached animals such as poultry, swine, horses, dogs, cats, fish, rats, mice, guinea pigs, monkeys, and some zoo animals. *Ruminant* nutritionists specialize in sheep, goats, dairy/beef cattle, and other wild or captive species that have a rumen. Most nutritionists tend to fall into one or the other of two additional camps: *basic nutritionists* or *applied/production nutritionists.*

Basic nutritionists study the metabolism of the animals and are most interested in the biochemical mechanisms of nutrient metabolism. They are sometimes interested in knowledge for its own sake rather than practical applications. What basic nutritionists learn is important to production nutritionists who take the basic nutritionists' findings and determine real-world applications. Applied or production nutritionists are more interested in maximal, cost-effective feeding of animals in a production setting. They explore the many possibilities for using feedstuffs in the best possible way to make money. Often thought of as "feed-um and weight-um" research, applied nutrition is really a complicated and often frustrating attempt to translate basic science into real-world application. We often find nutritionists who bridge several categories or even defy being categorized. Even a "specialist" must know something about several disciplines to be a good nutritionist. These disciplines include biochemistry, cytology, economics, marketing, endocrinology, genetics, inorganic chemistry, mathematics, microbiology, neurology, organic chemistry, physics, physiology, veterinary medicine, and waste management.

# THE NUTRIENTS AND THEIR USES

If we are to understand nutrition, we must understand the **nutrients.** "A nutrient is any chemical element or compound in the diet that supports normal reproduction, growth, lactation or maintenance of life processes" (Pond et al., 2005, p. 7). Nutrients are classified as either dietary essentials or dietary nonessentials. Dietary essential nutrients must be a part of the diet. The nutrient classifications are water, carbohydrates, vitamins, minerals, proteins, and fats (lipids). Energy is also needed by the body but is not a nutrient per se. Carbohydrates, proteins, and fats all provide energy to the animal through their breakdown products. Figure 5–1 lists the 50 dietary essential nutrients.

*Water* is frequently neglected in many discussions of nutrition. This seems odd, considering that the body will die quicker from water deprivation than from the deprivation of any other nutrient. Water is used by the body as a lubricant; as a regulator of body temperature; as a solvent for the body's solid components; as a transporting medium in body fluids such as blood, lymph, urine, and sweat; and as a necessary participant in the chemical reactions of the body. Animals meet their needs for water by drinking and eating feed that has water in it, and with metabolic water produced by the chemical reactions that break down protein, carbohydrates, and fats in the body. A direct, positive correlation exists between how much water an animal needs and how much of the other nutrients it consumes.

*Carbohydrates* are sugars, starches, and cellulose. The primary use of carbohydrates is to provide animals with energy. Energy is needed to do the body's work (chemical reactions). Grains are fed to livestock because their starch and sugar contents are easily digestible carbohydrates that animals readily use for energy. *Cellulose* is the major carbohydrate found in forages such as fresh pastures, silage, and hays. It is an important energy source for herbivores such as cattle, sheep, llamas, and horses. Carbohydrates account for the largest single percentage of nutrient content in most commonly fed feedstuffs.

*Proteins* are compounds made up of long chains of amino acids. Their primary uses in the body are as components of lean tissue, enzymes, hormones, and body metabolites. If excess proteins are present in the diet, they are used for energy. Animals typically need very different amounts of protein depending on the feeds used and the age and use of the particular animal. Young animals need protein to build their bodies (grow); mature animals must only replace exhausted proteins in the body. Pregnancy, milk production, and some other production functions require a substantial amount of protein. Oil seed meals such as soybean meal and cottonseed meal are common protein feedstuffs, as well as fishmeal, alfalfa meal, and dried skim milk. However, virtually all feeds contain some protein.

*Fats* are esters of fatty acids and glycerol. They are used by the body as a source of energy and as a source of essential fatty acids. Fats are high in energy. On average, they have 2.25 times more energy than carbohydrates on an equal weight basis. They are also important carriers of the fat-soluble vitamins. Common feedstuffs used for their fat content include tallow, lard, horse fat, and various vegetable oils.

*Vitamins* are organic compounds needed by the body in very small amounts. They are classified as either fat soluble or water soluble. The fat-soluble vitamins tend to be involved in regulating body functions such as vision, blood clotting, and tissue maintenance, and growth such as bone development. The water-soluble vitamins tend to be used more for body metabolic regulation. Many feeds are rich in vitamins, but some are very poor vitamin sources. We are able to isolate, synthesize, or otherwise gather

## Nutrient
A substance in the diet that supports the normal functions of the body.

## Carbohydrates
Chemically defined as polyhydroxy aldehydes or ketones, or substances that can be hydrolyzed to them.

## Cellulose
A carbohydrate composed of thousands of glucose molecules that forms the support structure of plants.

## Protein
Compounds composed of combinations of $\alpha$-amino acids.

## Fats
One of a class of biomolecules called lipids. Chemically, fats are triacylglycerides, which are composed of the alcohol glycerol, with three fatty acids attached.

## Vitamin
A term used to group a dissimilar set of organic substances required in very small quantities by the body.

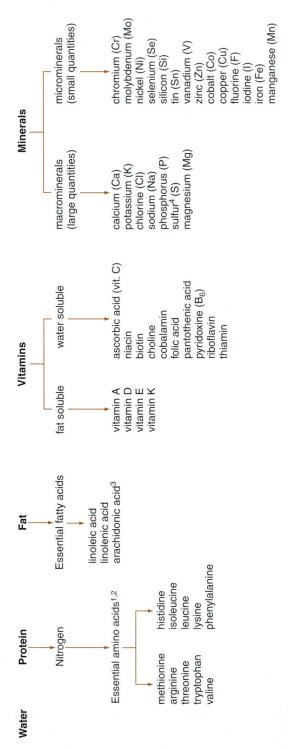

**FIGURE 5–1**   The essential nutrients.

[1]Mnemonic device for remembering essential amino acids = MATT HILL VP.

[2]For poultry, two additional amino acids are needed: glycine and proline.

[3]Arachidonic acid can be synthesized from linoleic acid if it is available so it is only essential if linoleic acid is absent or in short supply.

[4]Authors vary on whether or not to list sulfur as a macromineral or micromineral. The discrepancy arises because only a very small amount of inorganic S is needed, but the sulfur-containing amino acids (organic S) are needed in larger quantities.

vitamins into commercially prepared vitamin supplements that are heavily used in modern feed formulations and mixes.

*Minerals* are the inorganic constituents of bones and teeth and an important part of the body's enzyme systems. Virtually all feedstuffs contain some minerals. Supplemental minerals are provided in various forms including salt, trace mineralized salt, oyster shells, lime, bone meal, and a wide variety of other forms.

The nutritive value of feeds is quite variable. Feeds must be selected to provide an appropriate *diet* for the purpose intended. For example, some feeds are high in energy-furnishing nutrients and are used in finishing rations or for producing milk and eggs. When protein-rich feeds and vitamin and mineral mixes are added to the energy feeds, a *ration* of optimum production is born.

## BODY FUNCTIONS AND NUTRIENT NEEDS

Animals need nutrients to run the body's metabolic machinery for a wide variety of purposes including maintenance, production, and reproduction. How much of these essential nutrients does an animal need? The amount required depends on many different things. Definite species' differences are linked to the species' type of digestive tract as well as peculiarities of the species' metabolism. Other factors such as the level of production and the specific product, or combination of products, being produced (work, fetus, growth, milk, eggs, and fiber) are important. Nutrient needs are additive and must be fed in sufficient amounts to simultaneously meet all of the animal's current needs (Figure 5–2). Nutrients often have influence on each other owing to interactions that may increase or decrease the needs of the individual animal. Thus the job of the nutritionist is a complicated one. An understanding of the needs of the body is essential if nutritionists are to do their job. Individual discussions of the body's needs follow.

*Maintenance* in its simplest form refers to maintaining the body at a constant weight and temperature. The maintenance nutrient requirement is a combination of the nutrients needed for basal metabolism (activities such as heartbeat and breathing) and those for normal animal movement. Normal activity levels often vary

**Minerals**

In nutrition, the specific set of inorganic elements thus far established as necessary for life in one or more animal species.

**Diet**

All the feeds being consumed by an animal including water.

**Ration**

The specific feed allotment given an animal in a 24-hour period.

**Maintenance**

The needs of the animal exclusive of those required for a productive function such as growth, work, or milk production.

**FIGURE 5–2** Body functions dictate nutrient needs. Foals need additional nutrients specifically for growth, and mares need additional nutrients for lactation. (Photographer Alan D. Carey. Courtesy of Getty Images, Inc./PhotoDisc, Inc.)

**FIGURE 5–3** Finishing cattle for slaughter involves feeding for maintenance, growth, and fat deposition. (Photo courtesy of U.S. Department of Agriculture.)

tremendously. Imagine the difference in essential activity between a relatively sedentary riding horse standing knee-deep in good grass with a bucket of oats given to him twice a day, compared to a free-ranging mustang that has to search for every bite and avoid all kinds of hazards as well. It is rarely our goal to simply maintain animals in a production setting. However, the needs of maintenance must be met before the animal will produce any useful product. Thus maintenance is the first requirement that must be met in a production setting. Because energy is a big portion of maintenance, feeds relatively high in fats and carbohydrates are generally used for providing maintenance requirements.

*Growth* is the process of increasing the body weight by adding tissue like that already present. In growth, the body requires good-quality nutrients in relatively large amounts to build its structural units. All of the required nutrients are needed and balance is very important. Every system in the body is changing and developing, so the nutrient needs are high in the growth phase and nutrient deficiencies become readily apparent.

*Finishing* is the final growth and fattening phase of the production of meat animals. All nutrients are needed for the finishing phase, but energy is needed during this phase in much higher proportions. Energy feeds are required during finishing to provide the energy needed for the formation of lean tissue, bone, and fat. Less protein as a percentage of the total diet is needed during this phase than is needed for growth or production. We rely on feeds high in carbohydrates and fats during the finishing phase because their cost is usually lower than the cost of protein-rich foods (Figure 5–3).

*Production* refers to the output of products such as eggs, milk, and wool. The nutrients needed for production vary with the product. Lactation is the most demanding of all the production functions in mature animals. Milk contains water, protein, fat, vitamins, and minerals, and additional energy is required to produce it. Eggs contain protein, fats, minerals, vitamins, and water. Diets must be selected and rations must be balanced for these nutrients to accomplish cost-effective feeding. Other types of production require levels of nutrients specific to the product or combination of products being produced (Figure 5–4).

## Growth

The process of adding tissues similar to those already present in the body to increase the size of an organism toward the goal of maturity when growth stops.

## Finishing

Usually refers to the final feeding stage when animals are readied for market.

## Production

The general term used to describe the output of usable products and services by animals.

**FIGURE 5–4** Production of eggs, milk, and wool requires levels of nutrients specific to the product being produced. (Photo courtesy of Marty Traynor.)

**FIGURE 5–5** Work is a specialized production function that requires additional energy in the animal's diet.

*Work* is a specialized production function and the major product of some species such as working horses, working dogs, and packing llamas. The major nutrient increase needed for work is energy (Figure 5–5). Carbohydrates are generally the most cost-effective means of increasing energy in the diet, so the carbohydrate-rich feeds are generally used to meet this need.

*Reproduction* of live normal offspring is a basic biological necessity for a species to survive. For efficient, cost-efficient agriculture, we must provide for the arrival of a new generation at the same time we are receiving the products from the current one.

**Work**

Physical exertion as a production function.

**Reproduction**

The combined set of actions and biological functions of a living being directed at producing offspring.

**FIGURE 5–6**　　This cow must provide milk for her calf, rebreed, and gestate her next calf all at the same time.

**Estrous cycle**

The time from one period of sexual receptivity in the female (estrus or heat) to the next.

Providing adequate nutrition for the reproductive function must be accomplished in addition to maintenance and production. Poor nutrition can be manifested in abnormal or delayed *estrous cycles;* reduced calf and lamb crops; small and weak litters in swine, dogs, and other litter-bearing species; and poor egg production and/or hatchability in poultry. The high-producing animals are often the ones with whom we have the most difficulty. The third trimester of a pregnancy is the most critical for the reproducing female. Energy and protein must be increased, and the need for several minerals and vitamins is increased during this period as well (Figure 5–6).

## FEED ANALYSIS

**Digestibility**

A measure of the degree to which a feedstuff can be chemically simplified and absorbed by the digestive system of the body.

**Palatability**

The acceptability of a feedstuff or ration for consumption.

**Feed**

Foods used to feed animals.

Properly feeding livestock requires knowledge of the nutrients found in the feedstuffs available and balancing of these nutrients to meet the physiological needs for the species in question. A comprehensive evaluation procedure discovers nutrient composition, *digestibility,* productive value, *palatability,* and the physical or handling characteristics of *feeds.* Feeds analysis also provides useful information about feed harvesting and storing methods, thus helping to obtain the highest quality feed for the animals and the most profit to the producer.

Three basic types of analytical methods are commonly used to analyze feeds for nutrient content. *Chemical procedures* are standard chemistry applied to feeds. *Biological procedures* use animals to test the feeds. This is more time consuming, labor intensive, and expensive but gives a better estimate of how the animal will use the feed. We also use some species of animals as models for other species to save money (i.e., rats for pigs). *Microbiological procedures,* the third type, are similar to biological procedures but use bacteria in place of higher animals.

*Proximate analysis* is a set of chemical/analytical procedures designed to partition feedstuffs into water, ash, crude protein, ether extract, crude fiber, and nitrogen-free extract. Although proximate analysis is the most common set of chemical tests used on

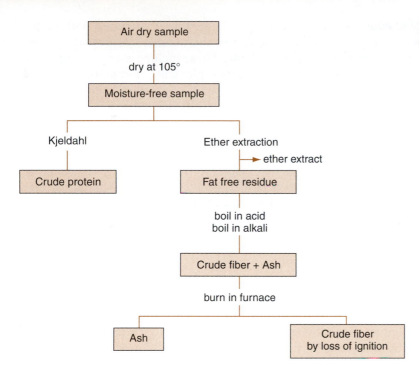

**FIGURE 5–7** Flow diagram for the proximate analysis. (*SOURCE:* Pond et al., 1995, p. 21. Reprinted by permission of John Wiley & Sons, Inc.)

feedstuffs, the information it provides is sometimes misleading or even inaccurate. However, proximate analysis is still used and will continue to be used. Thus good nutritionists must learn its limitations as well as its value. Figure 5–7 illustrates how a feed is analyzed for its proximate constituents.

*Dry matter* is determined by heating a feed sample until all water has evaporated. The percentage of the sample that is not water is then referred to as the dry matter of the sample. To balance a ration properly and know how much of each ingredient to mix into a ration, the dry matter must be determined.

*Ash* is considered the mineral content of the feedstuff. It is determined by burning a dry sample in a very hot oven (500–600°C) to burn off all of the organic matter. The actual value in knowing ash content is marginal. However, we must know how much ash is in the feedstuff to make other calculations. In some by-product feeds, the ash content can be quite high. More detailed mineral analyses are needed to help the nutritionist balance a ration for the various minerals.

*Crude protein* (CP) is determined by the Kjeldahl process, which isolates and measures the nitrogen in a feed. An average protein contains 16% nitrogen. Crude protein can then be calculated by multiplying total nitrogen by 6.25. As the name implies, this is a crude measure of nutritional value. Any nitrogen found in the feed is called crude protein. The crudeness of the measurement is not much of an issue for the functional ruminant, which can use nitrogen that is not in protein form (NPN) to make real protein in the rumen. For monogastric and immature ruminants, crude protein has less value because amino acid content is the information needed. Modern analytical techniques may soon make this laboratory procedure obsolete. (NPN means nonprotein nitrogen. Any nitrogen found in a feedstuff that is not part of a protein molecule falls in this category.)

**Dry matter**

Everything in a feed other than water.

**Ash**

The mineral content of a feed.

**Crude protein**

An estimate of protein content obtained by multiplying the nitrogen content of a substance by a factor, usually 6.25.

**Ether extract**

In proximate analysis, the portion of a sample removed by extraction with a fat solvent such as ethyl ether.

**Crude fiber**

In proximate analysis, the insoluble carbohydrates remaining in a feed after boiling in acid and alkali.

**Nitrogen-free extract (NFE)**

In proximate analysis, a measure of readily available carbohydrates calculated by subtracting all measured proximate components from 100.

**Bomb calorimeter**

A device into which a substance can be placed and ignited under a pressurized atmosphere of oxygen.

To determine ***ether extract*** (EE) (fat), a dry sample is extracted with diethyl ether. The ether extracts the fat from the rest of the sample. Unfortunately, some plant materials other than fat are ether soluble. These include many organic compounds such as chlorophyll, volatile oils, resins, pigments, and plant waxes that have little nutritional value. The goal of the test is to isolate the portion of the feed that has high-calorie density. The value of the results depends on the feed being analyzed.

*Carbohydrates (CHO)* are not determined by direct analysis. For the sake of analysis, carbohydrates are considered to be measured in two fractions: (1) ***crude fiber*** (CF) and (2) ***nitrogen-free extract (NFE)*** or nonfiber carbohydrates. The crude fiber procedure is an attempt to simulate digestion in the true stomach and the small intestine. Crude fiber is made up of cellulose, hemicellulose, and insoluble lignin. Nitrogen-free extract is determined by subtraction rather than by a direct chemical method. There is no "extract" per se, which makes this method confusing and sometimes hard to understand. The water, ash, crude protein, fiber, and fat found in the feed are added together and subtracted from the number 100. In theory, NFE should be the readily available carbohydrates, such as the sugars and starches. In reality, because it is derived from subtraction, it contains the errors of all the analyses and is generally considered to contain some hemicellulose and lignin. The error is especially true of forages that contain more of these materials. The most problematic limitations to the proximate analysis system are with crude fiber and NFE fractions. Sometimes NFE is less digestible than CF, especially for forages. This discrepancy would not happen if the individual tests were actually measuring what they should be. Because carbohydrates are such a large portion of animal diets, and the different species have very different capabilities of using the different carbohydrates, this shortcoming in the proximate analysis system is a major problem.

The Van Soest method is an alternative fiber analysis (Figure 5–8) that was developed as a means of better describing forages in response to the limitations of proximate analysis. The cell content is almost completely digested by the ruminant, but the cell wall is highly variable in its digestibility by the animal. Much of this variation can be explained by the species of plant and the stage of maturity. The Van Soest system has several uses. It can be used to predict the intake and digestibility of feedstuffs by animals and is a means of evaluating heat damage in forages.

*Vitamins* must be assayed individually. Biological assays are used for some; others are determined by chemical analysis.

The *determination of the energy content* of common feedstuffs is a very important part of nutrition. An instrument called a ***bomb calorimeter*** is used to determine the gross energy content of feedstuffs. Gross energy is expressed in calories, with a calorie defined as the amount of heat required to raise the temperature of 1 g water from 14.5° to 15.5°C. Gross energy itself has little value, but it is useful in calculating other values. For example, if we know how much gross energy is in both feed and feces, we can calculate how much was digested and absorbed.

Other methods of feed analysis are also used, and more are sure to be developed, to help the nutritionist. For example, infrared light rays are being used for feed analyses. Moisture, lipid, protein, and fiber contents are frequently determined this way, although calcium, phosphorus, salt, and a few other ingredients are also determined. This type of analysis has the major advantage of being much quicker than other methods. However, calibrating the machine requires a wide variety of samples of a wide variety of feeds. Cost is also a problem because instruments are expensive.

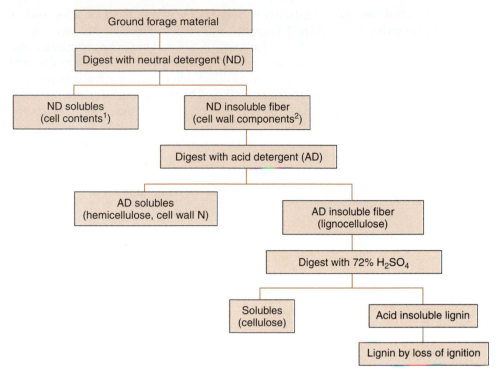

**FIGURE 5-8** *Flow diagram for Van Soest fiber analysis.*
[1]Cell contents include sugar, starch, soluble carbohydrates, pectin, protein, NPN, lipids, and miscellaneous vitamins.
[2]Cell wall components include cellulose, hemicellulose, lignin, silica, and heat-damaged protein.
(*SOURCE:* Jurgens, M. H. 1993, p. 58. Used with permission.)

## DETERMINING THE VALUE OF FEEDSTUFFS TO ANIMALS

Although chemical analysis of feed is important, it is only one step in determining the value of a *feedstuff* to animals. We must combine the chemical analysis with information about how feed affects the animal. To do this, we conduct trials in which the animal is fed the feed in question. Three major types of animal trials are used: *feeding trials, digestion trials,* and *metabolism trials.*

Information gathered from these types of trials allows us ultimately to develop feeding standards and recommendations for feeding animals at different levels of production. Many factors affect nutrient utilization of feeds by animals, including species of animal and type of digestive tract, age, physiological state, (pregnant, growing, lactating, and so on), physical form of the feed (pelleted, cracked, steamrolled, and so on), nutrient balance, and many more. In general, the types of trials used to determine the value of feeds are the same from species to species, with only minor changes in procedures needed.

A feeding trial is used to determine if animals will eat a feedstuff and how they will perform on the feedstuff. It doesn't tell us why things do or don't happen. It only tells us what does happen. Common types of feeding trials include growth trials, lactation trials, egg production, endurance (for horses), or even measurement of some specified body function (sperm count, bone levels of calcium, and so on).

**Feedstuff**

Any substance used as animal feed.

**Feeding trial**

A comparatively simple experimental tool in which animals are fed to determine their performance on specific feeds or substances added to feeds.

**Digestion trial**

An experimental tool used to determine the digestibility of a specific feedstuff, nutrient, or ration.

**Metabolism trial**

An advanced form of digestion trial that measures the body's use of nutrients.

Digestion trials are used specifically to discover the degree to which a feedstuff is digested and absorbed by the animal. Digestion trials can be conducted on any nutrient in a feed. In a digestion trial, we feed a feedstuff or diet that has been chemically analyzed to determine its nutrient makeup. During the feeding period we collect the fecal matter. The fecal material is then analyzed to determine chemically the nutrients that remain in it. We then calculate a digestion coefficient by taking the difference in the nutrients fed and then excreted and calculate a percent disappearance. Digestion coefficients can be calculated for all nutrients. The terms *digestion coefficients* and *percent digestibility* are identical in meaning. The following formula is used to calculate digestion coefficients:

$$\text{Nutrient digestibility (\%)} = \frac{\text{Nutrient intake} - \text{Nutrient in feces}}{\text{Nutrient intake}} \times 100$$

The basic procedure for calculating a digestion coefficient is given in the following example:

During a digestion trial, a steer ate 88 lbs of hay. The hay was 11.8% crude protein. The digestion trial was 7 days long. During the trial, the steer excreted 203 lbs of feces, which were analyzed and found to contain 1.3% protein.

**1.** Calculate a digestion coefficient for the protein in the hay.

$$\begin{aligned}
\text{Protein digestibility} &= \frac{\text{Total protein consumed} - \text{Total protein excreted}}{\text{Total protein consumed}} \times 100 \\[2mm]
&= \frac{(88 \text{ lbs} \times 11.8\%) - (203 \text{ lbs} \times 1.3\%)}{(88 \text{ lbs} \times 11.8\%)} \times 100 \\[2mm]
&= \frac{10.38 - 2.64}{10.38} \times 100 \\[2mm]
&= \frac{7.74}{10.38} \times 100 \\[2mm]
&= 74.6\%
\end{aligned}$$

**2.** Calculate the digestible protein content (DP) of the hay:

$$\text{DP} = \text{Crude protein content} \times \% \text{ Digestibility}$$

In this example, the crude protein content of the forage is 11.8% and the digestibility of the protein is 74.6%. Therefore, $11.8 \times 74.6\% = 8.8\%$ DP in the hay.

TDN (total digestible nutrients) is a measure that historically was used extensively to balance rations for livestock. It is considered to be the digestible energy content of a feed expressed in terms of a carbohydrate equivalent basis. TDN is calculated using digestion coefficients for the crude protein, ether extract, crude fiber, and NFE portions of a feed sample. Only the energy-containing components (not water or ash) are included. The basic procedure for calculating TDN is indicated by the following example:

| Fraction | Amount in Feed (%) | × | Percent Digestibility | = | Contribution to TDN (%) |
|----------|-------------------|---|----------------------|---|------------------------|
| Crude protein | 10 | × | 75 | = | 7.5 |
| Ether extract | 5 | × | 80 × 2.25 | = | 9.0 |
| Crude fiber | 5 | × | 60 | = | 3.0 |
| NFE | 70 | × | 90 | = | 63.0 |
| | | | % TDN in feed | = | 82.5 |

*Note:* One of the most confusing problems for some students is how to handle the decimal point. When you multiply a percent × a percent and you want your answer in a percent (as in the TDN example), move the decimal two places to the left on only one number (either one), and your decimal will then be where you want it to be in your answer.

## Expressing Feed Composition on a Dry-Matter (DM) Basis

The nutrient contents of a feed are commonly expressed on either an air-dry (as-is) basis or a dry-matter basis. To balance and feed rations, it is important to be able to convert them back and forth. The following example demonstrates how to make the conversion using a TDN as an example:

$$\frac{\text{TDN value on air-dry basis}}{\text{Dry-matter content of feed}} \times 100 = \text{TDN on dry-matter basis}$$

For hay, TDN was calculated to be 50.4% on an air-dry, "as-is," or "as-fed" basis. The following steps convert it to a dry-matter basis:

$$\frac{50.4}{0.91 \text{ (moisture content was 9\%)}} \times 100 = 55.4\% \text{ TDN}$$

The step can be repeated for the other nutrients, as demonstrated for crude protein (CP):

$$\frac{13.0 \text{ (protein content on air-dry basis)}}{0.91 \text{ (Dry-matter content of feed)}} \times 100 = 14.3\% \text{ CP on dry-matter basis}$$

## Measures of Energy

Energy is needed by different animals in very different amounts. Choosing the correct unit can help keep the math simple and the numbers easier to read. The units of energy measurement most frequently used are:

Calorie (cal) = Amount of heat or energy needed to raise 1 g of $H_2O$ 1°C from 14.5° to 15.5°C

Kilocalorie (kcal) = Energy required to raise 1,000 g of water from 14.5° to 15.5°C
= 1,000 calories

Megacalorie (Mcal) = 1,000 kcal or 1,000,000 calories.

In human nutrition, the term *kilocalorie* is used to measure and describe energy, although we often mistakenly use the term *calories*. For pet species, kcal are commonly used and referred to correctly. In livestock nutrition, Mcal are most frequently used because of the large amount of energy these species need. The figures 80 Mcal and 80,000 kcal represent the same amount of energy. The smaller number is more convenient to use when performing the math associated with ration formulation.

## Energy Content of Nutrients

Carbohydrates, proteins, and fats all provide energy to the animal. However, their energy contents (energy densities) are very different. Carbohydrates have approximately 4.0 kcal/g, whereas proteins have approximately 5.65 kcal/g on the average. However, about 1.65 kcal/g are lost as urea in the urine. The actual average gross energy value of most carbohydrates is about 4.4 kcal, but some energy is not digested, so we normally adjust the value to 4.0 kcal/g to compensate. The gross energy value for protein (5.65 kcal/g) is higher than for carbohydrates, but approximately 1.65 kcal/g is lost as urea in the urine when protein is used as an energy source, so the net value is about the same as for carbohydrates: 4.0 kcal/g. Therefore, carbohydrates and protein are given the same value in calculating TDN. Fats have approximately 9.0 kcal/g. Fat is given a value that is 2.25 times greater (9.00/4.00 = 2.25) than the values of the carbohydrate and protein because it does have a higher energy content per unit of measurement. This is the reason for the 2.25 correction factor for fat in a TDN calculation, and this is why TDN is considered an estimate of the digestible energy (DE) of a feed on a carbohydrate equivalent basis (Figure 5–9). The amount of energy lost from feed in the various compartments represented in Figure 5–9 depends on several factors, including the kind of feed, its digestibility, and the animal. The numbers in parentheses in Figure 5–9 represent the range of the losses. Losses in the feces are generally the largest and most variable. Highly digestible feeds have minimal losses in the feces; poorly digested feeds have much higher losses. These losses occur during digestion and metabolism and include heat of fermentation of the microbial population in ruminants and horses. Heat losses are usually the second largest and most variable losses.

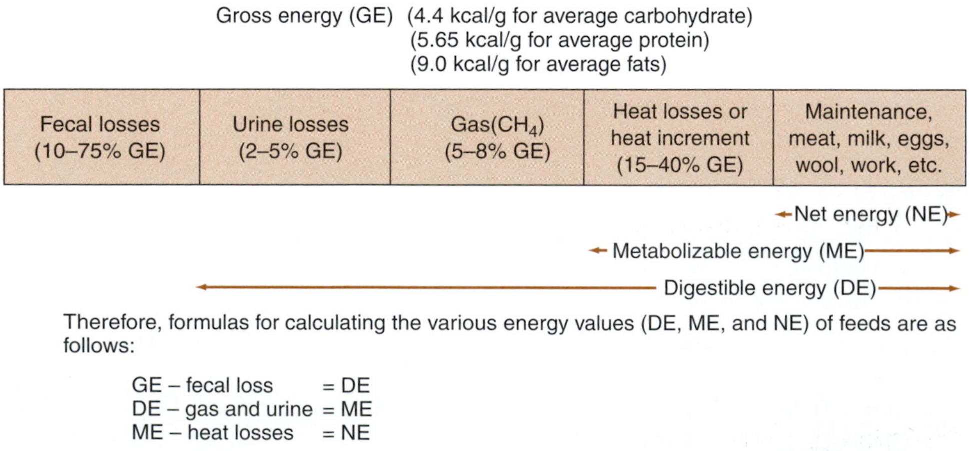

Gross energy (GE)  (4.4 kcal/g for average carbohydrate)
(5.65 kcal/g for average protein)
(9.0 kcal/g for average fats)

| Fecal losses (10–75% GE) | Urine losses (2–5% GE) | Gas($CH_4$) (5–8% GE) | Heat losses or heat increment (15–40% GE) | Maintenance, meat, milk, eggs, wool, work, etc. |
|---|---|---|---|---|

←Net energy (NE)→

← Metabolizable energy (ME)———————→

←———————— Digestible energy (DE)————→

Therefore, formulas for calculating the various energy values (DE, ME, and NE) of feeds are as follows:

GE – fecal loss       = DE
DE – gas and urine  = ME
ME – heat losses    = NE

**FIGURE 5–9**    Schematic diagram for partitioning energy values of feeds. (*SOURCE:* Adapted from Wagner, 1977. Used with permission.)

## Efficiency of Energy Use

The purpose of all the energy conversions in the body of animals is to convert energy into something useful, such as maintenance of the animals, and products such as meat, milk, work, wool, and eggs. Heat losses are also useful in the overall scheme because they help the animal maintain body temperature. The better the feed, the greater efficiency of net energy production from feeds. Metabolizable energy is used with different efficiency for various productive functions, some of which are more important in some species than in others. For this reason, we use different units of energy in ration formulation for different animal species. Figure 5–10 summarizes the varying efficiencies of energy utilization for the various productive functions of the body. This all relates back to TDN. The following formula is used to convert TDN to digestible energy and metabolizable energy:

$$1 \text{ lb TDN} = 2.0 \text{ Mcal DE or } 1.7 \text{ Mcal of ME}$$

*Metabolism trials* are more advanced digestion trials. If we include measures of such things as urine and hair loss, we have converted a digestion trial into a metabolism trial. One particular type of metabolism trial is a **balance trial.** The goal of a balance trial is to measure total intake and excretion so that retention in the animal's body can be calculated. Net retention of a nutrient within the body is positive balance and net loss is negative balance. Thus a metabolism trial also provides information on nutrient use after absorption by the animal. Balance trials are usually conducted to study protein, energy, and minerals. Collections include feces, urine, expired air, milk, eggs, sloughed skin, shed hair, dropped feathers, sweat, and measurement of heat loss.

**Balance trial**

A type of metabolism trial designed to determine the retention of a specific nutrient in the body.

## NUTRIENT REQUIREMENTS

The outcome of all the chemical tests and all the animal trials is that a set of nutrient requirements for the various species of animals commonly used in agriculture, research, and as companions has been developed. The National Research Council, a branch of the National Academy of Science, publishes these nutrient requirements and recommendations for feeding animals in various stages of production. The publications are updated and republished periodically to make the best information available. Included in these publications is a thorough treatment of the general nutrition and nutrient requirements of the various species as well as information on the feeding value of the commonly used feeds for the species. Examples of information found in these publications are given in Tables 5–1, 5–2, 5–3, and 5–4. These publications are the ultimate source of credible information about animal nutrition. The publications include *Nutrient Requirements of*

| | Productive function | Efficiency |
|---|---|---|
| ME ___ | Maintenance ⟶ | ~ 70% |
| ___ | Milk ⟶ | 70% |
| ___ | Growth ⟶ | 65–70% |
| ___ | Fattening (beef cattle) ⟶ | 35–58% |
| ___ | Fattening (simultaneously with milk production) ⟶ | 70% |
| ___ | Work ⟶ | 70% |

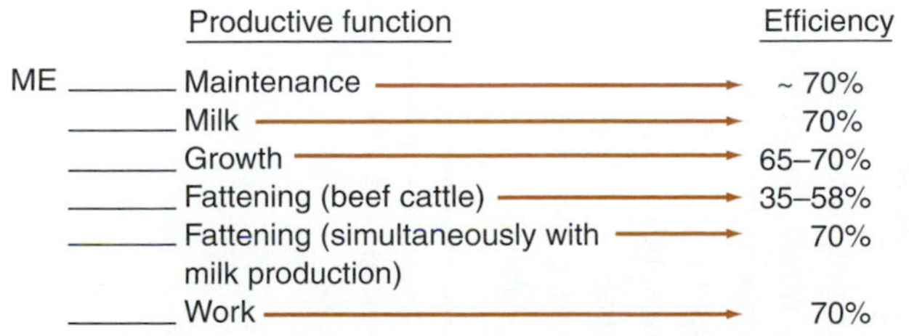

**FIGURE 5–10**  Efficiency of ME utilization. (*SOURCE:* Wagner, 1977. Used with permission.)

**TABLE 5-1   Selected Nutrient Content of Feeds Commonly Fed to Poultry and Swine on an As-Fed Basis**

| Feedstuff | (IFN) | DM (%) | CP (%) | Lysine (%) | Vitamin E (mg/kg) | Niacin (%) | Thiamine (%) | Ca (%) | P (%) | Fat (%) | ME Swine (kcal/kg) | ME Poultry (kcal/kg) |
|---|---|---|---|---|---|---|---|---|---|---|---|---|
| Alfalfa meal, dehydrated | 1-00-023 | 92 | 17.4 | 0.73 | 111.0 | 37.0 | 3.4 | 1.40 | 0.23 | 2.8 | 1,705 | 1,480 |
| Bakery waste, dehydrated | 4-00-466 | 92 | 9.8 | 0.3 | 41.0 | 26.0 | 2.9 | 0.13 | 0.24 | 11.7 | 3,600 | 3,630 |
| Corn, dent, yellow, grain | 4-02-935 | 88 | 8.5 | 0.54 | 20.9 | 23.0 | 3.7 | 0.03 | 0.28 | 3.6 | 3,420 | 3,350 |
| Corn, gluten meal, 60% | 5-28-242 | 90 | 61.2 | 1.0 | 23.4 | 60.0 | 0.3 | —[1] | 0.44 | 1.8 | 3,585 | — |
| Molasses, beet | 4-00-668 | 78 | 6.6 | — | 4.0 | 41.0 | — | 0.12 | 0.03 | 0.2 | 2,320 | 1,980 |
| Oats, grain | 4-03-309 | 89 | 11.8 | 0.4 | 14.9 | 14.0 | 6.0 | 0.08 | 0.34 | 4.7 | 2,735 | 2,543 |
| Sorghum, grain (milo) | 4-04-444 | 89 | 8.9 | 0.27 | 12.1 | 37.0 | 4.1 | 0.03 | 0.28 | 2.8 | 3,230 | 3,310 |
| Wheat, bran | 4-05-190 | 87 | 15.5 | 0.6 | 14.3 | 197.0 | 8.4 | 0.13 | 1.16 | 4.0 | 2,268 | 1,270 |

*SOURCE:* Various National Research Council publications.
[1] A — indicates that a feed does not have a significant amount of that nutrient.

**TABLE 5-2** Nutrient Content of Feeds Commonly Fed to Ruminants and Horses on a 100% Dry-Matter Basis

| Feedstuff | (IFN) | DM (%) | CP (%) | ADF (%) | Ca (%) | P (%) | TDN Beef Cattle (%) | NEm[1] Cattle (Mcal/kg) | NEg[1] Cattle (Mcal/kg) | NE$_L$ Dairy Cattle (Mcal/kg) | DE Horse (Mcal/kg) |
|---|---|---|---|---|---|---|---|---|---|---|---|
| Alfalfa, hay, midbloom | 1-00-063 | 90 | 17.0 | 35 | 1.37 | 0.22 | 58 | 1.24 | 0.68 | 1.30 | 2.3 |
| Brome, smooth, hay | 1-05-887 | 88 | 16.0 | 35 | 0.29 | 0.37 | 68 | 1.60 | 0.97 | 1.55 | 2.13 |
| Corn, silage | 3-28-250 | 33 | 8.1 | 28 | 0.25 | 0.22 | 70 | 1.60 | 1.03 | 1.60 | 2.68[2] |
| Cottonseed, meal, sol | 5-01-873 | 91 | 48.9 | 21 | 0.17 | 1.0 | 75 | 1.79 | 1.16 | 1.72 | 3.01[3] |
| Oats | 4-03-309 | 89 | 13.3 | 16 | 0.07 | 0.38 | 77 | 1.85 | 1.22 | 1.77 | 3.20 |
| Orchard grass, hay | 1-03-425 | 89 | 15.0 | 34 | 0.27 | 0.34 | 65 | 1.47 | 0.88 | 1.47 | 2.17 |

SOURCE: Various National Research Council publications.
[1] Net energy of maintenance and growth values for growing cattle
[2] Value for IFN# 3-02-912
[3] Value for IFN# 5-01-621

**TABLE 5 – 3**  **Daily Nutrient Intakes and Requirements of Swine Allowed Feed Ad Libidum**

| | Swine Liveweight (kg) | | | | |
|---|---|---|---|---|---|
| **INTAKE AND PERFORMANCE LEVELS** | **1–5** | **5–10** | **10–20** | **20–50** | **50–110** |
| Expected weight gain (g/day) | 200 | 250 | 450 | 700 | 820 |
| Expected feed intake (g/day) | 250 | 460 | 950 | 1,900 | 3,110 |
| Expected efficiency (gain/feed) | 0.800 | 0.543 | 0.474 | 0.368 | 0.264 |
| Expected efficiency (feed/gain) | 1.25 | 1.84 | 2.11 | 2.71 | 3.79 |
| Digestible energy intake (kcal/day) | 850 | 1,560 | 3,230 | 6,460 | 10,570 |
| Metabolizable energy intake (kcal/day) | 805 | 1,490 | 3,090 | 6,200 | 10,185 |
| Energy concentration (kcal ME/kg diet) | 3,220 | 3,240 | 3,250 | 3,260 | 3,275 |
| Protein (g/day) | 60 | 92 | 171 | 285 | 404 |
| **Nutrient** | **Requirement (Amount/Day)** | | | | |
| **INDISPENSABLE AMINO ACIDS (G)** | | | | | |
| Arginine | 1.5 | 2.3 | 3.8 | 4.8 | 3.1 |
| Histidine | 0.9 | 1.4 | 2.4 | 4.2 | 5.6 |
| Isoleucine | 1.9 | 3.0 | 5.0 | 8.7 | 11.8 |
| Leucine | 2.5 | 3.9 | 6.6 | 11.4 | 15.6 |
| Lysine | 3.5 | 5.3 | 9.0 | 14.3 | 18.7 |
| Methionine + cystine | 1.7 | 2.7 | 4.6 | 7.8 | 10.6 |
| Phenylalanine + tyrosine | 2.8 | 4.3 | 7.3 | 12.5 | 17.1 |
| Threonine | 2.0 | 3.1 | 5.3 | 9.1 | 12.4 |
| Tryptophan | 0.5 | 0.8 | 1.3 | 2.3 | 3.1 |
| Valine | 2.0 | 3.1 | 5.3 | 9.1 | 12.4 |
| Linoleic acid | 0.3 | 0.5 | 1.0 | 1.9 | 3.1 |
| **MINERAL ELEMENTS** | | | | | |
| Calcium (g) | 2.2 | 3.7 | 6.6 | 11.4 | 15.6 |
| Phosphorus, total (g) | 1.8 | 3.0 | 5.7 | 9.5 | 12.4 |
| Phosphorus, available (g) | 1.4 | 1.8 | 3.0 | 4.4 | 4.7 |
| Sodium (g) | 0.2 | 0.5 | 1.0 | 1.9 | 3.1 |
| Chlorine (g) | 0.2 | 0.4 | 0.8 | 1.5 | 2.5 |
| Magnesium (g) | 0.1 | 0.2 | 0.4 | 0.8 | 1.2 |
| Potassium (g) | 0.8 | 1.3 | 2.5 | 4.4 | 5.3 |
| Copper (mg) | 1.50 | 2.76 | 4.75 | 7.60 | 9.33 |
| Iodine (mg) | 0.04 | 0.06 | 0.13 | 0.27 | 0.44 |
| Iron (mg) | 25 | 46 | 76 | 114 | 124 |
| Manganese (mg) | 1.00 | 1.84 | 2.85 | 3.80 | 6.22 |
| Selenium (mg) | 0.08 | 0.14 | 0.24 | 0.28 | 0.31 |
| Zinc (mg) | 25 | 46 | 76 | 114 | 155 |
| **VITAMINS** | | | | | |
| Vitamin A (IU) | 550 | 1,012 | 1,662 | 2,470 | 4,043 |
| Vitamin D (IU) | 55 | 101 | 190 | 285 | 466 |
| Vitamin E (IU) | 4 | 7 | 10 | 21 | 34 |
| Vitamin K (menadione) (mg) | 0.02 | 0.02 | 0.05 | 0.10 | 0.16 |
| Biotin (mg) | 0.02 | 0.02 | 0.05 | 0.10 | 0.16 |
| Choline (g) | 0.15 | 0.23 | 0.38 | 0.57 | 0.93 |
| Folacin (mg) | 0.08 | 0.14 | 0.28 | 0.57 | 0.93 |
| Niacin, available (mg) | 5.00 | 6.90 | 11.88 | 19.00 | 21.77 |
| Pantothenic acid (mg) | 3.00 | 4.60 | 8.55 | 15.20 | 21.77 |
| Riboflavin (mg) | 1.00 | 1.61 | 2.85 | 4.75 | 6.22 |
| Thiamin (mg) | 0.38 | 0.46 | 0.95 | 1.90 | 3.11 |
| Vitamin $B_6$ (mg) | 0.50 | 0.69 | 1.42 | 1.90 | 3.11 |
| Vitamin $B_{12}$ (mg) | 5.00 | 8.05 | 14.25 | 19.00 | 15.55 |

SOURCE: *Nutrient Requirements of Swine*, 9th ed., 1988, p. 51, National Research Council.

**TABLE 5-4    Daily Nutrient Requirements of Horses**

| Animal | Weight (kg) | Daily Gain (kg) | DE (Mcal) | Crude Protein (g) | Lysine (g) | Calcium (g) | Phosphorus (g) | Magnesium (g) | Potassium (g) | Vitamin A (10³ IU) |
|---|---|---|---|---|---|---|---|---|---|---|
| Mature horses | | | | | | | | | | |
| Maintenance | 600 | | 19.4 | 776 | 27 | 24 | 17 | 9.0 | 30.0 | 18 |
| Stallions | 600 | | 24.3 | 970 | 34 | 30 | 21 | 11.2 | 36.9 | 27 |
| (breeding season) | | | | | | | | | | |
| Pregnant mares | | | | | | | | | | |
| 9 months | 600 | | 21.5 | 947 | 33 | 41 | 31 | 10.3 | 34.5 | 36 |
| 10 months | | | 21.9 | 965 | 34 | 42 | 32 | 10.5 | 35.1 | 36 |
| 11 months | | | 23.3 | 1,024 | 36 | 44 | 34 | 11.2 | 37.2 | 36 |
| Lactating mares | | | | | | | | | | |
| Foaling to 3 months | 600 | | 33.7 | 1,711 | 60 | 67 | 43 | 13.1 | 55.2 | 36 |
| 3 months to weaning | 600 | | 28.9 | 1,258 | 44 | 43 | 27 | 10.4 | 39.6 | 36 |
| Working horses | | | | | | | | | | |
| Light work | 600 | | 24.3 | 970 | 34 | 30 | 21 | 11.2 | 36.9 | 27 |
| Moderate work | 600 | | 29.1 | 1,164 | 41 | 36 | 25 | 13.4 | 44.2 | 27 |
| Intense work | 600 | | 38.8 | 1,552 | 54 | 47 | 34 | 17.8 | 59.0 | 27 |
| Growing horses | | | | | | | | | | |
| Weanling, 4 months | 200 | 1.00 | 16.5 | 825 | 35 | 40 | 22 | 4.3 | 13.0 | 9 |
| Weanling, 6 months | | | | | | | | | | |
| Moderate growth | 245 | 0.75 | 17.0 | 850 | 36 | 34 | 19 | 4.6 | 14.5 | 11 |
| Rapid growth | 245 | 0.95 | 19.2 | 960 | 40 | 40 | 22 | 4.9 | 15.1 | 11 |
| Yearling, 12 months | | | | | | | | | | |
| Moderate growth | 375 | 0.65 | 22.7 | 1,023 | 43 | 36 | 20 | 6.4 | 20.7 | 17 |
| Rapid growth | 375 | 0.80 | 25.1 | 1,127 | 48 | 41 | 22 | 6.6 | 21.2 | 17 |
| Long yearling, 18 months | | | | | | | | | | |
| Not in training | 475 | 0.45 | 23.9 | 1,077 | 45 | 33 | 18 | 7.7 | 25.1 | 21 |
| In training | 475 | 0.45 | 32.0 | 1,429 | 60 | 44 | 24 | 10.2 | 33.3 | 21 |
| Two-year-old, 24 months | | | | | | | | | | |
| Not in training | 540 | 0.30 | 23.5 | 998 | 40 | 31 | 17 | 8.5 | 27.9 | 24 |
| In training | 540 | 0.30 | 32.3 | 1,372 | 55 | 43 | 24 | 11.6 | 38.4 | 24 |

SOURCE: *Nutrient Requirements of Horses*, 5th ed., 1988, p. 44, National Research Council.

*Beef Cattle, Nutrient Requirements of Cats, Nutrient Requirements of Fish, Nutrient Requirements of Dairy Cattle, Nutrient Requirements of Dogs, Nutrient Requirements of Goats, Nutrient Requirements of Horses, Nutrient Requirements of Laboratory Animals, Nutrient Requirements of Mink and Foxes, Nutrient Requirements of Non-Human Primates, Nutrient Requirements of Poultry, Nutrient Requirements of Rabbits, Nutrient Requirements of Sheep*, and *Nutrient Requirements of Swine*.

## RATION FORMULATION

The amount of feed offered to an animal during a 24-hour period is called a *ration*. Providing the proper nutrients for a given animal to accomplish its specific production function is what we try to accomplish when we balance, mix, and subsequently feed a ration. If we accomplish this, we can say we are feeding a *balanced* ration. In addition to being balanced, a good ration must be palatable and free of damaging amounts of molds, diseases, and other quality-lowering factors. The ration should be economical and not have any negative effects on the general health and well-being of the animal or product quality. As an example, some substances in feeds and weeds give undesirable odors and flavors to the milk of dairy cows and "off" colors in eggs. Others cause undesirable carcass characteristics. Such feeds must be carefully managed if they are to be used at all. Tables 5–5, 5–6, 5–7, 5–8, 5–9, and 5–10 give examples of balanced rations for some classes of animals.

## SOURCES OF INFORMATION

Nutrition is a very dynamic field. New information and new procedures are being discovered constantly. Thus anyone interested in animals and/or nutrition should know about sources of information they can trust. *Feedstuffs* is a weekly newspaper that scientists and industry leaders alike consider a very credible source of information about nutrition and the feeds industry and thus about the animal industries. A yearly supplement that is an excellent source of current information about feeds and feeding is also published. *Feedstuffs* is available from the Miller Publishing Company (12400 Whitewater Drive, Suite 160, Minnetonka, MN 55343. Phone: 952–931–0211. Website: www.feedstuffs.com).

Scientists report their most significant research findings in scientific journals. The credibility of this information is very high because other scientists review and critique the articles in a process called **peer review.** Some of the major journals to consult for current research findings are *Animal Feed Science and Technology, Canadian Journal of Animal Science, Journal of Animal Science, Journal of Dairy Science, Journal of Nutrition*, and *Poultry Science*.

**Peer review**

A process in which those with expertise and experience similar to the author of an article (his or her peers) review and critique the article for scientific merit before it is published.

**TABLE 5–5    Balanced Creep Diets for Lambs**

|  | Diet 1 % | Diet 2 % | Diet 3 % | Diet 4 % |
|---|---|---|---|---|
| Corn grain | 48 | 48 | 29 | 24 |
| Wheat |  |  | 21 |  |
| Milo |  |  |  | 24 |
| Soybean meal | 9 |  |  |  |
| Cottonseed meal |  | 9 | 7 | 9 |
| Molasses | 4 | 4 | 4 | 3 |
| Limestone | 1 | 1 | 1 | 2 |
| Alfalfa hay or pellets | 38 | 38 | 38 | 38 |

SOURCE: Courtesy of Dr. Jerry Fitch, Oklahoma State University. Used with permission.

**TABLE 5–6    A Total Mixed Ration (TMR) for Lactating Dairy Cows**

|  | Per Cow Daily |
|---|---|
| Alfalfa hay | 10.0 lbs |
| Corn silage | 36.0 lbs |
| Shelled corn | 21.1 lbs |
| Soybean meal | 7.4 lbs |
| Dicalcium phosphate | 3.8 oz |
| Limestone | 5.2 oz |
| Sodium bicarbonate | 7.1 oz |
| Trace mineralized salt | 2.4 oz |
| Vitamin A | 59,200 U |
| Vitamin D | 29,600 U |

SOURCE: Adapted from Dunham and Call, 1989, p. 10.

**TABLE 5–7    Suggested Moderate Nutrient Density Diets for Growing Swine Using Sorghum Grain and/or Barley as Major Grain Sources[1]**

| Ingredient | Diet 1 (lb) | Diet 2 (lb) | Diet 3 (lb) | Diet 4 (lb) |
|---|---|---|---|---|
| Sorghum grain | 1,514 | 1,375 | 1,444 | — |
| Barley | — | — | — | 1,532 |
| Soybean meal, 44% | 437 | — | — | 319 |
| Soybean meal, 48% | — | — | 424 | — |
| Soybeans, full fat (cooked) | — | 577 | — | — |
| Fat | — | — | 80 | 100 |
| Calcium carbonate | 17 | 17 | 18 | 18 |
| Dicalcium phosphate | 22 | 21 | 24 | 21 |
| Salt | 7 | 7 | 7 | 7 |
| Trace mineral and vitamin mix | 3 | 3 | 3 | 3 |
| Totals | 2,000 | 2,000 | 2,000 | 2,000 |

SOURCE: Luce et al., 1995, p. 10.

[1]Suggested for average and high lean gain barrows 75–140 lbs, average gilts 75–140 lbs, and high lean gain gilts 140 lbs to market.

**TABLE 5–8** **Example Rations for Some Companion Species[1]**
**(percentage of ration)**

|  | Diet 1 | Diet 2 | Diet 3 | Diet 4 | Diet 5 |
|---|---|---|---|---|---|
| Ground yellow corn | 56.0 | 33.77 | — | — | 8.62 |
| Ground wheat | 5.0 | 9.0 | — | 28.9 | 13.47 |
| Ground oats | — | — | — | 17.75 | 7.54 |
| Wheat mill run | — | — | 36.5 | — | — |
| Cornstarch | — | — | — | — | 36.67 |
| Molasses | — | — | 3.0 | — | — |
| Corn gluten meal (60% CP) | 5.0 | 12.8 | — | — | — |
| Soybean meal (48% CP) | 15.0 | 10.0 | 6.0 | 13.25 | 10.77 |
| Meat and bonemeal | 10.0 | 3.0 | — | — | — |
| Fish meal | — | 1.0 | — | — | 4.85 |
| Poultry meal | — | 17.4 | — | — | — |
| Cellulose | — | — | — | — | 2.04 |
| Alfalfa meal | — | — | 54.0 | 38.15 | 2.15 |
| Animal digest | — | 1.0 | — | — | — |
| Animal fat | 7.0 | 6.0 | — | — | — |
| Soybean oil | — | — | — | — | 4.53 |
| Vitamin premix[2] | 0.8 | 0.8 | — | 0.05 | 1.0 |
| Trace mineral premix[2] | 0.5 | 0.5 | — | 0.05 | 1.9 |
| Dicalcium phosphate | 0.2 | — | — | 0.25 | — |
| Ground limestone | — | — | — | 1.10 | — |
| Calcium carbonate | — | 1.16 | — | — | — |
| Salt | 0.05 | 0.7 | 0.5[3] | 0.5[4] | — |
| Potassium chloride | — | 0.48 | — | — | — |
| Phosphoric acid solution | — | 2.3 | — | — | — |
| Citric acid | 0.01 | — | — | — | — |
| Taurine | — | 0.8 | — | — | — |
| Brewer's yeast | — | — | — | — | 6.46 |
| Total | 100.0 | 100.0 | 100.0 | 100.0 | 100.0 |

SOURCES: *Nutrient Requirements of Laboratory Animals,* 3rd ed., 1978, National Research Council; Pond et al., 1995, pp. 459, 537, 543.

[1]Diet 1 is a dry dog food. Diet 2 is a dry cat food. Diet 3 is a grower diet for weaned rabbits. Diet 4 is for guinea pigs. Diet 5 is for hamsters.
[2]The specific premixes are different for each species.
[3]Trace mineralized salt.
[4]Iodized salt.

**TABLE 5–9    Rations for Beef Calves[1]**

| | Receiving Protein Pellet | Growing Steers Gaining 1.5 lbs/day | Supplement for 400–500 lb Steers Fed Prairie Hay Free-Choice |
|---|---|---|---|
| Cottonseed hulls | 1.75 | | |
| Soybean meal | 90.875 | 5.0 | 22.0 |
| Salt | 3.0 | 0.3 | 0.8 |
| Calcium carbonate | 1.5 | — | — |
| Vitamin premix | 0.125 | — | — |
| Dicalcium phosphate | 2.75 | 0.5 | 1.1 |
| Limestone | — | 0.5 | 1.1 |
| Sudan hay | — | 55.7 | — |
| Wheat | — | 20.0 | — |
| Corn | — | 18.0 | 75.0 |

SOURCE: Hibberd et al., 1993, pp. 5–5, 5–15, 5–23. Used with permission.
[1]Values expressed as percentage of the total ration.

**TABLE 5–10    Rations for Mature Horses**

| | As-Fed Basis | | Dry-Matter Basis | |
|---|---|---|---|---|
| ALL FEEDS IN THE RATION | (LBS/DAY) | (%) | (LBS/DAY) | (%) |
| Bermuda grass, late | 13.001 | 70.000 | 12.091 | 70.898 |
| Oats, grain | 3.715 | 20.000 | 3.313 | 19.429 |
| Corn, dent, grain | 1.393 | 7.500 | 1.226 | 7.188 |
| Soybean meal 44% PRO | 0.371 | 2.000 | 0.331 | 1.941 |
| Sodium chloride | 0.046 | 0.250 | 0.046 | 0.272 |
| Calcium carbonate | 0.046 | 0.250 | 0.046 | 0.272 |
| Total ration | 18.573 | | 17.054 | |

SOURCE: Freeman, 1997. Used with permission.

## SUMMARY AND CONCLUSION

Nutrition is a dynamic and very complicated science. It is essential that we understand nutrition because feeds and the methods of feeding comprise a large portion of the costs of livestock production. This chapter will not make you a world-famous nutritionist. However, it does give you a broad working knowledge on which to build. The key to understanding the interconnectedness of nutrition is in reducing it to its essentials:

1. Nutrition is the study of how the body uses nutrients for maintenance and productive purposes.
2. Nutrients are the chemical entities that the body requires. We need to know which nutrients each animal needs and the amount of the nutrient needed to accomplish the production required of the animal.
3. Feeds contain nutrients. We must analyze the feeds to determine just how much of a particular nutrient is in each feed and then determine how much the animals can actually use.
4. When we know an animal's nutrient needs and a feed's nutrient availability for that animal, we can balance a ration from those ingredients to meet the animal's needs.

## STUDY QUESTIONS

1. Define nutrition and explain why one might choose to study it.

2. Why is the proper feeding of livestock such an important practical consideration?

3. Describe the disciplines of nutrition including the areas of specialty for nutritionists. Explain in detail the differences between basic and applied nutritionists.

4. Name and describe at least five disciplines involved in the study of nutrition.

5. Describe how you would set up an experiment to determine whether or not a nutrient is essential. How would you interpret the results?

6. List the nutrient categories. Why isn't energy a nutrient? It is required for life, correct?

7. List all of the essential amino acids. Are there any additional amino acids for poultry? Are there any deviations from this list?

8. List the fat-soluble vitamins.

9. List the water-soluble vitamins.

10. What is the difference between macro- and microminerals?

11. List all of the macrominerals and microminerals.

12. What are the overall body functions for which an animal uses nutrients?

13. What is the purpose of having feeds chemically analyzed?

14. What are the three general types of analytical methods used to evaluate feedstuffs? Briefly define each method.

15. What nutrients are measured by proximate analysis? What are the limitations of each individual test?

16. Describe in detail how carbohydrates are determined by proximate analysis. How is this process different from the way the other proximate constituents are determined?

17. What individual carbohydrate fractions are determined in each part of the proximate carbohydrate fractions?

18. What is the major problem with the proximate analysis?

19. Why was the Van Soest fiber determination system developed? What advantages does it offer over proximate analysis?

20. Why is there no standard test for determining the vitamin content of feeds?

21. Describe in detail the tests for determining the energy content of a feed. What are the units used to express energy content?

22. How do feeding trials help to generate feeding standards for animals?

23. List the factors that may have an effect on nutrient utilization.

24. What are the common types of feeding trials?

25. Describe how a feeding trial, a digestion trial, and a metabolism trial are different.

26. What is the difference between values for a feed expressed on an as-fed basis compared to a dry-matter basis?

27. Describe the differences in the energy content of an average carbohydrate compared to an average protein and an average fat.

28. What are the uses of the nutrient requirement information as published by the National Research Council?

29. What is the most important popular press publication about feeds and nutrition?

30. Where can you find reports on feed research?

## REFERENCES

Cheeke, P. R. 1991. *Applied animal nutrition*. 3rd ed. New York: Macmillan.

Church, D. C., and W. C. Pond. 1988. *Basic animal nutrition and feeding*. 3rd ed. New York: Wiley.

Dunham, J. R., and E. P. Call. 1989. *Feeding dairy cows*. MF-754 (Revised). Manhattan: Cooperative Extension Service, Kansas State University.

Fitch, G. Q. 2007. Personal communication. Oklahoma State University, Stillwater.

Freeman, D. W. 1997. Personal communication. Oklahoma State University, Stillwater.

Gillespie, J. R. 1987. *Animal nutrition and feeding*. Albany, NY: Delmar.

Hibberd, C. A., K. Lusby, and D. Gill. 1993. Mechanics of formulating stocker cattle rations and supplements. In *Oklahoma beef cattle manual*. Publication No. E-913. Stillwater: Cooperative Extension Service. Division of Agricultural

Sciences and Natural Resources, Oklahoma State University.

Jurgens, M. H. 1993. *Animal feeding and nutrition*. 8th ed. Dubuque, IA: Kendall/Hunt.

Luce, W. G., A. F. Harper, D. C. Mahan, and G. R. Hollis. 1995. Swine diets. In *Pork industry handbook*. Stillwater: Oklahoma Cooperative Extension Service. Division of Agricultural Sciences and Natural Resources, Oklahoma State University.

Perry, T. W., A. E. Cullison, and R. S. Lowery. 1999. *Feeds and feeding*. 5th ed. Upper Saddle River, NJ: Prentice-Hall.

Pond, W. G., D. C. Church, K. R. Pond, and P. A. Schoknect. 2005. *Basic animal nutrition and feeding*. 5th ed. New York: Wiley.

Wagner, D. G. 1977. *Livestock feeding*. Stillwater: Oklahoma State University.

# 6

# The Gastrointestinal Tract and Nutrition

## LEARNING OBJECTIVES

After you have studied this chapter, you should be able to:

1. Describe the methods of the breakdown of food.

2. Classify digestive systems according to stomach type and type of diet consumed.

3. Describe the steps of digestion.

4. Identify the differences and similarities in the digestive processes of animals.

5. Explain the importance of the complex stomach of the ruminant and its benefits to the animal.

## KEY TERMS

| | | |
|---|---|---|
| Abomasum | Denature | Papillae |
| Anaerobic | Digestion | Peristalsis |
| Bacteria | Duodenum | Prehension |
| Bolus | Enzymes | Protein |
| Carbohydrates | Eructation | Protozoa |
| Carnivore | Fermentation | Proventriculus |
| Cecotrophy | Forestomachs | Rumen |
| Cellulase | Herbivore | Ruminal bloat |
| Cellulose | Ileum | Ruminant |
| Chyme | Jejunum | Rumination |
| Colic | Mastication | Salivation |
| Coprophagy | Micturation | Symbiosis |
| Defecation | Monogastric | Vitamin |
| Deglutition | Omnivore | Volatile fatty acid (VFA) |

# INTRODUCTION

Feeds and feedstuffs are chemically complex mixtures of substances that contain the nutrients an animal needs. The digestive tract breaks down those complex materials to their constituent parts so the nutrients can be absorbed and metabolized by the body. Breakdown of food by the digestive system in preparation for absorption is called *digestion* and is accomplished in three ways: (1) the physical or mechanical actions of chewing (mastication) and muscular action of the digestive tract (peristalsis); (2) the chemical action of hydrochloric acid, which is used by the stomach to denature proteins and bile used in the small intestine to help digest fats; and (3) the action of enzymes, which increase the speed of the breakdown of the chemical bonds in foods by the addition of a water molecule (hydrolytic enzyme). *Enzymes* can be produced by the digestive tract and accessory organs (liver, pancreas), or by microorganisms living in symbiosis with the animal. Enzymes are biological catalysts that speed the rate at which a particular reaction reaches equilibrium. Many enzymes are found in the system and are needed for faster and more efficient digestion.

One of the intriguing things we know about nutrient metabolism and the metabolic pathways for different species is that they are essentially the same. *E. coli* and giraffes have much the same thing happening at the cellular level. However, they have different nutrient requirements, consume different types of feed, and digest food very differently. This difference in digestive types has allowed dissimilar species to fit into different places in the food chain and carve a niche for themselves. The type of digestive system an animal has dictates what the animal can successfully use as feed. The more complicated the feed (like forage), the more complicated the digestive tract. Thus the ruminant system is designed to retain feed for several days, which is a long time compared to the few hours that a feed is held in a carnivore's simple tract (Figure 6–1).

# CLASSIFICATION OF DIGESTIVE SYSTEMS

We classify the different digestive systems anatomically by the type of stomach an animal has. *Monogastrics,* or nonruminants, are one-stomached, or simple-stomach, animals. They can have a very simple tract, as in the mink and dog, or a cecal digestion, as in the horse, rabbit, or rat. Monogastrics that have a cecal fermentation can use much more fibrous feedstuffs than can a simple monogastric. Others, like the kangaroo, rely on a sacculated stomach to retain fibrous material for better digestion. The *ruminants*— cattle, sheep, goats, and pseudoruminants (llamas)—are more complex-stomached animals that have more than one stomach compartment. These compartments are located before the true stomach. A complicated fermentation takes place there to help the animal make use of fibrous feeds. These tracts usually also have a cecum, which helps to retain the feed even longer.

We also classify digestive systems according to the type of diet the animal normally consumes. There are three digestive categories based on diet. *Carnivores* are flesh-eating animals, for example, cats and birds of prey. *Omnivores* eat both animal and vegetable matter; examples include chickens, pigs, and humans. *Herbivores* eat vegetable-based feeds. They are able to digest this plant material with *cellulases* provided by bacteria during the fermentation process. Horses, cows, rabbits, guinea pigs, llamas, kangaroos, elephants, and many others are herbivores. All herbivores have some specialized way to help them digest *cellulose*.

## Digestion
The physical, chemical, and enzymatic means the body uses to render a feedstuff ready for absorption.

## Enzymes
Proteins capable of catalyzing reactions associated with a specific substrate.

## Monogastric
Having only one stomach. Also, nonruminant.

## Ruminant
Hooved animals that have a rumen and chew their cud.

## Carnivore
An animal that subsists on meat.

## Omnivore
An animal that selects a diet of both plant matter and meat.

## Herbivore
An animal that eats a diet of only plant material.

## Cellulase
An enzyme that specifically attacks and digests cellulose.

## Cellulose
A carbohydrate composed of thousands of glucose molecules that forms the support structure of plants.

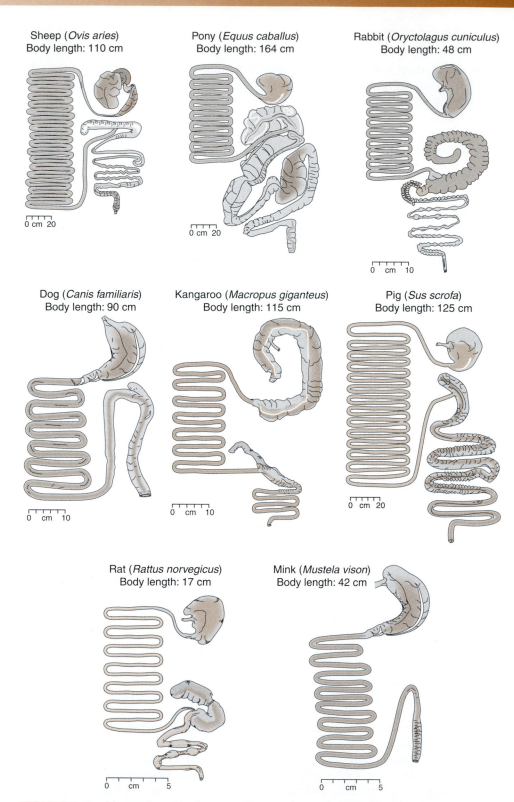

**FIGURE 6–1** Mammalian digestive tracts. (*SOURCE:* Swenson, 1984. Used by permission of the publisher, Cornell University Press.)

# AN OVERVIEW OF THE STEPS OF DIGESTION

*Prehension,* the means an animal uses to bring food into its mouth, is the first step of digestion. Animals use a variety of prehension methods, including use of their upper limbs, head, beak, and claws, and their mouth, teeth, and lips. Herbivores use the tongue, teeth, or lips to grasp forage. Ruminants have no upper incisor teeth. The cow uses its flexible tongue to bring forage into its mouth and cuts it off with the lower incisor teeth and upper dental pad. Sheep use their teeth and cleft upper lip to permit them to graze close to the ground, which can lead to overgrazing. Horses make use of their mobile, prehensile upper lip, as well as upper and lower incisor teeth.

*Mastication,* or chewing, involves the vertical and lateral action of the jaw and teeth to crush food. Carnivores chew only to the extent needed to reduce the size of their meat so it can be swallowed. All herbivores need thorough mastication of their feed, mostly forage, to allow bacterial enzymes access to the cellulose. Ruminants first form a bolus and then swallow it without much chewing. Later they regurgitate the feed and thoroughly chew it. This process is one of the key elements of *rumination.* A typical dairy cow "chews" between 40,000 and 50,000 times a day. Cecal fermentors and some simple monogastrics must chew thoroughly before swallowing because the food goes directly into the glandular stomach for digestion. They don't regurgitate their feed.

*Salivation* includes secretion and mixing of saliva with food. Three main paired salivary glands produce saliva (Figure 6–2). These are the parotid (below the ear), the submaxillary (mandibular) (at the base of the tongue), and sublingual (under the tongue). Saliva is mixed with food during chewing and has many functions. The first function is to lubricate food so the animal can chew and swallow it. All animals depend on this most irreplaceable salivary function. Both the water content and the protein mucin found in saliva are responsible for the lubricating function. Think how hard it would be for you to eat a cracker or for herbivores to eat forage without saliva. The solvent action of saliva dissolves small portions of food, allowing it to come into contact with taste buds, and gives food its taste. Saliva may also stimulate the taste nerves. The washing action of saliva cleanses the mouth and prevents decay of leftover food particles. The enzyme lysozyme supplies a disinfectant action that kills bacteria that could harm teeth and gums. Dogs and other carnivores have especially large amounts of lysozyme. Ruminants rely heavily on the bicarbonates in saliva to buffer the acids produced by the microorganisms in the rumen. The acids would soon damage the rumen wall if they weren't buffered. Saliva provides readily available nutrients for rumen microorganisms and helps maintain the proper pH for the microorganisms. Mucin, which

## Prehension
The act of seizing and grasping.

## Mastication
The process of chewing.

## Rumination
The process in ruminants where a cud or bolus of rumen contents is regurgitated, remasticated, and reswallowed for further digestion.

## Salivation
The elaboration of the mixed secretion (saliva) produced primarily by three bilateral pairs of glands in the mouth known as salivary glands.

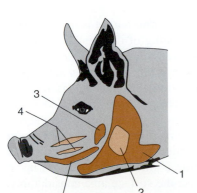

**FIGURE 6–2** Salivary glands of the pig.

Salivary glands
1. Parotid
2. Mandibular (submaxillary)
3. Palatine
4. Buccal
5. Sublingual

## Ruminal bloat

More correctly called *ruminal tympany.* An overdistention of the rumen and reticulum with the gases of fermentation. Commonly referred to as *bloat.*

## Deglutition

The act of swallowing. Passing material from the mouth through the esophagus to the stomach.

## Bolus

A rounded mass ready to swallow.

## Peristalsis

The progressive squeezing movements produced by the contraction of muscle fibers found in the wall of the digestive tract.

## Proventriculus

The glandular stomach in fowl.

## Abomasum

The true glandular stomach in the ruminant.

## Denature

In protein chemistry, to disrupt the structure of a native protein causing it to lose its ability to perform its function.

## Duodenum

The first segment of the small intestine.

## Jejunum

The second and longest portion of the small intestine.

## Ileum

The last short portion of the small intestine.

is a protein, is digested and used by the microorganisms once it reaches the lower tract. Urea is harvested from the blood by salivary glands and secreted with the saliva. The microorganisms use it to manufacture protein. Phosphorus (P) and sodium (Na) are also present and are used by the microorganisms. The antifrothing property of saliva helps prevent ***ruminal bloat*** from being a daily occurrence. It prevents formation of a stable foam that would interfere with eructation (belching). Saliva also has an excretory function.

Swallowing, or ***deglutition,*** is the passing of food and water (or anything else) from the mouth to the first stomach compartment via the esophagus. The swallowing reflex is involuntary and under neural control. It is caused by the presence of material in the back of the mouth. The tongue is responsible for forming a ***bolus*** of the food in the mouth and voluntarily pushing it to the back of the mouth. Muscle contractions (***peristalsis***) move the food to the stomach. The cardiac valve is a valve located at the end of the esophagus that prevents feed in the stomach from coming back into the esophagus in some species.

In the ruminant, the feed makes its first stop in the rumino-reticulum for fermentation. Fermentation is an important topic and discussed at length later in this chapter. In the monogastric, the first stop is in the true stomach, which represents the next stop after the forestomachs for the ruminants.

Significant chemical and enzymatic digestion begins in the glandular stomach, which is similar in most animals. In the chicken, the stomach is called the ***proventriculus,*** and in the cow, it is called the ***abomasum***. Physical breakdown of food occurs because of the churning action created by the contractions of the strong stomach muscles. The churning action also mixes the food (ingesta) with chemicals and enzymes. Chemical digestion is provided by hydrochloric acid (HCl), which is secreted by gastric glands in the stomach. HCl ***denatures*** proteins, a major step in their digestion. In the native state, proteins are poorly digested. In denatured form, enzymes can readily hydrolyze them. HCl activates the enzyme pepsin from its precursor pepsinogen. HCl provides an acidic pH in the stomach that is needed for gastric enzymes to work. HCl kills bacteria, which renders the stomach almost sterile. Several enzymes secreted in the stomach gastric juice provide enzymatic digestion of food. These include pepsin, gastricin, and rennin, all of which work on proteins. Rennin hydrolyzes the milk protein casein. This enzyme is not important in the adult animal but is especially important in young ruminants. Gastric lipase acts on fat.

In addition to digestion, the stomach stores food, another important function. This function allows the animal to eat at a much faster rate than if digestion had to occur as rapidly as the food was consumed. The food is then metered into the lower gut as it is capable of digesting it.

The small intestine is divided into three portions. The ***duodenum*** is the first part. It extends from the pylorus of the stomach to the beginning of the jejunum. In most species, the duodenum is in the form of a loop and is often referred to as the duodenal loop. It is generally only about 1 ft long. Bile and pancreatic secretions enter in this portion of the small intestine. Excluding microbial fermentation, the duodenum is the main site of food breakdown in the entire digestive system. The ***jejunum*** is the second and longest part of the small intestine. Digestion continues here, but absorption of the end products of digestion is its major function. The jejunum is several feet long. The ***ileum*** is the third part. Its major job is to form the connection to the large intestine, but absorption occurs here also. The entire small intestine is lined with a mucosa layer. There are folds within this layer that serve to increase the surface area considerably. The jejunum portion of many species is covered with microscopic villi, which may number in the millions. They increase the absorptive surface tremendously. Nutrients are absorbed into the villi and pass into the lymphatic system or the circulatory system. Carbohydrates, amino acids, short-chain fatty acids, water-soluble vitamins, and most minerals absorbed from the small intestines enter the bloodstream via the portal vein. They go first to the liver and then are distributed to the rest of the body. The rest of the lipids and fat-soluble vitamins are transported by the lymphatic system to the thoracic duct, and they empty into the vena cava to be further transported.

The small intestine is the chief site of food digestion and nutrient absorption for monogastrics. When *chyme* leaves the stomach, it is very acidic. Chyme is mixed in the duodenum with three alkaline secretions, all very important to digestion. These secretions are bile, pancreatic juice, and succus entericus. Bile is formed in the liver. It is concentrated (up to 20 times for some species) and stored in the gallbladder until needed for digestion. Some species, like the rat and horse, don't have a gallbladder. Because they are continual, rather than meal, eaters, they secrete bile continually. Bile salts, which are derivatives of cholesterol, assist in the digestion and absorption of fats.

Pancreatic juice, the second alkaline secretion of importance, contains the really important, very potent digestive (hydrolyzing) enzymes. In hydrolysis, a compound is split into two or more simpler compounds by the uptake of the H and OH parts of a water molecule on either side of the chemical bond cleaved. The main enzymes are trypsin, chymotrypsin, carboxypeptidase, aminopeptidase, intestinal lipase, and amylase.

The third secretion, succus entericus (intestinal juice), is secreted by glands in the small intestine itself in large amounts. For example, a human may secrete 20 liters per day. Its function is to lubricate, dilute, and increase the pH of the food mixtures.

Relative capacities of the large intestine and ceca vary greatly from species to species. Monogastric herbivores have the large intestine of greatest relative volume and carnivores have the smallest. The large intestine contains no villi; thus absorption is restricted. Mucous glands line the large intestine to provide lubrication. Note that no digestive enzymes are secreted in the large intestine by the animal. Any enzymes found there must be left over from secretions found earlier in the tract or are provided by microorganisms. The large intestine has three parts: colon, cecum, and rectum. The material that enters from the small intestines is liquid and contains cells, undigested foodstuffs, and digestive secretions. Water, electrolytes, vitamins, minerals, and volatile fatty acids (VFAs) are absorbed from this material, with emphasis on water and electrolytes. The contents of the large intestine are not sterile. The feces contains about 50% bacteria by weight. A variety of bacteria grow in the feces, the most common being coliform bacteria (fecal bacteria). These bacteria produce some vitamins. Large intestinal vitamin K is very important in chickens, for instance. Bacteria also produce gases, the most important of which are carbon dioxide and methane.

Monogastric herbivores like the horse, rabbit, guinea pig, and elephant have an extensive colon and cecum. The cecum is functional. Carnivores, by contrast, have a very short large intestine with nonfunctional ceca. The large intestine (colon and cecum) of the monogastric herbivores is comparable in size to the rumen of ruminants and has a large fermentation capacity. The large intestines and ceca in monogastric herbivores function much as the rumen does but with less advantage to the animal. The reason is twofold. Much of the good nutrient content has already been removed from the feed by the digestion and absorption of the digestive tract prior to the feed reaching the cecum. Thus the cecum has lower quality feed to work with than does the rumen. The second reason is that absorption is less from the cecum than the rumen. Very significant digestion does take place, however. Microorganisms break down cellulose in the cecum and large intestines, much as is done in the rumen. B-complex vitamins are produced.

## Chyme

The name given to the material consisting of food, saliva, and gastric secretions.

VFAs are produced, absorbed, and used in all monogastric herbivores as an energy source, similar to what happens in the ruminant. A large amount of bacterial protein is also manufactured in the cecum. However, animals must employ behavioral adaptation to be able to use this protein because there are no digestive secretions into the large intestine by the animal. The large intestine can only absorb free amino acids, not whole proteins, so the microbial protein remains unused. The adaptive mechanism is called *cecotrophy,* which is a form of pseudorumination. Animals who practice cecotrophy eat the contents of their cecum and digest it after it has been fermented in the cecum. Rabbits, for example, practice cecotrophy. They have two types of feces. One is the pellet type, which is what people usually see. The other is a soft feces that the rabbit eats. This soft feces is material that has been fermented in the cecum. The bacterial protein produced by the fermentation is digested the second time around by the upper gut. If a rabbit is denied the option of practicing cecotrophy (on a poor diet), it will die of malnutrition because it will be deficient in protein and B vitamins. Horses depend more on dietary protein and B vitamins than rabbits or ruminants. Thus they need more of each in their diet.

The omnivore large intestine is less capable than that of the herbivore, but more capable than that of the carnivore. Vitamin synthesis and water and electrolyte absorption are similar to that of the carnivores. Cellulose digestion in these animals is a function of retention time. The longer it is retained, the more likely some digestion will take place. Studies indicate that up to 18–20% crude fiber digestion in high-fiber diets fed to swine is possible. The usefulness of this is unclear, however, because the role of volatile fatty acid absorption and utilization in the pig is poorly understood. Because swine in modern swine production facilities are rarely fed much fibrous feed, the argument is moot.

*Defecation* is the discharge of excrement from the body via the rectum or cloaca. Internal and external anal sphincters control the exit of material from the body. Defecation is initiated by the defecation reflex, which is stimulated by the pressure of feces in the rectum. This reflex is assisted by parasympathetic nervous signals that intensify the peristaltic waves of the large intestines. Many animals also use the Valsalva maneuver. They breathe deeply, close the glottis, and then flex the abdominal muscles. This puts pressure on the fecal contents and helps expel them.

The contents of the fecal material include undigested feed, residues of digestive enzymes, sloughed cells, and bacteria. The quantity of feces is affected by how well the feed is digested. Feeds low in digestibility contribute more undigested material to the feces. In addition, the poorly digested material causes an increase in the amount of sloughed cells from the wall of the intestine.

*Micturition* is urination. The components of urine include the nitrogen compounds—urea in mammals and uric acid in birds and other species. Uric acid requires much less water from the body to excrete. Also included are minerals and water. The kidney is a major regulatory mechanism for keeping the body appropriately hydrated and for removing various wastes from the body.

## DIGESTION IN THE PIG

The pig is omnivorous, which means it eats and uses feed ingredients of both animal and plant origin. It is also monogastric and, as such, is discussed first to give us a "baseline" for discussing the other species (Figure 6–3). The mouth of the pig is used primarily for grinding feed by the teeth and mixing with saliva. The saliva, which has a pH of about 7.4, moistens feed and helps in the chewing and swallowing process. The pig is the only

**Cecotrophy**

The process by which mucus-covered soft fecal pellets are expelled from the intestine and consumed by the animal.

**Defecation**

The act of expelling fecal matter from the large intestine via the rectum or cloaca.

**Micturition**

The act of urinating.

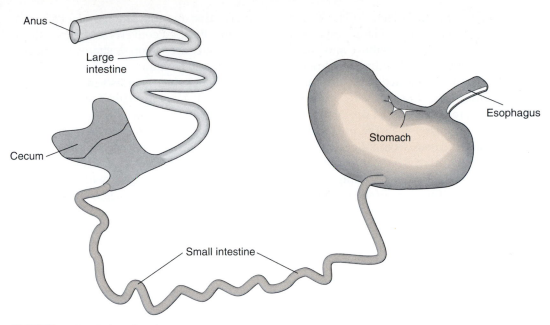

**FIGURE 6–3**   Swine digestive tract.

farm mammal in which any amount of the enzyme salivary amylase is secreted in the saliva. The salivary amylase is relatively weak compared to that in human saliva, which is about 100 times more powerful. Amylase begins to break down the starch in the feed. However, this is of very little nutritional importance because the feed does not stay in the mouth long enough for starch breakdown. The pH is too low (acidic) in the stomach for amylase to act on the starches. The stomach of the full-grown pig has a capacity of about 2 gals.

The enzymes contained in the gastric juices of the pig are those expected in a monogastric. One exception is that rennin, the enzyme that coagulates milk, is not found in the pig's gastric juice. Digestion of protein is completed in the intestine and most fat digestion occurs in the small intestine. The pig requires about 24 hours to empty a full stomach.

The small intestine of the full-grown pig is about 60 ft in length, has a capacity of about 2.5 gals, and is divided into three sections. The function of the small intestine is to continue the process of digestion by means of the pancreatic juice, bile, intestinal juice, and movements of the intestinal wall. A large amount of absorption of nutrients also occurs in the small intestine of the pig.

Pancreatic juice contains a number of enzymes that aid in the digestion of proteins, fats, and carbohydrates as well as sodium carbonate and sodium bicarbonate, which neutralizes acid from the stomach. The pancreas also produces insulin, which plays an important role in carbohydrate metabolism. Trypsin is initially secreted as trypsinogen and activated by calcium ions and the enzyme enterokinase, which is found in the intestine. Trypsin breaks down protein into amino acids and peptides. Chymotrypsin is secreted as chymotrypsinogen and activated by the action of trypsin in the intestine. It converts proteins into peptides and amino acids and has a coagulating action on milk. Carboxypeptidase acts on peptides and breaks them down to amino acids. All three of these enzymes continue the protein digestion that was started by pepsin in the gastric juice, and attack undigested proteins. Pancreatic lipase (steapsin) converts fats into fatty acids and glycerol. This action is most effective after the bile has emulsified the fats. Pancreatic amylase (amylopsin or pancreatic diastase) converts starch to maltose. Maltase changes maltose into glucose. Sucrase (invertase) changes sucrose to glucose and fructose. Lecithinase hydrolyzes the phospholipid lecithin.

Bile is secreted in the liver and stored in the gallbladder. It assists in digestion and absorption of fats and aids in absorption of fat-soluble vitamins. Bile may also activate pancreatic lipase and accelerate the action of pancreatic amylase.

The large intestine in the adult pig is about 16 ft in length, has a capacity of 2.5 gals, and consists of the cecum and the colon, which terminates as the rectum and anus. In the mature pig, the cecum is about 9.5 in. in length, with a capacity of about 0.5 gals. The colon in the mature pig is about 16 ft in length and has a capacity of 2.0 gals. The primary functions of the large intestine are to absorb water and to act as a reservoir for the waste materials that constitute the feces.

## DIGESTION IN THE RUMINANT

The most important difference between ruminants and nonruminants is the complex stomach of the ruminants and the very different digestion that takes place there. Because ruminants represent the largest percentage of the domestic herbivores, we discuss them in detail. The main function of the complex stomach of the ruminant is to allow the animal to use roughage (cellulose) as a source of energy. Microbial populations housed by the ruminant in the first three compartments ferment feed. The end products of the fermentation provide nutrients to the animal that would otherwise be unavailable for any productive use.

**Rumen**

The largest of the ruminant forestomachs. Contains microorganisms that degrade complex carbohydrates and produce volatile fatty acids, amino acids, and vitamins to the host animal.

A ruminant typically fills the **rumen** rapidly, taking little or no time to chew its meal, and then finds a place to rest and chew. During rumination, the animal regurgitates the food eaten earlier and spends time chewing it. The particle size of the feed is broken down mechanically in this way and its surface area is increased manyfold. This increased surface area allows the microorganisms many more places of access than would otherwise be available. When one mouthful is sufficiently dealt with, it is reswallowed, another is regurgitated, and the process is repeated. A ruminant may spend a third of its life in this process. The more fiber in a diet, the more the animal is forced to ruminate, and vice versa. Straw stays in the rumen much longer than feed of higher quality like corn silage, because of this need to ruminate and help the microorganisms.

Anatomy of a typical ruminant stomach is shown in Figure 6–4. The complex stomach is comprised of four compartments: the rumen, reticulum, omasum, and abomasum. Common names for these compartments are *paunch* for the rumen, *honeycomb* and *hardware stomach* for the reticulum, *manyplies* and *Stockman's Bible* for the omasum, and *true stomach* for the abomasum. The rumen and reticulum are often called the *rumino-reticulum* or *reticulo-rumen complex* because material passes freely between them and no wall separates them from each other.

The contents of the rumen generally equal about 20% of the body weight of the adult animal. This amounts to a very large fermentation capacity. For example, the combined capacity of the rumen, reticulum, and omasum is approximately 50 gals in an adult dairy cow. The relative volume of the rumen is 80%; the reticulum 5%, the omasum 7–8%, and the abomasum 8–9%. The relative amount of dry versus wet material in the rumen is variable depending on what the animal has consumed and what has been absorbed. This can range from 5–60 gals of liquid and from 5–50 lbs of solid.

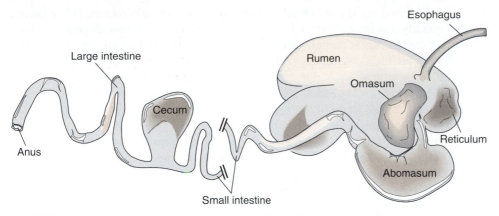

**FIGURE 6–4**    Digestive system of the ruminant. (*SOURCE:* Modeled on Jurgens, 1993, p. 44.)

The rumen, reticulum, and omasum are collectively known as the *forestomachs.* The lining of these organs is nonglandular and does not produce mucus or enzymes. The abomasum, or true stomach, is lined with a true mucous membrane and gastric juice is secreted there as in the stomach of all mammals. Generally speaking, from the abomasum to the anus, the ruminant tract is much the same as a monogastric tract.

The reticular groove is a groove that can contract and form a tube that acts as a by-pass of the rumen and empties into the omasum (Figure 6–5). This groove functions to keep milk out of the young ruminant's undeveloped rumen. If a calf drinks milk too quickly or if the groove does not close, milk can get into the rumen of the calf. The milk can't be adequately digested in the immature rumen and literally "rots." The calf develops a severe case of diarrhea (scours) owing to the proliferation of bacteria and production of toxins in the rumen.

The main function of the rumen is to act as a site of anaerobic bacterial fermentation. Undeveloped at birth, it begins developing in response to the young animal's eating solid food and can have some fermentation as early as 6–8 weeks of age. All of

**Forestomachs**

The name given to the three digestive compartments of the ruminant tract that are placed anatomically before the true stomach.

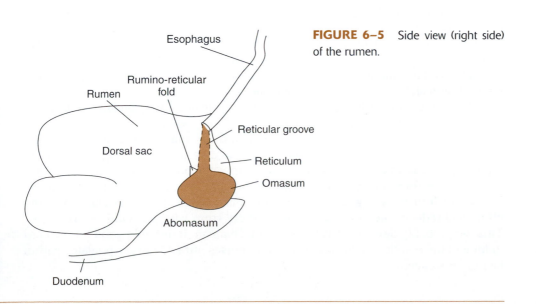

**FIGURE 6–5**    Side view (right side) of the rumen.

the food consumed by the ruminant animal enters the reticulo-rumen. Some passes through quickly and isn't affected, but 85–95% is fermented to some degree. The bacteria in the rumen digest carbohydrates (cellulose, starch, and so on) and other plant materials and produce volatile fatty acids (VFAs) as an excretory by-product. VFAs are absorbed by the animal across the rumen wall and supply about 50% or more (maybe up to 70% in some cases) of the energy requirement for the ruminant. The rumen wall is covered with small fingerlike projections called *papillae.* Papillae function to increase the absorptive surface of the rumen.

A major benefit of microbial fermentation is the protein manufactured by the microorganisms as they live and reproduce. Bacteria can even convert nonprotein nitrogen (NPN) such as urea into bacterial protein, which the animal subsequently uses. The bacteria pass out of the rumen and down the tract with the feed and are digested in the abomasum and small intestine. The amino acids released by the action of the animal's enzymatic digestion often meet the protein requirements of the animal. The microorganisms are about 50% crude protein with 3% lysine content. Their quality is similar to that of soybean meal, which is an excellent quality feedstuff. The bacteria also contain energy, vitamin K, and water-soluble vitamins that the animal is able to digest, absorb, and use. The formation of these vitamins by the rumen bacteria frees the ruminant from dietary water-soluble requirements under normal circumstances. The microorganisms do not synthesize vitamins A, D, and E. The bacteria also produce enough essential fatty acids for the animal.

Another important function of the rumen is to store food. This is probably how ruminants evolved—they would eat rapidly, run and hide in the woods, and chew their meal. This helped them avoid predators by minimizing their time in the open grasslands where they were vulnerable to attack.

The reticulum functions as a site of microbial action just the same as the rumen. The reticulum also acts as a pacemaker of rumen contractions. The contractions start with the reticulum and spread to the rumen. These rumen contractions are very strong and caused by contraction of the powerful rumen muscles. They function to mix the rumen contents and are essential for optimum microbial digestion. These contractions also serve to move the contents through the digestive tract. The heavier particles settle in the reticulum and are passed on through the omasum to the abomasum. The lighter particles float on the top of the rumen; these particles are then regurgitated and subjected to remastication.

The function of the omasum is unclear. It is probably involved in water and VFA absorption as well as reducing particle size of feed before it enters the omasum. Animals that have the omasum surgically removed suffer no ill effects and digestion proceeds as usual, so its value is questionable.

The abomasum of the ruminant is equivalent to the true glandular stomach in monogastrics and has the same functions.

*Eructation* is a very important mechanism for ruminants. The large quantities of gas produced by the rumen microorganisms as a by-product during fermentation (up to 600 L/day in a dairy cow) must be removed. Contractions of the upper sacs of the rumen force the gas toward the esophagus, which dilates, and the gas escapes. Much of the gas goes into the trachea and lungs. This provides a muffling effect and reduces the noise level considerably from what it would otherwise be. This too probably developed as a defense mechanism. It did little good for the herd to hide in the woods if a chorus of belching noises attracted every predator within hearing distance.

**Papillae**
Small fingerlike projections that greatly increase the surface area of the small intestine.

**Eructation**
Belching. Removing gas from the rumen via the esophagus.

## THE FERMENTATION PROCESS

Rumen microorganisms and the ruminant animal live in *symbiosis.* The animal benefits because the microorganisms digest feeds it could not otherwise use and generate nutrients it needs. This makes feeding the ruminant more complicated because feeding the ruminant actually involves feeding the microorganisms and feeding the animals. The ruminal environment is maintained by the animal to support the microorganisms, which are thus provided with a near ideal set of living conditions. These include warmth, moisture, a food supply, removal of the end products of digestion, darkness, and an *anaerobic* environment. In addition, ruminal contractions keep the contents mixed, and rumination reduces the particle size of the feed for easy access by the microorganisms.

Many different bacteria are found in the rumen. Concentrations can range from 15–50 billion/mL of rumen fluid. Nearly 40 species of ciliate protozoa also occupy the rumen. Typically, any individual animal has only a dozen or so different species. Many factors affect the concentrations of protozoa, which can number up to a half million per milliliter of rumen fluid. Protozoa dine on bacteria and eat food particles. Removing protozoa from the rumen (defaunation) may actually improve feed utilization. Protozoa are closely associated with the food particles and tend to stay in the rumen rather than passing on down the tract for digestion.

Virtually everything the ruminant consumes is affected by the rumen fermentation. Some feeds are changed tremendously; others less so. Both simple and complex carbohydrates are fermented in the rumen. VFAs are a major end product of ruminal fermentation. The VFAs commonly produced in the rumen are acetate, propionate, butyrate, isobutyrate, valerate, and isovalerate. The VFAs are very important. As an energy source, they may provide as much as 50–70% of the total energy needs of the animal. Bacterial cells provide 5–10% of the energy to the animal, and feed that is digested enzymatically amounts to 20–30% (Figure 6–6).

### Symbiosis
A relationship where dissimilar organisms live together or in close association.

### Anaerobic
Conditions that lack molecular oxygen.

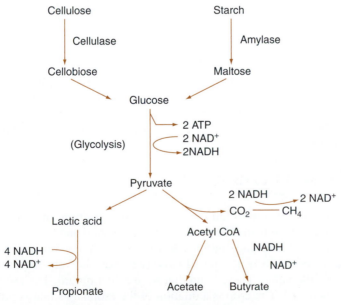

**FIGURE 6–6**  VFA synthesis in the rumen by microorganisms.

Much of the protein is broken down by microorganisms to ammonia and organic acids. The microorganisms then use the ammonia to manufacture amino acids for their own use. NPN sources can also be used by the microorganisms. This allows us to feed urea and other NPN sources to ruminants as a means of reducing costs. Most plant sources are low in lipids and those found in feeds aren't changed much by the fermentation process. High levels of fat are actually bad for the microbes and reduce the overall ability of the microorganisms to do their job.

Gas is a major product of fermentation and may amount to up to 600 L/day, as we mentioned earlier. Heat is also a major product of fermentation. The heat of fermentation is a part of the total heat increment of the animal, which is useful in winter to keep the body warm but can be detrimental in hot weather.

### Advantages and Disadvantages of the Ruminant System

There are both advantages and disadvantages to the ruminant digestive system when compared to that of the nonruminant animal. Microbial fermentation has the potential to digest feedstuffs that would not be digested by the animal's enzymatic digestion processes. This increases the feeding value of such feeds as prairie hay, corn cobs, and wheat straw. Essential nutrients not found in the animal's diet such as vitamins, amino acids, and fatty acids are also manufactured by the microbes. These are distinct advantages because nutrients can be absorbed and used by the animal. If microbial fermentation occurred only in the cecum, most of these nutrients would be excreted in the feces.

A disadvantage is that the fermentation processes in the reticulo-rumen have the potential to decrease the overall quality of the ruminant animal's diet by destroying essential nutrients such as vitamins, amino acids, and fatty acids that are found in the feed. Much current interest has been expressed in "bypassing" high-quality feedstuffs such as soybean meal and starch and not allowing the microbes to digest them. The fermentation process also requires and wastes a considerable quantity of energy. A significant portion of the dietary energy is lost as heat and gaseous products of incomplete bacterial metabolism. This explains in part why the feed consumption required to result in 1 lb of body weight gain is about 6–8 lbs for a growing steer and only about 3 lbs for a growing pig. Fortunately, we do not have to feed ruminant animals high-quality feeds throughout their production cycle. The ruminant animal provides us with an ideal mechanism to convert low-quality feeds not fit for human consumption into high-quality food products. Greater susceptibility to digestive upsets such as bloat (the accumulation of massive quantities of gas in the rumen) and acidosis (reduced rumen pH), which damages the rumen wall and leads to lowered productivity and liver abscesses, are other disadvantages.

# DIGESTION IN THE AVIAN

The digestive organs of fowl (Figure 6–7) are similar to those of other monogastrics except for the lack of teeth and the presence of the gizzard and the crop. The mouth is not sharply separated from the pharynx. Most birds have no soft palate or teeth, and only small numbers of poorly developed salivary glands. Salivary amylase exists in the saliva but is of little value. There is no digestion in the mouth. The esophagus in the avian tract is modified. The crop is a dilation of the esophagus that is present in most species, but not all. It functions as a food storage organ and as a moistening reservoir. The size of the crop varies with the eating habits of the bird and between species. In

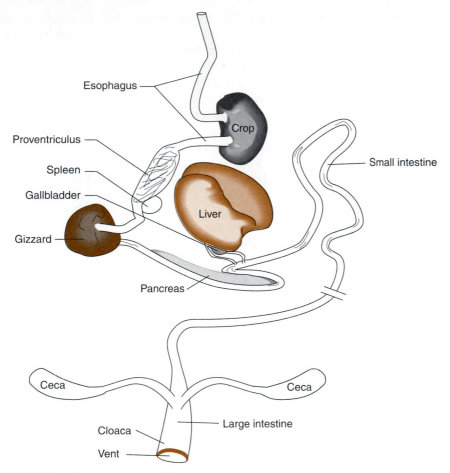

**FIGURE 6–7** Digestive system of the avian.

doves and pigeons, the crop produces "pigeon milk" or "crop milk," which is used to feed the young. Fermentation occurs in the crop in some species.

The proventriculus is equivalent to the glandular stomach in mammals or the abomasum in ruminants. It is small in some species, such as chickens or pigeons. It is large in some fish-eating species (i.e., those consuming high-protein diets). The proventriculus is the site of gastric juice production (HCl and pepsin) and has a pH of 4. Ingesta passes through in a matter of seconds, so virtually no digestion takes place. Carnivorous birds are an exception—there is digestion in their proventriculus.

The gizzard, or ventriculus, is a highly specialized grinding organ. It is very muscular but varies in muscularity depending on the type of food consumed. It may be large and well developed in wild animals and smaller and less developed in commercially raised species that eat preground rations. In free-ranging birds, the gizzard contains grit to aid in the grinding of feed to smaller particles; however, with modern preground rations grit is unnecessary. No enzymes are secreted; however, the enzymes from the proventriculus work here.

The small intestine functions in digestion and absorption of feed and nutrients just as in other monogastrics. The first part of the small intestine is the duodenum just as in other animals, but there is no specifically separated jejunum or ileum like that of mammals. The length of the duodenum varies among and between species, especially with eating habits. It is longer in herbivorous than in carnivorous birds. Generally the same enzymes found in mammalian species are present except for lactose. Milk by-products are not usually very good chicken feed for this reason.

The ceca are located at the junction of the small and large intestine. There are two ceca in the avian. The size is influenced by the type of diet—the ceca are larger when the animal consumes high fiber. They open into the large intestine via the muscular ileocecal valves. Water reabsorption occurs in the ceca along with some fiber digestion; perhaps as much as 18% of total fiber and water-soluble vitamin synthesis occurs in the ceca by bacteria. The modern chicken is not fed much fiber and therefore does not make use of its ceca.

The large intestine is relatively short in birds, only 2–4 in. It is not divided into a distinct rectum or colon as it is in mammals. From a digestive perspective, the large intestine is more important in water absorption than in any other function. However, vitamin K synthesis and absorption do occur here in the chicken. The cloaca is the common orifice for waste elimination (feces and urine), copulation, and egg laying in females.

## DIGESTION IN THE HORSE

Figure 6–8 illustrates the horse digestive system. The horse is a nonruminant herbivore. Horses, rabbits, and guinea pigs are all capable of using roughage because they have an active cecal bacterial population that digests fiber. In this section, we explore the digestion of the horse as a model for the cecal fermenters.

The horse accomplishes prehension with its teeth, a flexible upper lip, and tongue. The horse has both vertical and lateral jaw movements. In addition, the upper jaw is wider than the lower jaw. Because of this, horses chew on only one side of the mouth at a time. The saliva of the horse contains no enzymes but is important as a lubricant for the coarse material the horse eats. Total secretion is copious. A mature horse may secrete up to 10 gals per day if it is consuming hay or other similar material. The esophagus of the horse has only one-way peristaltic movements, which makes it almost impossible for the horse to regurgitate.

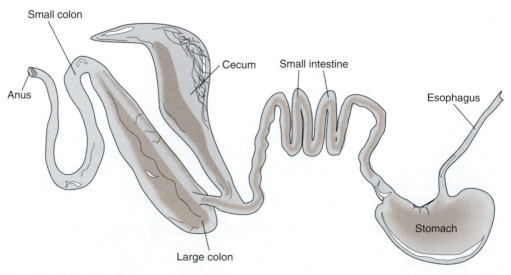

**FIGURE 6–8** Digestive system of the horse.

The stomach of the horse has some important distinctions compared to the stomachs of other mammals. Perhaps the most distinctive is that the stomach capacity is smaller. A pig less than half the size of a horse can have a stomach as large. A cow of equal size has an abomasum that is 50% larger. The horse's stomach provides only 8% of the capacity of the tract, or something on the order of 3–4 gals of capacity. The horse stomach is also unique in that it does not have equivalent muscular activity to that of other species. The feed tends to arrange itself in layers because of this lesser activity. These two factors result in the horse being more susceptible to stomach disorders than are other species. *Colic* and ruptured stomachs both occur, with colic very common. Horses should be fed small portions several times daily if possible. At a minimum, twice-a-day feeding is desirable, and sufficient hay should be fed to help avoid stomach disorders.

**Colic**

A broad term that means digestive disturbance, especially in horses.

The small intestine of the horse is similar to that of other monogastrics except for the absence of a gallbladder. The horse evolved as a continual eater and had no need for storage of bile to help with digestion of food consumed in meals. Instead, bile is secreted directly and fairly continually into the duodenum at a rate of approximately 300 mL/hr.

The large intestine of the horse is divided into the cecum, large colon, small colon, and rectum and is 25 ft long in an average-sized horse. It accounts for over 60% of the total gut capacity and is equal to about three fifths the size of the large intestine capacity in a comparably sized ruminant. The cecum and large colon contain a bacterial population similar to that of the rumino-reticulum of ruminants in both numbers and kinds of microorganisms. Bacterial fermentation produces VFAs similar to those of the ruminant, which are absorbed and used by the animal. The horse on a forage diet receives the majority of its energy in the form of these VFAs from the large intestine. These bacteria also synthesize water-soluble vitamins that are absorbed from the large intestine, but in limited amounts. Bacterial synthesis of protein occurs, but the horse is not able to absorb this product because of the lack of enzymes and absorptive mechanisms. Because this protein is not subject to action of the digestive juices, the horse doesn't benefit much, if any, from it unless it practices *coprophagy.* Sometimes, horses on a poor diet practice coprophagy in an attempt to balance their nutrient needs. The cecum is the primary area of water reabsorption from intestinal contents, although additional amounts are absorbed from the colon.

**Coprophagy**

The act of eating feces.

The soluble carbohydrates are digested and absorbed in the small intestine much as in any monogastric. The fiber fractions are fermented in the large intestine and colon. The end products are essentially the same as those of a rumen fermentation. The VFAs are even produced in almost identical molar concentrations as those in the rumen. The VFAs are absorbed from the large intestine and are used as an energy source by the cells of the horse. The horse is only about two thirds as effective at fiber digestion as the ruminant. This is generally believed to be because feed has a much more rapid rate of passage through the horse than the ruminant. Other reasons may yet be found.

The fats found in a horse's ration are digested and absorbed in the small intestine. The absence of a gallbladder does not seem to cause the horse any problem. Fats in rations of up to 20% fat are digested 90% or better. Proteins are digested in the horse just as in any other monogastric. A good deal of microbial protein is manufactured in the cecum, but the horse is not able to digest it. Even if the horse could digest the microbial protein, it would probably be unable to absorb it. Therefore, essential amino acids must actually be fed to horses. Horses should receive better feed than cattle receive if we expect them to perform to their potential.

## SUMMARY AND CONCLUSION

Different species have very different digestive tracts. These differences are both anatomical and functional and in many ways dictate how animals must be fed. However, for all their differences, there are also tremendous similarities. The overall goal of digestion is the same for all species: to release nutrients for the animal to absorb and metabolize in order to maintain and produce.

## STUDY QUESTIONS

1. Define digestion.
2. What are the three major methods of digestion? Describe them.
3. Why do different animals use very different kinds of feeds?
4. What are the stomach types in livestock?
5. Define carnivore, omnivore, and herbivore.
6. What is prehension?
7. Define mastication. How do different animals accomplish this?
8. What is salivation? How do mastication and salivation help the animal?
9. What do you think dictates the amount of saliva that various species produce?
10. What are the components and functions of saliva?
11. Define deglutition.
12. What is the role of the stomach in digestion?
13. What are the three portions of the small intestine, and what is the function of each?
14. What does bile do? Where does it come from?
15. What does pancreatic juice do? Include a description of the function of each enzyme found in pancreatic juice.
16. What is the function of the large intestine? What do bacteria do in the gut of a monogastric herbivore?
17. Why do rabbits practice cecotrophy?
18. Describe how an animal defecates.
19. What are the components of urine?
20. Describe how a bite of feed is digested by the pig from prehension to defecation.
21. What is the major anatomical difference in ruminants and nonruminants?
22. Describe the anatomy of the ruminant tract. What are the common names of the four ruminant digestive compartments found between the esophagus and the small intestine?
23. What is the relative volume of each compartment in the ruminant "stomach"?
24. What is the reticular groove and what does it do?
25. What is the function of the rumen?
26. Why are VFAs of importance in the rumen, and what does the animal do with them?
27. What is the relative quality of the protein manufactured in the rumen?

**28.** Name the nutrients produced as a result of rumen fermentation.

**29.** What is the function of the reticulum? The omasum? The abomasum?

**30.** Describe eructation and discuss the consequences if a ruminant is unable to eructate.

**31.** Describe symbiosis. How do ruminants and their microbes qualify for status as symbiotic partners?

**32.** What kinds of microorganisms are found in the rumen?

**33.** What is the relative importance of VFAs to the energy contribution of a ruminant animal?

**34.** What is the composition of the gas coming off a ruminant fermentation?

**35.** State clearly the advantages of the ruminant digestive system compared to that of the nonruminants.

**36.** Why don't ruminants practice cecotrophy?

**37.** What are the disadvantages of the ruminant digestive system?

**38.** Why is the feed conversion ratio for ruminants so much poorer than it is for monogastrics?

**39.** What are the major anatomical differences in the avian and the pig? Describe each along with its function.

**40.** How does digestion occur in the avian? What are the differences in the digestive processes of a chicken and of a pig?

**41.** Describe the significance to the horse of being a cecal fermenter.

**42.** How much saliva does the horse produce?

**43.** What are the unique features of a horse's stomach?

**44.** Describe in detail what happens in the large intestine and cecum of the horse.

**45.** Why must we feed horses their amino acids, and why don't we have to feed amino acids to cows?

# REFERENCES

Cheeke, P. R. 1999. *Applied animal nutrition: Feeds and feeding*. 2nd ed. Upper Saddle River, NJ: Prentice Hall.

Church, D. C., and W. G. Pond. 1988. *Basic animal nutrition and feeding*. 3rd ed. New York: Wiley.

Duke, G. E. 1984. Avian digestion. In *Duke's physiology of domestic animals*, 10th ed., Melvin J. Swenson, ed. Ithaca, NY: Cornell University Press.

Ensminger, M. E., J. E. Oldfield, and W. W. Heinemann. 1990. *Feeds and feeding digest*. Clovis, CA: Ensminger.

Evans, J. W., A. Borton, H. Hintz, and L. D. Van Vleck. 1990. *The horse*. New York: W. H. Freeman.

Gillespie, J. R. 1987. *Animal nutrition and feeding*. Albany, NY: Delmar.

Hibberd, C. A. 1990. *Animal nutrition manual*. Stillwater: Oklahoma State University, Animal Science Department.

Jurgens, M. H. 1982. *Animal feeding and nutrition*. 5th ed. Dubuque, IA: Kendall/Hunt.

Jurgens, M. H. 1993. *Animal feeding and nutrition*. 7th ed. Dubuque, IA: Kendall/Hunt.

Kellems, R. O., and D. C. Church. 1999. *Livestock feeds and feeding*. 4th ed. Upper Saddle River, NJ: Prentice Hall.

Maynard, L. A., J. K. Loosli, H. F. Hintz, and R. C. Warner. 1979. *Animal nutrition*. 7th ed. New York: McGraw-Hill.

McDonald, P., R. A. Edwards, and J. F. D. Greenhalgh. 1988. *Animal nutrition*. 4th ed. New York: Wiley.

Perry, T. W., A. E. Cullison, and R. S. Lowery. 1999. *Feeds and feeding*. 5th ed. Upper Saddle River, NJ: Prentice-Hall.

Pond, W. G., D. C. Church, K. R. Pond, and P. A. Schoknecht. 2005. *Basic animal nutrition and feeding*. 5th ed. New York: Wiley.

Swenson, M. J., ed. 1984. *Duke's physiology of domestic animals*. 10th ed. Ithaca, NY: Cornell University Press.

Teeter, R. G. 1988. *Animal nutrition manual*. Stillwater: Oklahoma State University, Animal Science Department.

# Feedstuffs Classification

## LEARNING OBJECTIVES

After you have studied this chapter, you should be able to:

1. Describe how feedstuffs are classified and identify the major categories of feedstuffs and their characteristics.

2. Identify the nutritive characteristics in various feedstuff categories.

## KEY TERMS

Anaerobic

Energy feed

Ensiling

Feed

First-limiting amino acid

Forage

Green forage

Lignin

Mineral supplement

National Research
   Council

Nonnutritive additive

Nutritive value

Pasture

Protein quality

Protein supplement

Range plants

Roughage

Silage

Vitamin supplement

Weathering

## INTRODUCTION

Feeds of many origins, qualities, and availabilities are used in animal diets in the United States and around the world. The nutritive content varies tremendously among them. Making sense of it all can be a daunting task. The National Research Council, a branch of the National Academy of Science, publishes a series of reports entitled *Nutrient Requirements of Domestic Animals*. Those publications use eight categories to group feedstuffs with others that have common characteristics. Feedstuffs within a group generally have similar nutritive values as well as other common characteristics. These categories help us organize types of feeds and give us a way to think about how we go about balancing a ration. In practical terms, most rations are balanced with the use of computers and databases that have information on every available feedstuff programmed into them. The category a feed falls into is not important to the computer. However, with the limits of the human mind, it is still useful to be able to put feeds into these categories: (1) dry forages and roughages; (2) pasture, range plants, and green forages; (3) silages; (4) energy feeds; (5) protein supplements; (6) mineral supplements; (7) vitamin supplements; and (8) nonnutritive additives.

## FEEDSTUFF CATEGORIES

### DRY FORAGES AND ROUGHAGES

Feeds placed in this category contain at least 18% crude fiber, with values ranging up to 50% crude fiber. Dry *forages* and *roughages* are high in cellulose, hemicellulose, and possibly lignin and low in readily digested carbohydrates such as starch and sugars. Consequently, they generally have a lower digestibility and therefore lower energy values than do concentrates. The protein content varies from nearly 30% for alfalfa to 2–3% for some straws. Because ruminants and cecal fermenters generally use these feeds, the quality of the protein is not usually a concern. It is hard to make general statements about the mineral and vitamin contents of these feeds because they vary so widely. Examples of feeds in this category are legume hay, grass hays, wheat straw, cornstalks, corncobs, cottonseed hulls, peanut hulls, and rice hulls (Figures 7–1, 7–2, and 7–3).

### PASTURE, RANGE PLANTS, AND GREEN FORAGES

Examples of feeds in this category are Bermuda grass pasture, sorghum-sudan grass, tall-grass prairie species, and wheat pasture (Figure 7–4). Many of these feeds could be harvested as dry feeds that would be classed in the previous category. The moisture content of these feeds is usually between 50–85% but can be quite variable. Winter range pasture in the range states may contain as little as 15–30% moisture, whereas wheat pasture can be as high as 90% moisture. The dry-matter nutritive content tends to be quite variable as well. Young, well-fertilized wheat pasture can have very high crude protein and can be very digestible, whereas late season prairie hay is the opposite.

**Forage**

Fiber-containing feeds like grass or hay. Contain at least 18% fiber but have high digestible energy (more than 70%).

**Roughage**

A bulky feedstuff with low weight per unit volume. Contains at least 18% fiber but can range up to 50%. Less digestible than forages.

**FIGURE 7–1**    Hay is an example of a dry forage and roughage.

**FIGURE 7–2** Beet pulp is a by-product of sugar production. It is often used in dairy rations and feeds for show cattle.

**FIGURE 7–3** Cottonseed hulls are a by-product of the cotton industry and an excellent fiber source.

## Silages

The process of *ensiling* plant material under *anaerobic* conditions produces silage (Figure 7–5). This is a common storage method for livestock feed. The plant material undergoes a controlled fermentation that produces acids. The acids then kill off the bacteria, molds, and other destructive organisms. As long as the silage is left undisturbed, it will keep for years. Many different materials can be ensiled. Corn silage is produced by chopping and ensiling the whole corn plant after the ears have formed. Other grain-producing species also produce good-quality silage, as do legume forage species, cannery waste, and roots and tubers. One common misconception is that ensiling improves the nutritive content of a feed. The opposite is actually true. The fermentation process uses nutrients and thus reduces the nutritive content of the material. This category of feedstuff also causes some confusion because the ensiling process can be used to preserve other

**Ensiling**

The process of producing silage from forage.

**Anaerobic**

Conditions that lack molecular oxygen.

**FIGURE 7–4**   Grazing is the world's most common use of growing forage.

**FIGURE 7–5**   Silage is plant material allowed to
ferment under anaerobic conditions in a silo such
as those pictured here.

products such as high-moisture corn. However, these other products are not automatically
classified as silage just because they are ensiled.

## Characteristics

The three categories previously discussed in this section have much in common in
terms of the plants found in each category, and thus their nutritive values are similar.
Most of the feeds from these three categories are commonly referred to as either
roughages or forages without further classification. A forage is generally considered to
be of higher quality than a roughage (Figure 7–6). Feeds in these categories provide the
bulk of the diets of the herbivorous species (ruminants and cecal fermenters) and as
such are the major feeds available for animal use in the United States and the rest of

(a)

(b)

**FIGURE 7–6** The hay (a) is a fine-stemmed forage of high nutritive value. The wheat straw (b) is a roughage of low nutritional value.

the world. For these reasons, a general discussion of them seems in order before discussing the remaining categories.

The characteristics of good-quality forage generally include being relatively immature when harvested by animals or by mechanical means; being green and leafy; having soft, pliable stems; being free from mold or mustiness; being palatable; and being free from foreign material. The further a feed gets from this ideal, the poorer the quality and the more likely it will be thought of more as a roughage than a forage. For example, under most range conditions an excess of high-quality forage is available during the growing season. However, animals need feed all year long. Thus a part of the growth must be retained to provide feed during the nongrowing months of the year. This feed is consumed as a mature, weathered, low-quality feed during the winter months and thought of as roughage.

It is common to divide forages and roughages into legumes (e.g., alfalfa, lespedeza, soybeans, and clovers) and grasses (e.g., prairie grasses, timothy, Bermuda grass, and wheat). Legumes are generally better quality feed than grasses because the former have a lower stem and a higher leaf content. Of course there are exceptions. Although wheat pasture is a grass, it may contain from 20–34% crude protein when it is in a young vegetative state. This protein level is higher than that of most legumes. For some nutrient parameters, there is actually little difference between legumes and grasses of equal maturity, but for other parameters, legumes are much higher in nutrient value. As a general rule of thumb, legumes and grasses have about the same energy content, but legumes have much higher protein, calcium, and carotene contents.

Many variables affect the nutritive content of forages and roughages. These include maturity at the time of harvesting, weather damage, soil fertility, plant species, and harvesting method. Maturity at the time of harvesting is perhaps the most important factor because all nutrients, except fiber, decrease in number with advancing maturity. Fiber increases with maturity. Young plants may contain only 20% crude fiber, but mature plants may have 40% or more. *Lignin* also increases as fiber increases.

The digestibility and palatability of a forage decrease with advancing maturity and increasing fiber level (Figures 7–7, 7–8). The rate of change is much greater for some plants than for others. For example, timothy, brome grass, and buffalo grass retain good palatability over a wide range of maturities. Orchard grass and lovegrass are very palatable and digestible when young, but they lose these characteristics quickly as they mature. The effects of maturity are more pronounced for grasses than for legumes. Nutrient changes with advancing maturity and *weathering* are illustrated in Table 7–1 and Figures 7–9, 7–10, 7–11, 7–12, 7–13, and 7–14. Table 7–2 gives the nutrient content of some representative forages and roughages.

### Lignin

Polymers of phenolic acids found in plants as part of the structural components of the plant.

### Weathering

Loss in nutritive value through exposure to the elements.

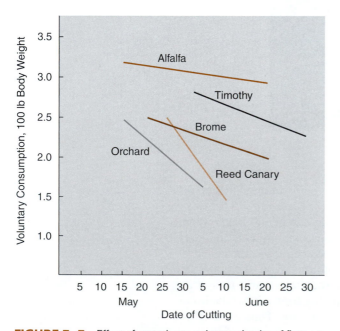

**FIGURE 7–7**   Effect of maturity on voluntary intake of first cutting forages by sheep. (*SOURCE:* Wagner, 1988, p. 50.)

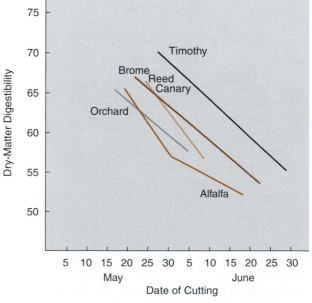

**FIGURE 7–8**   Effect of maturity on dry-matter digestibility of first cutting forages. (*SOURCE:* Wagner, 1988, p. 51.)

**TABLE 7–1** **Average Chemical Composition of Four Native Grasses during Different Months of the Year[1,2]**

| Month | Dry Matter (%) | Crude Proten (% on DM basis) | Crude Fiber (% on DM basis) | Calcium (% on DM basis) | Phosphorous (% on DM basis) | Carotene (mg/kg) |
|---|---|---|---|---|---|---|
| May | 44.0 | 10.01 | 30.69 | 0.30 | 0.13 | 174.8 |
| June | 45.5 | 7.84 | 32.23 | 0.30 | 0.99 | 195.6 |
| July | 48.3 | 6.04 | 33.11 | 0.35 | 0.09 | 145.2 |
| August | 56.5 | 4.92 | 34.83 | 0.31 | 0.08 | 81.4 |
| September | 60.8 | 3.99 | 35.51 | 0.28 | 0.07 | 40.4 |
| October | 66.5 | 3.79 | 35.80 | 0.32 | 0.06 | 20.2 |
| November | 81.4 | 2.55 | 38.56 | 0.25 | 0.03 | 3.6 |
| December | 94.4 | 2.63 | 38.07 | 0.30 | 0.04 | 0.9 |
| January | 93.4 | 2.49 | 38.45 | 0.34 | 0.09 | 0.4 |
| February | 93.6 | 2.52 | 37.21 | 0.29 | 0.04 | 0.3 |
| March | 96.0 | 1.95 | 43.19 | 0.21 | 0.03 | 0.0 |
| April | 92.6 | 2.91 | 39.97 | 0.29 | 0.04 | 0.6 |

[1]Data taken from *Chemical Composition of Native Grasses in Central Oklahoma from 1947 to 1962* by Waller, Morrison, and Nelson, Bulletin B-697, January 1972.
[2]Values represent averages for 15 years.
*SOURCE:* Wagner, 1988, p 47.

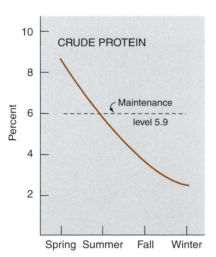

**FIGURE 7–9** Crude protein change with advancing age of the forage. (*SOURCE:* Wagner, 1988.)

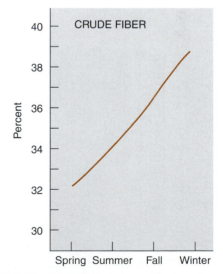

**FIGURE 7–10** Crude fiber change with advancing age of the forage. (*SOURCE:* Wagner, 1988.)

## ENERGY FEEDS

Energy feeds primarily include the cereal grains, by-product feeds made from cereal grains (e.g., corn hominy feed, wheat bran), and fruits and nuts. All are low in protein.

Feeds placed in this category contain less than 18% crude fiber or less than 35% cell wall and have a protein content of less than 20% (Figure 7–15). They are usually high in starch and NFE and are thus high in energy content. Protein supplements, which

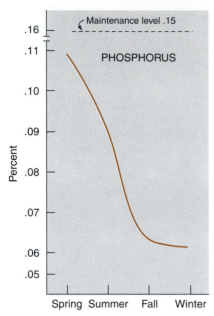

**FIGURE 7–11**   Phosphorus change with advancing maturity and weathering in forage. (*SOURCE:* Wagner, 1988.)

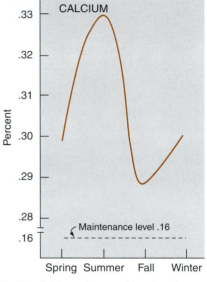

**FIGURE 7–12**   Calcium change with advancing maturity and weathering in forage. (*SOURCE:* Wagner, 1988.)

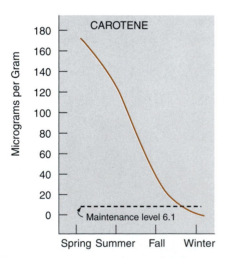

**FIGURE 7–13**   Carotene change with advancing maturity and weathering in forage. (*SOURCE:* Wagner, 1988.)

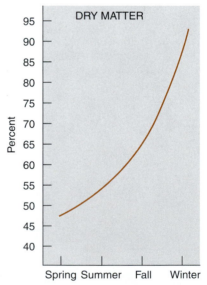

**FIGURE 7–14**   Dry-matter change with advancing maturity and weathering in forage. (*SOURCE:* Wagner, 1988.)

are discussed next, may have similar energy content but have greater than 20% crude protein. The cereal grains are very low in crude fiber, with the range being about 2–10%. Corn (Figure 7–16), sorghum (Figure 7–17), and wheat contain about 2%, barley about 6%, and oats 10–12%. Seeds that have a fibrous outer hull are higher in crude fiber. The lower the fiber levels, the higher the energy content tends to be because more readily digested carbohydrates such as starch and sugars will be present. The energy

**TABLE 7–2**     **Selected Nutrient Analysis of Some Forages and Roughages**

| Feedstuff | As Fed (% DM) | TDN (%) | NE$_m$ (Mcal/lb) | NE$_g$ (Mcal/lb) | CP (%) | EE (%) | CF (%) | ADF (%) | Ash (%) | Ca (%) | P (%) | K (%) | Mg (%) |
|---|---|---|---|---|---|---|---|---|---|---|---|---|---|
| Alfalfa, hay | 90.6 | 60.0 | 1.31 | 0.74 | 18.6 | 2.39 | 26.1 | 33.8 | 8.57 | 1.4 | 0.28 | 2.43 | 0.28 |
| Bermuda grass, fresh | 30.3 | 64.0 | 1.44 | 0.86 | 12.6 | 3.70 | 28.4 | 36.8 | 8.1 | 0.49 | 0.27 | 1.7 | 0.17 |
| Citrus pulp, silage | 21.0 | 78.0 | 0.86 | 0.57 | 7.3 | 9.7 | 15.6 | 25 | 5.5 | 2.04 | 0.15 | 0.62 | 0.16 |
| Corncobs, ground | 90.0 | 50.0 | 0.44 | 0.19 | 3.2 | 0.7 | 36.2 | 35 | 1.7 | 0.12 | 0.04 | 0.87 | 0.07 |
| Orchard grass, fresh, early bloom | 23.5 | 68.0 | 1.57 | 0.97 | 12.8 | 3.70 | 32.0 | 30.7 | 8.1 | 0.25 | 0.39 | 3.38 | 0.31 |
| Potato silage | 25.0 | 82.0 | 0.91 | 0.61 | 7.6 | 0.4 | 4.0 | 5 | 5.5 | 0.04 | 0.23 | 2.13 | 0.14 |
| Rice hulls | 92.0 | 12.0 | 0.00 | 0.00 | 3.3 | 0.8 | 42.9 | 72 | 20.6 | 0.10 | 0.08 | 0.57 | 0.83 |
| Sorghum silage, 30% DM | 30.0 | 60.0 | 0.60 | 0.34 | 7.5 | 3.0 | 27.9 | 38 | 8.7 | 0.35 | 0.21 | 1.37 | 0.29 |
| Wheat straw | 89.0 | 44.0 | 0.34 | 0.10 | 3.6 | 1.8 | 41.6 | 54 | 7.8 | 0.18 | 0.05 | 1.42 | 0.12 |

*SOURCE:* Bath et al., 1997; NRC, 1996.

TDN = total digestible nutrients
NEm = net energy for maintenance
NEg = net energy for gain
CP = crude protein
EE = ether extract
CF = crude fiber
ADF = acid detergent fiber
Ca = calcium
P = phosphorous
K = potassium
Mg = magnesium

**FIGURE 7–15** Energy feeds, generally the cereal grains or their by-products, contain less than 18% crude fiber (35% cell wall) and have a protein content of less than 20%. They are usually high in starch and NFE and are thus energy dense. Pictured here is ground wheat.

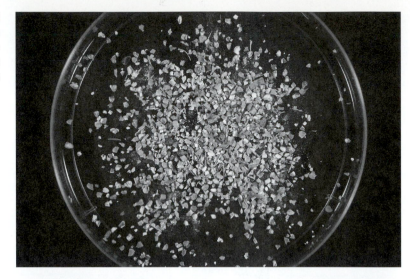

**FIGURE 7–16** Corn is the most common feed grain in the United States. This corn has been steam flaked.

**FIGURE 7–17** Grain sorghum is an excellent energy feed grown in drier parts of the United States.

**FIGURE 7–18** Wheat shorts is an energy feed that is also a by-product feed. By-product feeds usually have lower energy values and higher fiber levels than the original grain before processing.

value of grains is high, with the TDN as high as 90% on a dry-matter basis. These values are because of the high starch content (as high or higher than 70%), low fiber content, and high digestibility. The by-product feeds in this category usually have somewhat lower energy content because they contain more fiber and less starch as a result of processing (Figure 7–18). These feeds are fed to ruminants and cecal fermenters to increase the energy density of their rations, and to monogastrics as the primary source of energy for their diets.

Energy feeds are by definition below 20% in crude protein content. The cereal grains range between 8% and 12%. Some of the by-product feeds are higher. The protein digestibility ranges from 50–80% but the protein quality is generally poor. This is because the essential amino acid content is usually poor for grains. Lysine, methionine, and tryptophan are frequently the ***first-limiting amino acids*** in these feeds. If not first, they tend to be second or third limiting.

As a general statement, cereal grains are invariably very low in calcium (Ca), modest in phosphorus (P), and low in most trace minerals. Grains are very low in both vitamin D and carotene, which is a precursor for vitamin A. However, they are generally a good source of vitamin E and do contain some B vitamins. Table 7–3 shows the nutrient content of some representative energy feeds.

### PROTEIN SUPPLEMENTS

Protein supplements include feeds from three major sources. They are either of plant origin (e.g., soybean meal, cottonseed meal, and corn gluten meal), animal origin (e.g., fish meal, dried skim milk, and tankage), or nonprotein nitrogen (NPN) sources (e.g., urea, purified amino acids, and ammonium salts). Protein supplements are generally expensive feeds. Although balanced rations require less total protein than energy sources, the protein sources are expensive and represent the second largest source of expense in the ration. In some management and feeding situations, protein supplements are the largest out-of-pocket expense in the entire program.

Feeds placed in this category contain more than 20% crude protein. Some have high-energy contents as well. However, economics dictate that they be used to satisfy the protein needs of the animal. Because ruminants can convert the poorer quality

**First-limiting amino acid**

The first amino acid whose lack of availability restricts the performance of the animal.

**TABLE 7–3**   **Selected Nutrient Analysis of Some Common Energy Feeds**

| Feedstuff[1] | Dry Matter (%) | Crude Protein (%) | Ruminant TDN (%) | Poultry (kcal/kg) | ME (kcal/kg) | Swine TDN (%) | ME (kcal/kg) |
|---|---|---|---|---|---|---|---|
| Bakery products, dried | 91 | 10.0 | 82 | 1,650 | 3,630 | 1,635 | 3,600 |
| Barley, grain | 89 | 11.5 | 74 | 1,250 | 2,750 | 1,305 | 2,870 |
| Corn, yellow, grain | 86 | 7.9 | 80 | 1,540 | 3,390 | 1,520 | 3,350 |
| Hominy feed, corn expeller | 89 | 11.5 | 86 | 1,390 | 3,060 | 1,530 | 3,365 |
| Molasses, cane, dried | 91 | 7.0 | 80 | 1,080 | 2,375 | 1,165 | 2,560 |
| Oats, grain | 90 | 11.0 | 68 | 1,160 | 2,550 | 1,215 | 2,670 |
| Oat groats (dehulled oats) | 92 | 16.0 | 90 | 1,500 | 3,300 | 1,545 | 3,400 |
| Sorghum, milo, grain | 89 | 11.0 | 71 | 1,505 | 3,310 | 1,470 | 3,230 |
| Wheat, hard, grain | 88 | 13.5 | 76 | 1,440 | 3,170 | 1,465 | 3,220 |
| Wheat, soft, grain | 86 | 10.8 | 79 | 1,460 | 3,210 | 1,550 | 3,415 |
| Wheat bran | 89 | 14.8 | 62 | 590 | 1,300 | 1,055 | 2,320 |

[1]All table data are "as fed."

*SOURCE:* Dale, 1997. Used with permission.

**FIGURE 7–19**   Whole soybeans can be used as livestock feed. However, they are usually processed to remove their oil content for human consumption. The remaining soybean meal is a high-quality protein supplement.

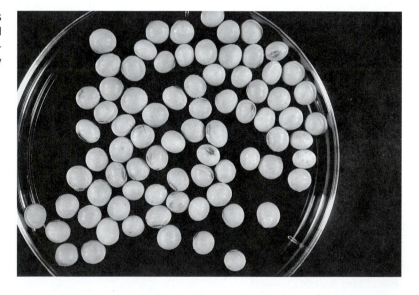

proteins to higher quality microbial protein, an effective cost-reduction strategy is to feed the NPN sources to ruminants and avoid the higher quality, and thus more expensive, of these. Very-high-quality, and thus very expensive, feeds must be used for some rations (e.g., baby pig rations, milk replacers).

The protein feeds of plant origin are primarily derived as an end product of the extraction of the oil from a group of seeds referred to as *oilseeds* because of their high fat content (Figure 7–19). These protein sources are thus referred to as *oilseed meals*. The most important of these sources are soybeans and cottonseed (Figure 7–20). However, significant amounts of flax, peanut, sunflower, sesame, and others are also available. The protein content is generally at least 40% and is highly digestible. The ***protein quality***

**Protein quality**

A measure of the presence and digestibility of the essential amino acids in a feedstuff.

| Feedstuff[1] | Swine TDN (%) | Carotene (mg/kg) | Vitamin A (IU/g) | Vitamin E (mg/kg) | Calcium (%) | Total Phosophorus (%) | Ash (%) |
|---|---|---|---|---|---|---|---|
| Bakery products, dried | 79 | 5 | 3.9 | 25 | 0.1 | 0.35 | 8.0 |
| Barley, grain | 70 | — | — | 36 | 0.08 | 0.42 | 2.5 |
| Corn, yellow, grain | 80 | 2 | 1.7 | 22 | 0.01 | 0.25 | 1.1 |
| Hominy feed, corn expeller | 82 | 9 | 15.3 | — | 0.05 | 0.5 | 3.0 |
| Molasses, cane, dried | 65 | — | — | 5.4 | 1.18 | 0.9 | 8.0 |
| Oats, grain | 65 | — | — | 20 | 0.3 | 0.35 | 4.0 |
| Oat groats (dehulled oats) | 84 | — | — | 15 | 0.07 | 0.45 | 2.2 |
| Sorghum, milo, grain | 78 | — | — | 12.2 | 0.04 | 0.29 | 1.7 |
| Wheat, hard, grain | 79 | — | — | 15.5 | 0.05 | 0.41 | 2.0 |
| Wheat, soft, grain | 83 | — | — | 15.5 | 0.05 | 0.3 | 2.0 |
| Wheat bran | 57 | — | — | 10.8 | 0.14 | 1.17 | 6.4 |

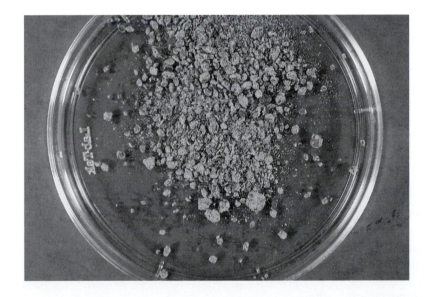

**FIGURE 7–20** Cottonseed is a by-product of cotton production. After the oil is extracted from the seeds for human use, cottonseed meal, a high-quality protein supplement, remains.

varies but is generally good. Lysine, cystine, and methionine levels are commonly low. Soybean meal is different in that its lysine level is usually higher. As you recall, the essential amino acid content is usually poor for grains with lysine, methionine, and tryptophan frequently being the first-, second-, and third-limiting amino acids in these feeds. These amino acids may need to be provided as purified amino acids, or animal-based protein supplements can be added to make up the deficiencies. The energy content varies depending on how much of the oil was removed in the extraction process. As a general statement, the oilseeds are low in calcium and high in phosphorus. Caution must be used when balancing rations for monogastrics because half or more of the phosphorus can be tied up as phytic-bound forms and is unavailable to the animals. The trace mineral content is variable but is generally considered to be low. Oilseeds are low

**FIGURE 7–21**    Blood meal is an example of a slaughter by-product used as a protein supplement.

in carotene, which is a precursor for vitamins A and E. They are low-to-moderate sources of B vitamins.

The protein feeds of animal origin are primarily derived as end products of the meat packing, dairy processing, and marine industries. The most important of these are meat meal, bonemeal, blood meal (Figure 7–21), feather meal, dried milk, and fish meal. The milk products are the highest quality of the end products and generally the most expensive. Good fish meals can rival milk products in quality. In addition, they generally contain much higher quantities of total protein. The lysine content of fish meal is considerably higher than that of other commonly available protein sources. Fish meals are usually good mineral and B vitamin sources. Milk and fish products are usually used for monogastric and young ruminant rations because of their high quality and subsequent cost. The quality of the protein in the meat products is usually lower than that of the milk and fish products. However, the quality is still good and their use is usually restricted to monogastrics. Some of the meat products have high mineral contents, depending on the percentage of bone they contain. The vitamin content is highly variable but generally low because of the types of processing required to make these feeds usable. Feather meal is a low-quality protein supplement, but it is very high in total protein content (over 90%). It is best used in ruminant rations, although it can be used as a part of monogastric rations.

The NPN sources technically include such a wide range of material that generalizations are impossible. Purified amino acids are technically NPN. However, because of the costs of such feeds, the practical use of the term NPN is for urea (Figure 7–22) and other similar products. Urea and similar products must be used with functional ruminants only, and then very carefully. The ruminant microbes are able to use substantial amounts of NPN and, because they are frequently of lower cost than proteins, they are often used to cheapen a ration. They are not a significant source of other nutrients as a rule. Table 7–4 shows the crude protein and amino acid contents of some commonly used protein supplements.

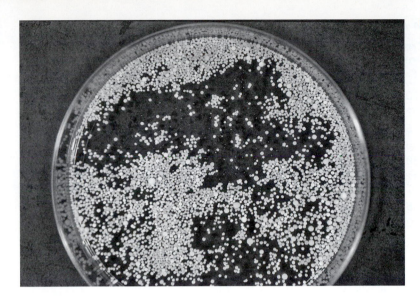

**FIGURE 7–22** Feed-grade urea is used as a nonprotein nitrogen source in ruminant rations.

## MINERAL AND VITAMIN SUPPLEMENTS

Virtually all feeds contain at least some vitamins and minerals. Animals need these nutrients in much smaller amounts than they do the other nutrients. Nevertheless, dietary needs must be met in a satisfactory manner to achieve good animal performance and economical production. Depending on the feeds used to balance the ration for the other nutrients, concentrated sources of vitamins and/or minerals may be needed.

Mineral supplements in common use include salt (often trace mineralized) (Figure 7–23), bonemeal, oyster shell (Figure 7–24), calcium carbonate, limestone, and fairly pure forms of other specific minerals such as selenium, cobalt, and others. In addition, a wide variety of premixed, complete, mineral supplements are generally available. The minerals of concern usually include sodium, calcium, phosphorus, magnesium, copper, iron, manganese, zinc, cobalt, and selenium. Table 7–5 shows the mineral content of some representative mineral supplements.

Vitamin supplements include such products as ensiled yeast, liver meal, fish oil, and wheat germ oil, as well as purified forms of individual vitamins. Although rations can be formulated and balanced for each individual vitamin easily with modern computer balancing programs, in practice, vitamin supplements are often added without regard for the amounts that may be found naturally in the feeds used in the ration. This is because of the highly variable availability of vitamins found in natural feedstuffs. Those vitamins usually supplemented in animal rations include vitamins A, D, and E for functional ruminants; and vitamin K, riboflavin, pantothenic acid, niacin, choline, folic acid, biotin, and cyanocobalamine for other species. Table 7–6 shows the vitamin content of a standard vitamin premix supplement (Figure 7–25).

## NONNUTRITIVE ADDITIVES

This is a catchall category for a large group of feed ingredients added to rations for some reason other than their nutritive value. They may be used to stimulate growth or

**TABLE 7–4**     **Crude Protein and Amino Acid Content of Some Protein Supplements[1]**

| Ingredients | Dry Matter (%) | Crude Protein (%) | Methionine (%) | Cystine (%) | Lysine (%) | Tryptophan (%) | Threonine (%) |
|---|---|---|---|---|---|---|---|
| Blood meal, animal | 89 | 80.0 | 1.0(91) | 1.4(76) | 6.9(86) | 1.0 | 3.8(87) |
| Brewers dried grain | 93 | 27.9 | 0.6 | 0.4 | 0.9 | 0.4 | 1.0 |
| Brewers dried yeast | 93 | 45.0 | 1.0 | 0.50 | 3.4 | 0.8 | 2.5 |
| Canola meal | 91 | 38.0 | 0.7(90) | 0.47(75) | 2.3(80) | 0.44 | 1.71(78) |
| Casein, dried | 90 | 80.0 | 2.7(99) | 0.3(84) | 7.0(97) | 1.0 | 3.8(98) |
| Cottonseed meal, 41%, direct solvent | 90 | 41.0 | 0.51 | 0.62 | 1.76 | 0.52 | 1.35 |
| Feather meal, poultry | 93 | 85.0 | 0.55(76) | 3.0(59) | 1.05(66) | 0.4 | 2.8(73) |
| Fish meal, herring, Atlantic | 93 | 72.0 | 2.2 | 0.72 | 5.7 | 0.8 | 2.88 |
| Meat and bonemeal, 45% | 92 | 45.0 | 0.53 | 0.26 | 2.2 | 0.18 | 1.8 |
| Milk, whole dried, feed grade | 96 | 25.5 | 0.62 | 0.4 | 2.26 | 0.41 | 1.03 |
| Peanut meal, solvent | 92 | 48.0 | 0.42 | 0.73 | 1.77 | 0.5 | 1.16 |
| Soybean meal, solvent | 90 | 44.0 | 0.65 | 0.67 | 2.9 | 0.60 | 1.7 |
| Yeast, Torula, dried | 93 | 48.5 | 0.80 | 0.6 | 3.8 | 0.5 | 2.6 |

[1]Numbers in parentheses represent percentage availability.

*SOURCE:* Dale, 1997. Used with permission.

**FIGURE 7–23**   NaCl and a variety of trace minerals are collectively referred to as trace mineral salt. It is used in loose form such as this for mixing with other feeds, or it may be fed in blocks that animals lick when they crave salt.

some other type of production, improve feed efficiency, enhance health, or alter metabolism. Feedstuffs in this category include antibiotics, coloring agents, flavors, hormones, and medicants. Examples of substances in this category include monensen sodium (makes rumen fermentation more efficient), butylated hydroxytolulene (antioxidant), aluminum sulfate (used as an anti-gelling agent for molasses), monosodium glutamate (flavor enhancer used in pet foods), propylene glycol (emulsifying agent),

| Ingredients | Isoleucine (%) | Histidine (%) | Valine (%) | Leucine (%) | Arginine (%) | Phenylalanine (%) |
|---|---|---|---|---|---|---|
| Blood meal, animal | 0.8(78) | 3.05(84) | 5.2(87) | 10.3(89) | 2.35(87) | 5.1(88) |
| Brewers dried grain | 2.0 | 0.47 | 1.69 | 3.2 | 1.3 | 1.82 |
| Brewers dried yeast | 2.2 | 1.3 | 2.37 | 3.2 | 2.2 | 1.86 |
| Canola meal | 1.51(83) | 1.10(85) | 1.94(82) | 2.6(87) | 2.3(90) | 1.5(87) |
| Casein, dried | 5.7(98) | 2.5(96) | 6.8(98) | 8.7(99) | 3.4(97) | 4.6(99) |
| Cottonseed meal, 41%, direct solvent | 1.33 | 1.1 | 1.82 | 2.4 | 4.66 | 2.23 |
| Feather meal, poultry | 2.66(85) | 0.28(72) | 4.55(82) | 7.8 (82) | 3.92(83) | 2.66(85) |
| Fish meal, herring, Atlantic | 3.0 | 1.91 | 5.7 | 5.1 | 5.64 | 2.56 |
| Meat and bonemeal, 45% | 1.7 | 1.5 | 2.4 | 2.9 | 2.7 | 1.8 |
| Milk, whole dried, feed grade | 1.33 | 0.77 | 1.74 | 2.57 | 0.92 | 1.33 |
| Peanut meal, solvent | 1.76 | 0.95 | 1.88 | 3.70 | 4.55 | 2.04 |
| Soybean meal, solvent | 2.5 | 1.1 | 2.4 | 3.4 | 3.4 | 2.2 |
| Yeast, Torula, dried | 2.9 | 1.4 | 2.90 | 3.5 | 2.6 | 3.0 |

**FIGURE 7–24** Oyster shell is a valuable source of calcium and frequently used in laying hen diets.

and aluminum potassium sulfate (color additive). Not all additives are fed. Some can be given to the animal as an injection or implant.

The list of nonnutritive additives changes over time. Those additives classified as drugs must be approved by the Food and Drug Administration (FDA). A good reference to become familiar with is the *Feed Additive Compendium*. Updated yearly, this is an invaluable resource for nutritionists.

**TABLE 7–5  Mineral Content of Some Representative Mineral Supplements**

| | Calcium (%) | Total Phos. (%) | Ash (%) | Sodium (%) | Potassium (%) | Magnesium (%) | Fluorine (%) | Manganese (ppm)[1] | Iron (ppm) | Copper (ppm) | Zinc (ppm) | Selenium (ppm) |
|---|---|---|---|---|---|---|---|---|---|---|---|---|
| Bonemeal (steamed) | 24.0 | 12.0 | 71.0 | 0.46 | n/a[1] | 0.64 | n/a | 30.4 | 840 | 16.3 | 424 | n/a |
| Calcium carbonate | 38.0 | —[1] | 95.8 | 0.06 | 0.06 | 0.5 | n/a | 279 | 336 | 24 | n/a | 0.07 |
| Diammonium phosphate (N-18%) | 0.5 | 20.0 | 34.5 | 0.04 | — | 0.45 | 0.2 | 500 | 15,000 | 80 | 300 | n/a |
| Deflourinated phosphate | 33.0 | 18.0 | 99.0 | 4.5 | 0.09 | — | 0.2 | 220 | 9,200 | 22 | 44 | 0.6 |
| Dicalcium phosphate | 20.0 | 18.5 | 85.6 | 0.08 | 0.07 | 0.6 | 0.18 | 300 | 10,000 | 80 | 220 | 0.6 |
| Phosphoric acid, 75% | — | 23.8 | n/a | n/a | n/a | n/a | — | n/a | 5 | n/a | n/a | n/a |

| | Potassium (%) | Magnesium (%) | Iron (%) | Copper (%) | Manganese (%) | Zinc (%) | Cobalt (%) | Sulfur (%) | Selenium (%) | Sodium (%) |
|---|---|---|---|---|---|---|---|---|---|---|
| Copper sulfate ($CuSO_4 5H_2O$) | — | n/a | n/a | 25.0 | n/a | n/a | — | — | — | — |
| Cobalt sulfate ($CuSO_4 7H_2O$) | — | 0.04 | 0.001 | 0.001 | 0.002 | — | 21.0 | — | — | — |
| Ferrous sulfate ($FeSO_4 7H_2O$) | — | 0.05 | 21.0 | 0.01 | 0.12 | 0.01 | — | 11.0 | — | — |
| Manganese sulfate ($MnSO_4 H_2O$) | — | 0.03 | 0.04 | — | 25.0 | — | — | 19.0 | — | — |
| Magnesium sulfate ($MgSO_4$) | — | 20.0 | n/a | n/a | n/a | n/a | n/a | 26.6 | — | — |
| Potassium sulfate ($K_2SO_4$) | 44.8 | — | — | — | — | — | — | 18.3 | — | — |
| Sodium selenite ($Na_2SeO_3$) | — | — | — | — | — | — | — | — | 45.6 | 26.6 |
| Zinc oxide (ZnO) | — | 0.5 | 0.8 | 0.07 | 0.01 | 73.0 | — | 1.0 | — | — |

[1]ppm = parts per million; n/a = data not available; — indicates that the ingredient does not contain a significant amount of nutrient.

*SOURCE*: Dale, 1997. Used with permission.

**TABLE 7-6**   **Vitamin Premix for Swine[1]**

| Vitamin | Amount/lb of Premix[1] | Suggested Source |
|---|---|---|
| Vitamin A | 2,000,000 IU[2] | Vitamin A palmitate-gelatine coated |
| Vitamin D | 200,000 IU | Vitamin $D_3$—stabilized |
| Vitamin E | 10,000 IU | dl-Tocopherol acetate |
| Vitamin K (menadione equivalent)[2] | 800 mg | Menadione sodium bisulfite |
| Riboflavin | 1,200 mg | Riboflavin |
| Pantothenic acid | 4,500 mg | Calcium pantothenate |
| Niacin | 9,000 mg | Nicotinamide |
| Choline | 20,000 mg | Choline chloride (60%) |
| Vitamin $B_{12}$ | 5 mg | Vitamin $B_{12}$ in mannitol (.1%) |
| Folic acid | 300 mg | Folic acid |
| Biotin | 40 mg | D-biotin |

[1]Premix is designed to be used at a rate of 5 lbs per ton of complete feed for sows and baby pigs, and 3 lbs per ton of complete feed for growing-finishing swine.
[2]A standard unit of potency. Defined by the International Conference for Unification of Formulae.
SOURCE: Luce et al., 1998. Used with permission.

**FIGURE 7–25** Vitamins are often mixed in appropriate individual quantities and blended with an inert carrier so they can be added to rations. Such *vitamin supplements* facilitate the balancing of rations and improve the accuracy of ration mixing.

## SUMMARY AND CONCLUSION

This chapter was designed just to give you an overview of feeds and their general uses and nutritive values. The system of categorizing feedstuffs used in this chapter is the one found in the NRC publications, which is generally used and accepted. Although the chapter has provided generalities about the feeds classifications, exceptions are easy to find. Research is continually telling us more about feeds and their nutrient availabilities. Even identical species of plants can produce different nutrient levels in a feedstuff because of the many environmental factors that affect nutrient content. The by-product feeds tend to change because the processes that generate them change. Plant breeders are continually developing new and nutritionally different crops. With the advent of accelerating recombinant DNA technology, even more rapidly changing nutritive values can be expected. All of these changes tend to cause the lines between nutrient classes to blur. It is important to keep up with changes in the value of the various feedstuffs. Consult current NRC publications and scientific journals. An additional important resource with high credibility with academics and industry leaders is the annual *Feedstuffs Reference Issue,* which can be purchased from Feedstuffs, Circulation Department, 191 S. Gary Ave., Carol Stream, IL 60188, (630)462–2883.

## STUDY QUESTIONS

1. What are the eight categories of feedstuffs as recognized by the NRC?
2. What are the similarities and differences in the first three categories described in this chapter?
3. What is the difference between a forage and a roughage? What are the characteristics of a good-quality forage?
4. What is the definition of an energy feed? What is the definition of a protein feed? What is the difference between an energy feed and a protein feed?
5. What are the differences and similarities of protein supplements from the three major sources of protein supplements?
6. What are the major similarities and differences in vitamin supplements and mineral supplements?
7. For what purposes are nonnutritive additives added to rations?
8. Why must one have a good, current source of nutrient composition for feedstuffs?

## REFERENCES

Bath, D., J. Dunbar, J. King, S. Berry, and S. Olbrich. 1997. Byproducts and unusual feeds. *Feedstuffs* 69 (30):32.

Dale, N. 1997. Ingredient analysis table: 1997 edition. *Feedstuffs* 69 (30): 24.

*Feed additive compendium.* 1998. Minnetonka, MN: Miller Publishing Co., and Alexandria, VA: The Animal Health Institute.

Luce, W. G., A. F. Harper, D. C. Mahan, and G. R. Hollis. 1998. Swine diets. *Pork industry handbook.* Lafayette, IN: Media Distribution Center.

NRC. 1996. *Nutrient requirements of beef cattle.* 7th ed. Washington, DC: National Academy Press.

Pond, W. G., D. C. Church, K. R. Pond, and P. A. Schokhecht. 2005. *Basic animal nutrition and feeding.* 5th ed. New York: Wiley.

Thaler, R. C., and R. C. Wahlstrom. 1998. Vitamins for swine. *Pork industry handbook.* Lafayette, IN: Media Distribution Center.

Wagner, D. G. 1988. *Livestock feeding.* 2nd ed. Stillwater: Oklahoma State University.

# Genetics

## LEARNING OBJECTIVES

After you have studied this chapter, you should be able to:

1. Explain the role that genetics plays in animal production.

2. Describe the location of genes within a cell.

3. Explain the process of cellular division with relation to the replication of cells containing a full complement of genetic information.

4. Explain the process of cellular division that ultimately produces cells containing only half of the genetic information.

5. Describe how variation in traits is passed from parent to offspring.

6. Describe how gene frequencies change in a population.

7. Explain the concept of relationship between individuals.

8. Describe several systems of mating individuals.

9. Summarize the implications of genetic engineering, the promise it holds for future animal production, and the opportunities that animals will have to provide even greater benefits to humanity.

## KEY TERMS

Additive gene action
Alleles
Aneuploidy
Animal breeding
Artificial selection
Autosomes
Bases
Biotechnology
Centromere
Chromosome

Codominance
Crossbreeding
Deoxyribonucleic
  acid (DNA)
Diploid
DNA polymerase
DNA replication
Dominant
Epistasis
Expression

Gametes
Gametogenesis
Gene
Gene frequency
Genetic drift
Genetic engineering
Genome
Genotype
Genotypic frequency
Haploid

Heritability
Heterosis
Heterozygous
Homologous
    chromosomes
Homozygous
Inbreeding
Inbreeding depression
Incomplete dominance
Inheritance
Linebreeding
Locus
Marker-assisted selection
Meiosis
Messenger RNA
Migration
Mitosis
Monosomy

Multiple alleles
Mutations
Natural selection
Nucleotide
Oocyte
Outbreeding
Phenotypic frequency
Polymerization
Polyploidy
Population genetics
Principle of independent
    assortment
Principle of segregation
Purines and pyrimidines
Qualitative traits
Quantitative traits
Recessive
Ribonucleic acid (RNA)

Ribosomes
Selection
Selection differential
Sex-influenced
    inheritance
Sex-limited traits
Sex-linked inheritance
Somatic cells
Sperm
Testcross
Traits
Transcription
Transfer RNA (tRNA)
Transgenic
Translation
Trisomy

## INTRODUCTION

*Genetics* could be termed the foundation of life, for without the ability to transfer genetic information from one generation to the next, existence would be impossible. In the nature versus nurture debate, genetics is the nature part. Inheritance takes place by the transmission of genes, in the form of chemical entities, from parent to offspring at the time of conception. During this transfer of molecular material, certain information is passed on to the offspring that is combined to form a blueprint of traits and characteristics that will describe both the physical appearance and the molecular composition of the animal. An animal's genetic makeup, or ***genotype***, sets the stage for disposition, coat type, coat color, speed, gait types, body composition, growth, reproduction, milk production, disease resistance, and other traits.

A large part of how efficiently animal products can be produced is related to the genetic composition of the animal or herd of animals. The ***expression*** of the genotype into traits of economic importance provides the basis for the animal's worth when marketed. Because there are many ways to market an animal, it is important to produce animals with the necessary genotype for maximum value in the target market.

Within each major animal species, producer and consumer preferences set the pace for type in the animals that are produced. From a livestock producer standpoint, efficiency of production might be the most important characteristic, with type, kind, or even disposition also considered. However, when viewing animals from a consumer's position, tenderness, flavor, color, and leanness might top the list of important characteristics. This is not to say that these characteristics don't overlap between the producer and the consumer, because they surely do. For instance, the consumer is interested in cost, which is related to efficiency of production. However, the priorities on each side are often different. In the companion species, there too may be differences in producer and consumer concerns. An elite breeder may strive to breed an international champion.

**Genotype**

The genetic makeup of an organism.

**Expression**

Manifestation of a characteristic that is specified by a gene.

(a)                                                    (b)

**FIGURE 8–1**    The application of animal breeding and selection techniques has led to remarkable changes, such as arranging the genes of the wild boar (a) into those of the modern meat-type hog (b).

The average "consumer" may just want a healthy dog with no major flaws and a good disposition that will be a good companion.

Applied genetics in animals is usually referred to as *animal breeding* (Figure 8–1). It is the science that helps in the quest to breed better animals. The practice of breeding and selection has led to remarkable changes in animal species. The wolf has been transformed into dogs as different as the Saint Bernard and the Chihuahua. The wild aurochs became the specialized milk, meat, or work breeds of modern cattle. The wild boar, the Red Jungle Fowl, the vicuña, the Siberian hamster, and several score of other wild animals have been converted to the modern domestic species we know today. Interestingly, the greatest changes in domestic species through the practice of animal breeding occurred before the science was named, and before the scientific basis of inheritance was discovered. Dogs have been dogs rather than wolves for at least 10,000 years. *Deoxyribonucleic acid (DNA)*, the stuff genetic codes are made of, was just determined to be the genetic material in 1953. However, research in *genetic engineering* has made possible advances that will come to dwarf those earlier accomplishments, and it will do so in the span of a few decades rather than a few millennia.

# THE GENE

The nucleus of the cell contains the *chromosomes.* Chromosomes are large molecules. Within these large molecules are smaller units called *genes*. The genes contain the information that controls all of the biochemical processes of the cell. By controlling the biochemical processes of the cell, genes control the life processes. A gene is a segment of DNA that codes for a specific protein. These DNA molecules are in the shape of a double helix. DNA comprises chromosomes, which are found in pairs. It is this arrangement of genes, DNA, and chromosomes that provides the basis for inheritance (Figure 8–2).

Because genes are segments of DNA, we should look at the structure of DNA to determine the existence of genetic material. Deoxyribonucleic acid (DNA) consists of two strands comprised of alternating sequences of the sugar deoxyribose and phosphate bonds. At each sugar, there is a bridge of nitrogen *bases* composed of chemical compounds called *purines and pyrimidines*. The purines present in a DNA molecule are adenine (A) and guanine (G); the pyrimidines are thymine (T) and cytosine (C). The bridges are always combined, with adenine attaching to thymine and guanine with

## Deoxyribonucleic acid (DNA)

Chemically, a complex molecule composed of nucleotides joined together with phosphate sugars. Chromosomes are large molecules of DNA.

## Genetic engineering

The term most frequently used to describe the technologies for moving genes from one species to another.

## Chromosome

The DNA-containing structures in cells. Composed of segments called genes.

## Gene

A short segment of a chromosome. Genes direct the synthesis of proteins or perform regulatory functions.

## Bases

One of the four chemical units on the DNA molecule that form combinations that code for protein manufacture. The four bases are adenine (A), cytosine (C), guanine (G), and thymine (T).

## Purines and pyrimidines

Organic ring structures made up of more than one kind of atom (heterocyclic compounds). Purines and pyrimidines contain nitrogen in addition to carbon.

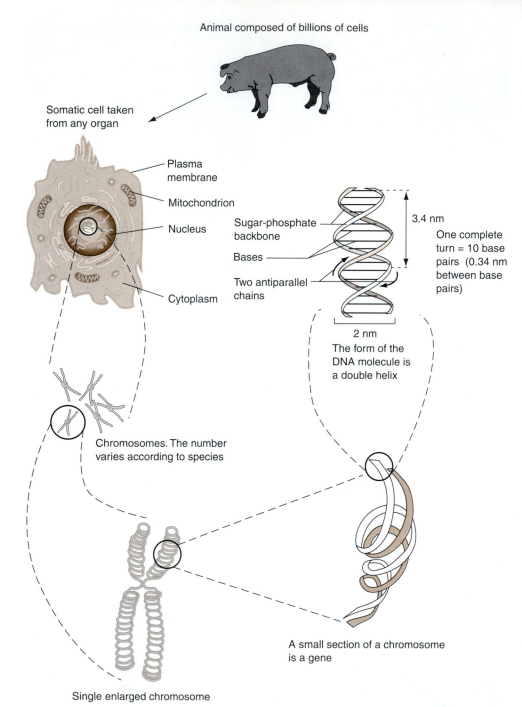

Animal composed of billions of cells

Somatic cell taken
from any organ

Plasma
membrane

Mitochondrion

Nucleus

Sugar-phosphate
backbone

Bases

Two antiparallel
chains

Cytoplasm

3.4 nm

One complete
turn = 10 base
pairs  (0.34 nm
between base
pairs)

2 nm
The form of the
DNA molecule is
a double helix

Chromosomes. The number
varies according to species

A small section of a chromosome
is a gene

Single enlarged chromosome

**FIGURE 8–2**   Location and structure of genetic material. (*SOURCE:* Adapted from Alcamo, 1996, p. 10.)

## Nucleotide

The building blocks of nucleic acids. Each nucleotide is composed of sugar, phosphate, and one of four nitrogen bases.

cytosine. The bases are attached with hydrogen bonds (Figure 8–3). A and T are attached with double bonds and G and C are held together with triple bonds, which give them a stronger hold. The segment of deoxyribose, phosphate, and one of the bases is called a ***nucleotide***. A gene is a segment of the double helix consisting of several nucleotides.

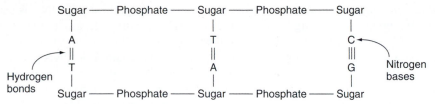

**FIGURE 8–3**    Chemical nature of DNA. This structure is the double helix form of DNA. A & T are joined with double hydrogen bonds, G & C are joined with triple hydrogen bonds.

These segments produce a genetic code that specifies the chemical composition of proteins, which ultimately are the end product of genetic expression. The entire genetic material of an animal is termed its ***genome***.

Genomes of organisms vary in size, with simpler organisms having genomes substantially smaller than those of complex multicellular organisms. Because chromosomes comprise the genome, there is an issue of how this genetic information is held in such small spaces. DNA segments can be of varying lengths, and this is important when they must be packed into small sections of the chromosome. DNA is supercoiled in such a way that it becomes very compact and is able to fit into extremely small sections of the chromosome. Imagine a DNA molecule that has ends attached in such a way as to produce a circle. Now imagine twisting this circle so it decreases in width by half, and now one circle is lying on top of the other. Picture a coiled garden hose that has been twisted in this manner. With each twist, the circumference of the circle decreases but gets deeper or thicker as the hose stacks up on top of itself. This is the same concept that is true of supercoiled DNA. As this sequence progresses, the DNA molecules in a chromosome become much more compact and the chromosome shortens and thickens, thus becoming smaller in overall size.

There is some potential for variation in chromosome numbers because of the many processes that must first take place for inheritance to be possible. One such variation in chromosome numbers is called ***polyploidy***. A polyploid individual has more than one full set of chromosomes. Polyploidy is very uncommon in vertebrates but quite widespread in plants. This is beneficial in some situations. Many times polyploidy causes failure of meiosis because there is more than one pairing partner for each chromosome and the result is just a few small seeds; for example, seedless watermelons.

Another example of variation is ***aneuploidy***, in which there is variation in chromosome number with respect to individual chromosomes. If an organism is missing a chromosome ($2n - 1$), it is said to be ***monosomic***. This is often a lethal situation. However, if an organism contains an extra chromosome ($2n + 1$), it is possible for development to be either normal or slightly abnormal. This is referred to as ***trisomy***. A common example of trisomy is Down syndrome in humans, which is caused by trisomy of autosome number 21. Trisomy is often related to failure of the chromosome to separate in meiosis, producing a gamete with two copies of the chromosome.

DNA is a sequence of base pairs (i.e., ATCG with TAGC) that represents the code for a specific gene. A sequence of four bases can be arranged in 256 different ways; therefore, it is easy to see how sequences of many more base pairs can be arranged in exponentially many more ways. This concept is what provides the complexity of DNA and thus the code for many different genes.

***DNA replication*** is the process of making a copy of a DNA molecule. Replication must occur accurately so the daughter cell (the cell being produced) inherits the same

**Genome**

The complete genetic material of an organism.

**Polyploidy**

Having more than two full sets of chromosomes.

**Aneuploidy**

A condition in which an organism has a chromosome number that is not an exact multiple of the monoploid (m) number.

**Monosomy**

The absence of one chromosome from an otherwise diploid cell.

**Trisomy**

The presence of one extra chromosome in an otherwise diploid cell.

**DNA replication**

The cellular process of making a copy of a DNA molecule.

## DNA polymerase

The enzyme that forms the sugar–phosphate bonds between adjacent nucleotides in a chain so that replication can occur.

## Polymerization

The process of building high molecular weight molecules by repeatedly chemically bonding the same compound to itself.

## Ribosomes

A component of cells that contain protein and tRNA. They synthesize proteins.

## Ribonucleic acid (RNA)

Long chains of phosphate, ribose sugar, and several bases.

## Transcription

In protein manufacture, the process of building RNA that is complementary to DNA.

## Messenger RNA (mRNA)

Nucleic acid that carries instructions to a ribosome for the synthesis of a particular protein.

## Transfer RNA (tRNA)

Molecules of RNA coded by DNA to bond with a specific amino acid. tRNA molecules "collect" the amino acids from the cytoplasm that the ribosomes use to manufacture proteins.

## Translation

In protein manufacture, the process of building an amino acid sequence according to the code specified by mRNA.

## Alleles

One of two or more alternative forms of a gene occupying corresponding sites (loci) on homologous chromosomes.

information contained in the parent cell. This process of synthesizing DNA is carried out by unzipping the existing DNA strand between base pairs (i.e., unzipping between A-T and C-G) to expose each base. *DNA polymerase* is the enzyme that forms the sugar–phosphate bond between adjacent nucleotides in a chain. For DNA polymerase to act, an RNA primer must be present to which the DNA polymerase attaches to begin the replication process. The process of *polymerization* is carried out by DNA polymerase. More specifically, this consists of deoxynucleotides being added to the existing single strand of DNA, thereby producing a complete DNA molecule with matching base pairs and sequences just like the initial molecule. Thinking through this process, it is easy to visualize the splitting of a DNA molecule to produce two strands of DNA. Then the addition of nucleotides that complement the existing bases (i.e., ATGC) forms a complete DNA molecule. By starting with one DNA molecule, this process ultimately produces two identical molecules.

The next step is to look at how this genetic information is used to specify the type of molecule to be produced. DNA is located in the nucleus of the cell, but proteins are produced by *ribosomes* located in the cytoplasm. Because the purpose of the DNA is to serve as a template for the construction of a protein, the genetic message found in the DNA must be transported from within the nucleus out to the ribosome. *Ribonucleic acid (RNA)* molecules are fitted especially for this process. RNA molecules are built to complement the DNA. This process is known as *transcription.* DNA serves as a template and codes for the manufacture of RNA. Once a complementary RNA has been synthesized, it is processed to remove sections of base pairs from the primary transcript that are not part of the coding sequence for the specific protein needed. The end product of this processing is a *messenger RNA (mRNA)* that contains only the sequences of base pairs used to code for the specific protein. The DNA also codes for a second RNA called *transfer RNA (tRNA),* which is used to collect the amino acids needed to build the protein. The mRNA leaves the nucleus and attaches to the ribosome where it is used as the template to manufacture the protein. The tRNA moves into the cytoplasm and attaches to the amino acid for which each is coded. Next, the ribosomes move along the length of the mRNA and align with the tRNA, which brings the amino acids into the chain. As they are aligned, the amino acids chemically bond to each other. A chain of amino acid sequences is thus constructed. This process is known as *translation* of the mRNA code for the protein being built. When the tRNA comes to a three-base sequence for which it has no match, the process is complete. The resulting chain of amino acids is a protein. The protein is now ready to do its work in the cell.

## PRINCIPLES OF INHERITANCE

In 1866, Gregor Mendel discovered the principles of inheritance while working with garden peas. He sought to understand why peas were consistent within lines but different between lines. Our understanding of how traits are inherited has sprung from the simple experiments of this monk.

The various forms of a given gene are called *alleles.* Alleles affect the same trait, but each allele causes the production of a different protein and thus differences in the way the trait is expressed. Genes are located on molecules called chromosomes. Chromosomes that have the same size and shape and occur in pairs are called *homologous chromosomes.* Homologous chromosomes have genes that affect the same traits. The number of chromosomes containing the genetic information of an individual differs among species (Table 8–1). An animal that has matching alleles at a given point on the

**TABLE 8–1    Number of Chromosomes by Species**

| Species | Number of Chromosomes (2n) |
|---------|---------------------------|
| Human | 46 |
| Cattle | 60 |
| Swine | 38 |
| Sheep | 54 |
| Goat | 60 |
| Horse | 64 |
| Chicken | 78 |
| Bison | 60 |
| Llama | 74 |
| Cat | 38 |
| Dog | 78 |

SOURCE: Complied from Bourdon, 2000, and Van Vleck et al., 1987

chromosome, or *locus,* is said to be *homozygous* (*AA*), and one with different alleles is *heterozygous* (*Aa*).

The method by which these alleles are passed on from one generation to the next is known as *inheritance.* Each parent produces reproductive cells called *gametes,* and within each gamete is a single allele for each gene. In the formation of these gametes, the parental alleles separate so that each gamete contains only half of the genetic code the parent possesses. Two important principles come into play at this point:

The *principle of segregation* states that alleles separate so that only one (randomly chosen) is found in any particular gamete.

The *principle of independent assortment* states that in the formation of gametes, separation of a pair of genes is independent of the separation of other pairs.

When the gametes combine to produce an individual, these alleles are brought together and coding for a protein begins. Any given gamete contains one allele for each gene in the genotype. A genotype is the entire genetic composition of the animal; however, genotype can also mean only the alleles of genes of interest to a particular situation.

The concept of sex determination is important in the formation of gametes. The male gametes are *sperm* and the female gametes are *eggs*. In mammals, female genotypes contain a pair of X chromosomes and males have an X and a Y chromosome. Thus a female can contribute only an X chromosome to her offspring; a male is capable of passing on either an X or a Y. The pairing of these sex chromosomes in the zygote ultimately determines the sex of an individual. In this situation, the male contributes the gamete that will determine the sex of the offspring. In the avian species, the female gamete contains pairs that do not match, making her the parent that passes on the chromosome that carries the information for sex differentiation; the male passes on gametes with only one type of sex chromosome.

Each normal body tissue cell, or *somatic cell,* of an individual has two sex chromosomes, or one pair. However, every somatic cell also has (2n – 2) *autosomes,* which are simply all chromosomes other than sex chromosomes. In other words, for each somatic cell of an organism, such as a human, that has 46 chromosomes (23 pairs), two chromosomes (1 pair) are sex chromosomes and the other 44 chromosomes (22 pairs)

## Homologous chromosomes

Chromosomes having the same size and shape, occurring in pairs, and affecting the same traits.

## Locus

The specific location of a gene on a chromosome.

## Homozygous

When two genes of a pair are the same.

## Heterozygous

When two genes in a pair are not the same.

## Inheritance

The transfer of gene-containing chromosomes from parent to offspring.

## Gametes

The sperm from the male parent and the egg from the female parent.

## Principle of segregation

Mendel's first law; often called the law of segregation. The law states that when gametes are formed, the genes at a given locus separate so that each is incorporated into different gametes.

## Principle of independent assortment

Mendel's second law. It says that in the formation of gametes, separation of a pair of genes is independent of the separation of other pairs.

## Somatic cells

All cells in the body other than gametes.

## Autosomes

All chromosomes other than the sex chromosomes.

**FIGURE 8–4**    A false-color light micrograph of a normal human female karyotype. The full complement of female chromosomes is arranged in numbered homologous pairs. The chromosomes are presented in this manner by matching up unpaired chromosomes photographed during the metaphase stage of cell division. (Photo by CNRI/Science Photo Library. Courtesy of Photo Researchers, Inc.)

are autosomes. Each gamete cell has one sex chromosome and $(n - 1)$ autosomes. In the human, a gamete contains 1 sex chromosome and 22 autosomes. A visual representation of the chromosomes of a species is called a *karyotype*. Karyotypes are put together by using pictures of individual chromosomes taken at metaphase (see following section). They are then arranged by chromosome number so that visual comparisons can be made (Figure 8–4).

The production of gametes is responsible for providing the means by which inheritance takes place. By applying the aforementioned principles, it is possible to understand how variation exists in a population. The principle of segregation, when combined with the principle of independent assortment, provides a means for randomization of alleles within the gametes.

Chromosomes occur in pairs in somatic cells. Thus a somatic cell contains a *diploid* ($2n$) number of chromosomes. The germ cells, sperm and egg, contain only a *haploid* number ($n$). A chromosome can be thought of as long strands of genes. Chromosome size depends on how many genes are located on each respective chromosome. Another feature of the chromosome is the ***centromere,*** which can be located anywhere along the chromosome. The centromere serves as the point of attachment for the spindle fibers during cell division. The location of the chromosome is another feature that can be used to identify chromosomes. To understand the method by which somatic cells and gametes obtain their respective number of chromosomes, it is important to identify the different types of cell division.

**Diploid**

Having two sets of chromosomes as opposed to the one set found in gametes.

**Haploid**

A cell with half the usual number of chromosomes. Sex cells are haploid.

**Centromere**

The region of a chromosome where spindle fibers attach.

**Mitosis**

The process of somatic cell division.

## MITOSIS AND MEIOSIS

### MITOSIS

*Mitosis* is the process of somatic cell division (Figure 8–5). It occurs in normal body tissues and is responsible for the everyday maintenance of the body and for growth in young animals. Mitosis is really just replication of cells. A diploid cell undergoes division that allows the production of two diploid cells. This is a replicational process whereby a ($2n$) cell has produced a pair of matching ($2n$) cells.

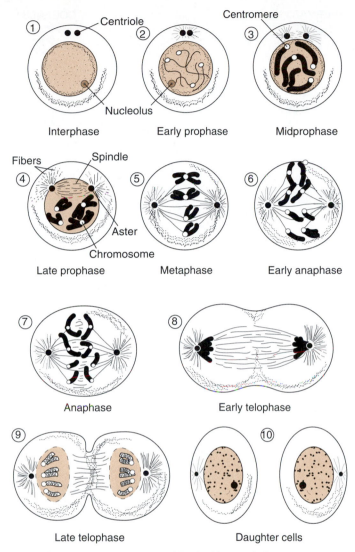

**FIGURE 8–5** Mitosis (*SOURCE:* Levine, 1980. Used with permission.)

## MEIOSIS

*Gametogenesis* is the development of the sex cells, (i.e, *sperm* and *oocyte*). This is a reductional process ($2n$ to $n$) responsible for forming cells that contain half of the genetic message. The cell division that occurs in gametogenesis is called *meiosis* and consists of two divisional procedures (Figure 8–6). During the first division, one diploid cell ($2n$) divides into two haploid cells ($n$). The second division consists of a replication of each of the two haploid cells to produce four haploids. This process of gametogenesis is efficient in that one diploid cell divides and replicates in such a way as to produce four haploid cells. These are the cells passed on in the form of sperm or oocytes that, when combined, produce a cell with the full genetic complement of DNA.

**Gametogenesis**
The formation of gametes.

**Sperm**
The gamete from the male.

**Oocyte**
The gamete from the female.

**Meiosis**
The process that forms sex cells. Cells formed through meiosis have half the chromosomes of the parent cells.

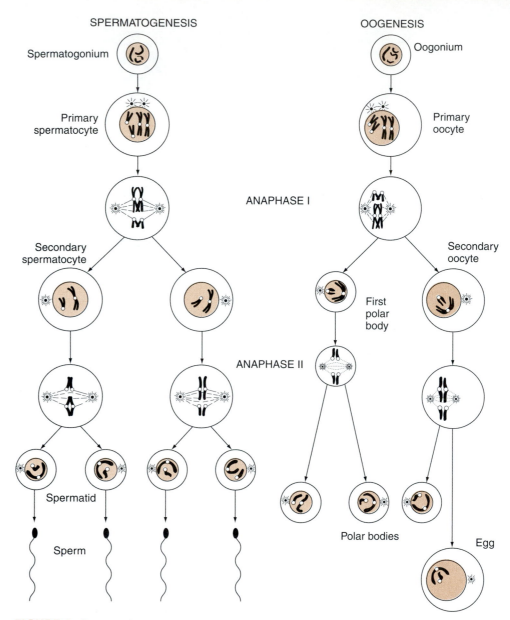

SPERMATOGENESIS

Spermatogonium

Primary spermatocyte

ANAPHASE I

Secondary spermatocyte

ANAPHASE II

Spermatid

Sperm

OOGENESIS

Oogonium

Primary oocyte

Secondary oocyte

First polar body

Polar bodies

Egg

**FIGURE 8–6**   Meiosis (*SOURCE:* Levine, 1980. Used with permission.)

# GENE EXPRESSION

## DOMINANT AND RECESSIVE EXPRESSION

Once the alleles have combined to determine the genetic makeup of an individual, the methods by which they become interpreted into traits, or are expressed, becomes important. *Dominant* alleles, signified by a capital letter (*A*, for example), express themselves over *recessive* alleles (*a*). For example, *R* stands for an allele that codes for black coat color, and *r* represents red coat color. If an individual receives *R* from

**Dominant**

One member of a gene pair is expressed to the exclusion of the other.

**Recessive**

The member of a gene pair that is only expressed when the dominant allele is absent from the animal's genome.

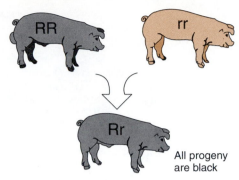

R represents black and r represents red. All progeny are black because
the R gene each received from the sire masks the expression of r.

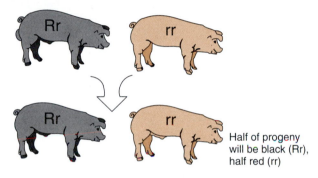

In this mating half the offspring receive an rr and half receive Rr.
Thus half the progeny are red and half are black.

RR is the homozygous dominant condition.
Rr is the heterozygous dominant condition.
rr is the homozygous recessive condition.

**FIGURE 8–7**    The behavior of simple dominant and recessive traits.

each parent, then its genotype is *RR* and the phenotype, or physical appearance, is
black. Likewise, an individual with an *Rr* genotype would also express a black coat
color. In this case, the dominant allele masks the recessive and the phenotype is representative of the dominant allele. However, if an individual inherited an *r* allele
from each parent, its genotype is *rr*, and the animal will express a red coat color.
For the individual to have a phenotype representative of the recessive allele, both of
the inherited alleles must be of the recessive form. This concept is illustrated in
Figure 8–7.

## CODOMINANCE

An exception to the rule of dominance occurs when neither allele masks the other and
both are expressed in the phenotype. This situation is known as *codominance*. An example would be that of coat color in Shorthorn cattle. An animal that inherits one of
each type of allele *RW* will express both red and white in its coat color to produce what
is called a *roan color pattern*. This is illustrated in Figure 8–8.

**Codominance**

Both alleles are expressed in
the phenotype when present
in the heterozygous state.

**FIGURE 8–8** Codominance.

Red MALE

| | R | R |
|---|---|---|
| **White FEMALE** W | RW (Roan) | RW (Roan) |
| W | RW (Roan) | RW (Roan) |

White MALE

| | W | W |
|---|---|---|
| **Roan FEMALE** R | RW (Roan) | RW (Roan) |
| W | WW (White) | WW (White) |

## INCOMPLETE DOMINANCE

**Incomplete dominance**

Condition in the heterozygote where both genes are expressed in a way different from either homozygous condition.

*Incomplete dominance* is a situation in which no dominance exists and a heterozygous individual will appear as an intermediate between the two alleles. We can illustrate this with an example of coat color. Assume that a particular breed expresses incomplete dominance such that a heterozygote will produce a color that is neither white nor black but rather a smutty gray that is intermediate between white and black. The terms *incomplete dominance* and *codominance* are easy to confuse. To help remember, use the Shorthorn cattle example above. The roan color in the Shorthorn is produced from codominance. If Shorthorn color were instead inherited by incomplete dominance, the coat color would be pink.

## EPISTASIS

**Epistasis**

Interaction among genes at different loci. The expression of genes at one locus depends on alleles present at one or more other loci.

The previous examples discussed inheritance of genes that are independent of each other to illustrate the concepts of dominance, codominance, and incomplete dominance. However, many genes are not actually inherited independently. In such cases, the expression of one gene is influenced by the presence of another. This is known as *epistasis*, which is capable of causing the appearance of a dramatic departure from the principles of Mendelian inheritance. A gene may express a typical action when another specific gene is not present, but when this gene does exist and epistasis takes place, the gene could have a totally different action. Coat color in many species is affected by epistasis. For example, horses either have black hair or they don't. This is controlled at the *E* locus. However, whether they have black hair all over their body or just on the points is controlled by the extension gene at another locus. Those with genotype *Ea* are black all over. Those with genotype *EA* are bay in color.

## MULTIPLE ALLELES

Remember that only two alleles can exist on each locus. However, there could be many alleles of a gene present in a population. This is known as a system of *multiple alleles.* A common example used to represent this concept is the A-B-O blood types in humans. Type A and type B are both dominant to type O, and types A and B are codominant to each other. In other words, if a person's genotype is *AA* or *AO*, he or she has the antigen A produced in his or her blood. Likewise, if the genotype is *BB* or *BO*, the antigen B is produced. If, however, the person has both the *A* and *B* alleles, codominance is in effect and the blood type is AB, in which both antigens are present in the blood. When the individual contains two alleles of the O blood group, the genotype is *OO*. In this case, no antigen is present in the blood.

**Multiple alleles**

Genes with three or more alleles.

## TESTCROSSING

It would be beneficial to be able to look at an animal and determine its genotype based on its phenotypic appearance. This would make the selection process much more efficient. It is not possible to do this when dominance is in effect. An animal that is homozygous recessive for a gene such as coat color could easily be identified because it has to have both alleles of the recessive gene. Therefore, it will express the recessive form in its phenotype. However, animals that phenotypically express the dominant form of the gene cannot be as easily identified genetically. An animal could be heterozygous or homozygous dominant, either of which will express the dominant phenotype. For instances such as this, a *testcross* can determine whether the animal is homozygous or heterozygous. A testcross can be conducted by mating this individual to one with a homozygous recessive genotype, which can only contribute a recessive allele. If an offspring is produced with the recessive phenotype, then it must be homozygous recessive, which means the test animal is heterozygous. If all offspring are produced with the dominant phenotype, we can conclude the test animal is homozygous dominant, and the animal is considered to be "true breeding."

**Testcross**

Mating with a fully recessive tester animal to determine if an individual is homozygous or heterozygous.

# SEX-RELATED INHERITANCE

## SEX-LINKED INHERITANCE

Some genes are located only on the X or Y chromosome and therefore are inherited only when that respective chromosome is passed on. This is referred to as *sex-linked inheritance.* X and Y chromosomes are homologous to each other in certain regions. This means they contain the same, or nearly the same, genetic sequence. Sex-linked genes are those that reside in the nonhomologous portions of the X and Y chromosomes. *X*-linked genes reside in the nonhomologous area of the X chromosome and can be inherited through the acquisition of that chromosome. *X*-linked genes can be passed on to either male or female offspring because each has at least one X chromosome. For males, it is easy to determine the genotype from the phenotype for *X*-linked genes. The male has only one *X* gene, so the phenotype representative of that gene is easily recognizable. For females, however, the genotype is more difficult to determine. Females have two X chromosomes, and the phenotype can be portrayed as either of those respective genes. *Y*-linked genes reside in the nonhomologous portion of the Y chromosome and are inherited only with that chromosome, meaning only males can have a phenotype representative of that gene.

**Sex-linked inheritance**

Traits inherited on the X or Y chromosome and therefore inherited only when that respective chromosome is passed on.

## SEX-INFLUENCED INHERITANCE

**Sex-influenced inheritance**

When the same genotype is expressed differently depending on the sex of the animal.

The genes for *sex-influenced inheritance* traits are carried on the autosomes. In sex-influenced inheritance, the phenotypes are not expressed in the same way in the two sexes. Let us examine an example of horns in sheep. In the male, the gene for horns (*H*) is dominant to the recessive form (*h*) for polledness. However, in females the *h* gene is dominant to *H*. In either case, when the genotype *HH* is present, the animals will have horns. Likewise, when either sex contains the *hh* genotype, no horns are present. In a case in which a male is heterozygous (*Hh*), horns will be present. However, if the heterozygous form of the gene is inherited in a female, she will be polled.

## SEX-LIMITED TRAITS

**Sex-limited traits**

Traits expressed in one sex or the other such as milk production in females. Both sexes carry genes for the trait.

Another example of sex-related inheritance is that of *sex-limited traits.* In this case, traits are unique to only one sex. Some examples are milk production, litter size, and egg production by females. Both sexes carry genes for these traits, but only one sex is capable of expression.

# POPULATION GENETICS

## GENE FREQUENCY

**Gene frequency**

The proportion of loci in a population that contain a particular allele.

**Genotypic frequency**

The frequency with which a particular genotype occurs in a population.

**Phenotypic frequency**

The proportion of individuals in a population that express a particular phenotype.

Knowing how often genes occur in a population of animals is important to make genetic change. *Gene frequency* is defined as the proportion of loci in a population that contain a particular allele. Thus the *genotypic frequency* can be defined as how often a particular genotype occurs in a population. The proportion of individuals in a population that express a particular phenotype is the *phenotypic frequency*. In a population of cattle in which half are horned and half are polled, the phenotypic frequency for polledness is 0.5, as is the frequency of horned cattle. Because many pairs of genes control most traits, it is important to expand these concepts of frequencies beyond just one pair of genes.

## ANIMAL BREEDING AND POPULATION GENETICS

**Population genetics**

The study of how gene and genotypic frequencies change, and thus change genetic merit in a population.

**Animal breeding**

The science of using the principles of genetics to make improvement in a livestock species.

The study of how gene and genotypic frequencies change, and thus change genetic merit in a population, is called *population genetics.* The principles of genetics discussed previously affect not only the proportion of genes in an individual, but also the occurrence of these genes in a population. The science of using the principles of genetics to make improvement in a livestock species is termed *animal breeding.* Robert Bakewell is given credit for being the first animal breeder. He lived in Dishley, England, from 1726 to 1795. In his productive life as a stockman, he made significant improvements in the English Longhorn breed of cattle, the Leicester breed of sheep, and the Shire breed of horses. He practiced "breeding the best to the best," even when the animals were blood relatives. This is said to have scandalized some of his neighbors. However, others saw the merits of what he was doing and joined him in his efforts. As a result, the discipline of animal breeding was born.

Animal breeders seek to influence population genetics. In a population of animals, represented either by an entire breed or just a set of animals within that breed, various factors play a role in the type of animals produced. Genetic merit of the population is influenced by many interactions among genes, as well as the frequency of genes in the population. Production animal breeders must incorporate an understanding of the principles of genetics into their production practices to make the necessary genetic changes within their herd.

## MUTATION AND GENETIC DRIFT

To understand how gene frequencies remain stable and how they change from one generation to the next, we must define some ways in which change is made. *Mutations* are changes in the chemical composition of a gene that alter DNA. This causes the production of new alleles that can affect gene and genotypic frequencies. However, the occurrence of mutations in a population is very small. Another method by which gene frequencies change is *genetic drift.* This is a change in gene frequency owing to chance that cannot be controlled in direction, but can be controlled in amount by the size of the population. As the population gets larger, the amount of genetic drift decreases. Mutations and genetic drift cannot be used to cause genetic improvement, but direct changes in gene frequency can be accomplished using migration and selection.

## MIGRATION AND SELECTION

*Migration* is the process of bringing new breeding stock into a population. Migration can be performed on a herd, a breed, or on an industrywide basis. Large changes in gene frequencies can be made quickly by this method. An important factor in migration is that the frequency of genes in the existing herd and the migrants must be different. Bringing a new sire into a population is a way to cause migration, but only if the sire's genotype is different from that of the existing population. When genotypes are very different, gene frequency can be changed quickly. *Selection* is the process of allowing some animals to be parents more than others. Selection can be expanded to include how many offspring each animal is allowed to produce and how long that animal stays in the breeding population. Two types of selection can occur.

*Natural selection* is based on the fact that some animals are more suited and/or have more natural opportunity to be parents than are others. This fact is controlled by nature and can be illustrated in domestic animals using the example of a herd of cows that run with several bulls. Some bulls will sire more offspring than others will because of their aggressiveness, age, and reproductive ability. Thus in managed populations, natural selection is still a very important force. Wild populations are selected through natural selection exclusively.

The other type of selection, *artificial selection,* is controlled by the herd manager. Artificial selection is based on management decisions to allow certain animals more opportunity to mate and produce offspring than others. (In common usage, the term *artificial selection* is shortened to *selection.*) Selection involves culling less desirable animals and choosing superior replacements. Selection is a very valuable tool to the animal producer, but it has limitations as well. The effect of selection is limited by the rate at which offspring are produced or by the generation interval (Figure 8–9).

Now that an understanding of possible methods of changing gene frequency has been established, we can apply an important principle to reinforce this concept:

The *Hardy-Weinberg law* states that in a large, random mating population where mutation, migration, and selection are nonexistent, gene and genotypic frequencies will remain stable from one generation to the next, and if there are two alleles with frequency of $p_A$ and $q_a$, the genotypic frequencies are $p_A^2$ *AA*, $2p_Aq_a$ *Aa*, and $q_a^2$ *aa*.

**Mutations**
Changes in the chemical composition of a gene.

**Genetic drift**
A change in gene frequency of a small breeding population owing to chance.

**Migration**
The process of bringing new breeding stock into a population.

**Selection**
The process of allowing some animals to be parents more than others.

**Natural selection**
Selection based on factors that favor individuals better suited to living and reproducing in a given environment.

**Artificial selection**
The practice of choosing the animals in a population that will be allowed to reproduce.

**FIGURE 8–9** Through artificial selection, animals within a species have been developed into breeds as diverse as the Great Dane and the Chihuahua. (Photographer David Allen Brandt. Courtesy of Getty Images, Inc.)

Practically speaking, if the desired genetic merit of the herd has been achieved, no methods of selection, migration, or mutation are occurring, and the herd is large, then the gene frequencies will remain constant within that herd from one generation to the next.

## QUANTITY VERSUS QUALITY TRAITS

### QUALITATIVE AND QUANTITATIVE TRAITS

There are different methods by which some traits are inherited in that varying numbers of genes are involved and different methods by which these genes are expressed. *Qualitative traits* are those for which phenotypes can be classified into groups rather than numerically measured. Examples of such traits are coat color and the presence of horns. *Quantitative traits* are those that are numerically measured and usually controlled by many genes, each having a small effect. One can rarely pinpoint the contribution of any particular gene to the phenotype of an animal. Most often, it is necessary to measure these traits with some kind of measuring tool. Evaluating the growth rate of pigs requires the use of scales, for example. Quantitative traits are influenced by the same types of gene action as qualitative traits are. If there is no dominance at a locus, it is referred to as *additive gene action*. Additive effects deal with individual genes, which allow for more efficient selection. For each individual gene, a representative effect on the trait occurs. For instance, if one $A^+$ represents a calf that is 2 lbs heavier at birth, then an animal with the genotype of $A^+A^+$ will be 4 lbs heavier than one with the alleles $AA$. Likewise, a calf with the genotype $A^+A$ would be 2 lbs heavier than a calf with the genotype $AA$.

Just as the type of inheritance affects a specific trait, the environment in which the animal is raised has an effect on the expression of quantitative traits. The genetic merit is very valuable in producing animals with certain characteristics or animals that perform to certain levels. However, factors such as climate, management practices, and health determine whether the animal performs to its genetic potential. This nongenetic source of variation is one that must be considered when evaluating animals based on quantitative traits. Using this understanding, it is easy to realize that the phenotype is a factor of the genotype and environmental interaction.

**Qualitative traits**

Those traits for which phenotypes such as coat color can be classified into groups.

**Quantitative traits**

Those traits that are numerically measured and are usually controlled by many genes, each having a small effect, such as milk or egg production.

**Additive gene action**

When the total phenotypic effect is the sum of the individual effects of the alleles.

## HERITABILITY

Differences in the phenotypes of animals are due to genetics and environment. Only the genetic effects are inherited. The proportion of the difference in individuals that is due to additive gene effects is known as *heritability.* Practically speaking, heritability is a measure of the proportion of phenotypic variation that can be passed from parent to offspring. Heritability is thus used as an indicator of the amount of genetic progress that can be achieved by choosing superior parents. The range of values for heritabilities is from zero to one, and they can be thought of as percentages or proportions. Heritability estimates have been calculated for most important traits for the different livestock species and can be used to give an indication of how much progress can be made in traits from generation to generation.

We can use an example of litter size here to demonstrate how heritability acts. Assume a group of sows is averaging 7 pigs per litter and a producer wishes to improve this number. For his next matings, he chooses boars and sows from litters averaging 11 pigs. The first step is to calculate a *selection differential.* This selection differential is the phenotypic advantage of those chosen to be parents. The selection differential here is 11 – 7, or 4 pigs. Now, assuming that the heritability of litter size is 0.10, we can expect only 10% of the 4-pig selection differential to be inherited, or 0.4 of a pig per litter. Thus we would expect the litters from these selected parents to be 0.4 pigs bigger, or 7.4 pigs per litter compared to 7 pigs, not much progress on a trait that is quite important from an economical standpoint. It is easy to see that traits with higher heritability estimates can be selected with much greater efficiency. Lowly heritable traits don't express much change from generation to generation from selection.

As a rule, carcass merit traits are considered to be highly heritable (0.4–0.6), which is encouraging because there is increased interest in, and demand for, specific carcass characteristics to meet today's market needs. This enthusiasm should be tempered by the knowledge that carcass traits are difficult to measure accurately in potential parents. Moderately heritable traits are those having heritabilities of 0.2–0.4. Growth traits are examples of moderately heritable traits and are selected for less progress when compared to carcass traits. Traits such as reproductive ability have low heritabilities (0–0.2), which is unfortunate considering they are among the most economically important traits in animal production.

Through the process of making genetic improvement in a population, animals begin to share similarities in their genetic composition. The relationship between two animals can be thought of as the proportion of genes they are expected to have in common. Siblings or offspring that have at least one parent in common inherit some of their genes from that parent, and thus the brothers and sisters have some of those gene pairs in common. The relationship coefficient can range from 0–1 and is most often by a factor of half. The following is a list of common relationships:

|  | *R* |
| --- | --- |
| Full-sibs | 0.5 |
| Half-sibs | 0.25 |
| Parent-offspring | 0.5 |
| Grandparent-offspring | 0.25 |
| Great-grandparent | 0.125 |
| Great-great-grandparent | 0.0625 |
| First cousin | 0.125 |

### Heritability

A measure of the proportion of the phenotypic variation that is due to additive gene effects.

### Selection differential

The phenotypic advantage of those chosen to be parents. The difference in the mean of those chosen to be parents and the mean of the population.

These values are an indication of how closely related the individuals are. Notice that the further back in the animal's pedigree a relative exists, the smaller the value of the relationship.

## SYSTEMS OF MATING

**Inbreeding**

The mating of closely related individuals.

**Inbreeding depression**

A loss or reduction in vigor, viability, or production that usually accompanies inbreeding.

**Linebreeding**

Mating system in which the relationship of an individual is kept close to an outstanding ancestor by having the ancestor appear multiple times on both sides of the pedigree.

**Outbreeding**

The process of mating less closely related individuals when compared to the average of the population.

**Heterosis**

The superiority of an outbred individual relative to the average performance of the parent populations included in the cross.

*Inbreeding* is the mating of closely related individuals (Figure 8–10). It is used to increase homozygosity for desired traits. When practiced in a herd, inbreeding decreases the variation in the genes existing in a herd or population. Related individuals begin to share more gene pairs in the homozygous state. As this happens, detrimental recessive genes also begin to express themselves because of the increase in homozygosity. Therefore, it is important to maintain control over direct inbreeding so that expression of bad genes can be minimized while allowing more expression of the good ones. Inbreeding causes a decline in performance that is called *inbreeding depression.*

*Linebreeding* is a form of inbreeding in which the purpose is to concentrate the genes of an outstanding ancestor in the linebred individuals. Linebreeding may result in mild inbreeding if the common ancestor appears at least three to four generations back in the pedigree. However, the inbreeding can be intense with parent offspring matings or after several generations of linebreeding to the same common ancestor. The adoption of modern genetic evaluations has replaced the practice of linebreeding in the livestock species. However, linebred pedigrees are relatively common in many companion animal species, especially in those bred for show ring type.

*Outbreeding* is the process of mating less closely related individuals when compared to the average of the population. This can be applied to animals in the population as well as animals as far out as another breed. The effect of this is directly opposite to that of inbreeding. This procedure produces individuals that have more heterozygous gene pairs. This increase in heterozygosity increases the vigor in the animals, which is termed *heterosis,* or hybrid vigor. Heterosis is defined as the superiority of an outbred individual relative to the average performance of the parent populations. Traits that are lowly heritable show high levels of heterosis. Reproductive traits are a good example of a lowly heritable trait that shows high levels of heterosis. Moderately heritable traits show moderate levels of heterosis, such as the growth traits. Highly heritable traits show little heterosis. Heterosis can either be increased or decreased by the system of mating used.

**FIGURE 8–10** Historically, inbreeding has been practiced to fix the traits associated with breeds such as color and markings, horns, and production traits. (Photo courtesy of U.S. Department of Agriculture.)

*Crossbreeding,* mating animals from different breeds, is a means of taking advantage of outbreeding. Each breed is generally more homozygous than the average of the population. By mating individuals from different breeds, a breeder can take advantage of the homozygocity of the parents to ensure heterozygocity in the offspring. The success of a crossbreeding program depends on the quality of the animals used in the system and whether or not their genetics complement each other. Therefore, it is essential to use breeds that complement each other well to strengthen the good traits and decrease expression of the bad ones. Reproductive traits benefit from crossbreeding systems. On average, crossbred females have more offspring (litter-bearing species) and have better mothering characteristics than the average of the two breeds used to produce them. Many times, breeders produce crossbred females and then mate them to a purebred sire of a third breed to take advantage of the traits in the purebred that have lower heritabilities. The crossbred females would be expected to have good mothering traits. The contribution of the sire could then be carcass merit. Elaborate methods of controlled crossbreeding have been developed to capitalize on the advantages breeds can offer in outcrossing. In general, crossbred individuals tend to be more vigorous, fertile, and healthy, and grow faster than the average of parental stock that make up the cross.

Several different methods can be used to evaluate how much genetic contribution an animal will have in any breeding scheme. Methods are being applied to evaluate parents and determine an estimate of their breeding value. This information can be utilized to get an idea of the additive genetic merit of an individual, and in selecting animals that have the opportunity to reproduce. Breeding values are based on records of the individual's own performance as well as the performance of all relatives. Relatives that are more closely related to the animal being evaluated are more beneficial because they are more likely to contain similar genetic makeup. However, all related animals are useful because they share at least some common genes. Breeding values are covered in more detail in the next chapter.

## BIOTECHNOLOGY AND GENETIC ENGINEERING

Genetics is an area of science that is contributing greatly to the overall advancements being made in *biotechnology.* In fact, the advances made in recombinant DNA technology, often called *genetic engineering*, have made it the most recognizable form of biotechnology being practiced today. It is common for people to think of biotechnology and genetic engineering as synonymous terms and not realize there are other areas of biotechnology.

Molecular biology has made it possible to identify the specific genes that control various characteristics. Worldwide, scientists are working to identify all of the genes of humans and animals. It is a mammoth undertaking that is nowhere near completion. However, scarcely a day passes without some significant contribution to the body of knowledge. This mapping of the genome is one of the finest examples of shared information and cooperation in the history of science. Scientists in laboratories scattered to the corners of the earth communicate, contribute information to shared databases, and work together to untangle the mysteries of the genetic code. The implications are truly mind boggling. Once genes are located on the chromosome and the functions they control are identified, then more precise animal breeding practices can be employed. For instance, lowly heritable traits like litter size can be precisely selected once the combination of genes responsible for large litters is identified. Much of the guesswork can be taken out of animal-breeding practices. When biochemical

**Crossbreeding**

Mating animals of diverse genetic backgrounds (breeds) within a species.

**Biotechnology**

A set of powerful tools that employ living organisms (or parts of organisms) to make or modify products, improve plants or animals, or develop microorganisms for specific uses.

## Marker-assisted selection

Selection for specific alleles using markers such as linked DNA sequences.

## Transgenic

An animal or plant that has had DNA from an external source inserted into its genetic code.

markers are identified for the genes that produce traits, then decisions about using the animal can be made. This is referred to as ***marker-assisted selection***. For example, Monsanto's swine genetics company has identified 6,000 specific markers they are using to make 40% faster genetic progress. Markers for genes for tenderness and marbling in beef cattle have been identified, and tests are commercially available for those and other traits in cattle and other species. A step beyond the use of traditional markers is the use of SNPS (single nucleotide polymorphisms; pronounced "snips"). USDA plans to use 50,000 SNPS to predict more accurately the breeding values of Holstein bulls prior to being progeny tested. In addition, inserting the genes from one animal into another can create new combinations of genes. This can even be done between species. This process of genetic modification creates a ***transgenic*** animal. These concepts are explored further in a later chapter.

The benefits of this new form of genetic manipulation over conventional forms are multiple. In traditional breeding practices, many of the genes passed to the next generation are unknown. In a mating of fully mapped individuals, all of the genetic material will be known. Precision breedings can be made with good genes passed on and undesirable ones excluded. Outcomes will be easier to predict. In transgenic animals, even less guesswork will be involved because the exact genetic information being transferred to create a transgenic animal will be known. The speed of genetic improvement will be increased because the genetics of a set of potential breeding animals can be mapped long before they even reach puberty. In traditional breeding schemes, identifying the genetics of an individual often has to wait for the birth and development of its offspring.

As exciting as the potentials are for improving livestock for production purposes, the area of greatest promise for gene manipulation and recombinant DNA technology is in a slightly different area. Genes can be inserted from animals of the same species. However, they can also be inserted from any species into another. Human genes can be inserted into bacteria or other species of animals. Fish genes can be inserted into pigs, and so on. Although several potential benefits can be envisioned from this procedure, one area holds incredible promise. Animals, microorganisms, and plants can be genetically manipulated to produce substances they otherwise could not produce. For example, bacteria now produce human insulin that is identical to the insulin produced by the human body. Prior to the commercial availability of this insulin, diabetics depended on the insulin of pigs and cattle, which was harvested as a slaughter by-product. Although life saving, this insulin was slightly different from that produced by humans. Now insulin-dependent diabetics can use the same insulin their bodies refuse to produce for them. Already, goats, sheep, and cows have been genetically manipulated to produce foreign proteins in their milk that have value in treating diseases. Most of these are protein products. The animals that produce these compounds are simply milked and the compound is purified. In this way, these transgenic animals can produce large quantities of therapeutic agents that are otherwise not available or are too expensive to produce. Many more of these applications are expected in the very near future. Transgenic chickens will probably be producing a range of therapeutic medicinals in their eggs in the near future.

Gene mapping and, subsequently, improved selection, combined with the use of transgenics for specific genes, will provide benefits in other areas as well. Specific gene therapies will be developed for diseases in animals and humans. Obviously, these are being developed with humans as the highest priority. One recently announced breakthrough causes hearts to grow new blood vessels. In the future, this process could take the place of some forms of open-heart surgery. Scientists working in this field confidently predict the availability of several dozen such gene therapies within the decade. Animals can be genetically manipulated to produce strains that will serve as models for the study of various human diseases. It is also likely that animals can be transgenically altered to produce organs that do not trigger rejection reactions for transplant into humans. Animals will be selected through the use of mapping technology to be resistant to specific diseases, reducing or eliminating the need for vaccines, antibiotics, chemicals, and other means of disease control and prevention. Biotechnology is explored in greater detail in a subsequent chapter.

## SUMMARY AND CONCLUSION

Genetics is the study of how DNA codes for the biochemical reactions of life. From a practical perspective, it is important to understand how to direct the genetics of the next generation so the genetic material of the animals produced causes them to have characteristics that are considered economically important. Cells in the body reproduce by two processes. Mitosis is the process used for the growth and maintenance of body tissue. Meiosis is the other process, which differs from mitosis in that the genetic material of an individual is halved before the cell reproduces itself. This process leads to a recombination of genetic material, half from each parent, when sperm and egg unite. Humans have been manipulating the genetic codes of some species for millennia. The organized effort called "animal breeding" began in the 18th century with the work of an English stockman named Robert Bakewell. Modern animal-breeding techniques have led us to organized efforts that help us manipulate the genetic code. These techniques include breeding relatives through inbreeding and linebreeding to "fix" certain genetic types, and outbreeding and crossbreeding to maximize heterozygosity. The tools of modern molecular biology are revolutionizing the science of genetics. What used to take generations to accomplish is now possible in one generation. In addition, animals, microorganisms, and plants are being genetically altered to provide a wide range of medicinal products for our benefit. This evolving area of science is perhaps the most promising area of scientific discovery being pursued today.

## STUDY QUESTIONS

1. Define the term *gene*.
2. Draw a DNA molecule with four base pairs.
3. List the two types of nitrogen bases in a DNA molecule.
4. What kind of bond holds these bases together?
5. Describe what is meant by supercoiling of DNA.
6. What is the variation in chromosome numbers when more than one full set is present in an individual?
7. Down syndrome is an example of what type of chromosomal variation?
8. What is the process that takes place within the nucleus of a cell whereby a copy of a DNA molecule is formed from another?
9. What is an allele?

10. What is the difference between an animal that is homozygous for a gene and one that is heterozygous?
11. Describe the relationship between a gamete and a somatic cell with regard to the number of chromosomes in each.
12. What is the entire genetic composition of an individual referred to as?
13. List the two types of sex chromosomes.
14. Each somatic cell contains _____ autosomes while the same cell contains _____ sex chromosomes.
15. Each gamete contains _____ autosomes and _____ sex chromosomes.
16. Describe in broad terms the process of mitosis.
17. Describe in broad terms the outcomes of meiosis.
18. Compare/contrast dominant and recessive alleles.
19. What is the physical appearance of an animal known as?
20. Describe an example of codominance and an example of incomplete dominance.
21. The blood type of humans serves as a good example of what kind of allelic situation?
22. Explain the concept of sex-linked inheritance with relation to the X and Y chromosomes.
23. When the sex of an individual determines how a gene is expressed in the phenotype, this is known as a _____ trait.
24. Milk production is an example of what kind of trait?
25. Define *population genetics*.
26. What does the term *gene frequency* describe?
27. What are four methods by which gene frequencies are changed in a population? Briefly describe each.
28. Compare natural selection with artificial selection.
29. What kind of gene action takes place in qualitative versus quantitative traits? Name an example trait for each type.
30. What is the major nongenetic source of variation discussed in this chapter? Give three examples.
31. Define *heritability*. What is the range for heritability values?
32. A trait that is lowly/highly (circle one) heritable for a given trait allows more genetic change for that trait from one generation to the next.
33. Give examples of high, moderate, and lowly heritable traits.
34. Describe the systems of mating discussed in this chapter.
35. What is biotechnology? Genetic engineering? How are they related?
36. Describe some of the potential benefits to humans of the manipulation of the genetic code.

# REFERENCES

*Author's Note:* For the second and third editions, Dr. David S. Buchanan, Professor, Animal Science, Oklahoma State University, reviewed this chapter. In addition, Dr. Buchanan contributed new material to the chapter. For the fourth edition, Dr. Tony Seykora, Professor, Dairy Genetics, University of Minnesota, reviewed this chapter. In addition, Dr. Seykora contributed new material. The author gratefully acknowledges these contributions.

Alcamo, I. E. 1996. *DNA technology: The awesome skill*. Dubuque, IA: Wm. C. Brown.

Bourdon, R. M. 2000. *Understanding animal breeding*. 2nd ed. Upper Saddle River, NJ: Prentice Hall.

Buchanan, D. S. 2001. Professor of Animal Science, Oklahoma State University, Stillwater, OK. Personal communication.

Buchanan, D. S., A. C. Clutter, S. L. Northcutt, and D. Pomp. 1993. *Animal breeding: Principles and applications*. 4th ed. Stillwater: Oklahoma State University.

Fowler, M. 1989. *Medicine and surgery of South American camelids*. 3rd ed. Ames: Iowa State University Press.

Hartl, D. L. 1994. *Genetics*. 3rd ed. Boston: Jones and Bartlett.

Klug, W. S., and M. R. Cummings. 2000. *Concepts of genetics*. 6th ed. Upper Saddle River, NJ: Prentice Hall.

Klug, W. S., and M. R. Cummings. 2002. *Essentials of genetics*. 4th ed. Upper Saddle River, NJ: Prentice Hall.

Lee, T. F. 1993. *Gene future*. New York: Plenum Press.

Levine, L. 1980. *Biology of the gene*. 3rd ed. St. Louis, MO: C. V. Mosby.

Lewin, B. 1983. *Genes*. New York: Wiley.

Mertens, T. T., and R. L. Hammersmith. 2001. *Genetics laboratory investigations*. Upper Saddle River, NJ: Prentice Hall.

Stufflebeam, C. E. 1983. *Principles of animal agriculture*. Upper Saddle River, NJ: Prentice-Hall.

Taylor, R. E. 1992. *Scientific farm animal production*. 4th ed. New York: Macmillan.

Van Vleck, D. L., J. E. Pollack, and B. E. A. Oltenacu. 1987. *Genetics for the animal sciences*. New York: W. H. Freeman.

# Animal Breeding

## LEARNING OBJECTIVES

After you have studied this chapter, you should be able to:

1. Define *animal breeding* and explain its contributions to animal science.
2. Describe the general principles of animal breeding as it applies to beef cattle.
3. Define *heritability* and *genetic correlations*.
4. Explain how to use EPDs in beef cattle breeding.
5. Describe the uses and benefits of a beef cattle sire summary.
6. Describe the general principles of animal breeding as it applies to dairy cattle.
7. Explain why associations among traits are so important to dairy cattle selection.
8. Identify goals on traits of emphasis in dairy cattle selection.
9. Describe the DHI system and explain its use in dairy cattle selection.
10. List the ways in which swine genetic improvement is similar to and different from the other major species.
11. Describe the difference in the way breeds influence the swine industry compared to the other industries.
12. Describe the general principles of animal breeding as it applies to sheep.

## KEY TERMS

Accuracy
Across-breed EPD
Animal breeding
Breed
Breeding soundness
  examination
Breeding value
Contemporary
  group

Expected progeny
  difference (EPD)
$F_1$
Feed efficiency
Generation interval
Genetic correlation
Heritability
Maternal effect
Parent average

Percentile
Phenotypic value
Porcine stress syndrome
Possible change
Predicted transmitting
  ability
Predicted transmitting
  ability net merit dollars
Quality grade

Rate of gain             Single-trait selection        Type production index
Reliability              Sire summary                  Ultrasonic scan measures

# INTRODUCTION

*Animal breeding* is the application of genetic principles in the selection of animals that will be the parents of the next generation. It is a field that has contributed enormously to animal agriculture. By applying the principles of animal breeding, the productivity of all the food-producing species has increased. Much of that progress occurred in the last half of the 20th century, primarily because the major tools of advanced animal breeding are complex mathematical models that must be applied to large data sets. Computers that are powerful enough to manage the mathematics have only recently become available. Scientists have made remarkable progress in providing the tools to make improvements logically and systematically in the major food-producing species. For many reasons, animal breeding work on horses, goats, and the companion species has not achieved the same level of understanding. The primary reason is economic justification. It is hard to justify the same level of research directed to a species that is of lesser economic importance. Most of this research has been done at the land grant universities of the United States. Their mission and limited resources have dictated that the research be aimed in the direction that will net the greater good.

Animal breeding is a field that is changing, just as all science is changing. The tools of genetic engineering, such as marker-assisted selection and transgenics are enhancing the work of animal breeders, helping make genetic progress in one generation that might otherwise have taken several generations to accomplish by traditional methods. However, the quantitative work of animal breeding will continue and will not become obsolete. Most economically important traits are brought about by the expression of several genes. More progress can be made in such traits through animal breeding than through the manipulation of individual genes. In addition, animal breeding also involves the measurement of how animals with certain genes respond to the environments in which they are placed. This evaluative work will always be needed.

This chapter is divided into sections based on species. Each section discusses the tools and goals of animal breeding as directed at that species. Only beef cattle, dairy cattle, swine, and sheep are included because these are the species for which the techniques and applications are the most advanced and useful. The chicken is the only major exception to this statement. However, the application of animal breeding principles to the poultry industry is almost completely in the hands of the major breeding companies that dominate that industry. On a commercial basis, virtually no producer is involved in making breeding decisions. For this reason, poultry is excluded. This does not imply that animal breeding principles are not used in that industry. Quite the contrary, they are used extensively and have helped the poultry industry attain the position in the animal industries that it enjoys today.

## **Animal breeding**

The use of biometry and genetics to improve farm animal production.

# BEEF CATTLE GENETIC IMPROVEMENT

The beef cow-calf producer is in business to produce beef as efficiently as possible. Modern breeding requires selecting for a balance of production performance (such as rate of gain) and end product merit (such as tenderness) to meet consumer expectations for eating satisfaction. Bull selection is a primary area in which producers can make directional change in their herd genetics. A wealth of performance information is available for what appears to be an endless list of traits.

The major areas of economic importance include mature size, calf growth, maternal performance, and carcass traits. The level of production must be matched with available feed resources and the production environment. The design of the herd is not an easy task because very rarely is a successful breeding program designed around selection for a single trait. Fortunately, studies in animal breeding have quantified the genetic variation in beef cattle traits, so that beef cattle producers may use this information in producing better beef through designed breeding programs.

The following sections address key areas of genetic improvement in the beef cattle industry. Animal breeding principles are related to specific beef examples and current genetic selection tools are discussed.

## HERITABILITY

An understanding of the principles of heritability and genetic correlations for beef cattle traits is needed to better use the variety of selection tools and **breed** trait information that is available. Differences in traits measured in animal populations are the sum of genetic and environmental factors associated with those traits. *Heritability* indicates the proportion of the differences between individuals that is genetic. Heritability is not constant. It varies from herd to herd and can vary within a herd if the management or the system of mating changes. Much research has been directed toward the study of heritability for various traits in beef cattle. Average heritability estimates from many studies for beef cattle are shown in Table 9–1. In general, reproductive traits tend to have low heritability (<0.20), growth traits tend to have moderate heritability (0.20–0.40), and carcass traits tend to have fairly high heritability (>0.40).

**Breed**

Animals with common ancestry. They have distinguishable characteristics, and when mated with others of the breed, produce offspring with the same characteristics.

**Heritability**

A measure of the amount of phenotypic variation that is due to additive gene effects. The proportion of differences between individuals that is genetic.

**TABLE 9–1**    **Heritability Estimates for Beef Cattle**

| Trait | $h^2$ |
| --- | --- |
| Birth weight | 0.35 |
| Weaning weight | 0.30 |
| Weaning score | 0.25 |
| Feedlot gain | 0.45 |
| Carcass grade | 0.40 |
| Fat thickness | 0.33 |
| Rib eye area | 0.58 |
| Marbling | 0.42 |
| Retail product % | 0.30 |
| Calving interval | 0.08 |
| Gestation length | 0.35 |
| Pasture gain | 0.30 |
| Yearling weight | 0.40 |
| Feed efficiency | 0.38 |
| Dressing % | 0.38 |
| Tenderness | 0.55 |
| Cancer eye | 0.30 |

*SOURCE:* Adapted from Buchanan et al., 1993, p. 86.

Probably the most practical use of heritability is that it indicates the ease with which we can make genetic improvement through selection. As we can see from the published estimates of heritability in Table 9–1, it is much easier to show selection progress for growth and carcass traits than for reproductive traits. This has led some beef cattle producers to decide that reproductive traits should not be included in a selection program. However, this idea overlooks the fact that reproduction is the most important factor in the efficiency of most beef enterprises. The importance of traits associated with reproduction makes up for the low heritability, so reproduction should be considered for most selection programs.

Also, heritability indicates the proportion of the superiority in an individual or in a group of individuals that can be passed on to the next generation. This property is used to estimate breeding value. *Breeding value* is the value of an individual as a parent. The actual breeding value of an individual is never known, but it can be estimated from the performance of the individual and its relatives. Common information on relatives includes progeny, sire, dam, and sibling records.

**Breeding value**
The worth of an individual as a parent.

## GENETIC CORRELATIONS

Very rarely are successful breeding programs based on *single-trait selection;* therefore, it is important to understand the genetic relationship between traits of interest. *Genetic correlation* refers to a situation in which the same or many of the same genes control two traits. The magnitude of genetic correlations may vary between $-1$ and $+1$. A genetic correlation of 0 indicates that different genes influence the two traits; thus the traits are uncorrelated. If the sign of the genetic correlation is positive, then the breeding values of the animals for the two traits tend to vary together. The reverse is true for a negative correlation.

**Single-trait selection**
Selection for only one trait or characteristic.

**Genetic correlation**
The situation in which the same or many of the same genes control two traits.

The absolute value of the correlation indicates the strength of the association between the two traits. When a genetic correlation exists between two traits, it means that the correlation does not equal 0. Rarely does this mean that the correlation is perfect at $+1$ or $-1$. For example, a genetic correlation of 0.10 is positive, but the magnitude of the correlation does not imply a strong genetic association between the two traits.

Knowledge of the magnitude of the genetic correlation between various traits is useful in a selection program. For example, *feed efficiency* is a difficult and expensive trait to measure. *Rate of gain* is a relatively easy and inexpensive trait to measure. A favorable genetic correlation exists between rate of gain and feed efficiency. Selection of sires can be directed toward rate of gain, which is easily measured. If the rate of gain is improved through selection, some improvement is expected in feed efficiency due to the favorable genetic relationship between the two traits. Genetic correlations are not always favorable. For example, selection for increased yearling weight has an adverse effect on calving difficulty. The fact that the genetic correlations are not perfect gives breeders the opportunity to try to identify sires that are exceptions to the unfavorable correlation.

**Feed efficiency**
Product (gain, milk, eggs, and so on) per unit of feed.

**Rate of gain**
Pounds of gain per day over a specified period.

Table 9–2 shows genetic correlations between growth and carcass traits. Some of these relationships may be beneficial if they are considered in a complete breeding program. As you can see, genetic correlations are seldom perfect. For example, many of the genes that control birth weight also control carcass weight in the same direction, as indicated by the positive genetic correlation of 0.60 in Table 9–2. Remember that the relationship is not perfect. Genetic correlations indicate what is likely to happen to one trait when selection is practiced for another trait.

**TABLE 9–2**     **Genetic Correlations between Growth and Carcass Traits**

| Trait | ADG^W | ADG^F | CAR | FAT | REA | MAR | SHR[1] |
|---|---|---|---|---|---|---|---|
| Birth weight (BW) | 0.28 | 0.61 | 0.60 | 20.27 | 0.31 | 0.31 | −0.01 |
| ADG to weaning (ADG^W) | | 0.49 | 0.73 | 0.04 | 0.49 | 0.31 | −0.05 |
| ADG feedlot (ADG^F) | | | 0.89 | 0.05 | 0.34 | 0.15 | 0.06 |
| Carcass weight (CAR) | | | | 0.08 | 0.44 | 0.25 | 0.00 |
| Fat thickness (FAT) | | | | | −0.44 | 0.16 | 0.26 |
| Rib eye area (REA) | | | | | | −0.14 | −0.28 |
| Marbling (MAR) | | | | | | | −0.25 |

[1]SHR 5 Warner-Bratzler shear.
SOURCE: Adapted from Benyshek, 1988.

To design the genetics of the beef animal for a particular production level, selection objectives must balance many traits of economic importance. Continued interest in carcass merit will result in an increase in multiple-trait selection practices in breeding programs. This makes information on the genetic correlations between carcass and other traits even more important.

## PERFORMANCE INFORMATION

To make genetic change in a desired direction, cow-calf producers have to know the current performance level of their herd. Through knowledge of beef cattle traits and their heritabilities, producers can use available selection tools to design a breeding program. The program should be designed with performance items in the plan that address breeding, calving, weaning, yearling, carcass, and maternal breeding objectives.

Performance programs come in many forms. Core programs begin at the ranch. Seedstock and commercial cow-calf producers have different needs. Seedstock producers sell breeding animals. Commercial cow-calf producers market calves to be finished for market. Also, herd size can influence the degree of detail that a producer is willing or able to assemble. Keep in mind that meaningful cow-calf records may be handwritten or computerized (Figure 9–1). The challenge is to choose performance records that are useful in making management decisions.

Seedstock breeders work closely with breed associations to develop extensive on-farm performance programs. Data collection and compilation are accomplished through strong ties between breeders and association personnel. Records across the country and, in some breeds, internationally, are used to generate genetic values for animals within a breed. Additional supplements to on-farm performance programs may include, but are not limited to, feedlot and carcass data collection programs and bull evaluation center data.

Commercial producers need an effective performance program that encourages the culling of inferior animals and selection of herd replacement breeding stock. Very rarely are effective selection programs based on single traits. Sire selection is the area in which commercial herds can place the greatest selection pressure. Commercial producers may rely on their seedstock "partners" to remove some of the guesswork in their bull selection and assist them in the use of specialized selection tools such as expected progeny differences (EPDs), discussed later in this chapter. Commercial cow herd and calf crop records are the nuts and bolts that assist producers in choosing the necessary bull power. Currently, a greater percentage of commercial and seedstock producers are

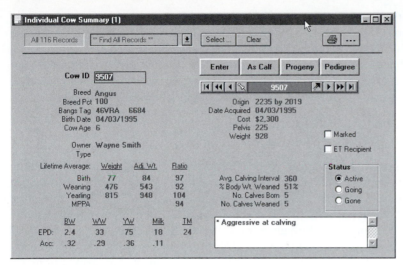

(a)

(b)

**FIGURE 9–1** (a) Cow summary report and (b) calf data entry form. (*SOURCE:* Cow Sense Herd Management Software, Midwest MicroSystems, Lincoln, NE. Used with permission.)

forming stronger links to better understand performance progress and genetic tools from conception to carcass.

Sometimes commercial producers think they need EPDs on their crossbred cows to use EPDs. This is not true. The choice of sires has a tremendous impact on the genetic improvement of the herd. The producer does not need to have registered cows, or straightbred cows of one breed, to use EPDs as a sire selection tool.

A good understanding of the herd performance level for reproductive, growth, and maternal merit, and even carcass merit, is the first priority. Calving and weaning percentages, pregnancy percentage, pounds weaned per exposed females, calf death loss, and average mature cow size are examples of decision-making tools. The sire selections are then made using EPDs to move the herd genetically in the desired direction. Throughout history, geneticists have studied methods for use in identifying superior individuals in beef cattle populations. Sire selection has tremendous value to the beef cow-calf operation. Choices of herd sires not only have an impact on the resulting calf crops, but also affect the performance of the cow herd if daughters of the sires are kept as replacement heifers. Ideally, beef producers like to select sires of desirable genetic merit for genetic improvement in economically important traits. Selection of desirable genetics to match with a cow herd is a challenging task. Fortunately, the concept of breeding value provides beef producers an avenue to make useful selection decisions. The background on breeding value estimation leads to a better understanding of the merit of EPDs.

## GENETIC EVALUATION, BREEDING VALUE, AND EXPECTED PROGENY DIFFERENCE

Breeding value is used as an estimate of the transmitting ability of an animal. Breeding value, or genetic merit, is calculated from information on an individual's performance and the performance of progeny, sibs, parents, grandparents, and so on. The information comes from purebred breeders who report data to the national herd improvement program for their breeds. The combined information from all contributors provides a national database for the breed. A mathematical model called an animal model (AM) is used to predict breeding values. These values are then used to calculate the expected progeny difference.

Half of the breeding value is equal to the ***expected progeny difference.*** The word *difference* implies a comparison. Thus EPDs let us compare or rank the superiority of individual animals. EPDs provide a prediction of future progeny performance of one individual compared to another individual within a breed for a specific trait. The EPDs are reported in plus or minus values in the units of measurement for the trait. For example, birth, weaning, and yearling weight EPDs are reported in pounds. In contrast, fat thickness EPDs are reported in inches.

Each individual member of a breed can have EPD values calculated for it. Age, sex of calf, or status as a parent are not limiting factors. Even a newborn calf can be assigned EPDs. It is possible to compare any two members of a breed, regardless of location. However, EPD values may be used to directly compare only those animals within a breed. The EPD values for a Hereford bull may not be compared against the EPDs for an Angus or Limousin bull.

Preferential mating of certain individuals does not bias the results. A genetically superior bull can be mated only to genetically superior cows and his EPD will not be inflated. This is accomplished by adjusting for the EPDs of the cows to which he is mated. Also, any genetic change within a breed for a particular trait is accounted for by adjusting for genetic trend. Thus comparisons may be made across generations of cattle. Young bulls with no progeny may be directly compared with older sires that have progeny.

Maternal genetic values, or milk EPDs, may be computed for the maternally influenced trait—weaning weight. EPDs are comparable across herds because mathematical models used to calculate EPDs account for differences in environment and management.

### National Cattle Evaluations

The first national sire summary was published in 1971 by a beef breed association. For the first time, bulls within a breed could be compared across herds, across regions of the United States, and across generations. All of the major beef breed associations today conduct national cattle evaluations (NCEs) and compute EPDs as a service to their breeders. These values are published and generally made available at least once a year.

### Contemporary Group

In the collection of beef cattle performance information, breed associations recognize that contemporary group definition is critical. A ***contemporary group*** is a group in

---

**Expected progeny difference (EPD)**

A prediction of the difference between the performance of an individual's progeny compared to all contemporaries for the progeny.

---

**Contemporary group**

A group in which animals of a given sex and age, having similar treatment, are given an equal opportunity to perform.

which animals of a given sex and age, having similar treatment, are given an equal opportunity to perform. The basis of sound performance testing relies on correct identification of contemporary groups. Accuracy in estimation of genetic differences within a group of animals depends on the accuracy of grouping.

## GROWTH TRAIT EPDS

EPD values are most useful when two individuals are compared directly. For example, consider the following birth weight EPD example of two sires. Assume both sires are from the same breed, and the EPDs have equal accuracies:

|  | Sire A | Sire B |
|---|---|---|
| Birth weight EPD, lbs | +5 | −2 |

The expected difference in the progeny of Sire A and Sire B for birth weight is 7 lbs. Sire A has an EPD of +5 and Sire B has an EPD of −2. On the average, we should expect the calves of Sire A to be 7 lbs heavier at birth than calves of Sire B, if all calves are managed the same and have dams of similar genetic merit. The predicted performance difference is 7 lbs. Because EPDs only allow the prediction of performance differences, not actual performance, it is not possible to estimate the actual birth weight average for these calves.

The following is an example for weaning weight. It describes a weaning weight difference in the progeny of two bulls:

|  | Sire A | Sire B |
|---|---|---|
| Weaning weight EPD, lbs | +25 | −10 |

The expected difference in the progeny of Sire A and Sire B for weaning weight is 35 lbs. Sire A has an EPD of +25 and Sire B has an EPD of −10. On the average, we should expect the calves of Sire A to be 35 lbs heavier at weaning than calves of Sire B, if all calves are exposed to the same environmental conditions and are out of cows of similar genetic merit.

An example for yearling weight follows. It describes a yearling weight difference in the progeny of two bulls:

|  | Sire A | Sire B |
|---|---|---|
| Yearling weight EPD, lbs | +50 | +10 |

The expected difference in the progeny of Sire A and Sire B for yearling weight is 40 lbs. Sire A has an EPD of +50 and Sire B has an EPD of +10. On the average, we should expect the calves of Sire A to be 40 lbs heavier as yearlings than calves of Sire B, if all calves are managed in a uniform manner and have dams of similar genetic merit.

## BREED AVERAGE EPD AND BASE YEAR

It is frequently said that an EPD is a comparison to an average bull. This, unfortunately, is not true. A zero EPD represents the average genetic merit of animals in the database at the time when sufficient information existed to calculate EPDs. Therefore, it represents a historic base point, or base year. Some breed associations now set the base year to a particular year. If the breed has made any genetic change for a trait, the average EPD for the trait will no longer be zero. Breed associations publish the average EPDs in the sire summaries made available to the public. Information printed in the summaries should be examined carefully before individual EPDs are studied.

## ACCURACY

**Accuracy** is the measure of reliability associated with an EPD. Each EPD value should have an accuracy assigned to it. Accuracy is expressed as a value between 0 and 1. A high accuracy ($>0.7$) means a higher degree of confidence may be placed on the EPD, and the EPD value is not expected to change much as further information is gathered. A low accuracy ($<0.4$) means that the EPD may change a great deal as additional information is gathered. Nonparent animals have lower accuracy values because no progeny information contributes to their EPD. From a practical standpoint, the EPDs are used to select bulls for use in the herd, and accuracies help determine how extensively to use them. Some sale catalogs do not list accuracies with the EPDs. For young animals with no progeny data, such as yearling bulls, accuracies are generally low. They improve as the animal has offspring to contribute to the data from which its EPDs are calculated.

## POSSIBLE CHANGE

**Possible change** is the measure of the potential error associated with EPD values. Many sire summaries are starting to include these values. Possible change is expressed as "+" or "−" pounds of EPD. These values quantify the amount a certain EPD may deviate from the "true" progeny difference. Accuracy and possible change values share a relationship. As more information is accumulated, accuracy increases and possible change diminishes. For a given accuracy, the "true" progeny differences of two thirds of all animals evaluated within a breed are expected to fall within the plus or minus possible change value. The following example illustrates this point:

Birth weight EPD $= +2.0$ lbs    Accuracy $= 0.90$    Possible change $= \pm 1.3$ lbs

Of all of the animals with this EPD and accuracy, two thirds are expected to have "true" progeny differences between $+0.7$ and $+3.3$. These "true" differences have a much greater chance of falling toward the center of the range defined by the possible change value than falling close to the extremes.

Also, one third of the individuals in the evaluation may have their "true" progeny difference values fall outside the range of $+0.7$ and $+3.3$. This means that one sixth of the individuals may have "true" values less than $+0.7$, and one sixth of the individuals may have "true" values more than $+3.3$.

## SIRE SUMMARIES

**Sire summaries** include a sampling of the available genetic material in each breed. Newly calculated and updated summaries for breeds that conduct national cattle evaluations come out at least once a year. Summaries include EPDs, accuracies, graphs of the average change in EPD for the particular breed, breed average EPDs, possible change values, and other useful materials. Descriptive material written at the beginning of each summary describes the format for reporting the EPDs.

Almost all sire summaries include birth weight, weaning weight, yearling weight, and milk EPDs (Figure 9–2). A few summaries currently include some characteristics that have

---

**Accuracy**

The measure of reliability associated with an EPD. If little or no information is available, accuracies may range as low as 0.01. A high accuracy would be 0.99.

**Possible change**

The measure of the potential error associated with EPD values.

**Sire summary**

Genetic information published on sires available within a breed.

### Sire Evaluation of Proven Sires

| ANIMAL IDENTIFICATION | OWNER | GRPS PROG | DTRS PROG/DTR | BW EPD ACC | WW EPD ACC | YW EPD ACC | MILK EPD ACC | TOTAL MAT. EPD | STAY EPD ACC | HPG EPD ACC | MARB EPD ACC | REA EPD ACC | FAT EPD ACC |
|---|---|---|---|---|---|---|---|---|---|---|---|---|---|
| **FORSTER LAKOTA 3100** REG.#: 405225 CATEGORY: A BIRTHDATE: 04/01/1993 SIRE: RED CENTURION C103 MATERNAL GRANDSIRE: LFS GALENA 132 | FORSTER RED ANGUS ALTA GENETICS, INC. WBD | 30 | 23 | 1.2 | 30 | 49 | 26 | 41 | 8 | | -0.06 | 0.03 | 0.00 |
| | | 89 | 1.9 | .82 | .75 | .72 | .65 | | .63 | | .22 | .22 | .19 |
| **FORSTER MAGNUM 6142** REG.#: 526563 CATEGORY: A BIRTHDATE: 03/30/1996 SIRE: RED BBCC PRESIDIO 5Z MATERNAL GRANDSIRE: BUFCRK COPPERTOP1628 | KINCHEN CATTLE CO FORSTER RED ANGUS | 19 | 2 | 1.6 | 17 | 28 | 22 | 30 | 8 | | | | |
| | | 79 | 1.0 | .80 | .68 | .58 | .33 | | .32 | | | | |
| **FORSTERS PAY DAY 376** REG.#: 192675 CATEGORY: A BIRTHDATE: 02/27/1986 SIRE: RAB ABE LINCOLN MATERNAL GRANDSIRE: ROCKY MT 19 | FORSTER RED ANGUS REDLAND RED ANGUS BOOT JACK RANCHES | 23 | 49 | 4.3 | 31 | 41 | 1 | 17 | 6 | 6.3 | -0.04 | 0.03 | -0.02 |
| | | 157 | 4.7 | .85 | .81 | .80 | .77 | | .75 | .19 | P | P | P |
| **FRITZ MONU 2X 211** REG.#: 373666 CATEGORY: A BIRTHDATE: 02/07/1992 SIRE: LEACHMAN MONU 2X 8096 MATERNAL GRANDSIRE: RCN DYNAMO 614 | KEITH & LINDA VANDE SANDT FRITZ RED ANGUS | 13 | 14 | -2.5 | 21 | 29 | 15 | 25 | 9 | | 0.17 | -0.15 | 0.00 |
| | | 29 | 1.8 | .71 | .64 | .62 | .57 | | .55 | | P | P | P |
| **FTF DOUBLE CHIEFN601** REG.#: 130241 CATEGORY: A BIRTHDATE: 03/29/1981 SIRE: FTF DOUBLE CHF L218 MATERNAL GRANDSIRE: FTF CHIEFTON 7068 | GREER RANCH | 59 | 93 | -1.5 | 18 | 31 | 20 | 29 | 14 | 22.0 | 0.16 | -0.10 | 0.00 |
| | | 210 | 4.7 | .88 | .86 | .85 | .83 | | .83 | .46 | .34 | .33 | .29 |
| **GENERAL VANGUARD** REG.#: 289875 CATEGORY: A BIRTHDATE: 11/12/1987 SIRE: FFC VANGUARD 638F MATERNAL GRANDSIRE: CV GENERAL LEE | MCKELVEY RED ANGUS CATTLE RANCH | 31 | 34 | 3.2 | 32 | 51 | 8 | 24 | 3 | | -0.04 | -0.07 | 0.01 |
| | | 64 | 2.3 | .81 | .76 | .74 | .71 | | .71 | | P | P | P |
| **GET-A-LONG LICORICE** REG.#: 128766 CATEGORY: A BIRTHDATE: 04/10/1979 SIRE: RED SR IMAGE 200F MATERNAL GRANDSIRE: RED VALLEY REVOLUTION | KENNETH FRAZER DOUBLE FORK RANCH | 167 | 236 | 4.4 | 37 | 51 | 8 | 27 | 14 | 8.7 | -0.03 | 0.04 | 0.00 |
| | | 610 | 5.0 | .93 | .91 | .90 | .89 | | .89 | .22 | .28 | .28 | .24 |
| **GILCHRIST CHIEF W515** REG.#: 273997 CATEGORY: A BIRTHDATE: 02/13/1987 SIRE: JKG CHIEFTON L303 MATERNAL GRANDSIRE: FTF RED RITO M428 | BUFFALO CREEK RED ANGUS J BAR K RANCH | 121 | 96 | -0.7 | 28 | 60 | 16 | 30 | 14 | 6.5 | 0.17 | -0.18 | -0.05 |
| | | 257 | 3.1 | .89 | .86 | .85 | .83 | | .82 | .19 | .61 | .59 | .51 |
| **GILCHRIST COYOTE B225** REG.#: 371018 CATEGORY: A BIRTHDATE: 03/22/1992 SIRE: BJR FIRE-MAN 0150 MATERNAL GRANDSIRE: JKG CHIEFTON L303 | STAR G RANCH GILL RED ANGUS BLUE RIDGE LAND & CATTLE | 76 | 55 | 2.4 | 33 | 50 | 23 | 39 | 3 | | 0.04 | -0.06 | -0.02 |
| | | 185 | 1.4 | .87 | .82 | .80 | .76 | | .76 | | .24 | .23 | .21 |
| **GILCHRIST HOWLER 95** REG.#: 477961 CATEGORY: A BIRTHDATE: 02/25/1995 SIRE: GILCHRIST COYOTE B225 MATERNAL GRANDSIRE: 741 PANHANDLER 189 | BRYLOR RANCH BLUE RIDGE LAND & CATTLE CODY E GRIFFIN | 29 | 7 | 2.0 | 20 | 34 | 15 | 25 | 4 | | 0.02 | -0.03 | -0.01 |
| | | 47 | 1.2 | .77 | .65 | .60 | .50 | | .49 | | .19 | .19 | .19 |
| **GLACIER ALPINE 658** REG.#: 514846 CATEGORY: A BIRTHDATE: 02/20/1996 SIRE: CREEK SIDE COPPER MATERNAL GRANDSIRE: LEACHMAN MONU 2X 8096 | GRILL CATTLE CO GLACIER RED ANGUS | 30 | 7 | 0.6 | 20 | 29 | 17 | 27 | 6 | | 0.29 | -0.05 | 0.00 |
| | | 78 | 1.0 | .78 | .68 | .64 | .50 | | .48 | | P | P | P |
| **GLACIER ARROW 664** REG.#: 514799 CATEGORY: A BIRTHDATE: 02/21/1996 SIRE: GLACIER STATURE 318 MATERNAL GRANDSIRE: GLACIER NATIONAL | SCHULER-OLSEN RANCH GLACIER RED ANGUS | 19 | 2 | -1.8 | 19 | 52 | 21 | 31 | 7 | 9.2 | -0.05 | -0.44 | 0.00 |
| | | 87 | 1.0 | .81 | .69 | .62 | .35 | | .33 | .23 | .51 | .48 | .41 |
| **GLACIER CAMAS 752** REG.#: 557327 CATEGORY: A BIRTHDATE: 02/21/1997 SIRE: GLACIER SUPER WEIGHT MATERNAL GRANDSIRE: GLACIER DYNARISE | GARY SONSTEGARD GLACIER RED ANGUS | 22 | 0 | -0.3 | 21 | 34 | 15 | 26 | 5 | | 0.03 | -0.41 | 0.00 |
| | | 70 | 0.0 | .78 | .66 | .55 | .27 | | .26 | | .26 | .25 | .22 |
| **GLACIER CASCADE 464** REG.#: 440319 CATEGORY: A BIRTHDATE: 09/26/1994 SIRE: GLACIER NATIONAL MATERNAL GRANDSIRE: GLACIER SUPER WEIGHT | KOLLE RED ANGUS GLACIER RED ANGUS | 26 | 8 | -3.1 | 17 | 39 | 24 | 32 | 6 | | 0.05 | -0.36 | 0.02 |
| | | 70 | 1.7 | .78 | .69 | .62 | .52 | | .51 | | .51 | .49 | .44 |

**FIGURE 9–2** Beef cattle sire summary. (*SOURCE:* Red Angus Association of America, Denton, TX. Used with permission.)

a role in reproduction such as calving ease, gestation length, heifer pregnancy, scrotal circumference, and stayability. Many of the associations are currently working to include some of these other characteristics into the summaries. There is a fairly large effort to incorporate more carcass information. In fact, several cattle breed associations publish EPDs for carcass weight, rib eye area, fat thickness, marbling score, and tenderness.

Many of the summaries contain two listings of bulls. The first section lists progeny-proven bulls. These are older bulls that have calves with performance records; therefore, the accuracies on the birth and weaning weight EPDs are generally at least 0.5. The second section is devoted to younger bulls that have lower accuracies (0.3 to 0.5 on weaning and birth weight). The criteria for listing varies among the breeds.

## MATERNAL TRAIT EPDS

Maternal effects are an important consideration in evaluating beef cattle performance. Extensive studies have been conducted to quantify maternal effects for a variety of traits, particularly those measured during the preweaning growth period. In beef cattle, the dam makes at least two contributions to the offspring phenotypic value. *Phenotypic value* is the physical expression of the genetic makeup of an animal, such as a weaning weight. These contributions are the sample half of her genes passed directly to the offspring and the maternal effect she provides her calf. A *maternal effect* is defined as any environmental influence that the dam contributes to the phenotype of her offspring. The contribution of the dam is environmental with respect to the calf (mothering ability, milk production environment, and maternal instinct). The genetics of the dam

**Phenotypic value**

A measure of individual performance for a specific trait.

**Maternal effect**

Any environmental influence that the dam contributes to the phenotype of her offspring.

allow her to create this environment for her calf. Maternal effects are important during the nursing period but have diminishing effects postweaning.

## MILK EPD

Weaning weight is influenced by the genes for growth in the calf and the genes for milk (mothering ability) in the cow. There are separate EPD values for these two components. The weaning weight EPD evaluates genetic merit for growth and the milk EPD evaluates genetic merit for mothering ability. The milk EPD that results from the separation of weaning weight into growth and milk segments is, like any other EPD, fairly simple to use. It is the expected difference in weaning weight of calves out of daughters of a particular sire, due to differences in mothering ability. For example, consider the following two bulls:

|  | *Sire A* | *Sire B* |
|---|---|---|
| Milk EPD, lbs | $+10$ | $-6$ |

Sire A has a milk EPD of $+10$; Sire B has a milk EPD of $-6$. The expected weaning weight difference, due to mothering ability alone, in calves out of daughters by the two bulls is 16 lbs. The 16 lbs are expressed in pounds of weaning weight, not pounds of milk.

## COMBINED MATERNAL EPD

The combined maternal EPD (sometimes called maternal weaning weight) reflects both the milking ability transmitted to daughters and the direct weaning growth transmitted through daughters to their calves. Here is an examples:

|  | *Weaning Weight EPD* | *Milk EPD* | *Weaning Weight Maternal Combined EPD* |
|---|---|---|---|
| Bull A | $+20$ | $+12$ | $+22$ |
| Bull B | $+4$ | $+6$ | $+8$ |

$$\text{Combined (Bull A)} = 1/2\ (20) + 12 = 22 \text{ lbs}$$

$$\text{Combined (Bull B)} = 1/2\ (4) + 6 = 8 \text{ lbs}$$

Bull A has a direct weaning weight EPD of $+20$ lbs, which expresses the ability of the bull to transmit weaning growth directly to his progeny. On the average, calves sired by Bull A should be 16 lbs heavier at weaning than calves sired by Bull B, assuming both bulls are mated to a comparable set of females and the calves are exposed to the same environmental conditions. The 16-lb difference in future progeny performance is due to genes for direct weaning growth.

The milk EPD for Bull A ($+12$) is the contribution to his daughter's calves solely through transmission of genes for mothering ability. Sire A has a milk EPD of $+12$; Sire B has a milk EPD of $+6$. The expected weaning weight difference, due to mothering ability alone, in calves out of daughters by the two bulls is $+6$ lbs.

The combined EPD for Bull A ($+22$) is computed by taking half the weaning weight EPD plus all of the milk EPD. The $+22$ lbs reflects both the milking ability transmitted to daughters and the direct weaning growth transmitted through the daughters to their calves. In a similar fashion, the combined EPD for Bull B is half times the weaning weight EPD plus the milk EPD, or $+8$ lbs. An average 14-lb difference between the combined EPDs for the two bulls ($22 - 8 = 14$) would be expected to be the difference in weaning weight of calves out of the daughters of the two bulls. The difference reflects the milking ability of the daughters and the direct weaning growth transmitted through daughters to their calves.

## Carcass EPD

Carcass trait EPDs are an additional performance tool that is becoming more available each year. Many of these EPDs are generated from progeny carcass data for various sires within a breed. Generating EPDs for sires within a breed can be an expensive project because carcass data on close relatives, particularly progeny, must be captured. When collecting data on progeny of sires for genetic evaluation, it is most desirable to have at least 25 progeny per sire, although this may be difficult to achieve. For carcass trait EPDs, a simple comparison of two bulls within a breed is conducted, as shown in Table 9–3. Look at the difference in EPDs between the bulls. On the average, future calves out of Bull A will have 50-lb heavier carcass weights, 0.40 sq. in. larger rib eye areas, and 1% greater retail product percentage than calves sired by Bull B. Future offspring of Bull B will have a fourth higher marbling score than calves out of Bull A.

Brangus was the first breed to generate rib eye area EPDs using ***ultrasonic scan measures.*** Ultrasonic data are relatively easy to collect on yearling bulls and heifers in seedstock herds. A bull, for example, may have his own ultrasonic measure contributing to the prediction of his ultrasonic rib eye area EPD. In 1998, the American Hereford Association released ultrasonic carcass EPDs. In spring 2001, the American Angus Association first published EPDs that incorporated ultrasound data in their sire summary. In the near future, most breeds will incorporate these scan measures into their genetic evaluations. Ultrasonic data will be useful in increasing the accuracy of prediction for young sires with few carcass progeny records available. National efforts are in progress to develop the database of ultrasonic measures and its implementation in genetic evaluation.

**Ultrasonic scan measures**

Measurements of body tissues taken with ultrasound waves.

## Mature Size

Mature size is an important issue in the beef industry today. Size is composed of closely related measures of weight and height; however, the relationship among these traits is not clearly understood. Studies of mature size in beef cows have estimated lifetime growth curves for weight through maturity. Other reports have considered the influence of body size on the biological efficiency of cows.

Genetic prediction of mature size may allow beef cattle breeders to make a directional change in the mature size of their cow herd or to emphasize uniformity of cow

**TABLE 9 – 3    Angus Carcass EPD Example**

| EPD | Breed Average | Bull A | Bull B | Difference |
|---|---|---|---|---|
| Carcass weight, lbs | +5.95 | 40 | −10 | 50 |
| Marbling score | +.05 | −.10 | .15 | .25 |
| Rib eye area, sq. in. | +.12 | .40 | .00 | .40 |
| Fat thickness, in. | −.003 | .00 | .00 | .00 |
| % Retail product | +.10 | .5 | −.5 | 1% |

SOURCE: Dolezal, 1999. Used with permission.

size for a particular production environment. Studies of the genetic components of mature size have addressed weight and height at maturity separately by trait, rather than as a composite measure.

Mature weight and height are highly heritable traits. For example, heritability estimates in Angus cattle are 0.49 for weight and 0.87 for height in mature cows. Heritabilities of this magnitude indicate that selection for these traits would be effective. In other words, if beef producers wanted to make changes in the height and weight of the cow herd, they would select sires of replacement heifers using some guidelines for desirable size for the herd production environment. Also, it is important to consider that the genetic correlation between mature weight and height is strong and positive. Large genetic association between these two traits suggests that selection for increased height would be associated with increased cow weight. Some beef breeds have incorporated cow size data into a genetic evaluation for the creation of mature cow size EPDs.

## USE OF EPDS

### Use of EPDs for Selection in Seedstock Herds

Purebred producers need to consider EPDs in their breeding programs. Competitors are using them and genetic change is happening. Care needs to be exercised when making selection decisions. Type fads have caused some problems in the past when single traits were emphasized. Similar, or worse, problems may arise if a single performance trait is emphasized. For example, if the members of one breed association begin to emphasize yearling weight and ignore all other characteristics, several concerns may result. Birth weight would be expected to increase, with the attendant calving difficulty. Mature size should also increase, perhaps to the point where the functionality of the cow herd would diminish. This could also lead to problems in reaching desirable *quality grade* at an acceptable weight. Each trait has a set of drawbacks, if changes are carried to an extreme. The availability of EPDs would make such extremes easier to reach, if breeders chose blindly to emphasize a single trait.

A more balanced selection program is certainly desirable. Some producers recommend choosing herd sires that have high yearling weight EPD, high milk EPD, and low birth weight EPD. Because these three characteristics are sufficiently different from one another, the difficulties from extreme changes in any one of them would be unlikely to result. It also needs to be recognized that there are still many important traits that are not included in the sire summaries. Careful monitoring of reproductive performance, including conception rates, calf mortality, libido in bulls, and regularity of calving, is still critically important. Carcass characteristics may have increased importance in the near future and breeders are certainly encouraged to obtain whatever carcass data is feasible and use it in making some selection decisions. Carcass EPDs should be available in several breeds soon, but more complete databases must be established.

Purebred breeders should certainly avail themselves of the opportunity to obtain EPDs on each member of their herd, if their association provides this service. Most associations have this ability at the present time. Even though the accuracies are sometimes low on these EPDs, they should be used when choosing replacements and, if possible, when culling cows.

Purebred producers are not only the users of EPDs, but they also provide the data used in calculating EPDs. Producers are strongly encouraged to provide complete, accurate records on all calves born each year. Complete, accurate recordkeeping is the only way that useful EPDs can be calculated.

---

**Quality grade**

Scale that indicates quality and value of the carcass such as *prime, choice,* and so on.

## Use of EPDs for Selection in Commercial Herds

Obviously, it will be the rare commercial producer who uses bulls that are listed in a breed association's sire summary. What then should the commercial producer do about EPDs? Many breed associations have a mechanism in place in which individual pure-bred producers can obtain EPDs on each individual in the herd, including calves. Commercial producers should certainly demand such information from the purebred sources of breeding stock.

A commercial producer has a first responsibility of choosing the appropriate breed, or breeds, for his or her program. Once breeds are chosen, examination of what is needed in replacement breeding stock is in order. Some recommendations for commercial scenarios are shown in Table 9–4.

Each of these recommendations should be followed with an awareness of the prevailing environmental conditions. Rougher conditions probably dictate avoidance of very high EPDs for growth or milk, and even more care to avoid high birth weights. Growth EPDs should be geared to the desires of the potential buyers. Again, traits for which there are no EPDs as yet can also be important. Traits associated with reproduction certainly fall into this category. Commercial producers should demand that bulls have passed a ***breeding soundness examination.*** The cow herd of the seller should be examined for regularity of calving.

EPDs within a breed are directly comparable between herds. Therefore, if a commercial producer has more than one source of breeding stock, he or she can compare the genetic merit of the different sources. Unfortunately, EPDs cannot be compared between breeds. A bull with a low birth weight EPD from a large mature size breed may sire calves that are heavier than those from a bull with a high birth weight EPD from a moderate sized breed. A low birth weight EPD does not guarantee a minimum of calving difficulty if the choice of breeds is incorrect.

**Breeding soundness examination**

Physical examination to determine the readiness of an individual for breeding purposes.

## Pedigree Estimated EPDs

After the first of each year, sale catalogs prepared for production sales and full of information on potential herd sires become available. Many sale catalogs contain EPDs for the bulls offered for sale. Data on some bulls appear in catalogs with limited or no EPD information. This may be particularly true for young bulls that have not had their performance information included in the breed genetic evaluation. Bull buyers may use a quick and easy procedure to compute "pedigree EPD" values for young bulls with no EPDs. Pedigree EPDs can be computed provided there is access to EPDs on the animals in the pedigree of the young bull. By using the EPDs on animals in the young

**TABLE 9–4    Recommendations for EPDs for Various Commercial Scenarios**

| Use of Individual | Breed | Birth | Weaning | Yearling | Milk[1] |
|---|---|---|---|---|---|
| Terminal sire on mature cows | Large carcass | Not too high | High | High | Not relevant |
| Bull to use with heifers | Small to medium size | Low | Moderate | Moderate | Consider if keeping heifers |
| Sire replacement heifers | Medium size maternal | Low to moderate | Moderate to high | Moderate to high | Varies |

[1]Selection decisions involving milk EPD should take into consideration the production environment and feed resources available for the cow herd.
*source:* Buchanan et al., 1993, p. 84.

bull's pedigree and the knowledge of how breeding value is transmitted from generation to generation, pedigree EPDs can be computed.

Every calf has received a random sample of half of the sire's genes and a random sample of half of the dam's genes to combine into its genetic makeup. Parents of the calf have received their genetic makeup in the same fashion, with half of their genetic makeup contributed by each of their parents. By understanding this halving nature of inheritance, the EPDs on parents and grandparents in the pedigree of a young bull may be used to compute pedigree EPDs.

**Procedure to calculate pedigree EPD.** The first step in calculating the pedigree EPD for a young bull of interest is to determine how much EPD information is available on the animals in the bull's pedigree. Many times, the breeder of the young bull will supply a performance pedigree including EPDs for the sire, maternal grandsire (MGS), maternal great-grandsire (MGGS), and maybe even the dam of the young bull. The next step is to calculate the pedigree EPD on the young bull of interest using the EPD information. For example:

1. If both sire and dam of the young bull have EPDs, take half the EPD of each parent:

$$\text{Ped. EPD} = 1/2 \text{ EPD of sire} + 1/2 \text{ EPD of dam}$$

2. If the EPD on the dam is missing, use EPDs on her sire (MGS):

$$\text{Ped. EPD} = 1/2 \text{ EPD of sire} + (1/2)^2 \text{ EPD of MGS}$$
$$= 1/2 \text{ EPD of sire} + 1/4 \text{ EPD of MGS}$$

3. Another option is to also use the maternal great-grandsire (MGGS) information:

$$\text{Ped. EPD} = 1/2 \text{ EPD of sire} + (1/2)^2 \text{ EPD of MGS} + (1/2)^3 \text{ of MGGS}$$
$$= 1/2 \text{ EPD of sire} + 1/4 \text{ EPD of MGS} + 1/8 \text{ EPD of MGGS}$$

*Note:* If the EPD of the dam is known, you *cannot* use the EPD information on the MGS and MGGS.

Some breed associations have an "interim EPD" program based on pedigree information to provide EPDs on young animals that have not had an opportunity to have their individual performance included in the most recent national cattle evaluation for the breed. Many sale catalogs may already provide the pedigree EPD for convenience.

## Across-Breed EPDs

Currently, most EPDs are used only on a within-breed basis. They are calculated for the specific breed; therefore, the EPDs are only useful for direct comparisons of future progeny performance for cattle within that breed. The across-breed EPD concept (AB-EPD) has been actively investigated since the late 1980s. Commercial bull buyers using more than one breed of bull are particularly interested in having this EPD option. The methodology for accomplishing AB-EPDs on a national scale is not yet perfected. To compare cattle of different breeds, additional information is required. This information includes (1) mean breed differences in the environments of interest; (2) the base year, or zero EPD point, for the breeds of interest; and (3) the expected effects of heterosis (or hybrid vigor) for matings between the breeds of interest. Breed comparison data from the U.S. Meat and Animal Research Center are the best resources available to date. Breed adjustments to make across-breed EPD comparisons are computed annually at this research station.

## EPDs and Crossbreeding

Planning a crossbreeding system first relies on the choices of breeds, followed by the use of within-breed EPDs as selection tools for performance traits. To assist beef producers in their choices of breeds, studies have tried to group or categorize breeds into general biological types. Perhaps the most famous and extensive of these studies involves the Germ Plasm Evaluation study conducted at the U.S. Meat and Animal Research Center (USMARC) at Clay Center, Nebraska. Table 9–5 lists 25 different sire breed groups that were evaluated in calves out of Hereford and Angus dams or calves out of the two-breed cross ($F_1$) dams. The breed groups illustrate relative differences (X = lowest, XXXXXX = highest) in growth rate and mature size, lean-to-fat ratio, age at puberty, and milk production. Increasing numbers of Xs indicate relatively higher performance levels and older age at puberty.

$F_1$

Two-breed cross animals.

**TABLE 9–5    Breeds Grouped into Biological Types for Four Criteria[1]**

| Breed Group | Growth Rate and Mature size | Lean-to-Fat Ratio | Age at Puberty | Milk Production |
|---|---|---|---|---|
| Jersey (J) | X | X | X | XXXXX |
| Longhorn (Lh) | X | XXX | XXX | XX |
| Hereford-Angus (HAx) | XXX | XX | XXX | XX |
| Red Poll (R) | XX | XX | XX | XXX |
| Devon (D) | XX | XX | XXX | XX |
| Shorthorn (Sh) | XXX | XX | XXX | XXX |
| Galloway (Gw) | XX | XXX | XXX | XX |
| South Devon (Sd) | XXX | XXX | XX | XXX |
| Tarentaise (T) | XXX | XXX | XX | XXX |
| Pinzgauer (P) | XXX | XXX | XX | XXX |
| Brangus (Bn) | XXX | XX | XXXX | XX |
| Santa Gert. (Sg) | XXX | XX | XXXX | XX |
| Sahiwal (Sw) | XX | XXX | XXXXX | XXX |
| Brahman (Bm) | XXXX | XXX | XXXXX | XXX |
| Nellore (N) | XXXX | XXX | XXXXX | XXX |
| Braunvieh (B) | XXXX | XXXX | XX | XXXX |
| Gelbvieh (G) | XXXX | XXXX | XX | XXXX |
| Holstein (Ho) | XXXX | XXXX | XX | XXXXX |
| Simmental (S) | XXXXX | XXXX | XXX | XXXX |
| Maine Anjou (M) | XXXXX | XXXX | XXX | XXX |
| Salers (Sa) | XXXXX | XXXX | XXX | XXX |
| Piedmontese (Pm) | XXX | XXXXXX | XX | XX |
| Limousin (L) | XXX | XXXXX | XXXX | X |
| Charolais (C) | XXXXX | XXXXX | XXXX | X |
| Chianina (Ci) | XXXXX | XXXXX | XXXX | X |

[1]Increasing number of Xs indicates relatively higher values.
SOURCE: Cundiff et al., Beef Improvement Research Federation Symposium, 1993, p. 130.

# DAIRY CATTLE GENETIC IMPROVEMENT

Dairy producers have been leaders in genetic improvement. Many commonly known techniques for evaluating genetic merit have been derived and tested initially on dairy cattle records. One advantage the dairy industry has is the focus on a limited number of economically important traits. Traditionally, milk yield has been the primary driver in trait emphasis for profitability. Yet very rarely are effective breeding programs based on a single trait. Dairy producers are challenged to balance traits of economic importance to address some of the following goals:

- Achieve profitable milk yield levels.
- Monitor milk composition.
- Generate profitable replacement animals that are productive under the stress of high production levels.
- Sustain and improve cow longevity in the herd.

The following sections address key areas of genetic improvement in the dairy cattle industry. Animal breeding principles are related to specific dairy examples and current genetic selection tools are discussed.

## HERITABILITY ESTIMATES

An understanding of the heritability and genetic correlations for dairy cattle traits is necessary to take advantage of the variety of selection tools and breed trait information available. Studies in animal breeding have quantified the genetic variation in dairy production, so producers may use this information to be more profitable through designed breeding programs. Although much emphasis is placed on milk yield and its impact on profitability, it is important to review the heritabilities of other commonly known dairy production traits. As discussed earlier, the most practical use of heritability is that it indicates how easily we can make genetic improvement through selection. However, one should not overlook more lowly heritable traits, such as the reproductive complex. These traits are highly influenced by environment and management practices.

Table 9–6 presents heritability estimates for dairy production parameters compiled from various research reports. In general, most reproductive traits tend to have low heritability (<0.20); yield traits tend to be moderately heritable (0.20–0.40); and composition traits and weights tend to have fairly high heritabilities (>0.40). Keeping in mind the earlier discussion, and viewing the heritabilities, one can see that it is relatively easy to change mature weight or wither height through selection. To improve reproductive efficiency or longevity, the dairy cattle breeder must take advantage of all the genetic tools available.

## ASSOCIATIONS AMONG TRAITS

Genetic correlation refers to a situation in which the same or many of the same genes control two traits. Table 9–7 presents phenotypic and genetic correlations between milk yield and other production characteristics. Phenotypic correlations are correlations between two traits that producers actually measure or see; thus a combination of genetics and environment plays a role in expression of the trait, such as weight or height. Genetic correlations are more difficult to visualize.

**TABLE 9–6** **Heritability of Various Traits in Dairy Cattle**

| Trait | Heritability | Trait | Heritability |
|---|---|---|---|
| Milk yield | 0.25 | Mature weight | 0.50 |
| Milk fat yield | 0.25 | Wither height | 0.50 |
| Protein yield | 0.25 | Conception rate | 0.05 |
| Total solids yield | 0.25 | Reproductive efficiency | 0.05 |
| Milk fat % | 0.50 | Calving interval | 0.10 |
| Protein % | 0.50 | Life span | 0.15 |
| Persistency | 0.40 | Feed efficiency | 0.35 |
| Peak milk yield | 0.30 | Mastitis resistance | 0.10 |
| Milking rate | 0.40 | Overall type score | 0.20 |
| Gestation length | 0.40 | Dairy character score | 0.20 |
| Birth weight | 0.40 | White coat color (Holsteins) | 0.90 |

SOURCE: Wilcox, 1992, p. 3. Used with permission.

**TABLE 9–7** **Genetic Correlations between Milk Yield and Other Traits**

| Trait | Correlation with Milk Yield | |
|---|---|---|
| | Phenotypic | Genetic |
| Fat yield | 0.85 | 0.70 |
| Solids not fat yield | 0.85 | 0.90 |
| Protein yield | 0.85 | 0.90 |
| Fat % | −0.35 | −0.35 |
| Solids not fat % | −0.30 | −0.25 |
| Protein % | −0.35 | −0.30 |
| Type score | 0.29 | 0.00 |
| Stature | 0.11 | −0.01 |
| Strength | 0.12 | 0.07 |
| Dairy character | 0.50 | 0.68 |
| Foot angle | 0.00 | −0.24 |
| Rear legs | 0.02 | 0.14 |
| Pelvic angle | 0.04 | 0.19 |
| Fore udder attachment | −0.09 | −0.47 |
| Rear udder height | 0.12 | −0.13 |
| Rear udder width | 0.16 | 0.09 |
| Udder depth | −0.27 | −0.64 |
| Suspensory ligament | 0.14 | 0.12 |
| Front teat placement | 0.02 | −0.12 |
| Productive life length | 0.25 | 0.75 |
| Mastitis susceptibility | −0.05 | 0.10 |

SOURCE: Adapted from Buchanan et al., 1993, p. 96.

Knowledge of the magnitude of the genetic correlation between various traits is useful in a selection program. The magnitude of genetic correlations may vary between $-1$ and $+1$. A genetic correlation of 0 indicates that different genes influence the two traits; thus, the traits are uncorrelated. For example, selection based on milk yield has little or no effect on front teat placement (Table 9–7).

The absolute value of the correlation indicates the strength of the association between the two traits. When a genetic correlation exists between two traits, it means that the correlation does not equal 0. Rarely does this mean that the correlation is perfect at $+1$ or $-1$. For example, a genetic correlation of 0.10 between mastitis susceptibility and milk yield is positive, but the magnitude of the correlation does not imply a strong genetic association between the two traits. Therefore, selection based on milk yield does not significantly increase the susceptibility to mastitis.

The sign $(+/-)$ of the genetic correlation indicates the relationship between traits (i.e., how selection for one affects the other). If the sign of the genetic correlation is positive, then the breeding values of the animals for the two traits tend to vary together. The reverse is true for a negative correlation. Therefore, if selection is for increased performance in one trait, performance in the other trait will likely decrease. For example, the percentage traits for composition (fat %, solids not fat SNF %, and protein %) tend to be negatively associated with milk yield. However, yield traits (fat, SNF, protein) tend to be highly correlated in a positive direction with milk yield, both phenotypically and genetically.

Consideration of the relationship between traits can be very beneficial if they are considered in a complete breeding program. Again, the tabular values reveal that genetic correlations are seldom perfect. For example, many of the genes that control productive life and dairy character also control milk yield in the same direction, as indicated by the positive genetic correlations of 0.75 and 0.68, respectively, in Table 9–7. The relationship is not perfect. The fact that genetic correlations are not perfect provides breeders with the opportunity to try to identify sires that are exceptions to the unfavorable correlation. Again, genetic correlations give us an indication of what is likely to happen to one trait when selection is practiced for another trait. The magnitude of the correlation suggests how closely traits will vary together.

## GOAL SETTING AND TRAIT EMPHASIS

The genetic improvement program for every dairy herd must have goals to design the cow herd with the genetics for making a profit. Milk yield and composition are important economic considerations. Within-herd genetics and production performance must be evaluated and scrutinized through effective recordkeeping. Also, access to genetically superior animals outside the herd through the use of artificial insemination is critical. Before deciding whether selection should be practiced for a particular trait, consider the following:

- Can the trait of interest be accurately measured?
- What is the heritability for the trait? Is genetic progress through selection possible?
- Will selection for this trait contribute to income (directly or indirectly)?

Most producers begin planning a well-founded breeding program through basic use of: (1) Dairy Herd Improvement Association (DHI) records and (2) semen purchase of bulls with genetic superiority for economically important traits.

Trait emphasis should be balanced with respect to the heritability of the trait, genetic correlations among traits, the reliability of the information, and economic importance of the trait. This is no small task considering there are national genetic evaluations for about 30 traits (Figure 9–3). U.S. dairy geneticists have attempted to simplify selection decisions for dairy farmers by properly weighting all the traits into the Net Merit Dollar Index (NM$). Similar indexes have been developed by breed associations, such as the Total Performance Index (TPI) by the Holstein Association. The traits in these indexes are properly weighted for the average or most typical dairy producer to maximize profit on a commercial dairy. Individual dairies may want to change emphases depending on their management systems and goals. Low-input grazing herds put more emphasis on reproductive traits and feet and legs and less emphasis on milk production. Also, smaller cows tend to make more efficient grazers than large cows.

Producers who sell breeding stock put more emphasis on fancy type traits. The so-called eye appeal of the animal seems to influence price in the sales ring much more than the animal's genetics for milk production. For dairy producers with aspirations of developing a great show cow, most emphasis needs to be placed on final score type, stature, dairy form, and the udder traits. Dairy producers must be careful in selecting their goals in that the genetics of a show animal is quite different from the genetics needed to produce milk most efficiently.

## GENETIC EVALUATION PROCEDURES

### DHI System

Much of the genetic improvement in milk production in the United States is attributable in part to good use of performance records through the Dairy Herd Improvement (DHI) system. Data obtained monthly through the DHI production testing and management system provide producers with detailed reports on the current status of their herd. These records are also compiled nationally through the U.S. Department of Agriculture (USDA) to calculate genetic evaluations for sires.

The DHI system of genetic evaluation consists of comparing sire daughters with contemporaries in the same herd. Superior sires are chosen based on their ability to pass specific traits to their offspring. Initially, young bulls are chosen based on superior pedigree value. These genetics are then randomly mated to cows across various locations for production of daughters. These daughters' records are compared with their contemporaries to allow the calculation of estimated genetic transmitting ability of the bull.

### Animal Model

The USDA-DHI Animal Model Genetic Evaluation compiles lactation yield information for milk, fat, protein, somatic cell score, productive life, and pedigree or relationships among animals. For a cow's lactation record, sources of variation such as management group, genetic merit, permanent environment, and herd by sire interactions are considered. Specific effects of age, length of lactation, and milkings per day are adjusted prior to analysis. The animal model procedure produces predictions of the breeding (genetic) value of an animal. Breeding value is defined as the value of an individual as a parent. Parents transfer a random sample of the genes to their offspring. Breeding value gives an estimate of the transmitting ability of the parent. Some basic values generated from the evaluation are as follows:

PTA—One-half the breeding value is equal to the ***predicted transmitting ability*** (PTA). The PTA implies a comparison. Thus, PTAs allow us to compare or rank

**Predicted transmitting ability**
Half the breeding value.

**Reliability**

A measure of accuracy in dairy records.

**Parent average**

The average PTA of the parents of a dairy cow.

**Predicted transmitting ability net merit dollars**

An economic index that measures relative lifetime profit of a dairy cow.

the superiority of individual animals. PTAs provide a prediction of future progeny performance of one individual compared to another individual within a breed for a specific trait.

REL—*Reliability* (%R) is the measure of accuracy, or the amount of information in an evaluation.

PA—*Parent average* (PA) is the average PTA of the sire and dam of the individual in question. If a parent is unknown, an unknown-parent group effect is used.

PTANM$—*Predicted transmitting ability net merit dollars* (PTANM$) is an economic index. PTANM$ combines evaluations for milk, fat, protein, somatic cell score, productive life, udder composite, feet and leg composite, size, daughter pregnancy rate, calving ease, and stillbirth rate. It is a measure of relative lifetime profit.

Producers benefit from the extensive herd summary reports provided by DHI. Herd analysis and management reports include production, reproduction, genetics, udder health, and feed cost information. Figure 9–4 illustrates the identification and genetic summary portion of a sample DHI report. This report is useful in verifying the number of replacement and producing animals in the operation. As one might expect, the usefulness of DHI records is enhanced by a higher percentage of identified animals.

The PTA$ and PA values are presented for cows within the herd, as well as their sires. Values are calculated using the USDA Animal Model Genetic Evaluation procedure. PTA$ in this report is the economic value of the PTAs for milk, fat, and protein. An increase in cow and sire PTA$ from younger to older cows is an indication of within-herd genetic progress.

### SIRE SELECTION

With sire selection playing an important role in genetic improvement, dairy producers spend a great deal of time studying bull proofs given in sire summaries. Figure 9–3 shows a sample data summary for a Holstein sire. Perhaps the best application of genetic evaluations involves the comparison among sires. An example comparison of data on two Holstein sires from a summary is shown in Table 9–8.

The TPI value is a *total performance index*. This multiple trait approach calculated by the Holstein Association combines PTAs (protein, fat, type), udder composite, feet and legs composite, somatic cell score, productive life, daughter pregnancy rate, dairy form, daughter calving ease, and daughter stillbirth rate. It provides a ranking of sires on their ability to transmit a balance of traits. Net merit dollars (NM$) is the economic index calculated by the USDA as an index of relative lifetime profit.

Table 9–8 shows that yield comparisons between the two sires favor the future offspring of Superior Brett over Average Jake. For example, if daughters of these bulls are housed in the same herd as contemporaries and are managed alike, the expected difference in milk, fat, and protein yield would be 495, 37, and 8, respectively, in favor of Superior Brett. However, protein percentage would tend to favor Average Jake, illustrating a negative association between these two traits in this example. From an economic standpoint, Sire Brett is still on top with respect to NM$ and TPI values. In this example, the reliabilities (%R) are similar for both bulls, indicating similar accuracies. If the %R values were largely different, decisions on how extensively to use a young sire (low %R) would be needed. A low accuracy bull is not bad; he is part of the new genetic information for the breed. Reliability values are expected to increase as more daughter records contribute to the bull's proof.

**Total performance index**

Index used by the Holstein Association to rank sires on their ability to transmit a balance of traits.

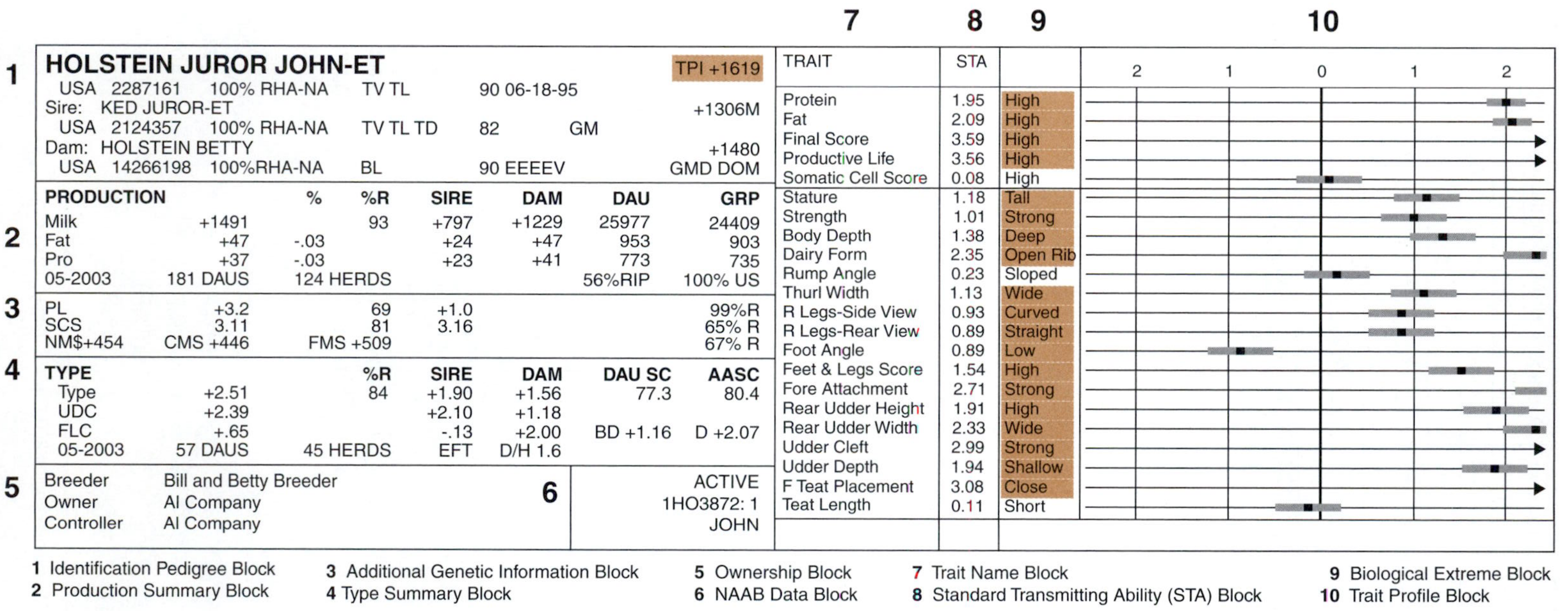

**1 HOLSTEIN JUROR JOHN-ET**

| | | | | |
|---|---|---|---|---|
| USA 2287161 100% RHA-NA TV TL | | 90 06-18-95 | | **TPI +1619** |
| Sire: KED JUROR-ET | | | | +1306M |
| USA 2124357 100% RHA-NA TV TL TD 82 GM | | | | +1480 |
| Dam: HOLSTEIN BETTY | | | | GMD DOM |
| USA 14266198 100%RHA-NA BL | 90 EEEEV | | | |

**2 PRODUCTION**

| PRODUCTION | | %R | % | SIRE | DAM | DAU | GRP |
|---|---|---|---|---|---|---|---|
| Milk | +1491 | 93 | | +797 | +1229 | 25977 | 24409 |
| Fat | +47 | | -.03 | +24 | +47 | 953 | 903 |
| Pro | +37 | | -.03 | +23 | +41 | 773 | 735 |
| 05-2003 | 181 DAUS | 124 HERDS | | | | 56%RIP | 100% US |

**3**

| | | %R | | SIRE | | |
|---|---|---|---|---|---|---|
| PL | +3.2 | 69 | | +1.0 | | 99%R |
| SCS | 3.11 | 81 | | 3.16 | | 65%R |
| NM$ +446 | CMS +446 | FMS +509 | | | | 67%R |

**4 TYPE**

| TYPE | | %R | | SIRE | DAM | DAU SC | AASC |
|---|---|---|---|---|---|---|---|
| Type | +2.51 | 84 | | +1.90 | +1.56 | 77.3 | 80.4 |
| UDC | +2.39 | | | +2.10 | +1.18 | | |
| FLC | +.65 | | | -.13 | +2.00 | BD +1.16 | D +2.07 |
| 05-2003 | 57 DAUS | 45 HERDS | | EFT | EFT | | D/H 1.6 |

**5**

| | | |
|---|---|---|
| Breeder | Bill and Betty Breeder | **6** |
| Owner | AI Company | ACTIVE |
| Controller | AI Company | 1HO3872: 1 JOHN |

**7  8  9  10**

| TRAIT | STA | |
|---|---|---|
| Protein | 1.95 | High |
| Fat | 2.09 | High |
| Final Score | 3.59 | High |
| Productive Life | 3.56 | High |
| Somatic Cell Score | 0.08 | High |
| Stature | 1.18 | Tall |
| Strength | 1.01 | Strong |
| Body Depth | 1.38 | Deep |
| Dairy Form | 2.35 | Open Rib |
| Rump Angle | 0.23 | Sloped |
| Thurl Width | 1.13 | Wide |
| R Legs-Side View | 0.93 | Curved |
| R Legs-Rear View | 0.89 | Straight |
| Foot Angle | 0.89 | Low |
| Feet & Legs Score | 1.54 | High |
| Fore Attachment | 2.71 | Strong |
| Rear Udder Height | 1.91 | High |
| Rear Udder Width | 2.33 | Wide |
| Udder Cleft | 2.99 | Strong |
| Udder Depth | 1.94 | Shallow |
| F Teat Placement | 3.08 | Close |
| Teat Length | 0.11 | Short |

1  Identification Pedigree Block
2  Production Summary Block
3  Additional Genetic Information Block
4  Type Summary Block
5  Ownership Block
6  NAAB Data Block
7  Trait Name Block
8  Standard Transmitting Ability (STA) Block
9  Biological Extreme Block
10 Trait Profile Block

**FIGURE 9–3** How to read Holstein sire information. (*SOURCE:* Holstein Foundation. Used with permission.)

**Identification and Genetic Summary**

| Age group | Number animals | Average age | NUM. identified by | | Number ID. changes | NO. animals with PTA$/PA$ | Average PTA$ / PA$ | |
|---|---|---|---|---|---|---|---|---|
| | | | Sire | Dam | | | Animal | Sire |
| 0-12 | 62 | 6 | 62 | 62 | | 58 | +108 | +187 |
| 13+ | 66 | 18 | 66 | 66 | | 61 | +97 | +163 |
| Replacements | 128 | 12 | 128 | 128 | | 119 | +103 | +175 |
| 1st lact | 34 | 23 | 34 | 30 | 2 | 34 | +84 | +140 |
| 2nd lact | 30 | 36 | 28 | 26 | | 28 | +73 | +122 |
| 3+ lacts | 58 | 59 | 51 | 44 | | 49 | +52 | +87 |
| All lacts | 122 | 43 | 113 | 100 | 2 | | +68 | +114 |
| % Identified (producing females) | | | 93 | 82 | | | | |

| Herd PTA$ option | Genetic profile of service sires | | |
|---|---|---|---|
| MFP | Proven A.I. sires | A.I. young sires | All other sires |
| % of herd bred to | 75 | 20 | 5 |
| Number of bulls used | 5 | 12 | 3 |
| Average PTA$ or PA$ | +200 | +216 | +10 |
| AV. percentile rank (net merit) | 83 | 90 | 0 |

**FIGURE 9–4** Sample DHI report. (*SOURCE:* Dairy Records Management Systems. Used with permission.)

**TABLE 9–8** **Comparison of Data on Two Holstein Sires**

| | Superior Brett | Average Jake | Difference |
|---|---|---|---|
| TPI™ | +1618 | +1430 | +188 |
| PTA | | | |
| M (milk) | +1850 | +1355 | +495 |
| F (fat) | +88 | +51 | +37 |
| P (protein) | +55 | +47 | +8 |
| PTA% | | | |
| F (fat %) | +0.09 | +0.09 | 0 |
| P (protein %) | −0.01 | +0.06 | −0.07 |
| % R (reliability of PTAM and PTAF) | 78% | 80% | −2% |
| PTA | | | |
| NM$ (net merit dollars) | +470 (69%R) | +407 (65%R) | +63 |
| SCS (somatic cell score) | +3.40 (57%R) | +3.10 (57%R) | +0.30 |
| PL (productive life) | +1.1 (42%R) | +0.2 (40%R) | +0.9 |
| T (type—final score) | +1.58 (75%R) | +1.30 (72%R) | +0.28 |
| %DBH (difficult births in heifers) | 9% (71%R) | 7% (72%R) | +2% |

*SOURCE:* Dolezal, 1999. Used with permission. Modified to August 2000 base change.

The somatic cell score (SCS) PTA is a tool that allows producers to select bulls based on their ability to sire daughters with lower rates of mastitis. Somatic cells are body cells. When found in milk, they indicate damage to the udder that is caused by mastitis. Research indicates that single-trait emphasis for higher milk yield is associated with increased incidence of mastitis. This is not a perfect relationship. Not all high production sires have associated rates of mastitis in daughters. Heritabilities used by the USDA are 0.10 for SCS and 0.30 for milk production. This suggests that genetic change to reduce mastitis is slow. In the previous example, the SCS of Sires Brett and Jake's daughters are expected to differ on the average by 0.3 (3.40 − 3.10 = 0.3). The PTASCS should be viewed as a selection tool, rather than as a sole selection criterion, to optimize total economic merit.

To continue to increase the genetic potential of the herd, follow a few basic rules:

- Use an index such as net merit dollars to weight properly the production and nonproduction traits to maximize total economic merit.
- Use 7 to 10 sires per herd per year.
- Select sires from the top 10% based on an index such as NM$ or TPI.
- Use sires with 70% reliability or higher.
- Consider calving-ease bulls for heifers.
- Young sires as a group may be used for up to 30% of semen purchases.
- Individually mate animals to lower average inbreeding in offspring.

## SWINE GENETIC IMPROVEMENT

Genetic improvement programs are a primary focal point for today's swine industry. The high reproductive rate and short *generation interval* in swine allow rapid genetic progress for economically important traits. In recent years, the swine industry has followed some of the patterns set by commercial poultry production. Much of the pork produced today originates from corporate swine production systems, which are vertically integrated from conception to consumer. This leads to an interesting structure for modern pork production, which contains seedstock breeders, commercial swine producers, and corporate production units.

The National Swine Improvement Federation (NSIF) and National Pork Producers Council (NPPC) are key organizations for documentation on swine genetic resources. The NSIF and NPPC, as well as other agencies, have historically sponsored "Guidelines for Uniform Swine Improvement Programs." Seedstock and commercial swine producers, corporate operations, researchers, and extension personnel utilize this publication. The guidelines give details on the use of uniform procedures for measuring and recording swine performance data.

The NPPC is a member organization of NSIF. The NPPC's mission is to make pork production successful and profitable from the production segment to the ultimate consumer. This council works closely with producers, researchers, and extension personnel, and has swine industry ties.

### PERFORMANCE INFORMATION

Efficient pork production relies on objective data collection for economically important traits, breeding value estimation, and planned selection decisions. Key areas include the

**Generation interval**

The average age of parents when their offspring are born.

reproductive complex, growth rate and efficiency, and carcass traits. With the high reproductive rate in swine, extensive evaluation of female reproduction is critical. Data include birth records, litter size (number farrowed alive and dead), farrowing ease scores, litter weight at weaning, and reproductive soundness. On the male side, reproductive soundness data are collected on boars for libido, mounting, mating ability, and semen evaluation. Herd reproductive measures include pigs per sow per year; pregnancy, farrowing, and weaning rate percentages; live pigs per litter; and mated female to service boar ratio.

Growth rate and feed efficiency are evaluated extensively in the swine production system. Economically important measures include days to 250 lbs, average daily gain (ADG), and feed efficiency. Body composition and carcass merit are important to producers as well as to the ultimate consumer eating experience. Data collected include backfat thickness (live), carcass fat depth, loin eye area, pounds of lean pork, and loin muscle color, firmness, and marbling.

Visual appraisal is important in swine breeding programs as it is in many other species. For swine, feet and leg soundness along with underline soundness may be scored; these areas affect production and reproduction success.

***Porcine stress syndrome*** (PSS) is tracked in swine populations. This condition has genetic control at a single locus and is identified as a homozygous-recessive genotype. Pigs under stressful conditions exhibit blotchy skin color and heavy breathing, and they can die from this condition. Phenotypic differences between normal and PSS pigs are that PSS individuals appear more muscular and shorter bodied. The ham area may appear more rounded and circular, along with prominent loins and rumps. Fortunately, the PSS animals can be identified by a blood test or Halothane anesthesia test. Those individuals with the condition should be culled.

**Porcine stress syndrome**

Genetic defect in which pigs are heavily muscled but have poor carcass quality and may die when subjected to stress.

## GENETIC PARAMETERS

Heritability indicates the proportion of the superiority in an individual or in a group of individuals that can be passed on to the next generation. This property is used to estimate breeding value. The actual breeding value of an individual is never known. It can be estimated from the performance of the individual and its relatives. Common information on relatives includes progeny, sire, and dam records. Table 9–9 presents heritability estimates and genetic/phenotypic correlations for some economically important traits.

Growth and carcass measures are moderately to highly heritable (e.g., ADG = 0.30; LEA = 0.47). The magnitude of these estimates indicates that selection for these traits will be effective. Also, the correlation between the traits is important to examine. In this case, ADG and LEA have a genetic correlation of −0.10. This low correlation indicates that selecting for either trait probably will not influence progress in the other trait. In contrast, LEA and backfat thickness are correlated at −0.35. Selection for increased LEA is associated with decreased backfat to some extent, although the relationship is not perfect. The phenotypic correlation is of similar magnitude. The relationship is not perfect. Genetic correlations give us an indication of what is likely to happen to one trait when selection is practiced for another trait.

## BREEDING VALUE AND EXPECTED PROGENY DIFFERENCE

Parents transfer a random sample of their genes to their offspring. Breeding value gives an estimate of the transmitting ability of the parent. Half of the breeding value is equal to the expected progeny difference (EPD). The word *difference* implies a comparison. Thus EPDs let us compare or rank the superiority of individual animals. These concepts

**TABLE 9–9** **Genetic Parameter Estimates for Swine**

| Trait | h² | Correlations[1] | | | | | | | |
|---|---|---|---|---|---|---|---|---|---|
| | | AP | OR | LS | S | R | NW | LBW | L21W |
| Age at puberty (AP) | 0.32 | | −0.10 | −0.01 | −0.25 | 0.14 | −0.01 | −0.24 | −0.24 |
| Ovulation rate (OR) | 0.39 | 0.12 | | 0.03 | | | −0.65 | −0.35 | −0.60 |
| Litter size (LS) | 0.10 | | 0.06 | | −0.25 | | 0.71 | 0.65 | 0.48 |
| Survival to weaning (S) | 0.05 | | | −0.23 | | | | −0.09 | 0.84 |
| Rebreeding interval (R) | 0.23 | | | | | | | | |
| Number weaned (NW) | 0.06 | −0.05 | 0.02 | 0.66 | | | | 0.67 | 0.93 |
| Litter birth weight (LBW) | 0.29 | −0.04 | 0.05 | 0.77 | −0.07 | | 0.70 | | 0.69 |
| 21-day litter weight (L21W) | 0.15 | −0.06 | 0.02 | 0.49 | 0.55 | | 0.86 | 0.66 | |

| | h² | ADG | AGE | BF | FE | LEA | DP | LEN | PL |
|---|---|---|---|---|---|---|---|---|---|
| Average daily gain (ADG) | 0.30 | | −0.93 | 0.22 | −0.70 | −0.10 | 0.0 | 0.10 | −0.15 |
| Days to 250 lbs (AGE) | 0.25 | −0.90 | | −0.20 | 0.65 | 0.05 | 0.0 | −0.10 | 0.10 |
| Backfat thickness (BF) | 0.41 | 0.20 | −0.18 | | 0.34 | −0.35 | 0.15 | −0.28 | −0.85 |
| Feed efficiency (FE) | 0.30 | −0.65 | 0.60 | 0.25 | | −0.35 | −0.43 | 0.0 | −0.07 |
| Loin eye area (LEA) | 0.47 | −0.06 | 0.03 | −0.25 | −0.20 | | 0.50 | −0.18 | 0.65 |
| Dressing % (DP) | 0.30 | −0.15 | 0.10 | 0.20 | 0.10 | 0.32 | | −0.32 | −0.10 |
| Carcass length (LEN) | 0.56 | 0.08 | −0.06 | −0.21 | −0.04 | −0.12 | −0.21 | | 0.18 |
| % Lean (PL) | 0.48 | −0.11 | 0.10 | −0.71 | −0.25 | 0.62 | 0.0 | 0.10 | |

[1]Phenotypic correlations are below diagonal; genetic correlations are above diagonal.
SOURCE: Lamberson and Cleveland, 1988, p. 3.

were explored in the beef cattle improvement section. The same principles and assumptions apply for swine genetic evaluations.

## SWINE BREEDS

Unlike the beef industry, fewer swine breeds have had a large impact on commercial swine production. Within these breeds, extensive evaluation of superior individuals has taken place. Specialized sire and dam lines have been developed using these evaluations. Subsequent commercial crossbreeding systems are designed for efficient pork production. Today's swine industry is strongly focused on genetic evaluation of performance data among and within breeds. For example, the NPPC has been instrumental in leading and supporting genetic evaluation programs. Examples include the Terminal Sire Line National Genetic Evaluation and the Maternal Line National Genetic Evaluation Programs.

An interesting angle to the swine industry is that many times the actual breed composition of a particular breeding line is not known. This approach was patterned similarly to commercial poultry production. Commercial units rely on the seedstock producer choices or corporate genetic selections to set the genetics of their animals. Private companies employ geneticists to carefully evaluate all production aspects of their base genetics. Hybrid boars and sows are developed with protected rights to the actual genetic makeup of breeds and individuals within breeds.

The swine industry capitalizes on the advantages of heterosis, particularly maternal heterosis benefits for reproduction. Crossbred sows are used in rotational crossbreeding systems, as well as the maternal side of the terminal cross programs. Research on hybrid boars has indicated that these sires have increased libido, structural soundness, and improved conception rates. Market offspring produced may express 100% of the individual heterosis, as long as the breeds are different for sire and dam lines in the crosses.

Selection index application is very common in the swine breeding programs. Index approach may be directed toward maternal, paternal, or general improvement strategies. Index equations allow the simultaneous evaluation of two or more traits. Traits are weighted based on their economic value and the overall selection objectives for the breeding population. The index accounts for economic value, but also heritability, genetic and phenotypic correlations, and the phenotypic variation for the respective traits. Specific indexes are designed for maternal lines, emphasizing reproductive performance as expressed in litter size and 21-day litter weight. Additional production traits may be included. On the sire, or paternal side, postweaning traits are important, as well as days to 250 lbs and backfat thickness. Feed conversion emphasis is included in the index through genetic correlations between the other traits.

Many times, maternal and paternal lines developed through index selection are combined into a terminal crossbreeding system in which distinct lines are crossed to produce market pigs. No replacements are generated from this system. In contrast, a general selection index is used in more rotational crossbreeding systems. Equal value is given to reproductive and production traits, since individuals in this system must serve as a sire and a dam. A rotational system generates its own replacements.

Breeds from other countries have been studied to determine if specialized genetics would benefit commercial hog production. Perhaps the most well-known quest is that of the Chinese breeds of swine. These breeds are of great interest because of their high reproductive rate (9–17 pigs born alive) as well as early puberty advantages. However, limiting factors associated with these breeds are low growth rate, poor conformation, and excessive fat deposition. Four main breeds are Meishan, Fengjing, Jiaxing Black, and Erhualian. Future developments in pig genome research may identify specific genetic material that these breeds may contribute to future reproductive advances.

## STAGES

A well-known performance resource in swine genetic evaluation is *STAGES* (Swine Testing and Genetic Evaluation System). This system evaluates the genetic superiority of swine using a statistical methodology similar to that of the beef and dairy industries. The U.S. swine breed associations use the program, which was developed jointly by Purdue University and USMARC. STAGES incorporates performance information on individuals, progeny, and collateral relatives, as well as the use of relationships among these animals, to generate breeding values (and ultimately within-herd EPDs). For example, postweaning and reproduction analyses are conducted on a within-herd basis. Breeders rely on performance information collected from on-farm programs, national breed tests, and progeny tests to generate within-herd EPDs for animals.

Also, a STAGES national evaluation is run for specific herds to generate across-herd EPDs. Across-herd runs are currently conducted for Yorkshire and Landrace breeds. Across-herd genetic evaluations to compare animals in different herds within a breed are used to identify the best genetic material nationally. The STAGES program began using this approach in 1990. Also, centrally tested boars have been evaluated nationally by this approach. For selection decisions, the seedstock breeders use these data in conjunction with their within-herd evaluation of breeding prospects. Commercial producers rely heavily on the progress of their seedstock suppliers to capture the value of these genetic evaluations.

An example of another index is the sow productivity index (SPI), used to select litters with future replacement gilt candidates. The index includes EPDs with reproductive emphasis. Weighting factors are placed on the EPDs for number of pigs born alive, litter weight, and number weaned relative to the economic value for the trait. Figure 9–5 illustrates the genetic change that has taken place in the Yorkshire breed for SPI (including reproductive traits only). Parent and nonparent lines are presented. The difference between the two lines shows that much selection pressure has been placed on the trait. When these nonparents have progeny in the future, more progress will take place in SPI.

Figure 9–6 shows the genetic trend lines for days to 250 lbs. In this case, negative EPD values are more desirable, so the change over time is in a negative direction. Thus the time it takes an animal to reach 250 lbs has been shortened genetically. The backfat genetic trend during this time period follows a similar pattern, with leaner animals being produced over the years.

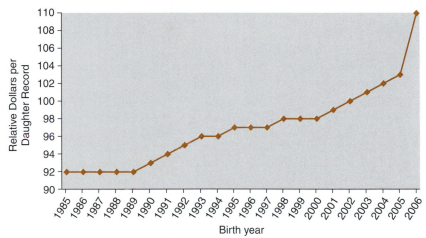

**FIGURE 9–5**  Genetic trend for Yorkshire sow productivity index (SPI). (*SOURCE:* Purdue University, 2007.)

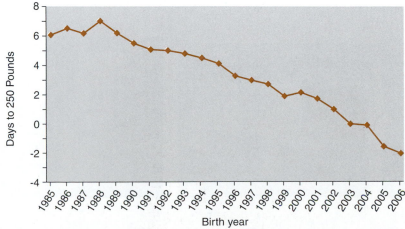

**FIGURE 9–6**  Genetic trend for days to 250 lbs in Yorkshire swine. (*SOURCE:* Purdue University, 2007.)

# SHEEP GENETIC IMPROVEMENT

The sheep producer is in the business of producing two products, lambs and wool, as efficiently as possible. The major areas of economic importance to the sheep producer are lamb growth, prolificacy, and, in some areas, wool quality and quantity. However, the tools are not the same for sheep breeding and genetic improvement as they are for the other major meat breeds. EPDs have only been available for sheep since 1986 and are not yet as useful as they are for the other species. In addition, artificial insemination is rarely used in the sheep industry. This makes gaining genetic progress through the widespread use of superior sires less of an influence on the industry. The following sections discuss the key areas of sheep breeding and genetic improvement available to the sheep producer.

## BREED AND BREED TYPES

To understand sheep genetic improvement and breeding systems, it is first necessary to understand something of the genetic diversity of sheep breeds. Breeds of sheep available in the United States can range from fine-wool breeds, to long-wool breeds, to hair breeds. These breeds can range in mature size from 100 lbs to over 400 lbs, and average from one lamb per ewe per year to over three lambs per ewe per year. Some breeds lamb year-round; most lamb only in the spring of the year. With this degree of genetic diversity available to the sheep producer, selection of the breeds utilized in a commercial operation is crucial to the success and profitability of the operation. The appropriate choice depends on the geographic location, feed conditions, weather conditions, and goals of the operation.

For simplification and ease of understanding, the breeds are grouped together and classified. They can be classified by face color (black face versus white face) or by wool type (fine wool versus long wool versus hair). However, the most common classification is by use, based on the major function of the breed in common mating systems: ewe breed, dual-purpose breed, and ram breed (Figure 9–7).

## Ewe Breeds

Ewe breeds are generally the fine-wool, white-faced breeds and those that were developed from crosses of the fine-wool breeds (Rambouillet, Merino) with long-wool breeds (Lincoln, Border Leicester), or the highly prolific breeds (Finnsheep, Ro-

(a)

(b)

(c)

**FIGURE 9–7**   Sheep breeds are often classified by use of the breed. The pictured breeds of sheep are examples of the different breed classes: (a) ewe breed, Rambouillet; (b) ram breed, Shropshire; and (c) dual-purpose breed, Corriedale.

manov). The breed most widely seen in commercial flocks in the United States is the Rambouillet or Rambouillet cross; the term most generally used for these range ewes is the Western ewe or Western White-face ewe.

## Ram Breeds

Ram breeds are the meat-type breeds used primarily as terminal sires on the ewe breeds or dual-purpose breed to increase lamb gain and carcass quality for market lamb production. The ram breeds are noted for size, fast growth, and carcass quality. The two most widely used ram breeds in the United States are the Suffolk and Hampshire breeds, with the Dorset, Shropshire, Oxford, and Southdown breeds used in some instances.

## Dual-Purpose Breeds

The breeds classified as dual purpose are those that can be used as either ewe breeds or ram breeds, depending on the environment and production goals of the operation. These are breeds that are not only noted for milk production, mothering ability, and twinning rate, but also exhibit some of the characteristics noted in the ram breeds such as better growth rate and carcass quality. Examples of dual-purpose breeds are the Dorset, Columbia, and Corriedale.

## SELECTION

### Heritability

An understanding of the heritability of important traits (low, moderate, high) in sheep production is needed to better understand the use of selection and mating systems in the sheep operation. Simply stated, the heritability of a trait describes how easily genetic improvement can be made. The variation in a trait is made up of genetic and environmental components. Heritability is the proportion of difference among animals for a trait (e.g., milk production) due to genetic difference, rather than environmental factors. This is important because geneticists are interested in the portion that is transmissible from parent to offspring. Strictly defined, heritability is the ratio of additive genetic variance to total phenotypic variance. This is important to genetic improvement because breeding value depends heavily on additive genetic variance. If variation does not exist, then progress through selection cannot be made.

Average heritability estimates for various traits are shown in Table 9–10. In general, reproductive traits tend to be low in heritability (<0.20), growth traits tend to be moderately heritable (0.20–0.40), and carcass and fleece traits tend to show high levels of heritability (>0.40). The most practical way to understand the use of heritability is in the selection process. The higher the heritability of a trait, the quicker improvements can be made through selection for that trait. Reproductive traits such as prolificacy are lowly heritable and, therefore, sheep breeders see very little progress when selecting for twinning rate. That is why other methods are used to increase twinning rate in the ewe flock. However, the carcass traits have very high heritabilities and selection progress can be noted from one generation to the next very quickly when selecting for the carcass traits. The growth traits are moderate in heritability and selection for these traits will show good genetic improvement from generation to generation.

Selection for growth traits.   Growth traits are of economic importance to sheep operations. Because growth traits show moderate levels of heritability, selection programs that emphasize growth traits can show good genetic improvement from generation to

**TABLE 9–10**   **Heritabilities of Various Traits**

| Reproductive Traits | Percent | Carcass Traits | |
|---|---|---|---|
| Ewe fertility | 5[1] | Carcass weight | 35 |
| Prolificacy[2] | 10 | Weight of trimmed retail cuts | 45 |
| Scrotal circumference | 35 | Percent trimmed retail cuts | 40 |
| Age at puberty | 25 | Loin eye area | 50 |
| Lamb survival | 5 | 12th rib fat thickness | 30 |
| Ewe productivity[3] | 20 | Dressing percent | 10 |
| **Growth Traits** | | **Fleece Traits** | |
| Birth weight | 15 | Grease fleece weight | 35 |
| 60-day weight | 20 | Clean fleece weight | 25 |
| 90-day weight | 25 | Yield (%) | 40 |
| 120-day weight | 30 | Staple length | 55 |
| 240-day weight | 40 | Fiber diameter | 40 |
| Preweaning gain: birth–60 days | 20 | Crimp | 45 |
| Postweaning gain: 60–120 days | 40 | Color | 45 |
| **Dairy Traits** | | | |
| Milk yield | 30 | | |
| Fat (%) | 30 | | |
| Protein (%) | 30 | | |
| Fat yield | 35 | | |
| Protein yield | 45 | | |

[1]May increase to 10% in ewe lambs and in ewes bred in spring.
[2]Lambs born per ewe lambing.
[3]Pounds of lamb weaned per ewe exposed.
SOURCE: *Sheep Production Handbook,* 1996, p. BRD-61. Used with permission.

generation. Selection for growth is most important in the ram breeds. These individuals are the predominant sires used for commercial market lamb production and their main contribution to the lamb crop is growth rate and carcass merit. Increased growth rate allows producers to market lambs at an earlier age or at heavier weights. In the sheep industry, the most important growth trait is weaning weight.

Weaning weights are influenced by not only growth rate of the lamb, but age of the ewe, type of birth, and method of rearing of the lamb. Selection for growth rate simply by using lamb weaning weight may not be effective unless some of these common nongenetic factors are adjusted for. Weaning weights should be adjusted for age of lamb, sex, type of birth, method of rearing the lamb, and age of the dam. These adjustments and an example of the calculations are shown in Table 9–11.

## NATIONAL SHEEP IMPROVEMENT PROGRAM (NSIP)

Expected progeny differences are widely used in the beef, dairy, and swine industries, but are just in the infancy stages in the sheep industry. The National Sheep Improvement Program (NSIP), initiated in 1986, was designed to provide both the purebred producer and the commercial producer with a performance recording and genetic evaluation program. The NSIP evaluates maternal traits, growth traits, wool traits, and is developing carcass traits.

**TABLE 9–11** Multiplicative Adjustment Factors for Adjusting Lamb Preweaning and Weaning Weights to a Common Age of Dam, Lamb Sex, and Lamb Type of Birth-Rearing

| Sex | Ewe Age | Type of Birth-Rearing | | | | | |
|-----|---------|-----|-----|-----|-----|-----|-----|
| | | 1–1 | 2–1 | 2–2 | 3–1 | 3–2 | 3–3 |
| Ewe | 1 | 1.13 | 1.29 | 1.38 | 1.40 | 1.51 | 1.80 |
| | 2 or over 6 | 1.08 | 1.19 | 1.29 | 1.28 | 1.38 | 1.54 |
| | 3–6 | 1.00 | 1.10 | 1.19 | 1.18 | 1.27 | 1.36 |
| Ram | 1 | 1.02 | 1.15 | 1.21 | 1.23 | 1.31 | 1.53 |
| | 2 or over 6 | 0.98 | 1.08 | 1.17 | 1.16 | 1.25 | 1.38 |
| | 3–6 | 0.91 | 1.00 | 1.08 | 1.07 | 1.15 | 1.23 |
| Wether | 1 | 1.10 | 1.25 | 1.33 | 1.36 | 1.45 | 1.72 |
| | 2 or over 6 | 1.05 | 1.16 | 1.26 | 1.25 | 1.35 | 1.50 |
| | 3–6 | .98 | 1.08 | 1.16 | 1.15 | 1.24 | 1.33 |

Adjustment factors are from the Report of the NSIP Technical Committee, 1986. Weights are adjusted to a single ewe lamb from a 3- to 6-year-old ewe equivalent. Adjustment factors are to be used on preweaning and weaning weights taken at approximately 30, 60, 90, or 120 days of age. Before applying adjustment factors, actual weight must be adjusted to the same age for each lamb using one of the following two equations:

1. If birth weight is available:

$$\text{Age adjusted wt.} = \left[ \frac{\text{actual wt.} - \text{birth wt.}}{\text{age when weighed}} \times \text{adjustment age (days)} \right] + \text{birth wt.}$$

2. If birth weight is not available:

$$\text{Age adjusted wt.} = \frac{\text{actual weight}}{\text{age when weighed}} \times \text{adjustment age (days)}$$

Final adjusted weight is given by multiplying the age adjusted weight by the appropriate adjustment factor. Example: A ewe lamb born as a triplet and reared as a single from a 7-year-old ewe weighed 7 lbs at birth and 66 lbs at weaning when 93 days of age. What is her adjusted 90-day weight?

$$\text{Age adjusted wt.} = \left[ \frac{66 - 7}{93} \times 90 \right] + 7 = 64 \text{ lbs}$$

Final adjusted 90 day wt. = 64 × 1.28 = 82 lbs

Lamb born in litters of greater than three should use the triplet adjustment factors. Lambs born as singles and reared as twins should use the twin–twin (2–2) adjustment factors and lambs born as singles or twins and reared as triplets should use the triplet–triplet (3–3) adjustment factors.

SOURCE: *Sheep Production Handbook*, 1996, p. BRD-63. Used with permission.

The NSIP uses the standard set of adjustment factors shown in Table 9–11 for the adjustment of weaning weights. In most breeds, the genetic evaluations are done on a within-flock basis until the database grows large enough to have sufficient ties to have across-flock genetic evaluations. Across-flock ties occur when related individuals are in two or more flocks. Genetic ties normally occur through the rams rather than the ewes because of the high number of progeny per ram and the movement of rams or sons of rams from flock to flock. The breed associations using across-flock genetic evaluations through NSIP are the Polypay, Suffolk, Columbia, Dorset, Hampshire, Katahdin, Romney, Rambouillet, and Targhee breeds.

## HETEROSIS IN SHEEP BREEDING

All of the meat-producing species rely on crossbreeding to improve productivity. However, the sheep industry has used crossbreeding systems very effectively for decades. Crossbreeding systems in sheep involve mating ewes and rams of different breed or breed crosses to produce offspring that are superior (due to heterosis) in performance to that of either of the parent stock. Systematic crossbreeding systems are advantageous because they utilize heterosis.

Heterosis, or hybrid vigor, for a trait is defined as the superiority of the crossbred individual relative to the average performance of the purebreds included in the cross. In general, crossbred individuals tend to be more vigorous, fertile, healthier, and grow faster than the average of parental stock that make up the cross. Traits that are lowly heritable show high levels of heterosis. Reproductive traits are a good example of a lowly heritable trait that shows high levels of heterosis. Moderately heritable traits show moderate levels of heterosis, such as the growth traits. Highly heritable traits such as fleece and carcass traits show little hybrid vigor. Average heterosis effects for the crossbred lamb and crossbred ewe are shown in Tables 9–12 and 9–13, respectively. The total effect of heterosis on the crossbred lamb is 17.8%; the effect of heterosis on the crossbred ewe is 18%. These advantages make it imperative for the sheep producer to use crossbreeding systems to improve the economic efficiency of the commercial sheep operation.

**TABLE 9–12      Average Heterosis Effects in the Crossbred Lamb[1]**

| Trait | Level of Heterosis (%) |
| --- | --- |
| Birth weight | 3.2 |
| Weaning weight | 5.0 |
| Preweaning daily gain | 5.3 |
| Postweaning daily gain | 6.6 |
| Yearling weight | 5.2 |
| Conception rate | 2.6 |
| Prolificacy of the dam[2] | 2.8 |
| Survival: birth to weaning | 9.8 |
| Carcass traits | approximately 0 |
| Lambs born per ewe exposed[1] | 5.3 |
| Lambs reared per ewe exposed[1] | 15.2 |
| Weight of lamb weaned per ewe exposed[2] | 17.8 |

[1]From the review by Nitter, G. 1978. Breed utilization for meat production in sheep. *Animal Breeding Abstracts* 46: 131–143.
[2]Purebred ewes mated to a different breed of ram to produce crossbred lambs.
*SOURCE: Sheep Production Handbook*, 1996, p. BRD-28. Used with permission.

**TABLE 9–13**   **Average Heterosis Effects in the Crossbred Ewe[1]**

| Trait | Level of Heterosis (%) |
| --- | --- |
| Fertility | 8.7 |
| Prolificacy | 3.2 |
| Body weight | 5.0 |
| Fleece weight | 5.0 |
| Lamb birth weight | 5.1 |
| Lamb weaning weight | 6.3 |
| Lamb survival: Birth to weaning | 2.7 |
| Lambs born per ewe exposed | 11.5 |
| Lambs reared per ewe exposed | 14.7 |
| Weight of lamb weaned per ewe exposed | 18.0 |

[1]From the review by Nitter, G. 1978. Breed utilization for meat production in sheep. *Animal Breeding Abstracts* 46: 131–143.
SOURCE: *Sheep Production Handbook,* 1996, p. BRD-29. Used with permission.

## SUMMARY AND CONCLUSION

Animal breeding is a discipline that takes the principles of genetics and applies them to practical selection and management systems. The goal is to produce the best animals for the conditions in which they will be produced. The cow-calf producer is in the business of producing beef as efficiently as possible. The techniques used and the selections made for breedings must balance production performance and end-product merit. EPDs are a tool to assist in this process. A wealth of information is available to use. The challenge is to identify the combinations of genetics and environment that is most profitable and competitive. Similar tools are readily available to the dairy producer. Extensive data collected through the DHI system has allowed rapid genetic progress to be made in economically important traits. Availability of PTAs provides a challenge for producers to evaluate the economic importance of each trait and to keep current with new technologies. A well-known performance resource for swine breeders to use in selection is the Swine Testing and Evaluation System. This system evaluates the genetic superiority of swine using mixed-model technology similar to that of the beef and dairy industries. Sheep producers have a similar program in place called NSIP. However, this is a new program compared to the other species. It is less developed because there are fewer sheep and fewer participating sheep producers. Sheep producers use selection and cross-breeding but have less information with which to work.

## STUDY QUESTIONS

1. Define animal breeding. Why is animal breeding more important as a discipline now than it was 50 years ago? Have all species of livestock benefited from animal breeding research? Why or why not?

2. Speculate on ways in which biotechnology and genetic engineering will make animal breeding a more useful science.

3. What are the major areas of economic importance in beef cattle breeding?

4. Define *heritability*. Describe the difference in heritability among growth traits, reproductive traits, and carcass traits in beef cattle.

5. Why is understanding genetic correlations important in beef cattle breeding? Describe such a relationship between two traits.

6. Why is it important for a beef cattle producer to know the performance of his or her herd? What are some types of performance programs? How do breed associations help?

7. What is an EPD? Define it in relation to breeding value. For what traits can EPDs be calculated? What animals within a breed can have EPDs calculated?

8. What is a contemporary group? Why is this important?

9. What does it mean for EPDs within a breed to be standardized? What is a base year?

10. Define and explain the value of *accuracy* and *possible change*.

11. What is a maternal effect in beef cattle? What are its components? What is milk EPD? What is weaning weight EPD? What is combined maternal EPD?

12. Describe the difference in the use of EPDs in the selection of seedstock herd compared to selection in a commercial herd.

13. In what situation might it be advantageous to calculate estimated EPDs from pedigree information? What would you expect the accuracy of such EPDs to be?

14. Speculate on the value of across-breed EPDs. What are the problems in calculating these values?

15. Name some major goals for dairy cattle genetic improvement. How much more or less complicated does the use of heritability estimates seem in dairy cattle compared to beef cattle? Is the definition of heritability the same?

16. What are some traits that a producer could expect his herd to make rapid improvement in through the application of sound animal-breeding techniques? Little progress?

17. What does it mean when we say that "genetic correlations are seldom perfect"?

18. Describe the value of DHI in dairy cattle breeding programs.

19. What are some traits that are difficult to measure and include with susceptible accuracy in breeding programs?

20. Define these terms: *predicted transmitting ability, reliability, parent average, predicted transmitting ability dollars,* and *percentile ranking*.

21. What is a somatic cell score (SCS)? Does available information suggest that the SCS is influenced much by genetics?

22. How does the short generation interval in the pig give swine breeders an advantage in making genetic progress?

23. What do NSIF and NPPC contribute to swine breeding excellence?

24. Based on the information provided, what are the areas of emphasis in swine breeding? How does visual appraisal contribute?

25. What is PSS? Why should individuals with this condition be culled?

26. What has been the influence of swine breeds on the swine production industry? What is their relationship to hybrid boar and sow breeding lines developed by commercial companies?

**27.** What is a selection index?

**28.** Presuming genetic engineering becomes widespread in livestock breeding, what genes would you select from the Chinese pig to contribute to a synthetic sow line? Boar line?

**29.** What is STAGES? What is its value? What is a sow productivity index and what is its value?

**30.** In the sheep section of this chapter, more emphasis was placed on breeds than in the other sections of this chapter. Why was this done?

**31.** Describe the importance of capitalizing on heterosis in sheep breeding. Integrate a discussion of ewe breeds, ram breeds, and dual-purpose breeds into your answer.

## REFERENCES

*Author's Note:* For previous editions Dr. Sally Dolezal, Dr. Daniel Waldner, Dr. Gerald Q. Fitch, Dr. David Buchanan, and Dr. John L. Evans of Oklahoma State University contributed to this chapter. For the second, third, and fourth editions, Dr. Tony Seykora of the University of Minnesota reviewed the chapter and contributed new material. The author gratefully acknowledges these contributions.

Beef Improvement Federation. 1996. *Guidelines*. 7th ed. Colby, KS: Beef Improvement Federation. http://www.beefimprovement.org.

Benyshek, L. L. 1988. Evaluating and reporting carcass traits. Proceedings of the Beef Improvement Federation 1988 Annual Convention, Albuquerque, NM.

Buchanan, D. S., A. C. Clutter, S. L. Northcutt, and D. Pomp. 1993. *Animal breeding: Principles and applications*. 4th ed. Stillwater: Oklahoma State University.

Cundiff, L. V., and K. E. Gregory. 1977. *Beef cattle breeding*. USDA Ag. Inf. Bull. 286.

Cundiff, L. V., F. Szabo, K. E. Gregory, R. M. Koch, M. E. Dideman, and J. D. Crouse. 1993. Breed comparisons in the germplasm evaluation program at MARC. Proceedings of the Beef Improvement Federation Research Symposium and Annual Meeting, May 1993, Asheville, NC.

DHIA. 1997. DHI-202 Herd Summary. Fact Sheet A-1. Columbus, OH: Dairy Herd Improvement Association.

Dickinson, F. N. 1985. *Genetic improvement of dairy cattle*. National Cooperative Dairy Herd Improvement Program Handbook, Fact Sheet 1–7.

Dolezal, S. L. 1999. Oklahoma State University, Stillwater. Personal communication.

Freeman, A. E. 1992. Integrating genetic evaluations into a breeding plan. In *Large dairy herd management*. Champaign, IL: Management Services, American Dairy Science Association.

Gregory, K. E., L. V. Cundiff, and R. M. Koch. 1999. *Composite breeds to use heterosis and breed differences to improve efficiency of beef production*. U.S. Department of Agriculture, Agricultural Research Service. Technical Bulletin 1875, pp. 1–175.

*Guidelines for Uniform Beef Improvement Programs*. Beef Improvement Federation. Ed. W. D. Hohenboken. 8th ed. 2002. pp. 1–161. (www .beefimprovement.org)

Kuehn, L. A., L. D. Van Vleck, R. M. Thallman, and L. V. Cundiff. 2007. Across breed EPD tables for the year 2007 adjusted to breed differences for birth year of 2005. Proceedings of the 2007 Beef Improvement Federation 39th Annual Meeting. Accessed online August 9, 2007, at http:// bifconference.com/bif2007/Symposium/ 074_Across_Breed_EPD.pdf.

Lamberson, W. R., and E. R. Cleveland. 1988. Genetic parameters and their use in swine breeding. Swine Genetics Fact Sheet Number 3, NSIF-F33. Raleigh, NC: National Swine Improvement Federation.

Lasley, J. F. 1978. *Genetics of livestock improvement*. Upper Saddle River, NJ: Prentice-Hall.

National Swine Improvement Federation. 1998. *Guidelines for uniform swine improvement programs*. Asheville, NC: http://mark.ansi.ncsu.edu/nsif/.

Northcutt, S. L., and D. S. Buchanan. 1993a. *Expected progeny difference: Part I, Background on breeding value estimation*. OSU Fact Sheet. F-3159. Stillwater: Oklahoma State University.

Northcutt, S. L., and D. S. Buchanan. 1993b. *Expected progeny difference: Part II, Growth trait EPDs*. OSU Fact Sheet. F-3160. Stillwater: Oklahoma State University.

Northcutt, S. L., and D. S. Buchanan. 1993c. *Expected progeny difference: Part III, Maternal trait EPDs*. OSU Fact Sheet. F-3161. Stillwater: Oklahoma State University.

Northcutt, S. L., and D. S. Buchanan. 1993d. *Expected progeny difference: Part IV, Use of EPDs*. OSU Fact Sheet. F-3162. Stillwater: Oklahoma State University.

Purdue University. 2007. *Swine testing and evaluation system (STAGES)*. West Lafayette, IN: Purdue University. http://www.ansc.purdue.edu/stages/welcome.html.

Rothschild, M. F., and G. S. Plastow. 1999. *Current advances in pig genomics and industry applications*. U.S. Genome Mapping Coordination Program. Ames, IA. http://www.genome.iastate.edu/~max/rev98/.

Rothschild, M. F., and A. Ruvinsky. 1998. *The genetics of the pig*. Wallingford, UK: CAB International. http://ansc.une.edu.au/genpub/genpig.html.

*Sheep production handbook*. 2002. Denver, CO: American Sheep Industry Association, Inc.

*Sire summaries supplement (SSS)*. 1998. Brattleboro, VT: Holstein Association.

USDA-DHIA. *Factors for standardizing 305-day lactation records for age and month of calving*. Columbus, OH: USDA-DHIA.

White, J. M. 1989. Characteristics of good dairy cattle. In *Guide to genetics*. Brattleboro, VT: Holstein Association.

Wiggans, G. R., and P. M. VanRaden. 1989. *USDA-DHIA animal model genetic evaluations*. Nat. Coop. Dairy Herd Imp. Prog. Fact Sheet H-2. Ames, IA.

Wilcox, C. J. 1992. Genetics: Basic concepts. In *Large dairy herd management*. Champaign, IL: Management Services, American Dairy Science Association.

Woodward, B. W., L. V. Cundiff, D. L. Notter, and D. L. Van Vleck. 1999. Understanding and using across breed expected progeny differences (EPDs). In *Beef Cattle Handbook*. Beef Cattle Resource Committee of the North Central Land Grant Universities. Accessed online August 3, 2007. http://www.iowabeefcenter.org/pdfs/bch/01310.pdf.

# Biotechnology and Genetic Engineering

## LEARNING OBJECTIVES

After you have studied this chapter, you should be able to:

1. Describe the magnitude of the biotechnology industry in the United States.

2. Define *biotechnology* and explain how genetic engineering is a part of biotechnology.

3. Describe in general terms how organisms are genetically engineered.

4. List the uses of such biotechnological innovations as monoclonal antibodies, gene therapy, cloning, and micropropagation.

5. Describe current and future uses of genetic engineering as it applies to field crop, food crop, and livestock production.

6. Describe developing uses of rDNA organisms as disease models, for organ donation, as bioreactors for pharmaceuticals, as nutraceutical producers, and as waste managers.

7. Explain the regulatory mechanism in place to control genetically engineered organisms.

8. Identify some of society's concerns about genetic engineering.

## KEY TERMS

Antibodies
Biotechnology
Bovine somatotropin
    (BST)
Cloning
Dolly
Founder
Gene enhancement
Gene map

Gene probe
Gene therapy
Genetic engineering
Genomics
Germ-line therapy
Hybridoma
Intergeneric
    microorganisms
Micropropagation

Monoclonal antibodies
Novel gene
Novel organism
Nutraceuticals
Pharming
Recombinant DNA
Tissue culture
Transgenic
Xenotransplantation

# INTRODUCTION

Biotechnology has and will continue to revolutionize modern life. The past and future contributions of biotechnology touch all areas of modern life, including medicine, agriculture, the environment, manufacturing, bioprocessing, and almost anything else one would care to mention. For this reason, this chapter includes discussions of a broad range of applications in areas other than just animal science. Any other approach would not do the science or you justice. Even so, this chapter is only meant to whet your appetite for a science that is producing miracles every day.

***Biotechnology*** is the development of products by a biological process. This may be done by using intact organisms, such as yeasts and bacteria, or by using natural substances (e.g., enzymes) from organisms. Biotechnology can also involve the use of plant and animal cells to produce products, especially products that could not previously be produced. A variety of biological processes are used, ranging from traditional fermentation to modern ***transgenic*** mammals that produce vaccines in their milk. The products of biotechnology include bread, cheese, wine, penicillin, plant and animal health diagnostic kits, vaccines, biopesticides, herbicide-resistant crops, nutrient-enhanced rice, cloned animals of several species, and the FlavrSavr tomato. Biotechnology integrates many disciplines, including agriculture, biology, genetics, molecular biology, biophysics, biochemistry, chemical engineering, and computer science and applies them to bring about the development of a practical and beneficial product. This may be a new product or it may be an enhancement of a traditional product. Those products may be used in disease research, food production, waste management, or a myriad of other areas of need.

Even though the word *biotechnology* is now part of the common vocabulary, biotechnology itself isn't really anything new. People have used it for millennia in traditional applications such as the production of beer, cheese, and bread (Figure 10–1). However, a new approach to biotechnology is causing a biotechnological revolution in the way that life can be lived. The modern era of biotechnology began in 1953 when

**Biotechnology**

The application to industry of advances made in the techniques and instruments from the biological sciences.

**Transgenic**

An animal or plant that has had DNA from an external source inserted into its genetic code.

**FIGURE 10–1** Biotechnology has been used for millennia to produce wine, cheese, bread, and fermented sausages. (Photo by Geostock. Courtesy of Getty Images, Inc./PhotoDisc, Inc.)

**FIGURE 10–2** Harry Klee, of the University of Florida, examines tomatoes used in his research at the university's Institute of Food and Agricultural Sciences on January 3, 1996, in Gainesville, Florida. Klee isolated a mutant gene that is responsible for thwarting the ripening of a tomato, opening the door on widespread manipulation of variety of economically important fruits, vegetables, and flowers. (Photo courtesy of AP/Wide World Photos.)

James Watson and Francis Crick announced the double-helix model for DNA. "New biotechnology" and its products are literally changing the way we function on this planet. "New biotechnology" owes its existence to the understanding of the cell and its components, and especially the genetic code. It is this "new biotechnology" that is stirring imaginations. It is "new biotechnology" that most people are referring to when they use the word *biotechnology*. Compare the previously mentioned products of alcohol, bread, and cheese as products of traditional biotechnology to "new biotechnology" that brings us tomatoes with an extended shelf life, tissue plasminogen activator to dissolve blood clots during heart attacks, bacteria that eat oil spills, goats and cows that produce milk that contains lifesaving drugs, animals with built-in disease resistance, biodegradable plastics, and gene therapy. These examples are only the tip of the iceberg (Figure 10–2).

## GENETIC ENGINEERING

Much, although certainly not all, of the new biotechnology has been made possible by the techniques of *recombinant DNA* (rDNA) technology. This term refers to the technologies used to transfer DNA (and the traits it codes for) from one organism to another and is commonly called *genetic engineering* (Figure 10–3). This technology was developed in the early 1970s by Paul Berg and Herbert Boyer at Stanford University and by Stanley Cohen of the University of California at Berkeley. The applications of this particular new biotechnology are the current stars of biotechnology. This status comes with good reason. The ability to recombine genetic material in some way other than the natural reproductive process was arguably the most important and exciting discovery of the last half of the 20th century. Genetic engineering has made possible the manipulation and transfer of genes from plants, animals, and microorganisms to and from each other. The cells of an organism that have new genes (transgenes) successfully incorporated into them pass the new genes on to the cells that spring from them when they divide. The implications of this stagger the imagination!

To many, the terms *genetic engineering* and *biotechnology* are synonymous. Biotechnology is actually a broader term that includes many things other than genetic

**Recombinant DNA**

DNA formed by combining pieces of DNA from different organisms.

**Genetic engineering**

The directed alteration of genetic material by intervention in the genetic processes.

**FIGURE 10–3** New biotechnology includes the techniques of genetic engineering. Compare the genetically engineered, transgenic Atlantic salmon with two sibling non-transgenic specimens. (Photo courtesy of Dr. Choy Hew, Hospital for Sick Children, Toronto, Ontario, and Dr. Garth Fletcher, Memorial University of Newfoundland, St. John's, Newfoundland.)

**Novel organism**

An organism with DNA from an outside source.

**Intergeneric microorganisms**

Microorganisms created to contain genetic material from organisms in more than one taxonomic genera. Another term for *transgenic*.

**Novel gene**

A previously nonexistent gene created in a laboratory.

**Gene map**

The locations of specific genes along a chromosome marked with probes.

**Gene probe**

A short segment of DNA used to identify and map an area of a specific chromosome or entire genome.

engineering. Microbial fermentation to produce foods like wine, beer, and cheese is a very traditional biotechnology with ancient roots. In addition, microorganisms have been put to many other uses in modern times without having their genetic code altered. However, some of those same processes are being done with microbes that have been enhanced with rDNA technology. Thus the lines tend to blur.

Because genetic engineering is done at the cellular level rather than with the whole organism as conventional breeding work is done, some new terms were coined to describe the results. Genetically altered plants, animals, and microorganisms are called ***novel organisms*** and are commonly referred to as *transgenic*. Novel microorganisms are also called ***intergeneric microorganisms.*** Ideally, novel organisms express the trait or traits that were inserted with the new gene(s). This technology creates a very different genetic code than natural breeding is able to create. Natural breeding cannot put elephant genes into corn or fish genes into lettuce. Genetic engineering can provide a distinct advantage over traditional breeding. Because the technology allows the genetic material to be altered before transfer, it may one day be possible to introduce ***novel genes***—genes that have been manufactured in a laboratory to do something that is now impossible. Genetic engineering also has the advantage of being more precise in the genetics it alters. In conventional breeding, many genes are affected with each production of a new offspring. In genetic engineering, as few as one gene can be altered, if that is what is desired, and the rest of the genetic code for the organism can be left intact.

How are organisms "genetically engineered"? Various methods of recombining DNA (hence the term *recombinant DNA*, or rDNA) have been developed. No doubt more will be developed, especially methods with greater efficiency than the current methods. Plants and microbes have proven fairly easy to genetically engineer, while mammals have been very difficult. Figure 10–4 demonstrates one of the earliest methods used to make transgenic bacteria.

## GENE MAP

The ***gene map*** of an organism is the location of specific genes on a chromosome that has been marked with ***gene probes*** as road signs, as it were. The code for the entire sequence of a species can vary from a few million for one-celled organisms to approximately 3 billion for humans. Thus behind all of the ability to genetically engineer is first the ability to map the genetic code. Techniques such as DNA sequencing, polymerase

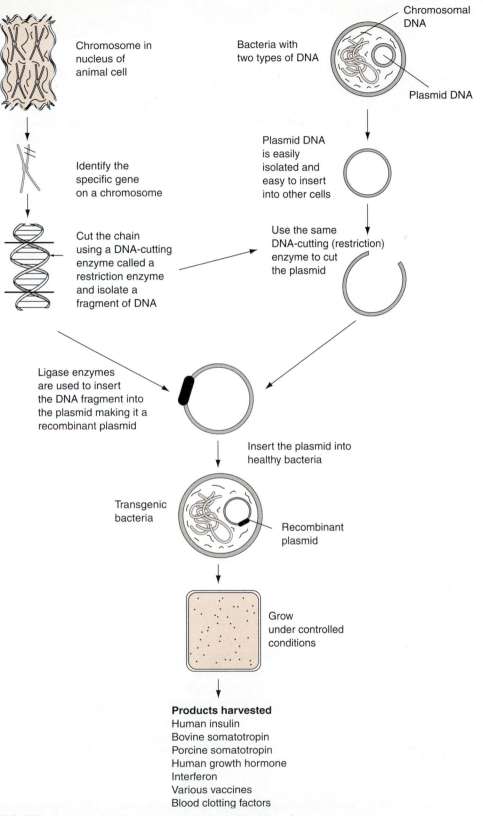

Chromosome in
nucleus of
animal cell

Bacteria with
two types of DNA

Chromosomal
DNA

Plasmid DNA

Identify the
specific gene
on a chromosome

Plasmid DNA
is easily
isolated and
easy to insert
into other cells

Cut the chain
using a DNA-cutting
enzyme called a
restriction enzyme
and isolate a
fragment of DNA

Use the same
DNA-cutting (restriction)
enzyme to cut
the plasmid

Ligase enzymes
are used to insert
the DNA fragment into
the plasmid making it a
recombinant plasmid

Insert the plasmid into
healthy bacteria

Transgenic
bacteria

Recombinant
plasmid

Grow
under controlled
conditions

**Products harvested**
Human insulin
Bovine somatotropin
Porcine somatotropin
Human growth hormone
Interferon
Various vaccines
Blood clotting factors

**FIGURE 10–4**   How to make transgenic bacteria and put it to use.

## Box 10–1   Genomics: A New Gene Pool of Potential

*Genomics.* It is one of the hottest buzzwords in science today. But it's far from being simply a new toy for making genetic improvements. Genomics refers to mapping and sequencing of the genetic material of a particular species and then associating the genes with the traits they express. "Genomic libraries" are constructed for each species to decipher the genetic information and associate it with traits. Such genomics libraries are either being built or are essentially complete for many important agronomic species. The Agricultural Research Service of the USDA has been a leader in developing genomic libraries for several animal, plant, and insect species as well as for microorganisms like the foodborne pathogen *Listeria*. For some species, the goal is to map the entire genome; with others, the aim is to concentrate on locating traits most important to breeders and producers.

As genomics data accumulates for a species from a wide range of production environments, scientists will not only understand how a trait operates in the real world, they will also know which specific genes control that trait. With genomics, a whole new horizon is opening up for conventional breeding, especially for complex traits previously beyond breeders' abilities to manipulate. The new genomics data are allowing researchers and breeders to identify genes that can add or improve a specific trait. Before, in conventional crosses, breeders had to depend on finding parents that exhibited the sought-after trait. But often the parent not only passes on the genes of interest but also other, sometimes undesirable, genes. Genomics data allow breeders to locate a specific gene sequence whether it is expressed or not. The precise genes can be moved directly into high-quality progeny without the need for many generations of backcrossing to produce consistent offspring that have the new trait. And there's no addition of undesirable genes. This will greatly speed up breeding. It will also allow breeders to tackle complex traits they haven't been able to introduce, amplify traits that are only faintly expressed, or suppress undesirable traits, and often without having to reach outside the species for new genes that would then create a transgenic organism. And when the source of the genes is a member of the same species, or a sexually compatible relative, the gene pool of the offspring would not be altered. This eliminates concerns that arise when genes are taken from wildly disparate species to achieve a new trait. Of course, genomics helps make genetic engineering from unrelated sources more precise. By narrowing the amount of genetic material to just those genes that need to be moved, potential problems and risks from unintended and unneeded gene transfer are minimized.

ARS has been a lead partner in creating genetic maps of chickens. Almost overnight, researchers went from having only about 2,000 genetic markers for traits to having potentially more than 3 million, which will help scientists make improvements that result in healthier, more nutritious, even tastier chickens. ARS scientists working with researchers from around the world have developed a physical bacterial artificial chromosome—or BAC—map of the cow and are participating in the international effort to sequence the entire cow genome. Traits being targeted for improvement include feed efficiency, reduced mastitis in dairy cows, and reduced susceptibility to diseases such as bovine spongiform encephalopathy, or mad cow disease. One result from the cow genome work has been development of a test for genes that code for the enzyme μ-calpain, which is a major reason for tenderness in beef. Breeding for tenderness will take a giant, but more precise, leap forward with the guidance of this DNA test.

Genomics research will markedly improve our ability to mine useful genes to enhance crops, livestock, insects, and microorganisms. A whole new pool of genetic potential for research and breeding is opening up.

This text was adapted from "Forum," published in the January 2007 issue of *Agricultural Research* (Volume 55, No. 1) magazine and may be accessed online at http://www.ars.usda.gov/is/AR/archive/2007.htm. It was authored by Judith B. St. John, Deputy Administrator, Crop Production and Protection, and Steven M. Kappes, Deputy Administrator, Animal Production and Protection. It is used with permission.

chain reaction (PCR), restriction fragment length polymorphism (RFLP), and statistical analysis allow the genes to be mapped. The mapping and sequencing of the genetic material of a particular species and associating specific traits with the genes is called *genomics* (Box 10–1).

Some applications of the science of biotechnology and genetic engineering do not involve manipulating the genetic code, but instead just identify it. In providing the map, they also provide an identity for anything being mapped. Very small differences in the genetic code between two individuals can be identified. These differences can be as little as one difference in a million. Thus disease organisms can be identified precisely and quickly. Family relationships can be identified, donor organ recipients and donors can be matched, and crime evidence can be provided. Maps can even serve as a modern pedigree for plants and animals. When still very new, this technology provided the information behind several well-publicized cases of switched babies being discovered and reunited with their birth parents. It routinely leads to convictions in criminal cases, and has also resulted in the release of individuals who had been wrongly convicted and imprisoned for crimes they did not commit. Some animal breed associations are beginning to require DNA testing as a condition of registration to prove the purity of bloodlines. Figure 10–5 demonstrates how to use DNA fingerprints to establish parentage.

## MONOCLONAL ANTIBODIES

*Antibodies* are proteins manufactured by the body in response to any foreign substance that finds its way into the body. If the foreign substance is a disease-causing microorganism, the body uses antibodies to fight the disease. Antibodies work by recognizing the foreign substances in the body and binding to them. If the foreign substance is a disease-causing microorganism, the antibodies binding to it render it harmless. Because antibodies attach to specific proteins, and thus specific disease-causing organisms, they can also be used outside the body to diagnose disease and in research on disease. Unfortunately, the antibodies manufactured by the body are relatively small in number. In addition, the body makes a range of antibodies to respond to the same substance. Thus collecting quantities of antibodies that respond to the exact same foreign substance in the exact same way is difficult and expensive. The use of *monoclonal antibodies* is a nongenetic biotechnology that solves these problems.

Through the techniques of biotechnology, antibody-producing cells are fused with tumor cells to create a *hybridoma.* Hybridomas make the same antibodies they made before fusion in unlimited quantities in the laboratory. These hybridoma-produced antibodies are identical, very specific, and bind to a specific site on a protein. A wide range of fast, inexpensive tests have been developed to diagnose viral and bacterial infections, screen for cancer, diagnose genetically caused diseases, and test for pregnancy. In addition, they are being used to locate tumors in the body in a very precise way with the use of modern computerized scanning and imaging processes. Monoclonal antibodies were first developed by Cesar Milstein and George Köhler in 1974. Monoclonal antibody use has revolutionized medicine.

## GENE THERAPY

Humans with defective, disease-causing genes can now receive newly engineered cells that are not defective in a process known as *gene therapy.* Gene therapy was first used successfully in 1990 to treat an immune deficiency in children called ADA deficiency. Scientists working in this pioneering field predict that gene therapy will be available for all genetically induced diseases by the year 2020. A more controversial application

### Antibodies

Proteins manufactured by the body in response to a foreign substance.

### Monoclonal antibodies

Antibodies produced by the daughter cells of a hybridoma that recognize, and thereby identify, only one antigen. Used to develop diagnostic tests and in research.

### Hybridoma

A cell line used to produce specific antibodies that are created by fusing an antibody-producing lymphocyte with a cancer-causing cell.

### Gene therapy

The process of replacing a missing or incorrectly functioning gene with a correct one to treat a disease. Also called *gene replacement therapy.*

DNA fingerprints

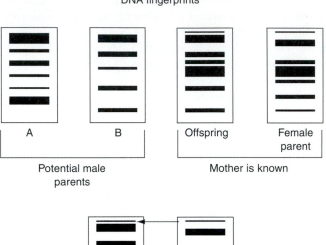

Collect tissue and make DNA fingerprints on offspring, female parent, and possible male parents.

A          B

**Potential male parents**

Offspring      Female parent

**Mother is known**

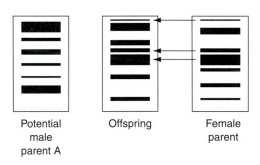

Match bands in offspring that correspond to female parent. NOTE: The offspring received half of its genetics from each parent.

Offspring      Female parent

The DNA fingerprint of possible male parent A does not match the remaining bands on the offspring. A is not the male parent.

Potential male parent A      Offspring      Female parent

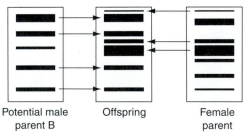

The DNA fingerprint of possible male parent B does match the bands unaccounted for by female. B is the male parent.

Potential male parent B      Offspring      Female parent

**FIGURE 10–5**   Using DNA fingerprints to establish parentage.

is called *gene enhancement,* which could be used in children to change their genetic potential for such traits as height. *Germ-line therapy* could be used to change reproductive cells so that a carrier for a genetic disease need not pass it on to his or her offspring.

## CLONING

A new biotechnology that does not include any recombining of the genetic code is *cloning.* Cloning is the process of producing a "twin" of an animal by transplanting the nucleus from a cell of the animal into an egg with the nucleus (and thus its original genetic material) removed. The egg with the new genetic material is allowed to develop in a female as would a naturally fertilized egg. The young animal bears no resemblance to the donor of the egg. All of its genetic material came from the donor of the nucleus. Thus it is genetically identical to the donor of the nucleus. The genetic material can come from an embryo (accomplished in the early 1980s) or, as the creators of the sheep *Dolly* have shown us, an adult animal. Dolly (Figure 10–6) was introduced to the world in 1997 as the first clone of an adult mammal. She was produced from a normal *somatic cell* taken from the udder of her "twin" Tracy. Of course, a genetically engineered animal could be cloned, but the genetic manipulation and the cloning are separate. Figure 10–7 demonstrates how cloning of animals is accomplished.

The potential advantages of cloned animals are many. Cloning should be useful in propagating transgenic animals with genes that cause them to produce important pharmaceuticals in their blood or milk. In addition, large numbers of genetically identical animals could be available for research purposes (Figure 10–8). For agriculture, outstanding animals could be cloned for production purposes, thereby making leaps in genetic progress in herds and reducing variability in production characteristics. Combining cloning with rDNA technology is considered to be one of the great potential benefits of cloning. Achieving a useful transgenic animal is complicated and expensive, and only a very small number of attempts succeed. Once a transgenic animal with a desired characteristic is achieved, cloning would allow the rapid development of a whole herd of these animals.

### Gene enhancement

The process of changing the genetic potential of an individual for a trait such as height or eye color.

### Germ-line therapy

The process of changing reproductive cells so an individual that carries a genetic defect need not pass it on to his or her offspring.

### Cloning

The process of producing a genetic copy of a gene, DNA segment, plant, animal, or embryo.

### Dolly

A normal Finn Dorset sheep who also happens to be the first clone of an adult mammal. She was born on July 5, 1996.

### Somatic cell

All cells in the body other than gametes.

**FIGURE 10–6** Dolly (right), the first cloned sheep produced through nuclear transfer from differentiated adult sheep cells, and Polly, the world's first transgenic lamb created at the Roslin Institute in Edinburgh, are seen in this December 1997 photograph. Dolly ignited international controversy about reproduction through cloning. Scientists at the Roslin Institute produced Polly with a human gene so that her milk will contain a blood-clotting protein that can be extracted for use in treating human hemophilia. Dolly died on February 14, 2003, euthanized because of an incurable lung infection. (Photographer John Chadwick. Courtesy of AP/ Wide World Photos.)

**FIGURE 10–7** How to clone an animal.

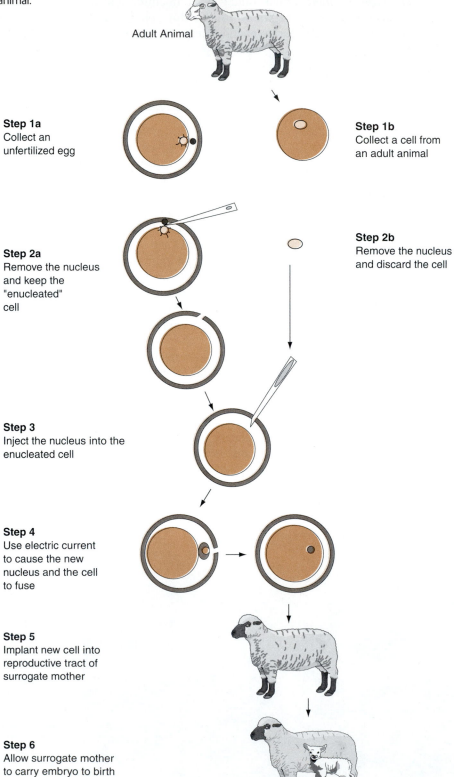

Adult Animal

**Step 1a**
Collect an
unfertilized egg

**Step 1b**
Collect a cell from
an adult animal

**Step 2a**
Remove the nucleus
and keep the
"enucleated"
cell

**Step 2b**
Remove the nucleus
and discard the cell

**Step 3**
Inject the nucleus into the
enucleated cell

**Step 4**
Use electric current
to cause the new
nucleus and the cell
to fuse

**Step 5**
Implant new cell into
reproductive tract of
surrogate mother

**Step 6**
Allow surrogate mother
to carry embryo to birth

**FIGURE 10–8** Newborn transgenic calves George and Charlie are shown in this University of Massachusetts and Advanced Cell Technologies, Inc. photo in January 1998. Cloning techniques combined with genetic engineering created the two healthy bull calves. The genetically modified cloned bull calves are the result of 4 years of research by James Robl, professor of Veterinary and Animal Sciences at the University of Massachusetts, and Steven Stice of Advanced Cell Technologies, Inc. of Worcester, Massachusetts. The science used to create the calves has potential applications for producing pharmaceutical proteins and cellular therapies to treat neurodegenerative disorders and diabetes. (Photo courtesy of AP/Wide World Photos.)

The technique that produced Dolly was extremely inefficient. This hurdle needs to be overcome before practical uses can be made of cloning. Progress is being made for several species on this front, but efficiencies are still low. Several species have now been cloned. However, cloning research is still in very basic stages and a long way from routine applications.

## MICROPROPAGATION

*Micropropagation* is a biotechnology that takes cells of a desired plant and uses them to generate another plant. This process is also called *tissue culture.* Because just a few cells can generate a whole plant, many identical plants can be made in a very short time from just a single leaf of a desirable plant. Micropropagation can be combined with genetic engineering to produce large numbers of a plant that has been engineered to do something considered of value.

**Micropropagation**

Individual cells of a plant used to generate another plant. Also called *tissue culture*.

**Tissue culture**

See *micropropagation*.

## CURRENT AND POTENTIAL USES OF BIOTECHNOLOGY AND/OR GENETICALLY ENGINEERED ORGANISMS

The potential number of genetically engineered plants, animals, and microorganisms is literally limitless. Genes are isolated daily from several species in research laboratories around the world. One day, the entire genome of any animal or plant of interest will be known. The human genome has been successfully mapped and sequenced. Maps of other species are well on their way to elucidation. The number of possibilities that will then be possible are effectively infinite. Many applications are and will be in agriculture and include novel foods, pesticides, feed additives, and animal drugs. Areas of interest and potential in agriculture include animals that produce leaner meat, plants with insect and herbicide tolerance, and bacteria that produce drugs for livestock. Some of the most exciting applications for genetically engineered products are in the realm of human medicine. These are tied to agriculture in some cases because transgenic animals or field crops are, or will be, used to produce the products. Some of the products and procedures developed for humans will eventually be available for animals. Others have important uses in food processing. Transgenic microorganisms are being developed to minimize pollution from livestock excretion, and for use in the pulpwood, ethanol, textile, detergent, compost, waste treatment, and pharmaceutical industries. Some of the applications of biotechnology are discussed in the following section.

### APPLICATIONS OF BIOTECHNOLOGY

In general, genetic engineering techniques have been developed that are considered *input traits* and *output traits,* many of which produce what are now known as genetically modified foods (GM foods). Although this terminology is most frequently used in relation to agronomic production, that is simply because most of the public debate has revolved around genetically manipulated plants. The principles are the same for animal products. Input traits are those designed for the convenience of the growers and include traits associated with pest, disease, and herbicide resistance, tolerance to temperatures, and changes designed to improve handling, distribution, and processing. Output traits are those designed to benefit the consumers and include improved nutrition, texture, uniformity, and appearance of the product and also enhanced safety and medicinal qualities. Modification of both the input and output traits can positively affect the environment through reduced herbicide and pesticide use and the reduced incidence of pollutants in animal wastes.

## Crops

Thus far, genetically modified crops outnumber all other genetically manipulated organisms. Already, several million acres of these crops are planted each year in the United States and around the world. Several thousand have been field-tested, with a much smaller number approved for commercial use. The number of transgenic plants that are field-tested in the United States alone roughly doubles annually. Corn, squash, tomatoes, cotton, canola, rice, and soybeans have gained the most attention to date. However, many other species and a multitude of traits are under study. Areas of interest include manipulating protein quality, vitamin content, and energy content. The most well known of these is Golden Rice, rice biofortified through genetic engineering to have increased amounts of beta-carotene (a pro-vitamin of vitamin A). Transgenic plants are being created that have resistance to various pests and diseases. Plants capable of withstanding adverse growing conditions such as extremes in temperatures and poor soil conditions are being developed. Plants resistant to herbicides have been developed and are widely planted around the world. Horticulturists are creating transgenic ornamentals. The possibilities seem endless: transgenic decaffeinated coffee, biodegradable polymer produced from switch grass, tropical plants growing in the snow, fruit that ripens on demand, and so on.

A very controversial area being explored is the possibility of creating edible vaccines with transgenic plants. Transgenic plants have already been successfully created to produce several plantbodies (antibodies produced in plants), and vaccine production by transgenic potatoes and bananas has been tested successfully. It also seems quite likely that transgenic plants will be developed to produce human insulin and other hormones at a fraction of the current cost.

## Transgenic Animals

Thus far, transgenic animals (Figure 10–9) are not nearly as plentiful as are transgenic microorganisms and plants. The techniques for producing transgenic animals are more

**FIGURE 10–9** One of Virginia Tech's valuable transgenic pigs is shown with one of her piglets in this 1997 photo. The pigs are the first in the world to produce human blood-clotting protein in milk. The protein could save human lives in the coming years. (Photo courtesy of AP/Wide World Photos.)

complicated and are not yet as successful. However, techniques are improving rapidly and more transgenic lines are being developed. Several hundred scientific papers are published each year describing techniques for producing transgenic animals. The transgenic animal that is subsequently used to establish the transgenic line is referred to as the *founder.*

Transgenic animals have been created since the early 1980s. Some of the areas of success in creating transgenic animals have been in enhancing growth rates, improved muscle-to-fat ratio of carcasses, improved resistance to disease, and enhanced feed nutrient use. Livestock species have been genetically engineered to produce human proteins in eggs, milk, blood, and urine. A few examples follow.

**Fish.** At least 35 species of fish and shellfish have been genetically engineered. This is at least in part because genetic manipulation is easier in fish than in mammals. Most have been developed for research. Targeted traits include growth rate, disease resistance, temperature tolerance, and flesh quality. A fast-growing Atlantic salmon has been engineered to contain a gene for growth hormone from Chinook salmon. The salmon is reported to grow as much as 400–600% faster than the original species. A fast-growing transgenic tilapia, with genetic material from rainbow trout, striped bass, and carp, has also been developed. The first transgenic pet, the GlowFish, is a zebrafish that carries the red fluorescent protein (DsRED) gene from a sea anemone. Originally engineered to detect pollutants, they have been on the pet market in the United States since 2003. "Electric Green" and "Sunburst Orange" were released on the market in 2006, with more likely to follow.

**Livestock.** Genetic engineering has produced several commercial applications for pigs including the enviropig developed at the University of Guelph that produces phytase in its saliva to utilize phosphorous from feed more efficiently and thus reduce the amount of phosphorous as a pollutant in manure. The University of Missouri has genetically engineered a line of pigs that has enhanced levels of omega-3 fatty acids, potentially leading to bacon and pork chops that might help your heart. Transgenic goats have been produced that produce silk in their milk. Both cows and goats have been genetically engineered to

**Founder**

A transgenic animal that is subsequently used to establish a transgenic line of animals.

produce human proteins in their milk. At the time of publication, no transgenic animal has been approved for human consumption.

**Animals as bioreactors to produce pharmaceuticals.** Transgenic bacteria now produce many pharmaceuticals. Although this is an amazing innovation, it does have drawbacks and limitations. It would be better if mammals could produce these same compounds. This method of producing pharmaceuticals is considered to be 2 to 3 times less expensive in start-up costs and 5 to 10 times more economical over the long term than microbial production of the same pharmaceuticals. Another reason this holds so much promise is that the complex mammalian organism is capable of much more complicated protein modifications than are bacteria. Transgenic mice, rats, rabbits, pigs, cows, goats, sheep, and crustaceans have been developed to secrete bioactive molecules into their blood, urine, or milk. The word *pharming* is used to describe the production of pharmaceuticals from livestock (pharmaceutical pharming). Products under active research include blood clotting factors to treat hemophilia, growth hormone, insulin, and biologicals to treat hereditary emphysema, cystic fibrosis, phenylketonuria, and several other diseases.

**Pharming**

The production of pharmaceuticals from livestock.

**Animals engineered as sources of transplant organs.** Companies are actively pursuing the development of transgenic animals that could donate organs to humans that would not be rejected by the human recipient. These organs would be used in *xenotransplantation*. The single largest hurdle is engineering a pig whose organs would not be rejected by a human recipient.

**Xenotransplantation**

The use of organs from genetically engineered animals to transplant into humans.

**Animals engineered to help researchers study and treat human diseases.** This is the use that most transgenic animals have been designed for thus far. Several hundred transgenic rodent lines have been developed to study cardiovascular disease, cancer, several autoimmune diseases, sickle-cell anemia, and neurological diseases. These transgenic laboratory animals develop diseases in a manner very similar to that of humans because they have been engineered to do so, allowing researchers a model to study the human disease.

**Disease resistance.** Transgenic chickens and turkeys have been developed that resist avian diseases, but they are not yet available commercially. This area of research should yield more applications as the genes responsible for disease resistance are discovered.

## Microorganisms and Invertebrates

*Bovine somatotropin (BST)* has been among the most successful products of genetic engineering made available to agriculture. It is produced commercially by transgenic bacteria. When administered to dairy cows, BST causes greater production of milk. Because of the high fixed costs and labor demands of a dairy, it is generally more profitable to increase average production of the cows in the herd. BST is discussed at length in a later chapter.

**Bovine somatotropin (BST)**

Hormone produced by the anterior pituitary gland that acts on various target tissues in the body.

**Invertebrates.** Various transgenic arthropods, insects, and nematodes have been developed and are being field-tested. Honeybees and other beneficial insects have been given increased tolerance to certain pesticides. Transgenic pink bollworm, pomace fly, spruce bollworm, and a nematode have been field-tested.

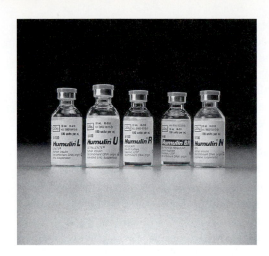

**FIGURE 10–10** Transgenic bacteria now produce many pharmaceuticals such as Humulin, manufactured by Eli Lilly and Company. It is identical to human-produced insulin. (Photo courtesy of Eli Lilly and Company.)

## Microorganisms used to produce drugs and vaccines.

Bacteria have been engineered to produce drugs for humans. Human insulin (Figure 10–10) produced by bacteria has been available since 1982 and has now almost completely replaced the cow and pig insulin diabetics once had to use. Growth hormone to treat some types of impaired growth in children, and tissue plasminogen activator (TPA) to treat heart attack victims, are also produced by fermentation of transgenic bacteria. The floodgates are about to open on this category of drugs, with several hundred new drugs tested each year. Certainly, they don't all reach market. However, if just a small fraction do, dozens of new drugs could be added each year. This research will escalate even more with time.

Other genetically engineered products are used in cancer treatment, to treat AIDS, to manufacture proteins that help hemophiliacs, and to stimulate red blood cell growth to help patients on dialysis. In addition, antibodies to help prevent organ rejection, Pulmozyme for cystic fibrosis patients, and beta-interferons to treat multiple sclerosis have been developed. Some of these products replace previous treatments. However, a significant number are completely new therapeutic agents with no predecessor.

Several biotechnology-produced vaccines have been developed for humans and animals. Rabies vaccines for wild animal populations have been approved and used. Vaccines for baby pig scours, footrot, sheep measles, and infectious bursal disease have also been developed. Several others are also expected to come on the market, including an HIV vaccine.

## Microorganisms used as pesticides.

Several bacteria have been genetically altered to improve their ability to kill or repel pests. Some have gained approval for commercial use and are being used as pesticides.

## Bacteria engineered for the foods industry.

Microorganisms have long been a part of food processing. Now products of transgenic bacteria are also in use. Genetically altered bacteria produce rennet, which is used in cheese making. Rennet was approved in 1990 and is the first food product from transgenic production to receive approval. Fast-acting yeasts have been developed that reduce the amount of time bread must rise before baking and also have better brewing characteristics. Among the valuable new tools of biotechnology are fast, easy tests that can be used on foods to detect disease-causing microorganisms or their toxins. These important tools will help provide a safer food supply. Both DNA probes and monoclonal antibodies are being used in this way.

**Nutraceuticals**

Products perceived to have both nutrient and pharmaceutical properties.

An emerging area of food biotechnology will include *nutraceuticals*, which are products that have both nutrient and pharmaceutical properties. Genetically manipulated soybeans may produce beneficial fish oils. Microorganisms used in a fermentation process may add increased levels of antioxidant vitamins. Transgenic microorganisms have also been developed to produce a safe, consistent supply of flavors for the food-processing industry. Many other uses are also being investigated.

## BIOTECHNOLOGY IN WASTE MANAGEMENT

Waste treatment facilities have long used microorganisms in traditional biotechnology applications to treat and process human waste. Modern applications of biotechnology include the development of new and more efficient ways of disposing of waste, providing new ways of producing products that do not make as much waste, and providing ways to convert wastes into useful products. Some of these applications include organisms genetically engineered to do a better job. Other applications involve a better way to use unaltered organisms, and still others include newly discovered microorganisms. Examples of waste management biotechnology include the following:

- Treating wastewater to remove phosphorus and nitrogen before they are discharged into rivers and streams, thereby reducing these contaminants in the water supply and in wildlife habitats.

- Removing chemical waste from the environment, including cleaning toxic dumps of various sorts.

- Using white rot fungi to clean up contaminated land and remove toxic waste. White rot fungi, which normally spend their time munching on wood, have been discovered to like organochlorine compounds such as DDT, dieldrin, aldrin, and polychlorinated biphenyls. This discovery is especially useful because these compounds are very stable and hard to dispose of when they are contaminants and pollutants. It is hoped this biotechnology will ultimately lead to the cleanup of contaminated land and the destruction of hundreds of thousands of drums of toxic waste stored at sites around the world.

- Treating industrial effluents in an economical and environmentally friendly way to remove pollutants before they are discharged into the environment. This has applications for pulp and paper mills, food processing, chemical manufacturing, textile manufacturing, brewing and distilling, and the foods industries.

- Releasing bacteria to break down oil and petroleum pollution. This practice has been in use for many years. Now genetically engineered microorganisms degrade oil slicks.

## REGULATION OF GENETICALLY ENGINEERED PRODUCTS

Three federal agencies currently regulate genetically engineered products: the Environmental Protection Agency (EPA), the Food and Drug Administration (FDA), and the U.S. Department of Agriculture (USDA). Products are regulated according to their intended use, with some being regulated under more than one agency. In addition, most states monitor development and testing of genetically monitored products within their borders.

The Animal and Plant Health Inspection Service (APHIS), a division of the USDA, regulates genetically engineered products under several statutes. Genetically engineered crops are regulated under the Federal Plant Pest Act. This legislation enables APHIS to regulate interstate movement, importation into the United States, and field-testing of altered crops.

The FDA has broad authority and primary responsibility for regulating the introduction of all new foods, including genetically engineered foods, under the Federal Food, Drug, and Cosmetic Act (FFDCA). Two different sets of provisions in the FFDCA pertain to genetically engineered foods. The first are the "adulteration" provisions, which give the FDA authority to remove unsafe foods from the market. The other pertinent section of the act requires premarket approval of food additives. In 1992, the FDA released guidelines for genetically engineered foods coming from plant sources. These guidelines can be found in the *Federal Register,* vol. 57, pages 22984 and 22986–88. Because the FDA has authority over drug approval, it also has authority over drugs produced through biotechnological means. In this case, the fact that a drug is produced through biotechnology is not the issue. The FDA already had authority over drugs. However, in a novel move, the FDA Center for Veterinary Medicine is regulating transgenesis in many cases, such as the transgenic salmon mentioned earlier, as a new animal drug. The logic is that transgenesis is being used as the means for delivering growth hormone to the animal tissues. This means that regulatory approval will follow closely the same procedures used to gain approval for a drug.

The EPA oversees genetically engineered microbial pesticides and certain crops that are genetically engineered to produce their own pesticides under the Federal Insecticide, Fungicide, and Rodenticide Act (FIFRA). The FFDCA authorized the FDA to set tolerances and establish exemptions for pesticide residues on and in food crops. Nonpesticidal, nonfood microbial products are regulated under the Toxic Substances Control Act (TSCA). The regulation under which the TSCA Biotechnology Program functions is titled "Microbial Products of Biotechnology; Final Regulation Under the Toxic Substances Control Act," which can be found in the *Federal Register,* vol. 62, No. 70, pages 17909–58. This rule was developed under TSCA because intergeneric microorganisms are considered new chemicals under the act.

Animal vaccines are regulated under the Virus, Serum, and Toxin Act; and engineered poultry and livestock fall under various meat inspection statutes. Transgenic animals other than poultry and livestock are not regulated for environmental risks. In December 2006, the FDA declared that products from cloned animals are safe to eat. At the time of the printing of this text, the FDA was actively working on regulations that would allow the meat and milk from cloned animals into the food supply.

## SOCIETAL CONCERNS

Biotechnology and genetic engineering have moral, ethical, and religious issues associated with them that society must reach conclusions about. It is outside the scope of this chapter to deal with these issues in any meaningful way. However, some questions that society must answer include these:

- How far is it acceptable to alter the genetic code? For instance, are featherless chickens acceptable?
- Is altering the genetic code of a human acceptable?
- Is cloning an animal acceptable for any and all purposes?

- Is cloning a human acceptable? Is cloning perhaps acceptable for some situations but not others?

- Is it acceptable to generate human embryos specifically for the treatment of disease with no intention of allowing them to develop to term?

- Is it acceptable to change the genes of a human if those genes can then be passed on to offspring? If it is determined that this is acceptable, should changes be restricted to correcting for diseases or expanded to such things as height, pattern baldness, and eye color?

- How far are we willing to go in mixing genetic information from species to species? Would you knowingly eat lettuce with dog genes inserted into its genetic code? How about genes from fish? What about genes from bacteria? What about genes from carrots?

- What will become of bioreactor animals when their productive life is over? Should they be used as food?

- Should genetically modified food be labeled?

GM foods are also under scrutiny. In the view of many, significant risks are posed by GM foods. Because no confirmed cases of illness or disease are associated with human consumption of GM foods, the risks are potential rather than actual. The companies that produce and market GM foods detect and eliminate plants with undesirable traits during development and before they are commercially available. However, critics argue that detection techniques are not foolproof and need improving. The risks generating the most concern about GM foods include direct risks to human health such as the production of allergenic or new toxic proteins or enhanced production of existing toxic proteins; indirect risks to human health such as reduced levels of nutrients, decreased efficacy of antibiotics, more rapid development of antibiotic-resistant bacteria, and greater human exposure to herbicides; direct risks to the environment and public welfare including turning crops into superweeds and the reduction in populations of beneficial nontarget species; and indirect risks to the environment and public welfare including increased development of pesticide-resistant pests, loss of valuable biological pesticides, turning related species into superweeds, greater environmental exposure to herbicides, and threats to biodiversity.

Policy debates over these and other related issues are sure to rage for the foreseeable future.

## SUMMARY AND CONCLUSION

Biotechnology is the development of products by a biological process using intact organisms or natural substances (e.g., enzymes) from organisms, or intact cells, plants, or animals to produce products that could not previously be produced. The use of rDNA through genetic engineering figures prominently in many of the modern or new applications of biotechnology. However, the term *agricultural biotechnology* is a broader term that describes the use of biological processes other than genetic engineering, such as micropropagation, plant and animal health diagnostics, vaccines, and biopesticides. Many different disciplines contribute to biotechnology including biology, genetics, molecular biology, biophysics, biochemistry, chemical engineering, and computer science. Consumers, agriculturists, and the agrifoods industry all have a stake in biotechnology. Safer foods, novel foods, cheaper foods, pharmaceuticals, diagnostics, and treatments all await the promise of biotechnology. Some of the more important applications include the following:

- Existing crops can be engineered for disease and insect resistance, drought tolerance, increased climatic tolerance, and the ability to produce their own fertilizers or be productive with less fertilizer. They can also be changed so they produce a different chemical makeup in their seed, such as more, less, or chemically different fats, proteins, and vitamins. Digestibilities can be improved, value enhanced, and new sources of products developed. Field crops with improved industrial applications are being developed. Food crops can be engineered to have greater nutrient content, improved taste, and better storage and processing properties.

- Microorganisms can be used to help increase crop yields, produce vaccines and other pharmaceutical agents, provide products for food processing, and so on.

- Livestock species can be improved for the useful traits, such as leanness, better growth, improved milk production, altered nutritional properties, and so on.

- Livestock can be altered to produce biologically active compounds in body fluids, including milk, for a variety of uses. These substances will be inexpensive compared to other methods of production.

- Diseases can be diagnosed and researched with the use of monoclonal antibodies and transgenic animals.

- Microorganisms can be used to help in the fight to reduce pollution and keep the environment clean.

The possibilities are as large as the size of the combined genome of all the world's species, and that is very big indeed.

## STUDY QUESTIONS

1. Define *biotechnology*. What are traditional biotechnologies and what are new biotechnologies? What are products of each?

2. Define *genetic engineering*. How does it fit into biotechnology? Describe which is the "broader" term and why. What is so biologically unique about genetic engineering?

3. In general terms, describe how the DNA of organisms gets recombined through genetic engineering. What are the processes used?

4. What do the terms *novel organism, transgenic organism, intergenic organism,* and *novel gene* mean?

5. Why is gene mapping important? What are the uses of a gene map? What is a genome? What is genomics?

6. What are antibodies? What do they do in the body? What is a monoclonal antibody? Do antibodies and monoclonal antibodies have the same purpose? If not, what are their functions? How are they made?

7. What is gene therapy? How does it differ from germ-line therapy?

8. What is a clone? What was unique about Dolly the clone? Has that feat been duplicated in any other species? What are the uses of clones?

9. Describe ways that genetically engineered plants are being put to use today. What are the major ways they have been engineered so far? What other changes do you think it would be useful to engineer into plants?

10. Why aren't there as many transgenic animals as there are transgenic microorganisms and plants?

**11.** What are the traits that would be useful to genetically engineer in food fish? Can you think of any traits of ornamental fish that would be beneficial to engineer into food fish?

**12.** What does it mean to refer to animals as bioreactors? How does this relate to pharming? What are the benefits of pharming as compared to microbial production of pharmaceuticals?

**13.** What is the use that most transgenic animals have been developed for to date? Why is this so valuable?

**14.** How is BST produced commercially? What is its value?

**15.** What are some of the uses that microorganisms have been genetically harnessed to perform in terms of producing drugs and vaccines?

**16.** How do genetically engineered organisms fit into the food-processing industry? What is a nutraceutical? Would it bother you to eat potatoes that produced fish oils?

**17.** How can biotechnology help in waste management? Are all of these uses connected to recombinant DNA technology?

**18.** Who has regulatory authority over transgenic organisms? Does this mechanism seem cumbersome to you? Do you have a suggestion for streamlining the process?

**19.** What are some of the concerns you have seen expressed in the media with reference to the ethics of genetic engineering? Is the list of social concerns in the text complete or do you think there are other issues to consider? Are there some in the text that you don't think are issues? Can you defend your position? In the end, what can one country do to stop genetic engineering of a particular type from happening?

# REFERENCES

Campbell, K. H. S., L. Pasqualino, P. J. Otaegui, and I. Wilmut. 1996. Cell cycle coordination in embryo cloning by nuclear transfer. *Journal of Reproduction and Fertility* 1 (1): 40.

International Food Information Council. 2006. Food biotechnology: A study of U.S. consumer attitudinal trends. Accessed online January 2008. http://www.ific.org/about/results.cfm.

McGarity, T. O., and P. I. Hansen. 2001. *Breeding distrust: An assessment of recommendations for improving the regulation of plant derived genetically modified foods.* Report prepared for the Food Policy Institute of the Consumer Federation of America, Washington, DC.

National Academy of Science, 2000. *Transgenic plants and world agriculture.* Washington, DC: The National Academies Press.

National Science and Technology Council (NSTC). 1995. *Biotechnology for the 21st century: New*

*horizons.* Washington, DC: Biotechnology Research Subcommittee, Committee on Fundamental Science, National Science and Technology Council, Office of Science and Technology Policy. http://www.nalusda.gov/bic/bio21/tablco.html.

North Carolina Association for Biomedical Research. 2006. *Issue brief: Animal Biotechnology*. Accessed online January 2008. http://www.aboutbioscience.org/pdfs/Animal_Biotechnology.pdf.

Rennenberger, R. 2008. *Biotechnology for beginners.* Amsterdam: Academic Press.

Ruane, J., and A. Sonnino. 2006. *Results from the FAO biotechnology forum.* Rome: Food and Agricultural Organization of the United Nations.

Union of Concerned Scientists. 1998. Various fact sheets. Cambridge, MA: Union of Concerned Scientists. http://www.ucsusa.org/agriculture/biotech.whatis.html.

CHAPTER **11**

# Animal Reproduction

## LEARNING OBJECTIVES

After you have studied this chapter, you should be able to:

1. Describe how the endocrine system drives the production of gametes.

2. Identify the various anatomical features of female and male reproductive systems.

3. Compare and contrast the functions of the male and female gonads.

4. State how conception, pregnancy, and parturition occur.

5. Discuss the considerable influence of the environment on reproductive function.

6. Describe the uses and advantages of the technologies recently employed in animal reproduction.

## KEY TERMS

Artificial vagina (AV)
Atresia
Blastocyst
Colostrum
Corpus luteum (CL)
Dystocia
Embryo transfer
Epididymis
Episodic
Estrous cycle
Estrus
Flushing
Follicle-stimulating hormone
Freemartin

Gametes
Generation interval
Gonads
Hypothalamus
In vitro
Libido
Lordosis
Luteinizing hormone
Luteolysis
Monoestrus
Morula
Oocyte
Ovulation
Parturition
Passive immunity

Pituitary gland
Placenta
Polyestrus
Postpartum
Postpartum interval
Pregnancy disease
Progesterone
Prostaglandin
Puberty
Recipients
Secondary sex characteristics
Semen
Testosterone
Zygote

# INTRODUCTION

Reproduction is required for propagation and continuation of a species, and as such, is an essential process in all species. Producers of domestic animals are particularly concerned with reproduction, as the production of young is the primary determinant of income for most livestock species. Even dairy producers, who generate the majority of their income through milk sales, require reproduction to occur to initiate lactation. In fact, increases in reproductive efficiency are considered to have a much greater impact on profitability than does progress in general production methods. A 3% improvement in birthrate would result in an additional 1 million beef calves born per year, 3.2 million pigs born per year, and 3.7 million gallons of milk produced per year. Considering these numbers, efforts to increase reproductive efficiency in domestic animal species are typically well rewarded financially. Likewise, selection against poor reproductive effectiveness is similarly rewarding.

Reproduction in all animals, both male and female, requires tremendous coordination between the endocrine system for the production of hormones, and the reproductive physiology and anatomy required to carry out the processes involved with germ cell development and maintenance, fertilization, pregnancy, and ***parturition***, the process of giving birth. These processes, although critical to produce offspring, are active only during certain phases of the life cycle. These phases are typically age dependent. In addition, many other factors come into play, including season (day length), presence of the opposite gender, and level of nutrition. Knowing which factors affect reproductive function and determining how to minimize the negative effects of those factors are critical to successful reproduction.

At the basis of the reproductive system is the ***gonad***. The female gonad is the *ovary* and the male gonad is the *testis*. The gonads have two primary functions: steroidogenesis, or the production of the sex steroids, and gametogenesis, or the production of ***gametes***. Both of these functions are hormonally controlled and require absolute coordination for proper activity to be expressed. The hormones responsible for proper function of the gonads are produced by the brain and ***pituitary gland***. Because the gonad is directly responsive to the action of the brain, environmental factors including nutritional status, length of daylight, and emotions have a profound influence on reproductive function. Some reproductive terms and other information of general value about some species are listed in Table 11–1.

# PUBERTY

Before an animal of either sex is capable of reproduction, it must go through the process of ***puberty***. The signals for puberty differ by species, but the most important factors influencing the onset of puberty are age and weight. However, nutritional stress, season of the year, and other factors can also be important. Puberty is simply the process of maturing from a nonfunctional endocrine and physiological reproductive state into a state of gamete and hormone production. After puberty, an animal is said to be *reproductively competent*. Puberty is associated with the ***secondary sex characteristics*** commonly associated with each sex. The transition through puberty is characterized by inconsistent reproductive competency.

Examples of secondary sex characteristics in males include such things as humps on the necks of bulls, beards on men, increased musculature in the male of most

---

**Parturition**

Process of giving birth.

**Gonads**

Sex organs; testis in male, ovary in female.

**Gametes**

Mature sperm in the male and the egg or ova in the female; the reproductive cells.

**Pituitary gland**

Gland sitting directly below the hypothalamus.

**Puberty**

Transitional state through which animals progress from an immature reproductive and hormonal state to a mature state.

**Secondary sex characteristics**

Characteristics that differentiate the sexes from each other; occur most profoundly during and after puberty.

**TABLE 11–1** Reproductive Terms by Species

| | Cats | Cattle | Dogs | Goats | Horses | Chickens | Sheep | Swine |
|---|---|---|---|---|---|---|---|---|
| Mature male | Tom | Bull | Dog | Buck | Stallion | Cock[1] | Ram | Boar |
| Mature female | Queen | Cow | Bitch | Doe | Mare | Hen | Ewe | Sow |
| Young male | — | Bullock | Puppy dog | Buck kid | Colt | Chick[2] | Ram lamb | Boar[3] |
| Young female | — | Heifer | Puppy bitch | Doe kid | Filly | Chick[2] | Ewe lamb | Gilt[3] |
| Newborn | Kitten | Calf | Pup | Kid | Foal | Chick[2] | Lamb | Pig |
| Unsexed male | Gib | Steer | Castrate | Wether | Gelding | Capon | Wether | Barrow |
| Groups | Bevy | Herd | Pack | Band | Herd | Flock | Flock | Herd, drove, or sounder |
| Genus | *Felis* | *Bos* | *Canis* | *Capra* | *Equus* | *Gallus*[4] | *Ovis* | *Sus* |
| Act of parturition | Littering | Calving | Whelping | Kidding | Foaling | NA | Lambing | Farrowing |
| Duration of heat | 6–7 days | 14 hrs | 2–21 days (6–12 avg) | 42 hrs | 6 days | NA | 30–35 hrs | 2–3 days |
| Length of estrous cycle (average; range) | 18; 14–21 days | 12; 18–24 days | 3 1/2–13 months; (6 months avg) | 21; 15–24 days | 21; 16–30 days | NA | 16; 14–20 days | 21; 18–24 days |
| Time of ovulation in (days) relation to heat | Stimulated by mating | 10–14 hrs after end of estrus | Usually 1–3 days after first acceptance of male | Near end of estrus | 1–2 days before end of estrus | NA | 1 hr before end of estrus | 18–60 hrs after estrus begins |
| Gestation period (average; range) (days) | 63; 62–64 | 281; 274–291 | 63; 58–68 | 151; 140–160 | 336; 310–350 | 21-day incubation | 150; 140–160 | 113; 111–115 |
| Age at puberty (months) | 4–18 (much breed variability) | 8–14 | 5–24 | 4–8 | 10–12 | 4–6 | 4–8 | 5–7 |

[1]Called a tom in turkeys.
[2]Called a poult in turkeys, a gosling in geese, and a duckling in ducks.
[3]Shoat refers to a young pig of either sex under one year of age.
[4]Genus for chicken.

**Hypothalamus**

Area of the brain responsible for many homeostatic functions.

**Episodic**

The pulsatile manner in which the gonadotropic hormones are secreted by the anterior pituitary gland. Controlled by the pulse-generating center of the brain.

**Luteinizing hormone**

Gonadotropic hormone primarily responsible for providing the signal to disrupt the mature follicle in females, and the production of testosterone by the Leydig cells of the testes in the male.

**Follicle-stimulating hormone**

Gonadotropic hormone responsible for growth, development, and maintenance of follicles in females, and the production of sperm in males.

species, and changes in the sound of vocalization (for instance, the voice change in boys that happens at puberty). For females this includes the many characteristics lumped together that we refer to as femininity: added body fat that creates curves where once angles were visible, mammary development, smoother hair coats, and so on. Behavioral characteristics such as "marking" territory (both sexes) or aggression in males are also part of the complex.

## ENDOCRINOLOGY

The endocrine functions involved in reproduction are initiated by the *hypothalamus*. The hypothalamus, a small area of the brain, is a critical part of the body's ability to adapt to the environment. The hypothalamus releases a hormone called *gonadotropin-releasing hormone* (GnRH). Release of this hormone is the first step in a cascade of hormonal events that must proceed in a coordinated manner to result in successful action by the gonads. Interestingly, GnRH is released in a pulsatile manner, causing all of the hormones and actions to be *episodic* (Figure 11–1). This action maintains a high degree of sensitivity in the system. The stimulation for GnRH release is controlled by a center referred to as the *pulse-generating center*. This center is under control of various parts of the brain and, in fact, integrates the many environmental signals to produce a driving force in the endocrine cascade.

GnRH travels a short distance to the anterior pituitary gland, which sits directly below the hypothalamus. The anterior pituitary gland, in turn, responds to GnRH by releasing two other hormones: *luteinizing hormone* (LH) and *follicle-stimulating hormone* (FSH). These two hormones enter the bloodstream and travel to the gonads

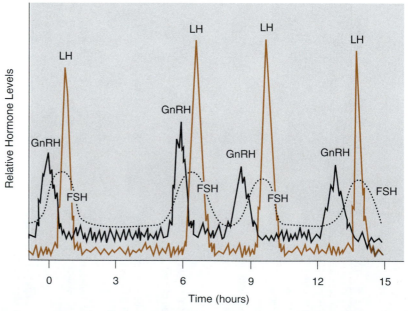

**FIGURE 11–1** Episodic release of the hormones of reproduction in response to the release of gonadotropin-releasing hormone (GnRH). LH = luteinizing hormone; FSH = follicle-stimulating hormone. (*SOURCE:* Senger, 1997, p. 170. Used with permission.)

of both males and females. Again, these hormones are released into the bloodstream in a pulsatile manner so that the gonads are only exposed to high levels of the gonadotropins (LH and FSH) intermittently.

*Testosterone*, the primary male hormone, is produced by the testes. In the male, release of LH is the signal to the testes to produce testosterone. If LH is not present, testosterone is not produced in adequate quantities for expression of the secondary sex characteristics associated with males. FSH in males is required for the production of sperm, the male gamete.

In the female, FSH is responsible for growth and maintenance of the developing follicle that is destined to produce the ova, which is the female gamete. Without adequate FSH support, the follicle undergoes death in a process referred to as *atresia*. As the follicle grows and develops, it produces estrogen (Figure 11–2). The increased level of estrogen causes a surge of LH to be released from the pituitary gland. This surge of LH results in a breakdown of the follicle wall, thereby releasing the ova from the ovary and making it available for fertilization (Figure 11–3). This release of the ova from the ovary is called *ovulation*.

After ovulation, the follicle turns into a *corpus luteum (CL)*. This structure's primary responsibility is to produce the hormone *progesterone*, which is required to support pregnancy. Progesterone inhibits LH and FSH release, prevents behavioral *estrus*, and decreases the motility of the muscles in the uterus. In some species (goat, rabbit, and sow), the CL is required throughout pregnancy. In other species (cow, mare, and ewe), the CL provides adequate progesterone during early pregnancy, but becomes unnecessary because the *placenta*, the organ that surrounds the fetus and unites it to the female during pregnancy, begins to produce enough progesterone to support pregnancy sufficiently. The placenta is the organ through which oxygen and nutrients are passed to the fetus from the female and waste is passed from the fetus to the female.

### Testosterone
Male steroid sex hormone.

### Atresia
The degeneration of follicles that do not make it to the mature stage, otherwise known as the Graafian stage.

### Ovulation
Release of the ova or egg from the ovary.

### Corpus luteum (CL)
Ovarian structure responsible for the production of progesterone for the support of pregnancy.

### Progesterone
Female sex steroid produced by the corpus luteum or the placenta.

### Estrus
The period when a female is receptive to mating. Synonymous with *heat*.

### Placenta
The organ that surrounds the fetus and unites it to the female while it develops in the uterus.

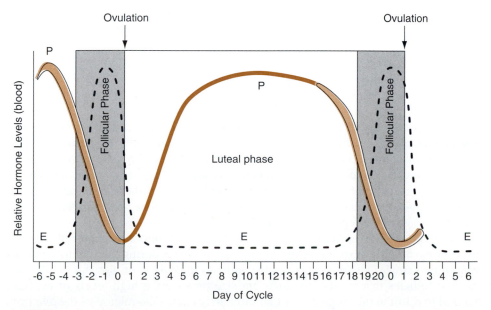

**FIGURE 11–2** Phases of the estrous cycle. P = progesterone; E = estrogen. (*SOURCE:* Senger, 1997, p. 119. Used with permission.)

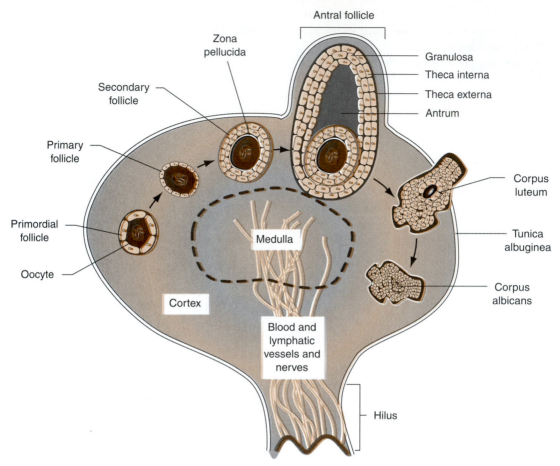

**FIGURE 11–3**   Illustration of the ovary. (*SOURCE:* Senger, 1997, p. 17. Used with permission.)

## ANATOMY

### FEMALE

The female reproductive tract consists of the ovaries (female gonad), oviducts (also called Fallopian tubes), uterus, cervix, vagina, and external genitalia (Figure 11–4). In most domestic animal species, the reproductive tract is suspended below the rectum by various ligaments. The ovaries occur in pairs, are attached to the ligament at the hilus, and are responsible for storage, development, and release of the ova. In addition, the ovaries produce the female sex steroid hormones progesterone and estrogen, depending on the stage of the reproductive cycle.

The ovaries of the newborn female contain all of the *oocytes* (gametes) the female will ever have. The oocytes develop into primary follicles during the monthly *estrous cycle* of the female. The primary follicle continues to develop or dies through atresia. The few follicles that survive to sufficient size produce the high levels of estrogen required to result in the surge of LH that ultimately causes the release of the ova from the follicle. It is not yet clear how the follicle that survives to ovulation is chosen among the 100 or so that begin to grow each month.

**Oocyte**
The gamete from the female.

**Estrous cycle**
The time from one estrus (heat) to the next.

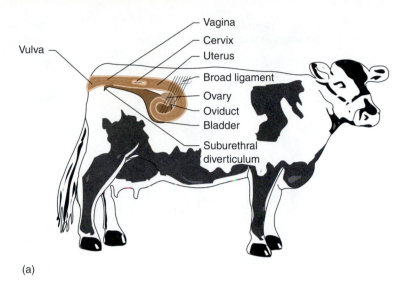

(a)

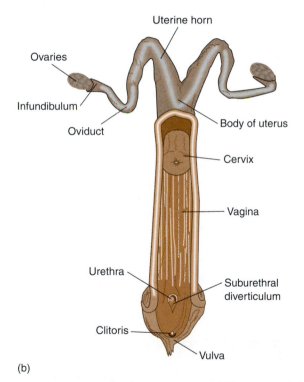

(b)

**FIGURE 11–4** Female reproductive tract. (*SOURCE:* (a) Bearden and Fuquay, 1997; (b) Texas A&M University. Used with permission.)

The ovum is released from the follicle and is "captured" by the *infundibulum*, a structure at the end of the oviduct that surrounds the ovary. The thin membrane directs the ovum into the oviduct, preventing it from entering the abdominal cavity. The intermediate portion of the oviduct is the ampulla, which connects the infundibulum with the isthmus, the final portion of the oviduct. The ampulla is generally the site of fertilization. It is the

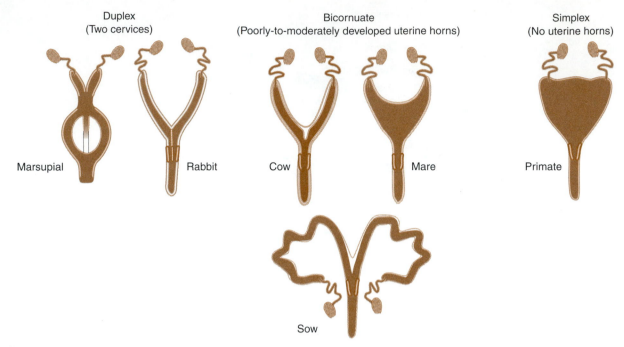

Duplex
(Two cervices)

Bicornuate
(Poorly-to-moderately developed uterine horns)

Simplex
(No uterine horns)

Marsupial            Rabbit            Cow            Mare            Primate

Sow

**FIGURE 11–5**   Types of uteri. (*SOURCE:* Senger, 1997. Used with permission.)

isthmus that makes the connection of the oviduct with the uterine horn. The oviduct is glandular, providing nutrients and a transport medium in the secretions. The eggs remain in the oviduct for approximately 3 to 6 days.

The oviduct is connected to the uterus. Species differ greatly in the type of uterus configuration present, as indicated in Figure 11–5. The primary difference between types of uteri is the presence of uterine horns. The uterus functions to provide a passageway for sperm cells from the cervix to the oviducts, to provide glandular secretions to nourish the embryo prior to development of the placenta, to provide nutrients and eliminate waste products for the developing fetus through the placental–uterus junction, and to expel the fetus during parturition. The uterus is a very muscular organ, and contractions aid in the expulsion of the fetus. However, this contractility of the uterus must be suppressed to allow the embryo to implant to the wall of the uterus. After ovulation, the development of a functional CL ensures that adequate progesterone is produced and secreted. Progesterone suppresses the contractility of the uterus, which allows the embryo to implant.

The uterus is connected to the cervix, which acts as a "gatekeeper" from the vagina into the uterus. The cervix has five primary functions:

1. To act as a passageway for sperm cells.
2. To act as a storage reservoir for sperm cells. In this way there can be a more consistent release of sperm into the uterus. This increases the chances that viable sperm will be present at the same time the ova is prepared for fertilization.
3. To act as the primary barrier between the external and internal environments.
4. To provide lubrication.
5. To act as a passageway for the fetus at parturition.

The cervix, a very thick-walled, sphincter-like organ, has a canal through which the sperm must travel (except for the mare and sow in which the sperm is deposited directly into the uterus). This canal becomes occluded, or shut off, from the vagina by viscous secretions that are produced under the influence of high progesterone. The viscous secretions are referred to as a *cervical seal* or *cervical plug*. Its function is to prevent the entrance of any contaminants when the embryo may be present. In fact, if the cervical seal is broken during pregnancy, spontaneous abortion generally follows. Under the influence of high estrogen, the cervix produces copious amounts of mucus to lubricate the vagina. This secretion also aids in preventing microorganisms from gaining entrance to the uterus by flushing the contaminants out.

The vagina serves as the copulatory organ in most species, and serves to expel the fetus. Unlike the uterus, the vagina is not a muscular organ, but it does have a well-developed mucosal layer. The vagina connects the cervix with the vulva, or the outside anatomical feature of the female. It consists of two labia, which, under normal circumstances, provide a closure protecting the female reproductive tract against entry by microorganisms.

## MALE

The primary structures of the male reproductive tract are the testes, penis, duct system, and accessory sex glands (Figure 11–6). The testes, which are analogous to the ovary in the female, are responsible for both gamete (sperm) production and production of the male sex steroids. Because the production of sperm is very temperature dependent, occurring at temperatures 4–6°C cooler than normal body temperature, the body has developed several mechanisms to maintain proper temperature control. The testes begin development in the abdominal cavity but descend from the abdomen to the scrotum,

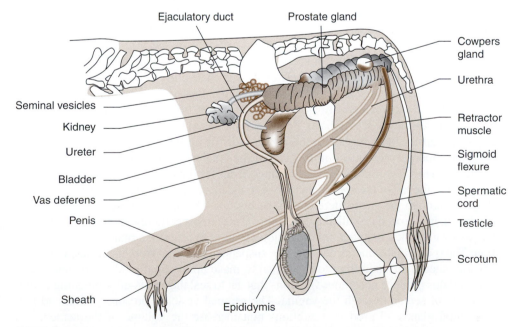

**FIGURE 11–6** Male reproductive tract. (*SOURCE:* Texas A&M University. Used with permission.)

usually during fetal development, through the inguinal canal. Infrequently, one or both of the testes may not descend and instead remain in the abdomen in a condition referred to as *cryptorchidism*. Because testosterone production can occur at body temperatures, the male will exhibit secondary sex characteristics of a male. However, sperm production requires cooler temperatures, so the affected testis is generally infertile. With unilateral cryptorchidism (one affected testis), the descended testis is fertile and often compensates for the lack of sperm production by the undescended testis. Therefore, the males are typically fertile. The condition appears to be hereditary, and, therefore, cryptorchid males should not be used for reproductive purposes.

The testes are suspended by the spermatic cord into the scrotum. The scrotum protects the testes and allows quick cooling for the testes to maintain proper temperature. The scrotum contains many thermosensors, which determine the outside temperature and cause several reactions, including scrotal sweating to dissipate the heat. In addition, neural connections from the scrotal thermosensors to the brain can affect the respiration of the male during heat stress. The increased respiration rate again dissipates the heat, thereby lowering testis temperature. The spermatic cord also connects the testis to the abdominal cavity through the inguinal canal. The spermatic cord contains the blood supply to the testes, neural connections, and a muscle that is capable of raising and lowering the testes. When the testes become cold, the cremaster muscle raises the testes closer to the body to warm them. The cremaster muscle works in concert with the tunica dartos, a smooth muscle layer underneath the skin of the scrotum. The tunica dartos is capable of holding the testes close to the body for sustained periods of time. When the testes become too warm, the cremaster muscle and tunica dartos relax, thereby allowing the testes to descend. In bulls and rams, the scrotum may become very pendulous during hot summer days. This is an example of how the body tries to compensate for environmental conditions to maintain processes such as sperm production. Another unique feature for temperature control is a specialized vascular system called the *pampiniform plexus*. This is a countercurrent blood supply that cools arterial blood as it travels from the body to the testes, and warms the blood traveling from the testes to the body. If the body is not able to disperse enough heat and the testes become too warm, sperm production may halt and pregnancy rates can be depressed for months.

The testes are responsible for sperm and testosterone production. Testosterone is responsible for the secondary sex characteristics in males. In contrast to females, which have limited ova production, the male has tremendous gamete production potential. Sperm is produced continuously, as opposed to females, who are born with the total number of gametes they will ever have. Sperm are produced by seminiferous tubules within the testis capsule, and migrate to the **epididymis** for further development, storage, and transport. Sperm collected from the head of the epididymis are typically immature and nonfertile. However, offspring have resulted from fertilization of eggs with sperm collected from the tail of the epididymis. Prior to ejaculation, the sperm travel from the epididymis through the ductus deferens to the urethra. The ductus deferens (vas deferens) can be cut in a procedure called a *vasectomy*. This almost certainly results in sterility but leaves sexual function intact. It is occasionally used to produce a male for estrus detection. The sperm in the ductus deferens are suspended in a fluid from the testes. However, several other organs add fluid to the sperm to make up the final product, **semen**. Boars and stallions produce gelatinous fractions that act to seal the cervix after breeding to prevent loss of semen through the vagina. The seminal vesicles, prostate gland, and bulbourethral glands all produce secretions that increase the volume of the semen, add nutrients to the semen, and aid in coagulation of the semen after ejaculation.

**Epididymis**

Duct connecting the testis with the ductus deferens. Responsible for sperm storage, transport, and maturation. It consists of a head, a body, and a tail.

**Semen**

Fluid from the male that contains sperm from the testis and secretions from several other reproductive organs.

The penis is the organ that deposits the semen in the vagina or cervix, depending on the species. The penis can be either vascular (stallions) or fibroelastic (bulls, boars, rams). The vascular penis enlarges during sexual excitement by retaining blood in specialized erectile tissue. The increased blood volume under high pressure causes erection. Following ejaculation, the blood is allowed to leave the organ, thereby decreasing blood pressure and volume in the penis.

With the exception of the stallion, the common farm animal species have a fibroelastic penis that exists in an S-shaped configuration inside the body until erection. During sexual excitement, the muscles responsible for retaining the penis in the sigmoidal flexure relax, allowing the penis to extend through the sheath. There is minimal increase in diameter of this type of penis. One modification of the penis is noted with the boar. The glans penis (the end of the penis) of the boar is corkscrew-shaped, such that it engages into the analogous corkscrew-shaped cervix of the sow. Therefore, the boar deposits the semen in the cervix rather than the vagina.

Because the penis is such a vascular organ, trauma can cause severe hemorrhaging. It is a fairly common injury for bulls to suffer a "broken penis," in which the penis is bent or kicked while extended. The blood from the penis leaks out and pools in the surrounding tissue. Many times the damage is irreparable and prevents the bull from ever mating naturally again. Semen can still be collected using an electroejaculator, which is explained later.

## PREGNANCY

Successful timing of ovulation and mating should result in pregnancy. During estrus, the female becomes receptive to the male, thereby encouraging copulation (the physical mating). The ova, recently released from the ovary and traveling down the oviduct, becomes fertilized in the ampulla region of the oviduct. The window of opportunity for fertilization is very narrow. If fertilization has not occurred within approximately 12 hours, the oocyte begins to degenerate, and the chance for successful fertilization decreases drastically. Fortunately, sperm travels to the ampulla very rapidly, as soon as several minutes after ejaculation. In addition, the cervix and uterus have numerous crypts and crevices that hold sperm. Thus sperm deposited for several hours or even days before ovulation have an opportunity to reach the oocyte.

Once the oocyte is fertilized, it becomes a one-celled embryo called a *zygote*. Cell division begins soon after fertilization and the zygote becomes a multicelled *morula* embryo. Further development transforms the morula embryo into a *blastocyst*. The blastocyst embryo is free-floating as it moves down the oviduct toward the uterine horns and body of the uterus. The day of implantation varies by species, but generally occurs between 14 and 40 days in farm animals. The blastocyst continues to develop, and cellular partitions become evident. The inner cell mass is the initial fetus, and the trophoblast partition develops into the placenta (Figure 11–7).

Relatively high levels of progesterone are required for maintenance of the pregnancy, and the CL supplies the progesterone during the early phases of pregnancy in all species. However, for pregnancy to be maintained, degradation of the CL must be prevented. The pregnancy must first be recognized by the female. If pregnancy is not recognized in time, the CL degrades, thereby removing the source of progesterone required to support the pregnancy. In the cow, ewe, and mare, the nonpregnant uterus produces *prostaglandin* $F_{2\alpha}$ ($PGF_{2\alpha}$), which travels to the ovary and causes *luteolysis*,

**Zygote**

Cell resulting from the fusion of the sperm and oocyte.

**Morula**

Early-stage embryo, after cell division multiplies cell numbers in the zygote.

**Blastocyst**

More differentiated embryo consisting of an inner cell mass, blastocoele, and a trophoblast.

**Prostaglandin**

A group of fatty acid hormones, one of which is prostaglandin $F_{2\alpha}$, which breaks down the corpus luteum allowing the female to return to estrus.

**Luteolysis**

Breakdown or degeneration of the corpus luteum. Occurs at the end of the luteal phase of the estrous cycle if pregnancy is not detected.

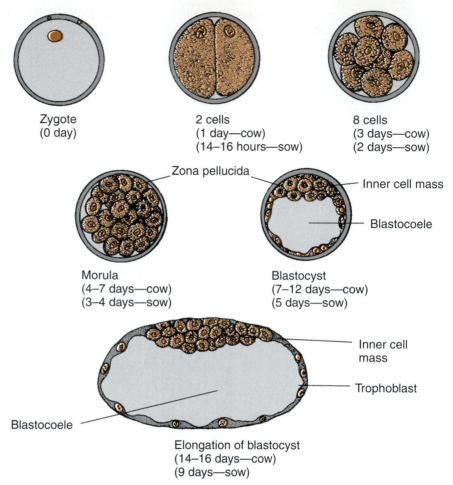

**FIGURE 11–7** Specific cleavage stages at given times after fertilization in the cow (281-day gestation) and the sow (114-day gestation). (*SOURCE:* Bearden and Fuquay, 1997, p. 91. Used with permission.)

or breakdown of the CL. The blastocyst of the species, however, produces proteins that block the production and release of $PGF_{2\alpha}$, thereby preventing luteolysis.

The trophoblastic cells of the blastocyst give rise to the various layers of the placenta. The fetal membranes are made up of the amnion, the chorion, the allantois, and the yolk sac (Figure 11–8). The amnion surrounds and cushions the fetus in amniotic fluid. In case of severe trauma, this amniotic fluid protects the fetus. The allanto-chorion, a fusion of the allantois and chorion, is in contact with the endometrium of the uterus. The yolk sac provides nutrients to the developing embryo/fetus. In some species such as birds, the yolk sac is prominent. In others, the yolk sac develops early but degenerates early in the pregnancy. In total, the placenta regulates the exchange of oxygen, nutrients, waste, and in some species, antibodies, between the mother and fetus. The actual connection between the uterus and placenta to facilitate these exchanges varies greatly by species, from a very diffuse connection as in the sow, to very localized connections seen in ruminants.

When more than one fetus is present, the membranes of the multiple fetuses typically fuse together. This is a normal occurrence and has no bearing on development, except in the cow. In cattle, the blood supplies for the multiple membranes fuse as

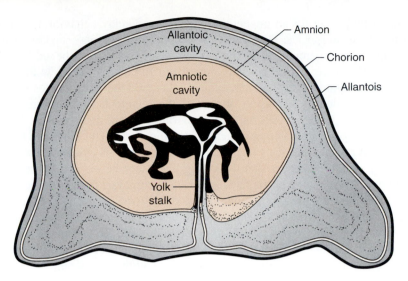

**FIGURE 11–8** Fetus with placenta. (*SOURCE:* Frandson and Spurgeon, 1992, p. 455.)

well, allowing hormones to transfer from one fetus to the other. If the twins are of the same sex, development continues normally. However, if the twin combination is a bull and heifer, the male hormones from the bull disrupt the development of the female. For this reason, about 10 out of 11 heifers born twin to a bull are sterile owing to incomplete reproductive tracts, a condition called freemartinism. A relatively inexpensive blood test has been developed to determine whether a newborn heifer is normal or a *freemartin*. The incidence of multiple births in beef cattle is relatively low but is about 6% in dairy cattle and has been increasing with higher milk production levels.

As a rule of thumb, approximately two thirds of fetal growth occurs in the last trimester of the pregnancy. Therefore, energetic demands of the female are greatest during this period and can create special nutritional requirements, which must be met. For example, ewes carrying more than a single fetus are susceptible to a condition called *pregnancy disease* in which the female is not capable of providing, or is not provided, enough nutrients to support the high demand of the fetuses. Likewise, cows and horses must be fed adequately such that energy reserves are appropriate at birthing. If not, the female is faced with high nutrient demands to support lactation without the necessary resources to draw on. Typically, these females are late to return to estrus, and subsequent breeding is delayed.

## PARTURITION

Parturition, the process of giving birth, is the culmination of pregnancy. Parturition is initiated by hormones secreted by the offspring. It is also possible to induce parturition using a cortisol-type drug such as dexamethasone. As the young move up the pelvic cavity and begin to push against the cervix, a neural connection from the cervix to the brain stimulates the release of oxytocin from the posterior pituitary gland. Oxytocin travels to the muscles of the uterus, causing contractions that aid in expelling the fetus. Most births occur naturally with no assistance. However, malpresentations and inappropriately sized offspring may necessitate intervention. Difficulty in birthing is called *dystocia*. The leading cause of dystocia is a fetus that is too large for the birth canal. This situation can be virtually eliminated by selecting for low-birth-weight sires and by using pelvic dimensions as a selection tool in the females.

**Freemartin**

Condition in cattle in which the female-born twin to a bull is infertile because of improper development of the female anatomy.

**Pregnancy disease**

Also referred to as *pregnancy toxemia*. A form of ketosis in females that occurs in late pregnancy because the female cannot eat enough of the feed she is provided or is not provided enough feed. Usually occurs in cases of multiple fetuses; common in sheep.

**Dystocia**

Birthing difficulty.

**Colostrum**

First milk given by the female after birth of the young.

Females preparing for an impending parturition display behavioral changes that vary from species to species. They generally become restless, may separate from other animals, and show extreme discomfort. The vulva swells and the udder becomes engorged. In the mare, *colostrum* may ooze from the teats in the few days prior to parturition. The ligaments around the tailhead relax, and the tail becomes relaxed.

Birth is preceded by expulsion of part of the allantochorion (referred to as the *waterbag*), which usually breaks as it releases fluid. As the fetus moves through the pelvic cavity, the placenta becomes detached from the uterus. This leads to a loss of connection for oxygen and nutrients to the fetus. Therefore, it is essential that labor progresses fairly rapidly to ensure that the young has adequate oxygen. In cows, mares, ewes, and does, the position of the fetus is normally front feet first, with the head lying between the two front legs. In pigs, either head- or tail-first presentation is normal.

The last phase of parturition is expulsion of the fetal membranes. This typically happens within a reasonable period of time after birth. In some circumstances, the placenta does not pass, resulting in a condition called *retained placenta*. Retained placenta in cattle used to be removed manually by a veterinarian, and the cow would be treated with antibiotics to prevent infection. However, the current dogma is to allow the placenta to slough off on its own, even up to a week after parturition. In horses, however, a retained placenta results in a serious health problem and should be treated as an emergency.

Following parturition, some species lick their young to clean and invigorate it, as well as to develop an identification bond with it. Others do not clean their young. Most farm animal species are relatively precocious and able to stand within minutes of birth. It is critical for the young to nurse as soon as possible after birth to obtain the antibodies contained in the milk. Colostrum, or first milk, differs from normal milk in that it is higher in protein, vitamin, mineral, and antibody concentrations. The antibodies, which are large protein molecules capable of fighting disease, are passively transferred to the young by being absorbed from the digestive tract. This confers *passive immunity* to the young. However, the gut begins to close to passage by the antibodies soon after birth. Therefore, if the young has not nursed soon after birth, it is incapable of taking advantage of the immunity of its mother. Many producers keep frozen colostrum in the event that a calf, lamb, or foal is born and has not nursed within 2 to 3 hours of birth. The frozen colostrum can be thawed and fed to the young through a tube placed in its stomach. Because the antibodies in the colostrum reflect the immunity of the mother, the more diseases the mother has been exposed to or vaccinated against, the more antibodies the young will gain. Therefore, it is better to collect colostrum from older, rather than young, animals.

The period between parturition and the onset of estrous activity is referred to as the *postpartum* period or *postpartum interval*. Management during this period is particularly important in cattle, which must become pregnant again within 80 days after calving to maintain the desired 12-month calving interval. In seasonal breeders, such as sheep and goats, this period has a much lower importance because season (actually, day length) has an overriding suppressive effect on reproduction. The females do not exhibit estrous activity until several months after parturition, regardless of other environmental factors. The postpartum interval is greatly affected by poor nutrition and by the presence and suckling of offspring. In combination, these two factors severely limit the onset of estrus. In swine, the suckling effect is so strong that producers wean the young early to stimulate the onset of estrus in a timely fashion. Beef cattle producers are now experimenting with short-term weaning to stimulate the onset of estrous activity, combined with increased feed intake to decrease the postpartum interval. Dairy cattle producers have a more difficult time managing the postpartum interval because they must keep the cow producing.

**Passive immunity**

Immunity conferred to an animal through preformed antibodies it receives from an outside source.

**Postpartum**

After parturition.

**Postpartum interval**

Period of time from parturition to first estrus in the female.

# ENVIRONMENTAL INFLUENCES ON REPRODUCTION

Many environmental conditions can have profound influences on reproductive function. The nutritional status of an animal greatly affects a female's ability to become pregnant, and it also influences a male's ability to exhibit the *libido* (sexual drive) and necessary sperm production to impregnate females. The nutritional effect can be divided into two categories:

1. Nutritional status: The long-term energy, protein, vitamin, and mineral regimen the animal has been exposed to; often reflected by how much body fat the animal is carrying.

2. Nutritional balance of an animal: The day-to-day consumption of proper nutrients in the proper amounts to support reproduction.

All nutrient groups can affect reproduction in animals; however, the primary nutritional limitation is energy. As a female loses body condition, or fat, her ability to become pregnant is limited. All species have a body condition level below which a female will not conceive. However, obesity can also become a limiting factor in reproductive function. Separation of nutritional status and nutritional balance is important, particularly in breeding females. For example, cows with questionable energy reserves (fat content) may be encouraged to ovulate and exhibit estrus by increasing energy intake for a short period prior to and during the breeding season. This process, called *flushing*, is commonly used in sheep, swine, and goats to increase ovulation rate.

Stress has a significant negative influence on fertility, although the mechanisms are not entirely understood. As increased knowledge of animal behavior is gained, increasing reproductivity by decreasing stress becomes an important concept for producers to adopt. Mixing unfamiliar animals together during the breeding season can present enough stress to decrease the incidence of estrus and reduce conception rates, particularly when combined with other environmental factors. (See Chapter 13, "Animal Behavior.")

Many species exhibit a very strong reproductive response to the length of day. This is not limited to the female. In sheep, the weight of the testes in the ram fluctuates throughout the year, being greatest in the fall during the breeding season, and the least in the nonbreeding season. However, the most significant effect of season is on the estrus of females during the breeding season and the lack of estrous cycles during the nonbreeding season. The mare is an example of a long-day seasonal breeder, so named because estrus is observed as days are becoming longer. Sheep, deer, elk, and goats are short-day breeders and onset of estrus is observed in the fall, as days become shorter. The exact mechanisms for seasonality are not known. However, the amount of sunlight present during the day is detected by the brain and transmitted to the pulse generator in the hypothalamus, which turns on the release of GnRH to start the endocrine cascade. Likewise, in the nonbreeding season, the pulse generator is shut down, and the ovary does not receive the gonadotropic support from the anterior pituitary gland for adequate development and maintenance of the follicles. Because seasonal breeders respond to length of light during the day, reproductive function can be manipulated by using artificial light to simulate the changing seasons. Therefore, it is possible to have ewes lambing in the fall, rather than the spring, and have mares foaling in late summer, rather than the spring. In addition, some breeds of season-breeding species are much less affected by day length than are other species. For example, most whiteface breeds of sheep have longer breeding

**Libido**
Sexual drive.

**Flushing**
Feeding extra feed to stimulate estrus and ovulation rates.

seasons than those of blackface breeds. Also, the closer to the equator one goes, the less seasonal the species become because of a more consistent day length throughout the year. Although cows are not recognized as being seasonal breeders, there is increasing evidence that fertility is higher at certain times of the year.

Species such as the sheep, goat, and mare can be described as being seasonally *polyestrus* because they exhibit more than one estrous cycle during a breeding season. However, some species such as the dog are seasonally *monoestrus* because they exhibit only one estrus during the breeding season.

## TECHNOLOGY AND REPRODUCTION

### ARTIFICIAL INSEMINATION

As early as the 1930s, it was discovered that semen collected from a male, frozen to extremely cold temperatures and thawed under proper conditions, was capable of impregnating a female. Inexpensive techniques have been developed, and artificial insemination (AI) (Figure 11–9) is now a common procedure in many species and is used, to some degree, in virtually all species. For example, in dairy cattle, approximately 80% of the calves born result from artificial insemination. This procedure allows dairy producers to use semen from the very best bulls in the country and throughout the world to increase the genetics of their herd. Artificial insemination is used on nearly 100% of the commercially grown turkeys because modern large-breasted turkey males are incapable of mating naturally. Several horse breed associations have recently changed registration rules so producers can take better advantage of this technology.

Successful insemination requires the acquisition of a quality sample of semen from a male, the detection of estrus in the female, and the ability to place the semen properly in the reproductive tract of the female. Semen is usually collected by causing the male to ejaculate into an *artificial vagina (AV)*, which has a receptacle to receive the sample (Figure 11–10). Another method of collecting semen in bulls, rams, and boars is to use an electroejaculator. The use of an electroejaculator is often preferred with males who refuse or are incapable of mounting a female naturally because of injury or age. This instrument consists of an electrical prod placed in the rectum of the male. The

**Polyestrus**

Exhibiting more than one estrous cycle.

**Monoestrus**

Exhibiting only one estrous cycle; for example, the bitch is seasonally monoestrus.

**Artificial vagina (AV)**

Device used to collect semen from a male. Following erection, the penis is directed into the artificial vagina and the male ejaculates, capturing the ejaculate in a reservoir of the AV.

**FIGURE 11–9** Rectovaginal method of artificial insemination in the cow. (*SOURCE:* From *Animal Agriculture: The Biology, Husbandry, and Use of Domestic Animals* by Cole and Garrett © 1974, 1980 by W. H. Freeman and Company. Used with permission.)

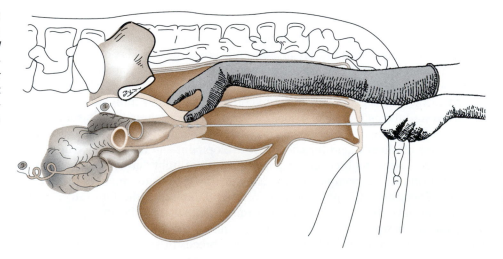

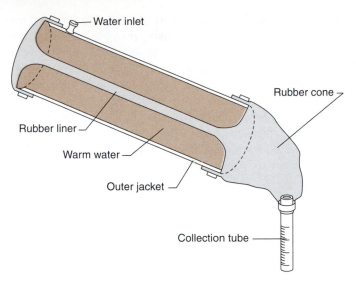

**FIGURE 11–10** Diagram of an artificial vagina. (*SOURCE: From Animal Agriculture: The Biology, Husbandry, and Use of Domestic Animals* by Cole and Garrett © 1974, 1980 by W. H. Freeman and Company. Used with permission.)

Labels on figure: Water inlet; Rubber cone; Rubber liner; Warm water; Outer jacket; Collection tube

**TABLE 11–2    Characteristics of Semen from Farm Animals**

| Characteristics | Cattle | Goats | Horses | Sheep | Swine |
|---|---|---|---|---|---|
| Volume of ejaculate (ml) | 5 | 0.8 | 30 | 1 | 225 |
| Sperm concentration ($10^9$/mL) | 1.1 | 2.4 | 0.15 | 3.0 | 0.2 |
| Motile sperm (%) | 70 | 80 | 70 | 75 | 60 |
| Ejaculates/week | 4 | 20 | 3 | 20 | 3 |
| Motile sperm (AI) | 10 | 60 | 100 | 120 | 1,200 |
| Females inseminated/ejaculate (AI) | 350 | 25 | 60 | 20 | 20 |

*SOURCE:* Compiled from Bearden and Fuquay, 1997, and Cole and Garrett, 1980.

electrical current stimulates contraction of muscles causing sperm to be expelled. By either method of collection, the sample is generally extended with a solution that contains proteins and buffers, and frozen in liquid nitrogen (for bulls and rams), or cooled sufficiently to prolong the life of the sperm (for boars and stallions). Frozen semen can be stored indefinitely. In fact, bull semen collected and frozen in the 1950s still retains enough fertile sperm to result in acceptable pregnancy rates. Characteristics of semen from various farm animal species are presented in Table 11–2.

Detection of estrus is one of the most difficult tasks in successful artificial insemination. Placement of the semen in the reproductive tract of the female must be coordinated with the timing of ovulation of the egg from the follicle on the ovary to ensure that fertilization can occur while the sperm is still alive and in the proper location. Fortunately, many species present specific behavioral activities that are characteristic of estrus. In the cow, a female allows other females, and males, to mount her approximately 12 hours prior to ovulation. Therefore, when a cow is standing to be mounted, she will be ready to be artificially inseminated in approximately 12 hours. The difference in timing between artificial and natural breeding is to accommodate for the placement of the semen. In natural breeding in the cow, semen is deposited in the vagina, and sperm must traverse through the cervix before entering the uterus. In artificial insemination, the sperm is deposited directly into the uterus, thereby bypassing the cervix. Mares have a characteristic "winking" of the vulva following urination when in estrus. Sows exhibit **lordosis**, a steadfast posture when pressure is applied to the back, and a swollen vulva that turns red in color. Therefore, to detect estrus in the sow, all one must do is press heavily on the back. If the sow walks away, she is not in estrus. If the sow not only stands but also refuses to move when pressure is applied, she is likely in estrus. The ewe presents a particular problem. Ewes do not exhibit overt signs of estrus and, therefore, are very difficult to identify for breeding. The female dog sloughs the lining of the uterus, losing blood and tissue for the duration of estrus. Cats typically exhibit marked behavioral changes.

### ESTROUS SYNCHRONIZATION

To combat the problem of estrous detection, and to minimize the labor involved in artificial insemination in cattle, drugs have been developed to synchronize estrous cycles in females. In this manner, several females can be given the proper injections to cause ovulation to occur simultaneously. There are three basic approaches to synchronizing estrus.

One approach is to give injections of prostaglandin $F_{2\alpha}$ ($PGF_{2\alpha}$) during the luteal phase causing regression of the corpus luteum. $PGF_{2\alpha}$ is a natural hormone produced and released by the uterus when pregnancy is not detected. The regression of the CL removes the inhibitory effect of progesterone on the release of LH and FSH. The levels of LH and FSH increase dramatically, allowing development of the follicles to occur. Therefore, females with a functional CL ovulate approximately 72 hours after the injection of $PGF_{2\alpha}$. If there was no functional CL present, the cow may be given another injection approximately 9 days after the first. The second injection is usually effective in evoking estrus.

A second method of estrous synchronization is the use of a combination of estrogen and progesterone, given as a slow-release pellet implanted in the ear of the female. The actions of the progesterone and estrogen block the release of LH and FSH, thereby interrupting the estrous cycle of the female. Upon removal of the implant, the inhibitory effects of progesterone and estrogen are removed, and the female exhibits estrus.

A more recent method incorporates the use of GnRH to time not only estrus, but ovulation as well. This method appears to be gaining popularity, particularly with dairy cattle producers.

### EMBRYO TRANSFER

**Embryo transfer** (ET) is the process of collecting fertilized embryos from one female and placing them in another for further development (Figure 11–11). This technology became available to livestock producers in the 1970s. The value of embryo transfer is that it allows for the production of many more offspring from superior females. ET is similar to, but less extensive than, the increased breeding potential from males using artificial insemination. Although used to some extent in several species, it has been most widely used in the cattle industry. There is less incentive for ET use with swine because they are litter producers and have a much shorter **generation interval**. With sheep and goats, there is less incentive for ET use because it is more difficult to make the procedure cost effective. The following discussion focuses on the cow for that reason.

**Lordosis**
Posture assumed by females in estrus such that they resist pressure applied to the back.

**Embryo transfer**
The process of transferring fertilized embryos from one female to another female.

**Generation interval**
The average age of animals within a species when they bear their first offspring.

**FIGURE 11–11** The dairy cow (upper right) is the genetic mother of the 10 calves. She was superovulated, and the embryos were recovered from her uterus 1 week after conception. After 3 to 10 hours of culture in vitro, the embryos were transferred to the uteri of the 10 recipient cows (left) for gestation to term. (Photo courtesy of George E. Seidel, Jr., Colorado State University.)

The process of embryo transfer begins with the superovulation of a superior female. Superovulation is the process of supplying exogenous FSH such that the ovary receives the gonadotropic support so more than the normal number of follicles reach the preovulatory stage. In fact, the FSH prevents follicles that would normally undergo atresia from dying. Therefore, the FSH does not increase the number of eggs being produced, it just keeps the follicles produced each month in a productive and developing state, therefore resulting in more ovulating, fertilizable eggs. Meanwhile, as the donor cow is being superovulated, the cows being used to receive the embryos, the *recipients*, receive hormone injections to synchronize their estrous cycles with the donor cow. When the follicles mature in the donor cow's ovaries, the donor and recipients release their eggs at approximately the same time. Insemination of the cow is timed to occur at approximately 12 hours after ovulation, thereby resulting in fertilization of the majority, if not all, of the eggs.

The embryos stay in the donor cow for about 7 days—the most viable stage for transfer—and are then flushed out of the oviducts, as they have not yet implanted. The embryos are identified under a microscope and placed, one by one, into a recipient female. If all goes well, the embryo develops into a fetus in the recipient, who produces a calf that carries the genetics of the sire and the donor female. Commercial embryo transfer companies boast greater than 60% pregnancy rates with good-quality embryos. In addition, if more embryos are collected than there are available recipients, it is possible to freeze the embryos indefinitely. The producer thereby has more flexibility in planning calvings, can hold on to frozen embryos to determine the quality of full siblings, and can maximize the use of recipients. Embryo transfer greatly increases the genetic impact that a cow can have on a herd by increasing the number of offspring born each year.

**Recipients**
Females used to carry the embryos of a donor animal throughout gestation.

## "IN VITRO" FERTILIZATION

Although embryo transfer is used extensively across the United States in cattle, it does have disadvantages. For example, ET can only be performed when a cow is open. Therefore, it is necessary to keep a superior cow nonpregnant for long periods of time to complete the procedure of superovulation, breeding, and flushing of the embryos. Second, some cows do not respond to the hormones as expected. When the ovaries are understimulated, no eggs are produced. When they are overstimulated, too many eggs are often produced and all may be nonfertile.

Recent technology allows the collection of the eggs directly from the ovary, before the follicles have fully developed. These ova are then fertilized *in vitro* and allowed to develop to stages appropriate for placement into recipients. This method has many advantages, including these:

**In vitro**

In a test tube or other environment outside the body.

- Collection of ova during pregnancy
- Use of different sires on the ova from each collection
- Frequent collections, as often as weekly

There are some drawbacks to collecting eggs directly from the ovary. The technology is not perfected. In addition, it is relatively expensive and time-consuming. However, in vitro pregnancy rates are approaching the success observed with embryo transfer. Increased pregnancy rates will be expected with advances in the technology and when more people become proficient at the procedure.

# REPRODUCTION IN POULTRY AND BIRDS

## HEN

The goal of reproducing young is the same in poultry as in mammalian species. However, the method varies in that the young are not carried inside the body. Rather, the eggs are laid outside the body where they are incubated until the young are ready to emerge (hatch) from the shell. Figure 11–12 shows the reproductive tract of the hen. The discussion focuses on the hen; however, the process is similar for other avian species.

Poultry females (hens) have only one functional ovary, unlike mammals, which have two. Only the left ovary in the hen is functional. When a female chick hatches, approximately 4,000 ova are attached to the ovary, each enclosed in a follicle. The follicle contains the blood vessels that will nourish the egg when it is time for it to mature. The ova begin maturing at sexual maturity. Each mature ovum is successively released from its follicle by rupture of the follicle wall. This release from the follicle is called *ovulation*. The rupture occurs along a line called the *stigma*. The ovum (yolk) moves into the oviduct where the additional parts of the egg are added.

In the oviduct, the egg moves from region to region where very specific steps in egg formation occur (Figure 11–13). The first section of the oviduct, the infundibulum, captures the yolk from the body cavity after ovulation. In the infundibulum, fertilization occurs if sperm are present. Sperm can remain viable for up to several weeks in some species. The total time the yolk spends in the infundibulum is 30 minutes or less. The yolk moves into the second portion of the oviduct, the magnum. The yolk stays in the magnum for 2 to 3 hours, during which time the thick portion of the albumen, or "white" of the egg, is added. The albumen is deposited around the yolk. In the next portion of the

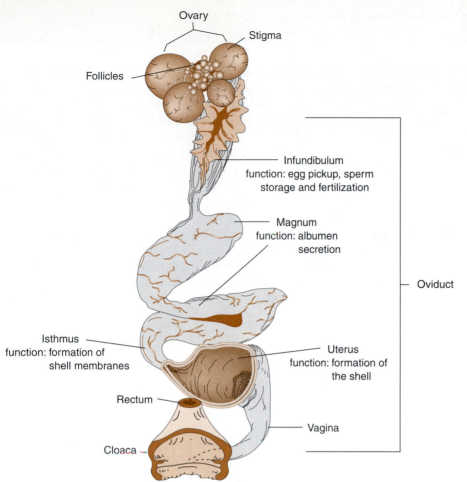

Ovary

Stigma

Follicles

Infundibulum
function: egg pickup, sperm
storage and fertilization

Magnum
function: albumen
secretion

Oviduct

Isthmus
function: formation of
shell membranes

Uterus
function: formation of
the shell

Rectum

Vagina

Cloaca

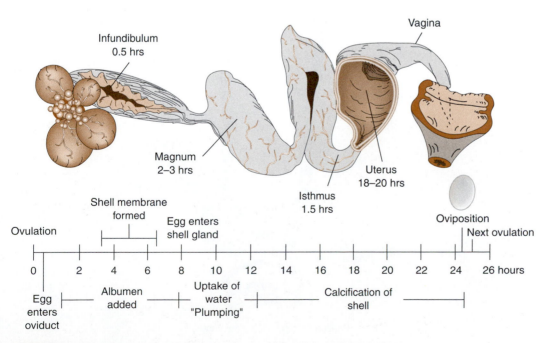

Vagina

Infundibulum
0.5 hrs

Magnum
2–3 hrs

Uterus
18–20 hrs

Isthmus
1.5 hrs

Shell membrane
formed

Egg enters
shell gland

Oviposition

Ovulation

Next ovulation

| 0 | 2 | 4 | 6 | 8 | 10 | 12 | 14 | 16 | 18 | 20 | 22 | 24 | 26 hours |

Egg
enters
oviduct

Albumen
added

Uptake of
water
"Plumping"

Calcification of
shell

FIGURE 11–13  Steps in egg production. (SOURCE: Adapted from Geisert, 1999.)

oviduct, the isthmus, the shell membranes are added. During the 1.5 hours that the egg is in the isthmus, it takes up water and mineral salts and the inner and outer membranes of the shell are added. Calcification of the shell occurs in the uterus, or shell gland, which is the next portion of the oviduct. In addition, the remainder of the albumen is added in the uterus. Early in the 18 to 20 hours the egg is in the uterus, it undergoes *plumping*, a process of adding water and minerals to the egg through the previously formed membranes. Once the plumping process is over, the calcification of the shell takes place and shell pigment is added. After the shell formation is complete, the egg moves from the uterus, through the vagina, where the bloom or cuticle is added, through the shell and then through the cloaca for oviposition, which is expulsion to the outside of the hen's body. This process is commonly referred to as "laying" the egg. The total time from ovulation to the laying of the egg is slightly more than 24 hours. Approximately 30 minutes after the egg is laid, another ovulation occurs and the process is repeated. Fertile eggs may be incubated by the hen or collected for artificial incubation. In the case of the vast majority of eggs laid by chickens, the eggs are not fertile and are processed as a food source.

## Cock

Figure 11–14 shows the reproductive organs in the male bird. Hens lay eggs in the absence of males, and it is not necessary for sperm to be present in the female's reproduc-

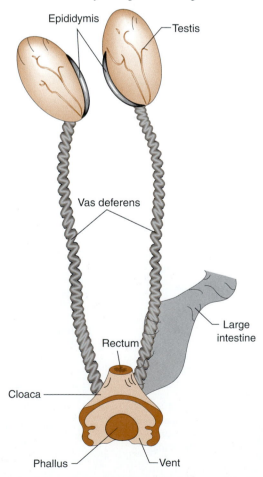

**FIGURE 11–14**   Reproductive organs in the male avian. (*SOURCE:* Adapted from Geisert, 1999.)

tive tract for an egg to form. For this reason, no males are found with chickens in commercial laying houses. However, if reproduction of the species is desired, sperm must be present in the infundibulum during the process of egg formation. As observed for the hen, there are substantial differences in the reproductive tract of male birds as compared to that of mammals. The differences begin with the testes, which are found inside the body cavity rather than outside it. The sperm cells move from the testes into the vas deferens, which leads to the cloaca and terminates in small papillae found in the wall of the cloaca. The male bird has a phallus, but it is rudimentary compared to that of the mammalian species. During copulation, the sperm are transferred from the papillae to the phallus and deposited in the oviduct via the cloaca of the female.

## SUMMARY AND CONCLUSION

Maintaining reproductive function is central to the success of livestock enterprises that rely on the production of young. The success of reproduction relies on absolute coordination among many systems of the body. Environmental factors may radically influence the expression of reproductive behavior, release of critical hormones, success of implantation of the embryo, and maintenance of pregnancy. Only when environmental factors are most favorable are animals able to express reproductive success fully.

Many tools of modern technology are improving our ability to manage the reproductive process. However, producers must also work diligently to increase the genetic potential for high reproductive function. A conflict can arise for the producer because of the advent of some of the emerging technologies. For example, it is not in the long-term best interest of the producer or the industry to use embryo transfer on a "good" cow that is not capable of maintaining a pregnancy. Likewise, using a stallion with a low sperm count may eventually decrease the fertility of numerous offspring, a defect that may take generations to correct. Opportunities for increasing production through advances in reproductive technology are tremendous. However, producers wishing to pursue this strategy have the responsibility to use it wisely.

## STUDY QUESTIONS

1. Why is reproduction so important to the producers of domestic livestock?
2. Describe the interplay among the many factors that affect reproductive function in the animal species.
3. What are gonads? What are their main categories of function? What does the brain have to do with the proper function of the gonads?
4. What is puberty? What does puberty have to do with secondary sex characteristics?
5. What does it mean for a hormone to be episodic? What are two examples of episodic hormones?
6. What is the hypothalamus? What does it have to do with reproduction? What is the anterior pituitary? What does it have to do with reproduction?
7. Make a table. Down the side of the table, list each of the hormones that has a function in reproduction of the female. Across the top, make categories for origin, target, and function. Fill in the table.

8. Make a table. Down the side, list each of the hormones that has a function in reproduction of the male. Across the top, make categories for origin, target, and function. Fill in the table.

9. Describe the different types of uteri of different species of animals.

10. List and briefly discuss the functions of the various portions of the female reproductive anatomy.

11. List and briefly discuss the functions of the various portions of the male reproductive anatomy.

12. What is the difference in the type of penis a horse has compared to that of a bull? What about a boar?

13. Why is it so important for the testicles to be outside the body cavity? Discuss temperature regulation of the testes.

14. What is cryptorchidism? What are the consequences to fertility of an animal that is a cryptorchid?

15. Why would it not be good for a bull to produce a gelatinous fraction in his semen like the boar and stallion do?

16. What can be the consequences of a male animal getting kicked during mating? What is the likely outcome of this? Is there any way to still retain the breeding services of the male, especially a bull?

17. Depending on the species, it is possible for mating to take place as much as 2 to 3 days before ovulation and a pregnancy still ensue. However, a natural service that takes place 2 to 3 days after ovulation will not result in a pregnancy. What is the difference?

18. Describe the role of the CL in pregnancy. Do you think it would be possible for a physical mishap to cause a disruption of the CL and therefore terminate a pregnancy? Can you think of what such a mishap might be?

19. Describe the role of the placenta. What happens to the placenta in multiple births? In what species does this cause a problem? What is the problem and what is it called?

20. Describe the reproductive and environmental interaction that leads to pregnancy disease, especially in sheep.

21. What are the steps in parturition?

22. Describe the value of colostrum to baby animals. Why must it be consumed soon after birth?

23. What is important about the management of the postpartum period in livestock? What are the major factors that affect it?

24. Describe the difference between nutritional status and nutritional balance. What effect does each have on reproduction?

25. What effect does stress have on reproduction in livestock? Do you think this could mean that handling systems for managing open females should be designed with stress reduction in mind?

26. What effect does season of the year have on reproduction? Is this more of an issue with some species than with others? If yes, which species?

**27.** What is the value of artificial insemination to a breeding program? How is semen collected for use in artificial insemination? How long will it last if properly processed, frozen, and stored? What are the challenges to successful AI?

**28.** What are some of the signs that various species display during estrus?

**29.** What is the value of estrous synchronization? Describe the three methods of estrous synchronization discussed in the text.

**30.** What is the value of embryo transfer? Can it be used on all the common livestock species? On what species is it most commonly practiced? Why?

**31.** Describe the procedure for embryo transfer in cattle.

**32.** What are the differences and similarities between embryo transfer and in vitro fertilization?

**33.** What is the caution that producers must exercise in order to protect the long-term integrity of the genetic pool of livestock if they use technology to enhance reproduction?

## REFERENCES

*Author's note:* This chapter was prepared in part by Dr. T. L. Beckett, California Polytechnic State University, San Luis Obispo, CA.

Bearden, H. J., and J. W. Fuquay. 1997. *Applied animal reproduction.* 4th ed. Upper Saddle River, NJ: Prentice Hall.

Bearden, H. J., J. W. Fuquay, and S. T. Willard. 2003. *Applied animal reproduction.* 6th ed. Upper Saddle River, NJ: Prentice Hall.

Bone, J. F. 1999. *Animal anatomy and physiology.* 3rd ed. Upper Saddle River, NJ: Prentice Hall.

Cole, H. H., and W. N. Garrett, eds. 1980. *Animal agriculture: The biology, husbandry, and use of domestic animals.* 2nd ed. New York: W. H. Freeman.

Frandson, R. D., and T. L. Spurgeon. 1992. *Anatomy and physiology of farm animals.* 5th ed. Philadelphia: Lea and Febiger.

Geisert, R. D. 1999. *Learning reproduction in farm animals.* A multimedia CD-ROM. Stillwater: Oklahoma State University.

Senger, P. L. 1997. *Pathways to pregnancy and parturition.* Pullman, WA: Current Conceptions, Inc.

# 12 CHAPTER

# Lactation

## LEARNING OBJECTIVES

After you have studied this chapter, you should be able to:

1. Describe the process of lactation.

2. List the components of milk.

3. Identify the major components of the mammary gland.

4. Identify and describe a typical lactation curve.

5. Explain the problem of lactose intolerance.

6. Describe the process by which various components are added to milk.

7. Compare and contrast the effect that species has on milk composition.

8. Explain the process by which BST increases milk production.

## KEY TERMS

Alveoli
Bovine somatotropin (BST)
Calcium
Casein
Colostrum
Involution
Lactation

Lactation curve
Lactogenesis
Lactose
Lactose intolerance
Lipoproteins
Mastitis
Milk
Myoepithelial cells

Oxytocin
Parturient paresis
Parturition
Pituitary gland
Saturated fats
Secretory cells
Unsaturated fats

# INTRODUCTION

*Lactation,* the process of producing milk, occurs in all mammalian species. In fact, the production of milk following *parturition*, the process of giving birth, is a major, if not sole, factor that defines the difference between mammals and other classes of animals. All mammalian species rely on milk for the nourishment of young. In addition, humans rely on milk throughout life as a readily available source of many nutrients including calcium, protein, fat, carbohydrates, and many trace minerals and vitamins. Thus the role of lactation in animal production is extremely important, not only for nourishment of young animals, but as a primary product marketed for human consumption. However, the production of milk is an extremely complicated process, and optimum production requires knowledge in all disciplines associated with animal production.

*Milk* is defined as the liquid—comprised of water, triglycerides, lactose, protein, minerals, and vitamins—that is produced and secreted by the mammary glands of the females in mammalian species. Milk varies in composition and consistency both among species and within species. The primary sources for milk consumed by humans include goats, sheep, water buffalos, and, of course, cows. However, around the world, many other species, including camels, llamas, and yaks, are utilized to produce milk for dairy products. The basic structures of the mammary gland in all species, both on a cellular and anatomical level, are remarkably similar.

# MAMMARY GLAND DEVELOPMENT, ANATOMY, AND FUNCTION

Development of the mammary gland, irrespective of the species, requires a coordinated effort of the endocrine, anatomical, and physiological systems of the female. In the prenatal period, the basic structures of the mammary gland develop. From birth to puberty, the mammary gland develops at about the same rate as the rest of the body, primarily under the influence of growth hormone. At puberty and with each subsequent estrus cycle, the female sex hormones progesterone and estradiol stimulate growth in the mammary gland at a rate more rapid than the rest of the body. In the pregnant female, progesterone stimulates the lobule-alveolar development, which is required to prepare the gland for milk synthesis. Close to parturition the hormone prolactin, necessary for initiation and maintaining lactation, is secreted. At parturition, the blood levels of progesterone and estrogen decrease abruptly and *lactogenesis* occurs. There are two stages of lactogenesis. In *stage I lactogenesis*, immunoglobulin uptake occurs and *colostrum* is formed, much of this prior to parturition. In *stage II lactogenesis*, copious milk secretion begins, and during the next few days colostrum production is shut down and milk of normal composition is produced.

The primary structures of the mammary gland responsible for producing milk are the *alveoli.* The alveoli are spherical (Figure 12–1) and capable of storing the milk produced by the *secretory cells* that surround the outside of the lumen, not unlike a water-filled balloon in which the water can be pushed out the opening. Milk is produced by these specialized secretory cells, which are housed in the alveoli and contain the necessary enzymes to produce the components unique to milk. These secretory cells have three primary functions: (1) to absorb the necessary precursors (nutrients) from the bloodstream; (2) to transform these nutrients into the lactate, fat, and protein in milk; and (3) to transfer the newly synthesized milk into the lumen of the alveolus. In addition, many minerals and vitamins are absorbed from the bloodstream and are combined

## Lactation
The process of producing milk.

## Parturition
The process of giving birth.

## Milk
The normal secretion of the mammary glands of female mammals.

## Lactogenesis
The series of cellular changes during which mammary epithelial cells convert from the nonsecretory state to the secretory state.

## Colostrum
Specialized milk produced in the early days following parturition to provide extra nutrients and immune function to the young.

## Alveoli
Spherical-shaped structures making up the primary component of the mammary gland. Consist of a lumen, which is surrounded by secretory cells that produce the milk, and myoepithelial cells, which contract to squeeze the milk out of the lumen into the mammary gland ducts.

## Secretory cells
The functional units of the alveoli that absorb nutrients, make the milk components, and transport the milk into the lumen of the alveoli.

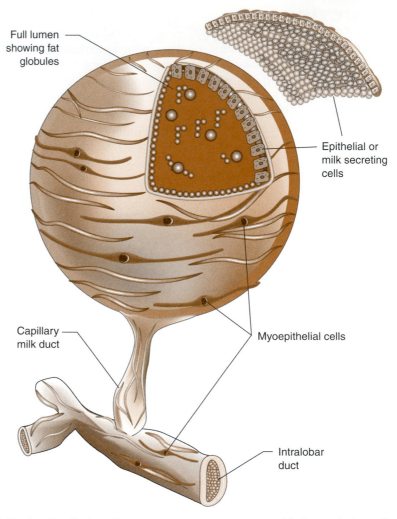

Full lumen showing fat globules

Epithelial or milk secreting cells

Myoepithelial cells

Capillary milk duct

Intralobar duct

**FIGURE 12–1**    Alveoli—the primary mammary structure responsible for producing milk. (*SOURCE:* Turner, 1969. Used with permission.)

### Oxytocin

A hormone produced in the hypothalamus and secreted by the posterior pituitary gland that stimulates contraction of muscle cells in the uterus to aid in parturition and the myoepithelial cells surrounding the alveoli to force milk out of the lumen into the mammary gland ducts.

### Pituitary gland

A small gland hanging just below the brain. Made up of the posterior pituitary gland and anterior pituitary gland.

### Myoepithelial cells

Specialized muscle cells that surround the secretory cells of the alveoli and cause them to contract when stimulated by oxytocin.

with the synthesized components prior to discharge into the alveolus. Fat, carbohydrate, protein, minerals, and vitamins are secreted in a liquid form. Therefore, the components are diluted in water and the milk remains in the alveolus until the milk is released.

Milk release, or letdown, is facilitated by the action of a hormone, *oxytocin*, released from the *pituitary gland*, which sits at the base of the brain. The oxytocin acts on specialized muscle cells called *myoepithelial cells*, which surround the secretory cells, causing them to contract. As the muscles contract, milk is squeezed from the lumen of the alveolus (Figure 12–2). Upon discharge of milk, the alveolus deflates, making room for more milk to be produced. The release of oxytocin depends on the female recognizing a stimulus, most often stimulation of the teats, such as the young beginning to nurse or the udder being washed. However, the female may be conditioned to respond to other factors associated with milk removal, such as the visual appearance of its offspring, or even the sound of a vacuum pump used to draw the milk from mammary glands at milking time.

The milk produced in the secretory cells is transferred directly into the lumen of the alveoli. The alveoli are arranged in lobules, and the contents are drained by a complex

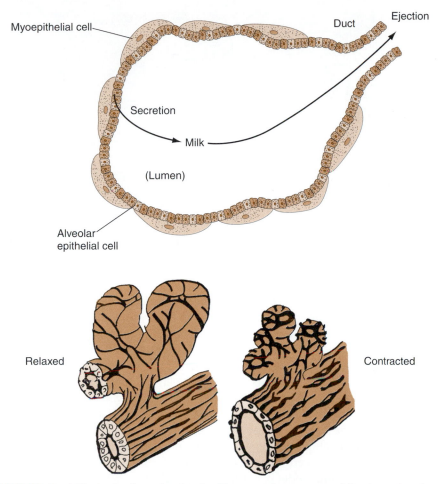

**FIGURE 12–2**  Milk ejection from the alveolus. The oxytocin acts on specialized muscle cells called myoepithelial cells, which surround the secretory cells, causing them to contract. As the muscles contract, milk is squeezed from the lumen of the alveolus. (*SOURCES:* [top] From HUMAN PHYSIOLOGY: FROM CELLS TO SYSTEMS, 2nd edition, by L. Sherwood. © 1993. Reprinted with permission of Brooks/Cole Publishing, a division of International Thomson Publishing. Fax 800 730–2215. [bottom] Mepham, 1987. Used with permission.)

ductwork system that carries the milk to the skin surface. The ducts terminate at the skin surface in structures referred to as *nipples* or *teats,* depending on the species. In cows and goats, the ducts terminate into a *gland cistern,* which is connected to a teat cistern and finally to a streak canal. It is the streak canal through which the milk ultimately leaves the female in these species. Surrounding the streak canal is a muscular ring, or teat sphincter, that is responsible for closing the streak canal after milk is released. In addition to holding the milk in the teat canal, this muscular ring, which is normally in a constricted configuration, prevents foreign materials from entering the mammary gland. This includes preventing the entrance of microbes, thereby providing the first line of defense in protecting the integrity of the mammary gland from infection. During milk release, the teat sphincter relaxes, allowing milk to pass through the canal. However, once the teat sphincter releases, it remains that way for 15 to 60 minutes. Therefore, the teat canal is susceptible to entry by microorganisms for an hour after each milking. For this reason,

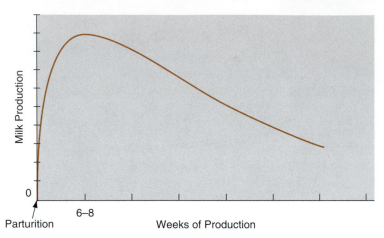

**FIGURE 12–3**   Lactation curve of a dairy cow. Volume of milk production generally increases for 6 to 8 weeks. After the peak, production declines until the end of the lactation.

cows must be kept standing immediately after milking to keep bacteria from entering the mammary gland. Most dairy producers accomplish this by providing feed after milking.

The prevention of bacterial entry is essential due to the susceptibility to ***mastitis***, an inflammation of the mammary gland most often attributed to bacterial infection. Mastitis is one of the leading causes of illness in dairy cattle and leads to the use of antibiotics that gain entrance to the milk. Because the law requires that antibiotics cannot be present in milk consumed by humans (referred to as *zero tolerance*), the use of antibiotics severely limits profitability by requiring milk from treated cows to be wasted and also increases the liability of the producer. Further, if not treated appropriately, mastitis can result in cellular damage to the mammary gland, reducing the animal's ability to produce milk. Severe cases can be fatal.

Although cows begin producing milk immediately after calving, the production of milk does not peak until several weeks after parturition. The reason for this is not known, but is likely owing to a physiological and biochemical adjustment period required by the animal for peak production. Generally, cows reach their peak production at 50 to 70 days postcalving. After that point, the secretory cells become less functional and a progressive decline in milk production is observed. The mammary gland undergoes a gradual ***involution***, decreasing in weight, volume, and productivity. However, this involution is reversed with the next pregnancy, and successive lactations often are progressive in production. This increase and decline of production during lactation is referred to as the ***lactation curve*** (Figure 12–3). The curves for different species have their own characteristic shapes. The rate at which a female's milk production declines after reaching its peak is referred to as *persistency*. A slower rate of decline, which produces more milk for longer periods, is referred to as *more persistent*.

## COMPONENTS OF MILK

### CARBOHYDRATES

***Lactose***, the primary carbohydrate found in the milk of virtually all species, is only produced by secretory cells in the mammary gland. Lactose is synthesized in the secretory cell by combining glucose with another 6-carbon carbohydrate—galactose. Because the

**Mastitis**

Inflammation of the mammary gland, most often caused by bacterial infection.

**Involution**

An organ's return to its normal state or normal size.

**Lactation curve**

A plot of milk production over the life of the lactation period.

**Lactose**

Disaccharide comprised of one glucose unit attached to a galactose sugar.

mammary gland is incapable of synthesizing glucose, all of the glucose used to produce lactose must be provided by the circulating blood. Therefore, one of the primary limiting factors in milk component synthesis is the availability of glucose. This requirement puts a tremendous metabolic demand on the cow, rendering her susceptible to several metabolic conditions during times that milk production is high.

As discussed in the chapter on nutrition, ruminant animals do not absorb much glucose. Instead, the microorganisms in the rumen break down most of the soluble carbohydrates into volatile fatty acids (VFAs). These VFAs (acetate, propionate, and butyrate) are absorbed through the rumen wall and travel through the bloodstream to the liver. The liver is able to make glucose out of propionate but is not able to make glucose out of acetate. Acetate and butyrate can be used to make milk fat or can be oxidized for energy to run milk synthesis. Therefore, the glucose that becomes available for the mammary gland depends greatly on what is supplied to the liver. Feeds (e.g., corn and other grains) broken down into propionate increase the production of glucose in the liver, thereby providing more glucose to the mammary gland to meet its potential to make lactose.

Because lactose is a disaccharide, it must be degraded in the small intestine after consumption. An enzyme, lactase, is present in the small intestine of most species to degrade the lactose into glucose and galactose, which are finally available for absorption. However, in individuals with less lactase, the lactose is not degraded, resulting in gastrointestinal distress known as *lactose intolerance*. Individuals who suffer from lactose intolerance generally experience cramps and diarrhea following milk consumption. Lactose intolerance is a genetic condition. Its incidence is lowest in people of Scandinavian and western European descent, in which only 2–8% of the population suffers from the condition. However, about 90% of Japanese, Thais, and Filipinos are afflicted to some extent. Estimates in the United States suggest that approximately 70% of the African American population is also affected. Fortunately, the condition can be alleviated through the consumption of cultured products such as yogurt. In addition, lactase pills can be taken as milk is consumed and most of the lactose is degraded in the small intestine before any deleterious effects are felt. Recent studies suggest that even individuals with severe lactose intolerance are able to consume small amounts of milk without symptoms.

**Lactose intolerance**

Condition in humans where the lack of the enzyme lactase leads to stomach distress following milk consumption.

## PROTEIN

The protein in milk exists in many different forms. The primary protein in milk is *casein*, which comprises more than 80% of the total protein in milk. As with lactose, the mammary gland is the only tissue capable of making casein. Casein exists in many different forms that can be further subdivided by biochemical characteristics. The casein proteins carry negative charges owing to the phosphate groups held in association with the proteins. To secrete these negatively charged proteins, the secretory cells draw calcium out of the bloodstream. Therefore, the negatively charged proteins are largely responsible for the high concentrations of calcium found in milk. Caseins are essential to the cheese-making process, and efforts are currently under way to select animals with casein fractions in their milk that will make better cheese.

In addition to casein are the milk serum proteins (lactoglobulin, lactalbumin, and immunoglobulins). The concentration of these proteins is much lower in milk compared with casein, but still comprises approximately 18% of the protein in milk from cows. When digested, milk serum proteins provide high concentrations of amino acids that are available for the consumer of the milk to use in synthesis of other proteins. As a critical component in the synthesis of casein, lactalbumin is frequently secreted into milk. Beta-lactoglobulin likely plays a role in the binding of proteins.

**Casein**

The major protein of milk.

The concentration of immunoglobulins varies greatly immediately after parturition. The first milk produced by mammals is referred to as colostrum. This milk is extremely high in protein concentration because of the large proportion of immunoglobulins (antibodies) secreted into the milk. These immunoglobulins increase the ability of the offspring to resist pathogens in the environment. However, the functionality of the immunoglobulins depends on the absorption of the antibodies from the small intestine. The young of species depending on this *passive* form of immunity have a short period of time immediately after birth during which the immunoglobulins are absorbed directly from the small intestine. However, gut closure occurs rapidly, most often within the first 24 hours. Therefore, the offspring must be provided this colostrum milk as soon after birth as possible. In addition, some species also confer immunity to their offspring through the production of specific antibodies in the mammary gland against antigens introduced by the young during nursing. The nursing young not only benefits from a generalized immunity, but also from a specific immunity for the particular pathogens present in the environment. Because these immunoglobulins are proteins, the protein concentration in colostrum milk is extremely high. However, by 3 to 4 days postparturition, the milk has returned to normal composition. This timing corresponds to the gut closure in the offspring.

### LIPIDS

The percentage and chemical makeup of fats in milk varies greatly by species. For example, milk may contain virtually no fat (in black rhinoceros) or up to 50% fat by weight (in seals). In addition, the amount of fat in milk is influenced by feed type and stage of lactation, among other factors. Of the primary lipids—triglycerides, cholesterol, and phospholipids—the vast majority of lipid in milk is made up of triglycerides (>90%). The fatty acids that make up triglycerides vary in length and degree of saturation. Fatty acids in milk derive from two sources: (1) fatty acids extracted from blood-borne *lipoproteins*, whose composition varies with diet, and (2) production in the mammary gland. Thus the source of the fatty acids in milk affects the characteristics of milk fat.

In ruminant species, the vast majority of **unsaturated fats** are saturated by the microorganisms in the rumen. Therefore, when the fats are absorbed in the small intestine, most are saturated. These fats are then packaged into lipoproteins and travel throughout the body. The mammary gland is then offered predominantly **saturated fats** to incorporate into milk. Because humans have a dietary essential requirement for at least one unsaturated fat—linoleic acid—some vegetable oil is traditionally added to human infant milk formulas based on cow's milk. By adding the vegetable oil, sufficient linoleic acid is incorporated into the formula to satisfy the needs of human babies.

Of the fats made in the mammary gland, the primary precursor for fatty acid synthesis in nonruminants is glucose. However, because ruminant species have very limited supplies of glucose, the primary precursor for fatty acid synthesis is one of the volatile fatty acids—acetate. Acetate is produced in the rumen when ruminants consume a high-fiber diet, so the amount of fat showing up in milk can be changed by manipulating the amount of fiber in the diet. Historically, we have selected feed to promote high milk-fat production in the dairy cow. The amount of milk fat is less emphasized today owing to consumers' preference for lower fat dairy products. More emphasis today is placed on the other components, like protein content. Dairy cows tend to be fed to produce as much as their genetics and environment will allow while

**Lipoproteins**

Compounds in the bloodstream that carry the lipids (fats) in the blood.

**Unsaturated fats**

Fatty acids that are not saturated with hydrogens, but instead have double bonds. The number of double bonds changes the physical, biochemical, and metabolic characteristics of the fats compared with saturated fats.

**Saturated fats**

Fatty acids that are completely saturated with hydrogens, and thus do not contain any hydrogen bonds.

maintaining an adequate milk-fat test. This is a good example of how knowledge of nutrition is supportive of the management for domestic animal production.

## CALCIUM

Milk serves as a very good source of minerals, particularly calcium. Human requirements for calcium are highest in the young because calcium and phosphate are used to build the matrix of bone. In our adult years, bone remodeling occurs. This process continues to require calcium, although at lower levels. However, as we age, bone breakdown often exceeds the replacement of the matrix, resulting in conditions such as osteomalacia and osteoporosis. Women are particularly susceptible to osteoporosis because the protective actions of estrogen are lost at menopause. Calcium consumption can be critical in preventing, or at least lessening, the devastating effects of these diseases. This is why the medical and nutritional communities so strongly encourage milk consumption (Figure 12–4).

As mentioned earlier, the reason that milk is high in calcium is the presence of strongly negative-charged proteins in milk. The secretory cells offset the negative charge by combining the protein with calcium that has been removed from the bloodstream. However, this process puts a stress on the cow. It is not uncommon for calcium levels in the blood of high-producing dairy cows to fall precipitously low when extraordinarily high levels of protein are produced. This condition, ***parturient paresis*** (otherwise known as milk fever), generally occurs near calving time when the production of colostrum milk requires calcium to be put into the milk faster than the cow is able to replace it from the diet. The cows exhibit paralytic symptoms, and the condition, if not treated quickly, can be fatal (Box 12–1).

As indicated in Table 12–1, tremendous variation is seen in the composition of milks among species. In dry-matter percentage alone, milk ranges from 12% to approximately 50% in whales. This increase in dry-matter percentage is related to a combination of necessity for delivery to young (whales nurse in an aquatic environment) and to a demand for a high proportion of protein and fat for fast-growing young.

**Parturient paresis**

Metabolic disorder generally occurring within 72 hours of calving. Caused by low blood serum calcium level.

**FIGURE 12–4** Dowager's hump in elderly women is linked to poor bone mineralization, especially calcium. Milk is an excellent dietary source of calcium. (Photographer Larry Mulvehill. Courtesy of Photo Researchers, Inc.)

## Box 12–1    Bovine Parturient Paresis

Parturient paresis, also known as milk fever, is a disease of mature dairy cattle that typically occurs in the 72-hour period following calving or parturition. Affected cows may initially be excitable and restless and are unstable on their feet. If the disease is not treated, cows become unable to stand and often lie with the head tucked into the flank. At this stage of the disease, cows show signs of inappetence, depression, and may have cold extremities. As the disease worsens, cows may bloat, and untreated animals gradually lose consciousness and die. Parturient paresis is caused by a sudden calcium loss associated with the onset of milk production, which leads to low blood serum levels of ionized calcium. Muscle contraction depends on adequate calcium levels, so when serum calcium falls below normal, muscle weakness occurs. The disease is most common in older, high-producing dairy cows; Jerseys may be affected more commonly than other breeds of dairy cows. Quick recognition of disease symptoms and rapid treatment with intravenous calcium gluconate solution are critical for saving the life of an affected cow. Nutritional modification leading up to the time of calving is often helpful in preventing the disease. One method involves feeding a low calcium diet prior to calving so cows begin to mobilize their own body stores of calcium and are therefore better able to respond to the sudden calcium loss associated with lactation. The addition of anionic salts to the diet, which enhances calcium absorption through the digestive tract as well as *resorption* of calcium from bone, is another nutritional modification that may be used to prevent milk fever.

### TABLE 12–1    Average Concentrations of Some Constituents in the Milk of Selected Species

| Species | Fat (%) | Protein (%) | Lactose (%) | Dry Matter (%) |
|---|---|---|---|---|
| Camel | 4.9 | 3.7 | 5.1 | 14.4 |
| Cat | 10.9 | 11.1 | 5.9 | 15.6 |
| Cow | 3.4 | 3.2 | 4.6 | 11.8 |
| Dog | 8.3 | 9.5 | 3.7 | 20.7 |
| Goat | 3.5 | 3.1 | 4.6 | 11.7 |
| Guinea pig | 3.9 | 8.1 | 3.0 | 15.8 |
| Llama | 4.7 | 4.2 | 5.9 | 15.6 |
| Pig | 9.6 | 6.1 | 4.6 | 21.2 |
| Human | 3.5 | 1.3 | 7.5 | 12.0 |
| Rabbit | 12.2 | 10.4 | 1.8 | 26.4 |
| Rat | 15.0 | 12.0 | 2.8 | 31.0 |
| Reindeer | 22.5 | 10.3 | 2.5 | 36.7 |
| Water buffalo | 10.4 | 5.9 | 4.3 | 21.5 |
| Whale | 36.6 | 10.6 | 2.1 | 50.0 |

SOURCES: Davies et al., 1983, Riek and Gerken, 2006, and Hurley, 2007.

## OTHER FACTORS IN MILK PRODUCTION

**Resorption**

The loss of tissue through normal physiological means or through a pathological process.

One of the hormones critically involved in the production of milk is growth hormone, otherwise known as *somatotropin*. This hormone is released from the pituitary gland and increases the production of milk by the secretory cells, either through increased activity of the enzymes involved in the production of milk components, through increased availability of the precursors to the secretory cells, or through a combination of the two effects. In fact, it is likely that one of the primary mechanisms by which genetic improvement in milk production has been so rapid in the dairy cattle industry is selection of those lines of cattle that are more responsive to the positive effects of somatotropin. In the 1930s, researchers at Cornell University observed increased milk production from cows injected with ground-up pituitary extracts. Years later, with the isolation of the somatotropin protein, this response was again observed, although to a much larger extent. In the 1980s, methods in biotechnology had advanced sufficiently that recombinant DNA methods could be employed to produce sufficient amounts of the bovine somatotropin such that it would be possible to explore its use to improve the production of milk from dairy cattle. In the early 1980s, drug companies became interested in the production and sale of a ***bovine somatotropin (BST)*** product that could be injected into cows to supplement their natural levels of somatotropin.

Of course, years of testing for drug effectiveness and safety for human consumption ensued. In 1985, the Food and Drug Administration (FDA) reported to the U.S. Congress that dairy products produced from milk from BST-treated cows were safe for human consumption. In addition, the National Institutes of Health concluded that milk and meat from BST-treated cows were just as safe for human consumption as were milk and meat from nontreated cows. These results have been supported by numerous studies, and no studies have indicated that the human consumption of milk produced from BST-treated cows is in any way harmful. Based on the results of numerous studies, the FDA approved the use of BST on November 5, 1993. Congress mandated a required 90-day moratorium on the sale of BST, and the commercially available product from Monsanto became available to dairy producers on February 3, 1994.

Bovine somatotropin has dramatic effects on the production of milk from dairy cows. An immediate increase of 10–20% in milk production is commonly observed, and the lactation persistency is improved. With increased milk produced per animal, the efficiency of milk production improves because the cost of maintaining the cow is less as a percentage of total requirements. However, the increase in milk requires increase in feed intake and appropriate nutritional management to support the increased production. There is also a cost for the BST.

It is important to note that BST is a naturally occurring hormone in cattle. Therefore, the addition of exogenous BST only supplements the hormone that is already present in the cow. Second, because BST is a protein, even if it did cross through the secretory cells and become discharged into the milk, it would be broken down in the stomach of the consumer, thereby rendering it inactive. Finally, the somatotropins are species specific. That is, the growth hormone in humans is different from the growth hormone in cows, which is different from the growth hormone in horses. Therefore, even if a person was injected with BST, the response would not be observable, because of the body's inability to recognize the foreign protein.

**Bovine somatotropin (BST)**

Hormone that acts on various target tissues in the body. Injections of BST increase milk production.

## SUMMARY AND CONCLUSION

Milk production requires tremendous coordination among the endocrine, physiological, biochemical, and anatomical functions in the female body. The nutritional requirement of the lactating animal is dramatically increased, not only because of the components transferred into the milk, but also owing to the increase in body functions required to produce the milk. Although levels of production and composition of milk vary greatly by species, the biosynthesis of the milk components is very similar among species. Levels of production have increased in those species primarily used for milk production. This increase in milk production can be attributed to several factors, including increased productive capacity of the mammary gland, increased knowledge of nutritional requirements of the female, better control of environmental conditions, and implementation of new technologies such as BST to improve milk synthesis capacity. Is there a physiological limit to the amount of milk that a cow or goat can produce? The answer to that question may never be determined. What is known, however, is that tremendous potential continues to be realized in the production and efficiency of milk production in the United States and around the world.

## STUDY QUESTIONS

1. What is lactation?

2. What is milk? What are the constituents of milk?

3. What are the alveoli? What do they do? What are the functions of the secretory glands in the alveoli?

4. What hormone facilitates milk letdown? What kinds of things can cause milk letdown?

5. What are the purposes of the teat sphincter? Does it provide protection from infection at all times? How can this problem be managed?

6. What is mastitis? Why is it such a physiological problem? Why is it such an economic problem?

7. What is the lactation curve? Describe what it illustrates.

8. What is milk sugar? Where is it manufactured? Where do the components come from?

9. What is lactose intolerance? Is there a "cure"?

10. What is casein? Where is it made? What is the link between casein and calcium in the mammary gland? How might biotechnology be used to determine the best bulls to produce superior cheese-making daughters?

11. In addition to casein, what are the other proteins found in milk? What does each do?

12. What is colostrum? What does it do? What is the difference between active and passive immunity?

13. Is there a difference from species to species in the fat content of milk? What is unique about the fat of a ruminant milk?

14. How does dietary fiber affect milk-fat production in the ruminant? How would the fat in milk be decreased by changing the ration? Increased?

15. What are osteoporosis and osteomalacia? How is calcium linked to these diseases? How is milk drinking a part of the solution?

16. What is milk fever? What mineral is it linked to?

17. What is somatotropin? BST? How do each of these work? What is the source of each?

18. Is BST safe? What are the negatives associated with its use?

# REFERENCES

*Author's note:* This chapter was prepared in part by Dr. T. L. Beckett, California Polytechnic State University, San Luis Obispo, CA. For the fourth edition, Dr. John P. Mc-Namara, Professor, Animal Science, Washington State University, reviewed this chapter. The author gratefully acknowledges his contributions. For the fourth edition, Melanie A. Breshears, DVM, PhD, Diplomate ACVP, Assistant Professor, Veterinary Pathobiology, Center for Veterinary Health Sciences, Oklahoma State University, contributed original material.

Akers, R. M. 2002. *Lactation and the mammary gland.* Ames: Iowa State Press.

Cunningham, M., M.A. Latour, and D. Acker. 2005. *Animal science and industry.* Upper Saddle River, NJ: Pearson Education.

Cupps, P. T. 1991. *Reproduction in domestic animals.* 4th ed. New York: Academic Press.

Davies, D. T., C. Holt, and W. W. Christie. 1983. The composition of milk. In *Biochemistry of lactation,* T. B. Mepham, ed. Oxford, UK: Elsevier Science.

Frandson, R. D., A. D. Fails, and W. L. Wilke. 2003. *Anatomy and physiology of farm animals.* 6th ed. New York: Blackwell.

Hurley, W. L. 2007. *Lactation biology website, milk composition and synthesis resource library.* Accessed online August 1, 2007. http://classes. ansci.uiuc.edu/ansc438/Milkcompsynth/ milkcomp_table.html.

Mepham, T. B. 1987. *Physiology of lactation.* Philadelphia: Open University Press.

*Merck veterinary manual.* 2005. 9th ed. Whitehouse Station, NJ: Merck.

Riek, A., and M. Gerken. 2006. Changes in llama (*Lama glama*) milk composition during lactation. *Journal of Dairy Science* 89:3484–3493.

Sherwood, L. 1993. *Human physiology: From cells to systems.* 2nd ed. Minneapolis, MN: West.

Turner, C. W. 1969. *Harvesting your milk crop.* Chicago: Babson.

# 13

# Animal Behavior

## LEARNING OBJECTIVES

After you have studied this chapter, you should be able to:

1. Describe the overall field of animal behavior and explain why it is important.
2. Explain the individual areas of study in animal behavior.
3. Cite the general effects of handling on livestock production.
4. Discuss how animal temperament and handling interact.
5. Describe the role of fear and fear memories in handling.
6. List the benefits of training animals to be handled and to accept restraint.
7. Cite the effects of novelty, vision, noise, and shadows on livestock movement.
8. Discuss the concept of flight zone.
9. Identify the role of genetics in handling.
10. Outline the basics of handling facility layout.

## KEY TERMS

Aggressive behavior
Animal welfare
Anorexia
Applied ethology
Artificial insemination
Auditory
Aversive event
Behavioral ecology
Charles Darwin
Classical conditioning
Comparative method of study

Comparative psychology
Conditioning
Cortisol
Cribbing
Critical period
Dichromat
Dominance
Electric prods
Ethogram
Ethology
Extensive rearing systems

Flight zone
Flighty
Flocking instinct
Habituation learning
Handling
Imprint learning
Nose tongs
Novelty
Olfactory
Operant or instrumental conditioning
Pacifier cow

Palatability
Phobia
Pica
Sensitive periods
Shelter-seeking behavior

Social structure
Sociobiology
Squeeze chute
Stereotyped behavior
Stress

Submissive behavior
Temperament
Wool chewing
Wool sucking

**Charles Darwin**

An English naturalist (1809–1882). Among other contributions, he proposed the theory of evolution by natural selection.

**Comparative method of study**

A systematic method of comparing the behavior of two or more species as a way of discovering the mechanism of behavior.

**Comparative psychology**

The study of the mechanisms controlling behavior, learning, sensation, perception, and behavior genetics in animals and making extrapolations to humans or other animals.

**Sociobiology**

The study of the biological basis of social behavior. Of particular interest is behavior that helps pass the gene pool on to the next generation.

**Behavioral ecology**

The study of the relationships between a species' behavior and its environment.

**Ethology**

The study of behavior of animals in their natural surroundings, focusing on instinctive or innate behavior.

**Ethogram**

A catalog or inventory of all the behaviors an animal exhibits in its natural environment.

**Applied ethology**

The term generally used to refer to the study of domestic animal behavior. Usually directed at companion species and livestock.

# INTRODUCTION

Humanity's interest in animal behavior no doubt dates back tens of thousands of years. Understanding the predator–prey relationship was important to early humans, who occupied both roles. Ample archeological evidence suggests that early humans had a good practical knowledge of the behavior of certain species. It is also clear that behaviors have been important in humans' development and selection of domestic animals. The flocking instinct of some breeds of sheep and the herding and hunting behaviors of different breeds of dogs serve to illustrate this. Countless other examples could be cited. However, animal behavior did not emerge as a science until the last half of the 19th century. The most notable stimulus was the work of **Charles Darwin** and others who proposed the theory of evolution by natural selection. This was followed by the development of the **comparative method of study,** which was a method of comparing the behavior of two or more species systematically. The final piece was the work of Gregor Mendel and others on genetics. From these roots have evolved four major approaches to the study of animal behavior:

1. **Comparative psychology** is the study of the mechanisms controlling behavior, learning, sensation, perception, and behavior genetics. Animal behavior has been studied for many decades by psychologists, physiologists, and cognitive scientists.

2. **Sociobiology** is the study of the biological basis of social behavior. Of special interest is behavior that helps pass on the gene pool to the next generation.

3. **Behavioral ecology** is the study of the relationships between a species' behavior and its environment.

4. **Ethology** is the study of behavior of animals in their natural surroundings, with its focus on instinctive or innate behavior. Originally ethology was the study of wild animals, but domestic species are now also studied in their surroundings. Associated with ethology is the **ethogram**, which is a cataloging of all the behaviors an animal exhibits in its natural environment.

Obviously these areas are not traditional areas of interest for production animal scientists or veterinarians. Rather, they have been studied by psychologists, zoologists, and cognitive scientists as a very basic science rather than an applied science. However, sometime around the middle of the 20th century, interest grew in the study of livestock behavior. Reproductive physiologists began studying reproductive behavior and nutritionists began documenting eating behaviors and patterns. The tools of ethology were adapted, and **applied ethology** was born as a field of study. The first major text on animal behavior, which contained substantial, credible information on livestock, was edited by E. S. E. Hafez and published in 1962. Until that time, most of the behavioral study had been directed at feeding and sexual behavior, and only for a few species.

In the preface to the second edition, published in 1969, Dr. Hafez commented that "progress in many areas of research is hindered by insufficient knowledge of animal behavior." He also bemoaned the fact that "literature on animal behavior is widely scattered in many different journals and known only to specialists." After the three editions of Dr. Hafez's text, the last in 1975, several other texts and journals came to prominence. The study of domestic animal behavior was launched as a branch of study, with some focusing on companion species and others on livestock.

Behavior of livestock, as a key to *handling* them, received added attention after the development of large intensive systems of concentrated livestock production. The topic received even more attention after emphasis was given to *animal welfare* issues. The decade of the 1990s saw a flurry of activity. Hundreds of scientific papers have been published, several good textbooks written, and research has escalated. The USDA opened the Livestock Behavior Research Unit at Purdue University in 1997 with the mission "to determine behavioral and physiological indicators of stress and/or well-being in food-producing animals and to develop management systems that maximize well-being in farm animals." This center has brought together ethologists, physiologists, immunologists, and others to work in cooperation. The work of Dr. Temple Grandin of Colorado State University has received wide attention. Several universities across the nation have added faculty positions in animal behavior and well-being. Perhaps yours is one of them. The veterinary profession has become more interested and involved, as evidenced by the American Veterinary Medical Association's creation of the American College of Veterinary Behaviorists. Much of the interest has been stimulated in the veterinary profession by problem behaviors in pets and companion animals.

The remainder of this chapter gives a brief overview of the general types of animal behavior and then focuses on behavior and handling as it pertains to the food species. This is not meant to imply that other species and other types of behavior are not important. However, this is a timely, relevant approach to behavior that has large economic importance to the livestock industries and it is often an overlooked area. A list of references and suggested readings is found at the end of the chapter. For those with a special interest in this area of study and who wish to explore it further, three of those references are recommended as the best place to start. These works have the status of being "classics" in the area. They are Lorenz (1981), Tinbergen (1951), and Skinner (1958).

## AREAS OF STUDY IN ANIMAL BEHAVIOR

### COMMUNICATION

Animals communicate in a variety of ways. They send a wide range of messages as well as receive communication and interpret signals in a variety of ways. Understanding all of this can be an important part of animal management. Perhaps you've known someone who had "dog sense," "cow sense," or "horse sense." When we apply these terms to people, we are just acknowledging their ability to correctly interpret the communication that animals send and to reply in a way that animals understand. Animals give and receive communication through visual, *auditory,* and *olfactory* means. A full study of the communication of a species necessarily means considering all three. Understanding animal communication means understanding such signals as the position of the ears and tail, general posture, hypervocalization, marking behavior, and the behaviors associated with the elimination of body wastes.

**Handling**
In this context, refers to any manipulation necessary to care for animals.

**Animal welfare**
Dealing with the animal's well-being and care.

**Auditory**
Related to hearing.

**Olfactory**
Relating to the sense of smell.

## AGGRESSION AND SOCIAL STRUCTURE

All animals have some form of **aggressive behaviors** whose purposes may revolve around obtaining food, acquiring mates, or securing a place in a social hierarchy. Thus, from a survival perspective, aggressive behavior can be the key to survival. In domestic species, this type of behavior can still have a place. Once a pecking order is established, serious aggression can be replaced with threats and injuries. Disruptions to the herd or flock are minimized. However, problems with aggression and **social structure** arise when humans and animals are forced to vie for **dominance,** such as becoming established as the "alpha" individual in a dog–human pack. Some dogs are definitely the "alpha" individual in their household, which leads to inappropriate aggression and an unacceptable and potentially dangerous social structure. Categories of aggression include dominance-related, territorial, pain-induced, fear-induced, maternal, and predatory aggression.

## BIOLOGICAL RHYTHMS AND SLEEP

Detecting abnormal sleep and activity levels in domestic animals can be difficult. This is especially so in light of the changes that confinement and modern nutrition have brought to animals. A stabled horse fed twice daily acts very differently from a free-ranging one. Understanding circadian (24-hour cycles) and other rhythms helps us understand animal activity, sexual cycles, and physiological responses. In addition to circadian rhythms, other areas of study include high-frequency (less than 30 minutes), ultradian (more frequent than 24 hours), infradian (less than 24 hours), annual, and circatrigentian (30-day) cycles. These cycles are influenced by light, barometric pressure, the endocrine system, drugs, feeding, and a variety of other factors. Obviously, knowing what is normal is the first step in knowing what is abnormal.

## SEXUAL BEHAVIOR

The study of sexual behavior in animals involves the wide range from development and maintenance of good mating behaviors in breeding animals to dealing with the unwanted sexual behaviors that may be exhibited in neutered animals, especially pets. Sexual behavior is greatly influenced in domestic animals by such factors as genetic selection for other traits, management practices, confinement rearing, and association with other species, especially humans. Study of sexual behavior requires a study of the physiological basis of sexual behavior, including endocrine and central nervous system influences. It also entails study of social and sexual experiences and environmental influences to promote normal mating behavior and to correct abnormal mating behavior.

## MATERNAL BEHAVIOR

Obviously the ability of a female to care for her young is an important part of successful production of most livestock and companion species. Factors that affect that success include environmental effects, previous experiences, endocrine and nervous system effects, heredity, and human intervention. Study of maternal behavior includes a study of bonding behavior between the female and her offspring, mutual recognition, negligence or neglect by the female, nest-building, nursing, weaning, and learned behavior. Study of maternal behavior also includes studying such aberrant behaviors as cannibalism, refusal to nurse, mis-mothering, and rejection of the offspring. Understanding maternal behavior is the key to preventing mother hamsters from eating their young, understanding why all the farm cats have their kittens in one big nest, and preventing a ewe from rejecting one of her twins.

---

### Aggressive behavior

Threatening or harmful behavior toward others of the same or different species.

---

### Social structure

The organization of a group; the patterns of the relationships to each other.

---

### Dominance

Refers to an animal's place in the social ranking. The most dominant animal in the group exerts the major influence over other animals. A term often used to describe this is *pecking order.*

## DEVELOPMENT OF BEHAVIOR

A particularly important area of study for those who wish to be veterinarians or work in some other area of animal science is the study of normal development of behavior. From a practical perspective, animals are only capable of responding to certain types of training at developmentally appropriate times. Issues of proper socialization with people and other animals become important in willingness and ability to mate as adults. From the study of behavioral development has come the concept of *sensitive periods* and *critical periods.* This concept refers to the developmental stage when experience, or lack of it, has an influence on later behavior. A good example is wild farm cats that cannot be petted and handled. To create tame farm cats, the kittens must be handled at a very early age. This study has led to better understandings of such factors as the critical nature of play and socialization with people and other animals. It has also led to the development of temperament tests for various species.

## LEARNING

The study of learning behavior has implications for all species. Later in the chapter, some implications of learned behaviors are discussed relative to handling livestock. Thus far, the greatest attention to learning behavior has been with the companion species and horses. Learning is broken down into different types. *Classical conditioning* is what Pavlov demonstrated when he showed how to make a dog salivate by ringing a bell. Understanding classical conditioning has implications for milk letdown in dairies and for dogs who hate going into a vet clinic. *Habituation learning* occurs when an animal learns to ignore something like a collar or a train that passes by the pasture. *Operant* or *instrumental conditioning* occurs when an animal can be taught to ring a bell for a food reward or open an automatic feeder door. This is often thought of as trial-and-error learning. *Imprint learning* helps young find their mothers and individuals recognize their own kind. Other areas of study of learning behavior in animals include discrimination learning, conceptual learning, imitation, and several other types. Some practical applications for each of these areas of learning can generally be found in all species. Anyone who wishes to be an animal trainer needs a keen understanding of animal learning.

## INGESTIVE BEHAVIOR: FOOD AND WATER INTAKE

Studying the way animals consume feed and water has implications in production systems and in day-to-day management of pets and companions. Many factors are at work, including hormonal, physiological, and psychological factors. In most livestock species, we are interested in maximizing consumption to improve rate of gain and increase efficiency. Animals that are fed ad libidum (i.e., freely) usually need less feed trough space compared to animals that are limit fed. Limit-fed animals must have sufficient space so dominant animals cannot chase the weaker animals away from their portion of the limited feed. However, that is not the case for a horse or a companion animal that will be kept for a much longer period of time and for a different purpose. Studying ingestive behavior requires study of what controls feed intake, such as the influence of herd or flock behavior, *palatability,* environment, hormones, and meal patterns. An understanding of ingestive behaviors can help in dealing with problems such

---

**Sensitive periods**

Times in an animal's life when certain types of learning are more easily accomplished.

**Critical period**

Similar to sensitive periods but with a more definite beginning and end.

**Classical conditioning**

The type of conditioning the Russian physiologist Ivan Pavlov demonstrated (hence the name Pavlovian conditioning), in which a reflex-like response can be stimulated by a neutral stimulus.

**Habituation learning**

A type of operant conditioning. An animal's ability to come to ignore something that occurs often enough.

**Operant or instrumental conditioning**

Learning that is primarily influenced by its effects.

**Imprint learning**

Learning that has restrictive conditions and time periods.

**Palatability**

A measure of the acceptability of feedstuffs.

as underperformance owing to insufficient feed consumption in livestock, obesity, *anorexia* and *pica* in all species, grass eating in dogs, and *cribbing* in horses.

## BEHAVIORAL DISORDERS

This is perhaps the area of behavioral study that has captured the greatest hold on public imagination. Many behavioral disorders are tied to the previously discussed types of behavior, but others are not. Helping owners and animals overcome behavioral disorders can be especially rewarding for veterinarians and trainers, and it often means the difference between keeping an animal or having it destroyed. The study of behavioral disorders includes determining the reasons why the disorder has developed or exists, and then instituting a treatment of the problem. Environment, early history, training, and related problems must all be assessed. Types of behavioral disorders in dogs include destructiveness, self-mutilation, tail chasing, *phobias,* car chasing, digging, jumping up on people, and vomiting. In cats, clawing, *wool sucking,* and plant eating most frequently need to be addressed. Horse behavioral problems frequently include *stereotyped behaviors* like stall kicking, trailering problems, head shyness, and phobias. Livestock behavioral disorders include kicking in cattle, bar biting in sows, and *wool chewing* in sheep.

In the big picture, finding the causes of behavioral disorders allows emphasis on preventing them. Socializing young animals with their own species enables animals to learn social interactions with other animals. Colts or puppies reared in isolation away from other animals are often vicious fighters against other animals. This occurs because they never learned that after they became dominant they no longer have to keep fighting to maintain their dominance. Many dog attacks on young children could be prevented by socializing young puppies to babies and toddlers so they learn the difference between children and prey. Young animals have to learn who they can socialize with and what they can attack.

# LIVESTOCK BEHAVIOR

An understanding of the behavior of livestock can facilitate handling and can improve both handler safety and animal welfare. Animals can seriously injure handlers and/or themselves if they become excited or agitated. Poor handling procedures can also cause animals to become stressed. Several studies have shown the adverse effects of *stress* on animals; reducing stress on animals has been demonstrated to improve productivity. Restraint, electric prods, transportation, inconsistent handling, other handling stresses, and abuse can cause lowered conception rates, reduced immune function, reduced digestive function, and early embryonic losses. However, studies with numerous species have shown that animals learn and adapt to stresses. Therefore, handling and management strategies can be implemented that increase productivity and maintain meat quality.

## TEMPERAMENT

An animal's *temperament* is one determinant of how it will react during handling. Temperament is determined by an interaction between a substantial genetic effect and environmental factors. In cattle, temperament is highly heritable. The heritability estimate

---

**Anorexia**

Inappetence or unwillingness to eat.

**Pica**

A craving for and willingness to eat unnatural feedstuffs.

**Cribbing**

A behavior in horses in which they bite or hold on to objects such as posts.

**Phobia**

Excessive and unwarranted fear.

**Wool sucking**

A *prolonged sucking syndrome* most frequently observed in cats where they continue to suck on objects and perhaps knead with the forepaws long after weaning.

**Stereotyped behavior**

A nonfunctional, repetitive, intentional, and often rhythmic behavior.

**Wool chewing**

Often referred to as *wool pulling*. Sheep nibble at their fleece and make bald spots.

**Stress**

A physical, emotional, or chemical factor causing body or mental strain or tension.

**Temperament**

Characteristic behavior or mode of response.

**Squeeze chute**

A restraining device used to handle livestock.

of temperament in cattle has been figured at 0.40, 0.53, and 0.45. Several studies have shown that cattle with Brahman genetics (*Bos indicus*) are more excitable than *Bos taurus* breeds when evaluated by observing their behavior in a ***squeeze chute.***

Cattle temperament can be rated on a numerical rating scale, measured while they are held in a squeeze chute, or by flight zone testing in a pen. To rate cattle temperament in a squeeze chute, the most common rating system is a four-digit scale. Cattle are scored as follows:

1. They stand calmly in the squeeze chute.
2. They are restless.
3. They vigorously shake the chute.
4. They violently shake the chute and try to escape (berserk).

**Aversive event**

A negative experience that may be painful, frightening, or nauseating.

Chute scoring may be more likely to assess the animal's genetic reactivity because restraint in a squeeze chute is a sudden ***aversive event,*** and the animal is forced to enter the squeeze chute. Another good method for measuring temperament is the speed that the animal exits from the squeeze chute. Animals with a faster exit speed are more flighty. The flight zone is the distance at which an animal will approach a person. Completely tame animals have no flight zone and allow people to touch them. When the flight zone test is used, a person stands in a pen and measures how closely individuals or groups of cattle approach them. Flight zone size is affected by both genetics and learning.

**Cortisol**

A hormone produced by the adrenal cortex. It is elevated during stress and has been used as a gauge for the degree of stress an animal is under.

A major component of temperament is fearfulness. Fear is a universal emotion that motivates animals to avoid predators. It is a natural reaction to being handled. Fear is a very strong stressor to the animal's body and as such can affect health and performance. Measuring blood ***cortisol*** levels is a means of measuring an animal's response to a stress. Fear caused by exposure to ***novelty*** elevates levels of cortisol higher than many husbandry procedures. For example, in beef cattle raised in ***extensive rearing systems*** not accustomed to being handled in a squeeze chute, the psychological stress of restraint raised cortisol levels almost as high as branding did. In sheep, fear stress can elevate cortisol levels higher than handling procedures such as shearing. Even nonpainful handling or restraint can induce very high cortisol levels in both sheep and cattle. The amount of stress caused by a handling procedure such as restraint in a squeeze chute is determined by how the animal perceives the event and how much it is frightened. Recent research indicates that when cattle become stressed, the whites of their eyes show.

**Novelty**

Anything new or sudden in an animal's environment.

**Extensive rearing systems**

Usually associated with range conditions in which hardy animals such as beef cattle receive little individual attention. They may be handled only once or twice per year.

Although temperament has a strong genetic component, it is also influenced by previous experiences and handling. All vertebrates can be fear conditioned. It follows that animals can also be desensitized to factors that would otherwise cause fear through training and habituating livestock to handling. Animals with a genetically ***flighty*** temperament can be trained to a handling procedure and may appear behaviorally calm. They can learn to behave calmly when they are with familiar people or are in a familiar handling facility, but they can suddenly panic when left alone in a strange place or exposed to the novelty of a noisy auction or a new farm.

**Flighty**

The tendency of an animal to take sudden flight when alarmed. Also called *mobile alarm.*

An animal's previous experience with handling affects its reaction to future handling. This can be especially true if the experience produced fear because animals can develop fear memories that are difficult to eradicate. Fear memories form a subcortical circuit in the brain that allows an animal to flee quickly if it sees or hears the same frightening stimulus. These memories can be suppressed by learning but are never completely erased from the brain's subcortical circuits. For ease and safety of handling, it is best to prevent fear memories from ever developing.

On farms, ranches, and feedlots, observations indicate that an animal's first experience with a handling facility, a new corral, a person, or pieces of equipment should be made as positive as possible. Farm animals can be very frightened in these novel situations. If a procedure is painful or very aversive the first time it is done, it may become difficult to persuade the animal to reenter the facility. First experiences are critical in how animals form future responses to similar situations. Some practical applications include the following:

- Train cattle by walking them through a squeeze chute a few times and giving them a feed reward, which should make future handling in the squeeze chute easier.
- Provide feed rewards to sheep to improve handling and movement through a handling facility.
- Accustom calves to regular gentle handling so no injuries occur during marketing.
- Avoid the use of dogs in a confined space where animals are unable to move away.
- Use *electric prods* sparingly on cattle and never use them on breeding pigs.
- Do not allow cattle to rush out of corrals back to pasture. Cattle should become accustomed to walking slowly past a handler when they exit corrals.

**Electric prods**

Small handheld devices designed to give a small electrical shock.

## TRAINING AND HABITUATING LIVESTOCK TO HANDLING AND RESTRAINT

The idea of training an animal to accept restraint voluntarily is a new concept to some people. Animals that are handled gently can be trained to accept restraint voluntarily in a comfortable device. This has many advantages for breeding animals or animals used in long-term research studies:

- Stress on both animals and people is reduced.
- One person can easily handle large animals that are trained to walk into a restraint device.
- Cooperative large animals are less likely to injure people or themselves.

Feed rewards can be used to facilitate animal movement through a facility. Sheep have been trained to enter a squeeze tilt table voluntarily for a grain reward in only one afternoon. The sheep were squeezed and tilted to a horizontal position nine times in one day. After being released from the squeeze tilt table, the animals ran rapidly into the crowd pen and lined up in the chute.

Training a group of cattle, which are somewhat more excitable than sheep, may take up to 10 days. Training sessions should be spaced 24 hours apart to give the animals an opportunity to calm down. A series of training trials in 1 day may result in increased agitation and excitement. Practical experience on ranches and feedlots shows that making cattle accustomed to people on foot and on horseback produces calmer and easier-to-handle cattle at the slaughter plant.

To train animals to accept restraint voluntarily, the restraint device must be introduced gradually and gently with feed rewards. At first, the animal is allowed to walk through the restrainer several times. The next step is to allow the animal to stand in the restrainer without being squeezed. On the fourth to fifth pass through, the squeeze is applied gently. During each step the animal is given a food reward. A relatively tame animal can be trained to voluntarily enter a restrainer in less than an hour. Animals do not voluntarily accept restraint if the restraint device causes pain. Selection of the right type of squeeze chute and headgate to fit the specific handling requirements is important.

Training animals to enter a restraint device voluntarily is easier and less stressful if the animal is tame and has little or no flight zone. If a wild animal is being trained, it is important to catch it correctly on the first attempt. Fumbling and failing to restrain an animal on the first attempt results in increased excitement. If an animal resists and struggles; it must not be released until it stops struggling; otherwise it will be rewarded for resisting. Animals that are released while resisting are more likely to resist in the future. The animal should be stroked and talked to gently until it calms down.

**Nose tongs**

Small clamp-like restraining device put in an animal's nose.

Cattle restrained with *nose tongs* become more difficult to restrain in the future. However, when a halter is used to hold the animal's head for blood testing, restraining the head becomes easier with successive tests. Cattle blood-tested with halter head restraint learn to turn their head and expose their jugular. Cattle that have had experience with nose tongs often fling their head about to avoid attachment of the tongs.

Some of the lean hybrid pig strains are more excitable and difficult to handle, and they are more likely to panic and pile up when driven through a high-speed slaughter plant. However, pigs balk less and drive more easily at the slaughter plant if the producer walks through the pens every day during finishing for as little as 10 to 15 seconds. This trains pigs to get up in an orderly manner and calmly move around the person. It is important to teach the pigs to flow around the person. If the handler stands still and allows the pigs to approach him and chew on his coveralls, they may become more difficult to drive at the slaughter plant because they tend to follow the handler, instead of allowing themselves to be driven. To avoid frightening the pigs, the handler must never kick or slap them. Walking in the pens with the pigs, or walking pigs in the aisles during finishing, helps produce calmer animals.

A review of many studies and practical experience has shown that animals with a more placid temperament habituate more easily to a forced, nonpainful handling procedure than animals with a flighty temperament. However, some animals do not habituate easily. In one group of cattle, some individuals violently shook the squeeze chute and never habituated to being restrained when they were handled every 30 days. Even though the cattle were handled quietly, they still struggled violently and became behaviorally agitated every time they were put in the squeeze chute. Research at Texas A & M University indicated that some pigs habituated when they were forced to swim. Other pigs remained fearful and never habituated during a series of swims.

Extremely flighty, excitable animals such as elk, bison, and antelope are less likely to habituate to a forced handling procedure. Bison are highly reactive and flighty and often injured during handling. Bison ranchers are concerned because it is difficult to handle their mature animals safely. These animals may be so excitable that the only way to handle them in a low-stress manner is to train them to cooperate voluntarily from an early age.

## EFFECTS OF NOVELTY

Novelty, anything new or sudden in an animal's environment, is a very strong stressor of animals. Examples of sudden novel stimuli include a stamping foot, a train passing a pen where newly arrived calves are received, or an auction ring. A sudden novel event, such as a person stamping his or her foot in a pen of commercial pigs, is one of the best tests for determining genetic differences in the reactivity of pigs reared under identical conditions. This test is superior to other tests, such as willingness to leave a pen or ease of movement through a hallway. The paradox of novelty is that it causes an intense behavioral and physiological reaction when suddenly introduced to an animal with a flighty, excitable temperament, but the same flighty animal may be the most attracted to a novel object when allowed to approach it voluntarily. In cattle, breeds with the largest flight zone generally have the greatest tendency to approach novel objects. Cattle approach and manipulate a piece of paper lying on the ground when allowed to approach it voluntarily, but balk and jump away if someone attempts to drive them over it. Numerous studies with many species have shown that animals raised in a variable environment are less likely to be stressed when confronted with novelty.

## VISION

Livestock have wide-angle vision. Cattle, pigs, and sheep have a visual field in excess of 300 degrees. This means that objects in over 80% of the space around them can distract them. Loading ramps and handling chutes should have solid side walls to prevent animals from seeing distractions outside the chute with their wide-angle vision. Moving objects and people seen through the sides of a chute can cause balking or can frighten livestock. Solid side walls are especially important if animals are not completely tame or if they are unaccustomed to the facility. Blocking vision stops escape attempts. This is why a solid portable panel is so effective for handling pigs. Sight restriction also lowers stress levels. The wildest cow remains calm in a darkened artificial insemination box that completely blocks vision.

Even though ruminant animals have depth perception, their ability to perceive depth at ground level while moving with their heads up is probably poor. This would explain why livestock often lower their heads and stop to look at strange things on the ground. Cattle, pigs, sheep, and horses often balk and refuse to walk over a drain grate, hose, puddle, shadow, or change in flooring surface or texture. For this reason, drains should be located outside of the areas where animals walk. Drains or metal plates running across alleys should be eliminated.

Lighting can help alleviate some problems. In areas where animals are handled, illumination should be uniform and diffuse. Shadows and bright spots should be minimized. Pigs, sheep, and cattle have a tendency to move from a dimly illuminated area to a more brightly illuminated area, provided the light does not glare in their eyes. A spotlight directed onto a ramp or other apparatus often facilitates entry. The light must not shine directly into the eyes of approaching animals. Moving or flapping objects can also disrupt handling. Fan blades or a flapping cloth or coat on a fence can cause balking. Animals may refuse to walk through a chute if they can see motion up ahead.

Numerous investigators have now confirmed that cattle, pigs, sheep, and goats possess color vision, but they are apparently *dichromats,* and the cones in their eyes are more sensitive to blue-green and yellowish green. Handling facilities should be painted one uniform color. All species of livestock are more likely to balk at a sudden change in color or texture.

**Dichromat**

Ability to perceive only two colors.

## NOISE

Cattle and sheep are more sensitive than people to high-frequency noises. The auditory sensitivity of cattle is greatest at 8,000 hertz (Hz), and sheep at 7,000 Hz. The human ear is most sensitive at 1,000 to 3,000 Hz. Unexpected loud or novel noises can be highly stressful to livestock. Continuous exposure to sounds over 100 decibels (dB) reduced daily weight gain in sheep and would likely affect other animals the same way. However, animals readily adapt to reasonable levels of continuous sound, such as white noise, instrumental music, and miscellaneous sounds. Continuous background sound has been shown to improve weight gain in some cases. Continuous playing of a radio with a variety of talk and music reduces the reaction of pigs to sudden noises and may help prevent weight loss caused by unexpected noises.

In facilities where livestock are handled, loud or novel noises should be avoided as much as possible. The sound of clanging metal can cause balking and agitation. Rubber stops on gates and squeeze chutes help reduce noise. The pump and motor on a hydraulic squeeze chute should be located away from the squeeze. Exhausts on pneumatic powered equipment should be piped away from the handling area. Noise can also be useful. Small amounts of noise can be used to move livestock. Cattle and sheep move away from a rustling piece of plastic. Observers of superior livestock handlers notice that even the sound of finger snapping can be used effectively to direct livestock. Shouting and yelling at animals is stressful to them and should be avoided if at all possible. Loud shouting may be as stressful as an electric prod.

## FLIGHT ZONE

**Flight zone**

The distance that an animal is caused to flee from an intruder.

An important concept of livestock handling is *flight zone.* Understanding the flight zone can reduce animal stress and can help prevent accidents to handlers. The flight zone is the animal's "safety zone." When a person enters an animal's flight zone, the animal moves away. The size of the flight zone varies depending on the tameness or wildness of the livestock. The flight zone of extensively raised cattle may be as much as 50 m (164 ft), whereas the flight zone of feedlot cattle may be 2 m (6 ft) to 8 m (26 ft).

Several factors can influence the flight zone:

- The size of the enclosure the livestock are confined in may affect flight zone size. Sheep experiments indicate that animals confined in a narrow alley had a smaller flight zone compared to animals confined in a wider alley.
- Approaching an animal at its head seems to increase flight zone size.
- The size of the flight zone slowly diminishes when animals receive frequent, gentle handling. Extremely tame livestock may be difficult or even impossible to drive because they no longer have a flight zone. These animals should be led with a feed bucket or halter. Excited animals have a larger flight zone.

Understanding the flight zone and how to use it is important in handling livestock. If the handler penetrates the flight zone too deeply, the animal either bolts and runs away, or turns back and runs past the person. The best place for the person to work is on the edge of the flight zone. This causes the animals to move away in an orderly manner. The animals stop moving when the handler retreats from the flight zone. To make an animal move forward, the handler should stand in the shaded area marked in the flight zone diagram (Figure 13–1). To cause the animal to back up, the handler should stand in front of the point of balance. A flag on the end of a stick can be used to sort cattle by moving it back and forth across the point of balance. Many people make the mistake of deeply invading the flight zone when cattle are being driven down an alley or into an enclosed area such as a crowd pen. If the handler deeply penetrates the flight zone, the cattle may turn back and run over him or her. If the cattle attempt to turn back, the person should back up and retreat from inside the flight zone.

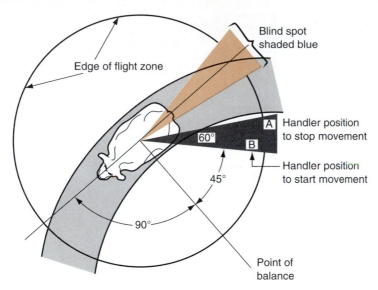

**FIGURE 13–1**    Cattle flight zone. Standing outside the flight zone causes the animal to stop moving (A). Standing in the shaded area within the flight zone causes the animal to start moving (B). Moving in front of the point of balance causes the animal to back up. (*SOURCE:* Grandin, 1989. Used with permission.)

The livestock attempt to turn back because they are trying to escape from the person who is deep inside their flight zone.

Cattle sometimes rear up and become agitated while waiting in a single-file chute. A common cause of this problem is a person leaning over the chute and deeply penetrating the flight zone. The animal usually settles back down if the person backs up and retreats from the flight zone. Inexperienced handlers sometimes make the mistake of attempting to push a rearing animal back down into a chute. The animal often reacts by becoming increasingly agitated, and both the handler and the animal have a greater likelihood of being injured. This also explains why livestock balk if they see people standing in front of the squeeze chute. The use of shields for handlers to stand behind improves animal movement.

## HERD ANIMALS

All livestock are herd animals, and they are likely to become highly agitated and stressed when they are separated from their herd mates. Productivity is affected by the physiological changes that occur during isolation. In addition, large animals that become agitated and excited are likely to injure handlers. If an isolated animal becomes agitated, other animals should be put in with it.

The desire to be with the herd can be used to help move animals. Cattle and sheep are motivated to maintain visual contact with each other and readily "follow the leader." If animals bunch up, handlers should concentrate on moving the leaders instead of pushing a group of animals from the rear. Trained animals can be used to lead others through a handling facility.

Groups of animals that have body contact remain calmer. A tame *pacifier cow* calms a wild cow during *artificial insemination.* The wild cow stands quietly while maintaining tactile contact with the tame cow. A loading ramp for pigs or sheep that has a "see-through"

**Pacifier cow**

A cow that has been trained to accept moving, restraint, and other types of management.

**Artificial insemination**

The procedure for placing semen in the reproductive tract of a female animal through means other than the natural mating act in the hopes of causing a pregnancy.

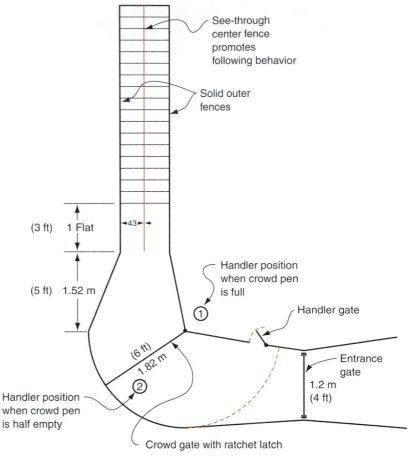

**FIGURE 13–2**    A loading ramp design for pigs and sheep. This design takes advantage of natural following behavior by allowing the animals to see each other through a see-through center fence. (*SOURCE:* Grandin, 1989. Used with permission.)

center partition takes advantage of natural following behavior (Figure 13–2). As the animals walk up the twin single-file chutes, they can see each other through the center partition. This works better if solid outer walls block outside distractions.

## GENETIC DIFFERENCES

Genetic factors affect an animal's reaction to handling. Brahman and Brahman cross cattle are more excitable and harder to handle than English breeds. Angus cattle are more excitable than Herefords, and Holsteins move more slowly than Angus or Herefords. When Brahman or Brahman cross cattle become excited, they are more difficult to block at fences. Visually substantial fences built with planks or a wide belly rail should be used with these breeds. Brahman cattle seldom run into a fence that appears to be a solid barrier. Highly excited Brahman cattle may lie down and become immobile if they are repeatedly prodded with an electric prod. Continuous electric prodding of Brahman or Brahman cross cattle can result in death. If the animal is left alone for a few minutes, it usually gets up. English or European cattle seldom become immobile. Even though Brahman cattle are more excitable than the British breeds, they can become extremely docile when they are handled gently. Brahmans are inquisitive, sensitive cattle that respond well to quiet, gentle handling, and they respond poorly and may become agitated if they are treated roughly. Because Brahman have a more reactive nervous system, they may become easily frightened when subjected to sudden novel experiences, for example, going through an auction ring. Brahmans often remain calm if they have a familiar person who can handle them when they are in a novel, strange situation. British cattle, however, are less likely to become fearful or agitated when subjected to sudden novelty.

In pigs, Yorkshires move more slowly during loading than Pietrians. Other breed differences likely exist. Observations at farms and slaughter plants indicate that certain types of hybrid pigs are difficult to drive. They have extreme *shelter-seeking behavior* (flocking together) and they refuse to move forward up a chute. They are also very excitable. This problem is most evident in some hybrid lines of pigs selected for high productivity. Pig breeders should select for temperament to avoid serious meat quality and animal welfare problems at the slaughter plant. Different breeds of sheep also react differently to handling. The *flocking instinct* of sheep is very evident when they are being handled. The Rambouillet breed and a few others, for example, tend to flock tightly together and remain in the group, whereas other breeds are more independent.

Observations of thousands of cattle and pigs in large slaughter plants indicate that some animals that have been bred for extreme leanness are very excitable and difficult to handle when they are brought to a new place. They become highly agitated when they are subjected to the noise and novelty of a large slaughter plant. It appears that the most excitable pigs, cattle, and dogs have long, slender, smooth bodies and fine bones. Animals bred for leanness with heavy bones and bulging muscles tend to be calmer. Genes are linked in ways that are not fully understood, as evidenced by the following example. In long-term selection experiments, Russian scientists selected foxes for temperament. For 20 years, they bred the calmest and easiest-to-handle foxes. Selection for the single trait of calm temperament resulted in a fox that looked and behaved like a Border Collie dog. Its coat color changed from gray to black and white. However, continued selection for the very calmest fox-dogs resulted in bitches that ate their puppies and in neurological problems such as epilepsy.

The previous example illustrates a challenge in breeding animals for temperament. To reduce stress and to improve both productivity and welfare, it is important to breed animals with a calm temperament. However, one must not make the mistake of overselecting for any single trait. Excessive selection for calmness may result in other problems, such as lack of mothering ability. To prevent handling and stress problems, it is advisable to cull the most flighty animals that become extremely frenzied and agitated when they are restrained, but it is probably a bad idea to select only for the very calmest animals.

## HANDLER DOMINANCE

Handlers can often control animals more efficiently if they exert dominance over the animal. Exerting dominance is not beating an animal into submission; it is using the animal's natural behavior to exert dominance. The handler becomes the "boss animal." Nomadic tribespeople in Africa control their cattle by entering the dominance hierarchy and becoming the dominant herd member.

Limited experience suggests that dominance can be achieved over a group of pigs. Slapping a dominant pig has little effect on its behavior. The aggressive behavior can be stopped by pushing the pig against a fence with a board pressed against its neck. The board against the neck simulates another pig to push and bite. Pigs exert dominance over each other by biting and pushing against the neck. It is often advisable to handle the dominant pig first. The odor of the dominant pig on the handler may make the other

---

**Shelter-seeking behavior**

Behaviors that an animal exhibits to escape from weather, insects, or danger.

**Flocking instinct**

A type of shelter-seeking behavior that has been selected for sheep. At the least hint of danger, they move close together and move as a group.

**Submissive behavior**
Behaviors a less-dominant animal exhibits toward a more-dominant animal to prevent being subjected to aggression.

pigs demonstrate more ***submissive behavior.*** More research is needed to develop simple methods of exerting dominance that will enable handlers to control boars and other large animals with a minimum of force and greater safety.

## HANDLING FACILITY LAYOUT

An animal's stress reaction to a handling procedure depends on genetics, individual differences, and previous experiences. Facility design can have a strong influence on the type of quality of the animal's previous experiences and can influence the ease with which the animal can be handled in the future. Handling facilities that utilize behavioral principles thus make handling easier in the present and future.

### Curved Chutes and Solid Fences

Curved single-file chutes are especially recommended for handling cattle (Figure 13–3). A curved chute is more efficient for two reasons. First, it prevents the animal from seeing what is at the other end of the chute until it is almost there. Second, it takes advantage of the animal's natural tendency to circle around a handler moving along the inner radius. A curved chute provides the greatest benefit when animals have to wait in line for vaccinations or

**Good Design Principles**

1. Cattle in crowd pen can see a minimum of 2 body lengths up the chute
2. Cattle make a 180° turn through the crowd pen and think they are going back to where they came from

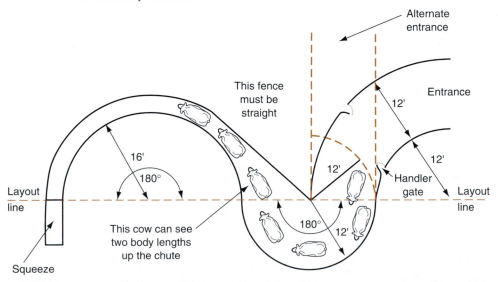

**FIGURE 13–3** Layout of a curved cattle handling facility. A curved chute is very effective for cattle because it prevents the animal from seeing what is at the other end of the chute until it is almost there, and it takes advantage of the animal's natural tendency to circle around a handler moving along the inner radius. A curved chute with an inside radius of 3.5 m (12 ft) to 5 m (16 ft) works well for handling cattle. If the chute is bent too sharply at the junction between the single-file chute and the crowd pen, it appears as a dead end and cows balk. If space is restricted, short 1.5-m (5-ft) bends can be used. If bends with a radius smaller than 3.5 m (12 ft) are used, there must be a 3-m (10-ft) straight, single-file section at the junction between the crowd pen and chute to prevent the chute from appearing to be a dead end. Handler walkways should run alongside the chute and crowd pen and never overhead. (*SOURCE:* Grandin. Used with permission.)

other procedures. A curved chute with an inside radius 3.5 m (12 ft) to 5 m (16 ft) works well for handling cattle. The curve must be laid out as shown in Figure 13–3. If the chute is bent too sharply at the junction between the single-file chute and the crowd pen, it appears as a dead end. This causes livestock to balk. If space is restricted, short 1.5-m (5-ft) bends can be used. If bends with a radius smaller than 3.5 m (12 ft) are used, there must be a 3-m (10-ft) straight single-file section at the junction between the crowd pen and chute to prevent the chute from appearing to be a dead end. Handler walkways should run alongside the chute and crowd pen. The use of overhead walkways should be avoided.

Livestock often balk when they have to move from an outdoor pen into a building. Animals enter a building more easily if they are lined up in a single-file chute before they enter the building. Conversely, pigs reared indoors are often reluctant to move out into bright daylight. A pig loading ramp should be designed so that the pigs are lined up in single file in an area where they cannot turn around before they leave the building (Figure 13–4). For all species, solid sides are recommended on both the chute and

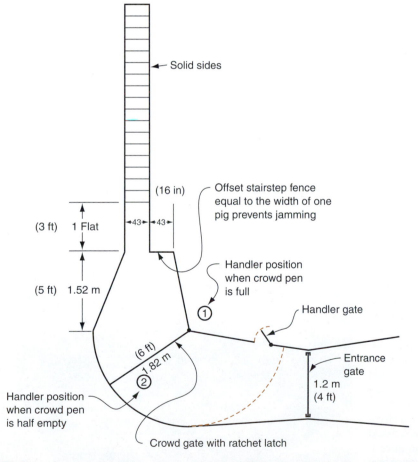

**FIGURE 13–4** Layout of a pig loading ramp with a single offset step to prevent jamming. The ramp is designed so that the pigs are lined up in single file, where they cannot turn around. For all species, solid sides are recommended on both the chute and the crowd pen, which leads to a squeeze chute or loading ramp. The solid crowd gate prevents animals from turning back. In holding pens, solid pen gates along the main drive alley facilitate animal movement. Handler gates allow people to escape charging animals. (*SOURCE:* Grandin, 1989. Used with permission.)

the crowd pen that leads to a squeeze chute or loading ramp. For operator safety, handler gates must be constructed so that people can escape charging animals. The crowd gate should also be solid to prevent animals from turning back. Wild animals tend to be calmer in facilities with solid sides. In holding pens, solid pen gates along the main drive alley facilitate animal movement.

## Crowd Pen Design

The crowd pen used to direct animals into a single-file or double-file chute must never be built on a ramp. A sloped crowd pen causes livestock to pile up against the crowd gate. Round crowd pens are very efficient for all species. In cattle facilities, a circular crowd pen and a curved chute reduced the time spent moving cattle by up to 50%. Practical experience has shown that the recommended radius for round crowd pens is 3.5 m (12 ft) for cattle, 1.83 to 2.5 m for pigs (6 to 8 ft), and 2.5 m (8 ft) for sheep.

Cattle and sheep crowd pens should have one straight fence, and the other fence should be on a 30-degree angle. This layout should not be used with pigs. They jam at the chute entrance, which is very stressful for pigs. A single offset step equal to the width of one pig should be used to prevent jamming at the entrance of a single-file ramp (Figure 13–4). Another good design is two single-file chutes side by side (Figure 13–2). Jamming can be further prevented by installing an entrance restricter at single-file race entrances. The entrance of the single-file chute should provide only 1/2 cm of space on each side of each pig.

The crowd pen for pigs or cattle should be only half to three-quarters full; half full is best. It is important to avoid using the crowd gate if possible. On a round crowd pen, the crowd gate should be closed and set on the first notch and left there. It should not be used to push animals. Cattle and pigs need room to turn and should be handled in small discrete bunches, with space in between the bunches. For sheep, the crowd pen may be filled completely, as long as the sheep are not too tightly packed. Sheep should be moved in one continuous stream, never breaking the flow, to maintain following behavior (Figure 13–5).

**FIGURE 13–5** Crowd pen design. This crowd pen design is efficient because the cattle go around the bend and think they are going back to where they came from. The handlers are quietly moving small groups by waving plastic flags. The crowd pen for pigs or cattle should be filled only half to three-quarters full. It is important to avoid using the crowd gate if possible. On a round crowd pen, the crowd gate should be closed and set on the first notch and left there if possible. It should not be used to push animals. Cattle and pigs need room to turn and should be handled in small discrete bunches, with space in between the bunches. For sheep, the crowd pen may be filled completely, as long as the sheep are not too tightly packed. Sheep should be moved in one continuous stream, never breaking the flow, to maintain following behavior. (Photo courtesy of Temple Grandin, Grandin Livestock Handling Systems, Inc.)

## Ramp Steepness and Flooring

Excessively steep ramps may injure animals. The maximum recommended steepness for a stationary cattle or pig ramp is 20 degrees for market-weight animals. If space permits, a 15-degree slope is recommended for pigs. Stair steps are recommended on concrete ramps because they provide good footing even when they are dirty or worn.

### IMPROVED HANDLING PRACTICES

Many present facilities are not necessarily designed well, but, of necessity, are still in use. But facilities are only part of the equation. How animals react to being handled depends largely on the methods of the handlers. Several strategies can be used to improve handling practices in all types of facilities:

**Move small bunches.** Move finishing pigs in small bunches of three to six during truck loading. On ranches and feedlots, move small bunches of cattle that can be easily handled. The staging alley leading to the truck loading ramp or processing area should be only half full.

**Eliminate electric prods.** Use other driving aids, such as plastic paddles or sticks with plastic streamers or flags tied on them. Use these devices to work the animal's flight zone and to turn the animals. These devices work better than plain sorting sticks because the animals can see them more easily (Figure 13–6).

**Open anti-back gates.** Many chute facilities have too many anti-back gates. Movement often improves if most are tied open. The only place an anti-backup gate may be needed is up close to the squeeze chute. Cattle handled calmly and quietly are less likely to back up. The anti-back gate at the single-file chute entrance can be equipped with a remote control rope so it can be held open by a person standing by the crowd pen. This facilitates entry of the cattle into the chute.

**Eliminate visual distractions.** Distractions and lighting problems may ruin the performance of even a well-designed facility and should be removed. To locate

**FIGURE 13–6** Dr. Temple Grandin demonstrates using a stick with plastic streamers to turn an animal. The streamers are gently moved alongside the animal's head to induce it to turn. (Photo courtesy of Temple Grandin, Grandin Livestock Handling Systems, Inc.)

distractions that impede animal movement and need removing, handlers should get in the chute and crouch down to visualize the area from the animal's eye level. If pigs or cattle balk or refuse to enter the single-file chute, look for distractions such as shiny reflections, a dangling end of a chain, water puddles, drain gratings, a coat hanging on a fence, or people visible up ahead. Pigs and cattle often refuse to enter a chute that is dark. When new feedlot processing areas are built, skylights are recommended to provide diffuse, shadow-free light because shadows that fall across a chute can make animals balk. However, animals will not approach blinding light and will not walk directly into the sun. Another distraction that may impede animal movement is air blowing in their faces.

**Reduce noise.** Avoid yelling at animals, whistling, or whip cracking. Clanging noises on steel should be silenced, and hydraulic systems should be quiet and designed to avoid the sound frequencies for which cattle have maximum sensitivity. On squeeze chutes, the clatter of the side bars should be quieted with rubber pads. Reducing the high-pitched whine in a hydraulic system can result in calmer cattle. In a pork slaughter plant, engineering conveyor equipment for reduced noise combined with quiet handling results in reduced squealing and pig pile-ups.

**Handler movement patterns.** Use the patterns shown in Figures 13–7 and 13–8 to move cattle and pigs through chutes. The use of these movement patterns enables handlers to eliminate electric prods in the processing area. Animals move forward in a chute when a handler walks past them in the direction opposite the desired movement. The handler must pass the point of balance at the shoulder to induce the animal to move away in the opposite direction. To make the animal move forward, the handler must be behind this point of balance. Animals speed up and move faster when a handler inside their flight zone walks in the direction opposite the desired movement. The same principles apply to other species of animals. Handlers should not put continuous pressure on an animal's flight zone. To induce a cow to walk into a squeeze chute, the handler should stand back out of her flight zone. The cow usually moves forward into the squeeze chute when the handler steps toward her and walks back past the point of balance at the shoulder (Figures 13–7 and 13–8).

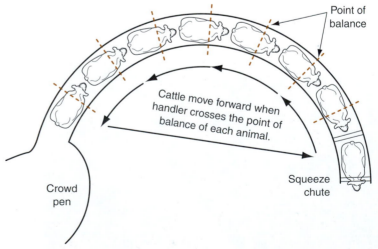

**FIGURE 13–7**   Handler movement pattern to keep cattle moving into the squeeze chute in a curved chute system. (*SOURCE:* Grandin, 1998b. Used with permission.)

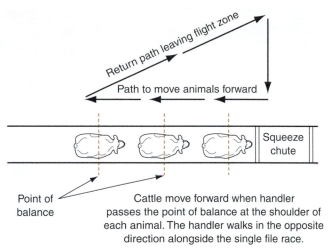

Return path leaving flight zone

Path to move animals forward

Squeeze chute

Point of balance

Cattle move forward when handler passes the point of balance at the shoulder of each animal. The handler walks in the opposite direction alongside the single file race.

**FIGURE 13–8**   Handler movement pattern to keep cattle moving into a squeeze chute or restrainer. (*SOURCE:* Grandin, 1998b. Used with permission.)

**Slow is faster.**  Move pigs and cattle at a slow walk. Fearful animals are more likely to balk and are more difficult to handle. Handlers should move slowly and deliberately. Sudden jerky motions frighten the animals. In the wild, sudden movements are associated with predators.

**Use following behavior.**  When handling cattle and pigs, do not fill the crowd pen until the single-file chute is partially empty because when there is space in the chute, a group of animals in the crowd pen will follow the leader into the chute. Cattle and pigs in the crowd pen turn around if the single-file chute is full. It is important to avoid overfilling the crowd pen. Cattle and pigs should be moved in small separate bunches, but sheep can be moved in a large bunch. When sheep are handled, the crowd pen should be continuously full so sheep will follow each other in a continuous flow. This behavior is a species difference between sheep and other hoofed animals.

**Behavioral principles of restraint.**  Four behavioral principles of restraint can be used to keep animals calm in a squeeze chute or similar restraint device. These are blocking vision; slow, steady movements of the restraint apparatus; optimum pressure; and providing secure footing so animals do not lose their balance and struggle because of slipping. On squeeze chutes, cover the open-barred sides or install angled rubber conveyor belt strips on the side bars to prevent cattle entering the squeeze chute from seeing the operator. Cattle often balk at the entrance to a squeeze chute because they see the operator deep in their flight zone (Figure 13–9). The crowd pen, the lead-up chute, and the squeeze chute should have solid sides. It is most important to cover the back half of the squeeze chute closest to the tailgate. Covering the sides of the squeeze chute also reduces sudden lunging at the headgate. Cattle should enter and exit the squeeze chute at a walk. During restraint, cattle remain calmer if the squeeze chute is closed with steady, strong pressure instead of suddenly bumping the animal. However, sufficient pressure must be applied to provide cattle the feeling of being held. Many people make the mistake of squeezing the animal tighter when it struggles. Remember that if an animal is squeezed too tightly, the pressure should be backed off slowly. A sudden release of pressure may scare the animal.

**FIGURE 13–9** To demonstrate the calming effects of solid sides, try fastening cardboard to the open barred sides of a squeeze chute as shown here. An opening can be left at the shoulder for giving injections. (Photo courtesy of Temple Grandin, Grandin Livestock Handling Systems, Inc.)

**Training handlers.** Quieter livestock handling techniques for loading and unloading trucks and handling animals in alleyways and chutes may take up to 2 weeks for handlers to learn fully. They may have to spend several days learning the most efficient handler movement patterns and making minor changes in the facility to improve livestock movement. Management has to be fully committed to changing handling procedures permanently on a farm, feedlot, or ranch. Top management has to implement the changes and impress on employees how serious they are about stopping rough handling. Most employees can be retrained. However, a few people who have been rough for so many years may not be able to change their ways, and they may need to be reassigned to jobs away from animals.

### OBJECTIVE SCORING OF HANDLING

In feedlots and slaughter plants, quiet handling has a tendency to become rough unless management maintains constant vigilance. Therefore, methods are needed to score handling procedures objectively. The simplest handling procedure to score is electric prod use. The lower the usage, the better. Trained handlers working in well-designed facilities can move large numbers of cattle without an electric prod. In large beef slaughter plants with line speeds of over 250 animals per hour, it is possible to move 95% of the cattle from the truck, unloading all the way through the stunning chute and restrainer without the use of electric prods. In most situations, a single 15-minute lesson in animal handling makes it possible to greatly reduce electric prod usage.

The percentage of cattle that vocalize can be used to assess handling stress. In both cattle and pigs, vocalizations are correlated with physiological stress measurements. In a slaughter plant survey, the percentage of bellowing cattle was significantly higher when a high percentage of cattle were prodded with an electric prod. Bellowing was associated with electric prodding, slipping and losing footing, and when excessive pressure was applied by a restraint device. Work with feedlot personnel has shown that reducing electric prod use also reduces bellowing. Some of Dr. Grandin's work has shown that when cattle are handled quietly in a squeeze chute for vaccinations and implanting, less than 3% of the animals vocalize during handling in the lead-up chute, when catching in the squeeze chute, or during vaccinations. Handling the ears and head for implanting and ear tagging causes some animals to vocalize. Handling of the head appears to be more aversive than carefully applied body restraint and driving cattle without electric prods. Some work suggests that blocking an animal's vision may reduce its resistance to having its head manipulated.

Squeeze chutes can be instrumented with load cells and strain gauges to measure the force exerted on them by the cattle. An instrumented squeeze chute can be connected directly to the feedlot computer system and used to monitor handling. In a feedlot in which cattle are individually identified, the computer could correlate the squeeze chute force scores to weight gains and sickness, and it could measure how hard the animal hits the headgate and the intensity of struggling. Through the use of special software, it would be possible to use the system for both temperament scoring and for assessing how employees are handling the animals.

## SUMMARY AND CONCLUSION

Genetics and experience interact to determine how an animal will behave during handling. Quiet, calm handling at an early age helps produce calmer, easier-to-handle adult animals. People working with animals need to understand the behavioral principles of handling. The use of behavioral principles should improve the efficiency of livestock handling and reduce stress on animals. Reducing stress also should help improve weight gain, reproductive performance, and animal health. To avoid agitation and facilitate handling, the handler should move small numbers of animals at a time, not overload the crowd pen, eliminate electric prods, open anti-back gates, eliminate visual distractions that make animals balk, reduce noise, and use flight zone and point of balance principles.

Restraint devices should be designed so they do not cause pain. In certain research situations, animals can be easily trained to enter a restraint device voluntarily. This practice helps reduce stress. The restraint device should be gradually introduced and should not cause pain. Feed rewards facilitate training. Training animals to submit to handling procedures voluntarily is especially useful for valuable breeding animals and animals used for research. All areas where animals are crowded such as chutes and crowd pens should have solid sides and diffused lighting with a minimum of shadows. Livestock have wide-angle vision and are easily frightened by shadows or moving distractions outside of chutes. Noise should be kept to a minimum because animals have sensitive hearing. Cattle, pigs, and sheep are herd animals, and isolation of a single individual should be avoided.

An animal's previous experience with handling will affect its reaction to handling in the future. Animals that have had frequent, gentle contact with people are less stressed during handling than animals that have had previous aversive treatment. An animal's first experience with a new corral, person, or pieces of equipment should be made as positive as possible. If a procedure is painful or very aversive the first time it is done, it may be difficult to persuade the animal to reenter the facility.

## STUDY QUESTIONS

1. Why has there been an increased interest in the study of animal behavior by veterinarians and animal scientists?
2. Describe the major areas of study in animal behavior and justify the importance of each.

3. What has caused the recent interest in animal handling behavior?

4. What are the benefits to better understanding and practicing improved animal handling techniques? What problems are associated with poor handling techniques?

5. Describe how temperament and handling practices interact. Name some methods for measuring temperament.

6. How does fear affect an animal and how does it influence the animal while it is being handled?

7. What are the advantages of training animals to accept handling and restraint?

8. Describe the general methods of training animals to accept restraint voluntarily.

9. What is the role of temperament in the habituation to handling?

10. What is novelty as it pertains to livestock? Give some examples. How do animals react to novelty?

11. Why is it hard for animals to forget fear memories? What are some examples of ways to prevent fear memories in animals?

12. What does it mean that animals have wide-angle vision? What are some ways to compensate for this in handling?

13. How does depth perception affect an animal's willingness to move? What can be done in a handling facility to compensate for poor depth perception?

14. How can manipulating lighting change an animal's willingness to be handled?

15. How do animals react to noise? How about continuous radio? What are some measures that can be taken to help animals deal better with sounds, or to eliminate them?

16. Describe flight zone. How does it affect an animal's handling? Is a small flight zone a good thing? Explain.

17. How can herd animal instincts be used to move animals and help prevent stress? Explain one way to design a loading chute to take advantage of the herd instinct.

18. Give some examples of how genetics interact with handling procedures to cause specific reactions to handling in livestock.

19. Think of some traits of various species of animals. Do you think selection for certain traits may have gone overboard and produced some traits that are not so desirable?

20. Describe the concept of handler dominance. How practical do you think the practice of handler dominance can become in modern management of livestock for high production?

21. Why is a curved chute effective in moving livestock? What other modifications can be made in chutes to facilitate handling?

22. Describe an effective design for a crowd pen. How is it best used by animal handlers to move livestock?

23. Why should ramps not be excessively steep? What angles are recommended for cattle? For hogs?

24. Describe some methods of improving handling practices.

25. What are the four behavioral principles of restraint? Give a practical use for each principle.

**26.** How can attention to handler movement patterns facilitate handling?

**27.** Describe several methods you would use to assess whether or not a group of livestock handlers is doing a good job in handling.

## REFERENCES

Beginning with the section on livestock behavior, this chapter is based on two papers published by Dr. Temple Grandin. The material was edited, condensed, and otherwise modified to fit the style and format of this book and used with permission. The articles are Grandin (1989) and Grandin (1997b), which are referenced here and available at www.grandin.com. These are refereed journal articles and thus fully referenced. The citations amount to nearly 200 individual articles by experts in the fields. Those of you interested in the more scientific approach of the original papers are encouraged to read them in their entirety. Dr. Grandin's website houses a wealth of information about livestock behavior, facilities design, and humane slaughter. It is highly recommended.

Dr. Grandin has reviewed and updated this chapter for all past editions and for the current edition. The author appreciates her contributions and support.

Beaver, B. V. 1994. *The veterinarians' encyclopedia of animal behavior.* Ames: Iowa State University Press.

Belyaev, D. K. 1979. Destabilizing selection as a factor in domestication. *Journal of Heredity* 70:301–308.

Boissy, A. 1995. Fear and fearfulness in animals. *Quarterly Review of Biology* 70(2): 165–191.

Brelands, K., and M. Brelands. 1961. The misbehavior of organisms. *American Psychologist* 16:681–684.

Dickson, D., G. Barr, L. Johnson, and D. A. Wickett. 1970. Social dominance and temperament in dairy cows. *Journal of Dairy Science* 53:904.

Drichamer, L. C., S. H. Vessey, and D. Meikle. 1996. *Animal behavior.* 4th ed. Dubuque, IA: William C. Brown.

Ewing, S. A., D. C. Lay, and E. Von Borell. 1999. *Farm animal well-being: Stress physiology, animal behavior and environmental design.* Upper Saddle River, NJ: Prentice Hall.

Grandin, T. 1989. Behavioral principles of livestock handling. *The Professional Animal Scientist* 5(2):1–11.

Grandin, T. 1993. Teaching principles of behavior and equipment design for handling livestock. *Journal of Animal Science* 71:1065–1070.

Grandin, T. 1997a. Assessment of stress during handling and transport. *Journal of Animal Science* 75:249–257.

Grandin, T. 1997b. The design and construction of facilities for handling cattle. *Livestock Production Science* 49:103–119.

Grandin, T. 1998a. Objective scoring of animal handling and stunning practices in slaughter plants. *Journal of the American Veterinary Medical Association* 212:36–39.

Grandin, T. 1998b. Livestock Handling Systems, Inc., Colorado State University, Fort Collins, CO. Personal communication.

Grandin, T. 1998c. Review: Reducing handling stress improves both productivity and welfare. *Professional Animal Scientist* 14(1):1–10.

Grandin, T. 2007. *Livestock handling and transport.* 3rd ed. Wallingford, Oxfordshire, UK: CAB International.

Grandin, T., and M. J. Deesing. 1998. *Behavioral genetics and animal science.* San Diego, CA: Academic Press.

Grandin T., and C. Johnson. 2005. *Animals in translation.* New York: Scribner.

Hafez, E. S. E. 1969. *The behavior of domestic animals.* 2nd ed. Baltimore, MD: Williams and Wilkins.

Hafez, E. S. E. 1975. *The behavior of domestic animals.* 3rd ed. Baltimore, MD: Williams and Wilkins.

Heffner, R. S., and H. E. Heffner. 1983. Hearing in large mammals: Horse (*Equus caballus*) and cattle (*Bos taurus*). *Behavioral Neuroscience* 97(2):299–309.

Houpt, K. A. 2005. *Domestic animal behavior for veterinarians and animal scientists.* 4th ed. Ames, IA: Blackwell Publishing Professional.

Jacobs, G. H., J. F. Deegan, and J. Neitz. 1998. Photopigment basis for dichromatic vision in cows, goats and sheep. *Visual Neuroscience* 15(3):581–584.

Jensen, P. 2002. *The ethology of domestic animals.* Wallingford, UK: CAB International.

Kilgour, R., and D. C. Dalton. 1984. *Livestock behavior.* Boulder, CO: Westview Press.

King D. A., C. E. Schuehle Pfeiffer, R. D. Randal, T. H. Welsh, R. A. Oliphant, B. E. Baird, Jr., K. O. Curley, R. C. Vann, D. S. Hale, and J. S. Savell. (2006). Influence of animal temperament and stress responsiveness on carcass quality and beef tenderness of feedlot cattle. *Meat Science* 74:546–556.

LeDoux, J. 1996. *The emotional brain.* New York: Simon & Schuster.

Lorenz, K. Z. 1981. *The foundations of ethology.* New York: Springer-Verlag.

Lucas, J. R., and L. W. Simmons. 2006. *Essays in animal behavior.* Burlington, MA: Elsevier Academic Press.

Mason G., and J. Rushen. 2006. *Stereotypic animal behavior.* 2nd ed. Wallingford, Oxfordshire, UK: CABI Publishing.

Price, E. O. 1987. Farm animal behavior. *Veterinary Clinics of North America* 3: 217–481.

Prince, J. H. 1977. The eye and vision. In *Duke's physiology of domestic animals,* M. J. Swenson, ed. Ithaca, NY: Cornell University Press.

Sandem A. J., A. M. Janczak, R. Salle, and B. O. Braastad. 2006. The use of diazepam as a pharmacological validation of eye white as an indication of the emotional state of dairy cows. *Applied Animal Behavior Science* 96:177–183.

Scott, J. P., and J. L. Fuller. 1965. *Genetics and social behavior of the dog.* Chicago: University of Chicago Press.

Skinner, B. F. 1958. *The behavior of organisms.* New York: Appleton-Century-Crofts.

Smith, B. 1999. *Moving 'em: A guide to low stress animal handling.* Kamuela, HI: The Graziers Hui.

Tinbergen, N. 1951. *The study of instinct.* New York: Oxford University Press.

# Animal Health

## LEARNING OBJECTIVES

After you have studied this chapter, you should be able to:

1. Explain the nature of disease.
2. Describe the causes of disease in general terms.
3. Outline a procedure for diagnosing disease.
4. Describe the body's defenses against disease.
5. Describe the elements of herd health.
6. Identify the effects of animal disease on human well-being.
7. Describe the elements of regulatory animal medicine.

## KEY TERMS

Active immunity
Acute disease
All-in, all-out animal management
Antibodies
Asymptomatic disease
Biopsy
Chronic disease
Clinical infection
Clinical sign
Contagious disease
Diagnosis

Diagnostician
Direct cause of disease
Disease
Etiology
Herd or flock health management program
Infectious diseases
Lesion
Necropsy
Passive immunity
Pathogen
Pathogenicity

Pathology
Predisposing cause of disease
Resistance
Stress
Subclinical infection
Toxin
Vector
Virulence
Zoonotic

# INTRODUCTION

It is a challenge, an obligation, and a necessity to animal stewardship that animals be maintained as near to a constant state of health as possible and/or feasible. It is a challenge because the area of disease prevention and treatment requires, at least for some species, constant vigilance. It is an obligation because appropriate stewardship of the animals in our care includes disease prevention and treatment. It is a necessity because the number of diseases that can bring illness and death, and with them economic ruin, is numerous for all species.

# DISEASE

In its broadest definition, *disease* is any state other than a state of complete health. In a state of disease, the normal function of the body, or some of its parts, is changed or disturbed. Microorganisms and parasites can cause states of disease. Disease includes injuries such as broken bones, cuts, and burns, and metabolic disorders such as grass tetany and milk fever. A *clinical sign* (Figure 14–1) is what the animal exhibits that is different from the normal function. Clinical signs include fever, weight loss, edema, and reduced performance. A *lesion* refers to changes in body organs. Changes in size, color, or shape of an organ are typical lesions, as are tumors or abscesses of organs.

Understanding disease is complicated by the possibility of several interactions between the causes of disease and the victim of it. *Pathology* is the study of the essential nature of diseases. *Etiology* refers to the cause of disease or the study of causes of disease. Pathology and etiology are pursued most frequently by individuals with training in medicine and/or various science disciplines. Biochemists, microbiologists, metabolic nutritionists, molecular geneticists, statisticians, and other scientists all contribute to the fields of pathology and etiology. These individuals can be associated with various disciplines in colleges and universities, with industry groups such as drug companies, in

## Disease
State of being other than that of complete health. Normal function of the body or its parts is disturbed.

## Clinical sign
Observable difference in an animal's normal function or state of health that indicates the presence of a bodily disorder.

## Lesion
Abnormal changes in body organs owing to injury or disease.

## Pathology
The branch of medicine that deals with the essential nature of disease.

## Etiology
The factor that causes a disease or the study of factors that cause disease.

**FIGURE 14–1**    Animals display clinical signs that indicate illness. Notice the characteristic "shipping fever" posture of this calf. (Photo courtesy of T. C. Stovall, Department of Animal Science, Oklahoma State University. Used with permission.)

local, state, or federal government, and in international organizations such as the Food and Agricultural Organization of the United Nations.

## CAUSES OF DISEASE

Animals have a natural defense system against disease involving a number of factors that can be loosely grouped under the term *resistance*. This includes the general state of health of the animal and several body mechanisms, including the immune system and behavioral adaptations, designed to repel attacks from disease-causing organisms. Obviously, these mechanisms fail from time to time. Some particularly virulent *pathogens* are so destructive because the body has virtually no defense of any kind against them. History is littered with great epidemics in both animals and humans, sometimes occurring at the same time, that have killed hundreds of thousands, even millions of victims. Other pathogens are perhaps just as deadly or incapacitating but are less destructive because the body has better defenses against them. Disease is usually caused by a combination of *predisposing causes* and *direct causes.*

Predisposing causes are often referred to as *stress* factors (Figure 14–2). Stress factors are varied, but have in common the fact that they place unusual or additional demands on the body. They make the animal more susceptible to disease. Poor nutrition can affect an animal's immune system and predispose it to infection (Figure 14–3). Poor mineral or vitamin balance during growth can predispose animals to joint problems. The genetic makeup of an animal may leave it vulnerable to many diseases, including a variety of metabolic diseases and certain cancers. Environmental factors such as chilling, crowding, inadequate ventilation, stress related to handling, improper use of feed additives or medications, and many more may predispose an animal or group of animals to disease. Poor leg structure can predispose a racehorse to fractures. Even the "normal" acts of life can predispose to disease. For example, milk fever almost never occurs except very close to parturition.

**Resistance**

The natural ability of an animal to remain unaffected by pathogens, toxins, irritants, or poisons.

**Pathogen**

Any living disease-producing agent.

**Predisposing cause of disease**

Any condition or state of health that confers a tendency and/or susceptibility to disease.

**Direct cause of disease**

Exposure to or contact with pathogens or other substances that cause a decrease in animal health.

**Stress**

A response by the body to a stimulus that interferes with the normal physiological equilibrium of the animal.

**FIGURE 14–2** Stress factors, such as the muddy conditions in this feedlot, predispose animals to disease. (Photo courtesy of Edward H. Coe, University of Missouri, Columbia.)

**FIGURE 14–3** Poorly nourished animals are vulnerable to infections and metabolic diseases. (Photographer B. C. McLean. Courtesy of USDA.)

(a)                                                                (b)

**FIGURE 14–4**   Contagious diseases are those that are transmitted from animal to animal. Most infectious diseases are contagious. (a) Several pigs in this herd are affected by a contagious disease. (b) In some instances, slaughter and burial of carcasses are necessary to control the spread of threatening contagious diseases. (Photos courtesy of USDA, APHIS.)

**Infectious diseases**

Diseases caused by living organisms, which invade and multiply in or on the body and result in damage to the body.

**Pathogenicity**

The capability of an organism to produce disease.

**Virulence**

Degree of pathogenicity.

**Contagious disease**

Disease capable of being transmitted from animal to animal.

Direct causes of disease include several categories of infectious etiologies: bacteria, viruses, *Rickettsia,* protozoa, *Mycoplasma,* external parasites, internal parasites, fungi, and the poorly understood prions associated with mad cow disease. These causes are distinguished by the fact that they are themselves living agents that cause disease by their presence in or on an animal's body. Diseases caused by living organisms are called *infectious diseases.* A living organism that causes disease in another living organism is a pathogen. Infectious diseases have four requirements for their perpetuation: (1) The pathogen must be able to gain entrance into the body, usually through the skin or a body orifice; (2) once there, it must be able to adapt to the host environment and multiply; (3) at some point it must be able to exit the host; and (4) then infect another host so that the cycle is perpetuated. The terms *pathogenicity* and *virulence* describe the ability of an organism to cause disease. A disease is *contagious* (Figure 14–4) if it is transmitted readily from animal to animal. Most, although not all, infectious diseases are contagious. Some pathogens infect only one species or a group of closely related species. Thus there are diseases only horses or chickens can get, diseases only ruminants can get, and so on. Some infectious agents are spread from animal to animal by a *vector*, usually an arthropod such as a mosquito or tick (Figure 14–5). Table 14–1 outlines the means by which infectious agents spread from one animal or herd to another.

Other direct causes of disease are noninfectious. Nutrient deficiency diseases result directly from improperly balanced and/or improperly fed rations. Perfectly balanced rations can cause disease if the animal overeats. Some diseases, such as hemophilia, are caused directly by genetic makeup. Genetic diseases are different

**Vector**

Animal, usually an arthropod, that transfers an infectious agent from one host to another.

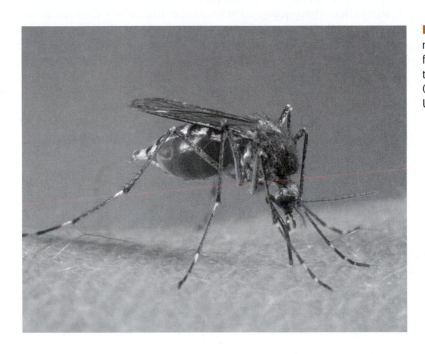

**FIGURE 14–5** Often arthropods, mammals, reptiles, and birds transmit diseases and parasites from one species to others. Mosquitoes, such as the one pictured here, transmit several diseases. (Photo courtesy of Agricultural Research Service, USDA.)

**TABLE 14–1    Ways of Introducing and Spreading Infectious Diseases**

1. Introduction of diseased animals.
2. Introduction of animals that have recovered from a disease but are carriers of the infectious agent and can transmit it to other animals.
3. Contact with inanimate objects such as trucks, trailers, feeders, waters, and so on, that are contaminated with infectious organisms.
4. Exposure to the carcass of an infected animal.
5. Water. This may be water a sick animal in the herd contaminated or impure surface water carried from another source.
6. Pathogens carried by rodents, varmints, and birds.
7. Vectors. Some diseases are carried by arthropods, especially blood-sucking and biting insects (Figure 14–5).
8. People. The shoes and clothing of a person can carry infectious organisms from one group of animals to another.
9. Contaminated feed or feed bags.
10. Contaminated premises. This could be soil, bedding, or litter.
11. Airborne organisms.

SOURCE: Adapted from Berry, 2007, p. 1. Used with permission.

from genetic predispositions to disease. One does not inherit cancer, for instance. However, certain genetic makeups predispose individuals to developing cancer. By contrast, diseases such as hemophilia and muscular dystrophy are inherited diseases caused by a specific genotype that is present from the moment of conception. Other direct causes of disease include those that may cause a traumatic injury, such as a nail, a sharp instrument, a falling object, or a wet and slick floor. A metabolic disturbance is a direct cause of many diseases. *Toxins* or chemical poisons can also be direct causes of disease.

## DESCRIBING DISEASE

The severity of diseases is described in several ways. If the disease is infectious in nature, then it may be either a *clinical* or *subclinical infection.* When clinical signs of a disease are present, it is clinical. Subclinical diseases don't have readily observable clinical signs. Sometimes the terms *inapparent* or *asymptomatic* are used to describe subclinical infections. Animals with subclinical infections are capable of shedding infectious agents, which can spread the disease to other animals. Such animals are often called *carriers.*

    *Acute diseases* have a sudden onset of clinical signs and a short duration of illness. They are also characterized by having a clear-cut termination of the disease, either the death of the animal or its recovery. The recovery should occur no longer than 2 to 3 weeks after onset of the disease. *Chronic diseases* have symptoms that develop slowly over a period of weeks or even months. Early symptoms are usually minor and often nonspecific and easy to overlook. Animals may be described as "off feed," "unthrifty," perhaps "lacking in bloom." Weight gains may be reduced, milk production "off," performance reduced, or activity levels may seem different. Often the clinical signs could be attributed to dozens of diseases, which gives them little diagnostic value. Chronic diseases have a lengthy duration compared to the 2- to 3-week period or less for acute diseases.

## DIAGNOSIS OF DISEASE

The combination of husbandry skills, scientific techniques, analytic ability, intuition, and luck that goes into the "art" of detection and identification of animal disease is collectively referred to as *diagnosis.* A person with expertise in diagnosing disease is a *diagnostician.* A good diagnostician can be worth his or her weight in gold. Obviously, the purpose of diagnosis is to identify the cause of the clinical signs that an animal is exhibiting so a treatment can be devised. However, other reasons are equally or even more important. In a herd situation, the extent to which the disease poses a threat to the entire herd must be assessed. Because many diseases can be transmitted from animals to humans, and vice versa, the health effects on the human population must also be considered.

    The animal or herd owner or manager will most likely be the first to observe that something is amiss. Experience has provided owners and managers with some uncanny abilities to detect illness in animals, and they often have a good idea as to the problem, its cause, and its cures. However, in many cases, a complete diagnosis requires the skills of a veterinarian. In addition to specialized training and continuing education requirements to help keep them current, licensed veterinarians have access to laboratories and other resources. A diagnostic procedure can be fairly simple or extremely complex, depending on the severity of the problem, and can include any and all of the following elements:

---

**Toxin**

One of several poisonous compounds produced by some microorganisms, plants, and animals.

**Clinical infection**

Infectious disease in which clinical signs are expressed. Allows identification of the disease.

**Subclinical infection**

Infectious disease in which no readily observable clinical signs exist.

**Acute disease**

Sudden or severe in onset and effect on the animal.

**Chronic disease**

Continuing over a long period or having a gradual effect.

**Diagnosis**

The process of determining the nature and severity of a disease; art of distinguishing one disease from another.

**Diagnostician**

An expert on diagnosing disease.

- History of the sick or dead animal and herd mates, if applicable.
- Clinical examination, including a record of visible signs, body temperature, observations of respiratory rate and character, and listening to the internal body sounds of lungs, rumen, intestine, and heart.
- Palpation of the body for swelling or other abnormalities.
- Rectal palpation to assess the state of various internal organs and to determine pregnancy.
- Collection of specimens such as feces, skin, blood, urine, semen, and other body fluids for various laboratory tests.
- A full diagnostic examination of a dead animal (*necropsy*).

Once the information is collected, the values are compared with normative information, such as the examples shown in Table 14–2. Abnormalities in these numerical values and other nonnumerical abnormalities or problems are detected and a diagnosis is reached through the process of *differential diagnosis* (Box 14–1).

**Necropsy**

The examination of a body after death.

**TABLE 14–2    Some "Normal" Body Values for Selected Species**

| Animal | Rectal Temperature (°F) | Heart Rate (beats/minute) | Respiratory Rate (breaths/minute) |
|---|---|---|---|
| Cow | 101.5 | 48–84 | 26–50 |
| Sheep | 102.3 | 70–80 | 16–34 |
| Goat | 102.3 | 70–80 | 15–30 |
| Pig | 102.5 | 70–120 | 32–58 |
| Dog | 102 | 70–120 | 18–34 |
| Cat | 101.5 | 120–140 | 16–40 |
| Rabbit | 103.1 | 180–350 | 35–60 |
| Horse | 100 | 28–40 | 10–14 |
| Chicken | 107.1 | 250–300 | 12–36 |
| Turkey | 105 | 160–175 | 28–49 |

SOURCE: Compiled from *Merck Veterinary Manual*, 2005, and Reece, 2004.

**Box 14–1    Differential Diagnosis**

Differential diagnosis is a systematic method used to identify the disease responsible for a patient's clinical signs. Diseases with similar characteristics can often be distinguished by comparing clinical signs, physical exam findings, and the results of laboratory tests or other diagnostic procedures. For example, a differential diagnosis list for potential causes of nasal discharge in cattle would include bacterial pneumonia, viral infection, lungworm infection, nasal trauma or foreign body, esophageal obstruction (choke), or pharyngeal abscess. A diagnostician (typically a veterinarian in this scenario) would use information such as the age and vaccination status of the animal, herd history of disease, presence of other affected animals, and so on, along with physical exam findings to determine which cause of nasal discharge is most likely. At that point, the diagnostician would either pursue further testing to verify the diagnosis or begin treatment for the most likely cause and reevaluate if the problem does not resolve.

# VETERINARY SERVICES AVAILABLE

## MOBILE

Mobile veterinary care has been a mainstay of agriculture for decades. It has been traditionally directed at emergency medicine in a rural setting. Such services are becoming more available in urban areas as well. Small-animal mobile clinics can meet a pet owner's need for both emergency and routine care. Many mobile veterinarians work out of a well-outfitted vehicle that makes available most of the equipment needed in everyday practice. Mobile veterinarians often provide more extensive services in affiliation with another veterinarian or at their own clinic, if the need arises (Figure 14–6).

## DRIVE-IN ANIMAL CLINICS/HOSPITALS

Traditional drive-in animal clinics and hospitals, in which one or more veterinarians practice medicine at a fixed location, are modeled on human medical clinics. These clinics are generally well equipped and often have the capacity to hospitalize patients for extended periods of time if necessary. Veterinarians may specialize in treating large or small animals, may focus on treating just one species of animal, or they may be in mixed practice. Small-animal owners frequently receive care for their pets at animal clinics or hospitals. Many drive-in clinics have working facilities to handle large animals efficiently. For example, horses may be brought in for procedures such as castrations. Working facilities for feeder calves or feeder pigs may be provided. A producer who purchases animals can stop and have the animals inspected, vaccinated, and dewormed before taking them home. In some parts of the country, these clinics are very popular.

## PROGRAMMED HEALTH MANAGEMENT

Programmed health management services are much more involved than simply treating sick animals. These programs are designed to be holistic and manage the

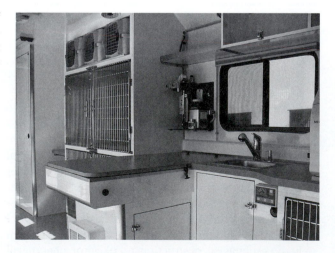

**FIGURE 14–6** A fully equipped mobile clinic for the care of small animals may include an examination area, cage space, a surgical suite, equipment for taking X-rays and laboratory equipment for blood analysis, all within a motor home-type vehicle. (Photo courtesy La Boit Inc. Used with permission.)

numerous aspects of a livestock operation that can affect animal health. The objective of a health management program is to reduce productivity losses owing to management errors or disease by ensuring the optimal care and well-being of the animals. Large kennels, horse and camelid stables, dairies, beef feedlots, and swine operations are among those most likely to have need for this type of program. Services include diagnosis and treatment of sick animals. In addition, the veterinarian may provide consultation on management, nutrition, housing, ventilation, milking technique, vaccination programs, and any other factor that affects animal health. Training is often provided to help those who are responsible for daily animal care to detect symptoms of disease and to accomplish routine management procedures effectively such as castration or vaccination. Medications and treatment guidelines may be provided for easy-to-diagnose ailments so that animals receive the care they need in a more cost-effective and timely manner. Regularly scheduled visits by the veterinarian are used for observation, consultation, treatment, pregnancy checks, and so on. Emergency veterinary care is also generally provided as a part of these programs.

## RESIDENT VETERINARIAN

Some operations are of such size and scale that they need and can afford a resident veterinarian; examples of such operations may include large horse farms, beef feedlots, integrated poultry and swine operations, and a few other isolated types of operations. Resident veterinarians provide virtually all veterinary services, including programmed health management for a single or small number of operations. The term *resident* may be taken very literally, as in the case of some horse operations where the animals are so valuable that the veterinarian lives on the premises and has access to a "clinic" where almost any procedure can be performed. The term may also be more figurative, as in the case of "resident" veterinarians who provide necessary services for large confinement operations for meat animals. Such an individual may be the resident veterinarian for a cluster of operations found in proximity to each other.

## DIAGNOSTIC SERVICES

Some types of diagnostic procedures are impractical to impossible for veterinary practitioners to perform because of prohibitive equipment costs, lack of specialized training, or a variety of other reasons. Veterinary diagnostic laboratories provide services that help practicing veterinarians diagnose diseases and thereby choose appropriate therapies for their patients. Examples of services provided by diagnostic labs include necropsies, examination of *biopsy* specimens, bacterial and viral cultures, serology testing, and toxicological assays. Some diseases pose such a threat to the health of animals and people that governments—local, state, and federal—require veterinarians to report suspect cases and provide diagnostic samples for verification of these diseases. States generally own and operate their own diagnostic laboratories that are staffed, in part, by veterinarians who specialize in diagnosis. These veterinarians may have advanced degrees in public health or advanced training in pathology, microbiology, toxicology, or other specialized areas of veterinary medicine. Private diagnostic facilities are also available and provide diagnostic

**Biopsy**

Surgical removal and examination of a tissue from a living body to reach a diagnosis.

**Antibodies**

Proteins produced by the body that attack infectious agents and neutralize them.

**Passive immunity**

The process of acquiring immunity by receiving preformed antibodies against an infectious agent, as from mother to offspring via colostrum.

**Active immunity**

Immunity to an infectious agent developed in response to exposure to the infectious agent or a vaccine for the disease.

services that focus on special niches, such as the testing of specimens from companion animals or zoo and wildlife species.

## THE BODY'S DEFENSE AGAINST DISEASE

In many instances, infectious agents are repelled by the body's complicated defense mechanisms. The body's general ability to ward off infection or disease is referred to as *resistance* or *immunity*. The body's defense against infection includes barriers that prevent an infectious agent's entry into the body as well as those mechanisms that attack infectious agents that have gained entrance to the body. These defenses, collectively known as the *immune system*, are the major way animals resist infection (Box 14–2 and Figure 14–7).

---

### Box 14–2    The Immune System

The immune system includes physical barriers as well as innate and adaptive types of immunity.

**Surface barriers**  Protective barriers include but are not limited to skin and mucous membranes, linings of the digestive and respiratory tracts, as well as protective mechanisms such as coughing and sneezing. Cilia are found on the cells lining the respiratory tract and help move inhaled infectious agents out of the body. Mucus within nasal passages and tears produced by tear glands also play a role in an animal's resistance to infection by entrapping, diluting, or washing away infectious agents.

**Innate immunity**  The innate immune system is the body's first defense against infectious agents that manage to enter the body despite surface barriers. The innate immune system is made up of cells and protective mechanisms that respond in a generic or nonspecific way to infectious threats. This branch of the immune system responds immediately to foreign invaders but does not provide long-lasting protection against specific pathogens to which the animal has been previously exposed.

**Adaptive immunity**  The adaptive immune system is a sophisticated defense that allows the body to recognize infectious agents it has previously encountered and therefore mount a quicker and stronger protective response against them. *Antibodies* are an important part of the adaptive immune system. They are proteins present in the blood and other bodily fluids and made by antibody-secreting plasma cells, which are special immune cells found primarily in the spleen and bone marrow. Antibodies are able to recognize infectious agents and attack them if they gain entrance to the body.

Maternal *passive immunity* is conferred by transfer of antibodies to an animal from its mother, either through the placenta before it is born or through colostrum shortly after it is born (Figure 14–7). These passively acquired antibodies provide protection against the infectious agents for which the mother made antibodies. They last a short time, around 3 to 4 weeks, and then deteriorate and are cleared from the blood. Another form of passive immunity that may be used to protect animals exposed to a disease to which they are not immune involves the infusion of blood serum (and the antibodies it contains) from an immune animal.

*Active immunity* is acquired by the animal on its own when antibodies are made by the animal's own immune system. Antibody production may be a result of exposure to an infectious agent or may be related to a vaccination designed to simulate exposure to an infectious agent. Because of the waning nature of maternal passive immunity, an animal needs to begin making its own antibodies shortly after birth to protect itself against infectious diseases.

**FIGURE 14–7** Passive immunity is conferred by transfer of antibodies to an animal. This newborn calf is receiving antibodies from the colostrum, or first milk, from its mother. (Photo courtesy of USDA.)

# HERD HEALTH

The demands of most types of animal production, especially horses and the food-producing animals, lend themselves to the absolute need for a coordinated *herd* or *flock health management program.* Some animals, although kept individually or in small groups, are so valuable that they need comprehensive health management programs as well. Even owners of small herds should have a plan for maintaining animals in the best of health. It may not be as elaborate or as necessary from a financial perspective, but the loss of even a single animal that could have been prevented through planning is one loss too many.

**Herd or flock health management program**

A comprehensive and herd-specific program of health management practices.

## MANAGEMENT

The manager is responsible for making the separate parts of a herd health program work together, as a whole. Management plans for small kennels, backyard rabbitries, and cow-calf or sheep operations will be very different than those designed and implemented for a large-scale intensive dairy or swine complex. Flexibility is important for smaller operations because their overall goals differ from those of larger, more intensive operations. In addition, they have much less monetary value at stake. In contrast, large, multimillion dollar operations with thousands of animals on the same site leave little room for flexibility.

## NUTRITION

As discussed previously, improperly balanced and improperly fed rations can be the direct cause of diseases. Certainly these problems must be prevented, and failing that, corrected by the proper mixing and feeding of feeds. Proper nutrition is also important from another perspective. Poorly nourished animals, either underfed or overfed, are more vulnerable to both infectious diseases and metabolic diseases. A sound nutritional program is an essential component of any herd health program.

## GENETICS

The elimination of genetic or hereditary diseases is largely within our ability. Many genetic faults have already been identified and documented using tools and techniques available to modern geneticists. Those tools, which include pedigree analysis, test mating, and laboratory profiling of DNA, provide the means to determine the genetics of individuals and the genetic lines of animals. As the field of molecular genetics continues to advance, more is understood each day; with that understanding, options for improving the genetic quality of livestock increase. The elimination of genetic faults in animals should be a goal of any sound and complete health program. A few specific examples of some undesirable genetic traits include ocular disorders such as retinal atrophy or dysplasia in certain dog breeds, bovine leukocyte adhesion deficiency in Holstein cattle, and hereditary skeletal muscle disorders in horses. Many genetic disorders have already been effectively removed from the gene pool, such as the dwarf gene in beef cattle. Advances in molecular genetics may soon give us tools to engineer enhanced disease resistance into animals or otherwise boost their productivity.

## PREVENTION

Adages are born from the need to pass well-founded advice along to others in a short, easy-to-remember fashion. The applicable and sound adage that applies here is "an ounce of prevention is worth a pound of cure." That short phrase contains volumes of wisdom. Small investments in time and money can reap huge rewards in a reduction of suffering and loss of profit. One important preventative measure is the provision of proper housing or other protection from the environment. Many diseases, whether serious or minor, can be prevented for pennies by the proper administration of vaccines. Early treatment of parasite infections reduces production losses and prevents other parasite-related diseases. Appropriate sanitation measures decrease the potential for exposure to parasites and other infectious agents.

Strategic preventive measures are those planned well in advance of disease and are based on a variety of factors, such as seasonal changes in weather, parasite life cycles, and patterns of infection. For example, grass tetany is a nutritionally related disease that often affects grazing beef cows who are nursing calves, usually in the early spring when the weather is still quite changeable and fertilization has promoted lush grass. Strategic prevention dictates that magnesium supplementation be appropriately carried out and begun early. This can greatly reduce or eliminate grass tetany.

Tactical measures, although they may be planned for in advance, are only put into action when necessary. A tactical approach to grass tetany would include more frequent supervision of susceptible cows during stormy and/or cold weather, and emergency treatment packs readily available to treat affected cows. A grass-tetany cow often doesn't have time to wait on the vet to arrive. She needs help immediately. Anyone trained to hit the jugular vein with a needle and administer the appropriate intravenous treatment can give that help.

Herd health programs have been shown in several studies to improve returns above the cost of the program. Every herd should have a comprehensive health program in place. Such programs should be specific for each operation and each species within the operation. There are just too many variables to have a one-plan-fits-all recommendation. However, the following practices are recommended starting points for the development of comprehensive herd or flock health management plans. The same general guidelines also apply to kennel owners, cattery owners, and other group management situations for animals.

- Consult a veterinarian. Make him or her aware of the objectives and goals for the herd. Hobby herds may have a different set of decision-making criteria applied to them than will commercial herds. Provide the veterinarian with all pertinent information.
- Have an operational procedure for the herd health plan. Make it detailed. Develop a calendar from the plan. Consult it often and do those things that the calendar says it is time to do!
- Identify animals in some permanent way and keep records of vaccination history, previous health problems, medications, reproductive information, and other health-

related information. A variety of preformatted record books and computer-assisted programs are available for most species. However, good useful records can be kept in longhand in a spiral notebook. The value of records generally depends more on the talent and conscientious nature of the record keeper than on the tools used to keep the records.

- Choose animals to be added to the herd from healthy, vigorous stock. Developing a closed herd of females as quickly as possible will reduce the exposure to outside animals because only males will need to be brought into the herd. Have a veterinarian examine any animals prior to bringing them into a herd. Reduce the need for introducing any outside animals by using artificial insemination if possible.

- Separate animals by such characteristics as age, source, intended function, and so on.

- Use *all-in, all-out animal management* whenever possible. This is more practical for such operations as broiler production than it may be for other situations. Even practicing this type of management on a limited scale, such as in a farrowing house, can prevent many infectious disease problems.

- Be sure rations are balanced, mixed properly, and fed as they should be. For small producers, hobbyists, and those working with specialty breeds, purchase high-quality, commercially prepared feeds from reputable companies and feed according to the recommendations of the company nutritionists.

- Provide clean water in a way that it will stay clean.

- Vaccinate for the diseases known to be a problem for your species in your area.

- Minimize the number of people to whom your animals/herd are exposed.

- Develop an "eye for disease." Observe animals frequently and regularly.

- Make all reasonable attempts to discover what caused the death of any animal from the herd. Have necropsy examinations done. In addition, have a plan for properly disposing of dead animals. Check with your local health authorities, your veterinarian, or the state veterinarian's office to determine any laws that apply, as well as the safest and most effective procedures.

**All-in, all-out animal management**

Adding all animals to a facility, such as a farrowing house, at the same time and then removing them at the same time.

# ANIMAL DISEASE AND HUMAN WELL-BEING

## EMOTIONAL LOSS

Although disease and/or death of almost any animal are accompanied by some economic loss, the emotional aspect of animal loss is also important. The very old and the very young are especially vulnerable, but few of us are immune to the emotional upheaval that death and disease of the animals in our care can bring to us. The loss of a valuable companion, a trusted mount, or even a prized breeding animal can inflict deep emotional wounds.

## ANIMAL DISEASE AND HUMAN HEALTH

Certain diseases are *zoonotic,* which means they may be passed to humans from birds, fowl, livestock, pets, or wild animals. Over 100 such diseases have been identified. Whereas some zoonotic infections, such as ringworm, are merely irritating in the discomfort they produce, others, such as leptospirosis or rabies, can be debilitating and/or life threatening. Some zoonotic infections are so problematic that their detection must be reported to one or more government agencies. Obviously, people in close contact with animals—farmers, veterinarians, slaughter plant workers, and so on—stand the greatest risk from the widest

**Zoonotic**

The ability to be passed from animals to humans under natural conditions.

variety of infections. However, the public is also at risk. The greatest public risk is from "food poisonings" by organisms like *Escherichia coli* 0157:H7, *Salmonella, Listeria,* and *Campylobacter.* Although these risks are small for any individual at any time, the very thought of the food supply contributing to disease is an emotional "hot button" issue with the public. The 2006 outbreak of *E. coli* in fresh spinach is a recent example of a food-associated zoonotic disease. In this instance, because the bacterial contamination was suspected to be from an animal source, the outbreak was considered zoonotic. Zoonoses can be transmitted to humans in the following major ways:

- Contamination of animal products with infectious agents or contamination of other food products with infectious agents from an animal source. Many safeguards are in place to avoid this means of disease transmission.

- Direct exposure to infected animals or to hay, water, or food contaminated by a diseased animal, or exposure to fetuses or tissues expelled in cases of infectious abortion.

- Animal or arthropod bites that can spread rabies, Venezuelan or Eastern equine encephalomyelitis virus, West Nile virus, Rocky Mountain spotted fever, Lyme disease, and others.

People are protected from exposure to zoonotic diseases in several ways. One way is to eradicate the disease from the animal population. The U.S. government, in cooperation with the states, has major eradication efforts under way to eradicate brucellosis, tuberculosis, and other zoonotic diseases. Interupting the cycle of transmission from animals to humans is another way to prevent zoonoses. Milk is pasteurized to prevent the possible spread of salmonellosis, brucellosis (undulant fever in humans), tuberculosis, and others. Meat is inspected to prevent the transmission of such diseases as trichinosis, brucellosis, and tapeworm infections. Sanitation procedures are taught to farm workers so they know how to protect themselves. Animals can be vaccinated against diseases like rabies so that, even if bitten by a rabid animal, they will neither get nor transmit the disease. Another method of controlling zoonoses is to destroy the infected population. This has been one of the strategies employed in Europe to try to bring mad cow disease under control (Figure 14–8).

**FIGURE 14–8** A worker at an animal crematorium northeast of London watches as a cow suspected of having bovine spongiform encephalopathy (BSE), known as "mad cow disease," is incinerated. Mad cow disease was first diagnosed in cattle in Europe in 1985. In 1996, the British government announced new evidence linking the disease to a deadly human sickness. The disease has had a major effect on the agricultural economy of Great Britain. (Photographer Stefan Rousseau. Courtesy of AP/Wide World Photos.)

In addition to the problem of zoonoses, the loss of an animal industry or industries because of disease decreases the supply of quality protein for the human diet. This increases prices. Although it affects everyone, the poor tend to suffer first and longest in these situations.

## ANIMAL DISEASE AND NATIONAL ECONOMIES

History is full of examples in which disease in the animal population has had profound effects on economies, nations, and peoples. Great Britain's war effort was seriously hurt in 1940 by an outbreak of foot-and-mouth disease, a disease that struck Europe again in 2001. In 1929, the Chicago livestock yards had to be closed because of the same disease. At that time, closing Chicago to livestock was equivalent to shutting down the meatpacking industry. More recently, the farm economy, and thus the national economy, of Europe has been under tremendous pressure because of the losses caused by another serious malady, mad cow disease. Confirmation of three cases of mad cow disease in the United States had a significant effect on the prices and trade of U.S. cattle and beef. The economic impact in the United States was much less severe than that in Great Britain, and as a result of vigilant surveillance and effective regulatory controls, U.S. beef exportation is on its way to recovery.

In many developing nations, trade suffers because diseases among the country's animals force other governments to prohibit shipment of products from entering their countries. Avian influenza, or bird flu, has recently had a tremendous economic impact on the poultry industry. Countries in Asia and Africa, many of which are poor, are among those hit hardest by mass poultry culling and import bans. Their economies falter and their people suffer as the livelihoods of millions are threatened and a critical source of dietary protein is lost. Additionally, a pandemic bird flu infection among humans, which is a realistic threat, could cost the global economy up to $2 trillion, according to a recent World Bank estimate.

A few examples of animal disease affecting our domestic economy include restrictions on the state-to-state transport of horses in 1971 because of Venezuelan equine encephalomyelitis and restrictions on horse transport in 1998 and 1999 owing to an outbreak of equine infectious anemia. Livestock diseases cost ranchers, farmers, and consumers approximately billions of dollars yearly.

Despite the serious impact animal disease can have on society, do not draw the conclusion that all of agriculture and, by extension, the civilized world is one disease away from collapse. This just isn't so. The state of our knowledge, the arsenal of weapons, and the resources available to marshal against animal disease are impressive and getting better. The general population is very well protected. We should all be comforted when we hear of a meat recall or detection of a reportable disease by a surveillance program. It means the system in place is working.

## ANIMAL DISEASE AND FARM INCOME

Disease can decrease income for operations that depend on either the production of individual animals, such as a horse farm, or profit from average or overall herd production, such as a beef feedlot or dairy farm. Operations that focus on the productivity of individual animals are vulnerable to the ravages of disease, which may cause losses in several forms. Death among livestock is an easily measured loss that can be accounted for on an individual basis. The profit from the remaining animals must pay for the lost individual before a profit can be declared. In most cases, the loss of a single animal robs the producer of the profit from several others. A less dramatic but sometimes greater economic drain to operations that depend on average rather than individual production within the herd is the loss of product. This can occur when meat or milk is condemned at the processing plant. The product may be deemed unwholesome owing to either the direct effects of the disease process or the antibiotic or other drugs used to treat the disease. Food safety inspection systems are in place to prevent such unwholesome product from entering the human food chain. Potentially, the greatest loss of all is the loss of gain or satisfactory production that results from unthriftiness caused by disease. An animal that has been ill and then recovers frequently performs at lower rates than if it had never been ill. A cow with subclinical mastitis has reduced milk production because of the infection. Reproductive failures and decreased numbers of offspring can be complications of some diseases. In addition to these losses, the costs of veterinary services, drugs, and labor drain the profit from the operation.

## BIOTERRORISM

Most thinking and planning regarding bioterrorism has focused largely on humans as the primary target. Certainly much human illness, death, and panic could be accomplished in this way. However, when economic and political vulnerabilities are considered, agricultural bioterrorism (the intentional targeting of a nation's livestock and crop resources) is at least as likely. It has also been suggested that an attack on the food supply or food economy might be more attractive to terrorists because of the secondary effects on humans and the potential for deniability that might make the response or retribution less likely. The General Accounting Office has concluded that intentional disease attacks against agricultural commodities, especially livestock, would be economically devastating.

U.S. agriculture is becoming more vulnerable to agricultural bioterrorism owing to the continuing trends of intensive production techniques, vertical integration, the increasing industrial dependence on the export market, and the lack of U.S. livestock resistance against many pathogens and pests. Added to this agricultural vulnerability is the relative ease of acquiring, producing, and disseminating animal pathogens. To better manage the increasing risks of agricultural bioterrorism, the Department of Homeland Security has led the development of a National Response Plan that spells out how the nation would work together in the event of a terrorist attack. Efforts to create stockpiles of important vaccines as well as create laboratory networks to enhance disease diagnosis and monitoring are under way.

It is beyond the scope of this text to do other than raise awareness on this topic. An excellent resource for current information on bioterrorism is www.bt.cdc.gov/bioterrorism, which offers an overview of bioterrorism and several links to more resources. For current information regarding the safety of the food supply, www.foodsafety.gov is a resource that brings together information from relevant websites.

## REGULATORY ANIMAL MEDICINE

Regulatory animal medicine is the sum of the activities directed by the government agencies charged with these tasks:

- Keep foreign animal diseases out of the United States.
- Stop or slow the spread of animal diseases across state lines.
- Eradicate selected animal diseases from the United States.
- Assist in the protection of the welfare of particular groups of animals.

Currently, some 40 diseases not present in the United States are considered a threat to U.S. poultry and livestock. A single outbreak of foot-and-mouth disease carries an estimated first-year price tag of $10 billion for containment alone. Clearly, preventing the introduction of foreign animal diseases needs to be, and is, a priority with government and private organizations.

## FEDERAL REGULATION

USDA's Animal and Plant Health Inspection Service (APHIS) is a multifaceted agency with a broad mission that includes protecting and promoting U.S. agricultural health, regulating genetically engineered organisms, administering the Animal Welfare Act, and carrying out wildlife damage management activities. APHIS is organized into six program units, and an Office for Emergency Management and Homeland Security, along with three management support units and the Office of Civil Rights. Five of the six program units as well as the Office for Emergency Management and Homeland Security are briefly described with relevance to regulatory animal medicine in Box 14–3. The Bureau of Customs and Border Protection, which is responsible for agricultural inspection (Figures 14–9 and 14–10), is briefly described in Box 14–4.

**FIGURE 14–9** Customs and Border Protection agricultural specialists detect, confiscate, and destroy items held by international travelers that could bring pests and diseases into the United States. Pictured here are confiscated meat and vegetables at Dulles International Airport, Sterling, Virginia. (Photographer Ken Hammond. Courtesy of USDA.)

**FIGURE 14–10** The Beagle Brigade is instrumental in helping Customs and Border Protection detect contraband at U.S. ports and airports. Undetected contraband could threaten U.S. agriculture, the food supply, and the economy. (Photo courtesy of USDA/APHIS; image BB 206.)

## Box 14–3   Animal and Plant Health Inspection Service

*Animal Care (AC)* is charged with the responsibility of inspecting animals in research, exhibition, and other regulated industries to ensure compliance with animal welfare legislation and to ensure the proper stewardship of these animals.

*International Services (IS)* works with foreign governments to protect U.S. agriculture and enhance agricultural trade. IS negotiates entry requirements for U.S. agricultural products to other countries and for agricultural products shipped to the United States. Preclearance in the country of origin is often easier, more practical, and a better means of excluding disease. IS also cooperates with foreign governments in agricultural pest and disease control or eradication programs. Cooperatiave eradication programs include efforts directed at screwworms and foot-and-mouth disease.

The *Plant Protection and Quarantine (PPQ) Service* was responsible for safeguarding U.S. agriculture and natural resources from damage related to the entry and spread of animal and plant pests and noxious weeds until March 1, 2003. At that time, The *Bureau of Customs and Border Protection (CBP)* was created within the Department of Homeland Security and made responsible for the inspection and prevention efforts necessary to keep prohibited agricultural items from entering the United States. PPQ works with the CBP to ensure the continued success of agricultural inspection by providing training and expertise in pest detection and identification. Box 14–4 provides a brief overview of agricultural inspection by the CBP. PPQ also oversees plant import and export and responds to the introduction of plant pests or plant health emergencies.

*Veterinary Services (VS)* works to prevent the introduction of animal diseases to the United States by regulating the importation of animals and animal products. If a disease is introduced, VS takes emergency action to contain it. Exotic Newcastle disease was accidentally introduced in the 1980s by pet birds, and then in 2002–2003 it was detected in backyard poultry flocks. VS led the effort to contain and eradicate the disease in both instances. VS also conducts disease eradication programs for existing animal diseases in cooperation with the states, provides health certification for exported animals and animal products, conducts diagnostic tests, and is the licensing authority for veterinary biological products and manufacturers. If private producers or practicing veterinarians detect or suspect an exotic disease, they contact one of over 300 specially trained state and federal veterinarians located throughout the United States to investigate the situation. If an exotic disease is diagnosed, VS can send a specially trained task force into the site of the outbreak to direct the eradication plan. The APHIS *Office of Emergency Management and Homeland Security* coordinates the response effort to ensure that available resources at federal, state, and local levels are used effectively and efficiently to contain and eradicate exotic disease threats.

*Wildlife Services (WS)* is responsible for reducing wildlife damage to agriculture and natural resources, minimizing potential wildlife threats to human health and safety, and protecting threatened and endangered species.

APHIS has responsibility for regulating the importation and exportation of plants, animals, and agricultural products. Regulations depend on the product and the country's product of origin. Animals must have a health certificate issued by veterinary health officials in the exporting country. Meat, some dairy products, and other animal products are restricted from countries that have a different disease status than that in the United States. Live animal imports are restricted, and many animals must be quarantined at one of three animal import centers located in Newburgh, New York; Miami, Florida; and Los Angeles, California. Cattle from countries affected with bovine spongiform encephalopathy (BSE), foot-and-mouth disease, or rinderpest cannot be imported into the United States. Personally owned pet birds and commercial shipments of pet birds enter through one of three USDA-operated bird quarantine facilities in New York City; Miami, Florida; and Los Angeles, California. Canada has sufficiently stringent import rules that birds may enter from Canada without quarantine.

## Box 14–4   U.S. Customs and Border Protection

The *Customs and Border Protection (CBP)* employs approximately 2,700 agricultural specialists that were formerly inspectors with the PPQ unit of USDA-APHIS. These agricultural specialists work at key U.S. ports of entry, international airports, border stations, and international mail facilities, where they inspect commercial cargo, luggage, and passengers/pedestrians. These inspectors enforce USDA regulations and seize prohibited meat, plant materials, and animal products (Figure 14–9). On a typical day in 2006, CBP inspectors were responsible for seizing 4,462 prohibited agricultural items. They use a variety of methods such as X-ray screening and trained detector dogs to find agricultural contraband. These dogs, known as the Beagle Brigade, are responsible for around 75,000 of the 2 million seizures of prohibited products each year (Figure 14–10).

*SOURCE:* Compiled from U.S. Customs and Border Protection, 2007.

The prevention of the entrance of mad cow disease, or bovine spongiform encephalopathy (BSE), to the United States is a good current example of APHIS's service. This disease has severely affected Europe's livestock industry. In 1989, APHIS began restricting the importation of live ruminants and ruminant products from Great Britain and other countries where BSE has been diagnosed. During the same year, it began a BSE surveillance program to monitor for the disease. This program included the direct examination of the brains of thousands of cattle. These and other measures were designed to prevent the entrance of this disease to the United States, and thereby protect the health of our livestock, our people, and our economy. However, in December 2003, it was announced by the USDA that a dairy cow in Washington state had tested positive for BSE. After an extensive investigation including the location and testing of animals that had entered the United States from Canada with the affected cow, as well as other animals in contact with the affected cow, it was determined that there were no collateral cases. As of June 1, 2004, an enhanced BSE testing program was implemented by the USDA. In June 2005, a second BSE-positive cow was detected in Texas and then in March of 2006 a third case of BSE in the United States was detected in a cow in Alabama. Based on data collected in the United States over the last seven years, including over a half million samples from the enhanced surveillance program, the USDA estimates that less than one in a million cattle in the United States is infected with BSE, with the total number of BSE infected cattle in the country estimated to be between 4 and 7. The number of BSE-infected cattle in the United States is expected to decline as long as current regulations maintain low risk for introduction and spread of the disease. For the latest on BSE, visit the USDA BSE information page at http://www.fas.usda.gov/dlp/BSE/bse.html.

Another example of the work done to protect the agricultural sector of the United States began in early 2001. At that time, the USDA began a coordinated effort among several government and private agencies, including APHIS, to guard against a North American outbreak of foot-and-mouth disease (FMD). FMD is a highly contagious viral disease of cattle, swine, sheep, goats, deer, and other cloven-hoofed ruminants. Although rarely transmissible to humans, FMD is devastating to livestock. An outbreak would potentially have severe economic consequences associated with losses in the production and marketing of meat and milk (Figure 14–11). The disease is difficult to control and has occurred in over 60% of the world. In today's highly mobile environment, FMD could be accidentally introduced and disseminated in the United States. A single infected animal or one contaminated sausage could carry the virus to American livestock. To protect against the occurrence of this disease, the U.S. government monitors FMD occurrence worldwide, evaluates the potential risk of foreign outbreaks to the United States, and reduces disease spread by assisting other nations with a rapid response when FMD or other devastating diseases are detected.

Avian influenza, or "bird flu," is another disease the USDA and other government agencies are working hard to safeguard the nation against. Avian influenza is a virus that infects wild birds and domestic poultry (Figure 14–12). High-pathogenicity strains of avian influenza (HPAI) have been rapidly spreading in some parts of the world and

**FIGURE 14–11**    Foot-and-mouth disease symptoms. Foot-and-mouth disease is characterized by fever and blister-like lesions followed by erosions on the tongue and lips, in the mouth, on the teats, and between the hooves. Many affected animals recover, but the disease leaves them debilitated. It causes severe losses in the production of meat and milk. (Photo courtesy of USDA.)

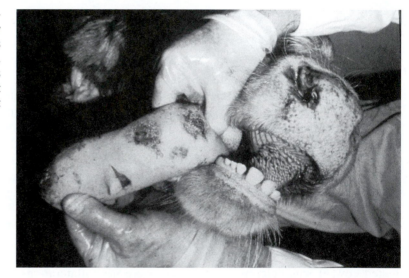

**FIGURE 14–12**    These turkeys are showing signs of diarrhea and depression related to infection with avian influenza. (Photo courtesy of USDA.)

have been responsible for three outbreaks in poultry in the United States, the most recent of which was in 2004. A rapid and coordinated response by the USDA and state, local, and industry leaders limited that outbreak to one flock and led to its rapid eradication from the country. To prevent the introduction of HPAI into the United States, the USDA quarantines and tests live birds that are imported into the country, maintains restrictions on poultry importation from regions with confirmed outbreaks of HPAI, and conducts surveillance of U.S. bird populations. Should an outbreak of HPAI occur within the United States, the USDA would lead an emergency response plan that would include quarantine, eradication, disinfection, continued monitoring, and testing of the affected region. In addition, USDA works with international organizations to assist countries affected by HPAI in disease prevention, management, and eradication efforts. At press time for this text, HPAI continues to be a problem in many countries throughout the world. For current information on avian influenza, visit the Department of Health and Human Services website at www.avianflu.gov.

## National Animal Identification System

The National Animal Identification System (NAIS) is a modern information system designed to help producers and animal health officials respond to U.S. animal health events quickly and effectively. A national animal identification system is expected to benefit disease control and eradication, disease surveillance and monitoring, emergency response to foreign animal diseases, consumer concerns over food safety, livestock production efficiency, and emergency management programs. Participation in the NAIS program, which is to be run cooperatively by state, federal, and industry partners, is voluntary and will guard producer confidentiality. NAIS has three voluntary components: premises registration, animal identification, and animal tracing. At press time for this text, the premises registration component of NAIS is currently available and animal identification is available for some species and in progress for others. Animal tracing is in progress for all species, and implementation of the entire system is projected for January 2009. Once fully in place, the NAIS should help in the rapid notification and containment of disease outbreaks, help keep markets open for unaffected producers, and prevent unnecessary restrictions on domestic and international transportation and trade. For additional up-to-date information about the NAIS, visit www.usda.gov/nais.

### State Regulation

Each state has a *state veterinarian*. These individuals usually serve under the direction of the head of the department of agriculture, although this may vary from state to state. State veterinarians and their offices are responsible for administering the legislation specific to their state—for example, interstate transport regulations and quarantine regulations for stock moving into or through the state. They are responsible for cooperating with APHIS on federally directed programs such as the brucellosis and hog cholera programs.

### Accredited Veterinarians

The Veterinary Accreditation program administered by the Veterinary Services division of APHIS qualifies private veterinarians to work cooperatively with federal and state veterinarians and animal health officials in preventing, controlling, and eradicating livestock and poultry diseases. More than 80% of all U.S. veterinarians are accredited. Accredited veterinarians assist producers in buying, selling, and transporting animals by examining animals for health and soundness, testing for specific diseases, and issuing certificates of inspection.

## UNITED STATES ANIMAL HEALTH ASSOCIATION (USAHA)

The United States Animal Health Association (USAHA) is a national nonprofit organization of approximately 1,400 members. It works with state and federal animal health officials, veterinarians, livestock producers, national livestock and poultry organizations, research scientists, the extension service, and seven foreign countries to control livestock diseases in the United States. The USAHA is not a government agency, but it does serve as an advisory body to the USDA. The USAHA informs state and federal authorities of current disease situations, methods of control, and other information of importance about animal disease. This group unofficially coordinates disease control work among states and the federal government and suggests new laws, regulations, and programs for disease control. The USAHA has 32 working committees set up to deal with the various aspects of disease control in domestic livestock. The USAHA holds an annual meeting each fall preceded by spring meetings of the four regional groups that make up the association.

## ORGANIZATIONS

The following resources are provided for those of you who wish to seek additional information.

**American Veterinary Medical Association**
1931 North Meacham Road, Suite 100
Schaumburg, IL 60173
Phone: (847)925-8070
Fax: (847)925-1329
E-mail: avmainfo@avma.org
http://www.avma.org/

**USDA, APHIS, Veterinary Services**
National Animal Health Programs
4700 River Road, Unit 43
Riverdale, MD 20737-1231
Phone: (301)734-4913
Fax: (301)734-7964
http://www.aphis.usda.gov

**United States Animal Health Association**
P.O. Box K227
Richmond, VA 23288
Phone: (804)285-3210
Fax: (804)285-3367
http://www.usaha.org/

## SUMMARY AND CONCLUSION

Disease is any state other than a state of complete health. In a state of disease, the normal function of the body, or some of its parts, is changed or disturbed. Maintaining animals as near to a constant state of health as possible or feasible is a challenge, an obligation, and a necessity to animal stewardship. Animals have a natural defense system against disease involving a number of factors that can be loosely grouped under the term *resistance*. When the defense mechanisms fail, disease occurs.

Disease is usually caused by a combination of predisposing causes and direct causes. Predisposing causes are often referred to as stress factors. They come in great variety, but have in common the fact that they place unusual or additional demands on the body. Direct causes of disease include several categories of infectious etiologies: bacteria, viruses, protozoa, external parasites, internal parasites, fungi, and the poorly understood prions associated with mad cow disease. These causes are distinguished by the fact that they are themselves living agents that cause disease by their presence in or on the body. Diseases caused by living organisms are called infectious diseases. Other direct causes of disease are noninfectious. Nutrient deficiency diseases result directly from improperly balanced and/or fed rations. Some diseases such as hemophilia are directly caused by genetic makeup. The direct cause of an injury may be a nail, a sharp instrument, a falling object, or a wet and slick floor. A metabolic disturbance is a direct cause of many diseases. Toxins or chemical poisons can also be direct causes of disease.

Detection and identification of animal disease is collectively referred to as diagnosis. One of several types of veterinary services can be called on to diagnose, treat, and prevent animal diseases. A comprehensive herd health plan is a good idea for any production unit. It should include the specifications for management, nutrition, genetics, and disease prevention. Several types of human losses are associated with animal loss related to diseases. As a means of minimizing those losses, the USDA, state governments, and private organizations work together to fight disease in animals.

# STUDY QUESTIONS

1. What are the motivating factors for keeping animals healthy? Are some of these factors different for companion animals than for livestock species?

2. What is a clinical sign? A lesion?

3. Define *pathology* and *etiology*.

4. Describe in detail the causes of disease. Include a discussion of the difference between direct and predisposing causes of disease.

5. What is the difference between a clinical and a subclinical infection? Acute and chronic disease?

6. In your own words, outline a diagnostic procedure for determining why an animal is "off his feed" and appears to be ill.

7. Describe the general types of veterinary services available to producers and other animal owners. Which would be the best choice, in your opinion, for a 100-dog kennel owner? A 10,000-sow farrowing house? A 50,000-sow farrowing unit? A horse owner who has three horses? A llama owner with 20 breeding females? Support your decision in each case.

8. What are the different parts of the immune system and types of immunity? How does an animal receive or develop immunity?

9. Why is a herd health plan so important? What are all the elements that must be considered?

10. Describe the effects of animal health on human health and well-being.

11. What is the purpose of regulatory animal medicine? Who are the major groups involved in regulatory medicine? What is the function of each?

## REFERENCES

For the fourth edition, Melanie A. Breshears, DVM, PhD, Diplomate ACVP, Assistant Professor, Veterinary Pathobiology, Center for Veterinary Health Sciences, Oklahoma State University, has assumed co-authorship of this chapter.

Berry, J. G. 1998. *Livestock disease and control.* Extension Facts No. F-3999. Stillwater: Oklahoma Cooperative Extension Service, Division of Agricultural Sciences and Natural Resources, Oklahoma State University.

Berry, J. G. 2007. *Livestock disease: Cause and control.* Extension Facts No. ANSI-3999. Stillwater: Oklahoma Cooperative Extension Service, Division of Agricultural Sciences and Natural Resources, Oklahoma State University.

Frandson, R. D., and T. L. Spurgeon. 1992. *Anatomy and physiology of farm animals.* 5th ed. Philadelphia: Lea & Febiger.

Noah, D. L., D. L. Noah, and H. R. Crowder. 2002. *Biological terrorism against animals and humans: A brief review and primer for action.* JAVMA, July 1, 2002. Available online at http://www.avma.org/public_health/biosecurity/zu_bioterrorism.asp.

Reece, W. O., ed. 2004. *Dukes' physiology of domestic animals.* 12th ed. Ithaca, NY: Cornell University Press.

Sainsbury, D. 1998. *Animal health: Health, disease and welfare of farm livestock.* 2nd ed. Oxford: Blackwell Science Ltd.

Smith, C. A. 1998. *Career choices for veterinarians.* Leavenworth, WA: Smith Veterinary Services.

*The Merck veterinary manual.* 2005. 9th ed. Whitehouse Station, NJ: Merck.

U.S. Customs and Border Protection. 2006. *Factsheet: On a typical day . . .* Accessed online July 6, 2007. www.cbp.gov/linkhandler/cgov/newsroom/fact_sheets/cbp_overview/typical_day.ctt/typical_day.pdf.

U.S. Customs and Border Protection. 2007. *U.S. Department of Agriculture, Animal and Plant Health Inspection Service—Protecting America's agricultural resources.* Accessed online July 6, 2007. www.cbp.gov/xp/cgov/toolbox/about/history/aqi_history.xml.

USDA-APHIS. 2006a. *Bovine spongioform encephalopathy (BSE).* Ongoing surveillance plan. Accessed online July 10, 2007. http://www.aphis.usda.gov/newsroom/hot_issues/bse/downloads/BSE_ongoing_surv_plan_final_71406%20.pdf.

USDA-APHIS. 2006b. *Strategic plan (2003–2008).* Accessed online July 3, 2007. www.aphis.usda.gov/about_aphis/downloads/APHIS_StrategicPlan-Feb2006.pdf.

USDA-APHIS. 2007a. *About APHIS.* Accessed online July 5, 2007. www.aphis.usda.gov/about_aphis.

USDA-APHIS. 2007b. *National Animal Identification System (NAIS).* Accessed online July 9, 2007. http://animalid.aphis.usda.gov/nais/index.shtml.

USDA. 1984. *Yearbook of agriculture, animal health, livestock and pets.* Washington, DC: USDA.

Van Kruiningen, H. J. 1999. *Daniels' health and disease management in animals.* 2nd ed. Storrs, CT: H. J. Van Kruiningen.

# PART THREE

# The Animal Industries

# Market Coordination in the Beef, Pork, and Poultry Industries

## LEARNING OBJECTIVES

After you have studied this chapter, you should be able to:

1. Describe market coordination and vertical integration and explain their significance in the beef, pork, and poultry industries.

2. Explain how the biological production cycle, the genetic base, industry stages, geographic concentration, and operation size contribute to or limit vertical integration.

3. List the motives and limitations to the industry for integrating an industry.

4. Explain the disadvantages to integrating an animal industry.

5. Compare the probability of increased integration in the three major meat-producing industries.

## KEY TERMS

Brand marketing
Capital requirements
Case-ready product
Certified Angus Beef
Contract integration

Differentiated product
Fresh-packaged product
Ownership integration
Strategic alliances
"The Other White Meat"

Value-added product
Vertical coordination
Vertical integration

## INTRODUCTION

The predominant organizational changes in the meat-producing industries in the last half of the 20th century were changes in the market structure and the development of vertical integration. It is essential for anyone working in the agricultural industries to understand how the markets are coordinated and integrated and how this affects the individual and collective industries.

## VERTICAL INTEGRATION DEFINITION

**Vertical coordination**
Organizing products flow from producers to consumers and information about the products from consumers to producers.

*Vertical coordination* is the process of organizing, synchronizing, or orchestrating the flow of products from producers to consumers and the reverse flow of information from consumers to producers. One extreme means of vertical coordination involves a totally open-market system where all coordination is accomplished by market prices. In an open-market system, market prices signal consumer preferences to producers and guide production decisions to fulfill consumer demands. At the other extreme is a totally vertically integrated operation where one firm owns and controls a commodity and the products processed from it through the entire producer-to-consumer chain. In this case, the integrating firm decides what, how, and how much to produce and process to meet consumer demands.

**Vertical integration**

The control of two adjacent stages in the vertical marketing channel from producers to consumers.

*Vertical integration* is the control of two adjacent stages in the vertical marketing channel from producers to consumers; for example, one firm engaged in both cattle feeding and meat packing. The two primary types of vertical integration are *contract integration* and *ownership integration*. There are many variations of contract integration. Some people include joint ventures as a form of contract integration; others may consider it a separate type of vertical integration.

**Contract integration**

When one firm from one industry phase contracts with a firm at an adjacent phase for products and/or services.

*Contract integration* involves a firm at one production–processing–distribution stage (such as meatpacking) contracting with a firm at an adjacent stage (such as cattle feeding) for specific services and/or products (such as fed cattle for slaughter). Both parties may own some, but not all, of the necessary resources. A contract, if written, would typically specify which party provides what resources, services, or products. Contracts also would likely include terms related to quality, quantity, time, and place of the services or products, how price is determined, and when payment is made.

**Ownership integration**

When an integrated operation is under one ownership.

*Ownership integration* differs in that the integrating firm owns most resources in both adjacent production–processing–distribution stages. An example would be a meatpacking firm owning a cattle feedlot and some or all of the cattle fed in the lot. Because most resources are owned by the integrating firm, alternative forms of ownership integration do not exist, only the extent to which resources are owned by the vertically integrated firm.

Most examples of vertical integration involving production agriculture are variations of contract integration. Vertical integration by definition involves two adjacent production–processing–distribution stages, but may include all stages. The former may be referred to as a *partially vertically integrated industry;* the latter may be called a *totally vertically integrated industry.* The poultry industry is essentially constructed of totally integrated operations (Figure 15–1). However, coordination changes have occurred in the beef and pork as well as poultry industries in conjunction with many other structural changes in each of these industries. Changes have been more noticeable for beef and pork because these industries both followed and responded to trends begun decades ago in the poultry industry, primarily for broiler chickens. Competitive pressures from poultry caused the beef and pork industries to seek greater efficiency and improved coordination. For example, several **strategic alliances** have been organized in the beef industry over the past decade to improve coordination. In the pork industry, contracting between pork-packing firms and larger hog operations increased sharply, along with packers integrating into hog production, both in an effort to increase production efficiency.

**Strategic alliances**

Partnerships between various independent segments of an industry to maximize cooperation, value, and return on investment.

## VERTICAL COORDINATION MOTIVES

Motives to improve market coordination in animal industries can stem from several sources. Purely open-market systems of coordination put tremendous pressure on market prices as the primary means of communicating consumer preferences to producers. Inadequacies of this process can end as market failures. However, market failures typically present opportunities for innovation and profit. In our capitalistic economy, profit

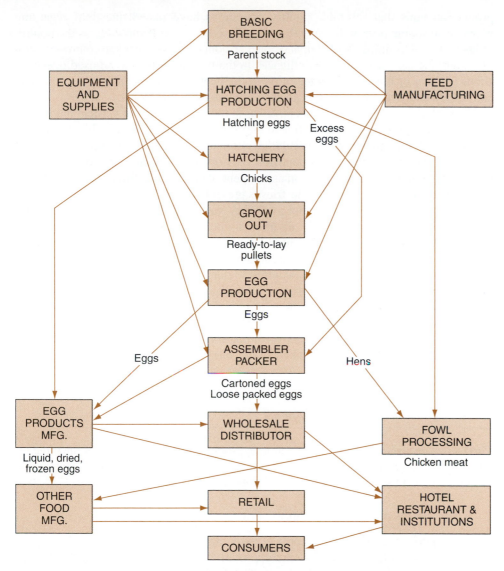

**FIGURE 15–1**   The organization of the egg subsector showing vertical market stages. (*SOURCE:* Adapted from Schrader et al., 1978, p. 23.)

opportunities are the ultimate economic incentive for many market changes. These profit opportunities may arise in response to inefficiencies in production, processing, or distribution, large transaction costs between stages in the producer-to-consumer channel, the application of new technology that may reduce costs or lead to new or improved products, or demand changes in the form of changing consumer preferences.

## CURRENT STATE OF VERTICAL INTEGRATION

Much discussion has taken place regarding the vertical structure and evolutionary changes in the beef and pork industries. Frequently those changes are compared with those of the poultry industry. The growth of contract integration and mega-sized hog production units that resemble poultry operations have raised questions regarding whether or not the pork industry will integrate vertically as completely as the poultry industry has. Declining beef demand, packer concentration, packer-controlled supplies, and the advent of strategic alliances in the beef industry have raised questions about whether or not vertical integration in the beef industry is desirable or possible and whether vertical integration is an answer to its many problems.

This chapter provides "*a perspective*" as to the likely extent of vertical integration in the beef and pork industries relative to the poultry industry. Specifically, differences that may contribute to or deter vertical integration are identified and discussed. Several incentives and disincentives are identified and compared among the three industries. Lastly, drawing on this discussion, an assessment of current and future vertical integration trends in the beef and pork industries is presented.

One caveat is noted. Integration in poultry vastly exceeds that of pork and beef. Therefore, in many cases, what is observed about poultry is the *result* of integration, not necessarily something that was foreseen prior to the industry becoming vertically integrated. To a much lesser extent, the same could be argued about pork. However, for beef, little vertical integration has occurred to date. Therefore, structural characteristics observed in beef production are primarily compared with vertical integration results or vertically integrated operations in poultry, and to a lesser extent with pork. Thus predictions for vertical integration in beef, and somewhat in pork, are based on similarities and differences compared with poultry that contribute to or impede moving toward more vertically coordinated and/or integrated industries.

## PRODUCTION CHARACTERISTICS: BEEF, PORK, AND POULTRY

There are some basic physical and economic production characteristics of the three industries that contribute to or limit vertical integration (Table 15–1). Many factors discussed in this and subsequent sections are interrelated.

### BIOLOGICAL PRODUCTION CYCLE

The conception-to-market period for beef, pork, and poultry varies widely. Time periods given here and in Table 15–1 are approximations. Variations in the production process can alter the time periods shown. Cattle have about a 24-month time period

**TABLE 15–1    Production Characteristics: Beef, Pork, and Poultry**

| Characteristics | Beef | Pork | Poultry |
|---|---|---|---|
| Biological production cycle | 24 months | 10 months | 5(4) months |
| Genetics base | Wide | Narrow | Narrow |
| Industry stages | Cow-calf, stocker, feeding | Farrowing, finishing | Hatching Growing |
| Geographic concentration in production | Dispersed; varies by production stage | Midwest, mid-Atlantic, southern Plains | Southeast |
| Operation size and specialization | Varies by production stage | Large and specialized | Large and specialized |

*SOURCE:* Clement E. Ward. 1997. Used with permission.

from conception to market: 10-month gestation phase, 6-month preweaning phase, 4-month growing phase, 4-month feeding phase, and 1-month processing/distribution phase. Some flexibility exists in altering these phases for all but the first phase. Hogs have about a 10-month time period from conception to market: 4-month gestation phase, 1-month preweaning phase, 4-month finishing phase, and 1-month processing/distribution phase. There is less flexibility and variability in absolute time from conception to market with hogs than with cattle. Poultry has about a 5-month period from conception to market: 1-month incubation phase, 3-month growing phase, and 1-month processing/distribution phase. The total time for broiler chickens is less than that for turkeys, reducing the 5-month figure by 1 month. The variability and flexibility from conception to market with poultry is even less than that of hogs.

The importance of the biological process to vertical integration is interrelated with factors that are discussed later in the chapter. Perhaps the primary factor involves the speed with which biological changes such as genetic improvements can be made. Although this factor is present both under open-market and vertically integrated systems in the same industry, it affects the incentives and disincentives for vertically integrating and the ease or difficulty for increasing coordination. For example, if a firm is considering improvements in product quality stemming from genetic or biological changes, there is more incentive to integrate vertically in an industry that has a shorter biological process and in which genetic changes can be made more quickly.

## GENETIC BASE

The genetic base for poultry is relatively narrow. Only a few breeds or genetic lines (i.e., fewer than 10) are used to provide the vast majority of final products. Genetic changes can be made more quickly in poultry because of the shorter biological process, ultimately leading to more consistent poultry products. Both the short biological process and more uniform animals resulting from a relatively narrow genetic base are important for managing the production process and production costs. They also affect managing costs in processing and in getting consistent products to consumers. Overall, they contribute to enhanced coordination in the vertical market channel. Genetic changes can also be made more quickly because one hen produces many more offspring in a short time than does either a cow or sow. This narrow genetic base may be a result of vertical integration in poultry and is an incentive for pork and beef.

The genetic base for hogs has narrowed considerably in recent years. Today, just a few specialized firms provide the breeding stock for nearly all large hog operations. Because of the shorter biological process, genetic changes in hogs can be made more quickly and, through larger hog litters, can influence more offspring in a single breeding cycle than with cattle. Making quicker genetic changes also affects efforts to reduce production costs and increase consistency of pork products for consumers.

In the beef industry, the genetic base is still quite wide. Some cattlemen are continuing to create or import new breeds. This results in further amalgamation or agglomeration of the genetic base. There are desirable genetics in every breed, but as yet, there is no easy, economical method of recognizing many of those desirable genetic traits in commercial cattle operations. The biological process is also a serious deterrent to changing the genetic base quickly because a cow produces only one calf per year and it takes about 24 months to learn whether or not the breeding process resulted in beef with more or less desirable eating characteristics. Therefore, making significant product quality improvements based on genetic changes is slow and a disincentive to vertical integration.

Technology can influence the speed of genetic changes. Artificial insemination and embryo transfers can speed the process somewhat for pork and beef. However, the costs are still high for using such technology on a large scale for beef. The high costs reduce the incentive for firms to integrate vertically and employ the available technology. However, work is progressing rapidly on identifying key genetic markers, which could could significantly speed genetic changes in beef and make other technologies worth the investment.

## INDUSTRY STAGES

The poultry industry has two primary production stages, hatching and growing, plus the processing and distribution stages common to all three industries. This contributes to the ease of managing a vertically integrated production process and reduces transaction costs between industry stages. Traditionally, the pork industry also has had two primary production stages, farrowing and finishing. The beef industry is at a relative disadvantage compared with the poultry and pork industries. The production process for cattle consists of cow-calf, stocker or growing, and feeding. Thus the beef industry has a third production stage, one more than for poultry, and production stages that are more variable than pork. The additional production stage increases the transaction costs for the industry. In addition, each stage has different resources and management needs. Combined, these differences increase the difficulty in managing a vertically integrated beef production unit.

## GEOGRAPHIC CONCENTRATION IN PRODUCTION

Poultry, pork, and beef are produced in every state. However, the concentration of production differs significantly by industry. Poultry production, especially broiler production, is more narrowly concentrated geographically in the southeastern United States. Turkey production is less concentrated in the southeastern United States than is broiler production, but the geographic concentration of turkey production is greater than that of pork or beef.

Traditionally, hog production has been geographically concentrated in Iowa and the surrounding Corn Belt states. However, pork production has increased sharply in North Carolina and the mid-Atlantic states, as well as in Oklahoma and the southern Plains states. Thus the geographic concentration in pork production has broadened. The growth areas in hog production are those areas that are more accepting of vertically integrated systems, culturally and legally, partly because of the presence of integrated poultry operations in those areas.

Cattle production is again distinctly different from poultry or pork production. A major reason is the significant land and forage base required for cattle. Beef and dairy cattle, which both contribute to the supply of beef, are geographically concentrated in different regions. Stocker or growing operations are quite diverse and frequently not concentrated in the same geographic regions as cow-calf production. Cattle feeding has increased in geographic concentration and involves some of the same states where there are numerous stocker and growing operations. However, because of the geographic dispersion combined with an added production stage, the beef industry incurs significant transaction costs moving animals from geographically dispersed cow-calf operations to more geographically concentrated stocker or growing areas and to still more geographically concentrated cattle-feeding areas.

## OPERATION SIZE AND SPECIALIZATION

Poultry operations are specialized units largely because of integration and contract production. Although operation size varies, many are relatively large, intensely managed operations. Hog production units have moved in the direction of poultry production. Owing in part to integration, hog production operations have become more specialized in farrowing, growing, and finishing operations. Contract production has increased significantly as has vertical integration of hog production by large packers. The size of operation has increased significantly to capture cost economies associated with larger units. Individual units range from a single farrowing or finishing unit to very large operations with several production units under the same management.

Cattle production is a mixture. A large number of cow herds are small, with fewer than 30 cows per operation, in part again because of the significant land and forage base required. Stocker or growing operations are larger, usually combining calves from several cow-calf operations into a larger production unit. Cattle feeding has moved predominantly to large specialized units. Additionally, increased consolidation among cattle-feeding companies has resulted in more feeding capacity in fewer and larger firms.

Implications for integration are interrelated with other factors. A large specialized production unit can be managed more efficiently than many smaller, diverse production operations. Specialization and larger size units in poultry are partly a cause and partly the result of enhanced coordination. Such units capitalize on more specialized management and economies of size. The pork industry has followed the poultry industry model to some degree and has trended toward increasingly larger and more specialized production operations. Too, contract production and vertical integration has led to improved coordination. Tighter vertical coordination in the beef industry will occur more slowly than it did for either poultry or pork, in part because of the difficulty of organizing and managing smaller, highly diverse production units. Incorporated with that are the disadvantages cited for the beef industry: longer biological process, diverse genetic base, an added production stage, and more geographically dispersed production. Thus there appear to be several obstacles to vertical integration in the beef industry.

## MARKET FACTORS ENHANCING COORDINATION

Other market-related characteristics of the beef, pork, and poultry industries lend themselves to improved vertical coordination. Enhanced coordination ideally enables firms in an industry to respond more quickly and correctly to changing consumer demands, especially changing tastes and preferences. Therefore, characteristics discussed in this section relate to how integrated firms or nonintegrated firms are able to meet consumer demands at the retail and food service levels, and how to capitalize on profit opportunities. Table 15–2 summarizes market characteristics affecting vertical coordination incentives. As in the previous section, the characteristics discussed are interrelated.

**TABLE 15–2  Market Characteristics: Beef, Pork, and Poultry**

| Category | Beef | Pork | Poultry |
|---|---|---|---|
| **Value-added products at retail** | Low, but increasing | Moderate and increasing | High |
| **New product development** | Moderately increasing | Moderately increasing | Slowly increasing |
| **Brand marketing** | Low, but increasing | Moderate | High |

## VALUE-ADDED PRODUCTS AT RETAIL

Greater profit opportunities exist with value-added, differentiated meat products than with commodity-type products sold in the traditional fresh form. (Note on terms: A pork chop sold as a *fresh-packaged product* is a traditional fresh form; a smoked pork chop is a *value-added product;* and a *Joe's Original Smoked Pork Chop* is a *differentiated product*. Smoking the chop allows the packer to add value and earn more on the chop. Differentiating the chop as *Joe's Original* gives the processor a means of making it different from other smoked chops, and thus allows advertising and promotion to the consumer. Differentiated products are typically brand-name products, again allowing advertising programs to create or emphasize product characteristic differences and differences in prices.)

Beginning in the 1970s, a concerted effort was made to develop more value-added poultry products. The space in the meat case for fresh whole birds or for fresh parts has declined as more products have appeared on the frozen food shelves. These frozen packaged products offered more opportunities for satisfying varied consumer demands such as for different package and serving sizes for varying size families, different flavors and styles for different ethnic and religious groups, and different degrees of convenience in meal preparation.

More recently, the emphasis has shifted somewhat to providing **case-ready products.** This entails improved packaging of fresh retail products. Case-ready products have probably affected the pork and beef industry more than poultry. The move to case-ready products has several sources. One is its responsiveness to consumer criticism regarding leaky, sticky, fresh meat packages at retail. This amounts to both a consumer satisfaction issue and a food safety issue. Second was the rapid expansion of Wal-Mart in food retailing and its emphasis on labor-saving handling in their stores. Case-ready products come into the store ready to be price-stamped and placed in the retail meat case. Improved packaging reduces meat case waste and cleanup also, in addition to enhancing food preservation and safety. In the pork and beef industries, new processing plants specifically geared to producing case-ready meat for one or a few retail supermarkets have become commonplace.

The pork industry has traditionally sold several processed value-added products. Consider the many traditional bacon, ham, and sausage products in the retail meat case. The remainder of the pork carcass was marketed in fresh form as chops, roasts, and other products. Some of those fresh pork products are now marketed as case-ready, value-added pork products. They are still fresh pork products; so the pork industry has not created as many frozen value-added products as the poultry industry. Pork industry efforts have focused on increased product quality and consistency in these case-ready products. Some of these quality and consistency gains have been achieved from a narrower genetic base as well as from new or improved processing methods. Several versions of newer, value-added products are offered to capitalize on varying consumer tastes and preferences.

Beef is primarily marketed in fresh form from the retail meat case. However, packaging improvements, including case-ready products, has likely benefited beef more than poultry. There are still relatively few value-added beef products throughout the retail supermarket, but new products are appearing (Figure 15–2). There are relatively few identifiable characteristics of fresh beef products that can be used as a basis for product differentiation apart from the primal cut source of the meat. As a result, the beef industry has had more difficulty developing value-added products. However, beef marketing is changing.

### Fresh-packaged product

Traditional fresh product sold with minimal processing.

### Value-added product

A product processed in some way that has enhanced its value.

### Differentiated product

A value-added product with a brand name.

### Case-ready product

A product of a business system in which centrally prepared meats and meat items are used to increase consumer satisfaction and related profits.

**FIGURE 15–2** Beef is primarily marketed in fresh form from the retail meat case. Few value-added beef products are available, although this is slowly changing. (Photographer Michael Littlejohn. Courtesy of Pearson Education/PH College.)

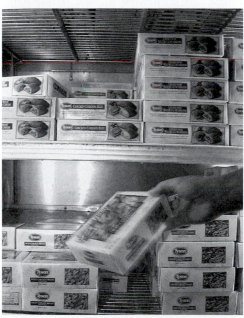

**FIGURE 15–3** The poultry industry, led by vertically integrated firms, has capitalized on the opportunities associated with product differentiation and new product development. (Photo courtesy of Tyson Foods, Inc.)

## NEW PRODUCT DEVELOPMENT

New product develop is tied closely to value-added products. One might argue the two are inextricably linked and should not be discussed separately. However, there is value in considering them separately.

Studies show that product differentiation allows firms to price products differently and receive premium prices for perceived or actual product differences from target market segments. The poultry industry capitalized on opportunities for new product development and product differentiation years ago (Figure 15–3). In Table 15–2, new product development is listed as slowly increasing. That description must be interpreted as meaning slowly increasing beyond the major gains achieved by the poultry industry over the past two or three decades. One of the notable gains in the past decade

**FIGURE 15–4**   *"The Other White Meat"* advertising campaign and aggressive new product development have changed the mix of pork products offered at retail and in restaurants. (Photo courtesy of National Pork Producers Council.)

has been what the poultry industry has done with lower-valued chicken cuts, such as wings. The success of "hot wings" or "drummies" attests to their success. These gains built on earlier success with chicken nuggets, strips, and sandwiches.

The pork industry capitalized on new processing techniques and case-ready technology to create several new products. Many relate to innovations in processing and packaging as discussed under value-added products. For many years, the industry aggressively used the "Pork, the Other White Meat" advertising campaign (Figure 15–4) to place more pork items on restaurant menus. Efforts continue to find new pork products that use lower valued pork cuts and meet consumer tastes and preferences. However, pork has clearly not achieved the degree of success poultry has experienced. In 2005, the National Pork Board changed the advertising campaign by adding the tag line "Don't be blah." to attempt to give the campaign a more contemporary message.

Considerable effort has been expended by the beef industry to better utilize lower valued beef primal cuts and create new consumer-accepted retail products. These efforts have met with some success, having developed a number of precooked, case-ready products. Additionally, the beef industry has attempted to create new products that might compete on restaurant menus with appetizers as well as entrees. There remains considerable dependence on burgers in the food service sector, but growth in deli-type restaurants has shifted some emphasis to deli-style beef products.

## BRAND MARKETING

It is similarly difficult to separate a discussion of brand marketing from the discussion of value-added products and new product development. Brand loyalty and perceived or actual product differentiation enables firms to extract premium prices at retail. Consumers are willing to pay a premium for consistent quality or perceived quality. Therefore, firms have an economic incentive to integrate vertically and to develop consumer brands and brand loyalty for differentiated products.

The poultry industry took a major step toward brand marketing in the 1960s when brands were developed successfully for fresh poultry. That success broadened as brands were placed on new value-added products. Most of the new products at retail are introduced by integrated firms, which own the brands and benefit most from brand-marketing success. The narrow genetic base of the animals and the vertically integrated production structure, which relies predominantly on contract production, enhances quality and consistency of the branded products.

**FIGURE 15–5** Consumers will pay a premium for consistent quality. Therefore, firms have an economic incentive to develop consumer brands and brand loyalty for differentiated products. Few brands exist for fresh beef. However, "Certified Angus Beef" has developed brand loyalty and the brand is an indicator of high-quality beef products. (Photo courtesy of Certified Angus Beef Program. CERTIFIED ANGUS BEEF™ is a registered trademark of the Certified Angus Beef Program and is used with permission.)

Numerous brands exist for traditional processed pork products such as bacon, hams, and sausage. Some firms have introduced case-ready pork products and capitalized on processor brand recognition and brand loyalty. In other cases, supermarket store brands are more important.

Several efforts have been made to develop branded fresh beef products. Some processors have experienced some success, but there are no overwhelming, industry-changing successes. One of the most recognized "brands" of beef products is "Certified Angus Beef" (Figure 15–5). In fact, several firms have attempted to capitalize on the consumer association of the Angus breed with quality, despite the fact that some beef marketed as Angus beef may not be from Angus cattle.

Premium prices and loyalty for retail brands offer incentives to enhance coordination and integration. However, brand loyalty demands consistent quality and eating satisfaction. Fresh beef products in particular have not had the necessary consistency owing to a broad genetic base and little or no control over the entire production process from selection of genetics to end-product distribution. Poultry integrators have capitalized on that production control capability and a narrower genetic base to produce, process, and distribute branded products. The same incentive for controlling production, developing new products, and targeting market segments with differentiated products exists with pork and beef. To date, the degree of success is lower and the probability of success for the large investment required is smaller. However, some gains have been made and efforts continue.

## VERTICAL COORDINATION LIMITATIONS

Many of the impediments to vertical coordination are interrelated with production and market characteristics. Some tend to be the opposite from economic factors that enhance coordination efforts. Table 15–3 summarizes the management characteristics that limit or make vertical coordination more difficult.

### CAPITAL

*Capital requirements* refer to the extent of capital needed by an individual firm to vertically integrate production, processing, and distribution. Capital requirements have two dimensions. First is the absolute capital needed to integrate vertically. Second is the capital needed to integrate vertically a sufficient volume to influence a large enough

---

**Capital requirements**

In this context, a combination of the capital required to integrate sufficiently to target a large enough market segment.

**TABLE 15–3    Management Characteristics: Beef, Pork, and Poultry**

| Category | Beef | Pork | Poultry |
|---|---|---|---|
| Capital | Varies by production stage | Moderate, but shared | Moderate, but shared |
| Risk | High | Moderate | Moderate |
| Control of Quantity, Quality, Consistency | Loose, but increasing | Moderate and increasing | Tight |
| Management Skills Needed | High | Moderate | Low |

target market segment. The capital needed to integrate vertically a small market niche may be small, but the profit potential from the small niche market may not make the venture attractive.

Capital requirements differ markedly between ownership integration and contract integration. The poultry subsector is predominantly organized in a manner that limits capital requirement by the integrator. Contract growers are required to provide part of the capital such as the buildings and equipment, thereby reducing capital requirements by the integrating firm. Along with a shift in capital requirements, some risks associated with production are effectively shifted to contract growers as well because risks follow the investment of capital. In addition, contract terms are written to limit the potential profitability of the contract growers. Growers can earn a reasonable return on investment, but significant returns above that accrue to the integrating firm.

One of the dominant forms of vertical integration in the pork industry has followed the poultry model. Contract growers (those engaged in farrowing, growing, and finishing) provide part of the capital for buildings and equipment and are allowed a reasonable but limited return on investment. The integrating firm provides the remainder of the capital, such as for hogs, and assumes the remainder of the risk but retains the potential for unlimited returns. With outright owner integration, the integrating firm provides virtually all the required capital, assumes virtually all the risk, but retains the potential for unlimited returns.

Vertical coordination in the beef industry has not followed a distinct model. Contract production is uncommon, although marketing contracts commonly exist between various stages.

One deterrent is the large amount of capital required. Again, two dimensions of capital needs are relevant. One is the capital required to integrate three production stages plus processing and distribution, even on a small scale. The second is the capital needed to integrate a sufficiently large operation to have a significant market influence. Because of economies of size in slaughtering–fabricating, an efficiently sized plant requires about 1 million fed cattle annually. That represents the output from about 10 feedlots, each with a 40,000-head onetime capacity. That, in turn, requires cattle from about 12,500 cow herds of 100 cows each. The capital required for the processing plant, feedlot, and cattle is immense, and excludes the land required for the cow herd. One means of reducing the capital outlay required is to develop a contract-integrated, capital-sharing operation, but this has not occurred.

## RISK

The absolute outlay of capital for a venture must be considered in light of the probability of success stemming from the investment. This introduces the dimension of risk and the typical trade-off between profits and risk. Higher risk ventures often have higher profit opportunities.

Some kinds of risk are less for poultry than for pork and beef. A type of risk faced by poultry in recent years might be categorized as geopolitical or trade-related risk. When poultry exports are disrupted, normal distribution is disrupted even for tightly coordinated industries. Poultry production risk seems to have increased from animal disease outbreaks both in the United States and abroad. Tightly coordinated systems, with shared capital structures, a shorter biological cycle, and less dependence on commodity marketing reduce risk. Adjustments to market interruptions are somewhat easier to make, risk is shared between capital owners, and consumers have increased loyalty to brands and value-added products.

Risks in pork production are similar to poultry but with some important differences. The biological process for pork is longer, coordination systems are not as tight, and there is more dependence on marketing commodity products (unbranded fresh pork). Therefore, market adjustments are made less easily or effectively, leaving firms with a greater exposure to market price risk.

Risk in the beef industry may be the greatest for the three industries. The longer biological process, lower degree of coordination, and more dependence on commodity marketing mean slower and less-effective adjustment to market interruptions. And the beef industry has suffered severe market interruptions for the past several years from sporadic bacteria contaminant events for beef. Then in 2003, the first known BSE (bovine spongiform encephalopathy) case in North America created another degree of risk and market disruption, closing several trade avenues with major trading partners whose business had benefited the beef industry over the past decade. These compounded the normal, high-risk nature of the beef industry.

## CONTROL OF QUANTITY, QUALITY, AND CONSISTENCY

Several factors come together in a discussion of controlling quantity, quality, and consistency. Quantity is tied directly to capital requirements. Quality and consistency are tied to the production characteristics discussed earlier, especially the genetic base, as well as the opportunities or difficulties in developing value-added branded products. All relate to how tightly or loosely coordinated the industry is.

The poultry industry, the most tightly coordinated, has demonstrated the ability to control the quantity of output in a vertical channel while simultaneously controlling quality and consistency. Narrow genetics, having only two production stages, capital-sharing and risk-sharing contracts, tight management specifications, the linkage between product differentiation and brand loyalty, and other related factors have all contributed to poultry's success.

The pork industry is following the poultry model generally, but differences limit the extent or success of vertical integration. Regulations on contract farming and environmental regulations in some states have limited the development of a tightly coordinated industry. Inconsistency in pork remains a problem but is diminishing. How much brand loyalty has developed for fresh value-added pork products is not yet clear. However, considerable brand loyalty exists for processed products. Integration in the pork industry trailed poultry by about two decades but occurred very rapidly in the 1990s.

Beef continues to face the biggest coordination challenges for several reasons. One of the primary impediments to improved coordination in the beef industry is the difficulty with controlling quantity, quality, and consistency. Quantity depends on the decisions of a large number of mostly smaller cow-calf producers geographically dispersed throughout the United States. For any one firm to control a sufficiently large quantity from production to consumption is difficult because of large capital requirements. Research is under way to find an economical technological test or method to predict and control end-product consistency, especially tenderness. Such a breakthrough might have a profound influence on coordination in the industry. The profit potential might be sufficient to provide the necessary incentive for organizing a more coordinated system from cowherd to consumer. A guarantee of beef's safety to consumers and increased quality and consistency would certainly provide an incentive to develop more tightly coordinated systems in the beef industry. These include identifying the proper genetics and narrowing the genetic base, more tightly linking the stages of production, and providing more incentive for new value-added products and brand marketing.

### MANAGEMENT SKILLS NEEDED

The biological characteristics of poultry, pork, and beef, the number of production stages, the geographic concentration, and the size and diversity of production units all affect the managerial skills required to manage a vertically integrated firm. The poultry industry first, and later the pork industry, found ways to manage each production stage, in part owing to narrower genetics, a shorter biological process, and specialized production units. Beef is distinctly different. The managerial skills needed to manage numerous small geographically dispersed cattle operations that have a broad genetic base are immense. Similarly, more managerial resources are needed at every step to control quantity, quality, and consistency of end products. Therefore, the extent of vertical integration in beef will continue to lag behind that of pork and poultry.

## SUMMARY AND CONCLUSION

The poultry industry is the most tightly coordinated system among the three major meat industries and is highly vertically integrated. It also involves the fewest firms responsible for the production to market coordination. There appears to be relatively little opportunity for further integration with the existing level of production, unless ownership integration replaces contract integration. Any further integration will likely occur at the processing–distribution stages in the form of more value-added, processed, branded poultry products.

Vertical integration in the pork industry has increased dramatically in recent years. Many of the large or mega-sized hog production units are already tied vertically to specific processors, and larger producers expect closer ties with processors in the future. Most integration is via contract, similar to the poultry industry, although there are cases of large ownership-integrated operations. Further changes may occur. The genetic base may continue to narrow, offering opportunities for more consistent pork products. This, in turn, provides an incentive for vertically integrated firms to capture the necessary control over quality and consistency. It also provides an economic incentive to develop processed, value-added, branded products for retail and food service. The pork industry is expected to continue focusing on new product development and market penetration to hold or enhance its market share among the three meat industries.

The beef industry has the most reliance on market prices or open market coordination and the lowest degree of coordination via contracts or vertical integration. Tightly controlled forms of vertical coordination in the beef industry will continue to trail poultry and pork. Several factors might reverse the trend or speed the move toward more tightly coordinated systems. One is an economical breakthrough in using gene markers to select for the genetics that produce beef having the eating quality consumers desire and being able to maintain identity of that beef from conception to consumer. Another is a breakthrough in processing or new product development to build a strong brand loyalty for value-added beef products. Lastly, there may need to be a means found to structure the industry in such a way as to share the capital requirements and risk of a more tightly coordinated industry.

## STUDY QUESTIONS

1. Why is it important to understand market coordination and vertical integration in the animal industries?

2. Define *vertical integration*. What are the two primary types of vertical integration? Explain why vertical integration and vertical coordination are not synonomous terms.

3. What is the current state of vertical coordination and integration in the beef, swine, and poultry industries?

4. What are the similarities and the differences in the biological production cycles of beef, pork, and poultry?

5. Describe the differences in the genetic base of the beef, pork, and poultry breeding stock in the United States.

6. How do the differences in the number of industry stages for beef, as compared to poultry and swine, affect the economics of integration?

7. Describe the differences and similarities in the geographic concentration of production for poultry, pork, and beef.

8. What is the trend in the animal industries with regard to operation size and specialization?

9. Describe the motives for an animal industry to coordinate vertically and integrate. What are the limitations?

10. Present a boiled-down version of how future integration will affect the industries.

## REFERENCES

The basis of this chapter is an article written and subsequently updated by Dr. Clement E. Ward, Professor and Extension Economist, Department of Agricultural Economics, Oklahoma State University. It is used with permission and appeared originally in *Current Farm Economics,* 1997, Vol. 79, No. 1, pp. 16–29. Oklahoma Agricultural Experiment Station, Oklahoma State University, Stillwater, OK 74078. Editorial changes and additional material were incorporated to accommodate the style and format of the text. It has been further updated with new material from Dr. Ward for each subsequent edition. With the fourth edition, Dr. Ward has assumed co-authorship of this chapter.

CAST. 2001. *Vertical coordination of agriculture in farming-dependent areas.* Report No. 137. Ames, IA: Council for Agricultural Science and Technology.

Schrader, L. F., H. E. Larzelere, G. B. Rogers, and O. D. Forker. 1978. *The egg subsector of U.S. agriculture: A review of organization and performance.* N.C. Project 117. Monograph #6. West Lafayette, IN: Purdue University.

Ward, Clement E. 2004. *Pork, beef, and poultry industry coordination.* Stillwater: Oklahoma Cooperative Extension Service, AGEC-552.

# 16

# Beef Cattle

## LEARNING OBJECTIVES

After you have studied this chapter, you should be able to:

1. Explain the place of beef cattle in U.S. agriculture.

2. Discuss the reasons why the United States has such a large beef industry.

3. Give a brief history of the cattle industry in the United States.

4. Describe the beef industry structure.

5. Give an accurate accounting of where the beef industry is physically located in the United States and explain why each region has the portion of the beef industry it does.

6. Discuss the role of genetics in the present and future of the beef industry.

7. Explain the breeds revolution in the U.S. beef industry, including the outcomes and ultimate implications to the industry.

8. Outline the basis of managing beef cattle for reproductive efficiency.

9. Describe the feed supply of beef cattle. Explain the purpose of beef cattle in the United States and other countries.

10. Outline and discuss nutritional benefits of beef to humans.

11. Discuss trends in the beef cattle industry, including factors that will influence the industry in the future.

## KEY TERMS

AI stud
Beef cycle
Breed
Breeding soundness exam
"Breeds Revolution"
British breeds
Composite breed
Estrous cycle
Estrus
Eutrophication

Expected progeny
　difference (EPD)
Feedlot
Finishing phase
Forage
Grain-fed beef
Hazard Analysis and
　Critical Control Points
　(HACCP)
Heritability

Least-cost ration
Net calf crop
Performance testing
Seedstock
Small grains
Stocker calf
Subtherapeutic
Vertical alliance
Withdrawal times

## SCIENTIFIC CLASSIFICATION OF CATTLE

| | |
|---|---|
| Phylum: | Chordata |
| Subphylum: | Vertebrata |
| Class: | Mammalia |
| Order: | Artiodactyla |
| Suborder: | Ruminata |
| Family: | Bovidae |
| Genus: | *Bos* |
| Species: | *taurus; indicus* |

## THE PLACE OF THE BEEF CATTLE INDUSTRY IN U.S. AGRICULTURE

The beef industry is the single largest money-generating commodity in all of agriculture in recent times, amounting to approximately a fifth of all yearly cash receipts (Figure 16–1). The gross annual income from beef in the United States is approximately $50 billion. About half of the value of the cash receipts from U.S. agriculture in any given year is generated from animal agriculture (Figure 16–1). On average, beef accounts for approximately 40% of animal agriculture's share (Figure 16–2). Another way to look at the importance of beef is to consider that cattle and calves rank in the top five commodities for 41 of the 50 states. There are approximately 970,000 beef cattle operations in the United States. Nine states record more than $1 billion each in gross income from marketing of cattle each year. An additional 12 states exceed $.5 billion. Although we have only 4.6% of the world's population, the United States raises 7% of

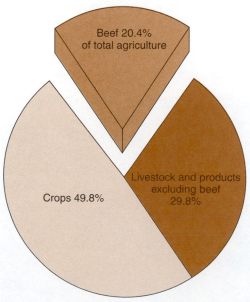

**FIGURE 16–1**  Beef farm cash receipts as a percentage of U.S. farm cash receipts, 2005–2007. (*SOURCE:* USDA-NASS, 2007a.)

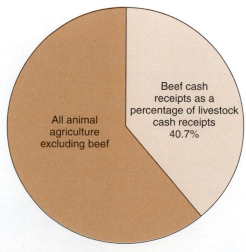

**FIGURE 16–2**  Beef yearly farm cash receipts as a percentage of total animal agriculture's cash receipts, 2005–2007. (*SOURCE:* USDA-NASS, 2007a.)

the world's cattle and produces nearly 20% of the world's beef and veal. By any standard, the beef industry is a major industry in the United States, and the United States is a major beef producer and consumer.

## PURPOSE OF THE U.S. BEEF CATTLE INDUSTRY

The purpose of the beef cattle industry in the United States is to make use of resources that would otherwise go to waste. The predominant resource is grass. As a ruminant, beef cattle convert grass that humans cannot use into high-quality food, by-products, and work we can use. This roughage conversion is cattle's chief contribution to human welfare (Figure 16–3). Much of the land of the United States is best suited to grass production. The 48 contiguous states have approximately 1,029 million acres of agricultural land, 57% of which is grazing land. The grazed land includes the vast expanses of dry lands in the American West, additional acreages that are unsuited to cultivation because of topography or some other reason, the swampy lands of the coasts, croplands currently used as pasture, forest that is grazed, and the high elevations found in the mountainous regions. In addition to these grazing resources, vast supplies of waste material from the agronomic crops of the nation's agriculture (e.g., cornstalks and wheat straw) and food by-products (e.g., corn cannery waste, brewer's grains, and sugar beet pulp) are also fed to beef cattle. All are resources that we would find difficult to use for productive purposes if not for beef cattle. Beef cattle are a rugged, adaptable domestic species that can be managed very extensively to harvest grass at low cost. Thus they fulfill a unique niche in resource utilization for the benefit of our whole society. In times of plentiful grain production, beef cattle can be used to help make the best use of the abundance from agronomic production. In this way they help moderate the fluctuations in grain prices that could otherwise disrupt agronomic practices and endanger the agricultural economy and the supply of grain for human consumption. Cattle are actually modest consumers of grain. This modest use of grain increases palatability of the product, improves efficiency, and helps provide for year-round production, thus avoiding great differences in the seasonal availability of beef.

**FIGURE 16–3**   The purpose of the beef cattle industry is to make use of resources, predominantly grass, that would otherwise go to waste.

Cattle are used extensively around the world by a wide variety of people for their grass conversion ability. With 1.37 billion head distributed globally, cattle are the most numerous and arguably the most important of all the domestic livestock species. They are also the most widely distributed. It is believed that every country in the world has some cattle. There is one cow in the world for every 4.8 people, and 26.9 head of cattle/square mile of earth land surface.

## HISTORICAL PERSPECTIVE

The word *cattle* was used at one time to mean all domestic species. It is derived from the Latin word *capitale,* meaning "wealth" or "property." Today the word is used only in relation to *Bovidae.* Cattle were probably domesticated by 6500 B.C.; however, don't be surprised if you see a different date elsewhere. Such milestones are hard to determine with certainty. Cattle spread with humans as we populated the globe, and they have had a wider range of uses by more people than any other domestic species.

Figure 16–4 gives a historical look at cattle numbers in the United States. The number of beef cattle in the United States fluctuates in a predictable manner; this fluctuation is described as the **beef cycle.** Based on roughly 10-year cycles, the number of cattle increases and decreases. The decisions to expand or reduce breeding animals are made by cow-calf producers all across the country in response to economics. Because not all producers are in the same economic circumstances, not all move in the same direction at the same time. The biological nature of beef production contributes to the length of the cycle. Once a heifer is added to a herd, over three years will pass before her calf becomes steak. A combination of these factors creates beef cycles that tend to be made up of approximately 5 years of expansion followed by approximately 5 years of reduction. Then the cycle repeats itself.

Christopher Columbus was the first cattle producer of the Western Hemisphere. He brought cattle to the West Indies on his second voyage in 1493. Cortez brought cattle to Mexico in 1519. Cattle were distributed across the American West by Spanish missionaries as a part of the mission culture they established at the beginning of the 17th century. The Spanish Longhorn was one of the livestock species used to stock the missions. In 1609, English settlers brought cattle to New England. Early uses of all of

**Beef cycle**

Historic fluctuations in beef cattle numbers that occur over roughly 10-year periods.

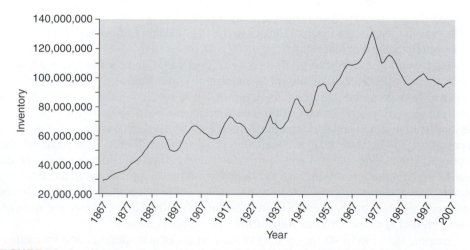

**FIGURE 16–4**   Historic cattle inventory in the United States. (*SOURCE:* USDA-NASS, 2007b.)

these cattle were as work animals and producers of milk. As the settlement of the continent proceeded, the grasslands became increasingly used as beef-producing areas.

By the middle of the 19th century, beef cattle in the West were being improved with newly imported Shorthorns. After the Civil War, many of the Texas Longhorns were rounded up and shipped east to feed the growing population. As Longhorn blood became increasingly scarce in the cattle population, the Shorthorns failed to meet the rigors of range life and were crossed with Herefords, who came to dominate the Western beef industry. This breed transformation was no doubt hastened by the devastating winter of 1886, when thousands and thousands of cattle were killed in the West owing to weather. Later, the Angus was used to cross on the Herefords, and the black baldie, which is still very much a part of the beef cattle industry, was introduced. Having learned the hard way that quality cattle needed better care than had previously been thought, the Western beef industry underwent a period that was to last two decades or so when greatly improved husbandry practices strengthened the position of that region as a major cattle-producing region in the United States. Quality cattle were reared and fattened on grass to meet the growing meat needs of an increasingly prosperous nation.

After World War II, the next major shift in the U.S. cattle industry occurred. Cattle had been fattened for many years as secondary enterprise operations on farms in grain-producing areas. These were small operations and only a small part of the cattle population was affected. This was destined to change. Technological and industrial innovation led to phenomenal increases in agricultural output. Huge grain surpluses developed in areas of the country so far removed from markets that shipping costs made realizing a profit impossible. Some enterprising cattle and grain producers began feeding increasing numbers of cattle on the surplus grain. Although more expensive, consumers quickly developed a taste for the improved flavor and tenderness of the grain-fed beef. A marketing revolution took place in which chain grocery stores began offering self-service meat counters with a wider variety of meat products readily available. Demand for the new ***grain-fed beef*** grew. The United States had ample pasture and grasslands available for the production of the feeder calves needed by this industry segment. A new industry segment developed that involved concentrating large numbers of cattle and feeding them grain in a ***finishing phase*** just prior to slaughter. This industry developed first in California and Arizona in the 1950s, and then in the Plains states of Colorado, Texas, Oklahoma, Kansas, and Nebraska in the early 1960s. Thus was born the beef feedlot industry. During the 1970s, this phase became established as the driving force in the U.S. beef cattle industry. Major meat processors moved to the feedlot area. Another reason contributing to the development of the feedlot industry in the Plains states was the dry climate that is important to the feedlot industry.

As a result of the shift to grain feeding of beef cattle, the beef industry began to increase in importance as a money generator in American agriculture. Beef cow numbers first exceeded dairy cow numbers in 1954, advancing the notion of beef as king in U.S. agriculture. The feedlot industry became dissatisfied with the type of cattle available to it and began demanding changes. The biggest concern was that cattle were getting too fat in the feedlot. Small-statured animals with fairly light adult weights produced calves that fattened well on grass. These cattle fattened too rapidly in the feedlot and did not grow to the size the industry felt made optimal use of the finishing phase. They also grew slowly. The industry needed cattle that could be fed on grain and not produce so much waste fat while growing more rapidly. This need ushered in an era often referred to as the ***"Breeds Revolution"*** in the beef industry.

Beginning in 1965 with the opening of the first major quarantine station for cattle in Canada, previously prohibited breeds of cattle could be imported into North America.

**Grain-fed beef**

Meat from cattle that have undergone a significant grain feeding.

**Finishing phase**

Grain-feeding period just prior to slaughter.

**"Breeds Revolution"**

Period of great expansion in numbers of breeds of beef cattle.

**FIGURE 16–5**    The Charolais was brought into the United States as part of the "Breeds Revolution" to help accommodate the new feedlot industry. (*SOURCE:* USDA photo/01cs1631/B. Tarpenning. Used with permission.)

Over the next 25 years, the number of beef breeds in the United States increased to more than 70 (Figure 16–5). Crossbreeding was reestablished as an integral part of the beef cattle industry. Cattle feeding changed to accommodate the new type of cattle. The rest of the country increasingly shifted its production style and techniques to accommodate the demand of the consumer for grain-fed beef. Total cattle numbers peaked in 1975 with 132 million head (cattle, calves, beef, and dairy). The current range is between 95 and 105 million head of cattle. This smaller number of cattle produces more meat in a year than was produced when cattle hit their peak. Genetics, nutrition, management, and shifts in the demands of packers have all contributed to this phenomenon.

Average annual per capita supplies (and consumption) of beef declined from 1976 until the early 1990s. Part of the decrease was a result of closer trimming (less fat content) of retail cuts. However, quality issues also plagued the industry with the change in the cattle, and consumers ate less beef. Today, the beef industry can be characterized as an industry struggling with consumers that have partially shifted their consuming habits away from beef to other lower-cost alternatives, predominantly poultry. However, as beef production increased and per capita supplies increased again, the precipitous decline in per capita beef consumption began to stabilize. Since the early 1990s, much of the yearly fluctuations in per capita consumption of beef appear to have been related to supplies. Pork and poultry products no doubt have taken a share of beef's consumer market, but consumer choices in the immediate future are expected to owe more to supplies and prices rather than to declining interest in beef.

## STRUCTURE OF THE BEEF INDUSTRY

Number, size, location, and major activity of the firms in an industry are the elements we refer to as the "structure of the industry." Beef cattle production differs from most other kinds of livestock production because structurally it is divided into more phases. The same animal is generally owned by several different owners, as compared with the other food-producing species. Animals change ownership when moving from one phase

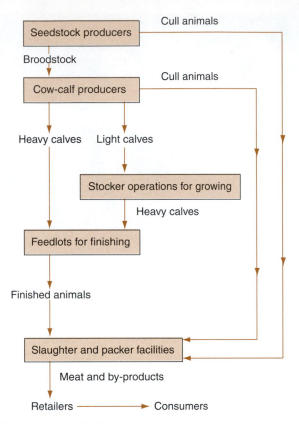

**FIGURE 16–6**   The movement of animals through the segments of the beef industry.

to the next. Thus the same animal tends to be owned by several people. There is also a distinct regionality to the beef industry in which different regions of the country tend to specialize in a single, or perhaps two, phases. The major segments of the beef cattle industry are seedstock producers, commercial cow-calf producers, yearling or stocker operators, and feedlot finishing operations. Figure 16–6 shows the general flow of animals through the various structure segments.

*Seedstock* production (Figure 16–7) is the only segment that doesn't produce animals whose primary use is for meat consumption. The vast majority of animals produced from this segment will ultimately be slaughtered for food, but only for salvage value at the end of a productive life. The primary goal of this industry segment is to produce breeding stock. Stock may be purebred or bred in controlled crossbreeding programs that produce the parent generation of cattle for commercial calf producers. The product of greatest demand from this segment is bulls. Usually these bulls are purebred animals whose pedigrees and qualitative information, such as *expected progeny difference (EPDs),* make selections easier and the results of the matings more predictable. Commercial cow-calf producers rely on these breeders for quality sires to pasture mate with their herds of commercial cows. The very best purebred seedstock producers will produce cattle of high enough quality that they can sell seedstock to other purebred breeders and help improve the overall quality of the breed. If the quality is high enough, they will also be able to market semen and embryos. These elite breeders are also able to sell or lease bulls to *AI* (artificial insemination) *studs* that sell semen to a variety of cattle breeders. Cull fe-

**Seedstock**

Broodstock intended for future production.

**Expected progeny difference (EPD)**

A prediction of the difference between the performance of an individual's progeny compared to all contemporaries for the progeny.

**AI stud**

Company that markets semen from high-quality males.

**FIGURE 16–7** Beef seedstock producers provide breeding stock to other seedstock producers and cow-calf producers. (Photo courtesy of American Angus Association. Used with permission.)

**FIGURE 16–8** Commercial cow-calf producers produce animals that are sold at weaning to feedlots or stocker operators. (Photo courtesy of American Brahman Breeders Association. Used with permission.)

males may be marketed as commercial cows. A substandard female by purebred standards can usually be a superior commercial cow as long as she is not suffering soundness problems. Controlled crossbreeding programs produce quality crossbred heifers to sell to cow-calf producers to become brood stock. These programs allow cow-calf producers to produce controlled crossbred offspring without having to own herds of different breeds.

Commercial cow-calf producers (Figure 16–8) represent the first phase of producing animals whose primary use is for the table. The product from this segment is 6- to

**Stocker calf**

Weaned calf being grown prior to placement in a feedlot for finishing.

10-month-old, 300- to 700-lb calves that are usually sold at weaning to either a feedlot or a ***stocker calf*** operator. Cow-calf producers generally produce crossbred calves bred for slaughter. Their goals are to produce the heaviest calves possible with the least cost. Margins are tight for these producers, and costs must be kept to a minimum. Cow-calf operations typically plan to have a spring or fall calving season. This is usually dictated by the type of feed available in the given region and other costs associated with rearing calves. Most of the U.S. cow herd calves in the spring. When the calves are ready to wean, the cow-calf producer has several options. The easiest option is simply to market the calves and begin making plans for next year's calf crop. Heavier calves with rapid growth potential are usually purchased and placed directly in a feedlot to ready them for slaughter. Lighter calves that need to grow are often purchased by a stocker operator to grow before they are placed in the feedlot. However, if the producer has feed or access to feed, he or she may choose to wean the calves and grow them to heavier weights before selling them. In this way, a cow-calf operator may also be a stocker operator.

Yearling or stocker operators (Figure 16–9) are in the business of growing calves to heavier weights on low-priced forage before the calves enter the feedlot. They typically purchase calves from cow-calf producers and grow them during a specific season and then ship them to feedlots. The improved potential for growth exhibited in today's cattle and the increased milk production of the cows have resulted in heavier calves at weaning. These heavier calves can go directly into the feedlot at weaning. However, in years of high grain prices, a percentage of these calves are also grazed in the stocker phase. Common feed resources on which this phase depends are very regional, but include small-grain pasture, cool-season grasses, summer pastures, summer range, crop residues, standing prairie hay, hay, or silage.

The feedlot phase (Figure 16–10) is referred to as the *finishing phase* of the industry. Here 600- to 850-lb cattle are finished to market weight and condition. Both steers and heifers are fed in feedlots. Feedlots vary in size and location. Some are fairly small operations that may hold only a few head on the farm of origin. These small operations represent a way for farmers to diversify and use their equipment and labor when they can't otherwise use them in the farming operation. Others are huge operations that

**FIGURE 16–9** Yearling or stocker operators grow calves to heavier weights on low-priced forage. The calves then enter the feedlot.

concentrate tens or even hundreds of thousands of cattle in highly specialized operations that do nothing but finish cattle for slaughter. The vast majority of the slaughter steers and heifers in the United States are fed on grain in feedlots before they are sent to slaughter. The amount of time an animal spends in the feedlot varies but is typically 120–150 days. Modern feedlot management is highly dependent on automation and mechanization to keep the cattle appropriately cared for. Cattle have such different genetic potential that it makes predicting an exact weight at slaughter difficult. They may range from 900–1,400 lbs at slaughter. Cattle feedlots and meatpacking plants that process the carcasses have both increased in size and decreased in number during the past 35 years. This is a trend that is expected to continue.

It is important to know that the beef industry has these parts. However, do not conclude that the various phases are cleanly and neatly divided. Nothing could be further from the truth. There is a growing trend for owners to retain the ownership of their cattle through one or more phases. A Tennessee cow-calf operator may stock his or her own calves after weaning and then ship them to a custom feedlot that will feed those cattle for a fee. A Texas Panhandle feedlot operator may purchase calves from Mississippi cow-calf producers and put them in a custom grazing program in the Oklahoma wheat belt and then select from them when he chooses to fill the pens of his feedlot. Such arrangements are destined to become more commonplace in the beef industry.

## GEOGRAPHIC LOCATION OF BEEF CATTLE IN THE UNITED STATES

Cow-calf production is found in every state. Figure 16–11 shows the actual numbers of beef and dairy cows found in the individual states. Dairy cows are included in the figure because they also contribute to beef supplies. The area in which beef cows are found generally depends on the availability of low-cost forage and roughage. In the Plains states, western states, and southwestern states, vast acreages of land grow grass but don't receive enough rainfall to grow crops. The use of beef cows on this land offers the best opportunity for the land to be productive. In Figure 16–11, notice the large number of cows in Montana, South Dakota, Nebraska, Kansas, Texas, and Oklahoma as examples. Other areas with significant cow populations can be found in the grain-producing areas of the country. A significant by-product of grain production is crop residues, including low-quality residues such as straw and cornstalks. The mature beef cow uses these low-quality, high-volume feeds better than any other livestock species. Much of the crop residue is found in the Corn Belt. Notice the large number of cows in Iowa and Missouri, for example. The southeastern region of the country has an abundance of grazing land and the advantages of mild winters and good rainfall that produce significant year-round grazing conditions. Year-round grazing is possible in the coastal areas (look at Florida) owing to mild climate, and also in the upper East South Central states of Tennessee and Kentucky because of their use of cool-season grasses. The only area of the United States that doesn't really contribute much to the beef industry is the northeastern region. Maine, New Hampshire, and Vermont don't have many cows, for example. However, even these three states have more than 200,000 beef and dairy cows combined.

Stocker operations are drawn to the same things that cow-calf operations are, except that the quality of the feed needs to be better. Thus the same areas that have cows and calves tend to have stocker calves also. The winter ***small-grain*** areas like Kansas, Texas, and Oklahoma generally import calves for winter grazing from areas that have lower-quality winter feed, such as Grain Belt states. The higher-quality winter wheat is an ideal feed for the stocker calf. Oklahoma uses its summer grazing for the cow-calf herd, and a portion of its calves then go to wheat pasture for the winter. Because there is so much wheat pasture,

### Small grains

Grains such as oats, wheat, and barley.

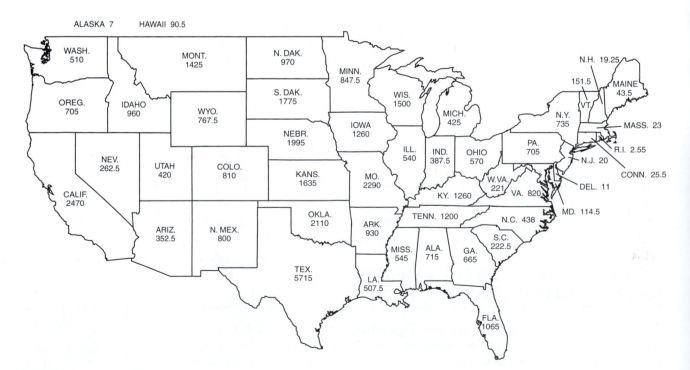

**FIGURE 16–11**    Distribution of total U.S. cows and heifers that have calved (beef and dairy) × 1,000. (*SOURCE:* USDA-NASS, 2007b.)

calves from other regions are shipped in to take advantage of the abundance. Many go to such states as Kansas to graze summer ranges before being sent to the feedyard.

The finishing of cattle in feedlots is one of the most important segments of the cattle industry. Cattle are finished for slaughter in many regions. As previously mentioned, secondary enterprise operations on farms in grain-producing areas have long fed cattle. These are generally small operations that now account for approximately 16% of total cattle fed. The majority of the cattle feeding is done in relatively few states. The first 12 states listed in Table 16–1 produce more than 88% of all finished cattle fed in the United States. Texas, Kansas, Nebraska, Colorado, and Iowa are responsible for marketing over 74% of the finished cattle in the United States. This segment of the industry underwent extensive structural changes during the last three decades of the 20th century. Both feedlots and meatpacking facilities have undergone reductions in numbers and increases in size.

Figure 16–12 shows all cattle and calves by state, averaged for some recent years to give an overall picture of where animals are located. Since 1997, the total cattle and calves in the United States has fluctuated between 94.9 million and 101.6 million head.

**TABLE 16 – 1**  **Cattle Marketed (× 1,000) Annually in the United States, 2006**

| State | Marketed: 1,000+ Capacity Feed Lot | Marketed: 1–999 Capacity Feed Lot | Total Marketed |
|---|---|---|---|
| Texas | 5,775 | | 5,775 |
| Kansas | 5,400 | | 5,400 |
| Nebraska | 4,635 | 300 | 4,935 |
| Colorado | 1,935 | | 1,935 |
| Iowa | 828 | 500 | 1,328 |
| California | 760 | | 760 |
| Oklahoma | 732 | | 732 |
| South Dakota | 425 | 290 | 715 |
| Idaho | 542 | | 542 |
| Wisconsin | | 350 | 350 |
| Arizona | 337 | | 337 |
| Washington | 315 | | 315 |
| Ohio | | 280 | 280 |
| Minnesota | | 270 | 270 |
| Illinois | | 240 | 240 |
| New Mexico | 226 | | 226 |
| Michigan | | 210 | 210 |
| Indiana | | 150 | 150 |
| Pennsylvania | | 120 | 120 |
| Missouri | | 90 | 90 |
| North Dakota | | 60 | 60 |
| Other States | 567 | 780 | 1,347 |
| United States | 22,477 | 3,640 | 26,117 |

SOURCE: USDA-NASS, 2007b.

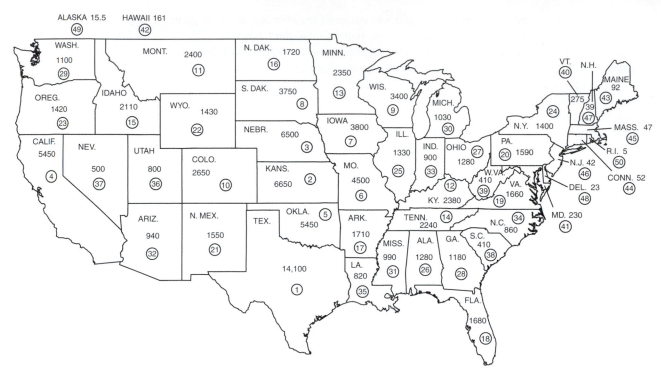

**FIGURE 16–12**    All cattle and calves × 1,000 for 2006 (circled numbers indicate state rank). (*SOURCE:* Based on USDA-NASS, 2007b.)

# GENETICS AND BREEDING PROGRAMS

The fundamentals of genetics and breeding are discussed in Chapters 8 and 9. See those chapters for more detailed information on these topics. In simplest terms, two factors affect all of the economic traits in cattle. These factors are the environment to which the animal is exposed (feeding, climate, and so on) and the genetics of the animal. *Heritability* is the difference in individuals that is related to genetic makeup. Obviously, this is the portion of the difference that can be passed on to the next generation. The pace of progress made in a herd to improve the genetic potential of the offspring depends on how heritable the trait is. Table 16–2 shows heritability estimates of some economically important traits in beef cattle. It is easy to see that selection helps in making good progress for some traits (e.g., feedlot gain and rib eye area) and that genetic progress is more difficult (e.g., calving interval and conception rate).

For those traits in which genetic progress can be made, a wide range of tools is available. Such practices as *performance testing,* sire summaries, and EPDs if purebred seedstock is used are examples of means by which the selection and breeding program of a herd can be enhanced. It is especially important to use all genetic tools available in selecting for the more lowly heritable traits. Again, seedstock producers commercial cow-calf producers vary tremendously in the tools they select and the degree to which they use the tools of choice.

For most commercial producers of meat animals, some system of crossbreeding could benefit the bottom line of the operation. Although measures of several types substantiate this, the simplest measurement is to look at how much crossbreeding increases

**Heritability**

A measure of the amount of phenotypic variation that is due to additive gene effects.

**Performance testing**

Evaluating an individual in terms of performance such as weight gain or milk production.

**TABLE 16–2** **Heritability Estimates of Some Economically Important Traits in Beef Cattle**

| Trait | Approximate Heritability |
|---|---|
| Conception rate | 0.05 |
| Calving interval | 0.08 |
| Birth weight | 0.35 |
| Weaning weight | 0.30 |
| Cow maternal ability | 0.30 |
| Feedlot gain | 0.45 |
| Pasture gain | 0.30 |
| Efficiency of gain | 0.45 |
| Yearling weight | 0.40 |
| Conformation score: | |
| Weaning | 0.28 |
| Slaughter | 0.38 |
| Carcass traits: | |
| Carcass grade | 0.40 |
| Rib eye area | 0.58 |
| Tenderness | 0.55 |
| Fat thickness | 0.33 |
| Retail product (%) | 0.30 |
| Retail product (lbs) | 0.65 |
| Cancer eye susceptibility | 0.30 |

SOURCE: Northcutt and Buchanan, 1992, p. 6–1. Used with permission.

weaning weights. Available information indicates that the pounds of calf weaned per cow in a herd can be increased by as much as 25% through the use of good systematic crossbreeding programs. As mentioned earlier, one of the problems with the nation's cow herd is that it is more mixed up than it is crossbred. The effective use of crossbreeding programs requires that a crossbreeding scheme be implemented and then followed. Figure 16–13 shows an example of one of several effective crossbreeding schemes for beef cattle. One word of caution: Crossbreeding is not a substitute for the other elements of herd management such as disease control, parasite control, and nutrition. However, it costs as much to deworm, feed, and vaccinate a genetically inferior brood cow as it does to perform those same functions for a genetically superior cow. The superior genetic program will produce better returns on the other costs.

## BREEDS

To have effective crossbreeding programs, purebred livestock must be available. *Breeds,* by definition, are animals that have been selected for a more uniform set of characteristics than their species shows as a whole. Survival of domestic species has depended on the animals becoming adapted to a specific set of conditions imposed by natural and human-made environments. Because humans occupy vastly different environments on this planet and impose a wide variety of conditions on livestock within those environments, it is only logical that many different genetic combinations of our favorite domestic livestock would be developed. Depending on the reference used, somewhere between 300 and 1,000 breeds and strains of cattle are recognized in the

**Breed**

Animals that have been selected for certain characteristics and that breed true for those characteristics.

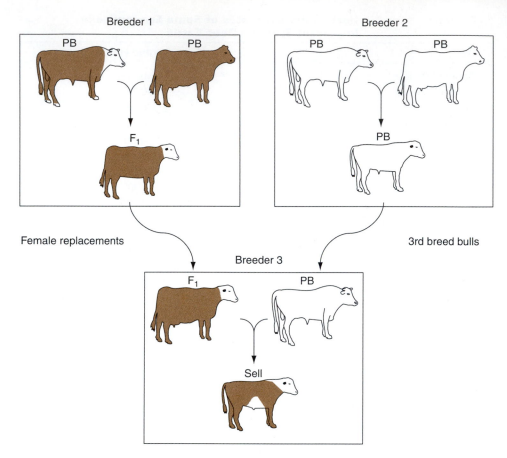

The cattle from Breeder 1 are termed $F_1$ hybrids and are the product of the mating of a purebred bull and a purebred cow from different breeds. Breeder 2 is producing purebreds of a 3rd breed. Breeder 3 is producing a calf that is $1/4$ breed 1, $1/4$ breed 2, and $1/2$ breed 3. Those calves are all fed for slaughter, hence the term terminal cross.

**FIGURE 16–13** Terminal crossbreeding system. (*SOURCE:* Northcutt and Buchanan, 1992. Used with permission.)

world. Most of those are found in small numbers in isolated geographic regions. Only a few of that number have ever been widely established in other areas of the world. Cattle are not indigenous to North America, so it is obvious our cattle must represent breeds that have managed to travel and become established beyond the confines of where they were developed.

Breed history was fairly simple in the United States until 1965. First there was the Texas Longhorn, which descended from the Spanish Longhorns (Figure 16–14). Shorthorns were imported to upgrade the Longhorns and effectively supplanted them. Herefords were imported and used to increase the hardiness of the Shorthorns and effectively supplanted them. Angus were then used to crossbreed with the Herefords and both became established as breeds in their own right. Brahman cattle were brought in for their superior abilities to handle heat, disease, and insects in certain parts of the country. Some ***composite breeds,*** Brangus and Santa Gertrudis, were developed. The

**Composite breed**

A breed developed from two or more previously established breeds.

**FIGURE 16–14** The Texas Longhorn, which descended from the Spanish Longhorns, dominated the beef cattle industry in the United States until after the Civil War. (Photographer Francois Gohier. Courtesy of Photo Researchers, Inc.)

Charolais entered the country just prior to World War II, and a few others found their way here in small numbers. All told, approximately 20 breeds were available until the mid-1960s. It is probable that additional breeds would have found their way into the United States if not for strict import restrictions placed on livestock to prevent the introduction of diseases from other countries. This situation changed in 1965 when a quarantine station was established by Canada on Grosse Isle. Cattle of many new and exotic breeds from Europe and other parts of the world soon began arriving. From 1965 until now, the number of breeds has been in flux. Somewhere around 70 are currently available. The number of breeds was increased through the importation of breeds from other countries as well as the development of additional composite breeds in the United States. Of that number, only 12–15 can be said to have a major influence on the U.S. cattle industry.

The development of the feedlot industry was the major stimulus for the increase in the number of breeds. The cattle being used before the development of the feedlot industry were small statured, early maturing, and bred to be fattened on grass. When fed grain in a feedlot, they got too fat to fit the needs of the beef-packing industry. Cattle were needed that could be fed to heavier weights and produce a leaner carcass. The cow-calf sector also recognized the need to increase its profits. Faster-growing cattle were one of the elements needed to accomplish this. Crossbreeding was a recognized way to add new genetics to the existing cattle. The introduction of new and different breeds was an easy and effective way to do this.

Several dozen breeds were ultimately imported. The value of many of those proved to be marginal. However, the importations did change the face of the American cattle industry. The ***British breeds*** had to make room for other breeds that have become important. In addition, the established breeds were selected to be more like the imports in traits. Commercial cows in the United States are now overwhelmingly crossbred rather than purebred or high-grade cattle as they had been before the influx of breeds. Unfortunately, fixing one problem invariably creates others. One of the major perceived problems with the current cow herd is that rather than being crossbred, it is just mixed up (Figure 16–15). This is causing problems with beef

**British breeds**

Hereford, Angus, and Shorthorn. Breeds that originated in England.

**FIGURE 16–15** One of the major problems with the current cow herd in the United States is that it is more mixed up than crossbred. This is causing quality and management problems. (Photo courtesy of Dr. J. Robert Kropp, Department of Animal Science, Oklahoma State University. Used with permission.)

quality in the retail outlet and problems of management. In response to these concerns, the genetic base of the U.S. cowherd is again narrowing. A look at Table 16–3 shows that the Angus breed is growing at a rapid pace, and most others are declining, or at best maintaining, their annual registration numbers. For information on cattle breeds, visit the breeds of livestock page at http://www.ansi.okstate.edu/breeds/. This is a comprehensive reference that includes pictures of livestock breeds. Color photos of representative beef cattle breeds are found in the color plate section of this text.

**TABLE 16 – 3**    **Major U.S. Beef Breed Association Registrations/Year for Selected Years**

| Cattle[1] Registrations | 1970 | 1990 | 2000 | 2005 | 2006 |
|---|---|---|---|---|---|
| Angus | 346,195 | 159,036 | 271,222 | 324,266 | 347,572 |
| Beefmaster | 2,800 | 46,469 | 30,416 | 19,524 | 18,202 |
| Brangus | 8,238 | 32,069 | 25,488 | 17,274 | 24,940 |
| Charolais | 158,132 | 46,713 | 45,354 | 45,944 | 45,740 |
| Gelbvieh | – | 22,668 | 29,260 | 27,877 | 36,222 |
| Hereford[2] | 421,853 | 169,746 | 75,019 | 71,054 | 69,935 |
| Limousin | – | 72,237 | 51,412 | 32,623 | 31,291 |
| Maine-Anjou | 2,843 | 5,000 | 12,219 | 11,115 | 12,437 |
| Red Angus | – | 15,315 | 41,900 | 47,064 | 47,011 |
| Shorthorn | 32,647 | 18,002 | 18,729 | 17,373 | 17,523 |
| Simmental | – | 78,456 | 41,499 | 48,274 | 50,495 |

[1]Different associations count their cattle at different times. Some use a calendar year and some use a fiscal year. This may result in some inconsistencies. This information is typically available from the National Pedigreed Livestock Council each September.

[2]The number of Herefords reflect both Polled Hereford and Horned Hereford. Their numbers have been added to reflect a merger of the two individual associations.

*SOURCE:* National Pedigreed Livestock Council and personal communication with breed association representatives.

# REPRODUCTIVE MANAGEMENT IN BEEF CATTLE

Chapter 11 is devoted to reproductive physiology. Refer there for the fundamentals of male and female reproductive anatomy, physiology, and function.

The goal of breeding herds of beef cattle is to produce and raise one calf per year from each of the reproductively mature females in the herd. Attainment of such a goal would give an operation a 100% calf crop for the year. The calf must be born within a 12-month interval and at the right time of the year to make best use of the operation's resources (mostly feed) and generate the most profit for the herd. This is a deceptively easy thing to say and not nearly so easy to accomplish. Each cow in the herd must be safely "in calf" within 85 days of calving or the 12-month calving interval cannot be maintained. Good management skills are essential if this goal is to be reached. The cow must recover from calving, start her heat cycle again, and conceive. She must do all of this while raising the calf she just bore. There is tremendous overlap among good reproductive management, good animal health management, and good nutritional management. Nothing on a functional animal unit occurs in a vacuum. Everything is related. Unfortunately, many producers underestimate the magnitude of the problems with the reproductive efficiency of their herds.

Replacement heifers must be managed so they can be mated to calve at 2 years of age. To do this, they must be mated at 15 months of age and need to be at least 65% of their adult weight at the time of breeding. Some producers find this difficult to do and opt for calving heifers first at 30–36 months. This works if the producer is running both spring and fall calving herds. If not, then the only alternative is to calve them at 3 years of age. The loss of a calf or a "half calf" in these systems makes it more difficult for the female to generate a profit in her early lifetime.

Diseases that affect reproductive efficiency directly include brucellosis (Bang's disease), vibriosis, leptospirosis, IBR/BVD complex, and trichomoniasis. Control of these and other diseases can be easily accomplished by developing and following a complete health program for the herd.

The normal season of birth for cattle can be year-round. However, for management purposes, either the spring or fall is preferred. The decision as to which season to choose is almost always made to accommodate the best use of the available feeds. The gestation period varies slightly from breed to breed, but 280–283 days is considered standard. The *estrous cycle* for a cow is 19–21 days and the duration of *estrus* (heat) is 13–17 hours.

Net calf crop as measured by the number of calves weaned/number of cows in the breeding herd is the simplest and best way to measure the effectiveness of a mating/reproductive health program herd. The normal calf crop for many of the nation's herds is about 90%. Obviously a greater percentage is desirable. Most of the nation's beef cows are mated by natural service in pasture or range conditions. Artificial insemination is used on a greater percentage of purebred cows than on commercial cows. The extensive nature of the management of much of the cow herd is a detriment to using artificial insemination cost effectively.

Bulls should be evaluated before they are used in a breeding program, including an evaluation of their genetic contribution and breeding soundness. The major factors that should be evaluated in a *breeding soundness exam* include testicular development, physical ability to breed females, semen quality, and libido. A bad bull can do tremendous damage to the profitability of an operation. A bull produces many more calves in a lifetime than does any single cow. In addition, his presence is felt for generations through his influence on replacement females chosen from his daughters. Thus bull evaluation is more important than the selection of any single female for the herd and should receive emphasis by the breeder.

**Estrous cycle**
Time from one estrus to the next.

**Estrus**
Period of sexual receptivity in the female.

**Breeding soundness exam**
Examination to determine the physical capacity of an individual to breed.

Good records are essential to managing reproductive issues in the cow herd. Because management practices and standards vary with conditions across the country, cattle breeders are encouraged to contact their local extension service or the state extension specialists for help in setting up a recordkeeping program for their area.

## NUTRITION IN BEEF CATTLE

Cattle are ruminants. The feeding niche they occupy is as forage and roughage users. Thus the nutrition and feeding of cattle in most of the production stages revolves around maximizing the use of forages as the base of economic production. Chapters 5, 6 and 7 give detailed information relating to nutrition, feeds, and feeding. See those chapters for more detailed information than is provided in this section.

Managing the nutrition of beef cattle is very dependent on the class of cattle to be fed (Figure 16–16). The breeding herd is always managed to take maximum benefit from forages and roughages. If supplemental feeds are fed, they are used to further extend the use of forage by supplementing missing nutrients. Mineral supplements are needed in all parts of the country. Individual mineral deficiencies specific to a region occur in the various geographic areas of the country and don't usually cause a problem in other areas. Salt, calcium, and phosphorus are likely to be needed everywhere. Protein supplementation is often needed when such feeds as late-season pasture, low-quality hays, crop by-products, and other waste by-products are used. Energy may need to be supplemented in drought, during the winter months, as a boost for 2-year-old heifers too thin to rebreed, for replacement heifers, and at other times in the production cycle. The goal is to supplement forages in a way that allows their maximum use because they are lower cost.

Stocker feeding programs also depend on maximizing the use of forages. Because stocker cattle are young and need to grow as well as be maintained, the plane of nutrition has to be somewhat higher for cattle in a stocker program than that for a brood cow herd. More energy and/or protein supplementation is usually necessary. Higher-quality forages are also needed to keep animals gaining weight rapidly enough to turn a profit. Although stocker programs vary somewhat from region to region, they generally fall

**FIGURE 16–16**   The nutrition and feeding of beef cattle varies according to the class of the animal. These feedlot cattle are finished for slaughter on high-quality energy and protein feeds. (Photo courtesy of American Angus Association. Used with permission.)

into two categories. The first is to full-feed calves a complete mixed ration based on forage as a low-cost energy source. The second is to graze calves or give them harvested forages free choice and then feed additional energy and protein as a mixed supplement. Table 16–4 gives examples of supplements that may be fed to grazing stocker calves. The stocker phase is also an excellent time to take advantage of the growth-enhancing abilities of growth implants and feed additives.

Feedlots are specialized finishing operations. High-quality feeds are needed to bring cattle to a suitable slaughter end point in as little time as possible. The goal here is to take calves from forage-fed cows and grow them further, if necessary, on forage and modest amounts of energy and protein supplements in a stocker phase, and then use a small volume of higher quality feeds to prepare the calves for slaughter. Several general types of feed programs will be used during the 120–150 days that a finishing beef animal is usually in a feedlot. These programs include a receiving ration given to the calves once they arrive at the feedlot and for a week or two after arrival. Even these rations take different forms. Hay and the supplement shown in Table 16–5 may be used or a complete milled ration might be used. As quickly as possible, cattle are switched to high-energy feeds designed to optimize gain and, therefore, returns on the finishing

**TABLE 16–4** **Supplement for Stocker Calves Grazing Late Summer or Early Fall Pastures**

| | COMPOSITION, % (AS FED BASIS) | |
| Ingredient | Oklahoma Gold | Oklahoma SuperGold |
| --- | --- | --- |
| Cottonseed | 86.0 | 17.0 |
| Soybean meal | — | 15.0 |
| Wheat middlings | 7.0 | 56.0 |
| Molasses (pellet binder) | 4.0 | 4.0 |
| Vitamin and mineral premix | 3.0 | 3.0 |
| Feed additive | Variable | Variable |
| Crude protein, % as fed | 38.0 | 25.0 |
| Feeding rate, lbs per day | 1.0 | 2.5 |

**TABLE 16–5** **Composition of Feed Supplement to be Fed with Hay for Cattle Receiving Ration**

| Ingredient | Percent as Fed |
| --- | --- |
| Soybean meal | 88.90 |
| Salt | 3.00 |
| Vitamin A—30,000 IU/g[1] | 0.11 |
| Vitamin E—222,800 IU/lb[2] | 0.09 |
| Premix[3] | 0.18 |
| Cottonseed meal | 5.00 |
| Dicalcium phosphate | 2.75 |

[1]To provide 13,140 IU vitamin A per pound.
[2]To provide 200 IU vitamin E per pound.
[3]To provide lasalocid, decoquinate, or monensin.
*SOURCE:* Gill, 1992, pp. 4–9. Used with permission.

**TABLE 16-6    Composition of Typical Beef Cattle Finishing Diets**

| Ingredient | Processed Corn and Dry Roughage | Whole Corn and Corn Silage (% of DM) | Dry-Rolled Corn and Wet Corn Gluten Feed |
|---|---|---|---|
| Roughages | | | |
|   Sudan grass hay | 4 | — | 8 |
|   Alfalfa hay | 6 | — | — |
|   Corn silage | — | 10 | — |
| Grain and grain by-products | | | |
|   Steam-flaked corn | 74.5 | — | — |
|   Dry-rolled corn | — | — | 52.5 |
|   Whole shelled corn | — | 71 | — |
|   Wet corn gluten feed | — | — | 35 |
| Liquid feeds | | | |
|   Molasses | 5 | — | — |
|   Condensed distiller solubles | — | 4 | — |
|   Fat | 3 | — | — |
| Supplement[1] | 7.5 | 15 | 4.5 |

[1]Supplement supplies calcium and phosphorus sources, urea and/or natural protein, trace minerals, vitamins, and feed additives.

*SOURCE:* Galyean and Duff, 1998, p. 276. Used with permission.

---

**Least-cost ration**

A ration formulated to meet the animal's nutritional needs at the lowest cost from the feeds available.

cattle. Any number of rations are used with success. Table 16–6 shows examples of some typical beef cattle finishing diets. Many other feeds can be used to mix rations that will finish cattle in a cost-effective manner. Most such rations are formulated with the use of *least-cost ration* formulation programs.

Seedstock producers are a small but important segment of the beef cattle industry. The feeds and feeding programs they use to produce replacement animals for the breeding herd are similar to those used by the commercial segments. However, they vary in one important way. Seedstock producers are more likely to use higher levels of supplementation at all phases of production than do the other sectors. The reason for this is simple. They have a vested interest in providing their animals with the nutrition necessary for them to demonstrate their full genetic potential to possible purchasers. Because seedstock producers command a higher price for their product than do the other sectors of the industry, they can afford to do this.

## BEEF'S NUTRITIONAL BENEFITS TO HUMANS

A 3-oz serving of cooked beef provides the following proportion of the recommended daily dietary allowance for a 25- to 50-year-old woman on a 2,200-calorie diet:

| | |
|---|---|
| Protein | 50% |
| Phosphorus | 25% |
| Iron | 17% |
| Zinc | 49% |
| Riboflavin | 16% |
| Thiamin | 7% |

**FIGURE 16–17** Beef is nutrient-dense healthful food that tastes good. (Photo courtesy of Certified Angus Beef Program. CERTIFIED ANGUS BEEF™ is a registered trademark of the Certified Angus Beef Program and is used with permission.)

| | |
|---|---|
| $B_{12}$ | 112% |
| Niacin | 23% |
| Calories | 8% |
| Fat | 14% |

Beef is a nutrient-dense food, which means it has lots of nutrition per calorie. Beef is a healthful food that can be part of the diet of virtually all people (Figure 16–17). A complete nutrient analysis for beef is given in Appendix I.

## TRENDS AND FACTORS THAT WILL INFLUENCE THE BEEF INDUSTRY IN THE FUTURE

The beef cattle industry is in a state of change. The following sections represent some of the challenges, changes, and areas of focus for the cattle industry. The beef cattle industry's future may well depend on how well it meets these challenges.

### CONSUMPTION

### Market Share

Annual per capita beef consumption in the United States declined from 1976 (highest year of consumption was 1976, one year after peak cattle numbers) until 1993 (Figure 16–18) while pork consumption stayed fairly stable and chicken and total poultry increased. Beef consumption has been fairly stable since the early 1990s. Table 16–7 shows the average per capita expenditures for meat. Beef expenditures as a percentage of total beef, pork, and chicken expenditures seems to have stabilized. Consumers decreased their beef purchases in large part because competing meats are lower priced. It follows that price will continue to be a factor in beef consumption.

### Nutrition

Nutrition and health consciousness will continue to play a role in food choices. Beef is good food that is nutritious. Lean beef is included in the dietary recommendations of all of the major health organizations. However, fat content is an issue with the consumer. The industry has already responded with leaner cuts by trimming and discarding fat

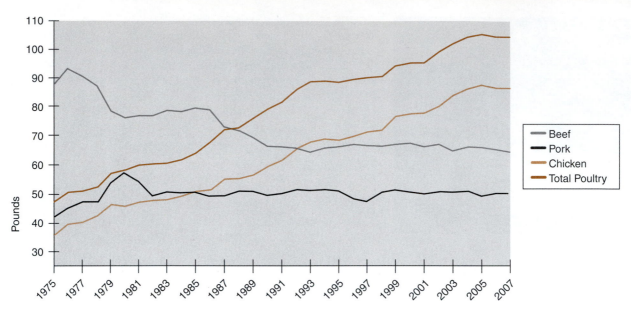

**FIGURE 16–18**   Beef per capita consumption compared to pork, chicken, and total poultry. Pounds retail weight. (*SOURCE:* USDA-ERS, 2007b.)

**TABLE 16–7**   **Average Annual Per Capita Expenditures for Meat**

|  | Beef ($) | Pork ($) | Broiler ($) | Beef as % of Total Beef, Pork, Broiler |
|---|---|---|---|---|
| 1980 | 178.8 | 84.5 | 44.4 | 58.1% |
| 1985 | 181.0 | 88.6 | 58.7 | 55.1% |
| 1990 | 190.3 | 111.8 | 86.6 | 49.0% |
| 1995 | 189.3 | 106.6 | 97.8 | 48.1% |
| 2000 | 207.7 | 132.3 | 119.5 | 45.2% |
| 2001 | 224.0 | 135.5 | 121.0 | 46.6% |
| 2002 | 224.4 | 136.9 | 130.5 | 45.6% |
| 2003 | 242.6 | 138.0 | 131.1 | 47.4% |
| 2004 | 268.7 | 143.2 | 145.6 | 48.2% |

Americans continue to spend more for beef than for pork or chicken.
*SOURCE:* USDA-NASS. 2007b.

from retail beef cuts and by improving the genetics of the animals. In addition, cattle are being sent to the market significantly leaner today than they were in the past. However, the beef industry has taken too much marbling out and has created a product that is inconsistent and undependable. Lack of consistency in beef products is now a problem. The real challenge is to have leaner animals that still produce a tasty product so we can stop this trim waste and still satisfy the consumer. That's quite a challenge!

At least a portion of the solution to this problem is likely to be technology based. One of the major concerns of consumers is beef tenderness. Because marbling is only one variable in meat tenderness, other methods of predicting tenderness are needed. The QualitySpec BT system developed by ASD Inc. in cooperation with USDA's Meat Animal Research Center is a potential technological solution. The goal is to provide a new standard for beef quality assurance (Box 16–1).

## Box 16–1    Visible/Near Infrared Spectroscopy

Visible/near infrared spectroscopy (Vis/NIR spectroscopy) is used in the food, beverage, and dairy industry for nondestructive and in situ analysis of crops, soils, raw materials, and finished products. Vis/NIR spectroscopy is also being used to develop nutritionally enhanced products. The foods industry routinely uses Vis/NIR spectroscopy for applications in processing cheese, eggs, butter, beer, wine, baked goods, cereal, produce, and jams/jellies. Properties typically measured in food, beverage, and dairy products include fat, protein, moisture, particle size, sugar content, acidity, and blend analysis. The goal is to decrease the chances of products that do not meet specifications, increase overall product quality control, and improve customer satisfaction and producer profitability. Other applications of the technology are being explored.

A recent major application of Vis/NIR spectroscopy was developed by the USDA's Meat Animal Research Center in Clay Center, Nebraska, for predicting beef tenderness nondestructively online at the grading stand. Major beef processors are using a Vis/NIR system, QualitySpec BT, developed by ASD Inc. The system reads a ribeye cut between the 12th and 13th rib during carcass processing. The system can be integrated into the processing plant's information systems so tenderness can be monitored and reported with all of the other parameters used in the facility. The system helps predict more accurately the overall tenderness of the whole carcass and permits processing of the specific carcass in a way that is most appropriate. Products from the carcass can be branded, along with tenderness guarantees, and priced accordingly.

## Convenient Foods

Beef, as a product, lacks the volume of good, home-prepared convenience foods that poultry and, increasingly, pork provide. Consumer surveys tell us that three fourths of the cooks in America are undecided at 4 P.M. about what they will make for dinner. Convenience is very important. Affordable, tasty, convenient, microwavable chicken and pork products are readily available. Beef must become more convenient and maintain quality in the process if it is to compete. However, beef marketing is changing. Many convenient products have become available just in the past few years. Some of these products have begun to affect demand positively (Figure 16–19). One bright spot is that beef is the single most ordered entree in the restaurant business. Eating out is a major form of convenience to modern consumers.

**FIGURE 16–19**  Convenient foods are a part of today's busy lifestyles. After years of lagging behind poultry and then pork, easy-to-prepare beef entrees are now affecting demand for beef. Here, the "heat 'n eat" section of the meat counter is being replenished after the late afternoon shoppers have depleted it.

## EXPORTS

The export market for beef looks bright for the long term. Exports have been increasing substantially since the early 1980s. The United States has been a net exporter of beef since 1992 on a value basis. The implementation of worldwide trade agreements has created a better trade environment and stimulated imports and exports with the United States. Burgeoning human populations, along with developing economies in some of those populations, are creating very favorable prospects for many agricultural exports from the United States. Although normal economic cycles, sporadic consumer concerns, and disease-related trade restrictions will surely affect year-to-year purchases by other countries, the overall trend is positive for the foreseeable future.

## TECHNOLOGY

The beef-producing sectors of the industry historically have not been progressive users of available technology. Cost competitiveness will give the edge to those who adopt and appropriately use the ever-increasing array of products and technological innovations. Expected aids include lower-cost sexed semen, "designer embryos" produced by recombinant DNA technology and perhaps multiplied through cloning, an expanded arsenal of drugs, growth enhancers and vaccines, and many other tools. Genetic engineering in particular will ultimately bring great dividends to this industry. The larger the operation, the more likely it is to use these tools. This relationship is circular in that the availability of these tools will help accelerate the trend to larger operations.

## National Animal Identification System

The National Animal Identification System (NAIS) has had a difficult beginning. However, NAIS is currently being implemented as a voluntary program under the control of the USDA with states deciding how to administer the program. The purpose of NAIS is to register livestock premises and livestock for the purpose of traceability. Although voluntary at the federal level, some states have made the program mandatory.

Traceability of livestock is emerging as an important political and economic issue. Increasingly, packers, foreign and domestic consumers, and exporters and importers want to know where food animal products come from (source verified) and how old the animal was when slaughtered (age verified). Further, all parties seem willing to pay premiums at all stages of the production system for those assurances. Although they can be accomplished without NAIS, it certainly could play a major role. Traceability is a key element of country of origin labeling (COOL).

Such a system and its potential benefits have been discussed for many years. After the September 11 terrorist attack on the World Trade Center, plans were developed, and the discovery of a dairy cow on U.S. soil with BSE in December 2003 provided the immediate impetus to move forward with the system. USDA announced plans for the National Animal Identification System (NAIS) in April 2004, provided the initial funding, and it continues to provide support to implement NAIS. The benefits of such a system as expressed by the Veterinary Services Division of APHIS include "disease control and eradication, disease surveillance and monitoring, emergency response to foreign animal diseases, regionalization, global trade, livestock production efficiency, consumer concerns over food safety, and emergency management programs."

Information on NAIS can be found on the USDA APHIS website, http://animalid .aphis.usda.gov/nais/index.shtml.

## FOOD SAFETY

The food supply in the United States is arguably the safest in the world. Meat, specifically, is closely monitored by the government prior to purchase. The Food Safety and Inspection Service of the Department of Agriculture is responsible for inspecting meat. The Food and Drug Administration regulates feeds, feed additives, and animal health products. Both agencies have enviable records of ensuring a safe food supply. Even so, safety of the food supply is a front burner issue with the consuming public. Examples of issues of concern to the public are feed additives and growth promotants of all kinds, antibiotic use, and foodborne pathogens. There is little doubt that most of the fears are unfounded. However, several well-publicized events, some with tragic outcomes, have heightened concern in the population. The beef industry must contend with both safety of beef and the consumer's *perception* of safety.

Many elements of the National Cattleman's Beef Association's Beef Quality Assurance program are designed to help producers ensure that domestic and international beef consumers enjoy ready access to a safe, wholesome, and healthy beef supply. Those guidelines can be accessed at http://www.bqa.org/.

## Foodborne Illnesses

Foodborne illnesses are especially volatile and emotional issues with consumers. Beef is an extremely safe food. However, events in the news easily shape public perception. A small number of unfortunate incidents have produced a general negative perception. Paradoxically, improper handling and preparation of foods by food handlers are the primary causes of such illnesses. Although a perfectly safe food supply is unachievable, several steps are being taken to reduce risks to consumers. For example, low-dose irradiation, high-temperature vacuuming, hot-water rinses, and organic acid rinses are being used to kill and/or remove pathogens from beef carcasses. Recent ground beef recalls because of concerns about pathogens were the result of newer, more sophisticated detection methods. The recalls made the food supply safer and, at the same time, the consuming public lost confidence in the food safety system to protect them. In 1998, the government initiated a *Hazard Analysis and Critical Control Points (HACCP)* plan to aid in detecting and preventing problems in packing plants. HACCP systems involve identifying points in a processing plant where contamination is most likely to occur and finding methods to combat it. In addition, the HACCP principles are being voluntarily implemented in production units, most notably feedlots, as a means of delivering an even safer and more wholesome product to consumers. It is probable that HACCP will be mandated to feedlots and other parts of the industry at some point as the government struggles to help make the food supply safer. Food safety concerns and HACCP are discussed in detail in Chapter 28.

**Hazard Analysis and Critical Control Points (HACCP)**

A process control system designed to identify and prevent microbial and other hazards in food production.

## Bovine Spongiform Encephalopathy (BSE)

BSE is a fatal disease in cattle commonly referred to as "mad cow disease" that was first diagnosed in Great Britain in 1986. In March 1996, it was announced that a new variant of Creutzfeldt–Jakob disease (vCJD) in humans might be linked to BSE and humans were contracting the disease from the infected animals. The disease is a degenerative disease of the central nervous system for which there is no cure. Ultimately, the disease has been confirmed in many European countries and others outside Europe as well. The cost of BSE to the affected countries (Great Britain has been affected the most) includes inestimable emotional distress, loss of consumer confidence, billions of dollars in lost revenues and lost animals, and the death of more than 150 people. Because it appears that the minimum incubation time of vCJD is 10 years with as much as 25 years possible, the loss in human life is certain to rise. The good news is that reported cases in both cattle and humans have been steadily declining. Tens of thousands of cattle died annually when the disease was at its worst. Now, only a few hundred die annually. Only five human cases were reported in 2005.

In December 2003, the USDA announced that a dairy cow in Washington state had tested positive for BSE. After an extensive investigation including the location and testing of animals that had entered the United States from Canada with the affected cow, as well as other animals in contact with the affected cow, it was determined that there were no collateral cases. As of June 1, 2004, an enhanced BSE testing program was implemented by USDA. As of press time for this text, three total animals have tested positive for BSE in the United States.

The beef industry was affected by the discovery of a BSE-infected animal on U.S. soil. Although U.S. consumers reacted with little more than short-term concerns, export markets were damaged and there are now costs to the industry that were not there before. Consumer surveys indicate that the potential of BSE to affect beef consumption dramatically is very real. What those costs will ultimately be is yet to be determined. For further information on BSE, visit http://www.fsis.usda.gov/Fact _Sheets/Bovine_ Spongiform_Encephalopathy_BSE/index.asp and http://www.fas .usda.gov/DLP/BSE/bse.html.

## Hormone Implants, Feed Additives, and Antibiotics

Hormone implants and feed additives are used to enhance animal performance and are not a safety risk to people who consume the meat. Many years of research have demonstrated over and over the safety of these products. Regardless of the science, however, a growing segment of the consuming public views the use of any of these products negatively.

When animals become infected with diseases, antibiotics are used to treat the disease. The Beef Quality Assurance Program helps producers and veterinarians follow *withdrawal times* when antibiotics are used. *Subtherapeutic* use of antibiotics is the practice of feeding low levels of approved antibiotics to promote better feed efficiencies and improve weight gains. Subtherapeutic levels of antibiotics are rarely used in beef cattle feeding anymore. The USDA's Food Safety and Inspection Service routinely checks meat for the presence of residues, and beef has an extremely good record for being free of residues.

### ENVIRONMENTAL CONCERNS

The effect of agriculture on the environment is a growing public concern. This is true of agronomic practices as well as animal agriculture. The animal industries are visible industries and considered major polluters. Manure and wastewater from concentrated animal feeding operations (CAFOs) have the potential to contribute pollutants such as nitrogen and phosphorus, organic matter, sediments, pathogens, heavy metals, hormones, antibiotics, and ammonia to the environment. Excess nutrients in water (i.e., nitrogen and phosphorus) can result in or contribute to low levels of dissolved oxygen (anoxia), *eutrophication,* and toxic algal blooms. Decomposing organic matter (i.e., animal waste) can reduce oxygen levels and cause fish kills. Pathogens in manure can also create a food

---

**Withdrawal times**

The period of time after a drug is administered to an animal that the milk or meat from the animal must not be used for human food.

**Subtherapeutic**

Quantities too low to treat disease.

**Eutrophication**

Promotion of excess growth of one organism to the disadvantage of other organisms in the ecosystem.

**FIGURE 16–20** Finding low-cost and effective methods of dealing with waste from feedlots is an industry priority. (Photo courtesy of Natural Resources Conservation Service/USDA.)

safety concern if manure is applied directly to crops at inappropriate times. In addition, pathogens of animal origin have been responsible for some shellfish bed closures. Nitrogen in the form of nitrate can contaminate drinking water supplies drawn from groundwater. One of the fundamental problems facing the animal industries is the recycling of potential pollutants back to cropland where they can again produce crops. This is important from a sustainable perspective as well as a pollution perspective.

Current environmental laws, regulations, policies, and guidance regarding CAFOs can be found at http://www.epa.gov/agriculture/anafolaw.html. Tougher waste disposal laws are inevitable and will add costs to production. Low-cost, effective means of managing animal waste will help blunt the added costs that better environmental stewardship will bring. Everyone involved understands that the air, soil, and water supply must be protected (Figure 16–20).

## ORGANIC AND NATURAL PRODUCTION

People are also becoming more aware of the environmental footprint of agriculture and its individual products. This is leading consumers to products that they perceive come from more ecologically friendly and sustainable production systems. In addition, many consumers are willing to pay a premium for organic or so-called natural meat from animals produced free of antibiotics, growth enhancers, and feed additives, believing them healthier to consume. In large measure, this is what has made organic farming the fastest growing segment of U.S. agriculture. The term *natural* has proliferated in product naming and marketing. A good deal of niche marketing has already developed around such products (Box 16–2), and more products are expected. In response, nearly 2 million acres of rangeland and pastureland have been certified for organic livestock production by livestock producers, with more expected. There is a resurging interest in grass-fed beef. Meat from cloned animals is likely to be rejected by many consumers on these grounds also.

## ETHANOL

The rapid increase in U.S. domestic ethanol production, part of a national initiative to decrease dependence on imported oil by increasing the availability of cleaner, domestic

## Box 16–2    Laura's Lean Beef

Laura's Lean Beef, the first naturally raised beef to be certified by the American Heart Association, is available in over 5,000 retail stores. At Laura's Lean Beef Company, we're proud to be both progressive *and* old-fashioned. In 1985, I made a commitment that we would raise our cattle on natural grains and grasses. Today we continue to maintain strict feeding guidelines that prohibit any animal protein—no bonemeal, no blood meal, no reprocessed animal tissue, no animal by-products. We also raise our cattle without the use of added growth hormones or antibiotics. It may be a little old-fashioned in these days of high-volume, high-speed production, but we are convinced it's the right thing to do. As a family farmer, I'm involved in every aspect of my farm; I know my animals and I know my land. The other farmers who raise Laura's Lean Beef are the same way. Farming isn't just a business for us; it's part of who we are. The result is that Laura's Lean Beef farmers produce a better beef. Many of our customers tell us that Laura's Lean Beef tastes the way beef used to—full of flavor and "clean."

alternatives to imported oil and gas, is affecting the animal industries because corn is the preferred grain in making fuel-grade ethanol. Corn prices have increased substantially in response to new ethanol-plant demand. Grain prices always affect the amount of grain fed to livestock.

Many factors will ultimately influence just how much ethanol is produced and the overall effect on agriculture and its component industries. Useful feed-grade by-products are available from ethanol production. A bushel of corn produces 2.8 gallons of ethanol and 17–18 lbs of distillers' grains. Generally, by-product feeds are best used if they can be fed close to the point of production. This could cause more cattle to be fed in the ethanol-producing regions of the country, which is most likely to be centered in the Corn Belt, and fewer cattle to be fed in the western states. Alternative feed sources and decreased grain use are possible responses. For example, cattle can be kept on forage to heavier weights before being placed in feedlots. The price of crude oil is also a major factor in how many ethanol plants are built and how much ethanol is ultimately produced.

### INDUSTRY STRUCTURE

## Industry Consolidation

The number of U.S. beef cattle operations has been declining for decades. In 2004, the number dropped to less than 1 million for the first time and continues to decline by more than 1.5% annually. The remaining operations will continue to get bigger. This is true of all phases of the industry. The greatest concentration has occurred so far with feedlots and beef packers. However, cow herds are also slowly increasing in size. This follows the overall trend in agriculture. Bigger operations can make better use of machinery, equipment, and management because they can more easily achieve economy of scale. Even so, part-time operators, people who are retired, people looking to diversify, and hobbyists will probably always play a role in the cow-calf segment and serve to keep herd size smaller than it might otherwise be.

## Industry Integration

One potential, highly probable, and desirable solution to many of the challenges to the beef industry is the increase in the amount of communication, cooperation, and coordination among beef industry segments. Terms like *alliance*, *vertical cooperation*, and *vertical coordination* describe the kinds of industry structures being discussed and tried. These are variations of, and alternatives to, vertical integration. All are expanding further into the beef industry. Alliances of many types are being forged in this historically segmented industry, both horizontally and vertically. What remains to be seen is which ones work, how far the restructuring will go, and what effects the consumer and producer will observe. One example of an alliance is Laura's Lean Beef Company, which has been in business since 1985, making it one of the oldest of the alliance groups (Box 16–2). See Chapter 15 for a complete discussion.

## INDUSTRY ORGANIZATIONS

### National Cattlemen's Beef Association

#### NCBA Chicago
444 North Michigan Ave., Suite 1800
Chicago IL, 60611
Phone: (312) 467–5520
Fax: (312) 670–9414

#### NCBA Denver
9110 East Nichols Ave., Suite 300
Centennial, CO 80112
Phone: (303) 694–0305
Fax: (303) 694–2851

#### NCBA Washington, D.C.
1301 Pennsylvania Ave. NW, Suite 300
Washington, DC 20004
Phone: (202) 347–0228
Fax: (202) 638–0607
http://www.beefusa.org/

NCBA is the result of a 1996 merger of the National Livestock and Meat Board with the National Cattlemen's Association. It is the marketing organization and trade association for the beef industry. The NCBA Checkoff Division is charged with promotion, research, information, and related activities as financed by the beef checkoff. The NCBA Dues Division is a trade association charged with non-checkoff-funded activities dealing with policy making, government affairs, and similar activities.

### Cattlemen's Beef Promotion and Research Board

9000 East Nichols Ave., Suite 215
Centennial, CO 80112–3450
Phone: (303)220–9890
Fax: (303)220–9280
http://www.beefboard.org

Often referred to as Cattlemen's Beef Board, or CBB, this organization approves the budget for beef checkoff money. Beef checkoff dollars are generated by a producer-approved program that assesses $1.00 per head for each cattle transaction to help market beef. They also certify Qualified State Beef Councils.

### United States Meat Export Federation

1050 17th St., Suite 2200
Denver, CO 80265
Phone: (303)623–6328
Fax: (303)623–0297
http://www.usmef.org/

This organization is a nonprofit, integrated trade organization representing a wide variety of groups including livestock producers, meatpackers, processors, farm organizations, grain promotional groups, agribusiness companies, and others. The organization works to develop foreign markets for U.S.-produced beef, pork, lamb, and veal.

## SUMMARY AND CONCLUSION

Beef cattle make use of resources that would otherwise go to waste. They convert grass to high-quality food, by-products, and work that humans can use. The United States is a major beef producer and consumer. The gross annual income from beef in the United States is approximately $50 billion. Beef cattle production differs from most other kinds of livestock production because it is divided into several distinct phases, and cattle are generally owned by different people in each phase. There is also a distinct regionality to the beef industry. There is a growing trend for owners to retain the own-

### Facts about Beef Cattle

| | |
|---|---|
| Birth weight: | Varies with breed and sex; 50–120 lbs |
| Mature weight: | Varies with breed, sex, and condition; male 1,400–3,000 lbs; female 900–1,800 lbs |
| Slaughter weight: | 1,000–1,500 lbs |
| Weaning age: | 5–8 months |
| Breeding age: | 14–19 months (female) |
| Normal season of birth: | Year-round, spring and fall seasons preferred |
| Gestation: | 280–283 days |
| Estrous cycle: | 19–21 days |
| Duration of estrus (heat): | 13–17 hours |
| Calving interval (months): | 12 desirable |
| Normal calf crop: | 90% |
| Names of various sex classes: | Calf, heifer, cow, bull, steer |
| Weight at weaning: | 400–800 lbs |
| Type of digestive system: | Ruminant |

ership of their cattle through one or more phases. Such arrangements are probably destined to become more commonplace in the beef industry as it reorganizes to remain competitive. The beef cattle industry is in a state of change. The beef cattle industry's future will depend on how well it meets current and future challenges.

## STUDY QUESTIONS

1. Cattle and calves amount to how much of the total cash receipts of animal agriculture? Describe the magnitude of this industry in other ways.

2. What is the purpose of the beef cattle industry in the United States? Describe it in terms of feed and use of resources.

3. How does the beef cycle describe cattle numbers in the United States?

4. When were the first cattle brought into the Western Hemisphere? Who brought them and what part of the continent were they on? Who is responsible for multiplying cattle across the American West?

5. When did beef cow numbers first exceed dairy cow numbers?

6. How does the structure of the beef industry differ from that of other animal industries? What are the major segments of the beef cattle industry?

7. Why are cattle located where they are in the United States? Why do different regions specialize in different segments?

8. Approximately how many total head of cattle are there in the United States?

9. What are some of the tools available to help cattle producers make sound decisions about the genetics they use?

10. What are the roles of the different breeds in the cattle industry?

11. Briefly discuss how crossbreeding has improved the beef industry and how it may have damaged it.

12. Briefly describe management for maximal reproductive efficiency.

13. Briefly discuss nutrition in beef cattle for each class of animal.

14. What are some important nutrients that a 3-oz serving of cooked beef provides? What does "nutrient dense" mean?

15. What has been happening to the market share of beef compared to pork and chicken?

16. What role will exports likely play in beef production in the future?

17. What role will technology play for the cattle industry in the future?

18. What are the food safety issues that concern people about beef?

19. How will the issue of nutrition and health consciousness affect the beef cattle industry in the future?

20. How will eating habits change in the future?

21. How will environmental concern affect the cattle industry in the future?

22. In the future, cattle operations will continue to grow larger. Why is this so?

23. Briefly discuss how and why vertical integration will affect the beef industry in the future.

## REFERENCES

Ensminger, M. E. 1987. *Beef cattle science.* 6th ed. Danville, IL: Interstate.

Ensminger, M. E. 1991. *Animal science.* 9th ed. Danville, IL: Interstate.

Evans, J., D. Buchanan, and S. Dolezal. 2004. Beef cattle breeding. In *Oklahoma beef cattle manual.* 4th ed. Stillwater: Oklahoma State University.

FAO. 2007. *FAOSTAT statistics database: Agricultural production and production indices data.* http://apps.fao.org/.

Galyean, M. L., and G. C. Duff. 1998. Feeding growing-finishing beef cattle. In *Livestock feeds and feeding,* 4th ed., R. O. Kellems and D. C. Church, eds. Upper Saddle River, NJ: Prentice Hall.

Gill, D. R. 1992. Nutrition, health and management on newly arrived stressed stocker cattle. In *Oklahoma beef cattle manual,* 3rd ed. Stillwater: Agricultural Experiment Station, Oklahoma State University.

Hibberd, C. A., K. Lusby, and D. Gill. 1992. Mechanics of formulating stocker cattle rations and supplements. In *Oklahoma beef cattle manual.* 3rd ed. Stillwater: Agricultural Experiment Station, Oklahoma State University.

Lalman, D., and D. Doye, eds. 2004. *Beef cattle manual.* 4th ed. Stillwater: Agricultural Experiment Station, Oklahoma State University.

National Cattlemen's Beef Association. 2007. *Cattle and beef industry statistics.* http://www.beef.org.

Neumann, A. L., and K. S. Lusby. 1986. *Beef cattle.* 8th ed. New York: John Wiley.

Northcutt, S. L., and D. S. Buchanan. 1992. Beef cattle breeding. In *Oklahoma beef cattle manual.* 3rd ed. Stillwater: Agricultural Experment Station, Oklahoma State University.

NRC. 2000. *Nutrient requirements of beef cattle.* 7th ed. Washington: National Academy Press.

Selk, G. E., and K. Lusby. 1992. The management of beef cattle for efficient reproduction. In *Oklahoma beef cattle manual.* 3rd ed. Stillwater: Agricultural Experiment Station, Oklahoma State University.

Taylor, R. E. 1984. *Beef production and the beef industry: A beef producer's perspective.* Minneapolis, MN: Burgess.

Taylor, R. E., and T. G. Field, 2007. *Beef production and management decisions.* 5th ed. Upper Saddle River, NJ: Prentice Hall.

USDA. 1997. *USDA nutrient database for standard reference.* Release 11–1. Nutrient Data Laboratory home page: http://www.nal.usda.gov.fnic/foodcomp.

USDA. 1998. *USDA nutrient database for standard reference.* Release 12. Nutrient Data Laboratory home page: http://www.nal.usda.gov.fnic/foodcomp.

USDA-ERS. 2007. *Red meat yearbook.* Accessed online September 2007. http://usda.mannlib.cornell.edu/MannUsda/viewDocumentInfo.do?documentID=1354.

USDA-FAS. 2006. *Livestock and poultry: World markets and trade.* Circular Series DL&P 2-06. October 2006.

USDA-NASS. 2007a. *Briefing room. Farm income and costs.* Accessed online September 2007. http://www.ers.usda.gov/Briefing/FarmIncome/.

USDA-NASS. 2007b. *Quick stats: Agricultural statistics data base.* Accessed online September 2007. http://www.nass.usda.gov/QuickStats/.

Ward, Clement E. 2004. *Pork, beef, and poultry industry coordination.* Stillwater: Oklahoma Cooperative Extension Service, AGEC-552.

Will, R. G., J. W. Ironside, M. Zeidler, S. N. Cousens, K. Estibeiro, A. Alperovitch, S. Poser, M. Pocchiari, A. Hofman, and P. G. Smith. 1996. A new variant of Creutzfeldt–Jakob disease in the UK. *Lancet* 347:921–925.

# Dairy Cattle

## LEARNING OBJECTIVES

After you have studied this chapter, you should be able to:

1. Explain the place of dairy cattle in U.S. agriculture.

2. Explain the reasons for the size of the U.S. dairy industry.

3. Give a brief history of the dairy industry in the United States.

4. Describe the structure of the U.S. dairy industry.

5. Give an accurate accounting of where the dairy industry is located geographically in the United States.

6. Give a brief synopsis of DHIA and its functions.

7. Identify and place in context the role of genetics in the dairy industry.

8. Describe the general basis of managing dairy cattle for reproductive efficiency.

9. Describe the feed supply of dairy cattle and explain how it affects dairy management.

10. Discuss bovine somatotropin and its use in the dairy industry.

11. Explain the nutritional benefits of milk to humans.

12. Discuss trends in the dairy industry including factors that will influence the industry in the future.

## KEY TERMS

| | | |
|---|---|---|
| Babcock Cream Test | Embryo transfer | Pasteurization |
| Bovine somatotropin (BST) | Genetic markers | Posilac |
| | Gomer bull | Silo |
| Condensed milk | Herd health | |
| Diversified farm | Milk | |

# SCIENTIFIC CLASSIFICATION OF CATTLE

| | |
|---|---|
| Phylum: | Chordata |
| Subphylum: | Vertebrata |
| Class: | Mammalia |
| Order: | Artiodactyla |
| Suborder: | Ruminata |
| Family: | Bovidae |
| Genus: | *Bos* |
| Species: | *taurus; indicus* |

# THE PLACE OF THE DAIRY CATTLE INDUSTRY IN U.S. AGRICULTURE

Dairy products provide approximately 10% of all yearly cash receipts from agriculture (Figure 17–1). This ranks dairy as third in animal industries behind beef and poultry and eggs. The annual cash receipts from dairy products in the United States is approximately $25 billion. Dairy products generally account for approximately 20% of animal agriculture's share of annual farm cash receipts (Figure 17–2). Dairy products rank in the top five commodities for 31 of the 50 states. Six states exceed $1 billion in yearly cash receipts from farm ***milk*** sales to milk-processing plants and dealers. An additional six states exceed $0.5 billion. The United States produces approximately 15% of the world's cow's milk, although we have only 4.6% of the world's population. U.S. dairy cattle account for less than 1% of the world's cows. Dairy cattle also produce beef from cull dairy cows and many Holstein calves that are fed in commercial feedlots (Figure 17–3). About 25% of

**Milk**

The normal secretion of the mammary glands of female mammals.

**FIGURE 17–1**    Dairy products farm cash receipts as a percentage of total U.S. farm cash receipts, 2005–2007. (*SOURCE:* USDA-NASS, 2007b.)

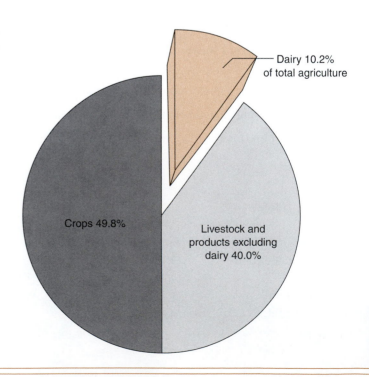

Dairy 10.2% of total agriculture

Crops 49.8%

Livestock and products excluding dairy 40.0%

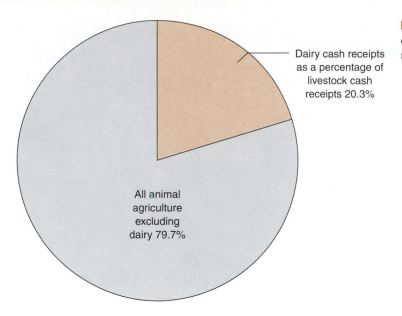

Dairy cash receipts as a percentage of livestock cash receipts 20.3%

All animal agriculture excluding dairy 79.7%

**FIGURE 17–2** Dairy products yearly farm cash receipts as a percentage of total animal agriculture's cash receipts, 2005–2007. (*SOURCE:* USDA-NASS, 2007b.)

**FIGURE 17–3** In addition to milk, much of the beef consumed in the United States each year comes from cull dairy cows and dairy steers, like these, which are finished for beef.

the milk cows in the United States are culled each year. Most veal comes from dairy calves that have been fed all-milk diets and slaughtered when weighing a few hundred pounds. By any standards, the dairy industry is a major industry in the United States, and the United States is a major dairy product producer and consumer.

## PURPOSE OF THE DAIRY CATTLE INDUSTRY IN THE UNITED STATES

The purpose of the dairy industry is to make use of resources that humans cannot use and to produce food that humans can use. As ruminants, most of the feed energy consumed by dairy cows is from forages. This forage conversion is their chief contribution

**FIGURE 17–4** The purpose of the dairy industry in the United States is to make use of forage and convert it to milk. (Photographer Todd Stone. Courtesy of AP/Wide World Photos.)

to human welfare (Figure 17–4). In addition to forage, vast supplies of waste material from the agronomic crops of the nation's agriculture (e.g., cottonseed) and food by-products (e.g., soybean meal, bakery by-products, distiller's grains, and citrus pulp) are also fed to dairy cattle. Dairy cows are very efficient at converting feed to usable product. Table 17–1 compares relative percentages of feed nutrients converted to edible products by livestock species. Milk production is the most efficient conversion of feed to food by all of the common food-producing species.

Dairy cows have also traditionally provided a means for farmers to add value to homegrown feeds. Farm-produced grain could provide more profit if fed to a cow and the milk sold than if the grain was sold. Dairying also provides a way for farmers to

**T A B L E  17 – 1**      **Estimate of Relative Percentages of Feed Nutrients Converted to Edible Products by Animal Species**

| Animal Product | Energy | Protein Gross Edible | |
| --- | --- | --- | --- |
| | CONVERSION (%) | CONVERSION (%) | OUTPUT AS PERCENTAGE FEED INTAKE (%) |
| Milk | 20 | 30 | 90 |
| Chicken (broiler) | 10 | 25 | 45 |
| Eggs | 15 | 20 | 33 |
| Pork | 15 | 20 | 30 |
| Turkey | 10 | 20 | 29 |
| Beef | 8 | 15 | 10 |
| Lamb | 6 | 10 | 7 |

Among farm animals, the most efficient way to convert feed nutrients to edible energy and protein is through milk production. To meet increased food needs, more attention will need to be directed to systems of crop and animal production that yield the most high-quality food for the least and best use of natural resources. Data support maintaining and improving dairy production through better management and application of new technology. In addition, the protein returned in the milk is much more biologically complete than that in the feed the cow eats, and it comprises a substantial component toward world food needs.

*SOURCE:* Voelker, 1978.

distribute labor needs across seasons in diversified operations. Fewer cows can be milked during the planting and growing season for crops, which allows time for the farmer to attend to the land. When the harvest is over, more cows can be milked to use available labor. In this way, dairy cattle help us maximize the use of available resources to the benefit of the whole society. In times of plentiful grain production, dairy cattle can be used to help us make best use of the abundance from agronomic production. In this way, they help to moderate the fluctuations in grain prices that could otherwise disrupt agronomic practices and endanger the agricultural economy and the supply of grain for human consumption.

Although cattle are milked around the world, most are multipurpose cows that produce milk, as well as meat and labor, for the people who own them. Only in developed countries such as the United States does an extensive dairy industry based on a specialized dairy cow exist. However, as countries develop, this is changing. Modern dairies are becoming more common around the world.

## HISTORICAL PERSPECTIVE

Until 1850 or so, the history of dairy cattle in the United States varied little from that of beef cattle because there were no specialized dairy breeds in the United States. The cows that were milked were simply the same animals used for all other purposes. Milk production was usually for the producer's own use or was produced by very small herds for local sale.

After the Civil War, the milk manufacturing system began developing and the dairy industry began evolving in earnest. The first cheese factory had actually been established in Oneida, New York, in 1851. However, several developments led to a rapid expansion of the fledgling industry. Gail Borden patented **condensed milk** in 1856 and established a condensery in 1857 in Burrville, Connecticut. The new product occupied less space and provided a use for surplus milk. Mechanical refrigeration was developed in 1861, which allowed for greater flexibility in shipping and storing fresh milk. In 1864, Louis Pasteur discovered the microbiological fundamentals that would lead to the process of **pasteurization**. The first commercial pasteurizing machines would be available in 1895. By the 1860s, there were specialty dairy centers near the larger Atlantic Coast cities. Cows that would support the industry were needed. Significant numbers of cows of various dairy breeds were imported, and between 1868 and 1880, breed associations were formed.

The refrigerated rail car was invented in 1871, which allowed shipment of products over greater distances. **Silos** were invented in 1875 at the University of Illinois, which allowed dairy farmers to store high-quality feed for the winter and gave dairying its year-round dimension. The cream separator was developed in 1878 by Dr. Gustav De Lavel. *Hoard's Dairyman* was first published in 1885 in Wisconsin, and by 1889 was recognized as the most important dairy journal in the country. Dr. S. M. Babcock introduced the test for milk fat (butterfat) that would bear his name, the **Babcock Cream Test**, in 1890. This test provided a basis for the pricing of milk. Testing for tuberculosis began in 1890. The fledgling industry got its first university curriculum when the University of Wisconsin established a dairy program in 1891. The first testing and recordkeeping association was formed on August 12, 1905, in Newaygo County, Michigan. Many similar organizations were formed and, in 1927, the Dairy Herd Improvement Association was instituted by the American Dairy Science Association with a rules committee for milk production testing. The first milking

**Condensed milk**
Milk with water removed and sugar added.

**Pasteurization**
Controlled heating to destroy microorganisms.

**Silo**
Structure in which silage is made and stored.

**Babcock Cream Test**
Test for determining the fat content in milk.

machines were assembled in 1903 and were soon manufactured by D. H. Burrell Company in Little Falls, New York. Milk was soon sold in glass bottles. Paper milk cartons were introduced in 1950, followed by plastic milk containers in 1964.

Tank trucks were first used to transport milk in 1914, although widespread use would not occur for several years. In the 1930s, bulk tank handling of milk on the farm began in California. Homogenized milk made its debut in 1919. Vitamin D fortification began in 1932. Artificial insemination began in dairy cattle in 1936 and was commercially available in 1939. It was not until after World War II that artificial insemination techniques were advanced enough to cause the genetic improvement of dairy cows to take a giant leap. Antibiotics were developed in the 1940s and 1950s, which allowed more animals to be kept in a small area.

During the early 1950s, the industry began restructuring. Operations got larger, which allowed for economies of scale. Milk production reached its peak in 1964 at 127 billion pounds. The all-time high number of dairy cows was reached in 1945 at 27.8 million. Numbers of cows and producers have been declining ever since. Production per cow has been increasing. Technological innovation and consumer demands are creating a very dynamic environment within the industry.

## STRUCTURE AND GEOGRAPHIC LOCATION OF THE DAIRY INDUSTRY

Dairying is among the least concentrated of all the major farm enterprises in the United States, with dairy cows found in every state in the union. In large part, this is because of the perishable nature of milk and the costs of shipping it over long distances. Thus dairy cows tend to be found near the human population. Figure 17–5 gives a full breakout of milk cows by state and region. Notice the numbers of cows in the individual states in the Mountain region and the Northern Plains compared to the most populous state in the union, California. Perhaps more graphic, notice the milk production for both the Northeast region and the Pacific region. The Pacific Coast and the Atlantic Coast are both densely populated regions. The other regions discussed in the figure are not as heavily populated. Among the 10 largest milk-producing states are the 6 most-populated states. An exception to this is Wisconsin, which has many more cows and produces much more milk than its population needs. However, Wisconsin has long had a strong milk manufacturing industry. Much of the nation's manufactured milk products is produced in Wisconsin, including over a fourth of the total cheese produced in the country. Idaho is another state whose milk-producing rank, at 4th, is quite different from its human population rank, at 39th. However, it ships much of its production to the West Coast states.

The dairy industry, like most of agriculture, has been restructuring since about 1950. The trend has been toward fewer operations (Figure 17–6) that have increasingly more cows per operation. As a result, over half of the U.S. milk supply is produced in herds of 500 or more cows (Figure 17–7) and that percentage increases yearly. In addition, the activities of the dairy have been changing. They have become more specialized and more likely to be the sole farm activity. Dairy herds have traditionally been self-contained family operations with as many cows as the family's resources would allow. The dairy was also an integral part of a complete, *diversified farming* program. Crops of various sorts were grown on the farm, most of the dairy's feed needs were met on the farm, and replacement animals were grown on the farm. The milk check came

**Diversified farm**

Farm with multiple income-generating enterprises.

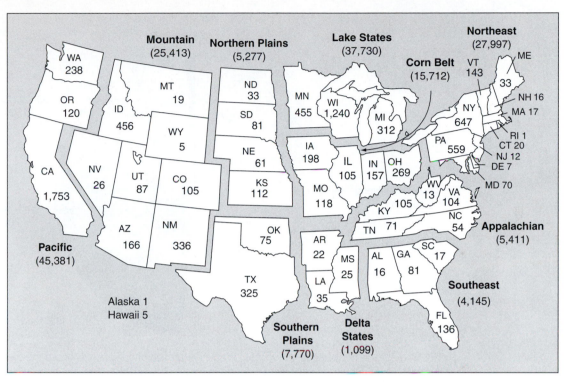

**FIGURE 17–5** Milk cows by state and total milk production by region. The number found within the state represents the number of dairy cows × 1,000. The number beside the region name represents the total milk produced in the region in millions of pounds of milk. All information represents averages for recent years. (*SOURCE:* USDA-NASS, 2007d.)

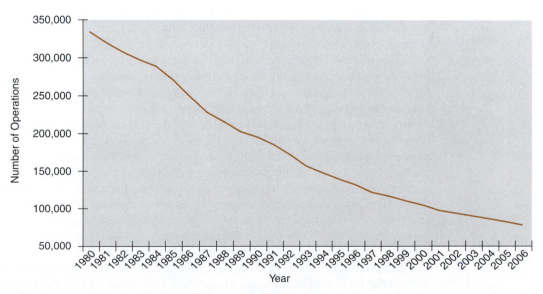

**FIGURE 17–6** Milk cow operations, 1980–2006. The trend since the 1950s has been to fewer total dairy cattle operations. At the same time, the average dairy has gotten larger. (*SOURCE:* USDA-NASS, 2007a.)

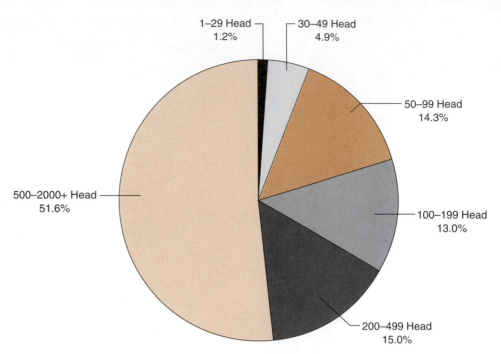

**FIGURE 17–7**   Percentage of U.S. milk production by size of operation. More of the cows are being found in larger herds, and fewer are found in smaller herds. This reflects the trend of small herds going out of the dairy business. As this is happening, fewer, larger herds are controlling a greater percentage of the cows. (*SOURCE:* USDA-NASS, 2007a.)

regularly, which helped meet the needs of the family and the cash flow for the remainder of the farming operation. In many ways, this epitomized the concept of "family farm" to much of the country. Much national farm policy has revolved around protecting this ideal. Although dispersed across the country, the dairy industry was historically centered in regions of the northeastern portion of the country.

Over the last decade, the evolution in the structure of the industry has reached a point at which the traditional dairy no longer predominates. Large-scale dairies with up to several thousand cows have increasingly been built. These large operations are able to take advantage of labor-saving technologies to expand that small operations cannot afford. Operations of 100 cows or less are still numerous, but the definition of *small* is changing as the overall average herd size increases. Production shifted west to take advantage of mild dry climates and to accommodate an increasing human population. From the beginning, these new, larger western producers adopted business organizations and management strategies that resulted in relatively low milk production costs, giving them a competitive edge. This further contributed to the westward shift in milk production, and it is still a driving factor. As it currently stands, the three major dairying areas in the United States are the Lake states, the Northeast, and the Pacific (Figure 17–5). The top-10 dairy states are listed in Table 17–2. These 10 states produce over 72% (2006) of U.S. milk production. The Mountain region also has a substantial portion of the nation's production. California, Texas, Washington, Idaho, and New Mexico are sometimes referred to as the West's "Big 5" in dairy production. California became the largest milk-producing state in 1994. Note that Wisconsin still produces almost twice as much milk as does New York, which is third (Table 17–2).

**TABLE 17–2** **Top-10 Dairy States in Various Categories**

| Total Milk (lbs) | | Total Cows (1,000 Head) | | Milk per Cow (lbs) | | Average Herd Size | |
|---|---|---|---|---|---|---|---|
| California | 38.83 bill. | California | 1780 | Colorado | 23,155 | New Mexico | 2,088 |
| Wisconsin | 23.40 bill. | Wisconsin | 1243 | Washington | 23,055 | Arizona | 1,331 |
| New York | 12.05 bill. | New York | 638 | Arizona | 22,855 | California | 908 |
| Pennsylvania | 10.74 bill. | Pennsylvania | 554 | Idaho | 22,326 | Nevada | 900 |
| Idaho | 10.90 bill. | Idaho | 488 | Michigan | 22,188 | Hawaiil | 860 |
| Minnesota | 8.36 bill. | Minnesota | 450 | California | 21,815 | Florida | 825 |
| New Mexico | 7.64 bill. | New Mexico | 355 | New Mexico | 21,515 | Idaho | 707 |
| Michigan | 7.10 bill. | Texas | 335 | Texas | 21,328 | Colorado | 647 |
| Texas | 7.15 bill. | Michigan | 320 | Kansas | 20,920 | Texas | 453 |
| Washington | 5.46 bill. | Ohio | 274 | Nevada | 20,667 | Washington | 389 |

SOURCE: USDA-NASS, 2007c.

The dairy cow population declined to less than 10 million cows in 1990 and has continued to decline (Figure 17–8). During that same time, the average amount of milk produced per cow has increased (Figure 17–9) and the total milk output from the nation's dairy industry has increased (Figure 17–10). Thus the increase in milk per cow is compensating for the decrease in the number of cows. The number of dairy farms in the United States fell below 100,000 in 2001.

Most U.S. dairies fall into one of two major categories. The first category consists of family dairy herds of around 50 to 200 cows, about as many as the average family could expect to handle and still do a good job. Most of the country's dairies remain in this category. However, a larger percentage of these dairies are located in the traditional dairy states than in the newly emerging dairy states. Dairies in California, New Mexico, Arizona, Texas, Idaho, and Florida, especially the newer ones, tend to have several

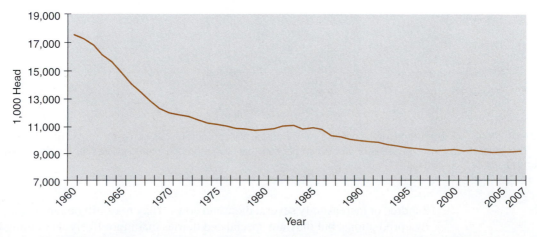

**FIGURE 17–8** The number of dairy cows in the United States. Dairy cow numbers have been declining for several decades. The decrease in cows has not caused a decrease in total milk production because of an increase in average productivity of the remaining cows. (SOURCE: USDA-NASS, 2007a.)

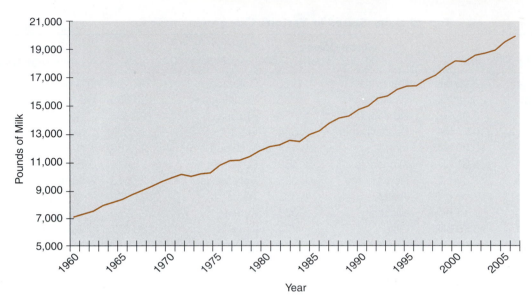

**FIGURE 17–9** Rate of milk production per cow. Improved genetics, management, and use of technological innovations have caused a sustained trend of more milk per average cow. This increase in production has more than compensated for the decrease in total dairy cows to result in more milk being produced from a substantially smaller number of cows. (*SOURCE:* USDA-NASS, 2007c.)

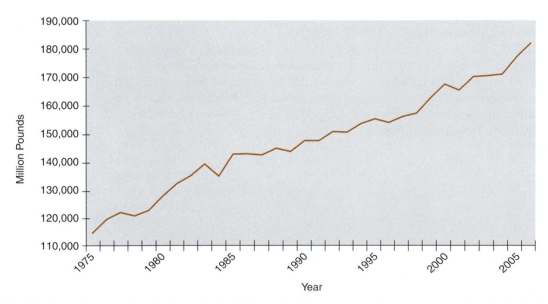

**FIGURE 17–10** Total milk production for the nation's dairy industry. Milk production has increased in spite of a decline in the number of cows. This is because of increased rates of production per cow that have more than compensated for the decreased number of cows. (*SOURCE:* USDA-NASS, 2007a.)

hundred or increasingly several thousand cows. They may still be family farms or family corporations, but they are specialized dairies that more likely do nothing but dairy. These large specialized operations tend to buy all of their feed and may even contract with someone to raise their replacement heifers. Several individual dairies in this category have over 10,000 head of milking cows.

In contrast to the other animal industries, there is very little vertical integration in the dairy sector, and few people are predicting that will change. The overwhelming majority of dairies, even the large ones, are either individual- or family-owned and operated, and many of the family-owned corporations and partnerships are restricted to family members.

## DAIRY HERD IMPROVEMENT ASSOCIATION (DHIA)

The National Cooperative Dairy Herd Improvement Program (NCDHIP) was established in 1965. It is administered at the state and local levels by the Dairy Herd Improvement Association (DHIA). The NCDHIP collects and processes information on dairy cows for about a third of the producers in the United States who own about half of the country's dairy cows. The program provides herd owners with management, production, and cost information. NDHIA, Inc., is the national DHIA organization made up of member state organizations that collect and record information on dairy herds. Information on DHIA records is collected on the dairy by a DHIA representative approximately 12 times each year. Milk weights and milk samples are collected on each cow. Once the information is processed, producers are provided with valuable information to help them determine the profitability of individual cows, make nutrition decisions, manage reproduction, control mastitis, and so on. It is arguably the best and most comprehensive management tool available to the dairy producer. The organization is mentioned here because it can be an integral tool in managing the cow for the needs outlined in the next few sections. The *National Dairy Herd Improvement Handbook* is available in print form or on the Internet.

Records compiled by this program are also used by the United States Department of Agriculture (USDA) for sire evaluation purposes. The USDA sire summaries are designed to provide accurate estimates on the genetic merit of sires. This is accomplished by comparing daughters of individual sires with daughters of other sires in the same herds. Any producer using artificial insemination to breed his or her dairy benefits from this service.

## GENETICS AND BREEDING PROGRAMS

The fundamentals of genetics and breeding are discussed in Chapters 8 and 9, which has a specific section on dairy cows. See those chapters for the fundamentals of dairy cattle breeding.

The dairy industry has made tremendous genetic progress in milk production over the last 42 years (Table 17–3). Milk, fat, and protein have increased significantly for all of the breeds. Holstein cows born in 1962 averaged 14,319 lbs of milk on a mature equivalent basis, whereas those born in 2004 are averaging 26,219 lbs. Of this approximately 12,000-lb increase in milk production, it is estimated that 7,161 lbs, or about 60% of the increase, is related to genetic progress and 40% of the increase to better management.

A disturbing statistic from Table 17–3 is the negative trend for daughter pregnancy rate (DPR). DPR is defined as the percentage of time that a cow would be expected to get pregnant during a three-week reproductive cycle during the breeding period of a lactation. It is equal to the percentage of cows inseminated times the percentage that conceive and maintain the pregnancy. The average DPR for Holsteins born in 1962 was

**TABLE 17–3** **Genetic Change in Selected Dairy Performance Traits for Different Breeds for Cows Born in 1962 Compared to 2004**

| | | | BREED | | | |
|---|---|---|---|---|---|---|
| | Ayrshire | Guernsey | Holstein | Jersey | Brown Swiss | Milking Shorthorn |
| Milk yield (lb) | 4,093 | 5,827 | 7,161 | 6,565 | 5,627 | 4,147 |
| Fat yield (lb) | 144 | 220 | 236 | 247 | 201 | 147 |
| Protein yield (lb)[1] | 54 | 89 | 119 | 109 | 87 | 50 |
| Somatic cell score[1] | +0.1 | +0.1 | +0.1 | +0.2 | +0.1 | +0.1 |
| Daughter pregnancy rate (%) | −4.9 | −6.6 | −6.5 | −4.1 | −6.7 | −2.9 |

[1]1986–2004.

*SOURCE:* Interpreted from www.aipl.arsusda.gov, May 2007.

30% compared to a DPR of only 23.4% for Holsteins born in 2004. The phenotypic decrease in DPR is about equal to the genetic decrease. Getting lactating cows pregnant has become one of the biggest challenges facing dairy farmers. Genetic evaluations for DPR were not officially published by the USDA until February 2003. It was often argued that because the heritabilities of reproduction traits were only 1–5%, it would not be worthwhile to try to improve fertility genetically. Ideally, with prudent use of these genetic evaluations for DPR, the negative trend for fertility can be reversed.

The USDA's Net Merit Index was modified in August 2006 to include traits known to be economically important and reflect more accurately the genetics that cows need for maximum lifetime net profit. Traits considered in the index, heritabilities, and genetic correlations are presented in Table 17–4. Interestingly, the current genetic correlation between milk and productive life is now only 0.08, contrasted with the 0.75 published from earlier work (Table 9–7). This means that in the past higher-producing cows stayed in the herd much longer than low-producing cows, but that strong relationship is no longer true today.

The relative index weights on the various traits in the Net Merit Index, the expected PTA gain/year and expected genetic trend per decade is reported in Table 17–5. All of the traits in the index are expected to change in a desirable direction. Even though no weight is put on PTA milk, milk is expected to increase 1,720 lbs over the next decade because of its positive genetic correlation with fat and protein. Somatic cell score should decrease significantly from the 2003 average of 3.1. The −0.80 in body size composite correlates to a reduction in about 15 lbs in weight per cow. Smaller cows tend to require fewer lbs of feed per lb of milk produced. The downward trend in daughter pregnancy rate should be reversed and significant increases in calving ease should be noted.

## BREEDS

Specialized dairy breeds were imported into the United States and breed associations were established between 1868 and 1880. The six dairy breeds used in the United States and their average DHIA performance levels are reported in Table 17–6. The most popular breed by far is the Holstein, accounting for almost 95% of the cows on DHIA test. Jersey is the second most popular breed with about 4% of the cows. The remaining four breeds account for only a very small percentage of the cows in the United States.

**TABLE 17–4**   **Heritabilities and Genetic Correlations of Traits Considered in the Net Merit Index**

| | | | | | PTA TRAIT | | | | | |
|---|---|---|---|---|---|---|---|---|---|---|
| PTA Trait | Milk | Fat | Protein | PL | SCS | Size | Udder | Feet/ Legs | DPR | CA$[1] |
| Milk | 0.30* | 0.45 | 0.81 | 0.08 | 0.20 | −0.10 | −0.20 | −0.02 | −0.32 | 0.15 |
| Fat | | 0.30 | 0.60 | 0.08 | 0.15 | −0.09 | −0.20 | 0.02 | −0.33 | 0.11 |
| Protein | | | 0.30 | 0.10 | 0.20 | −0.10 | −0.20 | −0.02 | −0.35 | 0.16 |
| Productive life | | | | 0.08 | −0.38 | −0.16 | 0.30 | 0.19 | 0.51 | 0.40 |
| Somatic cell score | | | | | 0.10 | −0.11 | −0.33 | −0.02 | −0.30 | −0.08 |
| Size | | | | | | 0.40 | 0.26 | 0.22 | −0.08 | −0.24 |
| Udder | | | | | | | 0.27 | 0.10 | 0.03 | 0.06 |
| Feet/legs composite | | | | | | | | 0.15 | −0.04 | −0.04 |
| Daughter pregnancy rate | | | | | | | | | 0.04 | 0.34 |
| Calving ability dollars | | | | | | | | | | 0.07 |

[1]CA$ (Calving Ability Dollars) is a function of calving ease and stillbirth rate.

**TABLE 17–5**   **Relative Net Merit Index Weights on the Various Traits and Expected Genetic Change**

| PTA Trait | Relative Weight (%) | Expected PT Gain/Year | Expected Genetic Trend/Decade |
|---|---|---|---|
| Milk (lb) | 0 | 86 | 1,720 |
| Fat (lb) | 23 | 3.8 | 76 |
| Protein (lb) | 23 | 2.6 | 52 |
| Productive life (mo) | 17 | 0.30 | 6.0 |
| Somatic cell score | −9 | −0.017 | −0.34 |
| Size composite | −4 | −0.04 | −0.80 |
| Udder composite | 6 | 0.04 | 0.80 |
| Feet/legs composite | 3 | 0.03 | 0.60 |
| Daughter pregnancy rate | 9 | 0.07 | 1.4 |
| Calving ability ($) | 6 | 1.3 | 25 |

In general, the number of registered dairy cattle is declining (Table 17–7) along with the declining number of total dairy cattle. Total registration of dairy cattle peaked in 1984 at 590,298 registrations and has declined ever since. Holsteins have dominated the registered cattle numbers in the United States since very early in the 20th century. For information on cattle breeds, visit the breeds of livestock page at http://www.ansi.okstate.edu/breeds/. This comprehensive reference includes pictures of livestock breeds. Color photos of the major dairy breeds are found in the color plate section of this text.

**TABLE 17 – 6    Production Level of U.S. Dairy Breeds, 2006**

| Breed | No. of Cows | Milk | % Fat | lbs. Fat | % Protein | lbs. Protein |
|---|---|---|---|---|---|---|
| Ayrshire | 5,796 | 15,515 | 3.9 | 601 | 3.2 | 491 |
| Brown Swiss | 15,007 | 18,152 | 4.0 | 716 | 3.4 | 595 |
| Guernsey | 7,862 | 15,402 | 4.5 | 665 | 3.4 | 494 |
| Holstein | 3,906,258 | 22,569 | 3.6 | 795 | 3.0 | 664 |
| Jersey | 186,581 | 16,127 | 4.6 | 728 | 3.6 | 563 |
| Milking Shorthorn | 2,641 | 14,676 | 3.7 | 522 | 3.1 | 443 |

SOURCE: DHI Report K3, USDA.

**TABLE 17 – 7    Registration of Animals by Dairy Breed Associations**

| Year | Ayrshire | Brown Swiss | Guernsey | Holstein | Jersey | Milking Shorthorn | Total |
|---|---|---|---|---|---|---|---|
| 1950 | 24,236 | 22,721 | 94,901 | 184,246 | 67,309 | 28,290 | 421,703 |
| 1960 | 16,831 | 23,949 | 62,891 | 265,861 | 54,695 | 9,525 | 433,752 |
| 1970 | 15,069 | 16,416 | 43,783 | 281,574 | 37,097 | 5,410 | 399,349 |
| 1980 | 10,977 | 12,871 | 20,907 | 353,949 | 60,975 | 4,924 | 464,603 |
| 1985 | 11,120 | 11,974 | 25,106 | 394,506 | 65,357 | 3,372 | 511,435 |
| 1990 | 7,752 | 11,756 | 13,930 | 395,906 | 53,547 | 2,596 | 485,487 |
| 1995 | 6,456 | 10,799 | 7,387 | 366,262 | 63,399 | 3,157 | 457,460 |
| 2000 | 6,046 | 10,648 | 6,151 | 317,567 | 63,776 | 2,595 | 406,783 |
| 2005 | 4,860 | 10,076 | 5,093 | 301,852 | 72,885 | 3,154 | 397,920 |

**Crossbreeding**

Mating animals from different breeds within a species.

Traditionally, there has been very little *crossbreeding* in dairy cattle in the United States. The major reason that dairy cattle were not crossbred was because Holsteins produced so much more milk than the other breeds that heterosis was not large enough to justify crossbreeding. However, in recent years, dairy producers have shown interest in crossbreeding. Reasons for the increase in crossbreeding include the decreased fertility in purebreds, a shift in payment away from milk volume and toward payment for lbs of fat and protein, and dairy producers wanting a more trouble-free cow.

It is difficult to determine what percentage of the cows in the United States are crossbreds, but the number is increasing. Many of the crossbred cows are in low input herds that are not on the DHIA test. The most popular crossbreds are Jersey-Holstein (Figure 17–11) and Brown Swiss-Holstein crosses.

There is growing interest in importing semen from four additional breeds from Europe for crossbreeding purposes: Swedish Red and Norwegian Red (collectively known as Scandinavian Red), Normande (France), and Montbeliarde (France). These breeds have had well-organized genetic improvement programs and have selected both for production and fitness traits (including fertility) for many years. Table 17–8 shows a comparison of these crosses with Holsteins in seven California dairies. On average, the pure Holsteins produced more milk in the first lactation. However, the crossbreds had fewer stillbirths, better fertility, and were less apt to be culled.

Crossbreeding studies are currently being conducted at several universities. Results may determine the extent of future crossbreeding in the U.S. dairy industry.

**FIGURE 17–11** It is difficult to determine what percentage of the U.S. dairy cows are crossbreds, but the number is increasing. Results of crossbreeding studies at several universities may determine the extent of future crossbreeding in the U.S. dairy industry. (Photo courtesy Dr. Tony Seykora. Used with permission.)

**TABLE 17–8** **Average Production of First Lactation of Crossbreds and Holsteins in Seven California Dairies**

|  | Holstein | Normande-Holstein | Montbeliarde-Holstein | Scandinavian Red-Holstein |
|---|---|---|---|---|
| Number of cows | 380 | 245 | 494 | 328 |
| Milk (lbs.) | 21,510 | 18,805 | 20,196 | 20,461 |
| Fat (lbs.) | 763 | 703 | 736 | 750 |
| Protein (lbs.) | 672 | 611 | 646 | 655 |
| % of Holsteins for combine fat plus protein |  | 92% | 96% | 98% |
| % stillbirths at first calving | 14% | 10% | 6% | 5% |
| Average days open | 150 | 123 | 131 | 129 |
| % culled in first 305 days | 14% | 7% | 8% | 7% |

SOURCE: *Journal of Dairy Science* 89:2799, 2805, 4944.

# REPRODUCTIVE MANAGEMENT IN DAIRY CATTLE

Chapter 11 is devoted to reproductive physiology. Refer there for the fundamentals of male and female reproductive anatomy, physiology, and function. Chapter 12 is devoted to lactation. Likewise, see that chapter for a look at the specialized function of lactation. Be aware that reproductive performance is closely tied to other factors, especially nutrition and level of management.

Reproductive problems are especially troublesome in the dairy industry. Sterility and delayed breeding cost the dairy industry hundreds of millions of dollars each year. A cow must have a calf to produce milk. In addition, because milk production declines over time, it is desirable to have the cow calve with relative frequency to initiate a new cycle of milk production. Other reasons for a frequent calving interval include the need for the daughters of the herd to become the replacements for their dams. Their genetics should be better than that of their dams, and they are needed to replace the cows that are culled in a given year. A realistic goal of breeding the dairy cow is to produce a calf at 13-month intervals. Because of the number of factors that influence the reproductive performance of a good dairy cow, this is a challenging goal. A cow must be pregnant within 115 days of calving or the 13-month calving interval cannot be maintained. The cow must recover from calving, start her heat cycle again, and conceive. She must do all of this while producing a sizable quantity of milk that is larger each year for the average cow. Good management is necessary or this simply won't happen.

Nutrition is central to reproductive efficiency in dairy herds. A cow generally loses weight for the first few weeks after calving because it is impossible for her to eat enough to increase production during early lactation and maintain body weight. Thus the body reacts by catabolizing stored tissue for energy and losing weight. This can create a problem because a dairy cow must maintain a certain body condition to begin preparing for heat cycles and impregnation. Thus good nutrition is necessary for both maximal milk production and reproductive performance.

**Gomer bull**

Bull rendered incapable of mating naturally.

Because most dairy cattle in the United States are artificially inseminated, one of the challenges to a successful breeding program is heat detection. Use of *gomer bulls* or hormone-treated cull cows to help spot cows in estrus (heat) can help. Learning the physical and behavioral signs of a cow in heat is also essential. Remember that a cow only displays the signs of heat for 12 to 18 hours on a 19- to 21-day cycle, which presents a narrow window of opportunity for heat detection. Increasingly, dairy producers use hormone programs to help control and detect estrus by making it more predictable. Use of heat-detection aids such as "HeatWatch," an electronic aid in heat detection, is increasingly common (Figure 17–12). Replacement heifers must be managed so they can be mated to calve at 2 years of age. To do this, they must be mated at 15 months of age and need to be at least 65% of their adult weight at the time of breeding. Dairy cows are freshened year-round. The gestation period varies slightly from breed to breed and ranges from 280 to 283 days.

Diseases that affect reproductive efficiency directly include brucellosis (Bang's disease), campylobacteriosis, leptospirosis, IBR/BVD complex, and trichomoniasis. Control of these and other diseases can be accomplished easily by developing and following a complete health program for the herd. Use of AI, which eliminates the possibility of a bull spreading disease, makes disease management in a dairy much easier.

**FIGURE 17–12**    HeatWatch Estrus Detection System is a computer-assisted system that determines when cows are in heat. (Photo courtesy of CowChips, LLC.)

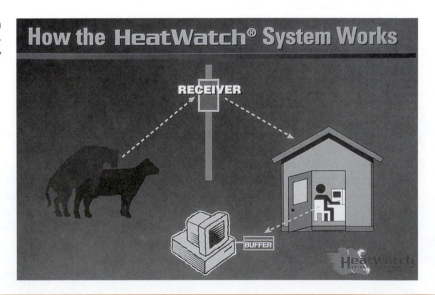

Good records are essential to managing reproductive issues in the dairy herd. A large number of programs exist to assist the producer in this regard. Both manual and computer programs are readily available. Most work as well as the person who is using them. Because management practices and standards vary with conditions across the country, cattle breeders are encouraged to contact their local extension service or their state extension specialists for help in setting up a recordkeeping program for their area. Many dairy producers use DHIA records as a tool in managing their breeding programs.

## NUTRITION IN DAIRY CATTLE

The current genetic potential in dairy cows is such that nutrition information and feeding systems are behind the curve. This is causing loss of opportunity for cost-effective feeding of the dairy herd, which is unfortunate because feed costs account for 45–65% of the costs of the dairy. Feeding is the single most important factor in the profitability of the dairy herd.

Dairy cattle are ruminants. They can use forages and roughages to produce a usable product for humans. The nutrition and feeding of dairy cattle is similar to that of beef cattle. Most of the production stages revolve around maximizing the use of forages as the feed base. However, higher-quality forages are required and dairy feeding has a greater dependence on concentrate feeds to supplement the forage. Chapters 5, 6, and 7 give detailed information relating to nutrition, feeds, and feeding. See those chapters for general nutrition information not provided in this section.

Managing the nutrition of dairy cattle is very dependent on the class of cattle to be fed. Different classes of dairy animals use different amounts of grain and other supplemental feeds to extend the use of forage by supplementing missing nutrients. The very young and the cow in production need the greatest amounts of supplementation of energy feeds. Mineral supplements critical in feeding the producing female include salt, calcium, and phosphorus. Additional individual mineral deficiencies occur in the various geographic regions of the country. Protein and energy feeds are often needed to keep heifers growing at an acceptable rate and to keep cows producing at optimal levels. The goal is to supplement forages in a way that allows us to maximize forage use.

Baby calves destined to be the next generation of producing cows are fed on a commercial calf milk replacer or on milk produced at the dairy. In some dairies, colostrum, milk being discarded because of antibiotic use on a cow, or otherwise-contaminated milk is used. It is economically advantageous to switch calves to dry concentrate feeds and forage as soon as possible to decrease the costs of producing the replacement heifers. Calves are usually fed their liquid feed twice a day, although systems are available that allow for free choice of a liquid diet. Calves are offered dry feed (Table 17–9) when they are only a few days old. They are weaned when their consumption of the dry feeds is adequate. They are switched to dry feeds to reduce the costs of feeding expensive liquid feeds and to reduce labor and housing needs.

Heifer-growing programs depend increasingly on maximizing the use of forages as the heifer gets larger and more able to use the forages effectively. Because replacement heifers should ideally enter the milking herd at 24 months of age, the plane of nutrition must be good to keep them growing at an appropriate rate. However, different programs must be developed for the animals at different ages, weights, and seasons. Energy and/or protein supplementation is usually necessary. Higher-quality forages are also needed to keep animals gaining rapidly enough to enter the milking string at age 2.

**TABLE 17–9** **Examples of Some Calf Starters**

| | GRAIN STARTERS[1] | | |
|---|---|---|---|
| | 1 | 2 | 3 |
| **Ingredients (air-dry basis)** | | | |
| Corn (cracked or coarse ground) % | 50 | 30 | |
| Ear corn (coarse ground) % | | | 50 |
| Oats (rolled or crushed) | 22 | 18 | |
| Barley (rolled or coarse ground) % | | 20 | 21 |
| Wheat bran % | | 8 | |
| Soybean meal % | 20 | 16 | 21 |
| Molasses % | 5 | 5 | 5 |
| Dicalcium phosphate % | .5 | .5 | .5 |
| Limestone % | 1.5 | 1.5 | 1.5 |
| TM salt and vitamins % | 1 | 1 | 1 |
| **Composition (dry-matter basis)** | | | |
| Crude protein % | 18.1 | 18.0 | 18.4 |
| TDN % | 80.0 | 78.8 | 78.0 |
| ADF % | 7.0 | 6.9 | 9.1 |
| Calcium % | .80 | .80 | .82 |
| Phosphorus % | .48 | .56 | .47 |
| Vitamin A, IU/lb | 1,000 | 1,000 | 1,000 |
| Vitamin D, IU/lb | 150 | 150 | 150 |
| Vitamin E, IU/lb | 11 | 11 | 11 |

**Examples of Some Complete Starters**

| | COMPLETE STARTERS | | |
|---|---|---|---|
| | 1 | 2 | 3 |
| **Ingredients (air-dry basis)** | | | |
| Corn (cracked or course ground) % | 40 | 25 | 30 |
| Oats (rolled or crushed) % | 14.5 | 8 | 18 |
| Beet pulp % | | 25 | 25 |
| Alfalfa hay (ground) % | | 10 | |
| Corn cobs (ground) % | 15 | | |
| Soybean meal % | 23 | 18 | 20 |
| Molasses % | 5 | 5 | 5 |
| Dried whey % | | 7 | |
| Dicalcium phosphate % | .5 | .5 | .5 |
| Limestone % | 1 | .5 | .5 |
| TM salt and vitamins % | 1 | 1 | 1 |
| **Composition (dry-matter basis)** | | | |
| Crude protein % | 18.3 | 18.0 | 18.2 |
| TDN % | 75.5 | 78.0 | 79.4 |
| ADF % | 13.3 | 15.8 | 14.2 |
| Calcium % | 13.3 | 15.8 | 14.2 |
| Phosphorus % | .45 | .44 | .43 |
| Vitamin A, IU/lb | 1,000 | 1,000 | 1,000 |
| Vitamin D, IU/lb | 150 | 150 | 150 |
| Vitamin E, IU/lb | 11 | 11 | 11 |

[1] Hay may be offered free choice with grain starters.
*SOURCE:* Linn et al., 1988.

**TABLE 17–10** **Example Rations for Large-Breed Dairy Heifers of Different Weights**

| Weight | Rate of Gain | Ration (lbs as fed) |
|--------|--------------|---------------------|
| 400 lbs | 1.7 lb/day | 8.5 lbs alfalfa hay, 20% CP<br>3.0 lbs grain mix (12.0% CP)<br><br>**Percentages**<br><br>71.0 coarse ground barley<br>23.0 rolled or ground oats<br>5.0 molasses<br>0.5 trace mineral salt<br>0.4 dicalcium phosphate<br>0.1 vitamin premix |
| 700 lbs | 1.7 lb/day | 18.0 lbs alfalfa hay, 20% CP<br>2.0 lbs grain mix<br><br>**Percentages**<br><br>95.3 coarse ground corn<br>1.5 trace mineral salt<br>3.0 dicalcium phosphate<br>0.2 vitamin premix |
| 1,000 lbs | 1.7 lb/day | 13.0 lbs alfalfa hay, 16% CP<br>35.0 lbs corn silage<br>0.5 lbs mineral-vitamin supplement<br><br>**Percentages**<br><br>80.0 soybean meal<br>12.0 trace mineral salt<br>6.0 dicalcium phosphate<br>2.0 vitamin premix |

SOURCE: Linn et al., 1988.

Some sample rations for replacement heifers are shown in Table 17–10. After 1 year of age, it may be possible to feed heifers good-quality forage and mineral supplements until just before calving. Table 17–11 shows desirable weights for heifers at various ages.

Dairies generally rely on one of two systems of feeding the producing cows. Cows can be allowed to graze on high-quality pastures and then fed their concentrate ration at a different time, often in the milking parlor. This has been the traditional method of dairying in the United States and is still used by many small dairies. Increasingly, however, dairy cows are handled in a dry-lot system. Cows are kept in confinement facilities, usually consisting of a barn or loafing shed and a small lot, often concrete, where they are fed and watered. The cows are often grouped by production level and fed a complete, mixed ration (total mixed ration, or TMR) from bunks. In this way, a cow uses fewer nutrients in procuring her feed and can divert more to producing milk. There is also less waste of feed in these systems. Forage is still the basis of the ration; however, it is mixed with the energy, protein, mineral, and vitamin concentrate portions of the ration prior to feeding. Thus each bite a cow takes is balanced for her needs. Cows are not fed in the milking parlor in this system. Most dairy rations are formulated with the use of least-cost ration formulation programs. Table 17–12 shows

**TABLE 17–11    Desirable Weights for Dairy Heifers**

| Age in Months | Brown Swiss or Holstein | Ayrshire, Shorthorn, or Guernsey | Jersey |
|---|---|---|---|
| Birth | 90–100 | 65–75 | 55–60 |
| 1 | 120 | 90–100 | 70–80 |
| 2 | 170 | 135–145 | 110–120 |
| 4 | 270 | 225–235 | 190–200 |
| 6 | 370 | 315–325 | 270–280 |
| 12 | 670–700 | 585–600 | 510–520 |
| 15[1] | 800–875 | 720–750 | 630–650 |
| 18 | 970–1,000 | 850–875 | 750–775 |
| 22 | 1,150–1,200 | 1,025–1,075 | 900–950 |

[1] Breed heifers in this weight range. Heifers should weigh about 60% of their mature weight when bred. With proper feeding, heifers should reach these weights and have good skeletal growth at 14–16 months of age.

SOURCE: Linn et al., 1988.

examples of complete, mixed rations for producing cows (Figure 17–13). Dry cows need a different feeding program than do lactating cows. Table 17–13 shows an example of a ration for a dry cow.

The following list shows the advantages and disadvantages of the TMR system. Cows should be grouped by production level to best take advantage of the complete ration system.

**Advantages:**

1. There is no parlor grain feeding.
   a. Reduced cost of parlor construction and maintenance of feeding equipment.
   b. Less dust and mess in the parlor.
   c. No delay in milking time waiting for cows to eat grain.
   d. More cows milked per person-hour and cows are in the parlor less time.
   e. Cows stand more quietly and defecate less during milking.
2. The dairy producer has more control over the total feeding program.
   a. Concentrates can be liberally fed to high producers without overfeeding the low producers.
   b. Silage tends to mask the taste and dustiness of feed ingredients, which allows for the use of more economical but less palatable feeds in the diet.
   c. Cows eat many times a day. Feed intake is greater and nutrients are used better, especially urea.
   d. Fewer cows have digestive upsets and go off-feed.
3. Labor is less for feeding the total herd.
   a. Equipment and rations for the lactating herd can be used for feeding dry cows, heifers, and calves from 2 months of age.
4. Cost of cow housing and feeding facilities is less.
   a. Feed bunks are simpler with no need for conveyers or augers.
   b. Less bunk space is needed, as little as 8 in. per cow.
   c. Free-stall numbers can be reduced to two stalls for every three cows.

**TABLE 17–12**  **Example Rations for Various Milk Production Phases[1]**

| Item | Phase 1 | Phase 2 | Phase 3 |
|---|---|---|---|
| Milk (lbs/day) | 90 | 80 | 50 |
| DM intake (lbs/day)[2] | 49 | 51 | 38 |
| **Ration 1** | | lbs/day (as fed) | |
| Alf hay (88% DM), 20% CP | 28 | 34 | 27 |
| Corn-oats[3] | 21 | 24 | 16 |
| SMB-44% | 5.0 | | |
| Dical-18% P | 0.5 | 0.45 | 0.30 |
| Salt, vitamins, TM | 0.30 | 0.25 | 0.25 |
| Weight change | −1.5 | — | +0.5 |
| **Ration 2 (corn silage limit fed)** | | | |
| Alf hay, 20% CP | 19 | 34 | 23 |
| Corn silage (35% DM) | 25 | 25 | 25 |
| Corn-oats | 18 | 12 | 10 |
| SMB-44% | 7.5 | 0.3 | − |
| Dical-18% P | 0.45 | 0.50 | 0.3 |
| Salt, vitamins, TM | 0.30 | 0.25 | 0.25 |
| Weight change | −1.2 | — | +0.5 |
| **Ration 3 (hay limit fed)[4]** | | | |
| Alf-grass hay, 16% CP | 10 | 10 | 10 |
| Corn silage | 41 | 70 | 57 |
| Corn-oats | 16 | 11 | 6 |
| SMB-44% | 11.5 | 8.2 | 4.5 |
| Dical-18% P | 0.40 | 0.30 | 0.25 |
| Limestone | 0.40 | 0.30 | 0.15 |
| Salt, vitamins, TM | 0.30 | 0.25 | 0.25 |
| Weight change | −1.4 | +0.7 | +0.5 |
| **Ration 4** | | | |
| Alf-grass hay, 16% CP | 23 | 32 | 24 |
| Corn-oats | 22 | 22 | 19 |
| SMB-44% | 8.5 | 3.5 | 1.1 |
| Dical-18% P | 0.45 | 0.40 | 0.25 |
| Limestone | 0.20 | | |
| Salt, vitamins, TM | 0.30 | 0.25 | 0.25 |
| Weight change | −1.9 | — | +0.5 |

[1] 1,350-lb cow, 3.8% fat test.
[2] Estimated average intake during the phase.
[3] 85% corn–15% oats mix.
[4] Feed amounts may have to be limited during phases 2 and 3 to avoid overconditioning.
*SOURCE:* Linn et al., 1988.

**FIGURE 17–13**   Complete mixed diets like this one can be fed in dry-lot systems. These diets have the advantage of providing balanced nutrition in every bite.

### TABLE 17–13   Examples of Forage-Based Dry Cow[1] Rations

|  | lbs/day (as fed) |
|---|---|
| **Grass Forage** | |
| Orchard grass-hay—12% CP | 25.0 |
| Corn | 3.0 |
| Soybean meal | 0.5 |
| Limestone | 0.15 |
| TM salt and vitamins | 0.1 |
| **Limited Legume Forage[2]** | |
| Alfalfa hay—20% CP | 12.0 |
| Corn silage | 43.0 |
| Monosodium phosphate | 0.1 |
| TM salt and vitamins | 0.1 |
| **Limited Corn Silage** | |
| Alf-grass hay—16% CP | 21 |
| Corn silage | 20 |
| Dical | 0.1 |
| TM salt and vitamins | 0.1 |

[1] 1,400-lb dry cow.
[2] Ration contains excess energy as formulated and may overcondition cows in some situations.
*SOURCE:* Linn et al., 1988.

### Disadvantages:

**5.** Special equipment is needed.

    **a.** The equipment must have the capability to blend the ingredients thoroughly.

    **b.** The mixer, preferably mobile, must have the capability for weighing each ingredient exactly.

   **c.** Once the ingredients are blended, separation should be avoided; this will occur with certain types of conveyers.

   **d.** A crowd gate or training gate may be needed initially to get cows into the parlor.

**6.** Cows should be grouped by production levels.

   **a.** If not grouped, cows in late lactation tend to get too fat.

   **b.** Dry cows must be removed from the lactating herd.

   **c.** Grouping cows is more difficult in small herds.

Forages commonly used in dairy production include legumes such as alfalfa or lespedeza, small-grain forage such as oats or barley, corn or sorghum forage, and various grasses. These forages may be harvested by the cow (grazing), harvested mechanically and fed immediately (often called "green chop"), or harvested and stored as hay, silage, and haylage. Commonly used grains and by-products are corn, sorghum, small grains, soybean meal, and cottonseed meal. Many by-product feeds are used to feed dairy cows. These by-products tend to be used in the region of the country where they are produced. Examples include brewers grains, distillers grains, beet pulp, citrus feeds, and many others.

## HERD HEALTH PROGRAM

In the dairy industry, ***herd health*** is a management tool concept that includes a total approach to management, nutrition, medicine, and environmental control for the dairy herd (Figure 17–14). Dairy production medicine includes herd health, but also factors in cow productivity and economic efficiency to improve dairy performance. In dairy areas, it is common to find veterinarians providing both. Herd health programs have proven to be a very valuable tool for the dairy producer. The concept and practice of herd health vary some from region to region; however, the general principles are fairly consistent and include (1) prevention rather than treatment; (2) planned health-related procedures and examinations; and (3) sound recordkeeping and record use in making herd management decisions. Herd health programs include the following:

**Reproductive health.** This includes individual health and reproductive records for each cow. Good records can be kept in a very simple fashion or in a very sophisticated computer program. The important thing is that they be kept and used. Routine exams are performed on all cows to help spot problems. Problem cows are checked as frequently as necessary.

**Mastitis control.** Checks on milking procedures and milking machine function are performed. Use of diagnostic tools such as somatic cell counts, California mastitis tests, and bacteriologic culturing are a means of keeping mastitis under control.

**Nutritional consultation.** Nutritionally balanced diets help prevent metabolic diseases as well as promote optimal production of milk. The goal is to meet but not exceed nutrient requirements. Proper feeding helps optimize health, decrease feed costs, and reduce the effects on the environment. Veterinarians, feed dealers, nutritionists, county agents, and producers may all have suggestions based on the nutritional analysis of the feeds available.

**Replacement health management.** This program of examinations includes blood and fecal samples to determine which vaccines and parasite control measures are

**Herd health**

A comprehensive and herd-specific program of health management practices.

**FIGURE 17–14** Dairy replacements are often reared in individual hutches until weaning age. This helps keep disease transmission at a minimum.

necessary to produce the best replacement animals possible. Nutrition and management programs for replacement females are important to promote optimal lifetime cow productivity.

**Overall management consultation.** Consultation can be provided from several sources, including the extension service, private consultants, and the veterinarian providing the service just discussed. The veterinarian can be a good source of information because he or she is a regular visitor to the operation and provides other services. For farms on DHIA testing, DHIA records would be the place to integrate that information.

## BOVINE SOMATOTROPIN (BST)

This section is included as a stand-alone section in this chapter because of the uniqueness of the topic to the dairy industry. However, the implications are broader and have both biological and social implications that deal with all of agriculture, medicine, and, indeed, anything biological that occurs on this planet.

**Bovine somatotropin
(BST)**

A harmone produced by the pituitary gland of the cow. Injections of BST increase milk production in most cows.

*Bovine somatotropin (BST)* is a hormone produced naturally by cattle in the pituitary gland, which is an endocrine gland located at the base of the brain. The effects of BST have been known since the 1930s. Each species produces its own unique version of this hormone. BST accomplishes several things in the body, including regulation of growth in the young. Injections of BST increase milk production in dairy cows. In and of itself, that is not earth-shaking news. What makes BST unique is that developments in recombinant DNA technology make it possible to produce BST on a large-scale basis. This product is referred to as rBST.

With any product that would be used on an animal, rBST required government approval. Herein lies the importance of rBST, which was the first such product to be approved for use in animals in the United States. As such, the process was long and painful. A tremendous amount of controversy surrounded the approval of BST, and its approval was a landmark accomplishment for the fledgling biotechnology industry. Most of the concern was raised by consumer watchdog organizations. It is important to understand that several consumer and medical groups had approved and endorsed the use of rBST. It had been in use in Europe and Mexico for some years before it was approved in the United States. Commercial use of rBST received approval in 1993 for use in 1994. Approximately 30% of U.S. dairy cows are in herds using BST. Actual herd records indicate that producers using rBST are realizing from 5 to 15 lbs of increased milk production per day per cow. Individual cows generally improve milk production by 10–15%.

Milk from rBST-treated cows is safe for humans to consume and has no ill effects on human health. A small amount of naturally occurring BST has always been found in milk. There are three major reasons why rBST use in cows is safe. First, BST is a species-specific protein that is active only in cattle. Second, over 90% of the BST found in milk is destroyed during the pasteurization process. Third, any BST found in milk is digested in the human digestive system as is any other protein. Slightly elevated levels of insulin-like growth factor (IGF-1) can be found in milk from cows treated with rBST. This is not a cause for concern for three reasons: First, these levels are not above normal ranges for cows. Second, human milk has higher levels of IGF-1 and it has not caused any problems. Third, IGF-1 is not biologically active when ingested by humans and is digested by the human digestive system just as BST is. Fourth, the amount of IGF-1 in a serving of milk is insignificant compared to the amount produced in the human body daily. A multitude of studies have demonstrated that BST, whether natural or synthetic, has no effect on humans.

*Posilac* (sterile sometribove zinc suspension) is the registered trade name of Monsanto Company's approved, commercially available rBST. The generic name of this product is recombinant DNA-derived methionyl bovine somatotropin. Posilac is administered as an injection every 14 days, beginning in the 9th week after calving. The injection is given in the postscapular region (behind the shoulders) or ischiorectal fossa (depression on either side of the tailhead). rBST has to be injected because it is a protein that would be digested by the cow if fed to her. Hormones like estrogen and progesterone (birth control pills) are smaller and can be absorbed intact from the gut.

For over a decade rBST use appeared to be a nonissue with most consumers. However, consumers' general concerns over health and wellness as well as interest in organic and natural foods led some producers to begin producing certified rBST-free milk. The availability of the rBST-free products seems to have resurfaced the issue with consumers and has led several retailers and several milk cooperatives to announce they are no longer willing to sell milk produced with the help of rBST, effective in 2008. This will effectively eliminate the use of rBST in many states. It is possible by the next edition of this text that rBST will no longer be in use in the United States.

**Posilac**

Monsanto Company's approved, commercially available BST.

## NUTRITIONAL BENEFITS OF MILK TO HUMANS

A 1-cup serving of whole milk provides the following proportions of the recommended daily dietary allowance for a 25- to 50-year-old woman on a 2,200-calorie diet:

| | |
|---|---|
| Protein | 16% |
| Phosphorus | 29% |
| Calcium | 36% |
| Zinc | 8% |
| Riboflavin | 30% |
| Thiamin | 9% |
| $B_{12}$ | 44% |
| Calories | 10% |

Milk is a very nutritious food. Recent trends in eating patterns have shown consumers decreasing their consumption of whole milk in favor of lower fat versions. The only thing that consumption of lower fat milk changes substantially in the preceding list is the calorie percentage, which goes down incrementally with the decreasing quantity of fat in the product, which improves the nutrient density.

Calcium is especially important in the average diet, and milk does an excellent job of providing it. A complete nutrient analysis of whole milk and nonfat milk is given in Appendix II.

## TRENDS IN THE DAIRY INDUSTRY AND FACTORS THAT WILL INFLUENCE THE INDUSTRY IN THE FUTURE

The dairy industry is a dynamic industry. Several factors are shaping the changes and direction of the industry. The following discussion is not all inclusive but does discuss some of the major issues and factors to watch.

### RESTRUCTURING

### Trends

The dairy industry has been restructuring since roughly 1950. Production is shifting to larger operations, total cow numbers are going down, and the number of dairy producers is declining (see Figures 17–6, 17–7, and 17–15). Large dairies have significant cost advantages over small ones. The cost of adopting new technologies into a dairy will continue to cause the minimum economically feasible size of a dairy operation to increase. Bigger operations can make better use of machinery, equipment, and management because they can more easily achieve economy of size. This increase in operation size fol-

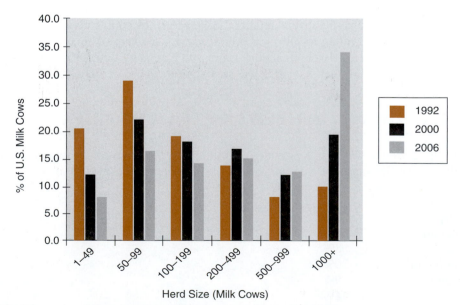

**FIGURE 17–15**  Milk production shifts to larger dairies. (*SOURCE:* McDonald et al., 2007.)

lows the overall trend in agriculture and is likely to continue. Technological innovations and continued increases in the amount of milk produced per cow will hasten this trend. Production will continue to shift to the West. Dairy farms will increasingly become more specialized as they increase in size and reduce in number and more mega dairies with more than 10,000 cows will emerge in both new and traditional dairy states. Fewer farmers will raise their own feeds; they will buy all or part of it. More dairy farmers will rely on specialized heifer-rearing operations to rear their replacements. Confinement dry-lot housing and automation with concomitant reduction in labor requirements will continue.

Dairying is among the least concentrated of the major farm enterprises in the United States. However, there is an increasing concentration of dairy producers in certain areas that may constitute a growing trend. Approximately 70 counties produce over half of the milk marketed in the United States. Granted, the counties are in several states, but nevertheless that is concentration. This is especially evident in the West.

## Managers and Labor

Dairying will need increasingly knowledgeable managers and high-quality labor, especially as the average size of a dairy increases and as the technological innovations become more and more sophisticated. Family-owned dairies will increasingly hire nonfamily members to manage their dairies.

## Family Farms

Most dairy farms are family enterprises. This will quite probably continue to be the case. However, more family operations will become legal partnerships or will incorporate. Large, corporate commercial dairies will increase in number but are unlikely to come to dominate the industry soon.

## Policy

The dairy industry has traditionally been the focus of a great deal of federal public policy and is heavily regulated. Increasingly, the argument is being made that U.S. dairy policy is too complex and a hindrance to market development, innovation in dairy product development, and development of export markets. A recent roundtable discussion (Bailey, 2007) concluded that "The dairy industry deserves a fresh new look at dairy policy to ensure every market opportunity is fully explored, profits are strengthened, and growth can continue." Another (Blayney et al., 2006) concluded, "The efforts of U.S. milk suppliers, processors, and product marketers to improve competitiveness depends more on innovation, flexibility, and investment than on policy support." There is every sign that U.S. dairy policy is in for an overhaul.

## Technology

Historically, dairy producers have been progressive users of available technology. Cost competitiveness will give the edge to those producers who adopt and appropriately use the ever-increasing array of products and technological innovations. Expected aids include improved vaccines, better diagnostic aids, lower-cost sexed semen, "designer embryos" produced by recombinant DNA technology, an expanded arsenal of drugs, better heat detection mechanisms, ways of controlling reproduction, and lactation curve extenders.

The use of *embryo transfer* will continue to help make genetic progress in the dairy herds of the United States. However, that progress will be increasingly restricted to the

**Embryo transfer**
Collecting the embryos from a female and transferring them to a surrogate for gestation.

progress realized through the bulls born as a result of embryo transfer, rather than the daughters. The reason is that the bulls will usually be "proven" and will either be used extensively or eliminated. The use of embryo transfer makes more bulls from good cows available for testing. The use of embryo transfer specifically for producing cows will increasingly fail to be cost effective unless costs per calf can be substantially reduced.

*Genetic markers* allow us to select the genetics of the next generation with greater accuracy. In the dairy industry, this will help us select young animals that do not yet have performance or progeny information available. There will be less value in selections on older animals because, once daughters are production tested, genetic markers don't provide much additional information.

---

**Genetic markers**

Biochemical labels used to identify specific alleles on a chromosome.

---

There is increasing interest in the U.S. dairy industry in robotic milking systems. Although several factors seem to motivate producers to consider robotic milking, labor availability and costs head the list. Finding, keeping, and paying for quality labor is a growing problem in the dairy industry. Robotic milking has potential as a solution. Some dairies that have adopted robotic milking systems report labor reductions of greater than 70%. Robotic systems are also reported to reduce stress on dairy cows. Very few robotic systems have been installed in the United States, but the number seems to be increasing. Only available in the United States since 2000, they have been commercially available internationally since the early 1990s. The technology is being used successfully in Europe, Australia, New Zealand, and Canada. One of the major benefits to producers who use certain types of robotic systems is that cows decide when they want to be milked. This eliminates the need for the dairy farmer to structure the day around milking times, the first of which is generally before sunrise.

## National Animal Identification System

The National Animal Identification System (NAIS) is not being driven by dairy producers. Other species with much larger total animals and total premises have more influence on the NAIS. However, as dairy farms continue to decline in numbers, participation in NAIS for the dairy industry becomes easier to manage. NAIS is currently being implemented as a voluntary program at the national level, but several states have already made premises registration mandatory. It is likely that more states will follow. The purpose of NAIS is to register livestock premises and livestock for the purpose of traceability. The benefits of such a system as expressed by the Veterinary Services Division of APHIS include "disease control and eradication, disease surveillance and monitoring, emergency response to foreign animal diseases, regionalization, global trade, livestock production efficiency, consumer concerns over food safety, and emergency management programs." Information on NAIS can be found on the USDA APHIS website, http://animalid.aphis.usda.gov/nais/index.shtml.

## CONSUMPTION

## Trends

Milk is consumed as a variety of products including fluid milk, cheese, ice cream, yogurt, cream, and many more. National consumption of several fluid milk products is shown in Table 17–14. Per capita fluid milk consumption has declined in the United States since the middle of the 20th century (Figure 17–16). This trend is probably destined to continue. Milk competes with other beverages for the consumer's dollar. Milk's market share is less than that of soft drinks and similar to that for coffee and beer. Industry groups identify several problems that contribute to the low market share:

**TABLE 17–14  U.S. Fluid Milk Sales by Product (Millions of Pounds)**

| Year | Whole Milk | Lower Fat Milk | Skim Milk | Flavored Whole Milk | Other Flavored Milk | Butter-Milk | Total Beverage Milk[1] | Total Cream Products[2] | Eggnog | Yogurt | Total All Products[1] |
|---|---|---|---|---|---|---|---|---|---|---|---|
| 1970 | 41,363 | 6,082 | 2,368 | 1,144 | 611 | 1,130 | 52,698 | 778 | 61 | 169 | 54,928 |
| 1975 | 36,188 | 11,468 | 2,480 | 1,366 | 719 | 1,011 | 53,232 | 1,070 | 76 | 425 | 54,803 |
| 1980 | 31,253 | 15,918 | 2,636 | 1,075 | 1,197 | 927 | 53,006 | 1,173 | 95 | 570 | 54,844 |
| 1985 | 27,760 | 19,812 | 3,009 | 882 | 1,430 | 1,046 | 53,939 | 1,586 | 121 | 940 | 56,586 |
| 1990 | 21,333 | 24,509 | 5,702 | 691 | 1,657 | 879 | 54,771 | 1,776 | 123 | 1,055 | 57,725 |
| 1995 | 18,662 | 24,202 | 8,359 | 704 | 1,914 | 739 | 54,580 | 2,095 | 112 | 1,646 | 58,433 |
| 2000 | 18,448 | 23,649 | 8,435 | 892 | 2,444 | 622 | 54,490 | 2,665 | 93 | 1,837 | 59,085 |
| 2001 | 18,007 | 23,630 | 8,225 | 973 | 2,553 | 592 | 53,980 | 2,933 | 105 | 2,003 | 59,021 |
| 2002 | 17,960 | 23,610 | 8,030 | 1,030 | 3,010 | 576 | 54,216 | 2,891 | 127 | 2,135 | 59,369 |
| 2003 | 17,832 | 23,559 | 7,789 | 1,049 | 3,141 | 547 | 54,364 | 3,307 | 134 | 2,387 | 60,192 |
| 2004 | 17,395 | 23,611 | 7,794 | 857 | 3,440 | 528 | 54,072 | 3,549 | 129 | 2,709 | 60,459 |
| 2005 | 16,760 | 23,882 | 7,984 | 753 | 3,549 | 512 | 53,865 | 3,661 | 130 | 3,058 | 60,714 |
| 2006[3] | 16,410 | 24,414 | 8,320 | 719 | 3,732 | 504 | 54,993 | 3,714 | 131 | 3,295 | 62,133 |

[1] Includes miscellaneous fluid milk products, beginning 2003.
[2] Light and heavy cream, half and half, sour cream, sour cream dips in CA, and sour cream used in dips elsewhere.
[3] Preliminary.
SOURCE: USDA-ERS, Livestock, Dairy and Poultry Outlook, June, 2007.

(1) health concerns; (2) milk is not a modern/contemporary drink; (3) adults think milk is only for children; (4) milk is usually associated with high-fat snack foods; (5) milk is not convenient or readily available for consumption outside the home; (6) milk packaging is boring, is not consumer friendly, and is not packaged for portable consumption; (7) milk does not provide consumers with the variety of taste options other beverages provide; (8) children have greater control over food choices; (9) eating out has increased; and (10) skim milk has a poor appearance and color. However, lower fat milk is more popular with consumers and has not suffered as much in recent consumption declines as has total fluid milk (Figure 17–16).

Cheese consumption (Figure 17–17) has increased dramatically. The amount of milk used to make cheese has been greater than fluid milk and cream use since the late 1980s and now accounts for over half of the end-product use of raw milk. Increased cheese consumption has provided for most of the growth in dairy product demand for decades. Factors contributing to increased cheese consumption include new cheese-containing products, wider availability of different kinds of cheese, demand for convenience foods, increased consumption of cheese-containing ethnic foods (especially Mexican and Italian foods), and cheese's ability to add rich flavor to a variety of foods. The increase in eating out and ordering in have also contributed to increased cheese consumption because it is a major ingredient in food manufacturing. Pizza and cheeseburgers both contribute significantly to overall cheese consumption. Resealable bags of shredded cheeses, cheese blends for use in Italian and Mexican recipes, and individually wrapped cheese products like cheese sticks and even small-package gourmet cheeses have contributed to increased at-home cheese consumption.

Other products such as yogurt have enjoyed increasing popularity. In addition to traditional yogurt products, yogurt is now in cereals, fast-food desserts, toothpaste, makeup, and pet food. Better-for-you yogurts are also entering the U.S. marketplace such as Dannon's Activia, which is advertised to improve digestive health.

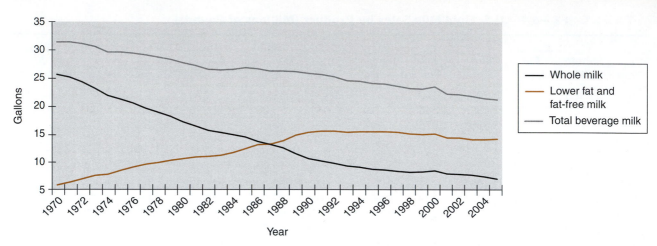

**FIGURE 17–16**    Whole, lower-fat, and total beverage milk per capita consumption, gallons, 1970–2003. (*SOURCE:* USDA-ERS. Food Availability (Per Capita) Data System. http://www.ers.usda.gov/data/foodconsumption/.)

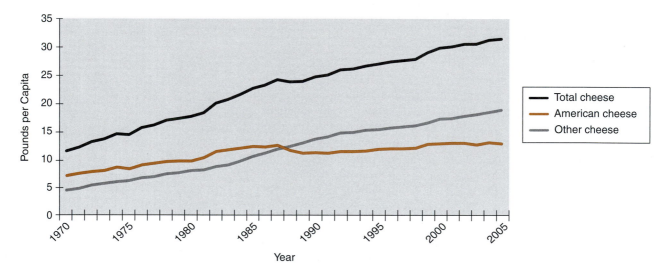

**FIGURE 17–17**    Per capita cheese consumption. (*SOURCE:* USDA-ERS, 2007b.)

Other changes are at work in U.S. dairy product consumption. Individual milk components are being marketed, such as individual proteins, skim solids, milk fat, and lactose. As both at-home and away-from-home eating patterns change, the demand for dairy products is shifting from retail sales to restaurant and food processor use, which now account for the majority of dairy product use. These uses are especially important for cheese, butter, and cream. The growing diversity of the U.S. consuming population is also a factor. Because of increased use of dairy products, per capita milk equivalent use is increasing modestly.

## Nutrition and Health Consciousness

Milk is good food that is nutritious (Figure 17–18). However, consumption has been affected by its nutrient content and will continue to be so affected. The most noticeable change in the consumer's habits has been the decline in whole-milk sales and the in-

**FIGURE 17–18** Milk is good food. The nutritive value of milk has long contributed to its reputation as the perfect food. While no such thing exists, milk comes close! (Photo courtesy of Dairy Management, Inc. Used with permission.)

crease in reduced-fat milk sales. Milk and other dairy products are included in the dietary recommendations of all of the major health organizations. However, milk faces a challenge in this regard. Fat-free milk is not as palatable as whole milk. Allowing the addition of solids other than fat to fat-free milk offers the advantage of improving taste as well as adding to the nutritive content of milk.

On the positive side, recent research studies have confirmed the positive role of calcium consumption in bone health, and milk is an excellent source of dietary calcium. Cheese already contributes over a quarter of the calcium in the U.S. diet. Other studies have suggested a positive role of milk consumption in a weight-loss diet and a positive role of milk consumption after exercise in muscle building. If these studies are confirmed, and the information becomes common knowledge, then milk consumption could get a boost with those relevant interest groups. Milk-based sports drinks, milk-based energy drinks, and cholesterol-cutting milk entered the marketplace in 2007. Other health-promoting products are being developed.

## Organic Milk Production

Dairy has been one of the fastest growing segments of the organic foods industry. Milk cows accounted for over half of the certified livestock animals from 1997 to 2003, and organic milk cows accounted for 1 and 2% of the total in California and Wisconsin, respectively, the two top dairy states for both organic and conventional production in 2003. Reasons consumers give for preferring organic milk include both environmental and animal rights concerns and health benefits. However, research has demonstrated no health benefits for organic milk over conventionally produced milk. In addition, organic milk is more expensive, selling for double the price of other milk in some parts of the country.

At this writing, the overall share that organic milk and milk products have of the market is less than 1% but is growing at a rate of 25% increase per year. This rate of growth will probably slow and then stabilize. However, for the near future, look for strong growth of the organic milk segment.

## FOOD SAFETY

### Consumer Concerns

The food supply in the United States is probably the safest in the world. Dairy products have an enviable record of safety. However, they cannot be perfectly safe and still be affordable. Consumers react to food safety concerns by changing consumption habits. This industry, and all food industries, must remain vigilant against threats to consumer confidence. Examples of issues of concern to the public are mycotoxins, naturally occurring allergens, chemical residues, hormones, feed additives of all kinds, antibiotic and other drug residues, and foodborne pathogens. Controlling these potential problems requires vigilance on the part of producers, processors, and distributors. It also requires the cooperation of the consumer. Many of the concerns that consumers have about the food supply are without basis. However, in the final analysis, perception is more important than reality when it comes to such issues. Chemical residues are considered the most important concern among the consuming public, but data from the Centers for Disease Control and Prevention indicate that pathogenic microorganisms are actually the greatest health threat.

Foodborne illnesses are especially volatile and emotional issues with consumers. Because of its perishable nature, milk requires a high level of processor care. It is an extremely safe food. However, public perception is easily shaped by events in the news. Any outbreak of *Salmonella*, *E. coli*, or any other disease linked to milk, as has occurred with much publicity in the last decade, erodes the consumer's confidence. Food safety concerns must begin with individual dairy producers. Although few food-linked illnesses can be traced to the farm, farmers must stay aware that perceptions are important. Good, dependable, quality practices with the cows and the milk are the first step in producing safe milk.

Bovine spongiform encephalopathy (BSE) should never become an issue in the U.S. dairy industry because milk is not a form of transmission of the disease.

### Environmental Concerns

Waste disposal is an increasing concern to dairies. Concentrated animal feeding operations (CAFOs) are considered a significant source of ground- and surface-water pollution because of the high levels of nitrates and phosphorus, harmful bacteria, and salt found in manure. Many modern dairies fit this category of operation (Figure 17–19).

Most analysts predict the enactment of increasingly tougher waste disposal laws that will add costs to production. Individual states have enacted environmental regulations that go beyond federal requirements, and more are expected to do so. New, low-cost, effective means of animal waste disposal should help blunt the added costs that environmental stewardship will bring. Quality of the air, soil, and water supply is an important issue with consumers.

## INDUSTRY ORGANIZATIONS

**American Dairy Science Association**
1111 N. Dunlap Ave.
Savoy, IL 61874
Phone: (217)356–5146
Fax: (217)398–4119
E-mail: adsa@assochq.org
http://www.adsa.org/

**FIGURE 17–19** Many modern confinement dairies incorporate "loafing" facilities and dry lots for housing during some or all of the year. These facilities will be under increasing pressure to control pollution.

**Dairy Management Inc.**

10255 West Higgins Rd., Suite 900
Rosemont, IL 60018–5616
Phone: 1-800-853-2479
http://www.dairyinfo.com/

Dairy Management Inc. (DMI) is the domestic and international planning and management organization that works to increase demand for U.S.-produced dairy products abroad. DMI manages the American Dairy Association, the National Dairy Council, and the U.S. Dairy Export Council.

**National DHIA (Mail Delivery)**

PO Box 930399
Verona, WI 53593-0399

**National DHIA (UPS, FedEx, etc.)**

421 S. Nine Mound Rd.
Verona, WI 53593
Phone: (608) 848-6455
Fax: (608) 848-7675
E-mail: dhia@dhia.org

**National Milk Producers Federation**

2101 Wilson Blvd., Suite 400
Arlington, VA 22201
Phone: (703)243–6111
Fax: (703)841–9328
E-mail: info@nmpf.org
http://www.nmpf.org/

The National Milk Producers Federation (NMPF) is a farm commodity organization representing most of the dairy marketing cooperatives in the United States. The NMPF provides a forum through which dairy farmers and their cooperatives formulate policy on national issues that affect milk production and marketing.

## SUMMARY AND CONCLUSION

Dairy products provide approximately 10% of all yearly cash receipts from agriculture in the United States, amounting to approximately $25 billion. The purpose of the U.S. dairy industry is to provide high-quality food from resources that the human population cannot use, such as forage and by-products.

The dairy industry is dispersed across the country because it is generally more cost effective to produce milk near the human population that will consume it. Dairy cattle fill a niche in resource utilization that could be filled by other ruminants, except that milk is such a valued product that it makes the higher inputs necessary for the dairy worth the effort because of higher returns.

As in most other areas of agriculture, the dairy industry is restructuring. The West is becoming more important, operations are becoming larger, and technology is increasingly important in driving the industry. Because of the dairy industry's importance to U.S. agriculture and to human nutrition, an extensive network of resources and information is available to the dairy producer. Organizations like the DHIA help in providing information for making management decisions. Good information is available to dairy producers, but increasingly good managers are needed to make best use of the information.

Herd health is an especially important management tool for the dairy producer. Dairy cows produce a high-volume product. They must stay healthy to continue producing in an economical fashion. A program of total herd health can help. Posilac, a recombinant DNA product, has been in use in the United States for over a decade. Although demonstrated as safe, it appears the product's use is destined to decline (perhaps stop altogether) because many processors and distributors are refusing to buy milk produced from herds where it is being used. This is widely considered a blow to the biotechnology industry. Trends and matters of importance to the dairy industry include continued restructuring, technological innovation, changing consumer preferences and demands, the fortunes of organic milk, food safety, and green concerns such as environmental impacts and animal welfare.

---

### Facts about Dairy Cattle

| | |
|---|---|
| Birth weight: | Varies with breed and sex; 50–110 lbs |
| Mature weight: | Varies with breed, sex, and condition; female 900–1,600 lbs; male 1,400–3,000 lbs |
| Weaning age: | Commonly removed from cow at 1–2 days of age and fed milk or milk replacer for 4–8 weeks |
| Breeding age: | 14–19 months (female) |
| Normal season of birth: | Year-round |
| Gestation: | 280–283 days |
| Estrous cycle: | 19–21 days |
| Duration of estrus: | 12–18 hours |
| Calving interval (months): | 11–15 (12 is preferable, but 13 is more realistic) |
| Normal calf crop: | 70–80% |
| Names of various sex classes: | Calf, heifer, cow, bull, steer |
| Type of digestive system: | Ruminant |

## STUDY QUESTIONS

1. Dairy amounts to how much of the total cash receipts of animal agriculture? Describe the magnitude of this industry in other ways. In addition to milk and its products, what other major product does the dairy industry provide?

2. What is the primary purpose of the dairy industry in the United States? What are the resources it uses? What is the dairy's role in using homegrown feeds?

3. Study the section on the history of the dairy industry. Select five elements that you feel are the most important and justify your list.

4. How geographically concentrated is the dairy industry compared to the other animal industries? Why is this?

5. What are the major dairy areas in the United States? Include the states in each area. Where are the dairy areas in a state most likely to be found? Name the 10 leading dairy states. What has been the historical role of the dairy in a farming enterprise?

6. How many dairy cattle are there in the United States? What is happening to the overall numbers? What is happening to dairy cow production per cow?

7. Describe the function and value of NCDHIP.

8. Name the major dairy breeds in the United States. Rank them on milk production and on popularity. What percentage of U.S. dairy cows is purebred? Explain.

9. Why are reproductive problems such an issue in the dairy industry? What role does nutrition play in reproduction? How can DHIA records be used to help reproductive efficiency?

10. Why is feeding the most important variable that needs to be controlled if a dairy is to be profitable? Why is good forage so important to dairy feeding and profitability?

11. Describe the basics of feeding the various classes of dairy animals.

12. Describe complete mixed ration feeding in dairy cattle. What are its advantages?

13. Herd health programs are important in a good dairy operation. What is the purpose of a herd health program? What are the elements of a good herd health program?

14. Describe the use of BST in dairy cattle. Does it work? Is it safe? Why or why not?

15. Describe the nutritional benefits of milk to the human diet. What nutrient is milk most famous for providing? After looking at the nutritional information, do you feel milk is being overlooked as an important source of other nutrients?

16. Describe the trends that are shaping the dairy industry in one sentence each.

## REFERENCES

For the second, third, and fourth editions, Dr. Tony Seykora, Professor, Animal Science, University of Minnesota, reviewed this chapter. In addition, Dr. Seykora contributed new material to the chapter. For the fourth edition, Dr. Noah Litherland, Professor of Animal Science, Oklahoma State University, reviewed the chapter and provided constructive comments. The author gratefully acknowledges these contributions.

Aitchison, T. E. 1992. Dairy sire evaluations in the U.S.A. *The National Dairy Database (1992) Collection; Genetics Improvement.*

Bailey, K. 1997. Dairy industry needs to reach today's consumers. *Feedstuffs* 37:11.

Bailey, K. 2007. Secretary Wolff's round table discussion on dairy policy (January 2007), revised February 15, 2007. Dairy Outlook Website. Accessed online at http://dairyoutlook.aers.psu.edu/.

Blayney, D., M. Gehlhar, C. H. Bolling, K. Jones, S. Langley, M. A. Normile, and A. Somwaru. 2006. *U.S. dairy at a global crossroads.* USDA-ERS report number 28, November 2006. Accessed online at www.ers.usda.gov.

Drache, H. M. 1996. *History of U.S. agriculture and its relevance to today.* Danville, IL: Interstate.

Ensminger, M. E. 1980. *Dairy cattle science.* 2nd ed. Danville, IL: Interstate.

Ensminger, M. E. 1991. *Animal science.* 9th ed. Danville, IL: Interstate.

FAO. 2007. *FAOSTAT statistics database: Agricultural production and production indices data.* http://apps.fao.org/.

Linn, J. G., M. F. Hutjens, W. T. Howard, L. H. Kilmer, and D. E. Otterby. 1988. Feeding the dairy herd: Collection, feeding, and nutrition. *The National Dairy Database (1992) Collection; Genetics Improvement.*

McDonald, J. M., W. D. McBride, and E. J. O'Donoghue. 2007. Low costs drive production to large dairy farms. *Amber Waves.* USDA-ERS. Accessed online September 2007 at http://www.ers.usda.gov/AmberWaves/.

Miller, J. J., and D. P. Blayney. 2006. *Dairy backgrounder.* USDA-ERS, LDP-M-145-01. Accessed online July 2006. www.ers.usda.gov.

Murley, W. R., and G. M. Jones. 1985. Complete rations for dairy herds. *The National Dairy Database (1992) Collection; Feeding and Nutrition.* http://www.Inform.Umd.Edu/Edres/Topic/Agrenv/Ndd/Feeding.

USDA-ERS. 2007. *Livestock, dairy, and poultry outlook.* Accessed online. September 2007. http://www.ers.usda.gov/Briefing/Dairy/.

USDA-ERS. 2004a. Characteristics and production costs of U.S. dairy operations. Statistical Bulletin Number 974-6. *Electronic Report from the Economic Research Service.* Accessed online August 2004. http://www.ers.usda.gov/Publications/sb974-6/.

USDA-ERS. 2007a. *Farm income data.* Accessed online September 2007. http://www.ers.usda.gov/Data/farmincome/finfidmu.htm.

USDA-ERS. 2007b. *Food availability (per capita) data system.* http://www.ers.usda.gov/data/foodconsumption/.

USDA-NASS. 2001. *Dairy products annual summary.* April 2004. Washington, DC: National Agricultural Statistics Service, USDA.

USDA-NASS, 2007a. 2007 Agricultural Statistics. Accessed online September 2007 at http://www.nass.usda.gov/Publications/Ag_Statistics/2007/index.asp.

USDA-NASS. 2007b. Briefing Room. Farm Income and Costs. Accessed online September 2007 at http:// www.ers.usda.gov/Briefing/FarmIncome/.

USDA-NASS. 2007c. Milk production. February, 2007. Accessed online September 2007 at http://usda.mannlib.cornell.edu/MannUsda/viewDocumentInfo.do?documentID=1103.

USDA-NASS. 2007d. Quick Stats: Agricultural Statistics Data Base. Accessed online September 2007 at http://www.nass.usda.gov/QuickStats/.

Varner, M. 1993. A workable herd health program. *The National Dairy Database Collection; Herd and Animal Health.*

Voelker, D. E. 1978. The dairy cow as a protein producer. *The National Dairy Database (1992) Collection; Business Management.*

Waldner, D. 2004. Professor of animal science, Oklahoma State University, Stillwater. Personal communication.

Weimer, M. R., and D. P. Blayney. 1994. *Landmarks in the U.S. dairy industry.* Agriculture Information Bulletin No. 694. Washington, DC: USDA-ERS.

# CHAPTER 18

# Poultry

## LEARNING OBJECTIVES

After you have studied this chapter, you should be able to:

1. Explain the place of poultry in U.S. agriculture.

2. Describe the purpose and the value of the different poultry industry segments.

3. Give a brief history of the poultry industry in the United States.

4. Describe the poultry industry segments and structure.

5. Give an accurate accounting of where the poultry industry is located in the United States.

6. Quantify the role of genetics in the poultry industry.

7. Explain the role of breeds in the U.S. poultry industry.

8. Describe the major breeding programs for poultry.

9. Give a basic outline for managing poultry to produce fertile eggs.

10. Describe the feeding practices for different classes of poultry.

11. Explain the concept of flock health management and give specifics regarding poultry house management.

12. Explain the nutritional benefits of various poultry products for humans.

13. Discuss trends in the poultry industry including factors that will influence the industry in the future.

## KEY TERMS

| | | |
|---|---|---|
| American Standard of Perfection | Blood spots | Brood |
| Avian | Breed | Brooder |
| Bantam | Broiler | Broodiness |
| Basic breeder | Broiler duckling or fryer duckling | Brooding |
| | | Candling |

Chick

Class

Cock

Cockerel

Contract grower

Dead germs

Defect

Delmarva

Dirties

Economy of size

Exotic fowl

Fowl

Game birds

Gander

General combining ability

Hatching

Hen

Heritability

Heterosis

Inbred line

Inbreeding

Incrossbred hybrid

Incubation

Incubator

Individual cage

Infertile

Layer

Laying

Litter

Mash

Mature duck or old duck

Mature goose or old goose

Molting

Omnivore

Outcrossing

Pelleting

Plumage

Poult

Poultry

Pullet

Quantitative traits

Roaster

Roaster duckling

Rooster

Setting

Sex-linked cross

Shelf-life

Specific combining ability

Strain

Strain cross

Tom

Trapnest

Variety

Vertical integration

Young goose or gosling

## SCIENTIFIC CLASSIFICATION OF POULTRY

| | |
|---|---|
| Phylum: | Chordata |
| Subphylum: | Vertebrata |
| Class: | Aves |
| Order: | Galliformes |
| Suborder: | Galli |
| Family: | Phasinanidae (chicken); Meleagrididae (turkey) |
| Genus: | *Gallus* (chicken); *Meleagris* (turkey) |
| Species: | *domestica* (chicken); *galiopavo* (turkey) |

## THE PLACE OF POULTRY IN U.S. AGRICULTURE

**Poultry**

Domestic birds raised for eggs and meat. However, this designation has some flexibility.

*Poultry* is a term that includes a wide variety of domestic birds of several species, and it refers to them whether they are alive or dressed. However, the important commercial species are chickens and turkeys and, to a much lesser degree, ducks and geese. The industry that revolves around these species is large. The combined gross annual income of the poultry segments amounts to almost $30 billion. The poultry industry has three major segments whose relative contributions to the total value of the poultry industry are shown in Figure 18–1. The chicken segment is the largest and is generally increasing its share (Figure 18–2). The poultry industry is responsible for over 11.6% of all U.S. farm cash receipts (Figure 18–3) and over 23% of animal agriculture's share of all U.S. farm cash receipts (Figure 18–4). The poultry sector is growing. It is the second-largest sector of animal agriculture.

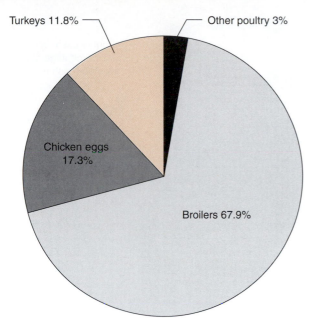

FIGURE 18–1    Relative contribution of poultry segments to total poultry industry value, 2005–2007. (*SOURCE:* USDA-NASS, 2007a)

FIGURE 18–2    Broiler chickens such as these account for the largest share of the poultry industry. Broiler production has steadily increased in the United States since 1975. (Photographer Norm Thomas. Courtesy of Photo Researchers, Inc.)

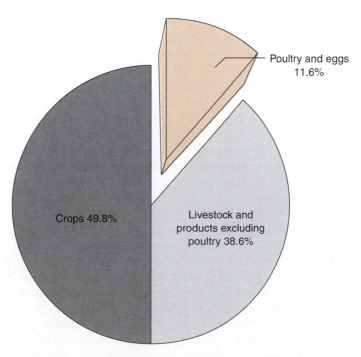

FIGURE 18–3    Poultry and eggs farm cash receipts as a percentage of total U.S. farm cash receipts, 2005–2007. (*SOURCE:* USDA-NASS, 2007a.)

**FIGURE 18–4**   All poultry and eggs yearly farm cash receipts as a percentage of total animal agriculture's cash receipts, 2005–2007. Includes broilers, eggs, turkeys, ducks, and other poultry. (*SOURCE:* USDA-NASS, 2007a.)

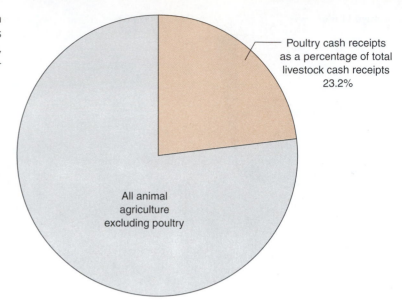

Poultry cash receipts as a percentage of total livestock cash receipts 23.2%

All animal agriculture excluding poultry

Six states have in excess of $1 billion yearly in gross income from broilers alone. An additional six have at least $0.5 billion. Chicken eggs are in the top-five agricultural commodities for four states, turkeys are in the top-five commodities for four states, and broilers are in the top-five commodities in 14 states. The United States produces approximately 8.8% of the world's chicken eggs and 22% of the chicken meat, and it is the largest poultry meat-producing country in the world. The United States is second to China in chicken egg production, produces 56% of the world's turkey meat, and is the largest producer of turkeys in the world.

The poultry industry in the United States has been growing at a steady rate. The broiler and turkey sectors have aggressively developed and incorporated technology, advanced animal breeding techniques, advanced nutritional information, and very savvy marketing and product development. The broiler industry especially has seized innovation as a way to generate a lower-cost product than the other meat-producing industries can produce. It has made the product convenient and good to eat. In so doing, the broiler industry has been the success story of the animal industries since the early 1970s. The egg industry is a major contributor to the nation's food supply. It supplies approximately 250 eggs per capita for the nation's population. Approximately 69% is consumed as fresh-shell eggs, and the remaining 31% is used in the manufacturing of products such as cakes, pies, pasta, and so on.

The duck segment of the poultry industry is the largest of the remaining segments (Figure 18–5). Ducks are raised primarily for meat. About 28 million ducks are slaughtered annually in the United States to produce approximately 86,000 metric tons of meat. Most are produced under confinement on specialized duck farms in a few commercially important duck-production areas. However, many farms still raise a few ducks primarily for family use or for local sale. Geese are raised in practically all parts of the United States, although they are almost exclusively a hobbyist or farm flock species. Some are marketed commercially.

**FIGURE 18–5** Ducks comprise the largest segment of the poultry industry after broiler chickens, laying hens, and turkeys. (Photo courtesy of Pearson Education Corporate Digital Archive.)

**FIGURE 18–6** The graceful and beautiful swan is a poultry species kept almost exclusively for ornamental reasons. (Photographer Edmond Van Hoorick. Courtesy of Getty Images, Inc./Photo Disc, Inc.)

*Game birds* and other *exotic fowl* are usually kept either for exhibition or ornamental reasons or for meat production (Figure 18–6). Game birds are also kept for release and subsequent hunting. Producing birds for hunting use is seasonal because birds are released for the fall hunting season. Demand for the meat is also somewhat seasonal and is associated loosely with the Thanksgiving and Christmas holidays. This is probably because these holidays fall within the traditional hunting seasons. Many families have traditions of game dinners during the holiday seasons. In earlier times, game meat was an important dinner fare for families in this country. The market for meat is generally local consumers or specialty processors, much of it for the hotel and restaurant trade.

**Game birds**

Fowl for which there is an established hunting season. Also refers to fighting chickens.

**Exotic fowl**

Nonindigenous, nonlivestock species often kept for ornamental reasons.

# PURPOSE OF THE POULTRY INDUSTRY IN THE UNITED STATES

The primary purpose of the poultry industry in the United States is to produce inexpensive sources of protein for human consumption. The industry takes grain and by-products and produces meat and eggs very efficiently. These products are among the best buys in the marketplace today for high-quality protein. Poultry are monogastric and can use only limited amounts of forage. In practice they are rarely fed any significant forage at all (except geese). However, they have the distinction of being the most efficient converters of grain to product of all the land-dwelling livestock species (Figure 18–7). Poultry are used to add value to the grain and help moderate the fluctuations in grain prices. Creating this secondary market for grain helps assure the supply of grain for human consumption. Because of their digestive system type (monogastric), poultry require high-energy feeds to be produced economically. Feed costs are a high fixed cost, so economical production depends on reducing labor demands and reducing time to market as a means of more efficient production. This is done with expensive confinement facilities and full feeding of high-energy rations.

Poultry are quite useful worldwide. Increasingly, U.S.-style confinement facilities are being built around the world for their production. In poor countries, poultry are often fed on wastes and allowed to range freely and forage for themselves. Left to their own devices, they eat a variety of feedstuffs from insects to seeds, even small rodents and reptiles. Worldwide, chicken-meat production surpassed beef production in 1999 but is considerably behind pork production, which is the world's most produced meat. However, chicken-meat production amounts to almost 70% as much as pork. Combined poultry-meat production equals over 31% of the world's total meat production. If the weight of the world's egg production is added to the previously mentioned poultry meats, then poultry accounts for approximately 1.4 times as much high-quality protein product as pork and nearly 2.5 times as much as beef. There are approximately 17 billion chickens in the world, easily more than the physically larger mammal species. There are 2.6 chickens per person in the world, and 325 chickens per square mile of earth land surface.

Another important group of individuals is interested and involved with poultry, who have purposes outside the industry. Although not large in terms of economic value, poultry hobbyists comprise a large and dedicated group of individuals who deserve acknowledgment. Their interests range from fancy (exhibition) chickens to game birds. Poultry keepers have a wide variety of species from which to choose, and virtually all are hardy and easy to raise. Space requirements and feed bills can be kept to a mini-

**FIGURE 18–7** Product protein output as a percentage of feed protein. (*SOURCE:* Compiled from several sources from the *Proceedings of the 8th World Conference on Animal Production,* Seoul, Korea, 1998.)

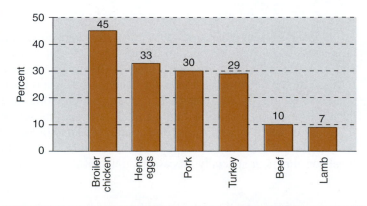

mum. Products can be harvested from the poultry for household use or sold to help defray costs. Pigeons can be raced. The ornamental species can grace a property or a body of water. A variety of exhibition opportunities exist for most species. Children in the family can be taught biology and responsibility, and they can learn to enjoy the animals and, by extension, other species as well. There is much to recommend poultry species for these roles, and many people keep them for these purposes. One can also make a powerful argument that without hobbyists, many breeds of poultry would effectively disappear because they have no place in the commercial industry.

In addition, several industrial uses of poultry give a somewhat different dimension to the purpose of the poultry industry and are worth mentioning. Fertile eggs are used in the preparation of vaccines for other animals and humans. Endocrine glands are used in making biological products. Eggs are used in the making of pharmaceuticals, and in paints, varnishes, adhesives, and printer's ink. Photography, book binding, wine clarification, leather tanning, and textile dyeing make use of eggs. Feathers are used in millinery goods, pillows, cushions, mattresses, dusters, and insulation material. The *chick* has also been a very important research animal. The chick is much more sensitive to the lack of several substances in the diet than are most other available species. Also, chicks are cheap and readily available, and large numbers of them can be hatched at the same time so as to provide accuracy and repeatability in research work.

**Chick**

A young chicken or game bird of either sex from 1 day to about 5 to 6 weeks of age.

## HISTORICAL PERSPECTIVE

### DOMESTICATION

Poultry have been with us for many thousands of years. DNA testing has demonstrated conclusively that modern chickens descended from the Red Jungle fowl *Gallus gallus,* which is native to Thailand and still found in the wild today. Today's Old English Game has a very similar appearance and temperament to the Red Jungle fowl. Gaming has been proposed and generally accepted as the primary motivation for the first domestication of the chicken. Cockfighting has been a favored pastime of many ancient civilizations. Domestic chickens were known in India over 3,400 years ago. Both Egypt and China had domestic poultry by 1400 B.C. Tombs in the Valley of the Kings at Luxor that date from around 1400 B.C. have drawings of jungle fowl in them. Coins from 700 B.C. have drawings of cocks on them. Plato (427–347 B.C.) complained about people watching cockfights instead of following more industrious pursuits. Cockfighting has not lost its appeal to all of the world's people. It is often cited as the largest spectator sport in the world.

### EARLY USE IN THE UNITED STATES

Columbus brought chickens along with a selection of other animals to this hemisphere on his second voyage in 1493. In 1607, the chicken was introduced to the North American continent by the settlers of the Jamestown Colony. The practice of small-flock keeping persisted until the middle of the 20th century (Figure 18–8). The flock provided meat and eggs for the family, and the extra was sold to provide cash to buy household staples such as sugar, flour, coffee, and so on. The practice began fading in earnest when agriculture began restructuring in the 1950s and the farm population began to decline. In 1840, the first census of poultry was taken in the United States, providing a means of tracking the industry. In 1844, the incubator was patented, although it would not be until 1892 that the first long-distance express shipment of

**FIGURE 18–8** Small flocks of chickens were commonly kept in this country from colonial times until the middle of the 20th century as an integral part of the sustenance for the family. The primary product was eggs. The meat was reserved for special occasions and considered a luxury. During the mid-1920s, meat production from chickens reached significant levels and the broiler industry was born. It has been growing ever since. (Photographer Arthur Rothstein. Courtesy of USDA.)

**FIGURE 18–9** Percentage of chicks supplied by hatcheries. (*SOURCE:* American Poultry Historical Society, 1973.)

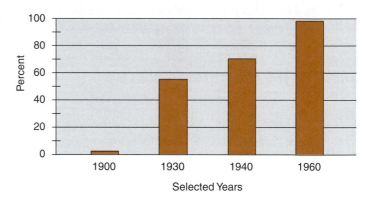

**Hen**

A mature female chicken or turkey.

baby chicks would take place. These chicks were shipped from Stockton, New Jersey, to Chicago, Illinois. This event marks the beginning of the commercial hatchery industry. Commercial hatcheries got another boost in 1918 when the U.S. Post Office allowed chicks to be shipped by mail. This allowed hatcheries to increase in size and improve their brood stock because they could ship anywhere in the country, even to small out-of-the-way places. The entire industry received another important device in 1923 when Ira Petersime developed the electrically heated incubator. At the beginning of the 20th century, virtually all chicks were hatched under **hens.** By 1960, virtually all were hatched in incubators and then shipped to producers (Figure 18–9).

All of animal agriculture got much-needed and far-reaching help in 1862 with the passage of the Morrill Agricultural Land Grant Act. From this act came the extensive land grant system of colleges of agriculture distributed across every state in the United States. The first state experiment station was established in 1875 by Connecticut. In 1887, the Hatch Act passed, which established agricultural experiment stations. This would be followed by the extensive system of experiment stations established by land grant universities across the different geographic areas of the country. These experiment stations allowed research to be conducted to help producers in specific geographic areas with the problems of those areas. The final piece of the system was put in place in 1914 with the passage of the Smith-Lever Agricultural Extension Act, which established the Cooperative Extension Service that was designed to distribute information to farmers. The poultry industry benefited tremendously from the research, education, and information dissemination of the land grant university system.

## EGGS

In 1872, a commodity exchange for the trading of cash eggs was chartered. In 1874, chicken wire, which made flocks easier to manage, was first introduced. In 1878, the first commercial egg-drying operation was established, giving eggs new uses and storage potential beyond the traditional spring–summer–fall *laying* season. In 1889, artificial light was first used to stimulate egg production in the winter months. This allowed producers to extend the season of lay beyond the traditional season. By 1899, frozen eggs were marketed, and the following year the first egg-breaking operation was established. In 1909, the electric candler, which allowed intact eggs to be inspected for deformities, was developed. This had positive *hatching* and egg-marketing implications. The USDA issued the first tentative classes, standards, and grades of eggs in 1923. These were followed in 1934 with legal standards for eggs, which provided uniformity for marketing. The egg-producing industry took a major leap in 1929 when *layers* were kept in *individual cages* at the Ohio Agricultural Experiment Station. In that same year, *pelleting* of poultry feed, which reduced waste and helped ensure that each animal received a balanced diet, began. In 1931, laying cages were first given general publicity and lights were kept on chicken hens all night at the Ohio Agricultural Experiment Station as a means of stimulating egg production. In 1932, forced *molting* for commercial egg production, which improved egg quality and provided an important management tool for egg production, started in Washington.

## GENETICS

Throughout the world, chickens evolved into different types based on the environments in which they were kept. Marco Polo (1254–1324), the famed Venetian traveler, described seeing chickens with "fur" that were most likely the ancestors of today's Silky breed. For centuries, travelers took poultry with them and brought chickens and other poultry home with them from exotic ports. Poultry are small and easy to transport. If food became scarce along the way, they were simply eaten by the travelers. Keeping small flocks on ships, especially on long voyages, was common practice in previous centuries. In this country, existing breeds and developed breeds were used across the many environments the landscape offered. A variety of breeds came into use. Poultry show divisions at county fairs helped ensure that breeds flourished and were kept in pure strains.

In 1828, the first Single Comb White Leghorns were imported into the United States. They would ultimately become the most important egg-producing breed in the United States after conditions were standardized in the egg-producing industry. Breeding programs involving red chickens began in 1830 near Narragansett Bay, Rhode Island. The dual-purpose Rhode Island Red was one of the outcomes of these programs. Later, the New Hampshire Red was developed from specialized selection from within the Rhode Island Red. In 1840, birds from China began entering the United States. In 1873, the American Poultry Association was formed. It was America's first livestock organization. In 1874, they issued the first American Standard of Excellence for poultry. It was known as the *American Standard of Perfection* and has been issued continuously since the first edition. The book provided a uniform set of standards for poultry breeders to breed toward.

### Laying
Technically, the expulsion of an egg (i.e., "That hen over there is *laying*"). However, commonly used to refer to hens in egg production (i.e., "Barn twelve is a *laying* hen barn").

### Hatching
The process of a chick leaving the egg and emerging into the world. Birth for fowl.

### Layer
A hen in the physiological state of producing eggs regularly.

### Individual cage
Small pens perhaps 8 to 12 in. wide and 18 in. long designed to hold one or more hens.

### Pelleting
A method of processing feed in which the feed is first ground and then forced through a die to give it a shape.

### Molting
The shedding of feathers by chickens.

### *American Standard of Perfection*
A standard published in book form by the American Poultry Association. It lists the recognized breeds and varieties of poultry and their characteristics.

### Trapnest

A nest that traps a hen while she is on the nest so that her production and egg quality can be recorded.

### Sex-linked cross

Sex-linked chicks can be sexed at birth by their color. Males are one color and females are another color.

### Incrossbred hybrid

Chickens developed by crossing inbred lines within the same breed. This technique is generally used to produce laying hens.

### Brooder

A device with controlled heat and light used to warm chicks from the day of hatch to approximately 5 weeks of age. The heat is usually contained in a large reflector or hover under which the birds congregate.

### Fowl

Any bird, but generally refers to the larger ones. In this context, it refers to poultry species only.

### Delmarva

Geographic region comprised of Delaware, Maryland, and Virginia.

### Vertical integration

A form of contract production in which all stages of production are owned by one entity, corporation, or individual.

In 1869, the first *trapnest* patent, which allowed a hen and her egg to be paired, was granted. Research had a valuable new tool. In 1870, the toe punch and a system of numbering based on the number and placement of toe punches were introduced for identification in breeding programs. In 1928, the first *sex-linked crosses* were advertised. In 1935, artificial insemination of poultry was introduced and the first successful *incrossbred-hybrid* chickens were produced for egg production. This led to an extremely important event in poultry breeding that occurred in 1940. Hy-Line Poultry Farms in Des Moines, Iowa, marketed its first Hy-Line layers. These birds were developed by applying inbred-hybrid corn-breeding principles to egg production. This was the effective beginning of the deemphasis of breeds, which were replaced by the use of lines and strains of birds. This is still common practice today.

## CHICKS

The problem of hatching large numbers of chicks had been dealt with by the development of the incubator, but there was still the problem of rearing large numbers of chicks without the help of a hen. In 1903, the Cornell gasoline *brooder* was developed by Professor James Rice of Cornell University. Large numbers of chicks could be kept warm and could grow in the brooder. By 1905, the first chickens were raised in batteries and the battery brooder was developed by Charles Cyphers. This allowed for better environmental control and space utilization. In 1913, the first oil brooder was marketed. It was more economical and safer than the battery brooder.

## INTEGRATION

In 1895, the commercial feed industry began in Chicago. This was the important first step of overall integration of the poultry industry. By 1898, battery fattening was introduced by Swift and Company. In 1918, the first USDA federal–state grading programs for poultry were established. During the 1920s, poultry production began on an industrial scale. In the early days, live poultry were transported from areas of production to areas of consumption. An outbreak of *fowl* plague at the New York Central railroad yards stimulated the New York Poultry Commission Association to organize an inspection program in 1924. This led to the establishment of the USDA's Poultry Inspection Service in 1926. In 1928, the live poultry inspection expanded to include dressed poultry and edible products. The *Delmarva* region of the United States began to be the center of commercial broiler production. The USDA recognized this activity as different from traditional farm rearing of birds and in 1934 began reporting commercial broilers separately. The production of broiler meat by the broiler industry since 1940 is shown in Figure 18–10. At this time, the chicken was still considered to be Sunday dinner rather than everyday fare. In 1940, mechanical poultry dressing was initiated, giving the meat-producing sector a giant boost. Improvements in diets, equipment, genetics, flock health, and processing came at a steady pace. Costs of production are estimated to have been cut in half from 1940 to 1960, and the industry spread across the Midwest. Prices declined and chicken became everyday food. The profit margin decreased for producers, who had to increase the size of their flocks to make a living. They soon began depending on the feed producer as their source of credit. As this practice grew, the feed dealers began to depend on the commercial feed manufacturers as their source of credit. To protect themselves financially, the feed suppliers and manufacturers began to acquire hatcheries and processors. Integration was born in the poultry industry. Consolidation began, and more *vertical integration* occurred, producing an industry that today is completely vertically integrated. In

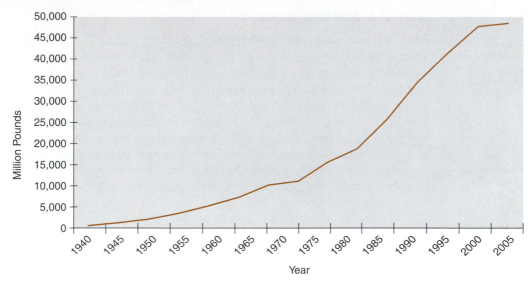

**FIGURE 18–10**   Production of broiler meat by the broiler industry. (*SOURCE:* Based on USDA, Meat Animals Production, Disposition & Income, 1998, and USDA-NASS, 2007b, Poultry Slaughter.)

1957, the Congress authorized the Poultry Products Inspection Act, and in 1968, the Wholesome Poultry Products Act.

In 1956, Colonel Harland Sanders began franchise operations of Kentucky Fried Chicken, which ultimately led to new life for chicken consumption. This was the same year that integration and contracting began in the egg business. By 1962, corporate egg farms had become a viable part of the poultry industry. In a span of just a few years, poultry production was composed of large integrated farms that contracted with individual producers to produce eggs, broilers, and turkeys. The companies supplied the animals, feed, *litter,* fuel, and medication, and the *contract growers* provided housing and labor. By 1970, it was estimated that vertical integration involved 95% of the nation's broilers, 85% of the turkeys, and 30% of the table eggs. In 1972, a vaccine for Marek's disease was approved, reducing some of the risk for concentrating large numbers of poultry in single sites. In 1981, combined cash receipts for poultry first exceeded those from hogs and in 1997 surpassed dairy. Today, per capita consumption of chicken and poultry is still increasing. This is because of the poultry industry's ability to contain costs and provide products at lower prices relative to alternatives, and its aggressive development of convenient and further-processed products. The modern mechanized poultry farm as we know it emerged in the late 1950s. The integrated poultry industry we know today emerged between 1955 and 1975. Integration brings all phases of the operation under centralized control. From 1975 until the present, tremendous consolidation occurred. Today, there is a relatively small number of integrated super companies.

**Litter**

Material such as wood shavings, straw, or sawdust used to bed the floor of a poultry house.

**Contract grower**

Producers of poultry and eggs who contract with an organization, usually integrated, to raise a product for a price determined through the contractual arrangement.

## STRUCTURE AND GEOGRAPHIC LOCATION OF THE POULTRY INDUSTRY

Vertical integration provides the structural framework of the commercial poultry industry.

### Broiler

A chicken of either sex produced and used for meat purposes. Generally slaughtered at 6 weeks of age or younger. The term *fryer* is often used interchangeably.

### Roaster

A young meat chicken, generally 12 to 16 weeks old weighing 4 to 6 lbs.

### Game hen

A standard meat-type chicken packaged at a smaller size. Also called Rock Cornish game hen or Cornish game hen to reflect the influence of the Cornish breed in most of the chickens sold this way.

### Economy of size

A relatively simple concept revolving around the maximization of the use of equipment, labor, and other costly items.

### Brood

A group of baby chickens. As a verb it refers to the growing of baby chicks.

## THE U.S. BROILER INDUSTRY

Meat chickens are marketed primarily as *broilers*, *roasters*, or *game hens*. All come from what is referred to as the broiler industry. Modern broiler production is concentrated in a relatively few large farms with large investments in facilities. In 1959, the average number of broilers raised per farm was 34,000 birds. Today, the average number of broilers sold annually per commercial operation is approximately 400,000. Several farms in the United States have more than 1 million birds per site. *Economies of size* favor large operations. In addition, gains in production efficiencies have allowed producers to rear more than six *broods* per house per year, so more birds can be grown in the same space. Before 1940, the broiler chicken was marketed at 16 weeks of age. Today, the broiler chicken is being reared to heavier weights and marketed at less than half that age.

The broiler industry is a highly integrated corporate industry rather than an industry of independent producers. A typical integrated broiler company owns or controls everything it needs but the consumer (Figure 18–11). All segments are either owned or controlled by the parent company. Well over 90% of the commercial broilers in the United States are grown under contract to an integrated broiler firm (Figure 18–12). A large percentage of what is left is grown on integrator-owned farms. Much of the product is sold with the producer's name on the label.

The broiler industry is centered in the southern and southeastern states (Figure 18–13). The industry located and developed in these areas initially for three reasons:

1. Favorable climate that reduced housing costs
2. Low-cost labor, which was often the small farmer who worked off the farm as well as on the farm
3. Nearby population centers that provided a demand for the product.

As the broiler industry grew in these areas, a large supporting infrastructure developed along with it. This infrastructure included processing plants, technical support companies,

**FIGURE 18–11**  Major segments of an integrated broiler company.

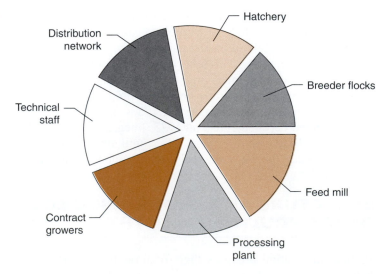

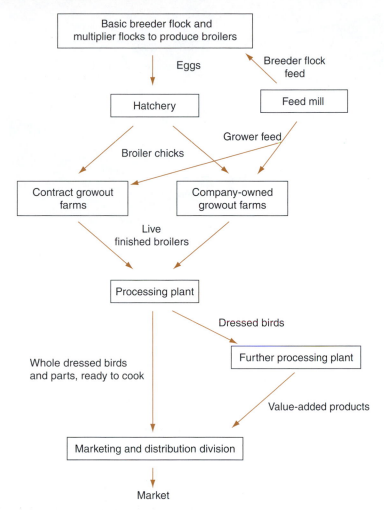

**FIGURE 18–12**    Structure of a typical integrated broiler company. (*SOURCE:* Adapted from Taylor and Field, 1998.)

and university research and extension groups. This infrastructure had and will continue to have a major influence on keeping the industry centered in this area. The top-12 broiler-producing states are relatively constant. The states are Georgia, Arkansas, Alabama, North Carolina, Mississippi, Texas, Delaware, Kentucky, Maryland, South Carolina, Virginia, and Oklahoma.

## THE U.S. EGG INDUSTRY

Egg production is also accomplished in large units that require large capital investments. Environmentally controlled housing and computer technology are common. Most eggs produced in commercial egg operations are never touched by human hands from the point of production until they are taken from their container to be used. The egg-producing industry is concentrated and integrated like the broiler industry for

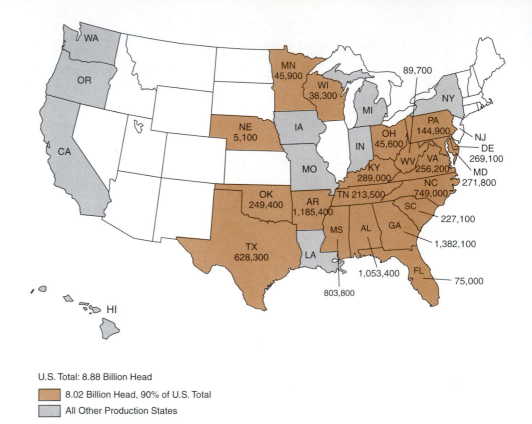

U.S. Total: 8.88 Billion Head

■ 8.02 Billion Head, 90% of U.S. Total

□ All Other Production States

**FIGURE 18–13**    The production of broilers by state, number raised (× 1,000), 2006. (*SOURCE:* USDA-NASS, 2007c.)

many of the same reasons. Most production is integrated from hatchery to marketing of the eggs (Figure 18–14).

Geographically, the poultry egg industry is distributed in a pattern much like that of the human population (Figure 18–15). Population is not the only factor that determines where layers are located, however. If that were so, then the distribution would perfectly mirror the human population, and it doesn't. However, egg production is much more dispersed across the country than is broiler production. There are several reasons for this, with these two being very important: (1) Eggs require less processing than broilers, which allows more flexibility in moving and marketing the product; and (2) locally produced fresh eggs can be promoted and sold at a premium.

The average number of laying hens and **pullets** in the United States is about 345 million birds. Total U.S. egg production is about 91 billion eggs. The number of laying

**Pullet**

A young female chicken.

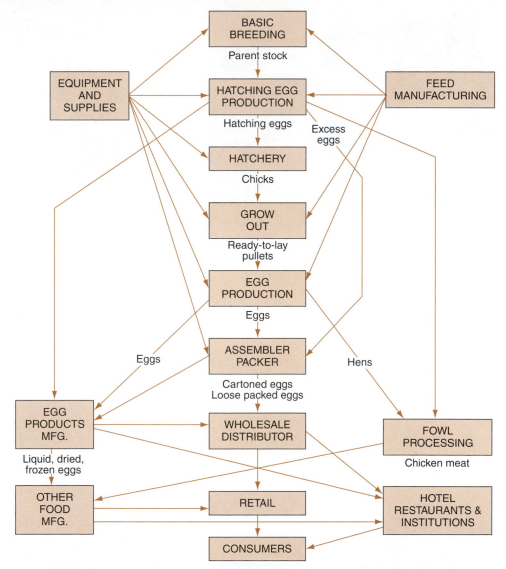

**FIGURE 18–14** The integrated egg-producing operation. (*SOURCE:* Schrader et al., 1978.)

hens and eggs increased until the early 1970s and then remained relatively constant, with some declines noted in the mid-1970s until the late 1980s. Since the late 1980s, egg production in the United States has increased (Figure 18–16). The number of eggs per layer has increased from 227 per hen in 1973 to 263 per hen in 2006, with an approximate increase of one egg per hen per year.

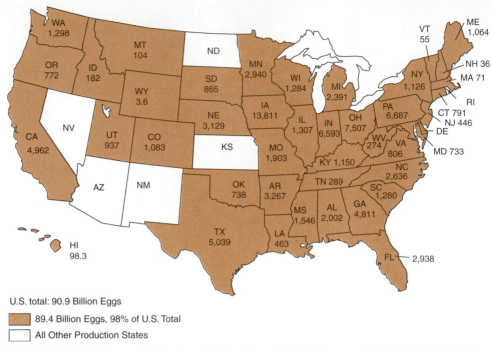

U.S. total: 90.9 Billion Eggs

▨ 89.4 Billion Eggs, 98% of U.S. Total

☐ All Other Production States

**FIGURE 18–15** Egg production by state, in millions, in 2006. (*SOURCE:* USDA-NASS, 2007c.)

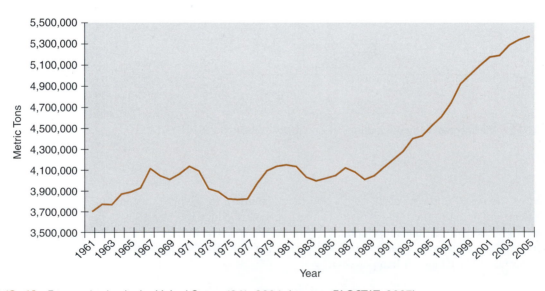

**FIGURE 18–16** Egg production in the United States, 1961–2006. (*SOURCE:* FAOSTAT, 2007.)

## THE U.S. TURKEY INDUSTRY

Most turkey production units grow from 50,000 to 75,000 birds and have approximately three and a half grow-out cycles per year (Figure 18–17). Large facilities often have a single **brooding** complex that serves two separate grow-out facilities. The brooder facility can brood seven groups each year and provides the **poults** for both growing facilities. Because turkeys are no longer produced just for seasonal consumption, turkey production is year-round. Virtually all turkeys are produced on contract to an integrator.

Turkey hens are marketed between 14 and 16 weeks of age and from 14–18 lbs. **Toms** are marketed between 17 and 20 weeks of age and from 26–32 lbs. Some variation exists based on whether the birds will be processed or sold whole and ready to cook. About 70% is further processed. Toms are preferred for further processing because of their larger weights as compared to hens. However, since only about a sixth of all turkey production is processed for the whole body market, many hens are also further processed. The remainder is processed and marketed as value-added products.

It is hard to see the same patterns in the turkey-producing industry as can be seen in the broiler industry (Figure 18–18). The reasons why particular states produce substantial amounts of turkey are varied. Virginia and South Carolina no doubt benefit from overlap from North Carolina, which is a top producer. Some states are located near consumer centers. Arkansas benefits from the corporate structure already in place for broilers and layers. Some states benefit from being located next to Arkansas. California has geographic isolation and population centers. The turkey industry has seen the same kinds of consolidation as the rest of the poultry industry, but the growth and the consolidation were generally slower. Over 7 billion pounds of turkey meat are produced in the United States each year on a live-weight basis. This requires approximately 260 million head of birds. Production of turkey underwent rapid increases during the 1980s and much of the 1990s but has had more variable production since (Figure 18–19).

**Brooding**

The act of raising young poultry under environmentally controlled conditions during the first few weeks of life.

**Poult**

Baby turkeys. Once sex can be determined, they are called young toms (males) or young hens (females).

**Tom**

A male turkey. Also called a *gobbler*.

**FIGURE 18–17** Turkeys are typically reared on productive units that grow 50,000 to 75,000 birds per year. Turkey production is the third-largest segment of the poultry industry behind broilers and layers. (Photo courtesy of Nicholas Turkey Breeding Farms.)

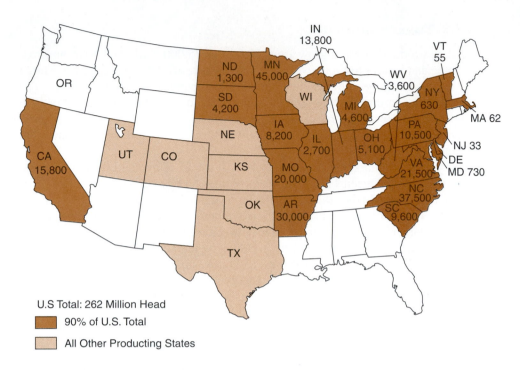

**FIGURE 18–18** Turkey-producing states, in thousands, in 2006. (*SOURCE:* USDA-NASS, 2007c.)

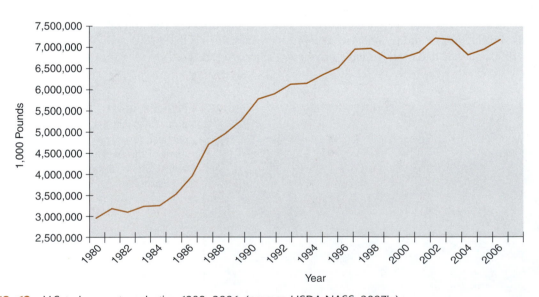

**FIGURE 18–19** U.S. turkey meat production 1980–2006. (*SOURCE:* USDA-NASS, 2007b.)

## THE U.S. DUCK, GOOSE, AND OTHER POULTRY INDUSTRIES

Table 18–1 gives statistics about several of the smaller poultry industries found in the United States. As can be seen from the table, none of these are large industries. Only the five most important states for each species are given in the table.

**TABLE 18-1** **Miscellaneous Poultry Inventory**

| DUCKS | | GEESE | | PIGEONS OR SQUAB | |
|---|---|---|---|---|---|
| Top 5 States | Inventory | Top 5 States | Inventory | Top 5 States | Inventory |
| Indiana | 1,143,160 | Texas | 17,813 | California | 168,532 |
| California | 956,606 | Indiana | 8,401 | Texas | 18,293 |
| New York | 431,867 | Wisconsin | 8,015 | Washington | 16,431 |
| Wisconsin | 417,522 | California | 7,641 | Kansas | 14,551 |
| Pennsylvania | 393,623 | Minnesota | 6,850 | Pennsylvania | 13,145 |
| U.S. Total | 3,823,629 | U.S. Total | 173,000 | U.S. Total | 449,255 |

| PHEASANTS | | QUAIL | | OTHER POULTRY | |
|---|---|---|---|---|---|
| Top 5 States | Inventory | Top 5 States | Inventory | Top 5 States | Inventory |
| Wisconsin | 370,737 | Georgia | 823,073 | Pennsylvania | 384,356 |
| Pennsylvania | 304,593 | South Carolina | 781,960 | Texas | 147,030 |
| California | 170,338 | Alabama | 575,764 | California | 168,029 |
| New Jersey | 155,168 | Texas | 480,993 | Colorado | 84,838 |
| Ohio | 121,289 | North Carolina | 451,139 | Tennessee | 70,827 |
| U.S. Total | 2,267,136 | U.S. Total | 4,888,196 | U.S. Total | 1,629,522 |

| OSTRICH | | EMU | |
|---|---|---|---|
| Top 5 States | Inventory | Top 5 States | Inventory |
| California | 3,388 | Texas | 11,616 |
| Texas | 2,782 | Alabama | 3,277 |
| Indiana | 975 | Oklahoma | 2,788 |
| Kansas | 906 | California | 2,501 |
| Wisconsin | 773 | Tennessee | 2,501 |
| U.S. Total | 20,560 | U.S. Total | 48,221 |

SOURCE: 2002 Census of Agriculture.

# GENETICS AND BREEDING PROGRAMS

The fundamentals of genetics and breeding are discussed in Chapters 8 and 9. See those chapters for more detailed information on these topics. Breeding and selection of poultry differs from breeding of the larger animals in three major ways: (1) It is more flexible because of short generation intervals and large numbers of offspring; (2) it has been the industry most subjected to modern animal breeding and selection techniques and has made the most progress; and (3) fewer people make all the decisions. Most of the traits of interest in poultry genetics are *quantitative traits*. These include traits such as egg production potential, egg size, growth rate, conformation, and so on. As a rule, these traits are more difficult to make progress in than are simpler traits. Traits with high *heritability* include body weight, feed consumption, egg weight, age at sexual maturity, egg shape, and egg color. Total egg production and feed efficiency are lower in heritability. Fertility and hatchability are very low in heritability.

*Heterosis* is very easy to demonstrate in poultry. Data from the 1950s indicate that heterosis can increase egg production by over 20 eggs per year. Other traits showing favorable heterosis include egg weight, body weight, and age at reproductive maturity (days to first egg). Many poultry-breeding systems specifically target heterosis because

## Quantitative traits

Traits inherited through the action of many genes. Most of the economically important traits are quantitative such as growth and carcass characteristics and egg production traits. Qualitative traits are those inherited through the action of a single pair of genes.

## Heritability

A measure of the amount of phenotypic variation that is related to additive gene effects.

## Heterosis

The superiority of an outbred individual relative to the average performance of the parent populations included in the cross.

of its benefits. The hybrids must be produced from pure strains, which can be continually improved to produce ever better heterosis results.

The most important traits that laying hen breeding programs target are rearing mortality, laying mortality, age at 50% production, feed per weight of eggs produced, egg weight, percentage of large and extra-large eggs, body weight, several egg quality measurements, and shell quality. Meat-producing chickens must be selected and bred for rate of growth, conformation, feed efficiency, structural soundness (fast-growing broilers often have leg problems), disease resistance, and skin and feather color. The parent stock of the broilers needs good reproductive characteristics so the eggs that will produce the broilers can be produced economically.

Poultry breeding programs make use of most of the animal breeder's tools in producing quality birds for production. *Inbreeding* is frequently used to produce strains of birds to use later by being crossed with a different *inbred line* for some specific purpose. These inbred lines are frequently only used by *basic breeders. Outcrossing* is the term used in poultry for crossbreeding. Outcrossing is also practiced through a variation called *strain crossing,* a process in which the inbred lines within the same breed might be used. When a specific strain usually contributes positively to a cross, it is said to have *general combining ability.* Other strains generally contribute only when crossed with other specific strains. These are said to have *specific combining ability.* Some strains are known to contribute to crossing more positively if always used as the male parent, and others when used as the female parent. Once good combinations are discovered, the pure lines are maintained and used to produce the crosses, which may be three- or four-way crosses involving different breeds and/or strains.

Today, poultry breeding is handled in an extremely controlled manner. The actual number of individuals who direct breeding programs is considerably fewer than 100. With the consolidation of the industry, control of the breeding programs is considered to be a major plus. It helps produce uniformity in the product. Companies invest large sums in breeding facilities, animals, labor, recordkeeping, and statistical analysis. For all practical purposes, all commercial poultry today are crosses of breeds, strains, or inbred lines. Meat chickens are marketed as broilers, roasters, or game hens.

## BREEDS, VARIETIES, AND STRAINS OF POULTRY

Generally speaking, chickens are one of two types: meat type or egg type. However, this generalization is too simple to explain poultry fully. Chickens exist in many colors, sizes, and shapes and in more than 350 combinations of these traits. Additional species have also been developed with different characteristics. To identify and classify the species, they are designated by *class, breed, variety,* and *strain.*

A class is a group of breeds originating in the same geographic area. Names are taken from the region where the breeds originated. Examples include Asiatic, Mediterranean, American, and English. A breed possesses a specific set of physical features, such as body shape or type, skin color, number of toes, and feathered or non-feathered shanks. Individuals within a breed, when mated to other individuals of the breed, pass these features to their offspring in a uniform way. Varieties are subdivision of breeds. Varieties are based on such characteristics as feather color, comb type, and presence of a beard and muffs. For example, the Plymouth Rock is a breed that has several color varieties including Barred, White, Buff, and Partridge. Their body shape and physical features are essentially identical, but each color is a separate variety. Strains are families or breeding populations that are more nearly alike than the

### Inbreeding
Mating individuals more closely related than the average of the population.

### Inbred line
An established line of chickens created by intensive inbreeding. They are usually mated to other inbred lines to produce commercial varieties.

### Basic breeder
Produces parent stock used for multiplication of poultry, either by outcrossing, inbreeding, or other methods. Also referred to as *primary breeder.*

### Outcrossing
The practice of mating unrelated breeds or strains in poultry. Used extensively in broiler production and to a lesser degree in other poultry.

### Strain cross
Mating different strains of the same breed and variety. Generally the strains have been inbred to some degree and selected for different strengths to get increased production in the offspring.

### General combining ability
A term that describes a strain that contributes positively to the genetic makeup of offspring resulting from mating it with several different strains.

### Specific combining ability
When a strain only contributes positively to a cross when mated with certain, specific other lines.

breed or variety they are a subdivision of. They may also be the products of systematic crossbreeding. However, a strain shows a closer relationship than that for others within the breed or variety. Today, the commercial poultry industry is based primarily on strains and strain crosses.

There are no effective breed registry associations for poultry that function in the same way as do the larger livestock breed associations. The American Poultry Association maintains a standard for recognized breeds and promotes exhibition of poultry. It publishes the *American Standard of Perfection*, which is a compilation of the breed standards. In this sense, the American Poultry Association functions as a breed registry. This book contains a complete description of each of the nearly 350 recognized breeds and varieties of chickens, ***bantams,*** ducks, geese, and turkeys. Size, shape, color, and physical features are described and illustrated. The terms *purebred* and *registered* are not used in poultry. Instead, poultry that meet the breed and variety descriptions found in the *American Standard of Perfection* are referred to as "standard bred." Some chicken and duck breeds have miniature versions (a fifth to a quarter the size of the large bird). Some breeds also only exist in the miniature forms. These small forms are referred to as *bantams* (the term *bantam* is an adjective). Thus there are bantam Leghorns and bantam Old English Game chickens. The organization that standardizes and promotes the bantam breeds is the American Bantam Association. It also publishes the *Bantam Standard*.

## POULTRY BREEDS

Descriptions of many of the world's poultry species (and pictures) can be viewed by visiting the poultry page of the Oklahoma State University Breeds of Livestock page at http://www.ansi.okstate.edu/poultry/. This is a comprehensive reference, which has information on chickens, turkeys, ducks, and geese. Photos of many breeds can also be seen in the color plate section of this text.

## CHICKEN BREEDS IN MODERN PRODUCTION

Relatively few of the breeds of poultry have any real place in the modern commercial poultry industry. Synthetic lines (strains) of poultry have gradually replaced the breeds in commercial poultry operations. They were developed originally from crossing pure breeds or even crossing within a breed by selecting from highly inbred lines within breeds. A brief discussion follows of breeds and varieties of chickens that are either still in use or were used to develop modern synthetic lines.

The Single Comb White Leghorn is one of several varieties of Leghorns (Figure 18–20). Many synthetic strains have been developed from this variety, and it is the most numerous breed in the United States today. Virtually all commercial white egg-producing flocks are strains of Leghorns. They are strictly egg type.

The Single Comb Rhode Island Red and Barred Plymouth Rock are used most often to produce sex-linked color differences in day-old chicks. In this way, chicks can be sexed easily and rapidly. Many commercial brown-egg layers are the result of crossing Rhode Island ***roosters*** with Barred Plymouth Rock hens. This cross is a good producer of large brown eggs.

The New Hampshire and White Plymouth Rock were, and still are, used to develop many of the synthetic lines of meat-type chickens. Most commercial meat-producing crosses have one or the other of both of these breeds somewhere on the female side.

The Cornish is an excellent meat-producing chicken, but it has poor reproductive characteristics. It is a very important part of the commercial meat-producing industry

**Class**

In poultry, a group of breeds originating in the same geographic area. Names are taken from the region where the breeds originated.

**Breed**

Birds having a common origin with specific characteristics, such as body shape, that distinguishes them from other groups within the same species and breeds that produce offspring with the same characteristics. In poultry, a breed may include several varieties different only in color or comb type.

**Variety**

A subdivision of a breed distinguished by color, pattern, comb, or some other physical characteristic.

**Strain**

Families or breeding populations within a breed. They have been more rigorously selected for some trait or set of traits than the average of the breed.

**Bantam**

Fowl that are miniatures of full-sized breeds. Some are distinct breeds, usually a fourth to a fifth the weight of standard birds. Considered ornamental.

**Rooster**

A mature male chicken. Also referred to as a *cock*.

**FIGURE 18–20**  Strains and strain crosses of the Single Comb White Leghorn are the most popular laying hen in the United States. (Photo courtesy of Dr. Joe G. Berry, Department of Animal Science, Oklahoma State University. Used with permission.)

because of its growth rates and carcass characteristics. Cornish males are crossed with females that are crosses of Barred Plymouth Rock, White Plymouth Rock, New Hampshire, or synthetic lines. It is safe to say that virtually all the commercial broilers in the United States contain some Cornish blood.

There is some concern today that the genetic base of poultry species is too narrow. The counterargument is that most breeds have no place in commercial production. The American Livestock Breeds Conservancy of Pittsburg, North Carolina, is an organization that works to preserve endangered breeds of livestock and is a source of interesting information on minor breeds of all livestock species.

## TURKEY BREEDS

The modern turkey is a descendant of the wild turkeys native to North and Central America and is North America's major species contribution to the livestock industry (Figure 18–21). The *American Standard of Perfection* lists only one breed of turkey: Turkey. There are eight varieties of turkeys: Bronze, Narragansett, White Holland, Black, Slate,

**FIGURE 18–21**  The wild turkeys of North and Central America were the wild progenitors of the domestic turkey.

Bourbon Red, Beltsville Small White, and Royal Palm. All of these varieties were developed in the United States except the White Holland. For all practical purposes, the term *breed* now refers to color and size types. The term *variety* now means different commercial brands offered for sale by poultry-breeding companies. Probably around 50 commercial turkey varieties are available from breeders and hatcheries in the United States. Large White, Medium White, Small White, and Bronze are generally offered. Because of the processing industry and the emphasis on size, the Large Whites have come to dominate. The white color varieties are the most popular because they withstand hot summer sun better than the colored varieties do. They are also easier to prepare for market because of the absence of dark pin feathers, which can leave a carcass looking dirty. The same is true for the other poultry species.

### DUCK AND GOOSE BREEDS

The Mallard is the most popular duck in the United States. The White Pekin duck, native to China, was brought to the United States in 1873. It is the major duck of commercial importance in the United States because it reaches market weight earlier than the other breeds, is a fairly good egg producer, and has white feathers. Pekins are also generally free of **broodiness.** The Rouen, a colored duck, is a popular farm flock breed. Although slower growing than the Pekin, it reaches the same weight under farm flock feeding and foraging. Its slower growth and colored **plumage** make it undesirable for commercial production. The Muscovy is also used in farm flocks because it is a good forager and the hens are good setters. It originated in Brazil and is the only one of the major breeds not originally developed from the wild Mallard. Meat production is generally the primary criterion in breed selection. Egg production for propagation and brooding tendency, and the white plumage that produces an attractive dressed carcass, are other characteristics that should be considered. There is growing hobbyist interest in bantam ducks such as White and Gray Calls, Black East Indias, Wood Ducks, Mandarins, and Teal ducks. Many poultry shows offer classes for these ducks.

The domestic goose, which was bred in ancient Egypt, China, and India, is said to have been in the United States since early colonial days. Most of the breed information and other references are obscure and difficult to authenticate. Most breeds arrived here via Europe, where they are much more popular. The White Emden (first believed brought to the United States in 1821) and Toulouse are the two most popular goose breeds. The African (said to be especially valuable in crossbreeding), Pilgrim, and White Chinese (said to be the best laying) are also raised in significant numbers.

**Broodiness**

When a hen stops laying eggs and prepares to sit on eggs to incubate them. Once the eggs hatch, the hen cares for them. Such hens are referred to as being *broody.*

**Plumage**

The total body feathering of poultry.

# REPRODUCTIVE MANAGEMENT IN POULTRY

In general, the management of poultry species for reproduction is very much alike for all poultry species. It is also similar in hobby and large commercial operations. When eggs fail to hatch properly, the reason may be the management of the breeder flock, the incubation procedures, or any step between the breeder flock and final hatch.

### BREEDER FLOCK MANAGEMENT

Proper management of the breeder flock is essential if healthy chicks are to be hatched. The breeder flock should be reared using proper management practices, and then selected to be healthy and free of **defects** that can interfere with proper mating and egg

**Defect**

Unacceptable deviation from perfection. Most defects are inherited.

production. They should be genetically superior and free of deformities and flaws that would interfere with normal eating, drinking, and maintenance of social stature in the flock. Males should be aggressive and willing to breed. Females should be selected for good egg-laying traits.

## MATING SYSTEMS

To produce hatching eggs, poultry can be mated by using one of the following mating systems:

1. Mass mating. Several males are allowed to run with a flock of females. This method is a practical means of obtaining the maximum number of hatching eggs.
2. Pen mating. One male is mated with a small flock of females. This system is used to keep track of ancestry.
3. Stud mating. One female is mated with one male. The females can be removed, and another female is then put with the male. Hens can be successively mated to different males. More females can be mated to a superior male.
4. Artificial insemination. This system is commonly used in turkey production and less so in other species. Birds of quite different sizes can be mated.

The breeder flock needs proper nesting facilities and ample, clean nesting material. This helps prevent damage such as breaking and contamination of the shells with dirt and manure, which spreads disease and reduces hatching. Excellent management including sanitation, vaccination programs, and pest control is essential. High-quality diets formulated for breeder birds should be fed.

## SELECTION AND CARE OF EGGS

Frequent collection of eggs—a minimum of once a day with more frequent collection if daily temperatures reach 85°F—is important. Commercial hatcheries frequently collect five or more times a day. Eggs laid on the floor should not be used because they spread disease. Nest eggs that are dirty should also be discarded. Hatching eggs cannot be washed because it removes the protective sealing substance from the shell, allowing bacteria to enter the egg. Oversized, undersized, and abnormally shaped eggs should be discarded because hatching is poor. Likewise, cracked eggs and eggs with thin shells should be discarded. Commercial operations fumigate eggs prior to *setting* to reduce the bacteria on the shells and increase hatchability.

**Setting**

Placing eggs to incubate. A setting hen is a broody hen incubating eggs.

## EGG STORAGE

Eggs should be placed in incubators as soon as it is convenient. Eggs held before incubation should be stored near 60°F and 75% humidity. Temperatures below 40°F reduce hatchability. The cool temperature delays embryonic growth until incubation begins, and the high humidity prevents moisture loss. Storage for less than 10 days is acceptable. After that time, hatch progressively declines to near zero for eggs stored for 3 weeks. Eggs that are not incubated within 3 or 4 days should be turned daily to prevent the yolk from touching the shell and injuring the embryo. Storage should be with the small end down and slanted at an angle of 30–45°. When eggs are to be placed in an incubator, they should be warmed slowly at first. Warming too rapidly causes moisture to condense on the shell, which may lead to mold and bacterial growth.

**TABLE 18–2   Incubation Times for Selected Poultry Species[1]**

| Item | Chicken | Turkey | Duck | Goose | Swan | Guinea | Quail |
|------|---------|--------|------|-------|------|--------|-------|
| Incubation time (days) | 21 | 28 | 28 | 28–32 | 35–40 | 26–28 | 17–24 |
| Temperature (°F) | 100 | 99 | 100 | 99 | 99 | 100 | 100 |

[1]Different incubators have different recommendations for temperature. There are also appropriate ranges for humidity and oxygen content of the air. For small incubators, follow the manufacturer's recommendations. Large commercial incubators are managed by skilled technicians and vary with the type and design.

## INCUBATORS

Eggs are incubated in devices aptly named *incubators.* They come in many shapes and sizes and with capacities from a dozen up to several thousand eggs. Table 18–2 shows *incubation* periods and incubation operation characteristics. Proper care must be given to relative humidity, temperature fluctuations, ventilation, and other factors to ensure a good hatch. Eggs must be regularly turned to prevent the embryos from sticking to the sides of the shells. Generally, eggs are not turned for the last 3 days before hatching because the embryos are moving into hatching position and should not be disturbed. After hatching, chicks are allowed to dry and "fluff up" before being moved to a brooder with feed and water.

## TESTING FOR FERTILITY

Eggs in incubators can be tested for fertility by *candling* them at 4–7 days of incubation. Infertile eggs can be discarded and the space used for additional eggs. Candling does not harm the young embryos. Candlers can be purchased or made by placing a light bulb and fixture inside a cardboard box. Cutting a 1/2- to 3/4-in. hole in the top or side of the box allows a narrow beam of light to escape. The internal features of the egg can be seen by placing it against the hole. A darkened room makes testing easier. Eggs with white shells are easier to candle and can be tested earlier than dark-shelled eggs. *Infertile* refers to an unfertilized egg or an egg that started developing but died before growth could be detected. *Dead germs* are embryos that died after growing large enough to be seen when candled. Both can be detected at this stage. An infertile egg looks clear except for a shadow from the yolk. In a live embryo, large blood vessels can be seen spreading out from the embryo. A dead germ can be determined by the presence of a blood ring around the embryo, which is caused by blood moving away from the dead embryo. A second test can be made after 14–16 days of incubation. If the embryo is living, only one or two small light spaces filled with blood vessels can be seen. The chick may even be observed moving inside the shell.

## NUTRITION IN POULTRY

Chapters 5, 6, and 7 deal with the intricacies of nutrition and feedstuffs in a much more detailed manner than this section. See those chapters for information on digestive anatomy, nutrition, feeds, and feeding. Poultry feeding has changed more than the feeding of any other species with the advent of modern production systems. The primary

**Incubator**

A machine that provides the environmental conditions to encourage embryonic development in fertile eggs.

**Incubation**

The process of sitting on eggs by a hen to warm them with body heat so that the eggs develop into young. Can also be done artificially in an incubator.

**Candling**

Inspection of the inside of an intact egg with a light to detect defects. Also, incubating eggs can be tested for dead germs and infertile eggs.

**Infertile**

An egg that is unfertilized or in which the embryo dies.

**Dead germs**

Embryos that have died.

reason for this is that most of the poultry is produced in large units where maximum technology is used. Poultry nutrition is also more critical, complicated, and thus a greater challenge to the producer because poultry have more rapid digestion, higher metabolic rates, faster respiration and circulation, and higher body temperature (107°F). Poultry are more active, more sensitive to environmental influences, grow more rapidly, and mature at earlier ages. Egg production is an all-or-none phenomenon. However, we probably know more about poultry nutrition than the nutrition of any other species. Economic production depends on economical rations.

Poultry species are monogastric and have a specialized *avian* tract. As *omnivores,* they seek a variety of plant and animal foods if left to scavenge. They have paired ceca, which will develop and digest some fiber if they are fed forages. The cecum is especially developed in geese. However, for the most part, modern production systems do not make use of this capacity. Poultry and swine generally compete for the same feedstuffs: concentrated feeds such as grains, soybean meal, and high-quality by-product feeds.

Feed is the largest cost in the production of the poultry species. Optimum growth and maximum growth are usually close to the same thing for growing animals. Thus the nutrition and feeding of most poultry species in most of the production stages revolves around optimizing growth. This is the best feeding strategy because the faster an animal reaches market, the smaller the percentage of total lifetime feed used for body maintenance. In addition, the faster an animal reaches market, the fewer days it requires housing, labor, and so on. Thus the fixed costs of production can be reduced on a per animal basis. Egg-producing animals are usually high enough producers that they find it difficult to overeat. For laying animals, the nutrition of the animal is quickly reflected in the size, quality, and number of eggs.

Feeding practices in poultry operations across the country tend to be fairly uniform. The industry is so integrated that very consistent technology and methods of production are employed. Feedstuffs vary somewhat, and thus rations are different. Commercial poultry rations are formulated with the use of least-cost ration formulation programs that factor in the cost of nutrients from the available feeds and calculate the lowest cost, balanced diet. Poultry species are fed almost exclusively on complete mixed diets that are offered in *mash* or pelleted form. A high percentage of poultry rations is pelleted because there is less waste. Table 18–3 gives examples of rations for different classes of poultry based on corn and soybean meal. These feeds are the most commonly used energy and protein feeds in poultry rations. However, regional differences exist. Lower-cost diets can sometimes be formulated using regionally available feedstuffs. Different rations are mixed for optimum performance of the birds at their particular stage of production. All species need good-quality clean water provided free choice and adequate watering space.

## FLOCK HEALTH MANAGEMENT

Disease control is absolutely essential to the poultry industry. Many diseases affect chickens, turkeys, and other types of poultry. Because of the concentrated nature of poultry production, a single virulent disease can cause millions of dead animals and millions in economic losses. Luckily, some of the more devastating diseases that affect poultry have been eradicated from the United States. Still others have been controlled with the use of vaccines (Figure 18–22). Nevertheless, poultry farmers must stay constantly on their guard against disease. Biosecurity measures that are routinely practiced are designed to (1) minimize the risk of disease transmission from sources outside the

**Avian**
Pertaining to poultry and/or fowl.

**Omnivore**
An animal that eats both plant and animal matter.

**Mash**
Finely ground and uniformly mixed feeds. Animals cannot separate feed ingredients; thus each bite provides all the nutrients in the diet.

## TABLE 18–3  Example Rations for Different Classes of Poultry

| Ingredient | Broiler Starter | Turkey Starter | Chicken Layer | Quail Grower | Duck Finisher |
|---|---|---|---|---|---|
|  | PERCENTAGE OF COMPLETE RATION | | | | |
| Yellow corn, #2 dent | 56.46 | 47.75 | 60.50 | 71.85 | 77.25 |
| Soybean meal (48% protein) | 27.33 | 38.83 | 21.50 | 18.7 | 16.13 |
| Meat and bonemeal (50% protein) | 7.0 | — | 5.09 | 5.00 | 5.0 |
| Meat meal (56% protein) | — | 9.50 | — | — | — |
| Bakery by-product | 6.00 | — | — | — | — |
| Alfalfa meal | — | — | — | 2.00 | — |
| Animal-vegetable fat | 1.82 | 0.31 | 3.00 | — | — |
| D,L-Methionine | 0.17 | 0.24 | 0.11 | 0.15 | 0.16 |
| L-Lysine HCl | — | 0.23 | — | 0.41 | — |
| Dicalcium phosphate | 0.13 | 1.54 | 0.49 | 0.70 | 0.15 |
| Ground limestone | 0.49 | 0.81 | 8.66 | 0.74 | 0.86 |
| Iodized salt | 0.10 | 0.09 | 0.20 | 0.25 | 0.25 |
| Sodium bicarbonate | 0.20 | 0.20 | 0.20 | — | — |
| Vitamin-mineral premix and feed additives | 0.3 | 0.50 | 0.25 | 0.20 | 0.20 |
| **Feed Analysis** | | | | | |
| % Protein | 22.50 | 28.00 | 18.00 | 18.0 | 17.0 |
| Metabolizable energy (kcal/lb) | 1,425 | 1,280 | 1,320 | 1,380 | 1,426 |
| % Calcium | 0.95 | 1.45 | 3.80 | 0.95 | 0.35 |
| % Available phosphorus | 0.48 | 0.83 | 0.45 | 0.45 | 0.35 |
| % Lysine | 1.21 | 1.80 | 0.94 | 1.2 | 0.80 |
| % Methionine + cystine | 0.92 | 1.10 | 0.71 | 0.70 | 0.70 |

SOURCE: Recommendations from multiple sources.

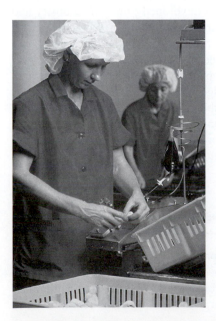

**FIGURE 18–22**  Technicians vaccinate chicks subcutaneously to prevent Newcastle disease. Vaccination is an important part of flock health management.
(Photo courtesy of Aviagen, Inc.)

production unit, and (2) reduce the transmission of diseases between groups of birds on the same farm. Here are some important biosecurity measures:

1. Poultry houses should be kept locked; fasten from the inside while inside the facility.

2. Street clothing should be kept separate from work clothing. A changing room should be provided so clothes can be changed prior to entering the facility. After caring for the flock, change clothes and wash hands and arms before leaving the premises.

3. Managers and other workers should not visit other flocks.

4. Visitors should be controlled. Nonessential visitors should not be allowed in poultry houses.

5. Essential visitors such as owners, meter readers, service personnel, delivery truck drivers, and poultry catchers and haulers must wear protective outer clothing, including boots and headgear, before being allowed near flocks.

6. Before entering the premises, vehicles should be scrubbed down and their undercarriage and tires should be spray disinfected.

7. All coops, crates, containers, and equipment should be cleaned and disinfected before and after use.

8. Laboratory diagnostics should be performed on sick or dying birds. Dead birds should be disposed of properly by burial or incineration.

9. People handling wild game must bathe and change clothes before being admitted to the premises.

10. "Restricted" signs should be posted at all entrances.

### THE NATIONAL POULTRY IMPROVEMENT PLAN

The purpose of the National Poultry Improvement Plan (NPIP) is to provide a cooperative industry–state–federal program for the application of technology to the improvement of poultry and poultry products. The plan was developed jointly by industry and government to establish standards for the evaluation of poultry-breeding stock and hatchery products with respect to freedom from hatchery-disseminated diseases. Products conforming to specific standards are identified by authorized terms that are uniformly applicable in all parts of the country. The plan is changed periodically to stay useful to the industry and incorporate new knowledge and techniques. Acceptance of the plan is optional with the states and individual members of the industry within the states. The plan is administered in each state by an official state agency cooperating with the USDA. The NPIP has active control programs for *Salmonella pullorum*, *Salmonella gallinarum*, *Salmonella enteritidis*, *Mycoplasma gallisepticum*, *Mycoplasma synoviae*, and *Mycoplasma meleagridis*.

## NUTRITIONAL BENEFITS OF POULTRY TO HUMANS

Table 18–4 presents the proportions of the recommended daily dietary allowance provided by eggs and chicken for a 25- to 50-year-old woman on a 2,200-calorie diet. Eggs have a very high nutritive value. They are considered to have a very high nutrient density, which simply means that we get a large amount of essential nutrients in

**TABLE 18–4** **Percentage of Daily Nutrients Provided by Poultry Products[1]**

| Nutrient | 3 oz Cooked Skinless Chicken (%) | 2 Eggs (%) |
|---|---|---|
| Protein | 49 | 28 |
| Phosphorus | 21 | 21.5 |
| Iron | 7 | 10 |
| Zinc | 15 | 9 |
| Riboflavin | 12 | 39 |
| Vitamin A | 6 | 79 |
| Thiamin | 5 | 6 |
| B$_{12}$ | 14 | 50 |
| Niacin | 52 | 0.5 |
| Folate | 6 | 26 |
| Calories | 7 | 7 |
| Fat | 9 | 14 |

[1]Calculated on the recommended daily dietary allowance for a 25- to 50- year-old woman on a 2,200-calorie diet.
*SOURCE:* USDA, 1997.

relation to the calories consumed. Eggs contain all essential amino acids needed by humans, many needed minerals, and all required vitamins except vitamin C. A large egg contains only 75 calories. The protein in meat products is also of high quality, contains all of the essential amino acids, and is easily digested. The fat content of *uncooked product* is lower than that of many other meat products. Poultry meat is also a good source of some vitamins and minerals. Poultry products are healthful foods that can be the part of the diet of virtually all people. A complete nutrient analysis of chicken, duck, goose, and eggs is shown in Appendix III.

# TRENDS IN THE POULTRY INDUSTRY AND FACTORS THAT WILL INFLUENCE THE INDUSTRY IN THE FUTURE

## TURKEY CONSUMPTION AND PRODUCTION

Turkey production was previously shown in Figure 18–19. Continued increase in production will probably be related to overall population increases and exports. Turkey has been promoted successfully as a nonholiday food in addition to its use as a traditional holiday food. Much of this success has been accomplished by further processing of the turkey into smaller portions and manufactured products. Per capita consumption appears to be stable at approximately 17 lbs.

## BROILER CONSUMPTION AND PRODUCTION

Broiler production continues to expand because of large consumer demand (Figure 18–10). Per capita consumption of broiler meat steadily increased until 2006 when it stalled for the first time since 1975 (Figure 18–23).

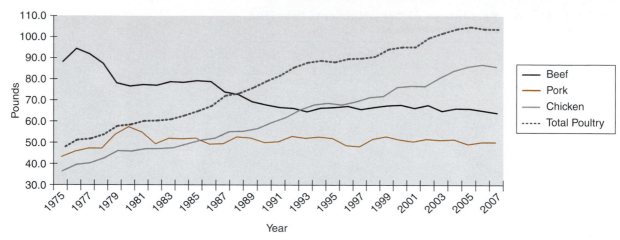

**FIGURE 18–23** Poultry meat per capita consumption, 1975–2007. (*SOURCE:* Based on USDA-ERS, 2007b.)

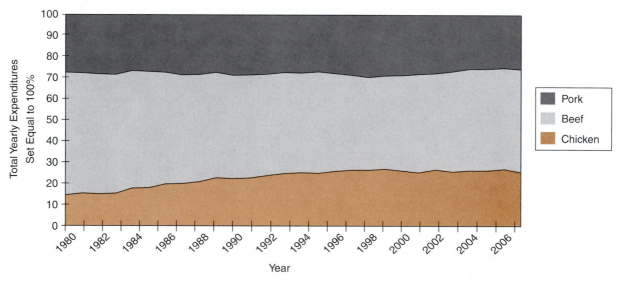

**FIGURE 18–24** Yearly chicken, beef, and pork expenditures as a percentage of total, 1980–2007. (*SOURCE:* Based on USDA-ERS, 2007c.)

## Cost Advantages

Broiler meat has advantages in production costs over competing meats. The industry has produced increased amounts of chicken for sale and, at the same time, lowered the price to consumers. Most market analysts suggest that this is a major reason why chicken has taken market share from beef. However, more is at work than simple price. If price was the only issue, then consumers would be expected to buy more total pounds because they could purchase chicken for less cost per pound. This would not explain how the product has managed to also gain an increasing percentage of the total yearly consumer expenditures for chicken, beef, and pork. Chicken has clearly done this, largely at the expense of beef (Figure 18–24).

## Convenient/Value-Added Products

Changes in social values, lifestyles, and demographic trends have caused an explosion in the convenience food market. Consumers with more disposable income but less time to prepare food are willing to pay for foods that they perceive are high in quality and in eating satisfaction, are healthy, and also convenient. The poultry industry has outdone the beef and pork industries in developing value-added products and has maintained an aggressive posture in this regard. They consider themselves in the food business, not the meat-production business. A look at the advertising material of integrated poultry companies finds all the bases covered with phrases and words such as "quality home meal replacements," "affordable yet simple-to-prepare entrees," "variety and appeal for kids," "portable foods for grab-and-go convenience," and always the word *new*: "New flavors! New textures! New forms! New products!" These people are excited about chicken. Perhaps the best phrase of all to illustrate the focus of these companies is "We convert chicken into customer satisfaction and loyalty." The poultry industry will continue to focus on price, packaging, *shelf life,* convenience, and taste. As long as they do these things, the industry will continue to grow.

**Shelf life**
The length of time fresh-dressed, iced, packed poultry can be held without freezing.

## Fast Food

Chicken is also a very important fast food. Harland Sanders probably had no inkling of the influence he would have on the poultry industry when he opened his first restaurant. However, before he died, the colonel witnessed multiple generations of Americans who had grown up deciding on " extra-crispy" or "original recipe." Kentucky Fried Chicken was the catalyst that revolutionized the fast-food industry where chicken is concerned. Fast-food chicken is available in virtually every fast-food restaurant and even in gas stations.

## Nutrition and Health Consciousness

Nutrition and health consciousness will continue to play a role in food choices. The early 1980s has been dubbed the beginning of the health-conscious era, although eating habits had been changing before that time. Poultry products are good food. They are nutritious, are usually leaner than alternative meats, and have gained the status of a health-conscious food. Poultry products found a place on the plate of the fat-conscious consumer, which has been cultivated jealously by competitors. A look at poultry consumption trends clearly shows a product that increased its share of per capita consumption during the same period when consumers were decreasing their consumption of red meat, whole milk, lamb, coffee, alcoholic beverages, and canned fruits and vegetables. Both chicken and turkey increased consumption at percentages that rival foods with a "health" connotation such as yogurt, fresh fruits and vegetables, rice, and low-calorie sweeteners.

### EGG CONSUMPTION

The high point for per capita egg consumption in the United States was in 1945 at 403 eggs per capita but began declining. Egg consumption declined somewhat because of health consciousness. Whether right or wrong, the cholesterol scare has hurt the egg-producing part of the poultry industry. However, changes in eating habits have probably been a more important factor. People either skip breakfast or choose a noncooked breakfast from a glass or a package because of busy lifestyles. Foods that can be held in the hand and eaten on the go are popular. A sunny-side-up egg isn't very good hand food. Another problem is that eggs are not traditionally eaten at meals other than breakfast. Also, there

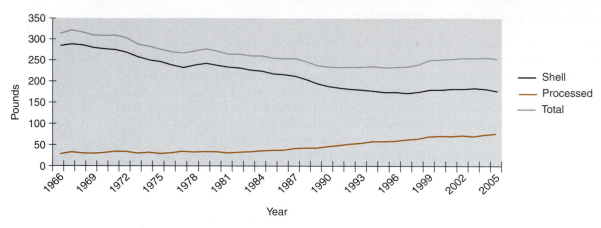

**FIGURE 18–25**   Per capita egg consumption in the United States. (*SOURCE:* USDA-ERS, 2007b.)

aren't many further-processed, convenient products that feature eggs. The low mark in consumption was in 1991 with 233 eggs. However, egg consumption stabilized in the 1990s and has actually increased in the new century. People on low-carbohydrate diets find the egg a very versatile and easy food to include in their diets. In addition, the egg is an important ingredient in many processed foods that are popular with consumers. The combined use of eggs in processed foods and a changing of attitudes about the nutritional value of the egg in the diet seem to be driving a consumption comeback for the egg. Combined per capita consumption for the egg in the United States is now around 255 eggs (Figure 18–25).

A changing ethnic mix in the U.S. population is also a driving force in the per capita consumption of eggs. The Mexican American population in the United States currently uses nearly a third of the shell eggs in the country, even though it represents only 17% of the population. As that population increases, it influences per capita egg consumption.

## Cholesterol

The health concerns about cholesterol have been a mainstream consumer issue for over 25 years. Early recommendations of some health-care professionals to reduce dietary cholesterol were based on the presumption that the level of circulating blood plasma cholesterol was directly tied to the quantity of cholesterol in the diet. Many scientists cautioned over and over that the link had not been established. Regardless of those cautions, eggs gained the status of a problem food because of their cholesterol content. A growing body of evidence has accumulated showing that dietary levels of cholesterol have very little to do with the plasma levels of cholesterol. Other factors such as total fat content of the diet, exercise, heredity, and several others are much more important. Based on research studies carried out for the past 40 years, it appears that the average plasma response of a normal, healthy person to a 100 mg/day change in dietary cholesterol intake is a 2.5 mg/dL change in plasma cholesterol levels. Approximately 15–20% of the population is somewhat more sensitive to dietary cholesterol, probably because of genetic factors. What all of this means is that reducing dietary cholesterol intake from 400 mg/day to 300 mg/day results in a plasma cholesterol reduction of 3.2 mg/dL in cholesterol-sensitive individuals, and as little as 1.6 mg/dL in cholesterol-insensitive individuals. Clearly other factors are more important. New studies have shown that egg eating has no effect on the small dense LDL-3 through LDL-7 particles that cause the greatest threat for cardiovascular disease. This is good news for egg producers and egg eaters!

## ALL-NATURAL AND ORGANIC PRODUCTION

The increasing interest by consumers in so-called natural products has prompted companies to begin producing and marketing chicken products such as Tyson's *100% All Natural, Raised Without Antibiotics*. In their press release Tyson stated, "The initiative is another example of Tyson's strategy of offering meaningful benefits to consumers that are focused on health, wellness and convenience" (http://www.tyson.com/Corporate/PressRoom/ViewArticle.aspx?id=27440). Tyson is the world's largest producer and marketer of chicken. (During the summer of 2008, Tyson withdrew the use of the label amid "uncertainty and controversy over product labeling regulations.") Coleman Natural Foods has also gained a great deal of industry and consumer attention with its products that are "antibotic-free, vegetarian-fed, and humanely handled." Gold Kist, Perdue Farms, and Foster Farms are other major poultry producers that have stopped using antibiotics for growth promotion. With major users of product such as the McDonald's Corporation refusing to buy chicken produced with the use of subtherapeutic antibiotics, other producers will follow suit.

Organic products are no longer simply a lifestyle choice for a small share of consumers. They are now being consumed at least occasionally by a majority of Americans. Reasons given for this include concerns about growth hormones and antibiotic use in conventional livestock, the environment, and humane care of livestock. Organic poultry and egg markets in the United States are expanding rapidly with products readily available in traditional and natural food supermarkets in addition to the more traditional venues for organic products such as farmers' markets and health food stores. Price comparisons between organic and conventional products show significant organic price premiums for both broilers and eggs. Organic production has increased so rapidly that it prompted the USDA-ERS to begin reporting monthly organic and conventional prices (http://www.ers.usda.gov/data/organicprices/).

Both the organic market and the organic production industry in the poultry sector are immature and developing. Consumer demand will dictate that development and in turn influence production, processing, and marketing of organic poultry meat and eggs. In the short term, the future of the organic poultry sector is one of growth as it struggles to meet very fast growing consumer demand.

Other interesting trends in egg production are developing in an effort to meet consumer demand such as vegetarian, cage-free, free-range, fertile, in-shell pasteurized, and nutrient-enhanced specialty eggs. Even blue-green eggs produced by the Araucana breed native to South America are available. All are produced with specific consumers in mind. Production costs are inevitably higher on specialty eggs, making them more expensive than generic shell eggs. Which, if any, will become significant parts of the industry remains to be seen.

## FOOD SAFETY CONCERNS

The Food Safety and Inspection Service (FSIS) of the Department of Agriculture monitors poultry-processing plants. The Food and Drug Administration regulates feeds, feed additives, and animal health products. The Environmental Protection Agency (EPA) regulates pesticide use. All agencies have enviable records of ensuring a safe food supply. However, several initiatives are under way to improve services. The safety of the food supply is an important issue. This industry must do what the other food industries are doing—it must provide the safest food supply that can be reasonably provided.

**FIGURE 18–26** Eggs are washed and graded for consumer quality and protection. (Photo courtesy of Dr. Joe G. Berry, Department of Animal Science, Oklahoma State University. Used with permission.)

## Eggs

Eggs are washed and graded for consumer quality and protection. Washing removes dirt and bacteria from the eggshells. Grading removes any cracks, ***blood spots,*** and ***dirties*** from the consumer market. Eggs are refrigerated on the farm and in transport vehicles. The USDA Egg Products Inspection Act does not define "fresh" eggs. However, most eggs are in the market within 1 to 7 days from the time of production. This speed helps provide a fresh, safe food product (Figure 18–26).

**Eggs and *salmonella.*** The inside of an egg was considered almost sterile until eggs contaminated with *Salmonella enteritidis* were found. It was discovered that *Salmonella enteritidis* could silently infect the ovaries of healthy-appearing hens and contaminate the eggs before the shells are formed. The number of affected eggs is small. Flocks associated with outbreaks of salmonellosis have been tested to have only one infected egg per 10,000. This translates to one egg per 20,000 eggs on a national basis that might be infected. The likelihood of finding an infected egg is about 0.005%. Even then, the numbers of microorganisms in a properly handled and refrigerated egg are so small that they cannot easily cause illness in a healthy person. However, when eggs are not kept properly refrigerated or cooked properly or perhaps eaten raw, problems can easily occur. This is especially true if an infected raw egg is used in a food that is not cooked and to further compound the problem is not then properly refrigerated. Whereas healthy adults and children are at risk for egg-associated salmonellosis, the elderly, infants, and persons with impaired immune systems are at increased risk for serious illness because a relatively small number of bacteria can cause severe illness. Most of the deaths caused by *Salmonella enteritidis* have occurred among the elderly in nursing homes. Egg-containing foods prepared for high-risk persons should be thoroughly cooked and promptly served.

## Poultry Meat

Poultry meat has not been unduly tainted by the foodborne illness scares associated with ground beef. Much of this has to do with the fact that poultry meat is rarely sold

---

**Blood spots**

Small spots of blood found inside an egg, probably caused by blood vessels breaking in the ovary or oviduct during egg formation.

**Dirties**

Eggs with dirt, fecal material, or other material on the shell.

as a ground product. Also, much of it is further processed and subject to strict quality control. The Hazard Analysis and Critical Control Points (HACCP) system helps all meat industries provide safer products.

## Integration and Consolidation

All phases of poultry production are and will continue to become more specialized, larger, concentrated on fewer farms, and more vertically integrated. More consolidation is inevitable, especially of the smaller companies. More integrators may opt to own the production units rather than to contract. Confinement rearing is here to stay (Figure 18–27).

## Technological Innovation and Standardization

The poultry industry has made the most dramatic advancements in both biology and technical aspects of any of the livestock industries. Since 1925, the time for broilers to reach market has gone from 15 to 6 weeks or less. The amount of feed required has been cut in half. It now takes less than 2 lbs of feed to produce 1 lb of meat. The number of eggs per hen has more than doubled. Much of this progress is related to the industry's willingness to take better advantage of modernization and technological innovation. It will continue to do so. Biotechnology will allow us to increase production and efficiency of all poultry segments. Mechanization will become a larger part of labor-saving (cost-saving) strategies.

## Waste Disposal

Poultry farms will face tougher and tougher laws regarding litter and dead bird disposal. A new set of rules pertaining to CAFOs were signed into effect, with the U.S. EPA as administrator, on December 15, 2002. The rule as well as explanatory material can be accessed at http://cfpub.epa.gov/npdes/afo/cafofinalrule.cfm or by writing the EPA. The EPA will no doubt be given more laws to administer that determine the way poultry waste disposal is managed. Current environmental laws, regulations, policies, and guidance regarding CAFOs can be found at http://www.epa.gov/agriculture/anafolaw.html.

One very exciting area of research has been in the area of nutritional effects on the content of poultry litter. Enzyme-modified diets have helped chickens digest more of the phosphorous in their feed, which allows for less phosphorous to be added to poultry diets. As a result of this and other nutrient management strategies, it is possible to reduce the phosphorous content of poultry litter by approximately a fourth. Other discoveries will continue to bring improvements and lessen the environmental footprint of the poultry industry.

## BIOTECHNOLOGY

The products of biotechnology that will most help poultry producers are in the areas of disease prevention and treatment. In addition, such tools as marker-assisted selection will help further the genetic progress. Transgenic strains of poultry species to produce specific substances in the egg will be developed. This will make it possible to have poultry species contribute to human welfare in new and exciting ways (Figure 18–28).

Genetic engineering work done by plant scientists also affects the animal industries. As an example, scientists at DuPont have discovered a gene in grain crops that can be silenced, resulting in greater availability of phosphorous from grain. This will allow reduced phosphorous supplementation and then reduced amounts of phosphorous in animal wastes. Corn genetically modified this way could be on the market within a decade.

## ANIMAL WELFARE

The layer industry has long been under attack from animal rights proponents who object to the practice of caging layer hens rather than allowing them to range freely. In 2002, as an answer to those concerns, the United Egg Producers (UEP) put in place welfare guidelines. Eggs produced from caged hens under the guidelines are referred to as *United Egg Producers Certified.* The program is billed as "the way U.S. egg farmers assure retailers, foodservice professionals, and consumers that their eggs originate from farms that follow responsible, science-based modern production methods in the care of their egg-laying

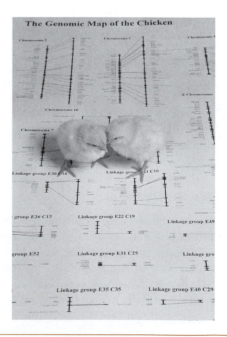

**FIGURE 18–28**    Chicks atop a picture of the genetic map of the chicken. The chicken genome has 39 pairs of chromosomes, whereas the human genome contains 23 pairs. Once the molecular geneticists have completed unraveling the "mystery genes" in the code, many production enhancements and transgenic applications will be found. (Photograph by Peggy Greb. Courtesy of Agricultural Research Service.)

flocks." Information on the program and its guidelines can be found at http://www .uepcertified.com/. (Similar guidelines have been developed for meat production by the National Chicken Council, http://www.nationalchickencouncil.com/aboutIndustry/ detail.cfm?id=19). Several restaurant chains put their own guidelines in place for producers from whom they buy eggs. However, animal rights groups have mounted a series of campaigns against caged layer production. U.S. city councils have passed resolutions urging their citizens not to buy eggs produced by caged hens. Major restaurant chains, food retailers, and food processors have announced policies on purchasing eggs from cage-free suppliers. According to the UEP, only about 5% of U.S. eggs are produced by cage-free suppliers. However, over a third are produced cage free in Europe, and the volume of cage-free eggs produced in the United States seems certain to increase. In recognition of this trend, the UEP adopted welfare guidelines for cage-free production systems in late 2007.

## AVIAN INFLUENZA (BIRD FLU)

Avian influenza H5N1 has been causing a great deal of concern in the United States and around the world. Millions of birds have died and so have well over 100 people. Trade disruptions have cost countries billions of dollars.

Two types of avian influenza (AI) are identified as H5N1. One is low pathogenic (LPAI), and the other is highly pathogenic (HPAI). The subtype HPAI H5N1, often referred to as "Asian" H5N1, is the type currently causing worldwide concern. LPAI H5N1, "North American" H5N1, is of less concern.

HPAI spreads rapidly and is often fatal to chickens and turkeys. In countries where HPAI H5N1 has struck, millions of birds have died. HPAI H5N1 has not been detected in the United States. However, other strains of HPAI have been detected and eradicated three times in the United States: in 1924, 1983, and 2004. No significant human illness resulted from these outbreaks.

HPAI H5N1 has also infected people, most of whom have had direct contact with infected birds. Although many people who have contracted HPAI H5N1 have died, many have no more than the symptoms of the common cold. Concerns about HPAI H5N1 revolve around the fear that the virus may mutate into a form that can pass easily from human to human. Such a mutation could create a virus capable of causing a worldwide pandemic.

Visit the USDA website Avian Influenza (Bird Flu) at http://www.usda.gov/ wps/portal/usdahome?navtype=SU&navid=AVIAN_INFLUENZA for a wealth of information on this disease threat.

# INDUSTRY ORGANIZATIONS

**American Association of Avian Pathologists**
953 College Station Road
Athens, GA 30602-4875
Phone: (706)542-5645
E-mail: AAAP@uga.edu
http://www.aaap.info/

**American Bantam Association**
Karen Unrath, Secretary
P.O. Box 127
Augusta, NJ 07822
Phone: (973)383-8633
http://www.bantamclub.com/

**American Egg Board**
1460 Renaissance Drive
Park Ridge, IL 60068
Phone: (847)296-7043
Fax: (847)296-7007
E-mail: aeb@aeb.org
http://www.aeb.org/

**American Livestock Breeds Conservancy**
P.O. Box 477
Pittsboro, NC 27312
Phone: (919)542-5704
Fax: (919)545-0022
E-mail: albc@albc-usa.org
http://www.albc-usa.org

**American Ostrich Association**
P.O. Box 163
Ranger, TX 76470
Phone: (254)647-1645
Fax: (254)647-1645
E-mail: aoa@ostriches.org
http://www.ostriches.org

**American Poultry Association**
Dave Anderson, President
1947 Grand Ave.
Filmore, CA 93015
Phone: (805)524-4046
E-mail: danderson@Keygroupinc.com
http://www.amerpoultryassn.com/

**American Poultry Historical Society**
Animal Sciences Building
1675 Observatory Drive, Room 260
Madison, WI 53706
Phone: (608)262-9764

**International Waterfowl Breeder's Association (IWBA)**
Box 212
Mayfield, NY 12117
Phone: (518)661-6645
http://www.home.alltel.net/md44721/

**World's Poultry Science Association**
World Secretary Dr. Ir P. C. M. Simons
World's Poultry Science Association, Centre for Applied Research
P.O. Box 31
7360 AA Beekbergen, The Netherlands
Fax: +31 55 506 4858

Guernsey

Shorthorn

Ayrshire

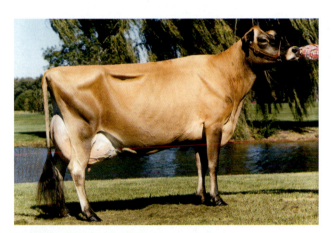

Jersey

Holstein

Brown Swiss

**Breeds of dairy cattle.**
Photos courtesy of *Hoard's Dairyman.* Used with permission.

Angus

Maine Anjou (Black 3/4 Blood)

Charolais

Polled Hereford

Limousin

Shorthorn

**Breeds of beef cattle.**
Photos courtesy of Christy Collins. Used with permission.

Beefmaster "Smooth Connection"[1]

Texas Longhorn[2]

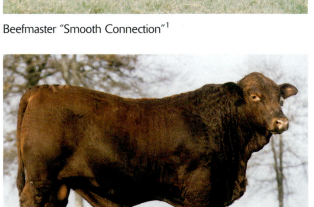

Red Brangus[2]

Brahman[2]

Braford[2]

Red Angus[3]

**Breeds of beef cattle.**

Photos courtesy of [1]Emmons Ranch–Fairfield, Texas; [2]Christy Collins; and [3]Red Angus Association of America. Used with permission.

White Holland Turkeys

Emden and Toulouse Goose

Muscovy Ducks

African Geese

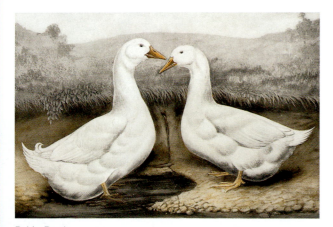
Pekin Duck

Jungle Fowl

**Poultry breeds and species.**
Courtesy of Watt Publishing Company. Used with permission.

White Leghorns

White Cornish

Mille Fleur Booted Bantams

Old English Games

Anconas

Rhode Island Reds

**Poultry breeds and species.**
Courtesy of Watt Publishing Company. Used with permission.

Yorkshire[1]

Landrace[1]

Duroc[1]

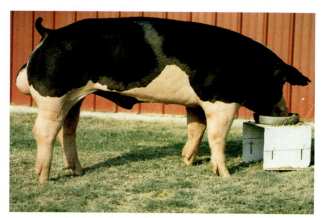

Spotted[2]

Hampshire[1]

Chester White[2]

**Breeds of swine.**

Photos courtesy of the [1]National Swine Registry; and the [2]Chester, Poland, and Spot Associations. Used with permission.

Alpine

Oberhasli

Lamancha

Saanen

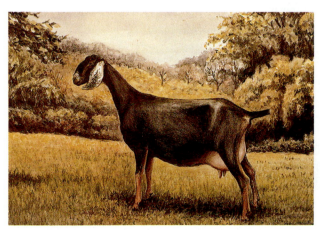

Nubian

Toggenburg

**Breeds of dairy goats.**
Courtesy of the American Dairy Goat Association. Used with permission.

Border Leicester

Cheviot

Corned

Debouillet

Dorset

Finnsheep

Lincoln

Montadale

**Breeds of sheep.**
Photos courtesy of U.S. Department of Agriculture.

Oxford

Polypay

Rambouillet

Romney

Shropshire

Southdown

Suffolk

Targhee

**Breeds of sheep.**
Photos courtesy of U.S. Department of Agriculture.

Quarter Horse[1]

Paint[4]

Appaloosa[2]

Palomino[5]

Arabian[3]

Tennessee Walking Horse[6]

**Breeds of horses.**
Photos courtesy of [1]American Quarter Horse Association; [2]Appaloosa Horse Club; [3]Judith Wagner, courtesy of Arabian Horse Trust; [4]American Paint Horse Association; [5]Palomino Horse Breeders of America, Inc.; [6]Stuart Vesty, courtesy of Tennessee Walking Horse Breeders' & Exhibitors' Association. Used with permission.

Cleveland Bay[1]

Standardbred[2]

Pony of the Americas[3]

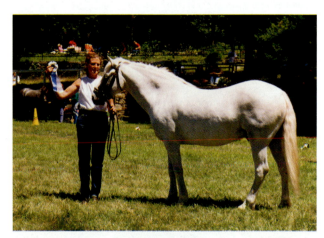

Connemara[4]

Standard Donkey[5]

Mule[5]

**Horses, ponies, donkey, and mule.**
Photos courtesy of [1]David Field; [2]Ed Keys, United States Trotting Association; [3]Pony of the Americas Club; [4]American Connemara Pony Society; [5]American Donkey and Mule Society. Used with permission.

Haflinger Horse

American Saddlebred[3]

Morgan Horse[1]

Percheron[4]

Classic American Shetland[2]

Modern American Shetland[2]

Miniature Horse mare[2]

**Breeds of horses.**
Photos courtesy of [1]American Morgan Horse Association; [2]American Shetland Pony Club, Inc. and Jamie Donaldson, courtesy of [3]American Saddlebred Horse Association (American Saddlebred); [4]Percheron Horse Association of America. Used with permission.

Scottish Fold Shorthair

Siamese

Cornish Rex

Bengal

Persian

Sphynx

Maine Coon

Manx

**Breeds of cats.**
Photos copyright of Chanan Photography. Used with permission.

Cirnecco Del Etna

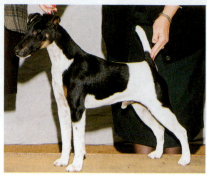

Fox Terrier

Doberman Pinscher

Bearded Collie

Boston Terrier

Irish Setter

Bloodhound

Italian Greyhound

**Breeds of dogs.**
Photos copyright of Melia Photography. Used with permission.

Australian Shepherd

Cavalier King Charles Spaniel

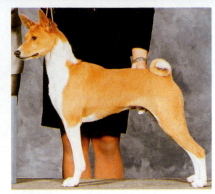

Basenji

Dalmatian

Shetland Sheepdog

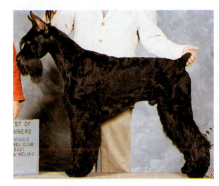

Giant Schnauser

Siberian Husky

Brittany

**Breeds of dogs.**
Photos copyright of Melia Photography. Used with permission.

Heavy-wooled llama

Some color variation and patterns of U.S. alpacas

White alpaca

Short-wooled pack-type llama

Guanaco llama cross-bred

Some color variations and patterns of U.S. llamas

**Breeds of lamoids.**
Photos courtesy of Susan L. Ley. Used with permission.

Silver Marten

Checkered Giant

New Zealand (Black)

Mini Rex

New Zealand (White)

English Angora

**Breeds of rabbits.**
Photos courtesy of American Rabbit Breeders Association, Inc. Used with permission.

Champagne d'Argent

Dwarf Hotot

Californian

Jersey Wooly

American Chinchilla

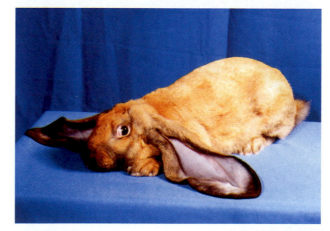

English Lop

**Breeds of rabbits.**
Photos courtesy of American Rabbit Breeders Association, Inc. Used with permission.

## SUMMARY AND CONCLUSION

The poultry industry is a large and thriving business based on meat- and egg-producing chickens and turkeys for meat production. A large contingent of hobby producers also use a number of other species such as ducks, geese, pigeons, and peafowl in addition to chickens and turkeys (Figure 18–29). The combined gross annual income of the poultry segments amounts to almost $30 billion. The primary purpose of the poultry industry in the United States is to produce inexpensive sources of protein for human consumption. It takes grain and by-products and produces meat and eggs very efficiently. These products are among the best buys available in the marketplace today for high-quality protein. Vertical integration provides the structural framework of the commercial poultry industry. Breeding and selection of poultry differs from breeding of the larger animals in three major ways: (1) it is more flexible because of short generation intervals and large numbers of offspring; (2) it has been the industry most subjected to

### Facts about Poultry

There are four major classes of chickens:

| | |
|---|---|
| *American Class:* | Plymouth Rock, Wyandotte, Rhode Island Red, New Hampshire, Jersey Black Giant, Rhode Island White, Java, Dominique, and Holland |
| *Mediterranean Class:* | Leghorn, Minorca, Ancora, Blue Andalusian, Buttercup, and Spanish |
| *English Class:* | Cornish, Australop, Orpington, Dorking, Sussex, and Red Cap |
| *Asiatic Class:* | Brahma, Cochin, and Langshans |
| Egg production takes 24.5 to 25.5 hours in the oviduct: | Infundibulum where fertilization occurs (15 min); magnum where white is laid down (3 hrs); isthmus where the two shell membranes are formed (1¼ hrs); uterus where eggshell is formed (20–21 hrs); vagina during laying (less than 1 min) |
| Average number of eggs/hen/yr: | 1937 = 120–130; 1990s = 250–260 |
| Hen/cock ratio: | Light breeds 15–20/cock, general-purpose breeds 10–15/cock, and heavy breeds 8–12/cock |
| Incubation and hatching: | Eggs incubated for 21 days at 98.6–100.4°F and turned 1–2 times daily |
| U.S. egg weight classes (oz/doz): | Jumbo (30), Extra Large (27), Large (24), Medium (21), Small (18), and Peewee (15) |
| U.S. grades of egg: | AA, A, B, depending on quality factors such as shell, air cell, white, and yolk |
| Names of various sex classes: | Rooster or cock, hen, capon, cockerel, pullet |
| Market classes of poultry: | Cornish game hen, broiler or fryer, roaster, capon, stag, hen or stewing chicken, cock or old rooster |
| Market classes of duck and goose: | Broiler duckling or fryer duckling, roaster duckling, mature duck or old duck, young goose or gosling, mature goose or old goose, gander |
| Comb: | The fleshy protuberance on top of the head of a fowl. The types of combs include: (1) buttercup, (2) cushion, (3) pea, (4) rose, (5) silkie, (6) single, (7) strawberry, and (8) V-shaped |

**Cock**
A rooster aged 1 year or older.

**Cockerel**
A rooster less than 1 year old.

**Broiler duckling or fryer duckling**
A young duck usually under 8 weeks of age of either sex. It must have a soft bill and weigh from 3 to 6½ lbs.

**Roaster duckling**
A young duck of either sex usually under 16 weeks of age. The bill must not have completely hardened. Generally weigh from 4 to 7½ lbs.

**Mature duck or old duck**
A duck of either sex with toughened flesh and a hardened bill. The meat is used in processed products.

**Young goose or gosling**
A young goose weighing from 12 to 14 lbs. A gosling weighs about 8 lbs.

**Mature goose or old goose**
A spent breeder whose meat is used in processed products.

**Gander**
A male goose.

**FIGURE 18–29** Geese comprise a very small percentage of the total poultry population. There is a small commercial segment, but most owners keep geese for home use or for ornamental reasons. (Photographer Sami Sarkis. Courtesy of Getty Images, Inc./PhotoDisc, Inc.)

modern animal breeding and selection techniques and has made the most progress; and (3) fewer people make all of the decisions. Most of the traits of interest in poultry genetics are quantitative traits, including egg production potential, egg size, growth rate, conformation, and so on. As a rule, these traits are more difficult to make progress in than are simpler traits. Generally speaking, chickens are one of two types: meat type or egg type. However, this generalization is too simple to explain poultry fully. Chickens exist in many colors, sizes, and shapes and in more than 350 combinations of these traits. Other species have also been developed with different characteristics. To identify and classify the species, they are designated by class, breed, variety, and strain. The management of the different poultry species for reproduction is very similar for all species. It is also of little matter if the birds are reared as a hobby or in a large commercial operation. When eggs fail to hatch properly, the reason may be the management of the breeder flock, the incubation procedures, or any step between the breeder flock and final hatch. Poultry feeding has changed more than the feeding of any other species with the advent of modern production systems. The primary reason for this change is that most of the poultry is produced in large units where maximum technology is used. Poultry nutrition is also more critical, complicated, and thus a greater challenge to the producer than the nutrition of other farm species. Eggs have recently increased in consumer demand. Turkey production has stabilized. Broiler meat continues to increase in its share of the consumer's dollar and per capita consumption.

## STUDY QUESTIONS

1. What is the monetary contribution of the poultry industry to U.S. agriculture? How do poultry species compare to other livestock species in monetary importance?

2. What species other than chickens and turkeys have a place in the poultry industry, and what is that place?

3. What is the purpose of the poultry industry in the United States? Describe this in terms of feed and use of resources. What is the role of the hobbyist?

4. Trace the development of the poultry industry from early colonial times to modern integration.

5. Briefly describe the structure of the modern broiler industry in the United States. Compare it to the turkey- and egg-producing sectors. Where is each major segment located?

6. Describe a typical integrated broiler firm and a layer firm.

7. Describe three major ways that poultry breeding and genetics is different from the breeding and genetics of the other livestock species.

8. What are the major breeding methods used in poultry breeding?

9. Generally, there are two types of chickens. What are they?

10. A system has been established to identify the different breeds and strains of chickens. Briefly describe class, breed, variety, and strain.

11. What is the *American Standard of Perfection*?

12. What is the importance of breeds to modern poultry production? What is the importance of breeds to hobbyists?

13. What are the major mating systems for the production of hatching eggs?

14. Describe how to collect, store, and incubate eggs in a way that maximizes fertility.

15. Why is poultry nutrition more complicated than nutrition for other species? Compare and contrast different types of poultry rations in terms of ingredients and amounts. What is the reason for the differences? What is an omnivore?

16. Describe important biosecurity measures for poultry flock keeping. Why was the National Poultry Improvement Plan developed?

17. Briefly discuss the nutritive value of eggs and the various poultry meats described in the chapter.

18. What is the per capita consumption of broiler meat in the United States? Why does it continue to increase?

19. Briefly discuss trends and factors affecting the poultry industry.

20. Why are chicks used for research?

## REFERENCES

Angulo F. J., and D. L. Swerdlow. 1998. *Salmonella enteritidis* infections in the United States. *Journal of the American Veterinary Medical Association* 213:1729–1731.

Carey, J. B., J. F. Prochaska, and J. S. Jeffery. 1999. *Poultry facility biosecurity.* Pub # L-5182. College Station: Texas Agricultural Extension Service, Texas A&M University System.

Ensminger, M. E. 1991. *Animal science.* 9th ed. Danville, IL: Interstate.

Ensminger, M. E. 1992. *Poultry science.* 3rd ed. Danville, IL: Interstate.

FAO. 2007. *FAOSTAT statistics database. Agricultural production and production indices data.* http://apps.fao.org/.

Ferket, P. R., and G. S. Davis. 1998. *Feeding ducks.* Publication PS Facts #2. Raleigh: North Carolina State University Extension. http://www.ces.ncsu.edu/depts/poulsci/.

Hanke, O. A., J. K. Skinner, and J. H. Florea. 1974. *American poultry history, 1823–1973.* Mount Morris, IL: American Poultry Historical Society.

Kellems, R. O., and D. C. Church. 1998. *Livestock feeds and feeding.* Upper Saddle River, NJ: Prentice Hall.

Moreng, R. E., and J. S. Avens. 1985. *Poultry science and production.* Reston, VA: Reston Publishing Company.

Oberholtzer, L., C. Greene, and E. Lopez. 2006. *Organic poultry and eggs capture high price premiums and growing share of specialty markets.* Outlook report from the Economic Research Service. LDP-M-150-1. December 2006.

Perez, A., and L. A. Christensen. May 1990. *Recent developments in the location and size of broiler growout operations.* USDA-ERS LPS-41.

Schrader, L. F., H. E. Larzelere, G. B. Rogers, and O. D. Forker. 1978. *The egg subsector of U.S. agriculture: A review of organization and performance.* N.C. Project 117. Monograph #6. West Lafayette, IN: Purdue University.

Skinner, J. 1978. *Chicken breeds and varieties bulletin.* A2880. Madison: University of Wisconsin Extension.

Smith, T. W. 1997. *Hatching quality chicks.* Publication 1182. Extension Service of Mississippi State University, cooperating with U.S. Department of Agriculture. http://www.msstate.edu/dept/poultry/exthatch.htm.

Taylor, R. E., and T. G. Field. 1998. *Scientific farm animal production.* 6th ed. Upper Saddle River, NJ: Prentice Hall.

USDA. 1997. *USDA nutrient database for standard reference.* Release 11-1. Nutrient Data Laboratory home page: http://www.nal.usda.gov/fnic/foodcomp.

USDA. 1998. *USDA nutrient database for standard reference.* Release 12. Nutrient Data Laboratory home page: http://www.nal.usda.gov/fnic/foodcomp.

USDA-ERS. 2007b. *Food consumption (per capita) data system.* Accessed online September 2007. http://www.ers.usda.gov/data/FoodConsumption/.

USDA-ERS. 2007c. *Red meat yearbook.* Table 114. Accessed online August 11, 2004. http://www.ers.usda.gov/data/sdp/view.asp?f=livestock/94006/.

USDA-NASS. 2004. *2002 Census of agriculture.* Washington, DC: National Agricultural Statistics Service, USDA.

USDA-NASS. 2007a. *Briefing room. Farm income and costs.* Accessed online September 2007. http://www.ers.usda.gov/Briefing/FarmIncome/.

USDA-NASS. 2007b. *Quick stats: Agricultural statistics data base.* Accessed online September 2007. http://www.nass.usda.gov/QuickStats/.

USDA-NASS 2007c. *Charts and maps.* http://www.nass.usda.gov/Charts_and_Maps/Poultry/brlmap.asp.

USDA-NASS. February 2007. *Chickens and eggs 2006 summary.* Bulletin # Pou 2-4 (07).

USDA-NASS. April 2007. *Poultry production and value.* Washington, DC: National Agricultural Statistics Service, USDA.

Voris, J. C. 1997. *California turkey production.* Poultry Fact Sheet No. 16c. Parlier: Cooperative Extension, Avian Sciences Department, University of California.

# Swine

## LEARNING OBJECTIVES

After you have studied this chapter, you should be able to:

1. Explain the place of swine in U.S. agriculture.

2. Describe the purpose and the value of the swine industry in the United States.

3. Give a brief history of the swine industry in the United States.

4. Describe the swine industry segments and structure and explain how that structure is changing.

5. Give an accurate accounting of where the swine industry is located in the United States and explain how that location is changing.

6. Quantify the role of genetics in the swine industry.

7. Explain the role of breeds in the U.S. swine industry and discuss how that role is changing.

8. Describe the traits of the ideal market hog.

9. Explain the major breeding programs for swine.

10. Give a basic outline for managing swine for reproductive efficiency.

11. Describe the feeding practices for different classes of swine.

12. Explain the concept of herd health management and give specifics regarding swine production.

13. Explain the nutritional benefits of pork to humans.

14. Discuss trends in the swine industry, including factors that will influence the industry in the future.

## KEY TERMS

| | | |
|---|---|---|
| Ad libidum | Barrow | Boar |
| Backfat | Biosecurity | Closed herd |

Complete diet
Creep
Cross-fostering
Farrow
Feed efficiency
Feeder pig
Finishing
Generation interval
Gestation

Gilt
Heterosis
Monogastric
Nursery pig
Nursing pig
Nutrient-dense food
Omnivore
Optimal growth
Palatability

Pig meat
Pork
Pork Quality Assurance
    Program
Show pigs
Sow
STAGES
Wean

## SCIENTIFIC CLASSIFICATION OF SWINE

Phylum:        Chordata
Subphylum:     Vertebrata
Class:         Mammalia
Order:         Artiodactyla
Suborder:      Suina
Family:        Suidae
Genus:         *Sus*
Species:       *domesticus*

## THE PLACE OF THE SWINE INDUSTRY IN U.S. AGRICULTURE

The swine industry is a large sector of U.S. agriculture. It did not suffer the declines of the beef industry. It would like to expand with the same vigor as the poultry industry, but it has had difficulty capturing additional market share. However, it has held its own. Hogs are the fourth most important money generator in food animal agriculture. The gross annual income from the swine industry is approximately $14.2 billion. Hogs currently generate 5.7% of all U.S. farm cash receipts (Figure 19–1) and over 11% of animal agriculture's share of all U.S. farm cash receipts (Figure 19–2). Three states have in excess of $1 billion yearly in gross income from hogs. An additional five states have at least $0.5 billion. A total of 13 states have in excess of $100 million. Hogs rank in the top five commodities in 14 states. The United States produces 9% of the world's **pig meat (pork)** with 6.2% of the world's hogs. The swine industry is an aggressive industry. It is very technologically driven and expansion minded.

## PURPOSE OF THE SWINE INDUSTRY IN THE UNITED STATES

The purpose of the swine industry in the United States is to use surplus grain production and high-quality by-product feeds to produce meat. The United States is capable of producing millions of tons of grain in excess of the needs of the human population and the export market. Swine are monogastric and, as such, can use only limited

**Pig meat**

The meat from a hog. Synonymous with *pork*.

**Pork**

The meat from a hog. In many parts of the world, the term *pig meat* is preferred.

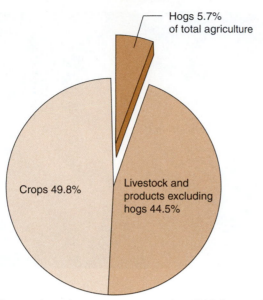

**FIGURE 19–1** Hog farm cash receipts as a percentage of total U.S. farm cash receipts, 2005–2007. (*SOURCE:* Based on USDA-NASS, 2007a.)

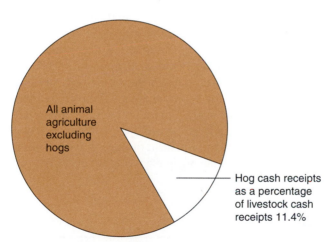

**FIGURE 19–2** Hog yearly farm cash receipts as a percentage of total animal agriculture's cash receipts, 2005–2007. (*SOURCE:* Based on USDA-NASS, 2007a.)

amounts of forage. However, they are the most efficient converters of grain to red meat of all the livestock species (Figure 19–3). In times of plentiful grain production, they can be used to add value to the grain and create an additional market for it. Like the other grain-using species, swine help moderate the fluctuations in grain prices, which could otherwise disrupt agronomic practices and endanger the agricultural economy and the supply of grain for human consumption. Because of their ***monogastric*** digestive tract, swine require high-energy feeds to be produced economically. Feed costs are a high fixed cost, and economical production depends on minimizing labor and maximizing efficiency of production by reducing time to market. Expensive confinement facilities and full feeding of high-energy rations are used to maximize weight gain. This requires a large investment in facilities and equipment.

**Monogastric**

Class of animals that do not have a rumen. Humans are monogastrics. Monogastrics require a better diet than ruminants do.

**FIGURE 19–3** The purpose of the swine industry is to turn surplus grain production and by-product feeds into meat. (Photo courtesy of National Pork Producers Council.)

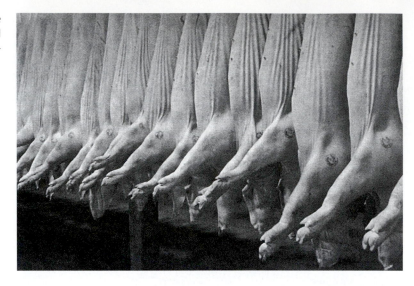

**Omnivore**

An animal that eats both plant- and animal-based foods.

In spite of the fact that they are monogastric, swine are distributed all around the world, including developing countries. When grain cannot be fed, they are fed on wastes and are often allowed to forage for themselves. As *omnivores*, they eat a wide variety of feedstuffs. They do not produce as efficiently this way but still produce high-quality food for their owners. The single most important swine-producing country in the world is China. Most of the pigs in China are in integrated systems, where they produce manure to be used as fertilizer, which is an important product. Swine also have the distinction of being the world's dominant meat-producing species. Pork production on a worldwide basis accounts for approximately 1.7 times as much meat as cattle produce and almost 39% of the world's meat production. Swine are used predominantly for meat and, therefore, a large percentage of them are slaughtered annually worldwide. There are nearly 990 million head of hogs in the world; they are the third most-numerous species. There is one pig in the world for every 6.6 people and 19 head of pigs per square mile of earth land surface.

## HISTORICAL PERSPECTIVE

Pigs were probably domesticated around 8000 B.C., roughly the same time the sheep and goat were domesticated. Columbus first brought the hog to this hemisphere on his second voyage in 1493. He brought eight head with him to the West Indies. In 1539, Hernando DeSoto brought pigs to the North American continent when he landed near what is now Tampa Bay, Florida. During DeSoto's three years of exploration, the original 13 hogs grew to 700. This occurred in spite of the fact that many were undoubtedly eaten, some were lost to predators, some escaped, and some were probably traded to the native people. This gives DeSoto the distinction of being the first swine producer in North America.

By the time swine were imported by settlers to North America, well-established breeds and types had been developed in different places around the world. The English settlers naturally brought animals from England and Ireland. The French brought different breeds to Canada and Louisiana. The Spanish brought even different breeds to Florida and Mexico. Breeds were also brought from Africa. All of these hogs contributed to the unique breeds developed in the United States.

In colonial times, the pig established itself as an important agricultural animal. By 1641, colonists had established a meatpacking plant and began to export salt pork and lard. (The term *meatpacker* dates to this time and came from the practice of colonists packing pork in salt as a means of preserving it.) Hogs were tended much as they had been in Europe. They mostly ran free and fended for themselves, feeding on roots, tubers, fruits, nuts, mushrooms, snakes, rodents, and so on. As omnivores, they were able to select from a wide range of foods, and they thrived. In the fall of the year, pigs were rounded up and slaughtered. Some were fattened on excess grain, table scraps, or skim milk.

The meat was less important than the body fat (lard) in the early days. Lard had many uses and was a valuable, tradeable product. Breeds of hogs were developed for their ability to lay down excessive body fat. By the mid-1800s, much of the corn grown in the Tennessee, Kentucky, and Ohio areas was being fed to hogs, which were then "walked" to market as live hogs. Cincinnati, Ohio, became the first pork-packing center and was even known as "Porkopolis." The center of activity changed to Chicago by 1860. Hogs were shipped there from throughout the country by railroad. Chicago became known as "hog butcher for the world."

The production and marketing of hogs became more decentralized after World War I. Slaughter facilities were built near where the animals were produced. This was largely in the grain-producing areas. Other products replaced lard, and hogs started to become leaner. The restructuring of the agricultural industries that became so evident in the 1950s started affecting the swine industry as well. Improvements in genetics and performance evaluation methods, crossbreeding systems, and nutrition improved efficiency and performance. New disease treatment and prevention strategies and tools allowed more animals to be concentrated in smaller spaces. All of this has led swine operations to fewer but larger units that are partially or totally confined, just as the other animal industries have done.

The swine industry of the 1980s and 1990s was characterized by the incorporation of increasingly sophisticated technologies into production. Not all producers took equal advantage of the available technologies. Producers in states other than traditional "hog states" became important producers because they aggressively incorporated the available advancements. This led to a further restructuring of the industry, and it brought such states as North Carolina to the second most-important swine state and other states to prominent positions from previous positions of obscurity. Oklahoma, for example, jumped from 23rd to 8th place in 7 years. This restructuring led to a distinct difference among the swine states, with some being dominated by more traditional approaches, and the "new swine states" being characterized by better genetics, better economies of size, and state-of-the-art facilities and management techniques. These new areas are dominated by large operations and a vertically integrated structure.

Figure 19–4 gives a historical look at swine numbers in the United States. Hog numbers have not really changed much in the United States since 1950, if you disregard cyclic ups and downs. However, increased production of red meat per animal has kept the supply of meat available, and per capita consumption has remained relatively stable during that period.

## STRUCTURE OF THE SWINE INDUSTRY

There are five primary types of swine operations:

1.  *Farrow* to wean. Consists of a breeding herd, which produces early-weaned pigs at 10–15 lbs or *feeder pigs* at 35–50 lbs (Figure 19–5). Pigs are generally early weaned in modern operations and placed in a nursery where they can receive specialized care until ready for *finishing* (Figure 19–6).

**Farrow**

In swine, the term used to indicate giving birth.

**Feeder pig**

Generally thought of as a pig weighing between 30 and 90 lbs. There is some regional difference in this range.

**Finishing**

The process of growing a pig to market weight.

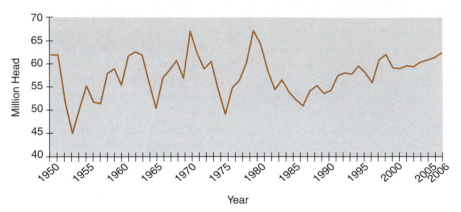

**FIGURE 19–4**    Historic swine inventory in the United States. (*SOURCE:* USDA-NASS, 2007b.)

**FIGURE 19–5**   Feeder pigs such as these are the primary product of the farrow-to-wean segment. They are often marketed to other producers who finish them for market.

**FIGURE 19–6**   The finishing segment grows pigs to market size. (Photo courtesy of National Pork Producers Council.)

2. **Finishing.** Feeder pigs are grown to market weight.

3. **Farrow-to-finish operation.** A breeding herd is maintained. Pigs are produced and finished for market on the same farm (Figure 19–7).

4. **Purebred or seedstock operations.** These are similar to farrow-to-finish except their salable product is primarily breeding *boars* and *gilts* or *show pigs,* which may be purebred or controlled crossbreeds (Figure 19–8).

5. **Integrated corporate production.** Integrated operations can generally be described as farrow-to-finish and often have their own seedstock production as well. The various phases of the operation are usually located on different sites. For instance, several brood *sow* facilities may be found in general proximity to each other but far enough apart that *biosecurity* measures between the facilities are easily handled. Nursery facilities for early-weaned pigs are commonly found on the same site or at a site close by. However, when pigs are ready for finishing, they are taken to another site, which may be a company-owned facility or a contractor (Figure 19–9). The boars that provide the semen to breed all the sows in a cluster of sow facilities are ordinarily housed in a centrally located facility. Semen is collected from them and transported to the sow facilities.

With the increased use of technology in the swine industry has come an increasingly integrated swine industry structure. Integrated, corporate swine production is causing several changes. First, the ownership of pigs is shifting to facilities in which more pigs are owned by fewer people. A contracting system has developed similar to that of the poultry industry, and with the general consolidation has come a consolidation of decision making. The most startling feature brought by these changes is the size of the operation and the total number of swine producers. In 1950, there were 2.2 million swine producers. In 1970, there were 871,200 swine operations in the United States. In 1980, the number had declined by 23% to 666,550. By 1990, this number had declined to 268,140, a reduction of nearly 60% for the decade. More dramatic yet, the number of producers declined to 87,470 by 2000, for a reduction of over 67% from 1990. The number of swine producers in 2000 was 10% of that in 1970. This trend is not over (Figure 19–10).

---

**Boar**

An intact male hog kept only for breeding purposes.

---

**Gilt**

Any female pig that has not yet given birth. Sometimes producers use the term "first-litter gilt" after the first litter is born.

---

**Show pigs**

Pigs bred for exhibition, usually by 4-H and FFA students.

---

**Sow**

Female pig that has given birth.

---

**Biosecurity**

Procedures designed to minimize disease transmission from outside and inside a production unit.

---

**FIGURE 19–7** Farrow-to-finish operations maintain a breeding herd and produce pigs that they then raise to market weight.

**FIGURE 19–8** Purebred and seedstock operators are specialized farrow-to-finish operations, except that their primary salable products are breeding boars, gilts, and show pigs.

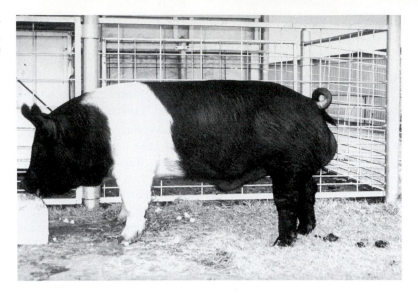

**FIGURE 19–9** Increased technology, improved housing, and the application of stringent biosecurity measures have allowed the concentration of swine on one site to be fed in facilities such as this one. (Photo courtesy of National Pork Producers Council.)

Along with the reduction in operations has come an increase in the size of operations. The trend in the swine industry since the 1970s has been a move toward larger operations and fewer producers. While the smaller herds have been disappearing, the larger herds have been getting even larger (Figure 19–11). Some of this has been facilitated by vertically integrated structuring. These changes have not occurred uniformly from region to region or from state to state. For instance, Iowa is relatively unchanged, whereas states like North Carolina, Mississippi, Alabama, Kentucky, Arkansas, and Oklahoma have changed tremendously. The industry now has a group of swine producers often referred to as *mega producers*.

Changes in the segmentation of the industry are also becoming evident. In areas of greatest structural change, specialization is much more likely, contract growing is more likely, and sales directly to a slaughter plant are more likely. In addition, feed use is 10% less, labor efficiency is 30% less, and death loss is lower at farrowing and while

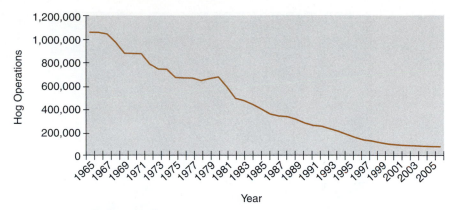

**FIGURE 19–10** Swine operations in the United States. In 1950, there were 2.2 million swine producers. The number of swine producers in 2000 was less than 10% of that in 1970. This trend is not over. (*SOURCE:* USDA-NASS, 2007b.)

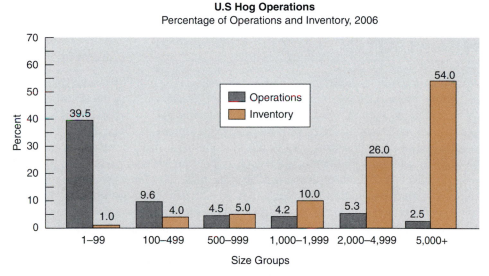

**FIGURE 19–11** U.S. hog operations and the percentage of inventory controlled by each size group. The trend in the swine industry since the 1970s has been moving toward larger operations and fewer producers. As shown here, as of the end of 2006, the largest 2.5% of the hog operations control 54% of the hogs in the United States. A staggering 90% are found in herds with 1,000 or more inventory. (*SOURCE:* USDA-NASS, 2007b.)

nursing, so more pigs per litter are **weaned**. Death loss from weaning to market is similar, but pigs are weaned earlier and lighter, which is a significant accomplishment. Much of this is probably owing to better facilities (total confinement) and use of the newest technologies. Newer facilities are less likely to grow their own feed, but costs are about the same because they save on fuel and repair. Overall technology usage is higher in these "new" structure areas (Figure 19–12).

The restructuring of the swine industry is not over. In fact, the swine industry is changing from an industry of independent producers to one that looks and acts very much like the consolidated, integrated poultry industry.

**Wean**

The process of removing pigs from the dam to prevent them from nursing.

**FIGURE 19–12** Adoption of the latest technology has allowed some states to greatly expand their swine industries. The "new" swine areas wean more pigs per litter than the "old" swine areas.

# GEOGRAPHIC LOCATION OF SWINE IN THE UNITED STATES

Swine are produced in all 50 states. Figure 19–13 shows the individual state statistics for total swine. The major concentration of hogs in the United States is in the Corn Belt. Iowa has had approximately 25% of the entire swine herd of the nation for a number of decades. Today, Iowa and its border states, plus Indiana, Ohio, Michigan, Oklahoma, and Kansas, contain 74% of the nation's hogs. Hogs are found in these states because they comprise the major area of grain production in the United States or are near grain. Feed prices are generally lower. With feed representing 65% of the total cost of hog production, low-cost feed is a powerful magnet to hog production. Also, most large hog-slaughter plants are located in this region, so the markets are usually better and are more conveniently located.

Even with the powerful lure of less-expensive grain, the traditional Corn Belt states have been losing their share of the total U.S. hog population. Although they are still major players in the swine industry, the Hog Belt has moved outside the confines of the Corn Belt. In recent years, North Carolina has been the major swine development story. It has come from relative obscurity as a swine state to have the second-largest swine numbers of any state. Other states like Oklahoma, which was never a major swine-producing state before 1991, and Colorado have developed swine industries of consequence. The development of the swine industry in some states has somewhat followed the development of the poultry industry in those same areas. In fact, some of the large poultry producers have significant involvement in the swine industry. Other factors that have contributed include the milder weather, which means that investment in facilities is often less than in the Corn Belt. Labor is usually less costly. Grain is not as plentiful, but significant amounts are produced in or near these states. The real backbone of the growth of these and several other states has been their willingness to adopt new technologies. Decreased costs of production and greater efficiency of production have more than offset the difference in grain price. Large swine producers who have moved to these areas have driven much of the increased production in these newer swine states.

The industry is still moving. The Hog Belt must now be considered to have decentralized somewhat from its previous Corn Belt shape and moved south, east, and west.

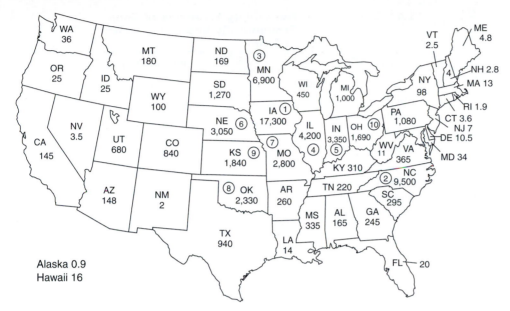

**FIGURE 19–13** Swine inventory (× 1,000) by state in 2006. The circled numbers represent the rank of the top 10 swine-producing states. Swine are produced in all 50 states. The Corn Belt has the major concentration of hogs. In recent years, the Hog Belt has moved outside the confines of the Corn Belt. North Carolina and states west and south of the Corn Belt have developed major swine industries. (*SOURCE:* USDA-NASS, 2007b.)

The most important direction for the long term seems to be west. The ultimate location of the swine industry will likely be determined by six major factors:

1. Availability of feed.
2. State regulations and restrictions on facilities.
3. Technological advances of the industry.
4. Availability of an adequate transportation infrastructure.
5. Labor availability.
6. Fossil fuel availability.

## GENETICS AND BREEDING PROGRAMS

The fundamentals of genetics and breeding are discussed in Chapters 8 and 9. See those chapters for more detailed information on these topics.

The economic traits are influenced by both the environment to which the animal is exposed (feeding, climate, and so on) and the genetics of the animal. Heritability is the portion of the difference that can be passed on to the next generation. The pace of progress in a herd to improve the genetic potential of the offspring depends on how heritable the trait is. Progress in swine genetics is easier to make than genetic progress for some species because the number of offspring from which to select in litter-bearing species is much higher than for species that bear only one offspring in a year. A sow can produce 30 pigs in a year compared to one calf for a cow. In addition, the *generation interval* for swine is fairly short. Pigs from a sow's litter can easily produce offspring within a year of their birth because of their rapid maturing rate and short *gestation* length.

**Generation interval**

In a herd, the average age of the parents when their offspring are born.

**Gestation**

The period when the female is pregnant.

**TABLE 19–1**    **Heritability Estimates of Some Economically Important Traits in Swine**

| Trait | Heritability (%) |
|---|---|
| Litter survival to weaning | 0 |
| Number farrowed | 10 |
| Number weaned | 10 |
| Birth weight | 20 |
| Weaning weight | 20 |
| Feed efficiency | 25 |
| Growth rate | 30 |
| Age at puberty | 35 |
| Backfat thickness | 40 |

SOURCE: Cleveland et al., 1999. Used with permission.

## Backfat

The fat on a pig's back. It is highly correlated to total body fat and often measured and used as a means of selecting lean brood stock.

## Feed efficiency

The amount of feed required to produce a unit of gain. A 3:1 ratio means that 3 units of feed were needed to produce 1 unit of gain. The smaller the number the better. Also referred to as *feed conversion ratio*.

## STAGES

Swine Testing and Genetic Evaluation System. A series of computer programs that analyze performance data of purebred swine and their crossbred offspring.

Table 19–1 gives heritability estimates of some economically important traits in swine. It is easy to see that some traits (e.g., **backfat** thickness and **feed efficiency**) are under more genetic control than others (e.g., litter survival and number farrowed). By using such tools as performance testing, sire summaries for boars available through artificial insemination, the **STAGES** program, and EPDs if purebred seedstock is used, the selection and breeding program of a herd can be enhanced. Seedstock producers and commercial producers will vary in the tools they select and the traits they emphasize.

Swine breeders have made remarkable progress in swine genetic improvement. Most notable is the decrease in fat content of the average pork carcass. Modern consumers demand lean pork, and today's hogs are much leaner than those of the past. Overall body fat has been reduced by more than 50%. Further progress is being made.

## BREEDS

Unlike other meat animal species, most of the commonly used breeds of swine in the United States were developed here between 1800 and 1880. Many of these breeds have effectively disappeared because they were so fat. However, this period left a purebred bias with swine breeders that hampered progressive crossbreeding strategies until well into the 1960s.

For many years, there were two recognizable categories of hogs in the United States. The first was the lard type. These hogs were developed and reared to get extremely fat so they would produce the maximum amount of lard. A fat carcass is also the best to preserve with salt. Salt-cured pork was in high demand in this country in the past. At one time, the Duroc, Chester White, Poland China, Spotted Poland China, and Ohio Improved Chester White were considered the major breeds in this category. The second type of hog was the bacon type. These animals were leaner, longer hogs. Because they were longer-sided animals, they produced more bacon. The major breeds in this category were the Tamworth and the Yorkshire. The Hampshire and the Berkshire were considered to be intermediate between these two types, although most classifications put them into the lard type. Today's hogs are considered to be meat-type hogs. They are something of an intermediate type between the previous types but are more muscular and much leaner. The emphasis on selecting and developing this meat-type animal can be dated to roughly 1950 (Figure 19–14).

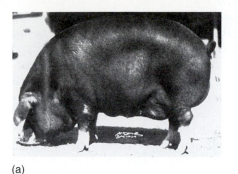
(a)

(b)

(c)

**FIGURE 19–14** Swine breeders have made tremendous progress turning lard-type (a) and bacon-type hogs (b) into the modern meat-type hog (c). (Photo [a] courtesy of National Pork Producers Council.)

Breeds of major importance in the United States today include Yorkshire, Duroc, Hampshire, Spott, Chester White, Landrace, Poland China, and Berkshire. Yorkshires, Landrace, and Chester White (all-white breeds) usually are considered to be mother breeds and are emphasized on the female side, although in recent years they have both improved in carcass merit and growth rate. The Hampshire is most often acknowledged to be the best carcass breed. It tends to be used more on the sire side. The Duroc is a good all-around breed with no major weaknesses. Durocs have good rates of gain, feed conversion, and carcass characteristics. It tends to be regarded more as a sire breed, but that is probably more of a perception than a truth. The swine industry is increasingly coming to depend on synthetic lines of hogs rather than purebreds. This, coupled with the decreased number of small producers who previously purchased purebred breeding stock for their operations, has caused a decline in the numbers of purebred swine being produced.

There is some concern today that the genetic base of swine is lacking the diversity of former years. However, with the continuing trend of large-scale confinement production, there is little demand for many breeds and several are nearly extinct. These breeds often did not have the traits desired in modern pork production and were better adapted for outside or open-range conditions. The American Livestock Breeds Conservancy of Pittsburg, North Carolina, an organization working to preserve endangered breeds of livestock, is a source of interesting information on minor breeds.

For information on swine breeds, visit the breeds of livestock page at http://www.ansi.okstate.edu/breeds/. This comprehensive reference includes pictures of livestock breeds. Photos of several breeds can also be seen in the color plate section of this text.

## SWINE BREEDING PROGRAMS

### Standards for the Ideal Market Hog

The National Pork Board has set forth standards for the ideal market hog to give producers a uniform goal. The standard for the ideal market hog, named Symbol III, includes a hog marketed at 270 lbs in 156 days for a *barrow* or 164 days for a gilt that produces a 205-lb carcass. Both barrows and gilts are to have a live-weight feed efficiency of 2.4 lbs of feed or less for each pound of gain. Loin eye area at 270 lbs is expected to be at least 6.5 sq. in. for barrows and 7.1 sq. in. for gilts. Barrows are expected to have a fat-free lean index of at least 53 and gilts should be 54.7 or better.

**Barrow**

A castrated male hog.

**Pork Quality Assurance Plus**

A voluntary educational program introduced by the National Pork Producers Council as a tool to enhance the quality of pork sold to the world's consumers.

**TQA**

Trucker Quality Assurance Program is an educational program for all involved in the transportation process of swine.

**Heterosis**

The superiority of the crossbred animal as compared to the parents' breeds. Commonly referred to as *hybrid vigor*.

Symbol III is expected to be free of the stress gene (Halothane 1843 mutation) and all other genetic mutations that have detrimental effects on pork quality and be the product of a systematic crossbreeding system emphasizing a maternal dam line and a terminal sire selected for growth, efficiency, and superior muscle quality. The maternal line should be greater than 25 pigs per year, after multiple parities. Symbol III should be produced under the *Pork Quality Assurance Plus* and the *Trucker Quality Assurance Program (TQA)*.

There are direct paybacks to producers who produce superior animals. The majority of the market hogs are sold on carcass merit pricing systems. These systems reward producers whose animals are heavily muscled and have low body fat. Further refinements in these systems are expected in the future.

## Crossbreeding

To produce the best hogs in the most economical manner, most commercial swine-breeding programs rely on crossbreeding. In excess of 95% of all market hogs in the United States are crossbred. Crossbred animals exhibit hybrid vigor, or *heterosis.* The advantages of crossbred hogs compared to purebred hogs represent a 20–30% improvement in productivity and efficiency. Swine producers started to practice and widely use crossbreeding programs effectively several decades ago. They were among the first livestock breeders to use systematic multibreed crossing programs.

Crossbreeding programs specify the breeds to be used and in what order. Figures 19–15, 19–16, and 19–17 show examples of some effective crossbreeding schemes for swine. Table 19–2 gives the percentage of heterosis that can be expected from several different crossbreeding schemes. The crossbreeding scheme chosen should be based on the size of the herd, availability of replacements, and several other management-related factors.

## Purebred Programs

Purebred operations have as their main function the production of seedstock for other purebred breeders and commercial producers. Purebreds should rarely, if ever, be used in commercial production except as the parents of crossbreeds. Crossbred hogs are 20–30% more efficient than purebreds. Less than 1% of all the hogs in the United States are registered purebreds. However, effective crossbreeding programs require the availability of superior purebred livestock. Breeds are by definition animals that have been selected for a more uniform set of characteristics than their species shows as a whole. Breeds in common use in the U.S. swine industry were discussed previously.

## Crossbred Seedstock

The use of purebred sires in swine breeding has had competition in recent decades from crossbred sires. The ability to use sophisticated computer models to predict the breeding value of animals has speeded the development of breeding programs that use controlled crossbreeding to select for and develop synthetic lines of breeding animals. These are commonly referred to as *maternal lines* and *sire lines*. Although purebred animals are needed to develop these breeding lines, fewer purebreds are needed than before. This is especially true because the crossbred boars from these programs are routinely used in breeding programs. Several swine seedstock companies in the United States have devel-

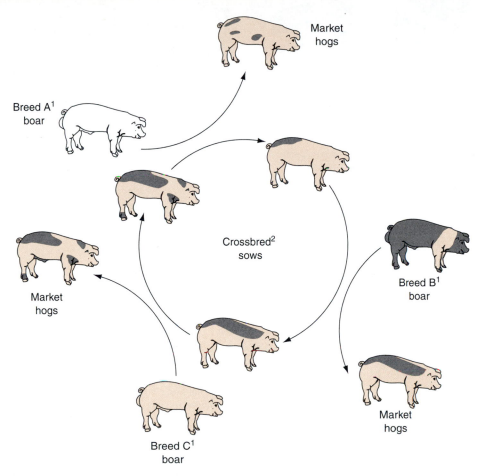

Market
hogs

Breed A[1]
boar

Crossbred[2]
sows

Breed B[1]
boar

Market
hogs

Market
hogs

Breed C[1]
boar

[1]The boars are alternated between breeds with different strengths. Breed A might be a boar from a breed known for maternal traits such as Yorkshire or Landrace; Breed B from a strong carcass and growth trait breed such as the Hampshire or Spotted; and Breed C from one known for being a more balanced breed such as the Duroc.

[2]The sow is always a crossbred once the rotation is stabilized.

**FIGURE 19–15** A three-breed rotational crossbreeding scheme for swine. Replacement gilts are selected from the market crosses and the breed of boar is changed every generation. In the initial cross, heterosis in the offspring is 100%. After the rotation has stabilized, the heterosis in the offspring and the sows is 86% for both. This general scheme can be extended to include additional breeds in the rotation. A six-breed rotation has a stable heterosis equilibrium of 98.4%. (*SOURCE:* Adapted from the *Pork Industry Handbook.*)

oped and marketed crossbred seedstock. The selection of the breeds and all other decisions within the companies relative to the breed composition of these animals is proprietary information that is not generally made available to producers. The replacement animals are developed according to whether the animals are sire lines or sow lines and are generally a mixture of three to four breeds. When a producer purchases seedstock, he or she also purchases the breeding program of the seedstock company. Several of these companies have been very successful in marketing their breeding programs along with their animals.

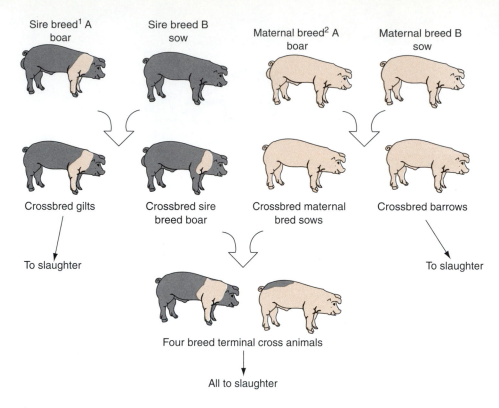

Sire breed[1] A boar

Sire breed B sow

Maternal breed[2] A boar

Maternal breed B sow

Crossbred gilts

Crossbred sire breed boar

Crossbred maternal bred sows

Crossbred barrows

To slaughter

To slaughter

Four breed terminal cross animals

All to slaughter

[1]The boar is from a group of breeds known for carcass merit and fast, efficient gain. The Hampshire, Duroc, Spotted, and Poland breeds are examples.

[2]The sow is a cross of white breeds known for their maternal traits. The Yorkshire, Landrace, and Chester White (all white breeds) are generally acknowledged the best for maternal traits.

**FIGURE 19–16** A four-breed terminal crossbreeding scheme for swine. Terminal crosses take advantage of breed differences. The sow is a cross of breeds known for their maternal traits. The boar is a cross of breeds known for carcass merit and fast, efficient gain. Breed differences are used to maximum advantage, and maximum heterosis is realized because there are no common breeds in the parents of the meat animals. (*SOURCE:* Adapted from the *Pork Industry Handbook.*)

**TABLE 19–2** **Percentage of the Maximum Heterosis Obtained from Various Crossbreeding Systems**

| | PERCENTAGE HETEROSIS OBTAINED | |
| System | Offspring | Maternal |
|---|---|---|
| $F_1$ cross; the initial cross A × B | 100 | 0 |
| Backcross, A × A − B | 50 | 100 |
| 2-Breed rotation | 67 | 67 |
| 3-Breed rotation | 86 | 86 |
| 4-Breed rotation | 93 | 93 |
| 5-Breed rotation | 97 | 97 |
| 6-Breed rotation | 98 | 98 |
| Terminal cross using $F_1$ sows | 100 | 100 |
| Rota-terminal cross using 2 breeds | 100 | 67 |
| Rota-terminal cross using 3 breeds | 100 | 86 |

*SOURCE:* Cleveland et al., 1999, and Ahlschwede et al., 1999. Used with permission.

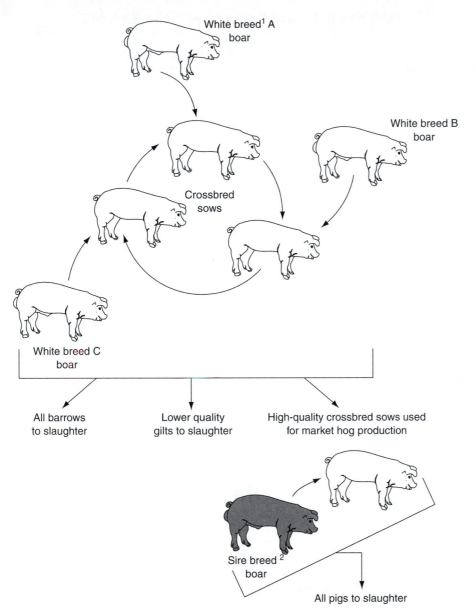

White breed[1] A
boar

White breed B
boar

Crossbred
sows

White breed C
boar

All barrows
to slaughter

Lower quality
gilts to slaughter

High-quality crossbred sows used
for market hog production

Sire breed [2]
boar

All pigs to slaughter

[1]The Yorkshire, Landrace, and Chester White (all white breeds) are generally acknowledged as the best for maternal traits.

[2]The boar is from a group of breeds known for carcass merit and fast, efficient gain. The Hampshire, Duroc, Spotted, and Poland breeds are examples.

**FIGURE 19–17**   A rota-terminal crossbreeding system for swine. This specialized crossing system is designed to make use of breed strengths while minimizing weaknesses. This system allows 100% heterosis in the offspring and 86% maternal heterosis. (*SOURCE:* Adapted from the *Pork Industry Handbook.*)

# REPRODUCTIVE MANAGEMENT IN SWINE

Chapter 11 is devoted to reproductive physiology and Chapter 12 is devoted to lactation. See these chapters for the fundamentals of reproduction and lactation.

Modern swine production systems demand high reproductive rates from swine. Litters need to be large. Sows need to either be lactating or gestating with as little non-productive time as possible. Use of facilities must be maximized. Expensive confinement facilities can't be paid for if they are empty. New gilts must be brought into the herd as early as they can be properly developed. Boars must maintain a high level of sperm production and have good libido. On the whole, reproductive management in a swine herd is a time-consuming task that has many facets and absolutely must be done well. There is tremendous overlap among good reproductive management, good animal health management, and good nutritional management. Nothing on a functional animal unit occurs in a vacuum. Everything is related.

## GILTS

New gilts must be brought into the breeding herd with regular frequency. They are needed to replace sows that must be culled for poor production. They are also expected to have better genetics than that of their parents' generation and, therefore, represent the introduction of the fruits of the selection program. Gilts should come from family lines of known mothering ability and should display no observable genetic defects. They should be grown to reach 250 lbs in their first 7 months of life. They may be bred at this weight. Gilts that do not meet these criteria should be culled. The exact timing of mating for gilts is dictated by the demands of the farrowing schedule for the whole herd, current costs, salvage values of cull breeding stock, and other management decisions.

## SOWS

Sows must be managed such that they produce ample milk for the litter and also to be ready to rebreed within 4 to 7 days of weaning the litter. For every 21-day delay in rebreeding, a sow must produce one to two additional pigs per litter just to pay for the feed and labor of the delay. Sows are easy to synchronize in heat. Within 3 to 7 days of weaning a litter, most sows come into heat. With today's commonly used practice of all-in-all-out, in which all litters born within a 7-day period are weaned together, the sows are automatically synchronized.

## ESTRUS DETECTION

Estrus detection in gilts and sows is critical to the success of a swine operation. Having a boar present can enhance heat detection. Sows and gilts in heat generally have a swollen vulva, respond to pressure on their back by standing solidly (hence the term *standing heat*), stiffen their ears (called "popping" their ears), and may display other behavioral signs.

## FARROWING MANAGEMENT

The presence of an experienced farrowing manager is acknowledged to be worth one additional pig per litter on the average. Intervals longer than 15–20 minutes between pigs may signal a problem that requires intervention. Baby pigs frequently succumb to being tangled in the afterbirth, and they suffer other mishaps like bleeding navels or inability to nurse in a timely manner. All of these problems can be alleviated. Many producers manage farrowings with the use of hormones to induce parturition. This helps keep litters closer to the same age and makes other management practices, such as equalizing litter sizes by *cross-fostering*, easier to accomplish.

**Cross-fostering**
Moving young from their dam and placing them with another female for rearing.

## BOARS

Boars should be evaluated for genetic contribution and breeding soundness before they are used in a breeding program. One in 10 untried boars is likely to have a fertility problem. The major factors that should be evaluated in a breeding soundness exam include testicular development, physical ability to breed females, semen quality, and libido. Most boars are sexually active by 7 months of age. Boars and females can be run together for breeding in pen-mating systems. Another method is hand-mating. In this method, the animals are only put together specifically to mate and are then separated. Increasingly, swine facilities are taking advantage of artificial insemination to accomplish the matings.

## ARTIFICIAL INSEMINATION IN SWINE

The greatest advantage of artificial insemination (AI) is the opportunity to use genetically superior boars. Currently, fresh boar semen is readily available from many different boar studs in the United States and anyone can purchase semen out of many outstanding boars. Pork producers who collect their own boars can reduce the number of boars needed in their herd and thus reduce the cost of breeding. Producers who have *closed herds* can use AI to reduce the risk of introducing new disease. They may purchase semen or collect their own boars at a different location. An additional advantage of artificial insemination is that it facilitates the mating of animals of different sizes. A 240-lb gilt can be safely mated to an 800-lb boar.

**Closed herd**
A herd into which no new animals are introduced.

Problems of swine AI include the storing of semen. Generally, the best success with AI is obtained by using fresh semen within 3 days of collection. However, long-term extenders can maintain semen viability for 10 to 14 days. Proper use of semen extenders and storage at correct temperature are necessary for the storage of fresh semen. Commercial frozen semen is available. Unfortunately, the average pregnancy rates are 25% lower and litter size is one to three fewer pigs per litter. Frozen semen is not commonly used in the United States. Fresh semen is used extensively by all segments of the industry including the commercial industry, seedstock industry, and show-pig producers.

## RECORD KEEPING

Good records are essential to managing reproductive issues in the swine herd. Because management practices and standards vary with conditions across the country and the specifics of the swine operation, producers are encouraged to contact their local extension service or state extension specialists for help in setting up a recordkeeping program. Several programs are available commercially as well. These range from simple record books to powerful and sophisticated computer programs. As the average size of swine units increases, it is hard to imagine that a hand-kept record book can be adequate. Keeping good records, interpreting the information in the records, and basing management decisions on those records are essential pieces of modern swine management (Figure 19–18).

In past years when swine were raised mostly in outdoor systems, sows were usually mated to farrow twice a year. The farrowings were timed to weather conditions as much as possible. Modern confinement and partial confinement systems represent such

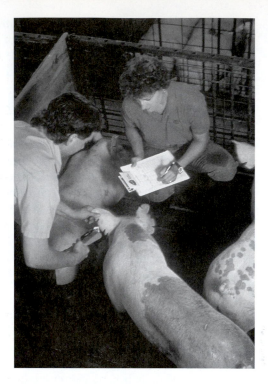

**FIGURE 19–18** Good record keeping is essential in managing production, reproduction, and health in a swine herd. (Photo courtesy of National Pork Producers Council.)

a capital outlay that they must be maximally used if they are to be profitable. Thus the modern approach to farrowing is to farrow year-round. The gestation period for a sow is 112–115 days. The estrous cycle for a sow is 18–24 days, with 20–21 days the average. The duration of estrus is 2–3 days. The average number of pigs reared per litter on a national basis is approximately 9. Those with more than 5,000 hogs average 8.9. Sows are capable of having 20 or more pigs per litter. However, such large litters rarely thrive. This is an area of swine production in which efficiency of production can be improved. The more pigs a sow can rear, the fewer sows will be needed. Obviously, a greater pigs-per-litter average is desired.

## NUTRITION IN SWINE

Swine are monogastric. They have a cecum, which develops and digests some fiber if they are fed forages. However, modern production systems do not make use of this capacity. The feeding niche that swine occupy is for concentrated feeds such as grains, soybean meal, and high-quality by-product feeds. Feed costs amount to approximately 60–65% of the cost of production in swine. The nutrition and feeding of swine in most of the production stages revolves around optimizing growth. This is the best feeding strategy because the faster an animal reaches market, the smaller the percentage of total lifetime feed is used for body maintenance. In addition, the faster an animal reaches market, the fewer days it requires housing, labor, and so on. Thus the fixed costs of production can be reduced on a per animal basis. Chapters 5, 6, and 7 give detailed information relating to digestive anatomy, nutrition, feeds, and feeding. See those chapters for more detailed information than is provided in this section.

Managing the nutrition of swine is very dependent on the intended purpose of the animal. Animals being readied for market are fed differently than the breeding herd. Likewise, different ages of growing pigs require different nutrient quantities and ratios. Monogastrics have exacting nutrient needs. Not only must they receive the absolute quantities of digestible nutrients to meet their nutritional needs, but the nutrients must also be in proper ratio to each other to ensure appropriate consumption and maximum utilization of the nutrients. *Palatability* of the feed is also an issue. If the animals do not consume the feed, it does no good to feed it. Following are some general comments relative to providing *complete diets* for swine:

- **Energy.** Age, activity level, level of production (for example, the number of pigs a sow is nursing), and environmental temperature all affect energy requirements. The energy density of the feed can affect how much of a ration the pigs will consume.

- **Protein.** Appropriate swine feeding is as much about quantity and quality of amino acids as it is about total protein. The essential amino acid requirements for the appropriate function must be met for cost-effective production. The younger the animal, the more exacting the requirements for protein and for specific amino acids. Young pigs are started on a 20–22% protein diet, and as they grow, the percentage is lowered until it is around 13–15% protein.

- **Essential fatty acids.** Most practical swine diets provide adequate essential fatty acids.

- **Minerals and vitamins.** Careful consideration of total amounts and ratios of minerals and vitamins is essential. Some minerals are toxic if fed in excess. Vitamins are easily damaged and may not be available to the animal.

- **Water.** Good-quality clean water should be provided free choice. It is important to provide adequate watering space for all animals.

## FEEDING PRACTICES

Feeding practices in swine operations across the country tend to have much in common. Uniformity is increasing in feeding practice as the swine industry adopts more consistent technology and methods of production. What is different from region to region is the availability of certain feedstuffs, which vary with regional agronomic practice and the types of by-product feeds available. Most hog rations are formulated with the use of least-cost ration formulation programs that factor in the cost of nutrients from the available feeds and calculate the lowest cost, balanced diet. Table 19–3 shows some examples of diets for various classes of hogs. All of the examples in the table are based on corn and soybean meal to show how different proportions of these same ingredients can produce diets intended for different purposes. However, economics dictate that other feeds are also used if they can "cheapen" the ration by making it less expensive. In general, growing and finishing pigs are fed for *ad libidum* intake. Other classes of hogs are usually fed a limited amount of feed consistent with the production function.

## Boars and Gestating Females

These animals can easily consume their daily needs in one feeding. The biggest problem is keeping these animals from becoming overly fat. Thus feed quantities frequently must be restricted. With restricted total intake, nutrient balance of the feed is critical. Generally, boars and gestating females are fed measured amounts one or two times a day. The amount of feed given varies with condition, size, and reproductive stage. For

**Palatability**
The acceptability of a feed or ration to livestock.

**Complete diet**
Diet formulated to meet all the nutritional needs of an animal.

**Ad libidum**
Having feed available at all times.

**TABLE 19–3    Example Diets for Four Different Classes of Hogs[1]**

| Ingredient | Gestation and Prefarrowing Diet | Lactation Diet | Growing High Lean Gain Barrows 45–75 lbs | Growing High Lean Gain Gilts 140 lbs to Market |
|---|---|---|---|---|
| Corn | 1,607 | 1,460 | 1,454 | 1,514 |
| Soybean meal, 44% | 315 | 465 | 492 | 435 |
| Calcium carbonate | 19 | 19 | 15 | 16 |
| Dicalcium phosphate | 44 | 41 | 29 | 25 |
| Salt | 10 | 10 | 7 | 7 |
| Vitamin/trace mineral mix | 5 | 5 | 3 | 3 |
| Total, lbs | 2,000 | 2,000 | 2,000 | 2,000 |
| **Nutrient Analysis** | | | | |
| Protein (%) | 13.76 | 16.44 | 17 | 16 |
| Lysine (%) | 0.65 | 0.85 | 0.89 | 0.81 |
| Tryptophan (%) | 0.17 | 0.21 | 0.22 | 0.21 |
| Threonine | 0.52 | 0.63 | 0.65 | 0.61 |
| Methionine + cystine (%) | 0.54 | 0.57 | 0.58 | 0.56 |
| Calcium | 0.91 | 0.9 | 0.70 | 0.65 |
| Phosphorus (%) | 0.70 | 0.7 | 0.60 | 0.55 |
| ME (kcal/lb) | 1,475 | 1,470 | 1,486 | 1,491 |

[1]Each of these formulations is based on corn and soybean meal to illustrate how amounts of the same ingredients can be changed to alter the nutrient content. A variety of ingredients is available and commonly used in swine diets. Generally the feedstuffs that produce the desired performance and the lowest feed cost is what is used.

SOURCE: Luce, 1998b. Used with permission.

group-housed sows, feeding space should be adequate to allow all sows access to feed. Individual feeding stalls can be used to ensure that individual animals receive the appropriate amount.

## Farrowing and Nursing Sows

Just before and directly after farrowing, sows are often fed laxative feeds or additives to minimize the fairly common problem of constipation. Modern sows with the large litters common today have difficulty eating enough to support the litter and keep adequate body condition. The sow must also be managed in such a way that she is nutritionally capable of rebreeding within a few days of weaning the litter. The general feeding strategy is to feed a balanced diet two or more times a day and minimize weight loss during lactation. Many are fed ad libidum.

## Nursing Pigs

It is common to provide *creep* feeds to *nursing pigs.* They ordinarily consume a small quantity. However, it is important to get the pigs started on solid feeds so that the weaning process is less stressful.

## Nursery Pigs

The balance of the feeds offered to *nursery pigs* is the most exacting of all the swine rations. Modern swine systems wean pigs as young as 14 days old. Nutritional balance

**Creep**

An area where young nursing animals can have access to starter feeds. Creep feeds are the high-quality feeds made available to the young animals.

**Nursing pig**

A pig still nursing the sow.

**Nursery pig**

An early-weaned pig of light weight that is housed in special environmentally controlled nursery facilities.

**FIGURE 19–19** Early weaned (nursery) pigs are pigs that have been weaned from the sow as early as 2–3 weeks of age. The nutritional requirements of these pigs is exacting and their rations the most difficult to balance, mix, and feed of all swine rations. (Photographer Ken Hammond. Courtesy of USDA.)

of the feed is critical to the success of these early-wean systems (Figure 19–19). Even in systems that wean at older ages (this is done with increasing rarity), the diet must meet exacting nutritional needs. The baby pig is not a very forgiving biological entity. Inadequate care and nutrition at this stage have health and performance implications in later growth stages and affect the ability of the animal to be profitable. Adequate feeder space and access to water is also critical in this class of pig.

### Growing and Finishing Pigs

The goal in the growing and finishing phases is to take pigs to slaughter weight as quickly as feasible. Most often, finishing pig diets are designed to achieve *optimal growth.* Such diets consider the influence of fixed costs of production, stage of production and weight of the animal, climatic conditions, and genetic potential of the pigs. Increasingly, the sex of the pig is considered as well. Barrows and gilts have slightly different nutrient requirements. Large swine operations generally place animals of the same sex in a group and feed slightly different rations to the male and female groups.

**Optimal growth**

When optimizing growth, such factors as cost of the ration, environment, labor, and other nonfeed inputs are considered.

## HERD HEALTH MANAGEMENT

Numerous diseases can bring havoc to a swine operation. The greater the number of animals on one site and the closer they are to other animal operations, the greater the chance of having an infectious disease outbreak. A discussion of the specifics of each disease is outside the scope of this chapter. Instead, we take a brief look at methods of prevention.

A preventive herd health management program should be designed specifically for the conditions and facilities of each herd. The key to swine health is to prevent problems. Not only are swine diseases difficult and expensive to treat, but many diseases so affect the survivors that they never perform well. It is essential to include a veterinarian in the planning process. A veterinarian's expertise is needed to provide the means of prevention, control, diagnosis, and treatment of diseases. Other areas to be included in the plan include biosecurity, management practices, standards of production efficiency, updates on new information, record keeping, and data analysis. Many producers and veterinarians use the Pork Quality Assurance Plus program as a guide in developing a comprehensive herd health plan.

### BIOSECURITY

Biosecurity refers to procedures designed to (1) minimize the risk of disease transmission from sources outside the production unit, and (2) reduce the transmission of diseases among groups of pigs on the same farm. Some important biosecurity measures include the following.

For new animals to be added to the herd:

- *Assess.* The health status of any animals to be added to the herd should be assessed in advance. This is best accomplished by a "let my veterinarian talk to your veterinarian" approach. Health status can thus be assessed and potential problems avoided or at least anticipated.
- *Isolate.* New animals should be isolated from the herd until their health status can be evaluated on site. People who come in contact with the isolated animals should shower and change clothes and boots before returning to the main herd.
- *Stabilize.* New animals should be stabilized by allowing isolation to last long enough for any incubating disease to manifest itself. Any detected diseases and parasites should be diagnosed and treated. The animals should be intentionally exposed to pathogens present in the recipient herd to allow them to develop immunity.

For day-to-day management of the herd:

- *People.* Minimum precautions of changing boots and coveralls when moving between houses or rooms should be observed. Anyone going off premises should shower and change into clean clothes and boots before returning to the facility. It is best to avoid contact with other hogs. When contact is necessary, a shower and change of clothes is a requisite precaution before returning. Visitors should be strictly controlled and provided clothing and footwear when allowed on site (Figure 19–20).

**FIGURE 19–20** Proper biosecurity on swine farms demands that visitors be screened. Many post signs such as the one observed at the entrance to a hog farm in North Carolina. (Photographer Ken Hammond. Courtesy of USDA.)

- ***Pigs.*** Pig groups should be segregated by age. This helps decrease disease transmission and facilitates cleaning and disinfecting facilities.

- ***Vehicles.*** Any vehicle is a potential source of pathogens. General transportation vehicles should be parked a safe distance from the facility. Feed trucks should either be washed before entering the premises or unloaded a safe distance away and the feed augured or hauled by a farm vehicle to the barns. Both trucks entering the farm to load and transport pigs and trucks returning to the farm after transport should be washed thoroughly. Rendering trucks should never be allowed on premises.

- ***Vermin.*** Flies, mosquitoes, rodents, skunks, starlings, pigeons, sparrows, stray dogs and cats, wild animals, and feral pigs can pass diseases to swine. Each should be controlled.

## NUTRITIONAL BENEFITS OF PORK TO HUMANS

A 3-oz serving of cooked, lean pork provides the following proportion of the recommended daily dietary allowance for a 25- to 50-year-old woman on a 2,200-calorie diet:

| | |
|---|---|
| Protein | 50% |
| Phosphorus | 25% |
| Iron | 6% |
| Zinc | 20% |
| Riboflavin | 22% |
| Thiamin | 67% |
| $B_{12}$ | 32% |
| Niacin | 30% |
| Calories | 8% |
| Fat | 11% |

Like other meat products, pork is a ***nutrient-dense food,*** which means it has lots of nutrition per calorie. Pork is also much lower in fat than it was just a decade ago. Pork is a healthful food that can be part of the diet of virtually all people. A complete nutrient analysis is shown in Appendix IV.

**Nutrient-dense food**
A food that has a variety of nutrients in significant amounts.

## TRENDS IN THE SWINE INDUSTRY AND FACTORS THAT WILL INFLUENCE THE INDUSTRY IN THE FUTURE

### PORK CONSUMPTION

For the past three decades per capita pork consumption has remained essentially flat while poultry continued to enjoy healthy increases in consumption and beef declined and then stabilized (Figure 19–21). The fluctuations in per capita consumption since the early 1980s have generally been caused by price fluctuations. Pork needs to gain market share if the industry is to grow. Total pork consumed in the U.S. should increase modestly because of population increases. However, some of the growing minority groups, especially Hispanics, tend to consume less pork than the overall average. Unless they change their eating habits, this could dampen the growth in the

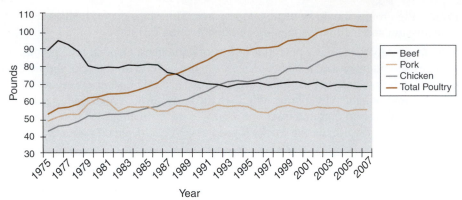

**FIGURE 19–21**    Per capita consumption of red meats and poultry. (*SOURCE:* USDA-ERS, 2007b.)

pork sector. Older people eat less pork than younger ones, and there is a graying population trend in the United States that could also affect overall pork consumption. Increased consumption of pork could be brought about through the production of a higher-quality, lower-cost product and additional convenient value-added products. Products that could significantly affect pork consumption include *functional foods* (omega-3 and selenium-enriched pork), natural pork, organic pork, vegetarian-fed pork, pork reared without the use of gestation and farrowing crates, and growth-enhancer-free pork. Also look for new pork cuts such as pork breast, petite tender, cap steak, and pocket roast (check them out at http://www.pork.org/newsandinformation/news/porkcheckoff24.aspx).

## NUTRITION AND HEALTH CONSCIOUSNESS

Nutrition and health consciousness will continue to play a role in food choices. Pork is good food that is nutritious. Lean pork is included in the dietary recommendations of all the major health organizations. The pork industry has made tremendous strides in producing a lean product that is lower in calories than it once was, and one that better fits into consumers' preferences for their diet. Today's hogs are genetically superior to those produced just a few years ago in terms of waste body fat. Fat content is, and will continue to be, an issue with consumers. Further progress needs to continue in this area. The educational efforts of the "Pork—The Other White Meat" campaign by the National Pork Board have apparently done a credible job of taking this message to consumers. A continuation of this effort and other informational campaigns should continue to improve the image of pork as a healthy food.

## CONVENIENT FOODS

The expanded integration of the pork industry is bringing about a shift in focus. Instead of considering itself to be in the pig-producing business, it is increasingly considering itself to be in the food business. With this change in focus (one the poultry industry adopted long ago) has come an expanded volume of good, convenience-oriented pork foods. Recognition that today's consumer has very different needs than those of the past is a key to maintaining or increasing consumption levels of any product. Surveys tell us that the average cook wishes to spend a total of 20 to 30 minutes in the kitchen preparing dinner. Children cook a big percentage of in-home meals. Younger people readily admit they don't really know how to cook. More and more,

**Functional foods**

Foods enriched with nutrients that may not be inherent to the food.

eating habits emphasize the need for less preparation time, ease of consumption, and good taste. Convenience at every level is very important. Affordable, tasty, convenient, microwavable pork products are now more available, but more innovation and progress in this area are needed.

## FOOD SAFETY CONCERNS

The safety of the food supply is a growing issue with consumers. It does little good to point out that the food supply in the United States is extremely safe. The consumer wants it perfectly safe—a goal that is simply unattainable. However, the industry must provide the safest food supply that can be reasonably provided. New initiatives are needed to improve the current system. Look for announcements and implementations of new plans and procedures.

In the past, the pork industry has had a problem with violative drug residues in meat. These residues can be avoided by observing proper withdrawal time for the products used before the pigs are slaughtered. A record of the product used, and the dose, duration of treatment, and period of withdrawal must be kept. Participation in the Pork Quality Assurance Plus Program (PQA Plus) is an effective mechanism to ensure that all producers follow residue avoidance guidelines.

Pork has escaped most of the unfortunate foodborne illness incidents that affected the beef industry during the 1990s. This is probably because most pork gets cooked to an internal temperature that virtually guarantees that pathogens are killed. The Hazard Analysis and Critical Control Points (HACCP) system was designed to aid in detecting and preventing problems in packing plants before the products ever reach consumers. With full implementation of this system, the safety of pork should be enhanced. The pork producer's responsibility under HACCP standards is to control violative tissue residues of antibiotics and chemicals. The PQA Plus provides excellent information to producers on how to accomplish this.

The swine industry is successfully controlling *E. coli* 0157:H7, which is a dangerous human pathogen. A study conducted by the U.S. Department of Agriculture, through its National Animal Health Monitoring System (NAHMS), determined the nation's swine herd to be free of the pathogen as of 1995. Periodic testing should be conducted to maintain the confidence of producers and consumers that this pathogen-free status is maintained.

## LAST STAGES OF RESTRUCTURING-CONSOLIDATION

The consolidation phase is the last phase of the restructuring of the hog industry. In 2006, 90% of the U.S. swine inventory was found in operations of 1,000 or more animals. These 1,000+ animal operations represented just 12% of the total swine operations. Conversely, 88% of the swine producers owned just 10% of the hogs. Small operations will continue to go out of business and the larger operations will continue to consolidate into even larger operations.

## TECHNOLOGICAL INNOVATION AND STANDARDIZATION

Producers who resist modernization and technological updates are being forced to leave the industry because they cannot compete. A so-called standard swine production unit is emerging and dominating the industry. The standard revolves around the economy of scale for the most efficient pork production. The standard includes the number of animals per unit, design for the facilities, genetics of the animals, feeds and feeding strategies, standard weights at marketing, and so on. The benefits of this standard include increased ease of adopting further technology. A more uniform product is produced and marketing strategies develop around value-added products that can be more easily designed with the standardized product being produced. This is the technology that has driven the emergence of large producers and it is the technology that will drive the production units that remain. Genetic engineering in particular will ultimately bring great dividends to this industry. The larger the operation, the more likely it is to use these tools. Conversely, the availability of these tools helps accelerate the trend to larger operations.

Confinement production will increase in the swine industry for several reasons. Confinement facilities lend themselves to the adoption of the latest technology. Swine are the best adapted of the red meat-producing animals to confinement rearing. Confinement production reduces labor, and facilitates manure handling and automatic feeding/watering. Maximum biological efficiency can be achieved in confinement rearing.

## WASTE DISPOSAL

As the prevalence of increasingly larger swine facilities has grown, manure disposal has become a major issue. Though brewing for many years, the issue finally came to a crisis point in the late 1990s and has boiled over into the new century. It was during these years that several well-publicized spills from swine lagoons occurred (Figure 19–22). Iowa and North Carolina were hardest hit by the adverse publicity, but other states had their share of problems. In addition to the physical waste, the public considers odor from swine operations to be a "waste," and they want it controlled. In the aftermath of the lagoon spills and the increased focus of attention that problem caused, a further focus of attention was brought to bear on the odor problem. State legislatures responded with moratoriums on the building of swine facilities, new and tougher rules for lagoon construction and safety, and, in some states, very explicit laws about where facilities could be located in relation to public roads, existing human housing, and so on.

**FIGURE 19–22**    The swine industry has to control the wastes from its facilities. Careful attention to every detail is a must if the industry is to regain the trust of the public on this sensitive issue. Here, a swine producer in North Carolina checks the operation of his lagoon pump. (Photographer Ken Hammond. Courtesy of USDA.)

There is no doubt that a portion of the problem has been dealt with through these measures. However, problems still remain. Most of these new laws apply only to new construction. Existing facilities still face problems that must be dealt with. Public perceptions about the offensiveness of swine odors are not going to change. For the industry's long-term well-being, the odor issue must be addressed. Industry groups must be willing to sponsor research studies to treat, reduce, and eliminate the chemicals that cause odors at swine farms. Effective methods of containing and disposing of solid and liquid wastes must be developed so that no spills occur and no surface or groundwater contamination occurs. The accidents and contaminations that have occurred have received such publicity that the industry should consider itself in a time of zero tolerance for errors on this issue.

Much progress has already been made in this area. Alternative strategies of feeding protein are being developed that reduce the amount of nitrogen in pig waste. Nitrogen is a problem because it is converted to ammonia, which is one of the major components of the odor from a swine facility. Likewise, much progress has been made in developing methods for reducing the phosphorous content of swine waste. Phosphorus is a major environmental contaminant from swine waste. Means of controlling odor are actively being explored, with several showing promise. Look for further improved methods and implementation of these and other strategies.

The U.S. EPA will continue to have a say in the way swine waste is managed in the future. Current environmental laws, regulations, policies, and guidance regarding CAFOs can be found at http://www.epa.gov/agriculture/anafolaw.html.

## BIOTECHNOLOGY

Numerous tools will ultimately come to assist the swine producer as a result of biotechnology. Biotechnological tools already developed or in various developmental stages for swine include:

- Innovative and faster techniques for the diagnosis of swine infectious diseases, gene-deleted or genetically engineered vaccines, more sophisticated diagnostic tests that differentiate vaccinated animals from infected animals, tests for the identification of infectious agents, and differentiation of related infectious agents, DNA fingerprinting of infectious agents, detection of carrier animals that are often difficult to diagnose by conventional techniques.

- Methods for the control of reproductive functions, such as induced estrus, induced ovulation, synchronization of prepubertal and mature gilts, superovulation, artificial insemination, embryo transfer, and sexed semen.

- Changes in body composition such as pork with significant amounts of omega-3 fatty acids—the kind believed to stave off heart disease.

Marker-assisted selection promises to assist swine breeders greatly in the selection of superior breeding stock. Of major interest are the genes that control growth, carcass composition, and resistance to disease. Marker-assisted selection will allow swine geneticists to predict the inheritance of economically important traits in specific animals and thus determine the genetic merit of animals more accurately and at younger ages. Animals do not have to be mated to determine the genetics they will pass on. Marker-assisted selection allows much more rapid genetic progress. It effectively allows geneticists to bypass generations in the selection process. Marker-assisted selection will allow for the selection of "designer pigs" to produce the next generation.

## FOREIGN COMPETITION/TRADE

Enhanced trade agreements and a general opening of world markets have created, and will continue to create, import–export opportunities for the United States and other countries. Several European countries are pork-exporting nations. Some of the South American countries are potential pork-producing and pork-exporting nations. Canada is making great progress in swine production, in both quality and quantity. Competition from foreign imports to the United States, and competition against the United States for other markets, is inevitable. However, U.S. pork exports have increased. In 1995, the United States became a net exporter of pork. The value of U.S. pork and pork by-product annual exports exceeds $2.75 billion (2006) and continues to increase yearly. This export volume represents a value of over $25 per hog produced in the United States and has allowed the industry to grow outside of domestic demand. The export market for pork shows promise for the long term. Burgeoning human populations, along with developing economies in some of those populations, are creating very favorable prospects for many agricultural exports from the United States. Although normal economic cycles and sporadic consumer concerns will surely affect year-to-year purchases by other countries, the overall trend should be positive for the foreseeable future.

## ANIMAL WELFARE/ANIMAL RIGHTS

The confinement systems of the swine industry have long been criticized by animal rights groups. Gestation and farrowing crates have been considered especially irksome. The European Union is phasing in a gestation crate ban to be complete in 2013. Recent state referenda (Florida and Arizona) and legislative action (Oregon) banning sow crates brought the issue to new attention. A cascade of announcements by major pork purchasers (Burger King, McDonald's, Chipotle Mexican Grill, Panera Bread, and Oregon-based New Seasons Market) altering their buying preferences based on whether the supplier used gestation crates was followed by a similar set of announcements by major pork producers (Smithfield Foods, Cargill Pork, Maple Leaf Foods in Canada) that they were phasing out gestation crates or "transitioning to group housing" for their sows. In 2007, respected animal system expert Dr. Temple Grandin was quoted as saying, "I think sow stalls will be gone 10 years from now" (Arnot & Gauldin, 2007). Grandin attributes the shift to "the organic and natural farming movement" rather than to the animal rights movement per se.

In recognition of the continuing need to improve animal welfare in pig production, and with an understanding that consumers increasingly want assurances about the way that animals are raised and cared for, the National Pork Promotion and Research Board officially launched the Pork Quality Assurance Plus program in 2007. It combines food safety, animal health, and welfare elements into one program. Details are available at http://www.pork.org/Producers/PQA/PQAPlus.aspx.

## WORK FORCE

Developing and retaining a workforce has become a major problem for the swine industry. Over the next decade (at least), the labor force, both skilled and unskilled, will tighten dramatically. This problem will be further complicated by such issues as group housing of sows, which requires more labor and management skills than gestation crate management, and other such initiatives.

# INDUSTRY ORGANIZATIONS

**American Berkshire Association**
P.O. Box 2436
West Lafayette, IN 47996
Phone: (765)497–3618
Fax: (765)497–2959
E-mail: berkshire@nationalswine.com
http://www.americanberkshire.com

**American Livestock Breeds Conservancy**
P.O. Box 477
Pittsboro, NC 27312
Phone: (919)542–5704
Fax: (919)545–0022
E-mail: albc@albc-usa.org
http://www.albc-usa.org/

**Certified Pedigreed Swine**
P.O. Box 9758
Peoria, IL 61612–9758
Phone: (309)691–0151
Fax: (309)691–0168
E-mail: cpspeoria@mindspring.com
http://www.cpsswine.com

**National Pork Board**
1776 NW 114th St.
Des Moines, IA 50325
Phone: (515)223-2600; 1-800-456-PORK
E-mail: info@pork.org
http://www.pork.org
Composed of members who are nominated by producers and appointed by the Secretary of Agriculture. They are responsible for administering the funds from the pork checkoff. The checkoff was created by the Pork Promotion and Research Act of 1985. This act provided that all producers contribute a portion of their sales (0.4) to be used for pork promotion, research, and consumer education.

**National Pork Producers Council**
122 C Street
Washington, DC 20001
Phone: (202)347–3600
Fax: (202)-347-5265
E-mail: pork@nppc.org
http://www.nppc.org/
Composed of 44 affiliated state pork-producer organizations. State organizations select delegates who collectively set policy. Their purpose is to "increase the quality, production, distribution, and sales of pork and pork products."

**National Swine Registry**
P.O. Box 2417
West Lafayette, IN 47996
Phone: (765)463–3594;
Fax: (765)467–2959
E-mail: nsr@nationalswine.com
http://www.nationalswine.com

**U.S. Meat Export Federation**

1050 17th St., Suite 2200

Denver, CO 80265

Phone: (303)623-MEAT (6328)

Fax: (303)623-0297

E-mail: info@usmef.org.

http://usmef.org/

A nonprofit integrated trade organization representing a wide variety of groups including livestock producers, meatpackers, processors, farm organizations, grain promotional groups, agribusiness companies, and others. The organization works to develop foreign markets for U.S.-produced beef, pork, lamb, and veal. The headquarters is in Denver, Colorado. However, offices are also located in several foreign countries to facilitate the work of the organization.

## SUMMARY AND CONCLUSION

The U.S. gross annual income from pork is approximately $14.2 billion. Pigs currently generate 5.7% of all U.S. farm cash receipts. The swine industry is not only a major industry, but it is also an aggressive, technologically driven, expansion-minded industry. The purpose of the swine industry in the United States is to produce meat from millions of tons of excess grain and available by-products. Swine are monogastric and use only limited amounts of forage. However, they are the most efficient livestock converters of grain to red meat. They help moderate the fluctuations in grain prices, which could otherwise disrupt the economy and the supply of grain for humans.

Feed costs are a high fixed cost, so economical production depends on minimizing labor and maximizing efficiency by reducing time to market. This requires a large investment in facilities, equipment, and technology. This has caused an increasingly integrated and consolidated swine industry with larger operations. Distinct regional shifts have occurred in the swine industry. A group of megaproducers have driven expansion east, south, and west. The Hog Belt has let out a few notches. The industry is currently in the last stages of a consolidation phase.

Swine breeders have made remarkable progress in genetic improvement, especially in reducing the fat content of the average pork carcass. Modern consumers demand lean pork, and today's hogs are much leaner than those of the past. Further progress is also being made. Modern swine production systems demand high reproductive rates from swine. Expensive confinement facilities must be kept full. Reproductive management in a swine herd is a time-consuming task that must be done well. There is tremendous overlap among good reproductive management, good animal health management, and good nutritional management.

Feeding practices in swine operations across the country are becoming more uniform. Most hog rations are formulated with the use of least-cost ration formulation programs. In general, growing and finishing pigs are fed for ad libidum intake and other classes of limited intake consistent with level of production. Herd health programs should be designed for each herd. Prevention is much more important than treatment. Swine diseases are difficult and expensive to treat and affect the survivors to a degree that prevents them from ever performing well. Many producers and veterinarians use the Pork Quality Assurance Plus Program as a guide in developing a comprehensive herd health plan.

Like other meat products, pork is a nutrient-dense food, which means it has lots of nutrition per calorie. Increased consumption of pork could be brought about through the production of a higher quality, lower cost product, more convenient foods, and emphasis on pork's excellent record in food safety.

### Facts about Swine

| | |
|---|---|
| Birth weight: | 2–3 lbs |
| Weaning weight: | 10–12 lbs at 2–4 weeks; 30–40 lbs at 6 weeks |
| Mature weight: | Male 500–800 lbs; female 400–700 lbs; miniatures 140–170 lbs |
| Slaughter weight: | 230–260 lbs |
| Weaning age: | 2–6 weeks |
| Breeding age: | 6–8 months |
| Normal season of birth: | Traditional, spring and fall; Modern, year-round |
| Gestation: | 112–115 days |
| Estrous cycle: | 19–21 days |
| Duration of estrus (heat): | 2–3 days |
| Boar/sows services: | Pasture 1/15; hand-mating 1/20 (limit to 1–2 per day); AI 6–10 sows per service |
| Litter size: | 7–15, although larger litters are common |
| Names of various sex classes: | Sow, gilt, boar, barrow |
| Digestive system: | Nonruminant, monogastric |

## STUDY QUESTIONS

1. What is the economic value of the swine industry to U.S. agriculture? Describe the magnitude of this industry in other ways.

2. What is the purpose of the swine industry in the United States? Describe it in terms of feed and use of resources.

3. What is the importance of the pig to worldwide meat production?

4. When were the first hogs brought into the Western Hemisphere? Who else brought them and what part of the continent were they on? What was the most important use of the pig until well into the 20th century? How do today's swine numbers in the United States compare to those of the past?

5. What are the major types of swine production practiced in the United States today? Describe each. Compare the number of swine producers today to the past. How is the industry restructuring? What are the factors that will ultimately determine the location of the U.S. swine industry?

6. Why are swine located where they are in the United States? Describe what is happening to the swine industry in terms of its geographical location.

7. What are some of the tools available to help swine producers make sound decisions about the genetics they use? What are the roles of the different breeds in the swine industry? How are synthetic lines being used in the breeding of swine? Briefly discuss the value of crossbreeding in swine production.

8. Describe the ideal market hog according to the National Pork Board.

9. Briefly describe swine management for maximal reproductive efficiency. What is the role of artificial insemination in the swine industry?

10. Briefly discuss nutrition in hogs for each class of animal. What are the similarities and what are the differences?

11. Why is having and following a herd health plan so important in raising hogs?

12. Describe methods of addressing the concept of biosecurity in a swine unit. Be specific. Can you think of any others that might be used?

13. What does a 3-oz serving of cooked pork provide nutritionally? What does "nutrient dense" mean? Are other animal-based foods nutrient dense?

14. How much pork does the U.S. population consume on a per capita basis? Is that amount changing? What has happened to the market share of pork compared to that of beef and poultry?

15. What changes have been made in pork to make it a more health-conscious food? How do you think the issue of nutrition and health consciousness will affect the swine industry in the future?

16. Why will swine operations continue to grow bigger? Briefly discuss how and why vertical integration is involved in swine industry consolidation.

17. Why is it important for swine producers to use the technological innovations available to them? How do confinement operations and the adoption of technology go together? Speculate about the role that technology will play for the swine industry in the future.

18. Why is the issue of waste disposal so important to the swine industry? What do you think will happen with laws and regulations regarding waste disposal in the swine industry? Does the issue of odor differ from the issue of solid waste? If so, how does it differ?

19. What role will exports likely play in the growth of the future swine industry?

## REFERENCES

Ahlschwede, W. T., C. J. Christians, R. K. Johnson, and O. W. Robison. 1999. *Crossbreeding systems for commercial pork production*. PIH-39. Pork Industry Handbook.

Arnot, C. and C. Gauldin. 2007. Sow stall debate is at a "tipping point." *Feedstuffs* 79 (15).

Cleveland, E. R., W. T. Ahlshwede, C. J. Christians, R. K. Johnson, and A. P. Schinkel. 1999. *Genetic principles and their applications*. PIH-106. Pork Industry Handbook.

Ensminger, M. E. 1991. *Animal science*. 9th ed. Danville, IL: Interstate.

FAO. 2007. *FAOSTAT statistics database. Agricultural production and production indices data*. http://apps.fao.org/.

Jones, R. D. 1989. History of the pig: Swine over 40 million years old. *Livestock Newsletter*. Athens: University of Georgia.

Luce, W. G. 1998b. Regents professor, Oklahoma State University, Stillwater, OK. Personal communication.

NPPC. 2007. *Pork facts*. Des Moines, IA: National Pork Producers Council.

USDA. 1997. *USDA nutrient database for standard reference*. Release 11–1. Nutrient Data Laboratory home page, http://www.nal.usda.gov/fnic/foodcomp.

USDA. 1998. *USDA nutrient database for standard reference*. Release 12. Nutrient Data Laboratory Home Page, http://www.nal.usda.gov/fnic/foodcomp.

USDA-ERS. 2007. *Red meat yearbook*. Table 114. Accessed online Oct 4, 2007. http://www.ers.usda.gov/data/sdp/view.asp?f=livestock/94006/.

USDA-NASS. 2007a. *Briefing room. Farm income and costs*. Accessed online September 2007. http://www.ers.usda.gov/Briefing/FarmIncome/.

USDA-NASS. 2007b. *Quick Stats: Agricultural Statistics Database*. http://www.nass.usda.gov/QuickStats/.

# Sheep and Goats

## LEARNING OBJECTIVES

After you have studied this chapter, you should be able to:

1. Explain the place of sheep and goats in U.S. agriculture.

2. Explain the reasons for the size of the U.S. sheep and goat industries and describe how they are changing.

3. Give a brief history of the sheep and goat industries in the United States.

4. Describe the structure of the sheep and goat industries.

5. Give an accurate accounting of where the sheep and goat industries are located in the United States.

6. Identify the elements of a practical approach to breeding and genetics in the sheep and goat industries.

7. Explain the significance of breeds in the U.S. sheep and goat industries and describe the general classifications of sheep and goat breeds.

8. Outline the basis of managing sheep and goats for reproductive efficiency.

9. Describe the feed supply of sheep and goats and explain why that feed supply is so important to the purpose of sheep and goats in the United States and other countries.

10. Explain the nutritional benefits of sheep and goat milk to humans.

11. Discuss trends in the sheep and goat industries including factors that will influence the industries in the future.

## KEY TERMS

| | | |
|---|---|---|
| Anestrus | Doe | Flushing |
| Angora | Estrous cycle | Forage |
| Breed complementarity | Estrus | Heterosis |
| Browse | Ewe | Intensive systems |
| Conception | Flock | Kidding |

Lamb

Lambing

Mohair

Palpation

Roughage

Seasonal market

Secondary enterprise

Sheep

Singleton

Specialty market

Survival of the fittest

Trimester

Wool

## SCIENTIFIC CLASSIFICATION OF SHEEP AND GOATS

| | |
|---|---|
| Phylum: | Chordata |
| Subphylum: | Vertebrata |
| Class: | Mammalia |
| Order: | Artiodactyla |
| Suborder: | Ruminata |
| Family: | Bovidae |
| Genus: | *Ovis* (sheep); *Capra* (goat) |
| Species: | *aries* (sheep); *hires* (goat) |

## THE PLACE OF THE SHEEP AND GOAT INDUSTRIES IN U.S. AGRICULTURE

**Sheep**

An animal of the genus *Ovis* that is over 1 year of age.

**Lamb**

A sheep under 1 year of age. Also, the meat from a sheep under 1 year of age.

The sheep industry has declined in the United States and is now approximately two tenths of 1% of the total U.S. farm revenue from livestock and products (Figure 20–1). The gross annual income from *sheep* and *lambs* in the United States is approximately $500 million and declining. It comprises only four tenths of 1% of animal agriculture's share of cash receipts (Figure 20–2). The U.S. population consumes very small amounts

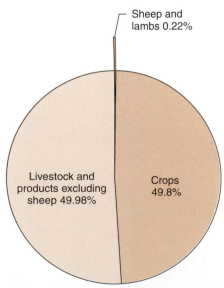

**FIGURE 20–1**  Sheep, lambs, and wool combined farm cash receipts as a percentage of total U.S. farm cash receipts, 2004–2007. (*SOURCE:* Based on USDA statistics.)

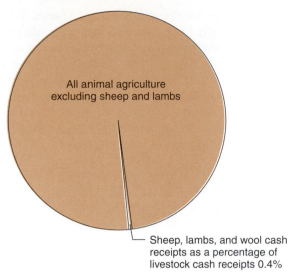

All animal agriculture
excluding sheep and lambs

Sheep, lambs, and wool cash
receipts as a percentage of
livestock cash receipts 0.4%

**FIGURE 20–2**    Sheep, lamb, and wool farm cash receipts as a percentage of animal agriculture's cash receipts, 2004–2007. (*SOURCE:* Based on USDA statistics.)

of lamb meat, less than 1 lb per capita on a boneless weight basis. The sheep industry was once a mighty industry in the United States, but those days are long gone and are not likely to return. If the United States stopped growing sheep altogether, there would be ample imports to supply the needs of the whole country. Clearly, the current status of this industry now places it in the ranks of a specialty industry. However, it is an important industry to the more than 69,000 sheep producers in the United States who make their livelihood, or part of it, from sheep. Sheep are also an important 4-H and FFA club species. Many young people who may not have the size, skill, space, or money to raise a steer or horse can raise a lamb and have a meaningful project (Figure 20–3). It is also a species that produces a specialty meat product in demand in the hotel and restaurant trade and, in certain parts of the country, in grocery stores as well. The dairy sheep industry shows some potential to become an industry segment of significant value to some states.

**FIGURE 20–3**    Once a major industry, the sheep industry has declined to a specialty industry. Many 4-H and FFA club projects include rearing and showing lambs.

### Angora

A specialized fiber-producing breed of goat.

The goat industry has never been large in the United States. With the exception of the **Angora,** goats have generally been dispersed in smallholdings and hobby operations scattered across the country. However, Angora goats have had a place in the agriculture of a small number of southern states. Dairy goats have been a small but noticeable industry in several states. Meat goat production has been on the increase in the United States over the past decade or so and has emerged as an alternative agriculture for some of the nation's farmers. Goat production has grown most actively in the South and in a few eastern states. Gross annual income for the goat industry is estimated to be approximately a third that of the sheep industry.

## PURPOSE OF THE SHEEP AND GOAT INDUSTRIES IN THE UNITED STATES

### Forage

Fiber-containing feeds like grass or hay. Can be grazed or harvested for feeding. Contain at least 18% fiber but are have high digestible energy (>70%).

### Roughage

A bulky feedstuff with low weight per unit volume. Contains at least 18% fiber but can range up to 50%. Less digestible than forages.

### Flock

A group or band of sheep.

The purpose of sheep and goats in the United States is much the same as the purpose of other ruminant species: to take advantage of *forage* and *roughage* to produce products that humans can use (Figure 20–4). The products include milk, meat, and fiber. Historically, sheep have been used as part of a mixed farm in the eastern states and in large *flocks* that graze rangelands in the western part of the country. Range production, in turn, differs depending on whether the range is dry or wet. In most situations in which sheep are kept as part of a diversified farm, meat production is more profitable than wool production. Therefore, when forage conditions are good enough to permit weaning a high percentage of lambs fat enough for slaughter, meat production is the primary goal. Wool production is secondary and is more of a by-product. Wool production is likely to be more profitable than meat production if forage conditions are poor. In this case, wool production is emphasized and meat production is the secondary product. This is the situation in the dry range areas, but not in the wet range areas. In wet range conditions, the nutritional level is high enough that both meat and wool can be produced satisfactorily. The value of wool has declined to the point that raising sheep for wool in the United States is rarely profitable (Figure 20–5). Thus most systems focus on meat production.

**FIGURE 20–4**  Sheep are excellent converters of forage into food and fiber. (Photo courtesy of Dr. Jerry Fitch, Department of Animal Science, Oklahoma State University. Used with permission.)

**FIGURE 20–5** Wool production from sheep was once vital to the comfort and economy of the country. Competition from synthetics and availability of wool on the world market have reduced the importance of its production. (Photo courtesy of Union Pacific Railroad Museum.)

Sheep and goats have always been able to help us better use our forage resources when used in conjunction with cattle. When sheep and/or goats use the land along with cattle, they make better use of the resources because they graze slightly different forage. Sheep are better able to select their diet than cattle and can thus pick a better diet. Goats like to eat *browse* and select it first if it is available. In this way, goats and cattle complement each other. It is generally accepted that one *ewe* or *doe* can be added per each existing cow unit and no additional forage will be needed. In wooded or scrub conditions, goats can be the preferred species because they like the browse. Both goats and sheep can actually improve pastures and other grazing resources because they eat many species, including weeds, that cattle leave behind. For this reason, sheep farms frequently have cattle on the same operation. However, the use of mixed species to make better use of our grazing land has not been practiced as widely as it might be in this country. This is so in spite of the fact that research has demonstrated the economic and biological value of grazing mixed species. Goats have been used even less widely as part of mixed grazing systems. Much of this is because sheep require more care and are subject to greater predation losses than cattle. Because of their liking for browse, *mobs* of goats are often used as "brush clearers" in lands that need to be reclaimed. They are also often used in brush control under power lines and other similar areas in some states.

*Wool* and *mohair* were subsidized by the federal government from 1954 until 1995, which helped these industries. The importance of the wool and mohair incentive program to the sheep industry is easy enough to see. Since the program ended, there has been a precipitous decline in sheep and Angora goat numbers (Figure 20–6).

Although the United States is not a major sheep- and goat-producing country, much of the rest of the world has sizable sheep and goat interests. Sheep and goats are used extensively around the world for their grass conversion ability by a wide variety of people. With over 1.1 billion head of sheep and approximately 840 million head of goats distributed globally, they are the second and fourth most numerous agricultural animals, respectively, excluding the poultry species. Because of their smaller size, five sheep or seven goats can be kept on the same amount of feed as one cow. Also, because of their smaller size, a goat or sheep carcass can be consumed more rapidly than the

**Browse**

The tender twigs and leaves from brush and trees.

**Ewe**

A female sheep.

**Doe**

A female goat.

**Mob**

A group or herd of goats.

**Wool**

The fiber that grows instead of hair on the bodies of sheep.

**Mohair**

The fiber produced by the Angora goat.

**FIGURE 20–6** Angora goats produce mohair. Although Angora production was never a large industry, the phase-out of the federal wool and mohair subsidy in 1995 has caused a precipitous decline in Angora numbers. (Photographer S. Alden. Courtesy of Getty Images, Inc./PhotoDisc, Inc.)

carcass of cattle and is thus less likely to spoil. This is important for people who do not have cold storage available for meat. Goats and sheep are thus a better choice for people in less-developed economies. On a worldwide basis, goats and sheep are increasing modestly. They are also widely distributed. It is believed that every country in the world has some sheep and some goats. There is one sheep in the world for every 5 people and 22 head of sheep per square mile of earth land surface. There is one goat in the world for every 8 people and 16 head of goats per square mile of earth land surface. However, the United States has only 0.34% of the world's goats and 0.56% of the world's sheep. If estimates are correct, the United States has a greater percentage of the world's llamas and alpacas.

## HISTORICAL PERSPECTIVE

Sheep and goats were probably both domesticated by 8000 B.C. The sheep was probably domesticated first. It is generally accepted that sheep, goats, and pigs were all domesticated around the same time, with the pig the last of the three. This makes sheep and goats the first of the food-producing animals to be domesticated. Both sheep and goats went with humans as we populated the globe and have been used virtually across the planet (Figure 20–7).

Christopher Columbus has the distinction of being the first sheep and goat producer in the Western Hemisphere, just as he was for cattle. He brought sheep and goats to the West Indies on his second voyage in 1493. Cortez brought sheep and goats, as well as cattle, to Mexico with him in 1519. Sheep were brought to the East Coast in 1609 by the English who settled in New England. Early on, the value of sheep as a fiber producer was recognized. Importations improved the value of the wool until the sheep industry eventually supported a thriving wool industry. As the settlement of the continent proceeded, the grasslands became increasingly used as sheep-producing areas. Eventually, the sheep industry became located predominantly in the western part of the country.

Figure 20–8 gives a historical look at sheep numbers in the United States. The numbers of sheep have undergone wide swings to reach the industry's *specialty market* status. During the early colonial period, they were especially important for their wool

**Specialty market**

A term that suggests a product generally aimed at a specific segment of the overall market.

**FIGURE 20–7** An established and useful animal in Europe, sheep were brought to the New World by European settlers. They quickly became important to the economy. (Photo courtesy of Corbis.)

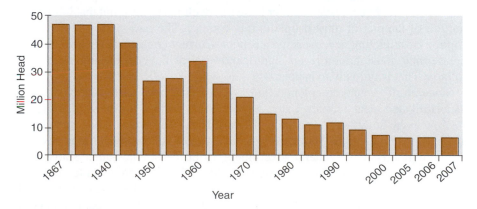

**FIGURE 20–8**    Sheep and lamb inventory. (*SOURCE:* USDA-NASS, 2007a.)

production. This was really the case until the early 1900s. The production of wool was of primary importance and the production of slaughter animals was incidental. Starting in the late 1800s this began to change and the production of lambs for slaughter became increasingly important. This is also when U.S. sheep production shifted to the western states.

The total count of sheep and lambs on January 1, 1867, just after the Civil War, was 46.3 million head. From 1867 until now, sheep numbers have passed through many cyclical phases. The all-time high was in 1942, when sheep numbers totaled 56.2 million head. The numbers declined after World War II to less than 30 million head in 1950 and dropped to less than 15 million by the early 1970s. Since the early 1970s, the numbers have cycled to successively lower levels. Numbers may decline further. The declining importance of sheep and lambs has been attributed to a number of causes, including:

1. Less demand for wool. As more synthetics and cheaper substitute materials were introduced, wool wasn't as competitive.

2. The declining demand for lamb in consumer diets and the relatively high price of lamb relative to other meats.

3. Increased difficulty in obtaining and keeping reliable herders to manage and care for range flocks.

4. Increased competition for public-owned rangeland and increasing grazing fees.

5. An increasing problem of predators in many range and farm flock-producing states.

6. Decreased government support, especially the demise of the wool support program.

7. Farmer diversification into other enterprises.

8. Seasonal nature of lamb production and consumption.

9. Inadequate profit to keep producers producing.

Although large decreases in numbers have left the sheep industry a much smaller part of U.S agriculture, there does seem to be a new optimism about the sheep industry. The declines in total numbers have become less drastic, with the numbers for breeding animals actually showing some increases starting in 2005. Lambs per 100 ewes have also increased modestly, which has given a larger lamb crop. The inventory value for sheep increased by nearly 50% from 2000 to 2007. Toward the end of the decade, sheep operations actually showed a modest increase. There are still problems to overcome, however. Per capita demand for lamb has declined to the point that it is no longer a portion of the diet of most people in this country. The outlook for improving consumption by attracting new consumers is poor. Current consumers are also unlikely to increase their consumption levels significantly unless the cost of lamb declines substantially. That is unlikely to happen.

Since North American colonial times, goats have been used in very small numbers as milk and meat animals. Later, Angora goats became useful as fiber producers for a time in the Southwest, especially in Texas. However, goat numbers and goat meat consumption in the United States has historically been so low that it has not even been reported regularly by the federal government. Prior to 2005 no annual statistics were reported for goats except for Angora goats, although data were included in the Census of Agriculture. Increasing interest in goats and demand for goat meat led USDA-NASS to publish its first annual goat survey in 2005. Since then, goat numbers in the United States have been increasing 3–5% annually with almost all of the growth related to increased meat-type goat production (Figure 20–9). Much of the increased demand for goat meat is attributed to growing immigrant and ethnic populations. Additionally, many states offered incentives to farmers who participated in the tobacco buyout program if they would get into other areas of agriculture. The Southeast

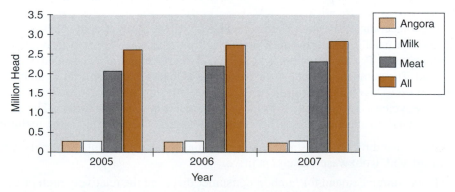

**FIGURE 20–9**   U.S. goat inventory by class. (*SOURCE:* USDA-NASS, 2007b.)

is where the majority of the tobacco was grown and has also seen the largest increases in goat numbers. The increase in number of goats is not showing any signs of slowing down. Goat production appeals to people who have limited access to land and want an animal enterprise. Goat production is also appealing to women and children and requires little start-up costs in terms of equipment and buildings. A substantial youth exhibition industry has also developed around the goat, with some states having more students enrolled in 4-H and FFA goat projects than sheep projects.

## STRUCTURE OF THE SHEEP AND GOAT INDUSTRIES

### SHEEP

The infrastructure of the sheep industry has consolidated in response to declining sheep inventories. More slaughter plants have located in the lamb-feeding areas. Range sheep production, lamb feeding, and the purebred sector are the three major segments of the U.S. sheep industry. Purebreds are a very small part of a very small industry. Purebred producers produce breeding stock and are located all over the country.

Range sheep producers keep grazing flocks on pasture and/or range. Dry range often offers few alternatives for productive use of the land. Even then, cattle compete and often win as the enterprise that will make more money with less fuss. The grazing flocks of sheep are referred to as *stock sheep*. They provide a portion of their lamb production directly to slaughter and a portion to feedlots. They also provide wool from the ewe flock to sell (Figure 20–10).

Lambs not ready for slaughter directly from range are placed in feedlots at a wide range of weights from 40–90 lbs and are fed grain until they are ready for market (finished). In the past, most lambs were finished on grass. Now, most are finished in a feedlot. In general, this is done because it is more cost effective to feed lambs in a feedlot than to grow them on forage. There are exceptions to this, however. For example, many lambs are finished on forage in California and produce a high-quality product ready for slaughter. However, the conditions in which many of

**FIGURE 20–10** Range sheep production is often practiced on dry range with little alternative use. Flocks are often large with several thousand stock sheep in them. (Photo courtesy of USDA.)

**FIGURE 20–11** Most sheep producers keep small flocks as a secondary enterprise or hobby. However, most sheep are kept in large range flocks.

the lambs are raised do not provide high-quality feed for finishing and don't produce a very high-quality product. Producing a uniform product is important to have satisfied customers. Lambs in these areas are generally fed on concentrated feeds in a feedlot prior to marketing.

Most U.S. sheep producers have small flocks and raise sheep as a **secondary enterprise** or as a hobby (Figure 20–11). Most large producers are found in 17 western states, where the bulk of the sheep are kept in large range flocks. Most small flocks and, therefore, most sheep producers are found in states other than where most of the sheep are found (Figure 20–12).

**Secondary enterprise**

In diversified farming operations, the enterprise secondary to the one providing the bulk of the income.

### GOATS

The Angora goat sector is shrinking because of the demise of government payments for mohair. The meat goat industry is the largest segment (Figure 20–9, Figure 20–13). Primarily located in the South, considerable interest in the meat goat as a potential expansion industry has been observed in several other states since the 1990s. There seems to be room for this industry to grow.

The dairy goat sector has had a few large-scale commercial operations and numerous small-scale holders who milk a few goats for small returns or as a hobby or part of a diversified operation. Traditional dairy cow states have also tended to have larger numbers of dairy goats. As an economic force, the dairy goat industry does not contribute substantially to agricultural income. Viable commercial enterprises exist only in a few states. The few commercial dairies that do exist market their milk as fresh or evaporated or as cheese. Typically, dairy goats have had such poor returns that farmers wanting to dairy have been better off doing so with cattle.

## DAIRY SHEEP

In many countries of the world, particularly those surrounding the Mediterranean, sheep dairying has a rich history and is a viable, modern agricultural enterprise. Sheep milk is especially valued for producing some of the world's best and most expensive cheeses for

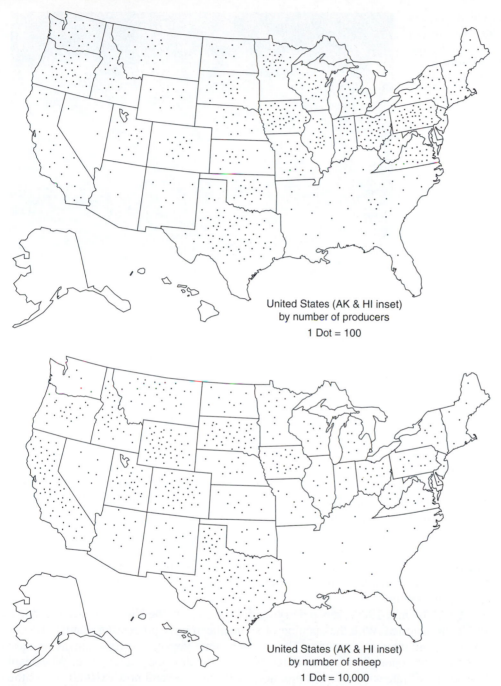

United States (AK & HI inset)
by number of producers
1 Dot = 100

United States (AK & HI inset)
by number of sheep
1 Dot = 10,000

**FIGURE 20–12**   The top map is a distribution of sheep producers. The bottom map shows the distribution of sheep in the United States. (*SOURCE:* USDA-NASS, 2007a.)

**FIGURE 20–13**   Most meat goats are kept in small flocks. These meat-type kids reflect several different breeds. However, as this industry has grown, more meat-type breeds are being used. (Photo courtesy of Dr. Rebecca L. Damron. Used with permission.)

the gourmet market. Sheep milk is valuable for producing cheese because the solids are higher in sheep milk (18%) than they are in cow or goat milk (12–13%). A pound of sheep milk yields almost twice as much cheese, and the flavor is different as well. Worldwide, at least 50 cheeses are made from sheep milk, totaling approximately 636,000 metric tons in 1998. Some of them include feta, ricotta, Brebicet, Fiore Sardo, pecorino, and Roquefort. The United States imports more than 30 million pounds of sheep cheeses annually. It is estimated that five times more sheep than cows are milked in the world, although they produce only a small fraction of the total milk (Figure 20–14).

Worldwide, there are many systems in place for milking sheep. Some of these systems involve allowing the lambs to be reared by the ewe for periods ranging from 1 to 4 months. The lambs are then weaned and the ewe is milked for whatever remains of a 5-month period. Nomads have long kept ewes and lambs separated at night and milked the ewes in the morning before allowing the lambs to nurse. Some *intensive systems* have developed similar milking strategies. In these systems, the lambs are weaned at 8 weeks of age and the ewe is milked twice a day for an additional 3 months. More intensive systems of milking usually wean lambs at 24 hours of age and milk the ewes twice daily after that for as many as 10 months. This system is similar to the intensive dairy cow system.

Until the mid-1980s, sheep dairying did not exist in the United States or the rest of North America. With the opening of some sheep milk-processing plants, this has changed, and interest is growing. Dairy sheep is not yet an industry segment because it is not large enough. It is considered to be in the developmental stage. Wisconsin has been the state with the most producers, although several now exist all across the country (Figure 20–15). If this is to develop into a full-scale animal industry segment, several things will need to happen. Additional processing plants will need to open at various locations around the country. The genetics for milk production will need to improve in the U.S. flock. Milking parlor design and equipment need to improve and be made available at reasonable prices as a means of decreasing labor. Current labor needs for sheep dairies are approximately eight times more than those for cow dairies.

**Intensive systems**

Animals produced in intensive systems are kept in more concentrated conditions, fed higher-quality feeds, and have much more labor, technology, and overall attention directed to them.

**FIGURE 20–14**   Both goat's milk and sheep's milk are used in the making of specialty and gourmet cheeses. (Photographer Michael Busselle. Courtesy of Corbis.)

**FIGURE 20–15**   Milking sheep at the Spooner Agricultural Research Station, University of Wisconsin at Madison. Sheep dairying has long been practiced outside the United States and is a developing industry in the United States. (Photographer Wolfgang Hoffmann. Courtesy of Spooner Agricultural Research Station, University of Wisconsin-Madison.)

## GEOGRAPHIC LOCATION OF SHEEP AND GOATS IN THE UNITED STATES

### SHEEP

U.S. sheep inventory is shown in Table 20–1. The western states have the majority of all breeding sheep, predominantly because of the arid rangeland found there. Sheep are ruminants, and they can use forages. Sheep are often found on operations that have cattle holdings as well. A mixture of sheep and cattle generally make better use of available forage because they select different diets and the native range species therefore do better. Rented Bureau of Land Management and Forest Service lands have traditionally supplied the grazing for 30–40% of the sheep in the West. With the increase in user

**TABLE 20-1** **All Sheep and Lambs: Number by Class, State, and United States**

| State | Total Sheep & Lambs 1,000 Head | Total Breeding Sheep 1,000 Head | Total Market Sheep & Lambs 1,000 Head |
|---|---|---|---|
| Arizona | 105 | 70 | 35 |
| California | 650 | 325 | 325 |
| Colorado | 390 | 190 | 200 |
| Idaho | 260 | 220 | 40 |
| Illinois | 69 | 61 | 8 |
| Indiana | 50 | 46 | 4 |
| Iowa | 235 | 170 | 65 |
| Kansas | 100 | 65 | 35 |
| Kentucky | 35 | 29 | 6 |
| Maryland | 22 | 16 | 6 |
| Michigan | 83 | 61 | 22 |
| Minnesota | 155 | 110 | 45 |
| Missouri | 75 | 65.5 | 9.5 |
| Montana | 295 | 270 | 25 |
| Nebraska | 106 | 80 | 26 |
| Nevada | 74 | 64 | 10 |
| New Mexico | 155 | 130 | 25 |
| New York | 70 | 54 | 16 |
| North Carolina | 18 | 14 | 4 |
| North Dakota | 104 | 76 | 28 |
| Ohio | 141 | 117 | 24 |
| Oklahoma | 80 | 62 | 18 |
| Oregon | 220 | 145 | 75 |
| Pennsylvania | 110 | 94 | 16 |
| South Dakota | 385 | 280 | 105 |
| Tennessee | 27 | 22 | 5 |
| Texas | 1,090 | 870 | 220 |
| Utah | 280 | 260 | 20 |
| Virginia | 67 | 51 | 16 |
| Washington | 50 | 41 | 9 |
| West Virginia | 32 | 27 | 5 |
| Wisconsin | 89 | 73 | 16 |
| Wyoming | 450 | 350 | 100 |
| New England | 47 | 40.5 | 6.5 |
| Other States | 111 | 91 | 20 |
| United States | 6,230 | 4,640 | 1,590 |

SOURCE: *USDA-NASS* 2007a.

fees for these lands, some sheep producers are finding themselves unable to make a profit. This is contributing to the decline in western sheep numbers.

The Great Plains and California are the areas of greatest concentration of lamb feeding. Colorado is the primary lamb feedlot state. Other important lamb-feeding states are California, Texas, and Wyoming. Many lambs are also finished for market on winter alfalfa pastures in the Imperial Valley of California.

## GOATS

Goats are distributed across the United States (Figure 20–16). Texas has over three fourths of the Angora goats in the United States (Figure 20–17), with most located in the Edwards Plateau region. Figure 20–18 shows the top-ten milk goat states. Wisconsin has increased its milk goats in recent years. California and Texas are also significant dairy goat states. It is probably no coincidence that the top-10 dairy goat states are also important dairy cow states. Most goat dairies are smallholdings that are part of a hobby or part-time operation, and are comprised of 5 to 20 animals. The family probably uses most of the product, with some excess sold or given to neighbors and friends (Figure 20–19).

**FIGURE 20–16**   U.S. goat inventory (× 1,000 head) by type and state. (*SOURCE:* USDA-NASS, 2007c.)

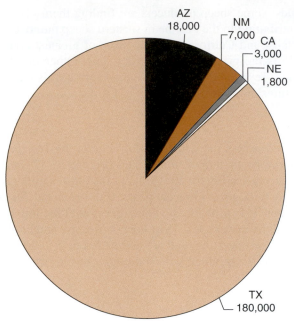

**FIGURE 20–17**    Top-five Angora goat states. (*SOURCE:* USDA-NASS, 2007c.)

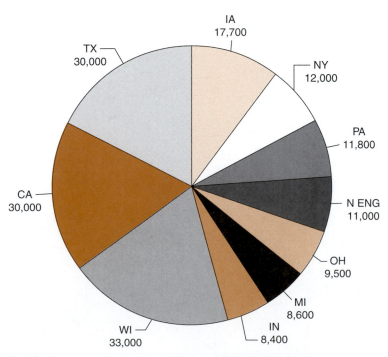

**FIGURE 20–18**    Top-ten milk goat states. (*SOURCE:* USDA-NASS, 2007c.)

**FIGURE 20–19** Almost all of the dairy goats in the United States are found in small herds with the product being used predominantly by the family. To many families, the animals are also considered family pets. (*SOURCE:* Photographer Larry Rana. Courtesy of USDA.)

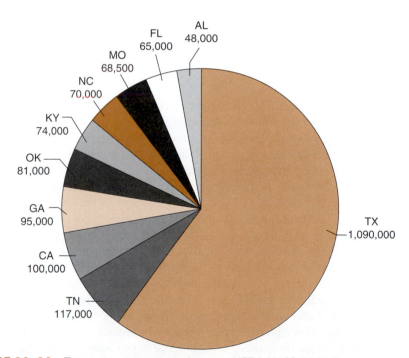

**FIGURE 20–20** Top-ten meat-goat states. (*SOURCE:* USDA-NASS, 2007c.)

The meat-producing sector of the goat industry has been expanding. Texas has long been the most important goat meat-producing state (Figure 20–20). However, other states have been cultivating this industry and have developed significant meat-goat populations. Goats are browsers. This makes them useful for controlling invasive brush species (Figure 20–21). This browsing habit has been exploited very well by the meat-producing sector to get double use from the animals.

**FIGURE 20–21** Goats prefer to browse on woody and weedy plant species rather than graze. This makes them a good choice for marginal areas containing brush species. They are often used to help control or eliminate invasive plant species. (*SOURCE:* Photographer Scott Bauer. Courtesy of USDA.)

# SELECTION AND BREEDING PROGRAMS

### SHEEP

The fundamentals of genetics and breeding are discussed in Chapters 8 and 9. See those chapters for more detailed information on these topics.

It is the job of the person in charge of the selection program (the breeder) for a flock of sheep to combine the best set of genetics available for the environment in which the animal will be producing. When the breeder makes decisions determining which individuals will become the parents of the next generation, he or she is practicing the science and art of *selection*. The challenge of selection is to improve the genetic potential of the next generation. Of course, selection will take place according to the principles of ***survival of the fittest*** if nature is allowed to run its course. However, natural selection rarely leads to improvement in the economically important traits in any species. Improving the ability of a species to produce a needed product in an economically beneficial way takes a more studied approach. In sheep production, breeding programs revolve around the goals of the individual producer. These can be as varied as producing high-quality wool for the hand spinning and weaving segment, producing high-quality lambs for the 4-H and FFA club market, or producing lambs for the slaughter market. In part, much diversity exists in sheep breeds because this adaptable animal has been selected for different uses in different environments.

The producers of seedstock animals generally practice purebreeding in their flocks and produce registered animals. The various breed associations that offer member services to help their breeders in recordkeeping and breeding decisions facilitate this. Some commercial lamb and wool producers may practice purebreeding techniques even if their animals are not registered. Examples in which this is justified include those whose production environment is uniquely suited to the specific set of characteristics possessed by a certain breed or those that produce for a specialty market.

For most commercial sheep producers, crossbreeding is the method of choice for a breeding program. The reason is simply that crossbred sheep perform better than purebred sheep for meat production. Crossbreeding programs for sheep are fairly easy to

**Survival of the fittest**

Natural selection. A process whereby those best equipped for the conditions in which they are found survive and pass their genetic material on to the next generation.

design, and a good variety of productive, adaptable breeds of sheep from which to select is available. Crossbreeding is practiced when the benefits of *heterosis* are desired in the offspring. Crossbreeding in sheep produces beneficial effects in many economically important traits including, but not limited to, birth weight, weaning weight, daily gain, yearling weight, body weight, prolificacy, survival rate, fertility, pounds of lamb weaned per ewe exposed, and fleece weight. Most crossbreeding is done within the confines of a system designed to maximize diverse and complementary traits in the individual parent breeds. These systems rely not only on heterosis for their benefits, but also on breed selection to take advantage of *breed complementarity.* This simply means that we select breeds based on what they have to offer and place them in the appropriate place in the crossbreeding system. For example, breeds whose strengths are in maternal traits are used on the dam side and carcass-trait breeds are used on the sire side. Most sheep produced in the United States are a product of a crossbreeding system. This is because these systems produce the best slaughter lambs, and producing slaughter lambs is the primary purpose of the majority of ewes in the United States. Crossbreeding can be effectively accomplished using any of the systems shown in Figures 20–22 and 20–23.

The National Sheep Improvement Program (NSIP) was established in 1986 to assist sheep breeders in making genetic decisions. It is a computerized genetic evaluation that can be used to estimate the breeding value of every sheep in a flock for most commercially important traits. NSIP calculates genetic values for the traits from producer-provided information. Traits for which estimates of genetic value can be calculated include number of lambs born, total litter weight of all lambs at 60 days of age for each ewe that is *lambing,* individual body weight at 30, 60, 90, 120, 180, or 360 days of age (up to three weights can be analyzed for each flock), wool weight, fiber diameter, and staple length. A variety of reports derived from the data are available to the producer to help make breeding decisions. Enrollment forms and further information on NSIP are available from the NSIP Processing Center, 6911 South Yosemite Street, Suite 200, Englewood, CO 80112-1414; phone: (303)771-5717; fax: (303)771-8200; Website: http://www.nsip.org; e-mail: info@nsip.org.

**Heterosis**

Hybrid vigor. The performance improvement of a crossbred animal above the average of the parents' breeds.

**Breed complementarity**

When the characteristics of different breeds complement each other in crossbreeding systems.

**Lambing**

Parturition in sheep.

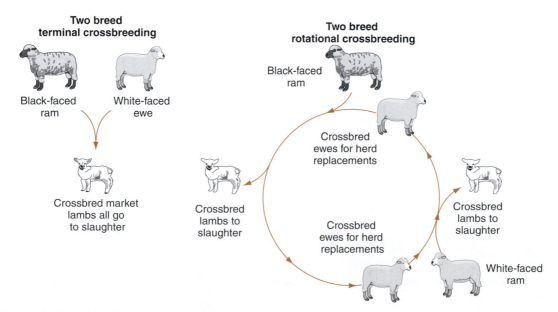

**FIGURE 20–22** Crossbreeding systems in sheep using two breeds.

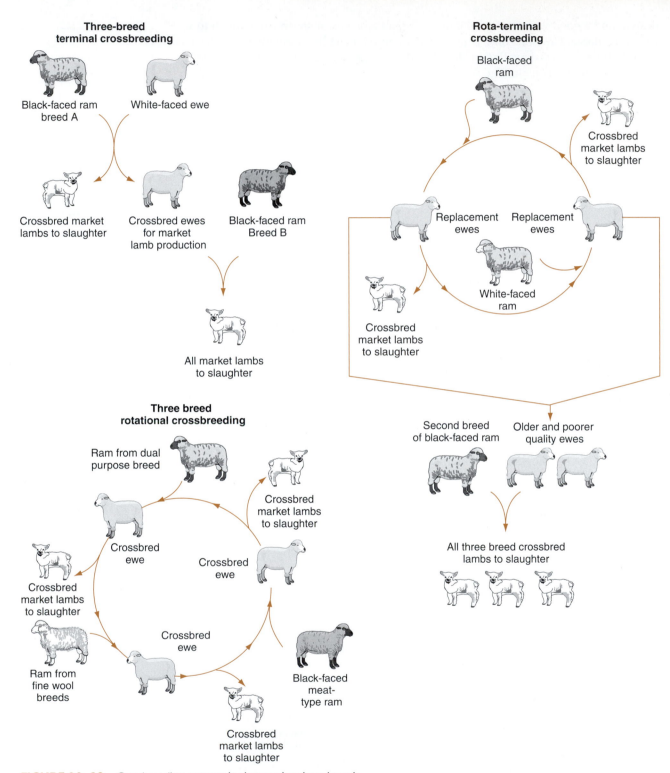

**Three-breed terminal crossbreeding**

Black-faced ram breed A

White-faced ewe

Crossbred market lambs to slaughter

Crossbred ewes for market lamb production

Black-faced ram Breed B

All market lambs to slaughter

**Rota-terminal crossbreeding**

Black-faced ram

Crossbred market lambs to slaughter

Replacement ewes

Replacement ewes

White-faced ram

Crossbred market lambs to slaughter

Second breed of black-faced ram

Older and poorer quality ewes

All three breed crossbred lambs to slaughter

**Three breed rotational crossbreeding**

Ram from dual purpose breed

Crossbred market lambs to slaughter

Crossbred ewe

Crossbred ewe

Crossbred market lambs to slaughter

Crossbred ewe

Ram from fine wool breeds

Black-faced meat-type ram

Crossbred market lambs to slaughter

**FIGURE 20–23** Crossbreeding systems in sheep using three breeds.

## GOATS

The state of knowledge about genetics and breeding for goats of all types is in a very rudimentary state compared to the state of the knowledge for any other traditional livestock species. However, there are USDA genetic evaluations for dairy goats. They can be found at http://www.aipl.arsusda.gov/. In 2003, the American Boar Goat Association initiated the Boer Goat Improvement Network, or "B-GIN" (pronounced "begin"), a genetic improvement program developed with assistance from the Texas Agricultural Experiment Station. The International Kiko Goat Association initiated a similar program in 2005. These programs should dramatically improve the ability of meat-goat producers to select breeding stock for specific measurable traits.

When choosing and breeding goats, the best line of approach is to combine a general understanding of the principles of their genetic improvement with any available research information on how animals perform in specific environments. That information should be applied to the selection of whatever trait is important in the goats in question, whether it is milk, meat, or fiber. If the meat goat industry is to emerge and thrive, a concerted effort aimed at the genetics of the goat is essential.

# BREEDS

## SHEEP

Historically, sheep have been bred for different uses and purposes. Thus a variety of characteristics are found in different breeds (Figure 20–24). Some have been bred for quality wool production (Box 20–1). In some areas, meat production has been more emphasized and breeds have been developed that emphasize growth and carcass characteristics. Modern dual-purpose breeds have been developed to produce both quality meat and quality wool. Other characteristics vary also. In mature adults, size varies among breeds, ranging from 100 to 400 pounds. Some ewes routinely have single offspring, others routinely twin, and still others are capable of litters of three to six offspring. The breeds in common

**FIGURE 20–24** Historically, sheep were bred for quality and quantity of wool, such as the wool of this ram. Today, wool is of decreasing value and receiving less emphasis in breeding programs. (Photo courtesy of UN/DPI.)

## Box 20–1    Wool Grades

Wool is graded by three systems based on the fineness of the fiber. The American, or blood, system was developed in the United States using the fleece of the Merino as the standard. Merino wool was considered the standard and is referred to as fine wool. The standard was actually based on the quality of fleece generally produced by sheep with different percentages of Merino breeding. The Bradford system is a system developed in England. It is based on the length of yarn that can be spun from 1 pound of clean wool. Very fine wool yields 120 hanks (560 yards) of wool. Its Bradford grade is 120's. A Bradford grade this high is very rare. The third system is based on the measurement of the diameters of the fibers. It is measured in microns. One micron is 1/25,400 of an inch.

### Comparison of Wool Grading Systems

| Blood or American System | Bradford or English System | Micron Count System |
|---|---|---|
| Fine | Finer than 80's | Under 17.70 |
| Fine | 80's | 17.70–19.14 |
| Fine | 70's | 19.15–20.59 |
| Fine | 64's | 20.60–22.04 |
| 1/2 Blood | 62's | 22.05–23.49 |
| 1/2 Blood | 60's | 23.50–24.94 |
| 3/8 Blood | 58's | 24.95–26.39 |
| 3/8 Blood | 56's | 26.40–27.84 |
| 1/4 Blood | 54's | 27.85–29.29 |
| 1/4 Blood | 50's | 29.30–30.99 |
| Low 1/4 blood | 48's | 31.00–32.69 |
| Low 1/4 blood | 46's | 32.70–34.39 |
| Common | 44's | 34.40–36.19 |
| Braid | 40's | 36.20–38.09 |
| Braid | 36's | 38.10–40.20 |
| Braid | Coarser than 36's | Over 40.20 |

use in the United States have been selected largely from those developed in Europe. They tend to be much more similar than different when compared to the great diversity of breeds available in the world. Breeds of sheep can be classified in many ways. Examples include wool type, face color, productive function, or commercial use.

It is estimated that approximately 200 breeds of sheep exist in the world today. A visit to the Oklahoma State University Breeds of Livestock website (http://www.ansi.okstate .edu/breeds) will give you the opportunity to view information about virtually all of those breeds and see pictures of most of them. The color insert of this text provides pictures of some of the more commonly used breeds in the United States. Because of their widespread use and distribution, sheep breeds have been developed with quite a range of genetic variation. Certainly not all of those breeds have found use in the U.S. sheep industry. The Seedstock Committee of the American Sheep Industry Association (ASI) lists 47 breeds of sheep in the United States. However, only 8 are used in any significant number. Table 20–2 lists these breeds and characterizes them according to traits of interest in sheep production. In general, the breeds of sheep in the United States are classified by their major commercial use. The categories are all-purpose (dual-purpose or general) breeds, sire (ram) breeds, dam (ewe) breeds, hair breeds, and milk breeds. Brief descriptions follow.

## All-Purpose or Dual-Purpose Breeds

These breeds are those that have utility as both meat and wool producers and are adapted to diverse environments. Examples of these breeds include Clun Forest, Dorset, Montadale, North Country Cheviot, Polypay, and Texel. Some people include the Columbia, Targhee, Romney, and Corriedale breeds in this group. Breeds in this category can be used as either ewe breeds or ram breeds in crossbreeding schemes for different production situations. For example, the Columbia is a popular ewe breed in the West, but it is a ram breed in the Midwest. These breeds are often chosen for small flocks in which crossbreeding is difficult to maintain. They can be used effectively in three-way crossbreeding systems.

## Dam or Ewe Breeds

These breeds include the Rambouillet, Merino, Targhee, Columbia, Corriedale, Border Leicester, Coopworth, and Romney. The Finnsheep and Romanov are two additional dam breeds that are highly prolific. These are typically white-faced breeds of fine-wool type or crosses of fine-wool types. The Rambouillet cross is the most important of these breeds. The majority of the sheep in the United States are producing females in stock sheep operations. It thus stands to reason that these breeds are the most numerous in total numbers in the United States. Ewes that carry a predominance of breeding from this group are mated to rams from the sire breed group to produce lambs for meat. These breeds contribute traits for good mothering ability, hardiness, good fleece characteristics, and good volume of wool. They also need adequate meat-producing characteristics.

## Sire or Ram Breeds

These breeds are selected for the growth and meat qualities of their offspring because they are used as terminal sires in crossbreeding programs. They are further grouped by the target weight of lambs at slaughter. The breeds favored for heavyweight lamb production are the Suffolk, Hampshire, and Oxford. These breeds are the most commonly used sire breeds in the United States, with the Suffolk having the distinction of being the single most common sire breed; the Hampshire is next. The Shropshire and Texel are used if medium weights at finishing are the goal, and the Cheviot and Southdown produce lightweight lambs. Crossbreeding in sheep is used extensively in the U.S. sheep flock, and this is the major use of the ram breeds.

## Hair Sheep Breeds

Not all sheep have wool. In hot, often underdeveloped parts of the world, wool is not especially useful or needed and has not been selected for in the animals that developed there. To the untrained eye, hair sheep often look more like goats than sheep. Considerable interest developed in these sheep in the United States because they can have superior fertility, livability, and parasite resistance, plus the added bonus of having an extended breeding season when compared to wool sheep. The greatest interest in hair

**TABLE 20–2** **General Classification of U.S. Breeds of Sheep[1]**

| Breed | Wool Type | Hardiness[2] | Mature Size[3] | Growth Rate[2] | Prolificacy[2] | Avg. Breeding Season[4] | Fiber Diameter Microns | Ewe Grease Fleece (wt./lb.) |
|---|---|---|---|---|---|---|---|---|
| **More Common Breeds** | | | | | | | | |
| Border Leicester | Long | M– | L– | M+ | M+ | S | 30–38 | 8–12 |
| Cheviot | Medium | M+ | S+ | L+ | M | S | 26–33 | 5–8 |
| Columbia | Medium | M+ | L | H | M– | M | 23–30 | 9–14 |
| Coopworth | Long | M | M | M | M+ | S | 30–36 | 8–12 |
| Corriedale | Medium | M+ | M | M | M– | M | 24–31 | 9–14 |
| Dorset | Medium | M– | M | M | M | L | 27–33 | 5–8 |
| Finnsheep | Medium/Long | L+ | S+ | L+ | H+ | L | 24–31 | 3–7 |
| Hampshire | Medium | M– | L | H | M | M | 25–33 | 5–8 |
| Montadale | Medium | M | L– | M+ | M | M | 25–30 | 5–9 |
| Oxford | Medium | M | L | H– | M | S | 30–34 | 5–8 |
| Polypay | Medium | M | M+ | M+ | H– | L | 24–33 | 6–10 |
| Rambouillet | Fine | H | L– | M+ | M– | L | 19–24 | 9–14 |
| Romney | Long | M– | M+ | M | L | S | 32–39 | 8–14 |
| Shropshire | Medium | M– | L– | M+ | M | M | 25–33 | 5–8 |
| Southdown | Medium | M– | M– | L+ | M– | M | 24–29 | 5–8 |
| Suffolk | Medium | L | L+ | H+ | M+ | M | 26–33 | 3–7 |
| Targhee | Medium/Fine | M+ | L– | M+ | M | L | 21–25 | 8–14 |
| **Less Common Breeds** | | | | | | | | |
| Am. Cormo | Fine | H | M | M– | L | L | 19–22 | 10–14 |
| Barbados Blackbelly | Hair | M+ | S | L | M+ | L | — | — |
| Black Welch Mountain | Medium | H | S | L | L+ | S | 29–36 | 3–4 |
| Bluefaced Leicester | Long | M– | L | M+ | H– | S | 26–31 | 6–8 |
| Booroola Merino | Fine | M+ | S | L | H+ | L | 18–23 | 9–15 |
| CA Red | Medium | M | S+ | L+ | M– | L | 26–31 | 4–7 |
| CA Var Mutant | Medium/Fine | M | M | M | M | M | 22–25 | 6–12 |
| Clun Forest | Medium | M | M– | M | M+ | S | 28–33 | 5–8 |
| Cotswold | Long | M– | L | M+ | M | S | 33–40 | 11–15 |

*(continued)*

breeds has been for hot climates. The most common breeds available are the Barbados Blackbelly and St. Croix. The advantages of these breeds have not far outweighed their disadvantages, which include lighter weights, slow growth, and inferior carcass merit as compared to the more commonly used breeds.

**TABLE 20-2** General Classification of U.S. Breeds of Sheep[1]—*continued*

| Breed | Wool Type | Hardiness[2] | Mature Size[3] | Growth Rate[2] | Prolificacy[2] | Avg. Breeding Season[4] | Fiber Diameter Microns | Ewe Grease Fleece (wt./lb.) |
|---|---|---|---|---|---|---|---|---|
| **Less Common Breeds** *(continued)* | | | | | | | | |
| Debouillet | Fine | H | M | M— | L+ | L | 18–23 | 9–14 |
| Delaine Merino | Fine | H | M— | M— | L+ | L | 17–22 | 9–14 |
| E. Friesian | Medium | L | L— | H— | H | S | 27–31 | 8–12 |
| Gulf Coast Native | Medium | H | S | L | L | L | 26–32 | 3–5 |
| Icelandic | Long | H | M— | L+ | M+ | S | 20–28 | 3–5 |
| Jacob's | Medium | M— | S | L | L | S | 27–35 | 3–6 |
| Karakul | Carpet | H | S+ | L | L | M | 24–36 | 4–8 |
| Katahdin | Hair | M+ | M— | M | M+ | L | — | — |
| Lincoln | Long | M— | L | M | L | S | 34–41 | 10–14 |
| Navajo-Churro | Carpet | H | S | L | L | L | 28–40 | 3–7 |
| North Country Cheviot | Medium | M | M+ | M+ | M+ | S | 27–33 | 5–8 |
| Panama | Medium | M+ | M+ | M+ | M | M | 24–30 | 9–14 |
| Perendale | Long | H | M— | M— | M— | S | 30–38 | 7–10 |
| Rideau Arcott | Medium | M+ | M | M | H | L | 25–31 | 6–8 |
| Romanov | Medium/Hair | H | S+ | L+ | H+ | L | 28–35 | 3–5 |
| St. Croix | Hair | M | S+ | L | M+ | L | — | — |
| Scottish Blackface | Carpet | H | S+ | L+ | L+ | S | 28–38 | 5–6 |
| Shetland | Medium | H | S— | L— | L | S | 19–29 | 2–4 |
| Texel | Medium | M | M+ | M+ | M+ | M | 28–33 | 7–9 |
| Tunis | Medium | M+ | M— | M— | L | M | 26–31 | 4–8 |
| Wiltshire Horned | Shedding | M | M | M | L | S | — | — |

[1]The evaluations of the breeds for hardiness, mature size, growth rate, and prolificacy are subjective to varying degrees and assume all breeds are performing in a common environment.

[2]Hardiness, growth rate, prolificacy: H = high; M = moderate; L = low.

[3]Mature size: L = large; M = medium; S = small.

[4]Breeding season: L = long (6–8+ mos.); M = medium (4–6 mos.); S = short (less than 4 months). In most cases, long breeding season implies early onset; Finnsheep have a late onset (Aug./Sept.) but not a long season.

SOURCE: *Sheep Production Handbook,* 1996. Used with permission.

## Dairy Breeds

It is only recently that an interest in dairy sheep has developed in the United States. Dairy breeds include East Friesian, Lacaune, Sarda, Manchega, Chios, Awassi, and Assaf. The extent to which any of these become common in the United States will depend on the success of those trying to develop this new industry segment.

### GOATS

Goats are generally grouped as milk breeds, meat breeds, dual-purpose breeds, or fiber breeds, reflecting the uses for which they have been bred. There is great diversity in the breeds. The few that are in use in the United States have come from several parts of the world. A visit to the Oklahoma State University Breeds of Livestock website (http://www.ansi.okstate.edu/breeds) provides information about different breeds. Goat breeds vary tremendously from dwarf goats with a 20-lb mature female body weight, to large meat breeds with *bucks* weighing in at 340 lbs. The males in some breeds can be 50 in. at the withers; the dwarfs may be less than 20 in. Normal birth weights range from 3 to 9 lbs. Multiple births are so normal that flocks may have more than a 200% kid crop.

**Buck**

An intact male goat.

## Dairy Breeds

Goats of Swiss origin are the world's leaders in milk production. The most important are Saanen, Toggenburg, and Alpine. The LaMancha is a dairy breed developed in the United States. The most popular U.S. breed is the Anglo-Nubian, a breed developed in England. The color insert of this text provides pictures of some of the more commonly used dairy goat breeds in the United States.

## Meat Breeds

The meat breeds include the South African Boer (introduced into the United States in 1993), Kiko, Indian Beetal, Black Bengal, Latin American Criollo, and the U.S. Spanish goats (Figure 20–25). The South African goats are best known for meat-producing ability. However, many meat-type goats are quite variable in appearance and can be best described as nondescript. It is only since the 1990s that there has been enough demand for goat meat to cause breed development work and/or extensive importation and upgrading of meat-type goats and their breeds.

## Dual Purpose

Indian- and Nubian-derived goat breeds are dual-purpose meat and milk producers. The Anglo-Nubian, or Nubian as it is often called in the United States, is generally con-

**FIGURE 20–25** The Boer goat is a meat-type goat developed in South Africa using performance testing. It has become increasingly popular in the United States.

sidered to be a dual-purpose breed. In addition, Pygmy goats from Western Africa are of increasing interest as laboratory and pet animals.

## Fiber Producers

The Turkish Angora, Asian Cashmere, and Russian Don goats are kept for fiber production. The long upper coat (mohair) is the valuable product in the Angora, whereas fine underwool is the product from the Cashmere.

## Pygmy

In their native Africa, these are multipurpose goats known for their superior disease resistance, especially to *Trypanosoma*, which limits the use of other livestock. They have been used in the United States as a laboratory species. However, their diminutive size and all-around cuteness have earned them a spot as a livestock pet breed in the United States.

# REPRODUCTIVE MANAGEMENT

### SHEEP

Chapter 11 is devoted to reproductive physiology. Refer there for the fundamentals of male and female reproductive anatomy, physiology, and function.

Sheep are capable of routinely producing more than one offspring per pregnancy (Figure 20–26). This gives them an advantage in efficiency of reproduction over cattle, the major species with which they compete for resources. However, it also creates a different set of physiological circumstances for the ewe, which must be managed well if the sheep flock is to be profitable. Good management of the reproductive process is

**FIGURE 20–26** Twins holding twins. Whereas multiple offspring births like these girls happen infrequently in humans, sheep routinely produce more than one lamb per birth. In some breeds the lambing percentage can be over 200%. (Photographer Jack Schneider. Courtesy of USDA).

critical because reproductive efficiency, or the overall percentage of lamb crop raised and marketed per ewe, is the most important variable affecting profitability of a sheep flock. Optimal reproductive efficiency may not be the same as maximal reproductive efficiency. Optimal efficiency is that which is maximal for the set of production constraints in an individual operation. A hobbyist who has little market for lambs and is rather interested in colored fleece production would consider an optimal reproductive efficiency to be one that produced just enough replacements for the flock. A dry-range commercial flock could be at optimal reproductive efficiency at 100–110% if the nutritional plane of the range doesn't support enough milk production for twins. A purebred seedstock producer with good grass might prefer a 200% lamb crop. Rarely is greater than 200% lamb crop desirable because of ewes only having two teats. Reproductive efficiency must be optimized within the context of the operation and its goal. However, for the majority of the sheep in production in the United States, the production goal is to maximize output from the most limiting resource, usually feed or land. In the majority of these situations, increasing reproductive rate is the best way to accomplish this because the maintenance costs of the ewe are spread over more and/or larger lambs. Several reproductive strategies can improve reproductive efficiency. The most obvious, perhaps, is to produce more lambs per lambing by increasing the number of multiple births and decreasing the number of ewes that do not lamb. However, other strategies also result in improved lamb production. Because the gestation length for sheep is only 148 days, time would allow more than one lamb crop per year. However, most of the popular breeds of sheep in the United States are seasonal breeders. Sexual activity is controlled by the ratio of light to darkness. The sheep typically breed when the days are short. Overcoming this tendency requires the use of different breeds, hormonal control of estrus, or manipulation of the light-to-dark ratio with the use of artificial lights. Age at first breeding also influences lifetime production of a ewe. Ewes that lamb as yearlings have better lifetime productivity but require careful feeding and management. Improvements in average longevity of the ewes in the flock allow more of the total lamb crop to be marketed by decreasing the number needed for replacement. Careful attention at lambing reduces the death loss of lambs.

A substantial portion of good reproductive management can be attributed to good animal health management and good nutritional management. Replacements must be economically reared and managed with optimal reproductive efficiency in mind. The nutritional plane of both the ewe and the ram affects reproductive rate. Managing the ewe flock to minimize the negative effects of high temperature and humidity during breeding and shortly after is critical in some regions. Disease prevention programs are different from flock to flock. However, all sheep producers should be aware that sheep are especially sensitive to the effects of internal parasites. A regular program of parasite control will pay for itself. Any commonly occurring disease can reduce reproductive efficiency by reducing the general thriftiness of the flock and should be controlled. Diseases that affect reproductive efficiency directly include enzootic abortion of ewes, toxoplasmosis, *Brucella ovis* abortion, *Vibrio* campylobacteriosis (vibriosis), salmonellosis, leptospirosis, vaginal prolapse, hypocalcemia, and pregnancy toxemia. Control of these and other diseases can be easily accomplished by developing and following a complete health program for the flock.

Common domestic breeds of sheep exhibit a sexually inactive period referred to as **anestrus**. They have a breeding season of 5 to 7 months, usually beginning in the fall or, for some, the late summer. Breeding season varies with the length of daylight and is thus not the same from region to region. The normal season of birth for sheep in the United States is the spring. How early or late largely depends on facilities, labor, and

**Anestrus**

Any period of time when a nonpregnant adult female is not having regular heat cycles. In sheep it is most frequently caused by sensitivity to the length of the photoperiod.

when grazing becomes available. The gestation period varies slightly from breed to breed, averaging 148 days with a range of 144 to 150 days. The ***estrous cycle*** for a ewe averages 16 to 17 days and the duration of ***estrus*** is 24 to 36 hours. Most of the nation's ewes are mated by natural service in pasture or range conditions. However, artificial insemination techniques are available and are being used, especially by progressive purebred breeders.

Rams are also affected by day length. They produce the greatest volume of high-quality semen during the same period that ewes of their breed are in the breeding season. To maximize the value of high-quality rams and give each ewe the best chance at early ***conception***, each ram used in a breeding program should have a breeding soundness exam before being placed with the ewes. The major factors that should be evaluated in a breeding soundness exam by ***palpation*** and observation include testicular, epididymis, and penis development and soundness. Feet, legs, eyes, and jaws should be examined. The ram should also be screened for common diseases. Observations should be made on physical ability to breed females and libido. Semen evaluations are useful but difficult because of the difficulty in collecting semen from untrained rams.

### GOATS

In the goat, scent glands are located around the base of the horn. They help stimulate estrus and improve conception. These glands are the source of the goat odor, which can be very unpleasant to the human nose! The smell can also be taken up by the milk and can give it a terrible off-flavor. For this reason, bucks are generally kept separate from the milking does if the milk is to be used for human consumption. Like sheep, many breeds of goat breed in response to length of daylight and are thus seasonal breeders. Also like sheep, goats reach sexual maturity at approximately 6 months of age and can be mated to produce their first offspring at 1 year of age, although for longevity in the herd they should not be bred until 1 year of age or older. The normal estrous cycle of a doe is 21 days, and estrus lasts 1 to 2 days. Length of gestation is about 150 days. In most of the continental United States, the normal time of breeding for goats is August through February. Some breed twice per year; this is more prevalent the farther south the animal is found.

# NUTRITION

## SHEEP

Feed is the single most expensive part of sheep production. Thus a cost-effective program of feeding and nutrition is essential. Sheep have the ability to use forages and roughages because they are ruminants. In fact, only limited amounts of grain are used in sheep production. Some exceptions occur in creep feeding and feedlots; for show, club, and purebred animals; and sometimes for ewes just before breeding and then again just before lambing. Thus most feeding strategies for sheep revolve around how to get the most from forage. Chapters 5, 6, and 7 give detailed information relating to nutrition, feeds, and feeding. See those chapters for more detailed information than is provided in this section.

The cost of feeding the ewe is approximately half of the entire production cost of a lamb-producing operation. The nutrients need to be met in a way that minimizes feed costs and optimizes production to optimize returns. It is easy to overfeed ewes for part of the year and to underfeed them for the other part. The ewe must be fed with the goals of the specific program in mind and with appropriate consideration for the specific

**Estrous cycle**

The time from one estrus to the next. Occurs at a regular, periodic rate.

**Estrus**

Heat. The period when the female is receptive to mating.

**Conception**

When the sperm fertilizes the ovum.

**Palpation**

Physically touching and examining with one's hand.

environment in which she is producing. For instance, nutritional requirements are different for a purebred ewe producing rams for breeding stock than they would be for a dry-range ewe producing lambs for the feedlot. Yet there are certainly similarities between the two. The first of three crucial times in a ewe's productive cycle are approximately 2 weeks before breeding when she should be *flushed* for breeding. The second is the last *trimester* of gestation when it may be difficult for her to eat enough low-quality feeds to appropriately nourish the fetus. This is especially true for ewes having more than a *singleton.* A third crucial time of feeding the ewe is the first 6 weeks of lactation. Commonly used feeds for ewe flocks include grasses, legumes, by-products, and crop residues. These may be grazed by the animal or harvested, stored, and hand fed. Sheep are very adaptable. They can effectively use rangelands, pastures, and forage crops to meet their nutrient needs.

Sometimes it is advantageous to get lambs to market as soon as possible, either to capture a *seasonal market* or simply to take advantage of available labor that will have other, more pressing uses later on. In these situations, lambs may be creep-fed a high-energy ration from a very young age. Examples of lamb creep rations are given in Table 20–3.

Feedlots are specialized finishing operations. High-quality feeds are needed to bring lambs to a suitable slaughter end point as cost-effectively as possible. Different rations and feeding regimes can be used to effectively finish lambs. Table 20–4 shows examples of specialized feedlot diets for lambs. Lambs may also be finished on high-quality pastures with only the addition of appropriate mineral supplementation and perhaps a small amount of grain. This type of pasture finishing is commonly done in California.

## GOATS

One of the values of goat meat production is that they can be produced without intensive feeding strategies or systems. Although some goats are no doubt given supplemental feed as a means of making them market ready, there is no market for a fattened goat, and heavy grain feeding is not practiced and not likely to ever be done. In periods of drought and perhaps in winter, supplemental feedstuffs are used, but they tend to be roughages and forages. Similar feeding strategies are employed for Angoras. Neither the profit margins nor the production level support the use of much expensive feed.

Dairy goats require higher-quality feed in general. Good-quality forage supplemented with commercially available dairy goat feeds is probably the best approach for

**Flushing**

The practice of increasing feed to a female just before and during the breeding season.

**Trimester**

A third of a pregnancy. The last trimester in a ewe is approximately the last 50 days.

**Singleton**

An offspring born singly.

**Seasonal market**

A time of the year when there is increased demand for a product.

**TABLE 20–3**    **Creep Feeding Rations for Lambs**

| Feedstuff | 20% Protein Total Grain Ration to Be Fed with High-Quality Hay | 20% Protein Total Grain Ration to Be Fed Without Hay | 16% Protein Grain Ration to Be Fed with High-Quality Hay |
|---|---|---|---|
| Corn, no. 2, yellow, rolled[1] | 53.7% | — | 59.0% |
| Oats, whole or crimped | 13.4% | 63.6% | 14.8% |
| Soybean meal, 44% | 30.4% | 27.6% | 19.7% |
| Molasses | — | 6.36% | 3.93% |
| Limestone | 1.07% | 1.06% | 1.18% |
| Salt | 1.07% | 1.06% | 1.18% |
| Vitamin pre-mix and feed additives | 0.36% | 0.32% | 0.21% |

[1]Wheat can be substituted for corn on an equal basis up to 50% of the ration. Any more can cause digestive problems.

**TABLE 20–4**   **Feedlot Diets for Finishing Lambs**

| Ingredient | Diet 1 | Diet 2 |
|---|---|---|
| | PERCENTAGE OF DIET | |
| Corn, no. 2, yellow, unprocessed | 37.32 | — |
| Corn, no. 2, yellow, cracked | 37.33 | — |
| Corn, no. 2, yellow, whole ears, ground | | 76.85 |
| Cottonseed hulls | 10.00 | — |
| Soybean meal (48% CP) | 12.80 | 15.60 |
| Liquid molasses | — | 5.00 |
| Limestone | 1.00 | 1.00 |
| Trace mineral salt | 1.00 | 1.00 |
| Vitamin premix and feed additives | .55 | .55 |

**TABLE 20–5**   **Sample Lactation Rations for Milk Goats[1]**

| Ingredients | 16% Protein Ration | 20% Protein Ration |
|---|---|---|
| | PERCENTAGE OF RATION | |
| Corn, no. 2, yellow, steam rolled | 40.0 | 32.8 |
| Oats, whole or crimped | 15.0 | 10.0 |
| Soybean meal | 17.8 | 30.0 |
| Beet pulp | 10.0 | 10.0 |
| Dried brewers grain | 7.5 | 10.0 |
| Molasses | 7.5 | 30.0 |
| Trace mineral salt | 1.0 | 1.0 |
| Dicalcium phosphate | 0.5 | 1.0 |
| Monosodium phosphate | 0.5 | — |
| Magnesium oxide | 0.2 | 0.2 |

[1]These feeds to be fed in addition to forage and/or browse.

the vast majority of goat owners. Examples of supplement feeds for dairy goats to be fed in addition to good-quality forages are shown in Table 20–5.

## NUTRITIONAL BENEFITS OF LAMB AND GOAT MEAT TO HUMANS

The proportion of the recommended daily dietary allowance for a 25- to 50-year-old woman on a 2,200-calorie diet that a 3-oz serving of cooked, lean lamb, and a serving of goat meat provides is as follows:

| | *Lamb (%)* | *Goat Meat (%)* |
|---|---|---|
| Protein | 48 | 52 |
| Calories | 8 | 6 |
| Phosphorus | 22 | 21 |

|             | Lamb (%) | Goat Meat (%) |
|-------------|----------|---------------|
| Iron        | 12       | 21            |
| Zinc        | 37       | 37            |
| Riboflavin  | 18       | 40            |
| Thiamin     | 9        | 7             |
| $B_{12}$    | 111      | 51            |
| Niacin      | 36       | 22            |
| Folate      | 11       | 2             |

Lamb and goat are nutrient-dense foods, just like other animal products. It is doubtful that nutritional benefit has anything to do with the level of consumption of lamb in this country. People who eat lamb eat it as a specialty product, and we tend not to care about nutritional value in such situations. Nevertheless, it is nice to know that lamb is good food. A complete nutrient analysis of lamb is shown in Appendix V. It is hard to assess whether those who routinely eat goat meat take its nutrient content into consideration when doing so. Given that most consume it as a part of their traditional diet, nutrient content is probably not a consideration. However, goat meat has the advantage of generally being a lean product. A complete nutrient analysis of goat meat is given in Appendix VI.

## NUTRITIONAL BENEFITS OF GOAT MILK TO HUMANS

A 1-cup serving of goat milk provides the following proportion of the recommended daily dietary allowance for a 25- to 50-year-old woman on a 2,200-calorie diet:

| | |
|-------------|---------|
| Protein     | 19.7%   |
| Calories    | 7.6%    |
| Phosphorus  | 34%     |
| Calcium     | 41%     |
| Zinc        | 6%      |
| Riboflavin  | 26%     |
| Thiamin     | 11%     |
| $B_{12}$    | 8%      |

Goat's milk is very similar to cow's milk and is also a nutrient-dense food. A complete nutrient analysis of goat's milk is given in Appendix VII.

## TRENDS IN THE SHEEP AND GOAT INDUSTRIES AND FACTORS THAT WILL INFLUENCE THE INDUSTRIES IN THE FUTURE

### CONSUMPTION

Lamb consumption is approximately 1 lb per capita. The sheep industry has declined to the point that it is a specialty industry. Any increase in the industry will depend on the industry attracting new consumers willing to pay enough for the meat to support lamb production. Given the small percentage of the U.S. population that consumes lamb, there is certainly an opportunity to find new customers. Opportunities exist with the more diverse ethnic mix developing in the United States, some of whom are more accustomed to consuming lamb. The quality, availability, and convenience of lamb must be improved if consumption is to increase. Fat is an issue with the consuming public. A portion of demand for meat products can be traced to consumers desiring a leaner product. Lamb is still too fat. This not only leads to consumer dissatisfaction with the product but also causes the product to be higher priced than it need be, which further drives away customers. In addition, the availability of convenient, cook-friendly lamb product is essentially nonexistent. Fresh product is not even available year-round in most of the country. Relatively few people in the United States today know what to do with a leg of lamb. They are simply not going to buy it.

A portion of the expanding immigrant population of the United States tends to be goat meat-eating and is creating a demand for increased goat meat production. The largest groups of these immigrants are Hispanics, Asian, Caribbean, and various Muslim groups. Predominantly, these populations are settling in California, Texas and other southwestern states, New York City, Atlanta, and Miami. Any development in the industry will need to cater to these peoples as a base for expanded production. Detriments to the development of an increased demand for domestically produced goat meat include the seasonal nature of consumption (often around holidays), and the preference of each of these groups for a different type of goat carcass. Figure 20–27 shows recent goat slaughter rates. Part of the observed increases in goat slaughter have been related to Angora goats that were slaughtered because it was no longer profitable to keep producing fiber with them. However, the goat-meat industry is in a growth phase. Certainly the number of meat goats is increasing in the United States and there are almost a third as many goats slaughtered in USDA-inspected slaughter plants in the United States as sheep. In addition, a substantial number of goats are slaughtered outside of USDA-inspected facilities, including on-farm slaughter and state-inspected slaughter facilities. Data from these slaughter channels are unavailable. Therefore, knowing exactly how many goats are slaughtered in the United States annually is not possible. Thus these numbers underestimate to some unknown extent the actual slaughter. Direct and niche markets need to expand if a substantial increase in goat-meat production is to take place. The cultural market must be satisfied with the product and have access to it. The goat-meat industry also stands to gain from interest of an increasing portion of the general public in so-called ethnic foods, goat products, lean meats, farm-fresh product, and natural and organic products. Should the current trends continue, goats could well become a more numerous species in the United States than sheep.

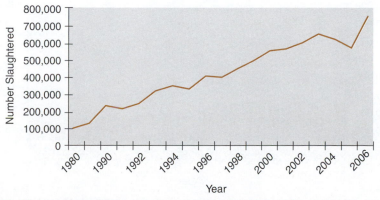

**FIGURE 20–27**   Number of goats slaughtered. (*SOURCE:* USDA-NASS, 2007d.)

Domestic consumption of wool and mohair is essentially a nonfactor in sheep and goat production now and will stay that way. Synthetics are more versatile, cheaper, and have a higher appeal to the modern consumer than do wool and mohair. That portion of the fiber trade that depends on these products can easily purchase the amounts they need on the international market.

### INDUSTRY SIZE AND STRUCTURE

The number of sheep in the United States has been declining. Figure 20–28 shows both the numbers of operations and total inventory for recent years. The downward trend in each is evident. However, a recent slight upturn in both the number of producers and total sheep is cause for cautious optimism in this industry. No one knows how many sheep are necessary to sustain this industry.

Some shift in the geography of production will probably occur. The western range bands of sheep are probably destined to be reduced in numbers even further. Some of those animals could move into smaller farm flocks in the more eastern states.

### ENVIRONMENTAL CONCERNS

Environmental concerns will be an increasing consideration to all concentrated animal feeding units like lamb feedlots. Finding low-cost, effective means of disposing of animal waste will help offset the costs of complying with new rules.

### TECHNOLOGY

Science and technology are underused in the sheep and goat industries. Some of the tools of biotechnology could be especially useful. Accelerated, or out-of-season, lambing and **kidding** could help the sheep and goat industries be more productive and thus more competitive. Ewes and does that produce once per year are being maintained for a significant portion of the year while they are doing nothing. Under good management, offspring could be produced twice a year, if not for the physiological limitation that the period of anestrus imposes. Successful strategies have been developed, but more research and work are needed.

Predators are a potential problem for both goats and sheep virtually anyplace they are found in the country. Surveys indicate that combined losses to producers of the two species are between $15 and $20 million annually, amounting to nearly 40% of all

**Kidding**

Parturition in goats.

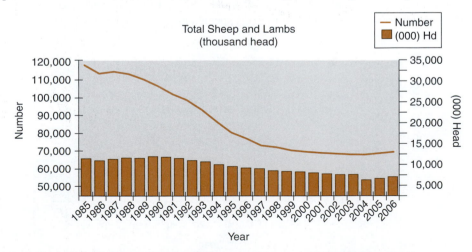

**FIGURE 20–28**  Sheep operations and total sheep. (*SOURCE:* USDA-NASS, 2007a.)

**FIGURE 20–29** Predators plague the sheep and goat industries. Guard dogs, such as these Akbash Dog pups being imprinted on a herd of goats, are used to ward off predators. (Photo courtesy of Dr. Rebecca L. Damron. Used with permission.)

annual losses. Another $10 million is spent on predator control measures. Predators include coyotes, dogs, mountain lions, bears, foxes, eagles, bobcats, and other animals, with coyotes the most common. Many of the approaches to dealing with predators such as poisons and traps are increasingly unacceptable to certain elements of society. Other approaches such as guard animals have helped but have not been totally effective. Any goat or sheep operation needs an effective predator plan in place. The costs and worry of predator control is a deterrent to producers (Figure 20–29).

## INDUSTRY ORGANIZATIONS

**American Boer Goat Association**
1207 South Bryant, Suite C
San Angelo, TX 76903
Phone: (325)486-2242
Fax: (325)486-2637
E-mail: info@abga.org
www.abga.org

**American Dairy Goat Association**
P.O. Box 865
Spindale, NC 28160
Phone: (828)286-3801
Fax: (828)287-0476
E-mail: info@adga.org
www. ADGA.org

**American Kiko Goat Association,** Jean Thomure, Secretary
295 Cardinal Lane
Trenton, GA 30752
Phone: (706)657-8649
E-mail:2meracre@tvn.net
www.2mereacres.com

**American Meat Goat Association**

P.O. Box 676,
Sonora, TX 76950
Phone: (915)387-6100;
Fax: (915)387-5814
http://www.meatgoats.com/

**American Sheep Industry Association, Inc. (ASI)**

9785 Maroon Circle, Suite 360
Centennial, CO 80112
Phone: (303)771-3500
Fax: (303)771-8200
E-mail: info@sheepusa.org
http://www.sheepusa.org/
ASI distributes the *Sheep Production Handbook*, which is an excellent resource for sheep producers.

**E (Kika) de la Garza Institute for Goat Research**

Langston University, Agricultural Research and Extension Programs
P.O. Box 730
Langston, OK 73050
Phone: (405)466-3836
http://www2.luresext.edu/index.htm

**National Lamb Feeders Association (NLFA)**

1270 Chemeketa St. NE
Salem, OR 97301–4145
Phone/fax: (503)370-7024
E-mail: info@nlfa-sheep.org
http://www.nlfa-sheep.org/

**National Sheep Improvement Program (NSIP)**

6911 South Yosemite St., Suite 200
Englewood, CO 80112-1414
Phone: (303)771-5717
Fax: (303)771-8200
E-mail: info@nsip.org
http://www.nsip.org

**North American Dairy Sheep Association**

Route 3, Box 27
Hayward, MN 56043
Phone/fax: (612)384-6612
www.dsana.org

**United States Meat Export Federation**

1050 17th Street, Suite 2200
Denver, CO 80265
Phone: (303)623-0297
http://www.usmef.org/
A nonprofit, integrated trade organization representing a wide variety of groups including livestock producers, meat packers, processors, farm organizations, grain promotional groups, agribusiness companies, and others. The organization works to develop foreign markets for U.S.-produced beef, pork, lamb, and veal.

## SUMMARY AND CONCLUSION

The sheep industry has become a specialty industry in the United States. This is related to several factors, all of which add up to poor profit margins, the fact that profits have not been great enough to attract or keep new producers. Sheep are able to make use of extensive (especially dry) range and resources, but they must compete with other ruminant species for those same resources. Although sheep can work well in multispecies grazing systems with other ruminants, even that practice is failing to help the sheep industry hold its place in production. The United States is no longer a major lamb producer or consumer.

### Facts about Sheep

| | |
|---|---|
| Birth weight: | 5–15 lbs (varies with breed and number born) |
| Mature weight: | Varies with breed, sex, and condition; Female 100–200 lbs; Male 150–300 lbs |
| Slaughter weight: | 90–110 lbs |
| Normal season of birth: | Spring, fall—some breeds |
| Recommended breeding age: | As yearlings (female) |
| Gestation: | 145–150 days |
| Estrous cycle: | 16–17 days |
| Duration of estrus: | 24–36 hours |
| Ram/ewes: | Range 1/25; pen mating 1/50 |
| Normal lamb crop: | 100–160% (range usually lower than farm flock) |
| Age weaned: | 3–6 months |
| Weight at weaning: | 60–100 pounds |
| Names of various sex classes: | Ewe, ram, wether |
| Fleece weight: | 6–14 lbs |

### Facts about Goats

| | |
|---|---|
| Breeds: | Alpine (French, Swiss, Rock), American La Mancha, Nubian, Saanen, Toggenburg, Angora, Spanish-type |
| Birth weight: | 6 lbs |
| Mature weight: | Varies with breed, sex, and condition; doe 110–135 lbs; buck 120–150 lbs |
| Weaning age: | 8 weeks (completely off milk); take from doe after 3 days |
| Recommended breeding age: | 9 months or weighing at least 75–80 lbs |
| Normal season of birth: | Tend to be seasonal breeders (bred late August–March) |
| Gestation: | 145–155 days |
| Estrous cycle: | 17–21 days |
| Duration of estrus: | 1–2 days |
| Buck/doe: | Buck 6–8 months = 6–8 does, 18–20 months = 25–30 does; mature 50–60 does per season. |
| Normal kid crops: | High conception rate. Does over 18 months of age normally average 1.5 kids per birth; frequently have three and sometimes four kids. |
| Average milk production per doe: | Good producing doe averages 1,800 lbs in 10-month lactation. This is equal to 3 quarts per day. Can get 3,000 lbs/year. If the bred doe is milking, she should be dried off (not milked) 6–8 weeks before kidding. |
| Names of various sex classes: | Kid, doe, buck |

There is a distinct regionality to the sheep industry. Most of the sheep are found in the western states in large flocks. Fewer sheep are found in the rest of the country, and those are generally in small flocks. The net result is that many producers are scattered across the country but not so many sheep. The sheep industry is struggling to retain enough numbers to even be considered a viable agricultural enterprise in the United States. Its survival will depend largely on whether or not a coordinated effort by the various segments of the industry can be mounted. Recent increases in sheep numbers and inventory value may signal better prospects for the sheep industry.

The goat industry has always been a specialty industry in the United States. Although there have certainly been numerous individuals who have kept goats, those who have done so as a bona fide commercial enterprise have been few and are restricted geographically. Texas has been the largest goat-producing state and will probably continue to be so. However, the production of goat meat on a larger scale should not be ruled out as an expanding industry because the number of consumers is increasing because of the current demographics of the immigrant population. Producers are entering this industry, especially the meat industry, and it is a growing industry.

## STUDY QUESTIONS

1. Describe the relative size of the sheep and goat industry segments compared to other animal industries.

2. What is the purpose of sheep and goats in the agricultural industries of the United States? What are the resources they use and what are the products they return?

3. What is the value of mixed-species grazing? Why does it work?

4. How do the sheep and goat industries in this country compare to those in other countries?

5. Give a brief historical account of the sheep and the goat in U.S. agriculture.

6. What is a specialty market and how does it pertain to the sheep and goat industries?

7. What are the reasons for the long decline in sheep numbers and thus the sheep industry in the United States?

8. The sheep and goat industries have similar structures in many ways and different structures in other ways. Compare and contrast them to each other and to the beef industry.

9. Compare the location of the sheep in the United States to the location of the sheep producers. Why is the distribution uneven?

10. Why is there currently so much interest in dairy sheep in the United States?

11. Where are the geographic locations of sheep and goat industry segments in the United States? Why aren't they distributed together?

12. Compare and contrast the relative information available about selection and breeding programs for sheep and for goats.

13. What is breed complementarity? How does it work in crossbreeding?

14. Describe crossbreeding systems for sheep.

15. How is the quality of wool measured?

16. Give a brief accounting of the breed classification for both sheep and goats. What are the similarities? The differences? How do these compare to the breed classification of other species?

**17.** Give a brief overview of reproduction in the sheep and goat.

**18.** What role does day length play in sheep reproduction? Goat reproduction?

**19.** Why are ewes flushed?

**20.** What are the similarities and differences in sheep and goat feeds and feeding programs?

**21.** Look at the rations described in the chapter for the various classes of animals. What differences and what similarities do you see?

**22.** What does a 3-oz serving of cooked lamb provide nutritionally to a human?

**23.** What does a 3-oz serving of cooked goat meat provide nutritionally to a human?

**24.** What does a 3-oz cup of goat milk provide nutritionally to a human?

**25.** Describe the trends that are influencing the direction of the sheep and goat industry segments.

**26.** What is your opinion about the future of meat goats in the United States? Dairy goats? Dairy sheep? Meat sheep?

## REFERENCES

Ensminger, M. E. 1991. *Animal science*. 9th ed. Danville, IL: Interstate.

FAO. 2007. *FAOSTAT statistics database. Agricultural production and production indices data*. http://apps.fao.org/.

Gibson, T. A. 2007. *Demand for goat meat: Implications for the future of the industry*. Langston University Field Day Proceedings. Accessed online October 2007. http://www2 .luresext.edu/goats/library/field/goat_meat _demand99.htm.

Jones, K. G. 2004. *Trends in the U.S. sheep industry*. USDA-ERS, Agricultural Information Bulletin Number 787. Electronic report accessed online. http://jan.mannlib.cornell.edu/reports/general/aib/ aib787.pdf.

Packham, J. H., and J. C. Teuscher. 1992. *Sheep dairying*. National Sheep Database. 1997.

Pinkerton, B., ed. 2007. *Meat goat production and marketing handbook*. Raleigh, North Carolina and Mid-Carolina Council of Governors: Rural Economic Development Center. Accessed online October 2007. http://www.clemson.edu/agronomy/ goats/handbook/cover.html.

Ross, C. V. 1989. *Sheep production and management*. Upper Saddle River, NJ: Prentice Hall.

*Sheep production handbook*. 2002. Englewood, CO: Sheep Industry Development Program.

Solaiman, S. G. 2005. *Meat goat industry outlook for small farms in Alabama and surrounding states*.

George Washington Carver Agricultural Experiment Station, Tuskegee, AL: Tuskegee University.

USDA. 1997. *USDA nutrient database for standard reference*. Release 11–1. Nutrient Data Laboratory home page: http://www.nal.usda.gov/fnic/ foodcomp.

USDA. 1998. *USDA nutrient database for standard reference*. Release 12. Nutrient Data Laboratory home page: http://www.nal.usda.gov/fnic/ foodcomp.

USDA-ERS. 2007. *Farm income data*. Accessed online March 7, 2008. http://www.ers.usda.gov/ Data/farmincome/finfidmu.htm.

USDA-NASS. 2007a. *Agricultural statistics data base*. Accessed online October 2007. http://www .nass.usda.gov/Data_and_Statistics/index.asp.

USDA-NASS. 2007b. *Sheep and goats*. Accessed online October 2007. http://usda.mannlib.cornell.edu/MannUsda/view DocumentInfo.do?documentID=1145.

USDA-NASS. 2007c. *Overview of the U.S. sheep and goat industry*. Accessed online October 2007. http://usda.mannlib.cornell.edu/usda/nass/ ShpGtInd//2000s/2007/ShpGtInd-09-28-2007.pdf.

USDA-NASS. 2007d. *Livestock slaughter, 2003 summary*. Accessed online October 2007. http://usda.mannlib.cornell.edu/MannUsda/view DocumentInfo.do?documentID=1097.

# 21 CHAPTER

# Horses

## LEARNING OBJECTIVES

After you have studied this chapter, you should be able to:

1. Put the horse industry in economic context.

2. Compare historical and current uses of the horse and understand the very unique features that make the horse industry different from all the rest of the commercial livestock industries.

3. Give a historical perspective on horses in North America and the United States.

4. Describe the change in the purpose of the horse in the developed world over the course of the 20th century.

5. Describe the structure and geographic location of the horse industry as far as available information will allow.

6. Explain the basics of horse genetics, especially color genetics.

7. Classify the types of horses.

8. Discuss the important tasks to be accomplished if a mare is to reproduce.

9. Cite the basics of how to feed a horse.

10. Identify and discuss areas of concern for the horse industry.

11. Discuss some of the opportunities for growth in the horse industry.

## KEY TERMS

| | | |
|---|---|---|
| Alleles | Continuous eater strategy | Equine |
| Anestrus | Cubed hay | Estrous cycle |
| Artificial vagina | Cutting | Filly |
| Body condition | Diet | Foal |
| Colic | Diluter gene | Foaling; to foal |
| Colt | Dorsal | Full-time equivalent |
| Conformation events | Dressage | Gait |

Gelding
Gene
Hand
Horsepower
Horse slaughter
Hunter under saddle
Impacted intestine
Incompletely dominant
Laminitis
Long hay
Mare
Modifier gene
Mustang

National Animal
    Identification System
Pellets
Points
Prepotent
Ration
Recreational horses
Reining
Saddle seat pleasure
Sclera
Seasonally polyestrous
Stadium jumping
Stallion

Stock horse
Stud
Stud book
Stud fee
Teasing
Three-day eventing
Trail
Ultrasonography
Unsoundness
Western pleasure
Western riding
Working cow horse

## SCIENTIFIC CLASSIFICATION OF HORSES

| | |
|---|---|
| Phylum: | Chordata |
| Subphylum: | Vertebrata |
| Class: | Mammalia |
| Order: | Perissodactyla |
| Family: | Equidae |
| Genus: | *Equus* |
| Species: | *caballus* (horse); *asinus* (ass) |

## THE PLACE OF HORSES IN THE UNITED STATES

It is impossible to present facts and figures for the horse that are comparable to those of the other livestock species. The federal government does not collect and summarize information on the horse as it does for other species. After the horse was replaced by the petroleum-fueled engine on the nation's farms, and before it emerged as an important leisure and recreation species, the government decided it was no longer a good use of taxpayers' money to collect and publish extensive information about horses and stopped doing so in 1960, when it estimated there were 3 million horses in the United States. However, information is collected periodically by private agencies and organizations and for the Census of Agriculture. The most authoritative information currently available on the *equine* industry is found in studies commissioned by the American Horse Council, the latest released in 2005. The 2005 study, *The Economic Impact of the U.S. Horse Industry on the United States* (EIHI), found 9.2 million horses in the United States that produce $29 billion in direct economic impacts to the U.S. economy. More than 1.96 million people are horse owners, with an additional 2 million involved as supportive family members and volunteers. The study did not include people under the age of 18 in these numbers or each would have been much higher. In total, the industry directly provides 1.4 million *full-time equivalent* (FTE) jobs each year and pays $1.9 billion in taxes. Racing and showing each contribute 27% of the employment generated, with recreation contributing approximately 31%. The remaining activities account for 15% of the employment generated. Based on

**Equine**
Pertaining to horses.

**Full-time equivalent**
In referring to employment, a 40-hour equivalent position.

**FIGURE 21–1** Horse/mule yearly farm cash receipts as a percentage of U.S. farm cash receipts, 2005–2007. (*SOURCE:* Based on USDA statistics.)

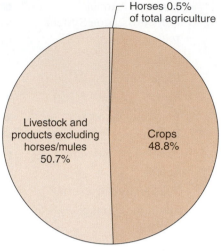

Horses 0.5%
of total agriculture

Livestock and
products excluding
horses/mules
50.7%

Crops
48.8%

**FIGURE 21–2** Horse/mule yearly farm cash receipts as a percentage of total animal agriculture's cash receipts, 2005–2007. (*SOURCE:* Based on USDA statistics.)

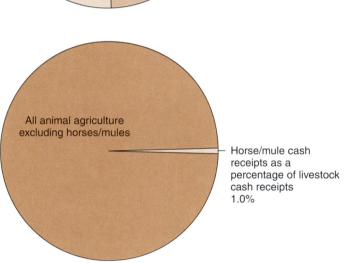

All animal agriculture
excluding horses/mules

Horse/mule cash
receipts as a
percentage of livestock
cash receipts
1.0%

limited statistics collected by the USDA, horse sales account for approximately 0.5% of total U.S. farm cash receipts (Figure 21–1) and approximately 1.0% of animal agriculture's share of all U.S. farm cash receipts (Figure 21–2).

## PURPOSE OF THE HORSE INDUSTRY IN THE UNITED STATES

The horse industry is unlike the industries that surround other livestock species. It is a hybrid industry with interests and segments in agriculture, and also in the sports, recreation, and entertainment industries. The feed for all horses, as well as most of the horses themselves, come from the agricultural sector, but most people who own horses are not in agriculture. Also, a wide variety of people who never owned a horse have interests in the horse industry. This makes the people of the horse industry very diverse (Figure 21–3). They range from those who spend their leisure time with horses to those who own horses as a business. The major purposes for keeping horses can be broken down into four categories: racing, showing, recreation, and other activities, which includes

**FIGURE 21–3** Although considered a livestock species, the horse industry is a unique hybrid. Certain of its interests are firmly rooted in agriculture; however, entertainment, sports, and recreation segments are also important parts of the horse industry.

**TABLE 21–1   Number of Horses by Activity**

| Activity | Number of Horses |
|----------|------------------|
| Racing | 844,531 |
| Showing | 2,718,954 |
| Recreation | 3,906,923 |
| Other | 1,752,439 |
| Total | 9,222,847 |

SOURCE: The American Horse Council Foundation, 2005.

rodeo, polo, ranch use, police work, and breeding. Table 21–1 summarizes the number of horses and people involved in each activity.

People tend to be in the horse industry for one or more of three general reasons: competition, leisure, and youth education. All can be considered as tools that improve the quality of human life. Many people are willing to spend large amounts of time and money for the satisfaction of winning a competition such as showing and racing. The horse industry produces one of the few leisure activities that allows for rapid recognition of excellence for relative newcomers in the industry. Many of those people recognized as leaders in various competitions today were not even involved 5 years ago. The variety of competition events include English events such as *hunter under saddle*, *saddle seat pleasure*, *dressage*, *stadium jumping*, and *three-day eventing*, and Western events such as *reining*, *cutting*, *working cow horse*, *western pleasure*, *trail*, *western riding*, and *conformation* judging, which are broken down into age and sex divisions. The reason most people have interest in the horse industry is recreation. People like horses and enjoy owning and caring for them. Also, many parents consider owning and caring for horses, and competitions that involve horses, to be wholesome and educational experiences and activities for their children (Figure 21–4).

The horse industry is unique, as compared to the cattle, swine, sheep, and aquaculture industries, in that the horse is not primarily kept to provide food or fiber, or to convert otherwise unusable material to something useful. On the contrary, horses compete with food-producing livestock for high-quality feedstuffs and return little food to

### Hunter under saddle

An English division class in which movement and mannerisms are judged with the intent of a pleasurable ride, which is to depict a horse on the hunt chasing the hounds.

### Saddle seat pleasure

An English event in which the horse's movement and mannerisms are judged.

### Dressage

A competition in which horses are required to perform highly advanced maneuvers in a specified pattern. This competition may be held alone or may be part of a three-day event.

### Stadium jumping

A competition in which horses jump a course of fences (jumps) in a specified order.

### Three-day eventing

An event of the Olympic Games since 1912. It is comprised of three parts—stadium jumping, dressage, and cross-country.

### Reining

A competition in which a horse and rider perform a specified pattern of advanced maneuvers. These include long sliding stops; spins (up to four 360° turn-arounds at high speeds); rollbacks (180° turns after a sliding stop); and circles of different sizes.

**FIGURE 21–4**   The single largest reason for interest in the horse industry is recreation. (Photo courtesy Emily Cooper.)

### Cutting

A competition in which a horse cuts one cow from the herd and holds the cow away from the herd. The horse must work the cow without assistance from the rider for 2½ minutes.

### Working cow horse

A western event broken into two scored performances with those scores combined for a grand score to determine the winner. The two categories include "dry work," which is the reining portion, and "cow work," in which the horse shows its ability to control the cow.

### Western pleasure

A western event in which the manners and movement of the horse are judged. A good western pleasure horse is quiet, responsive, and gives a very smooth ride.

### Trail

A western event in which the horse is scored on its ability to pick cleanly through a set course that mimics outdoor trail riding.

### Western riding

A western event in which the horse is scored on lead changes through one of three potential patterns as well as its manners.

### Conformation events

A competition in which the horse's conformation is judged.

### Horsepower

Term that originated as a measure of the pulling power exerted by a horse. Technically equal to the rate of moving 33,000 lbs a distance of 1 foot in 1 minute.

humans at all. Their contributions to humans are not as practical as that, but are just as important. Horses contribute quality to our lives. Horses bring a satisfaction that defies description to those who are involved with them. Most livestock feed our bodies; the horse feeds our being (Figure 21–5).

Another unique feature of the horse industry when compared to other livestock industries is that many horse owners have a decidedly nonpragmatic attitude toward horses. Many have bonded with their animals. In this regard, horses have much more in common with pets and companion animals than with livestock. Thus a different set of human values enters the horse industry than is associated with the livestock species. The ramifications of this are enormous. Certainly, this view of the horse has not always held sway. The horse has been a very practical animal to own in the past serving humans as a beast of burden, a means of transportation, and a tool of war (Figure 21–6). The horse is still doing those things somewhere around the world. However, during the 20th century, recreation, sport, and companion became the most important uses of the horse, especially in the United States and other developed countries. No longer is the horse needed to provide *horsepower*—machines do that. The horse has been given a higher purpose. This purpose sets the horse apart from the other livestock species (Figure 21–7).

## HISTORICAL PERSPECTIVE

Although the horse was among the last of the livestock species to be domesticated, this hasn't deterred it from having a fascinating history. The domestic horse's fate has been inexorably linked to the rise and fall of many of the world's great civilizations. The ancestors of the horse developed about 1 million years ago in North America and spread throughout South America and the Old World via land bridges. They then became extinct in North America but luckily flourished in other parts of the world. The exact date and place of domestication is as difficult to pinpoint for the horse as it is for all the other species. Evidence suggests that it probably occurred at more than one place at approx-

**FIGURE 21–5** "There's nothing so good for the inside of a man as the outside of a horse."* That could well be the motto of therapeutic riding. Here young Matthew Sitton takes part in one of his regular outings aboard Bucky. (Courtesy Shelly R. Sitton. Used with permission.) *The quote is attributed to Henry John Temple, Viscount Palmerston (1784–1865). Teddy Roosevelt, Sir Winston Churchill, Will Rogers, Ronald Reagan, and a host of others have used it in different forms.

**FIGURE 21–6** The rise and fall of many of the world's great civilizations have been inexorably linked to the horse. Albrecht Durer, 1471–1528, "The Knight, Death and the Devil." Engraving, 1513, $9\frac{5}{8} \times 7\frac{3}{8}$ inches. (Photo courtesy of the Metropolitan Museum of Art, Harris Brisbane Dick Fund, 1943. [43.106.2].)

imately the same time. China and Mesopotamia are among the earliest places of domestication, probably somewhere close to 3000 B.C.

Columbus is credited with introducing horses to the Western Hemisphere in 1493 by bringing them to the West Indies along with several other species of livestock. The first horses to be reintroduced to North America were those of the Spanish conquistadors, beginning with those of Cortez in 1519 when he invaded Mexico, and then to what is now the United States by de Soto in 1539. Myth has it that the ***mustang*** herds of the American West were descendants of strays from these expeditions. They might have contributed, but if so, only in a minor way. The Spanish missionaries, who established missions in the late 1500s and early 1600s, brought horses and other livestock in significant numbers. At these missions, Native Americans learned the ways and

**Mustang**

The term used to describe the feral horse of the American West.

**FIGURE 21–7**   In the developed countries, and to a degree in the entire world, the horse underwent a transformation in use during the 20th century. Its utilitarian roles were largely replaced by its role as a sport and recreational animal. Pictured here is Oklahoma State University's Spirit Rider and Spirit Horse, Bullet.

value of the horse. From these early exposures to the horse until roughly 1750, Native Americans all across the Western Plains acquired and spread the horse, which led to the development of the great horse culture of the Native American. From 1750 to 1850, the "wild" horse adapted and flourished along with the Longhorn cattle, setting the stage for the most colorful and romanticized period of the history of the United States, "The Wild, Wild West," which lasted roughly from the end of the Civil War to the turn of the century. Other horses were brought by Franciscan missionaries to the Southeast at the same time the missions were established in the West. These became the base for what would become known as the Chickasaw horse. Over the course of colonization, various breeds and types of horses were brought to the continent by the settlers. From the 1890s to the late 1920s, the horse was used in this country mostly for draft. From 1920 to 1960, horse numbers declined steadily in the United States. In 1920, there were 25 million horses in the United States. By 1960, there were only 3 million. Thanks to the 40-hour workweek and a booming U.S. economy, the horse began a comeback as a recreational animal and numbers increased to a probable 10 million head in the early 1980s. Then, because of the economic turmoil of the 1980s and changes in the tax code, which removed some of the tax incentives of horse ownership, horse numbers declined to 6.9 million head in 1996. Horse numbers increased to 9.2 million by 2005.

## STRUCTURE AND GEOGRAPHIC LOCATION OF THE HORSE INDUSTRY

Horses are found in all 50 states (Figure 21–8). There is both a rural and an urban segment to the industry. Breeding, rearing, and training are generally rural activities. Racetracks, shows, and so on, are generally urban. Table 21–2 shows that a substantial number of horses are owned by people in all income brackets.

Several classification schemes can be used to describe the horse industry. The EIHI divides them into the categories shown in Table 21–3.

In general, the *recreational horses* follow the population and the racehorses follow the racetracks, which also follow the population. Racing produces big revenues from relatively few animals (Figure 21–9). Heavily populated states like California,

**Recreational horses**

Horses of the light breeds kept for riding, driving, or nonprofessional racing and show.

**FIGURE 21–8** Horses by state. (*SOURCE:* American Horse Council Foundation, 2005.)

**TABLE 21–2** **Horse Ownership by Household Income**

| Household Income | Horses Owned | Percentage |
|---|---|---|
| $0–$24,999 | 209,879 | 11% |
| $25,000–$49,999 | 453,511 | 23% |
| $50,000–$74,999 | 435,930 | 22% |
| $75,000–$99,999 | 306,797 | 16% |
| $100,000–$124,999 | 199,646 | 10% |
| $125,000–$149,999 | 94,672 | 5% |
| $150,000+ | 179,268 | 9% |
| Not Reported | 76,124 | 4% |
| **TOTAL** | **1,955,827** | **100%** |

*SOURCE:* American Horse Council Foundation, 2005.

Florida, and Texas have more horses because their human population is so large. Some states that have strong show horse industries, like Oklahoma, as well as substantial use of horses for other reasons, gain significant economic impact on their state's economy and provide employment (Figure 21–10). Regional and economic factors also affect horse distribution.

**TABLE 21–3**  Horse Industry Participants by Form of Participation[1]

| Type of Participation | Number of Participants | Percentage of Total Participation |
|---|---|---|
| **Horse Owners** | **1,955,827** | **41.97%** |
| Primary Activity, Breeding | 237,868 | 5.10% |
| Primary Activity, Competing | 481,238 | 10.33% |
| Primary Activity, Other | 1,117,330 | 23.98% |
| Primary Activity, Service Provider | 119,392 | 2.56% |
| **Employees** | **701,946** | **15.06%** |
| of Owners | 598,398 | 12.84% |
| of Racetracks | 70,382 | 1.51% |
| of Shows | 33,166 | 0.71% |
| **Family Members and Volunteers** | **2,001,946** | **42.96%** |
| **TOTAL** | **4,659,719** | **100.00%** |

[1]Excludes horse owners under the age of 18.
SOURCE: American Horse Council Foundation, 2005.

**FIGURE 21–9**  Event facilities such as racetracks comprise a large segment of the horse industry. Racing produces big revenues from relatively few animals.

**FIGURE 21–10**  An important segment of the horse industry is the employees. Shown here are ranch workers training young horses at Oklahoma's Lazy E Ranch. Other segments of the industry also generate significant employment.

# HORSE GENETICS

Basic information on genetics and breeding is found in Chapters 8 and 9. Refer to those chapters to gain a basic understanding of genetics.

Scientific approaches to the study of horse genetics are not nearly as extensive as those for other livestock species because less research work has been done on horses. There are several reasons for this. First, horses are more important to recreation than to food and fiber production, and research on them is thus viewed as being less important. Second, many of the traits considered important in horses are difficult to measure and, therefore, are difficult to evaluate in a research study. Other hindrances to horse research include the high cost of animals and the horse's long generation interval. Recently, however, more attention has been placed on horse genetics, and this lack of information is changing. Many breed associations are now sponsoring this work.

There are a number of heritability estimates for some quantitative traits in horses. A list is shown in Table 21–4.

One area that has received a good deal of attention is coat color. Some colors are considered more desirable than others, to such an extent that several breed associations have been formed on the basis of color. Horses of certain colors can command higher prices than equal quality animals of less desirable color. This has created a demand for knowledge about color genetics. It is also possible to discover much of the genetics of color by simply observing the results of many matings. Purebred registries have helped in this study by keeping records on color along with parentage. Analysis of these records has produced good information about the genetics of color. The tools used in genetic engineering are also helping. Many DNA tests are now available to determine a horse's genetic code. Table 21–5 shows the various *alleles* and actions of horse color genetics. Table 21–6 gives the genetic formulas for each of several common color types.

**Alleles**

The alternative forms of any given gene.

**TABLE 21–4**   **Heritability Estimates for Certain Traits in Horses**

|  | Average Heritability Estimate |
|---|---|
| Height at withers | 45–50 |
| Body weight | 25–30 |
| Body length | 35–40 |
| Heart girth circumference | 20–25 |
| Cannon bone circumference | 20–25 |
| Pulling power | 20–30 |
| Running speed | 35–40 |
| Walking speed | 40–45 |
| Trotting speed | 35–45 |
| Movement | 40–50 |
| Temperament | 25–30 |
| Cow sense | Moderate to high |
| Type and conformation | Moderate |
| Reproductive traits | Low |
| Intelligence | Moderate to high |

SOURCE: Johnson, 1993. Used with permission.

**TABLE 21–5**    **Alleles and Actions of Horse Coat Color Genes**

| Color | Gene | Alleles | Observed Effect of Alleles in Homozygous and Heterozygous Conditions |
|---|---|---|---|
| White | White (*W*) | *W*<br>*w* | *WW:* Lethal<br><br>*Ww:* Born white. Horse typically lacks pigment in skin and hair. Eyes are dark.<br><br>*ww:* Horse is fully pigmented. |
| Gray | Gray(*G*) | *G*<br>*g* | *GG:* Horse shows progressive silvering with age to white or flea-bitten, but is born in any nongray color. Pigment is always present in skin and eyes at all stages of silvering.<br><br>*Gg:* Same as *GG*.<br><br>*gg:* Horse does not show progressive silvering with age. |
| Chestnut | Extension (*E*) | *E*<br>*e* | *EE:* Horse has ability to form black pigment in skin and hair. Black pigment in hair may be either in a points pattern or distributed overall.<br><br>*Ee:* Same as *EE*.<br><br>*ee:* Horse has black pigment in skin, but hair pigment appears red/yellow. |
| Bay/black | Agouti(*A*) | *A*<br>*a* | *AA:* If horse has black hair (*E*), then that black hair is in points pattern. *A* has no effect on red (*ee*) pigment.<br><br>*Aa:* Same as *AA*.<br><br>*aa:* If horse has black hair (*E*), then that black hair is uniformly distributed over body and points. *A* has no effect on red (*ee*) pigment. |
| Palomino/<br>buckskin/<br>cremello/<br>perlino | Cream (*C*) | *C*<br>*C$^{cr}$* | *CC:* Horse is fully pigmented.<br><br>*CC$^{cr}$:* Red pigment is diluted to yellow; black pigment is unaffected.<br><br>*C$^{cr}$C$^{cr}$:* Both red and black pigments are diluted to ivory. Skin and eye color are also diluted. |
| Dun | Dun(*D*) | *D*<br>*d* | *DD:* Horse shows a diluted body color to pinkish red, yellow-red, yellow, or mouse gray and has dark points including dorsal stripe, shoulder stripe, and leg barring.<br><br>*Dd:* Same as *DD*.<br><br>*dd:* Horse has undiluted coat color. |
| Tobiano | Tobiano (*TO*) | *TO*<br>*to* | *TOTO:* Horse is characterized by white spotting pattern known as tobiano. Legs are usually white. White crosses the dorsal line.<br><br>*TOto:* Same as *TOTO*.<br><br>*toto:* No tobiano pattern present. |
| Overo | Overo (*O*) | *O*<br>*o* | *OO:* Homozygous lethal (white).<br><br>*Oo:* White spotting characterized by horizontal pattern (usually) dark legs, white not crossing dorsal line.<br><br>*oo:* No white spotting. |
| Leopard spotting | Leopard spotting (*LP*) | *LP*<br>*lp* | *LPLP:* Variable pattern of roaning and spotting with mottled skin, eyes showing white sclera. Also known as appaloosa or tiger spotting. Homozygotes have more white than heterozygotes.<br><br>*LPlp:* Same as *LPLP* but less white.<br><br>*lplp:* No white spotting. |

*(continued)*

**TABLE 21–5**   **Alleles and Actions of Horse Coat Color Genes—*continued***

| Color | Gene | Alleles | Observed Effect of Alleles in Homozygous and Heterozygous Conditions |
|---|---|---|---|
| Champagne | Champagne (*CH*) | *CH* <br> *ch* | *CHCH:* Red pigment diluted to yellow, black pigment to brown or olive. Both have a metallic sheen. <br><br> *CHch:* Same as *CHCH*. <br><br> *chch:* No color dilution. |
| Silver | Silver (*Z*) | *Z* <br> *z* | *ZZ:* Black pigment diluted to chocolate; minimal effect on red pigment. <br><br> *Zz:* Same as *ZZ*. <br><br> *zz:* No color dilution. |
| Roan | Roan (*RN*) | *RN* <br> *rn* | *RNRN:* Hair is mixture of white and any other color. Points usually dark. <br><br> *RNrn:* Same as *RNRN*. <br><br> *rnrn:* Full color. |

*SOURCE:* Bowling, 1999. Used with permission. Modified according to Bowling & Ruvinski, 2000.

**TABLE 21–6**   **Genetic Formulas and Color Definitions**

| Genetic Formula | Color |
|---|---|
| *W* | White |
| *G* | Gray |
| *E, A, CC, dd, gg, ww, toto* | Bay |
| *E, aa, CC, dd, gg, ww, toto* | Black (includes brown) |
| *ee, aa, CC, dd, gg, ww, toto* | Red (chestnut, sorrel, and so on) |
| *E, A, CC$^{cr}$, dd, gg, ww, toto* | Buckskin |
| *ee, CC$^{cr}$, dd, gg, ww, toto* | Palomino |
| *C$^{cr}$C$^{cr}$* | Cremello |
| *E, A, CC, D, gg, ww, toto* | Buckskin dun |
| *E, aa, CC, D, gg, ww, toto* | Mouse dun |
| *ee, CC, D, gg, ww, toto* | Red dun |
| *E, A, CC, dd, gg, ww, TO* | Bay tobiano |
| *ee, CC, D, gg, ww, TO* | Red dun tobiano |

*SOURCE:* Bowling, 1999. Used with permission.

## BASIC COAT COLORS

Bay horses have the combination of a red-brown body with black legs, mane, and tail (the ***points*** of a horse). Black horses are black all over—black body as well as black points. Chestnut horses have red bodies, manes, tails, and legs with no black anywhere. Each of these basic colors has several shades ranging from dark to light. The mechanism for just which shade of bay, black, or chestnut will be exhibited has not been worked out. Adding to the confusion is the fact that the names of the three basic colors differ among the breeds and the regions of the country. All other colors are produced by the action of other ***genes*** called ***modifier genes***. For instance, gray horses are not born gray. Rather they are born one of the colors just mentioned, and they change to gray in response to a modifier gene. They still carry the genes for color they were born

**Points**

The legs, mane, and tail of a horse.

**Gene**

A segment of a chromosome.

**Modifier gene**

Gene that influences the expression of another gene or genes.

**Diluter gene**

A type of modifier gene that changes a base color to a lighter color.

with. Sorrel, chestnut, and liver chestnut are just shades of basic red. Brown horses may be blacks that have faded in the sun; palominos are basic chestnut but get changed by a separate *diluter gene,* and so on. Understanding that this is complicated is the first step in understanding the genetics of color in horses.

### Gene *G*

Gene *G* causes horses to be gray. Gray horses are usually born with color and gray as they age. A horse with a *G* allele may become solid gray or gray with red or black flecks. During the process, they may be dappled for a time, even for years (Figure 21–11). A *G* horse keeps its skin pigmentation, which is how it is differentiated from a *W* horse (discussed later).

### Gene *E*

Gene *E*, the Extension gene, controls black hair. Pure recessives (*ee*) are some shade of red (dark chestnut to light sorrel) with no black anywhere. Mating red to red always produces red because these horses do not carry the black gene. *E* animals will either have black hair that is restricted to the points or black hair that covers the whole body. Whether that black is restricted to the points or covers the whole body is determined by the action of gene *A* (discussed later). Gene *E* just controls black hair. Both a bay and a black have the same genotype for this gene (*E*) and both have black hair over at least part of their body. Black is dominant to red. This is confusing to some because bay horses have a body color that most of us call red or brown. Just remember that this gene controls the presence or absence of black hair on either the whole body or the points. Bay horses have black points.

### Gene *A*

Gene *A*, the Agouti gene, is the determining gene that controls the distribution pattern of black hair. This determines if an *E* horse is a bay or a black. If the dominant allele *A* is present, and is combined with *E*, then only the points of the horse will be black and the body will be some shade of red-brown (i.e., a bay). A bay horse must carry genes *A* and *E*. If a horse carries *aa*, then black hair is not restricted to the points and will

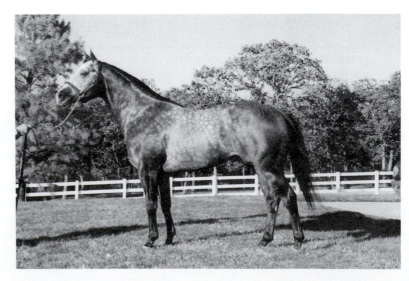

**FIGURE 21–11** Gene *G* is the gene responsible for gray. Gray horses start out as colored, and the expression of gene *G* causes them gradually to lose their original color and become gray. During the process, they often become dappled like the horse in this picture. Most will eventually become gray over most or all of the body. These horses are different from *W* horses in that they keep their body pigment.

cover the whole body (except for stockings and blazes and other patterns to be discussed later). Thus solid black is recessive to bay. If a horse is genetically red (*ee*), then the *A* gene has no effect on the horse. The animal will carry the genetics and pass them to its offspring but will not express them because *A* only affects distribution of black hair, and the animal does not have any black hair.

## Gene *W*

Gene *W* makes a horse unable to form pigment in skin and hair. If a horse carries this gene, it will have color according to one set of alleles but will be unable to manufacture the pigments necessary to display the color because of a different set of alleles. A horse carrying the dominant allele *W* will have pink skin, white hair, and brown or blue eyes. These horses are often called *albino*. White horses always carry the *Ww* genotype. The *WW* genotype is lethal and kills the embryo early in pregnancy.

### DILUTING GENES

Having established the genetic makeup of white, gray, bay, black, and chestnut horses, it is now time to see how the action of diluter genes changes the colors.

## Gene *C*

Gene *C* is the cream gene. This gene causes pigment dilution and is an ***incompletely dominant*** gene. If a horse carries the genotype *CC*, then nothing happens to its color. However, chestnut horses that are heterozygous (*CC$^{cr}$*) have their red pigment diluted to yellow and become palomino. Palominos have white manes and tails. Black is generally unaffected by the heterozygous state and stays black, although some horses have a slightly modified color referred to as smoky black. The bay (*E, A*) color is diluted to buckskin, which is yellow with black points. The body color changes but the black points stay the same. A horse of any color that carries the homozygous-recessive condition (*C$^{cr}$C$^{cr}$*) has its color diluted to very pale cream with pink skin and blue eyes. These horses are called *cremello* if the base color is red, *perlino* if the base color is bay, and *smoky cream* if the base color is black. Cremello can sometimes be difficult to distinguish from white. Some breed associations use the designation *albino* for all.

**Incompletely dominant**

Neither allele is dominant to the other. Both influence the trait.

## Gene *D*

Gene *D* is also a dilution gene that produces dun coloring. It is different from *C* in that it dilutes the body color only; the points are not diluted. In addition, this gene produces a uniquely recognizable pattern of dark points consisting of a ***dorsal*** stripe, a shoulder stripe, and leg barring often referred to as *zebra stripes*. Black dilutes to mouse-gray with black points referred to as *mouse dun* or *grulla*. Bay is diluted to yellow or tan with the zebra stripes, and it is called *zebra dun* or *buckskin dun*. Chestnut is diluted to yellowish red with darker red points and zebra stripes and referred to as *red dun* or *claybank dun*. One additional difference is that homozygous *DD* does not produce the extreme dilution to cream seen in *C$^{cr}$*. Homozygous *dd* has no effect on color.

**Dorsal**

Refers to the back on an animal.

## Gene *CH*

Gene *CH*, the Champagne gene, has only recently been described. This gene produces a color dilution plus mottled gray skin, a metallic sheen to the hair, and eyes blue at birth that change to hazel as the horse ages. Horses can appear to be palomino, buckskin, or cremello and have been called these colors prior to the discovery of the Champagne gene. Champagne in combination with black produces what is called classis champagne, an olive-hued metallic color. Many breed registries have not yet made provisions to include colors caused by the Champagne gene in their selection of colors for their breeds.

## Gene Z

Gene $Z$ (proposed) is the Silver gene. Actually described over 100 years ago, it has been difficult to understand and often ignored. In a black horse ($aaE$-), the presence of $Z$ causes the color to be diluted to a black-chocolate or chocolate and the mane and tail become silver gray or flaxen. A bay horse with the $Z$ gene can appear to be a chestnut. Chestnut horses appear to be changed only in that they get a silver (flaxen) mane and tail and can sometimes look like a palomino. In interaction with Gray, Silver can cause a "white-born" gray. This gene does not appear to be found in all breeds, or at least is found in much greater frequency in some compared to others, and many breeds do not recognize the color variation with an official designation as a color for their breed.

### ROAN

Gene $RN$ controls roan, when white hair is mixed with colored hair over the animal's body. The points generally retain the color with no white. Mixtures of white and any shade of red hair give a red roan. With a black horse, roan produces what is called a *blue roan*. Roan is a simple dominant-recessive gene, with the roan expression dominant. The homozygous-dominant condition $RNRN$ has been proposed to be lethal. Animals of this genotype are thought by some to die in early development and are never born. This would mean that all roan horses are heterozygous, $RNrn$. This is one of the unknowns of color breeding.

### LEOPARD

The gene $LP$ controls whether or not a horse has the complex of spotting and diffuse roan patterns referred to as leopard, appaloosa, and tiger spotting. It is believed the gene controlling all these patterns is a single incomplete dominant gene. The genetics of Leopard coloring leaves many unanswered questions. The patterns included in Leopard coloring include leopard, blanket, snowflake, varnish roan, and others. These different patterns are caused by various modifier genes in combination with $LP$, including some probably not yet identified. Leopard markings can occur on any color. Striped hooves, mottled skin (most evident around the muzzle and eyes), and prominent white *sclera* can also identify Leopards.

### SPOTTING

Paint and pinto are horse color patterns characterized by some body color interspersed with white. Think of this pattern as being white spots superimposed over the basic colors. If the horse is black and white, it is called a *piebald*. A horse that is any color other than black with white is called *skewbald*. Paints and pintos are produced in two different color patterns called *tobiano* and *overo*. These patterns are inherited separately and both can occur on the same horse. A horse with both patterns is called *tovero*. Other patterns include *sabino* and *splashed white*.

## Tobianos

Tobianos generally have white on the legs below the hocks and white across the back. The skin under the white is pink. The head of a tobiano is usually solid except for

---

**Sclera**

The tough white outer coat of the eyeball.

**FIGURE 21–12** Spotting in horses can be controlled by several genes. This baby is considered a tobiano. A simple dominant gene, *TO*, is believed to control this pattern.

facial markings. The pattern of white on the animal's body is arranged vertically. A single gene, *TO*, controls this color. It is a simple dominant-recessive trait. Thus pure dominants, *TOTO*, always produce a tobiano (Figure 21–12).

## Overos

The overo pattern seldom produces white across the back. Overo heads have more white as a general rule, displayed as bald faces or what is called a *bonnet* or medicine hat pattern. All four legs are usually colored, although the pattern sometimes produces as few as one colored leg. Overo can be difficult to determine because other genes that control white markings may be present in an overo or any other type of horse. These genes can cause a white leg. When two overo horses are bred together, the product can be a white foal that dies shortly after birth because it cannot absorb food from the digestive tract. The precise genetics of overo markings are unknown. Until recently, it was assumed overo coloring was recessive because it sometimes occurs in the offspring of two solid horses. However, work by the late Dr. Ann Bowling of the Equine Research Laboratory, School of Veterinary Medicine, University of California, Davis, suggests it is actually a dominant trait that mutates more frequently than most genes. This would explain how two solid horses could produce an overo paint. Complicating the picture is the fact that the color pattern called overo is variable enough to be produced by additional genes. Sabino and splashed white are other variations that may be controlled differently, because an analysis of the breeding records of overo horses has yet to establish the pattern of a simple dominant-recessive trait. This is one area where there is not yet a definitive answer to the genetic puzzle.

## FLAXEN MANE AND TAIL

Gene *F* produces a normal red mane and tail on chestnut (*ee*) horses. The *f* causes a flaxen mane and tail on chestnut or sorrel horses.

**FIGURE 21–13** Perhaps as many as 10 different genes control the presence or absence of stars, stripes, snips, and other white markings. This is one of many pieces of the horse genetic puzzle left to be completed.

### ADDITIONAL COMMON MARKINGS

When white markings such as stars, stripes, and snips on the face and white stockings are added to horses, the effects can be very appealing to the eye. However, the explanations of how these markings come about are incomplete at this time. It is suspected that at least 10, and probably more, different genes control white markings (Figure 21–13).

### GENETIC DISEASES IN HORSES

Various genetic diseases and abnormalities occur in horses. Some occur with relative infrequency and others with much higher frequency. Table 21–7 gives some examples of common genetic diseases in horses. The key to controlling these conditions is being aware of the ones that affect the breed or breeds you are interested in breeding, and making knowledgeable breeding decisions. Until the last few years, it was often difficult to know the genotype of a horse with certainty where many of these diseases were concerned. However, great strides have been and are continually being made in the genetic mapping of the horse. Already, fairly inexpensive tests have been developed to detect several of these disorders. In the not-too-distant future, a horse breeder will be able to know the exact genetic makeup of breeding stock and can make decisions accordingly.

## BREEDS OF HORSES

**Hand**
One hand equals 4 inches. Horses are measured for height at the withers in hands.

Horses are generally classified as light horses, draft horses, or ponies. Most horses are in the light horse class. Each of these divisions may be broken down by height, build, weight, use, or other variables. Most authorities classify horses as being over 14.2 **hands** at the withers, and ponies as being under 14.2 hands. Some breed associations take exception to this philosophy, and so do some individuals. The problem arises from the fact that some small horses of established light breeds can be shorter

**TABLE 21-7** **Examples of Genetic Diseases Caused by a Single or a Few Genes**

| Genetic Disease | Clinical Description |
|---|---|
| CID | Failure of immune system to form; animals die of infections |
| HyPP | Defect in movement of sodium and potassium in and out of muscle; animals intermittently have attacks of muscle weakness, tremors, collapse |
| Myotonic dystrophy | Spasms occur in various muscles |
| Hemophilia A | Failure to produce blood clotting factor; bleeding into joints; development of hematomas |
| Hereditary multiple exostosis | Bony lumps develop on various bones throughout the body |
| Parrot mouth | Lower jaw is shorter than upper jaw; incisor teeth improperly aligned |
| Lethal white foal syndrome | Failure to form certain types of nerves in the intestinal tract; foals die of colic within several days of birth |
| Laryngeal hemiplegia | Paralysis of the muscles that move cartilages in the larynx; results in noise production in the throat with exercise and exercise intolerance |
| Cerebellar ataxia | Degeneration of specific cells in the part of the brain called the cerebellum, resulting in incoordination |
| Hydrocephalus | Accumulation of fluid within compartments of the brain, resulting in crushing of normal brain tissue |
| Umbilical hernias | Opening in the body wall of the navel does not close normally, resulting in the presence of a sack into which intestines may fall |
| Inguinal hernias | Opening through which the testicles descend into the scrotum is too large and intestines can escape into the scrotum, sometimes causing colic |
| Hereditary equine regional dermal asthenia (HERDA) | Dysfunctional collagen bundles within the dermis, resulting in loss of strength and durability of the skin |
| Epitheliogenesis imperfecta | Skin fails to form over parts of the body or in the mouth |
| Cataracts | Cloudiness of the lens in the eye, resulting in blindness |

*SOURCE:* McClure, 1993. Used with permission (modified).

than 14.2 hands. Calling their horse a pony generally upsets breeders and owners of those animals. So how does one distinguish between ponies and horses? The breed associations make this decision. One way of knowing the association's decision is that a horse's height is referred to in hands, whereas a pony's height is expressed in inches.

The various horse uses dictated that several types of light horses and ponies were developed from the classes to perform specialized functions. Some were developed primarily for riding, some for work, some for racing, some for driving, and others to be capable of many tasks. Different horses have natural tendencies toward the use of certain *gaits*. Riding horses are generally three-gaited horses. The walk, trot, and canter are the three basic gaits of horses. Five-gaited horses add a slow-gait and rack to the basic three.

**Gait**

Forward movement of a horse. The three natural gaits for most horses are the walk, trot, and canter or gallop. Some breeds have additional or different gaits such as the pace, foxtrot, running walk, rack, and others, and they are considered five-gaited.

The horse industry revolves around breeds and breed associations. Several representative breeds are pictured in the color insert of this text. For a complete look at horse breeds from around the world, visit the Breeds of Livestock page at http://www.ansi.okstate.edu/breeds/. All of the breeds now popular in the United States are light horse and used primarily for riding. Most of these breeds were developed here, usually to meet a specific need in a specific region. There are more than 150 breed organizations in the United States.

## DRAFT HORSES

During the peak years of horse numbers in the United States (1910–1920), 75–85% of the total horses in the United States were the heavy draft horses that originated in Europe. The two most popular breeds of draft horses were the Percheron and Belgian, with the Clydesdale and Shire a distant third and fourth. There were some Suffolks, but very few. The Clydesdale was never very popular in this country. This is ironic in light of the TV ads of a well-known beer company, plus the personal appearances of its team of Clydesdales—this is the only draft breed that millions of people have ever seen.

## IMPORTED LIGHT BREEDS

Only two breeds of light horses imported to the United States have maintained significant importance in today's horse industry. They are the Arabian and the Thoroughbred.

The Arabian is known for its intelligence, durability, and stamina. Its origin is unknown, but it is very likely that Arab peoples had been breeding and improving this breed since the beginning of the Christian era, and probably long before. Whatever the exact date of origin, there is no question it is the oldest breed of horse, perhaps the oldest of any class of agricultural animal. The Arabian has contributed genes to the foundation of most of the breeds of light horses in the world.

The Thoroughbred was developed in England in the late 17th and early 18th centuries using Arabian breeding. It came to the United States in 1730. It has had the greatest direct influence of any breed on the development of the American breeds of light horses. No breed can best the Thoroughbred at distances of three-fourths to 2 miles. Although the Thoroughbred is most known as a racehorse, it is valuable for many other uses as well, including hunting, jumping, and polo.

## BREEDS OF HORSES DEVELOPED IN THE UNITED STATES

Modern American breeds of light horses in use in the United States were developed from two different groups of Native American light horses plus horses imported later (Figure 21–14). One group developed on the Eastern Seaboard. The foundation horses were breeds and strains of light horses brought by the early settlers from Europe. The Thoroughbred was used quite heavily in developing these horses in Europe. The other groups of Native American light horses were those that descended from the Spanish horses that became the mustang and the horse of the Native American.

## Morgan

**Stallion**

A mature male horse that is not castrated. Castrated male horses are *geldings*.

The Morgan breed was developed in the Northeast as a versatile animal for light draft and driving as well as for riding. This breed is unusual not only among horses, but among all classes of livestock because it traces to a single foundation animal, the *stallion* "Justin Morgan" who was foaled about 1789. Most breed historians believe his sire was a Thoroughbred and his dam carried considerable Thoroughbred breeding.

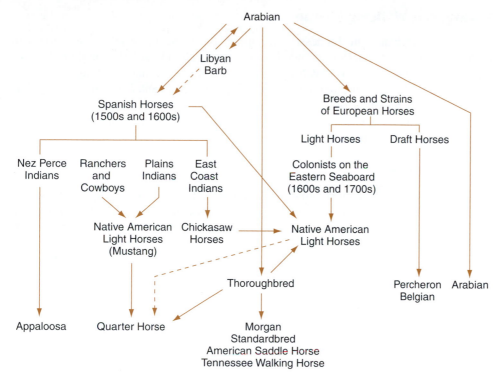

**FIGURE 21–14** Outline of the development of breeds and strains of horses that have been or are of major importance in the United States. (*SOURCE:* Turman, 1986.)

Justin Morgan was an outstanding horse and a ***prepotent*** sire. His descendants became very popular for both riding and driving. Today, the breed is used primarily for riding, driving, and showing and is most popular in the New England states.

## Standardbred

This breed was developed initially to meet the need for a carriage horse and also to be used in harness racing. Its foundation included trotters and pacers among the Native American horses, Canadian Trotters, and the infusion of Thoroughbred and Morgan breeding. It is now bred almost exclusively for harness racing. The name "Standardbred" comes from its ability to trot or pace a mile in under "standard" time. "Standard" time was 2:30 for trotters and 2:25 for pacers. Today the standard is 2:20 for 2-year-olds and 2:15 for mature animals. There is no difference for trotters and pacers—most are pacers anyway.

## American Saddlebred Horse

This breed was developed in the South, especially in Kentucky, with emphasis more on riding comfort than on speed. The foundation included Native American horses, the Thoroughbred, and Standardbred, plus a limited infusion of Arabian and Morgan. The American Saddle Horse can be either three- or five-gaited, and some may be used as fine harness horses. The breed is used primarily as a pleasure and show horse. Many consider it the premier show horse because of its great beauty and very attractive way of moving.

**Prepotent**

An animal that transmits its characteristics to its offspring in a consistent fashion.

## Tennessee Walking Horse

This breed originated in Tennessee, for much the same purpose as the American Saddle Horse. Its foundation included American Saddle Horse, Thoroughbred, Standardbred, and Morgan as well as Native American Horses. Tennessee Walking Horses are three-gaited. The *gaits* are the walk, running walk, and canter. The running walk is peculiar to this breed. When performing the running walk, the horse overstrides by placing a back hoof significantly ahead of the print of the forehoof. This breed is known for its comfortable gaits and pleasureful ride.

## Quarter Horse

**Stock horse**

Any horse of the light breeds trained and used for working livestock, mainly cattle.

The Quarter Horse is the most numerous breed in the United States. The Quarter Horse is still an important work animal that is used on American ranches as a ***stock horse***. No other breed can really have this said about it. Its use as a work animal is far overshadowed by its use as a racing, rodeo, show, and pleasure horse. The Quarter Horse originated in Virginia with the Quarter Pather, but the bulk of its improvement occurred in the Southwest. The foundation included Native American horses, the Mustang, and the Thoroughbred. The Quarter Horse is unexcelled for two purposes: as a working cow horse and for speed at short distances.

## Appaloosa

**Stud book**

The set of records a breed association keeps on the animals registered with it.

The Nez Perce tribe of the Northwest developed these horses. It takes its name from the Palouse River area where the tribe lived. Appaloosas were selected for their very distinctive color pattern. Because of an open ***stud book*** policy, the modern Appaloosa has varying amounts of Quarter Horse, Thoroughbred, and Arabian heritage.

### BREED POPULARITY

Table 21–8 lists historic and current horse breed registration numbers. The Quarter Horse registers by far the largest number of new horses each year (Figure 21–15). In many years, as many or more Quarter Horses were registered as all the others combined. However, several breeds have been growing in numbers.

**TABLE 21–8    Horse Breed Registration Figures**

|  | 1960 | 1975 | 1980 | 1985 | 1990 | 1995 | 2000 | 2005 | 2006 |
|---|---|---|---|---|---|---|---|---|---|
| Appaloosa | 4,052 | 20,175 | 25,384 | 16,189 | 10,669 | 10,903 | 10,906 | 7,055 | 6,749 |
| Arabian | 1,610 | 15,000 | 19,725 | 30,004 | 17,676 | 12,398 | 9,660 | 6,359 | 7,003 |
| Morgan | 1,069 | 3,400 | 4,537 | 4,538 | 3,618 | 3,053 | 3,624 | 3,156 | 3,461 |
| Paint | NA | 5,287 | 9,654 | 12,692 | 16,153 | 34,846 | 62,511 | 42,557 | 39,357 |
| Quarter Horse | 35,507 | 97,179 | 137,090 | 157,360 | 110,597 | 107,332 | 127,763 | 144,955 | 144,236 |
| Saddlebred | 2,329 | 4,064 | 3,879 | 4,353 | 3,569 | 3,239 | 2,908 |  |  |
| Standardbred | 6,413 | 12,830 | 15,219 | 18,384 | 16,576 | 10,918 | 13,846 | 10,457 | 9,791 |
| Tennessee Walking Horse | 2,623 | 6,591 | 6,847 | 7,633 | 7,609 | 10,020 | 15,000 | 13,366 | 9,345 |
| Thoroughbred | 12,901 | 29,225 | 39,367 | 50,382 | 44,143 | 34,958 | 36,700 | 34,070 | 36,317 |

SOURCE: Breed associations.

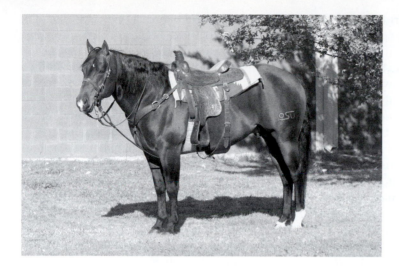

**FIGURE 21–15** The Quarter Horse is the most popular breed in the United States.

# REPRODUCTIVE MANAGEMENT

As few as 50–60% of the **mares** bred **to foal** the following year actually have a foal. This has earned the horse the reputation of being a fairly infertile animal. It is probably true that horses have more problems with reproduction than other species that have been specifically selected for their reproductive efficiencies. Even though reproductive rates are generally lowly heritable, substantial improvement has been made with many of the other species because it is a trait that is selected for. In horses, it is not usually one of the traits that is even considered. However, the horse is generally more fertile than it is given credit for. The way that many people breed their horses is destined to a high percentage of failure from the onset. The typical way many approach the task is as follows:

1. Spend months of decision making over just which of the available stallions to use.
2. Scrimp and save to afford the stud fee.
3. Take the mare for an early spring ride and notice her showing signs of heat.
4. Rush back home and call the stallion owner and arrange to bring her right over.
5. Breed the mare, load her up, and bring her home to avoid paying board.
6. Turn her out to pasture and wait 11 months for the blessed event.
7. Be disappointed when no foal appears.

To be fair, people who make their living breeding horses could not survive in the business with such poor management. However, this scenario is all too common and has contributed to the horse's reputation as being infertile. In truth, a high percentage of mares can settle and bring a new foal into the world. An understanding of how to make this happen is all that is needed. The horse is much less frequently the problem than the people managing the breeding.

   The mare has a **seasonally polyestrous** estrous cycle and generally conceives only during certain times of the year. Mares respond to the hour of daylight by initiating the estrous season during lengthening days. The common way to think of this is that mares are "long-day breeders." In North America, **anestrus** is usually from mid-November until mid-February. For some mares and even whole breeds, the fertile

**Mare**

A mature female horse.

**Foaling; to foal**

Parturition in the horse.

**Seasonally polyestrous**

When an animal has repeated estrous cycles but only in response to some environmental factor associated with the seasons, frequently the photoperiod.

**Anestrus**

Period of time when a female is not having estrous cycles.

**Estrous cycle**

The period of time from one estrus to the next.

**Stud**

A unit of male animals kept for breeding. Also a term for a stallion.

**Teasing**

Placing a stallion and a mare in proximity to each other and observing the mare's actions. Mares that are coming into heat display specific behaviors.

**Ultrasonography**

Using ultrasound waves to visualize the deep tissues of the body.

period tends to start later in the spring. During the rest of the year, the mare generally has an *estrous cycle*, can mate, and can conceive offspring (Figure 21–16). For mares that are to be bred, it is critical to manage breeding so the mare is mated when an egg is present to be fertilized. Because a mare may very well be receptive to the stallion at times other than when she is ovulating, this can be a bit tricky. For most owners of just a mare or two, the best course of action is to choose a stallion that is managed by a knowledgeable *stud* manager, and to trust he or she will get the mare safely in foal. For those who will be standing stallions to the public or breeding larger numbers of mares, the only course of action is to familiarize themselves thoroughly with the hormonal and environmental control mechanisms of reproduction in both the mare and the stallion. This knowledge, put to use with good insemination techniques and management for both mare and stallion, is the only course of action. Breeding managers must effectively learn to manipulate the estrous cycle, detect estrus, mate at the proper time, and confirm pregnancy.

Heat detection is usually accomplished through the use of *teasing* with a stallion and palpation. Combining *ultrasonography* with rectal palpation helps determine the time of ovulation. Generally, mares are teased with a stallion to determine if they are showing signs of estrus (Figure 21–17). The ones that are showing signs of estrus are palpated. Signs of estrus in mares include winking of the vulva, urination, squatting,

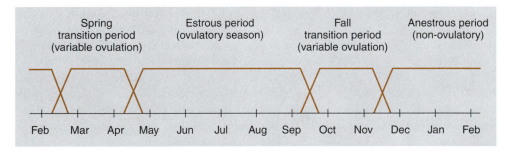

**FIGURE 21–16**    Mare seasonality. (*SOURCE:* Slusher et al., 1998, p. 3974.1. Used with permission.)

**FIGURE 21–17**    Observing a mare's behavior when in the proximity of a stallion is one useful tool in heat detection. In the pictured arrangement, a stallion is confined to the pipe enclosure and several mares are allowed into the surrounding corral. Trained personnel observe from a distance and keep a daily record of the mares' actions. Analyzing this record of observations helps determine when to start taking ultrasound readings on each mare to detect ovulation.

and seeking the stallion. Ultrasound can help determine many of the physical changes that accompany ovulation, and can thus help time the mating (Figure 21–18). Conception rates are best when insemination occurs within the 36 hours directly preceding ovulation. Ovulation most frequently occurs 24 to 48 hours before the end of estrus (Figure 21–19).

Pregnancy is detected in various ways. A good indicator of pregnancy is to wait and see if the mare returns to heat. Generally, a nonpregnant mare should once again come into estrus 18 to 20 days from the time she last ovulated. Rectal palpation can detect a pregnancy in as few as 18 days following insemination. Ultrasonography can be used to detect pregnancy in as few as 10 days. Either of these methods in the hands of a skilled person can take the guesswork out of pregnancy diagnosis.

The estrous cycle can be manipulated in several ways in horses, to varying degrees of success. Because a mare's estrous cycles begin in response to length of day, artificial lights can be used to "extend" the day from 1 to 16 hours. To be successful, this technique must be started 60 to 90 days before the first ovulation is desired.

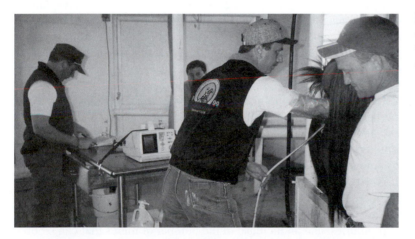

**FIGURE 21–18** Timing of breeding is critical to ensure conception. Combining ultrasonography with rectal palpation helps determine the time of ovulation.

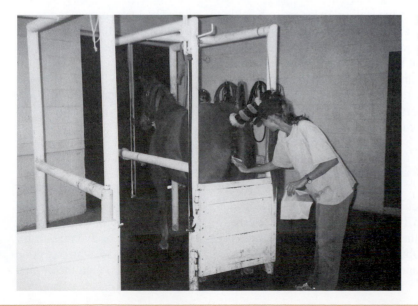

**FIGURE 21–19** Use of a palpation stock such as this one helps ensure the safety of the horse and the people involved in breeding horses.

This must be carefully done. The days must be "extended" by the same amount each day. Light intensity must be a minimum of 2 foot-candles. Hormonal treatments are also available to alter the length of estrus, change its duration, and even stimulate ovulation.

Good record keeping is an important part of breeding management. Figure 21–20 shows an example of a comprehensive teasing, breeding, and palpation record for a mare.

The stallion should not be overlooked in this process. Breeding soundness examinations and semen evaluation should be carried out on the stallion before the breeding season begins and periodically during the breeding season. Stallions used to breed mares naturally should be checked for transmissible diseases. Stallions used in artificial insemination programs have their semen evaluated much more frequently as part of the processes of extending and dividing the semen for multiple inseminations.

Various methods are used to inseminate mares. At some breeding farms, mares are simply turned into the pasture with a stallion (pasture breeding) and allowed to stay there for a period of time long enough to ensure mating. Such mating systems can result in good conception rates, but risks of injury are higher. This is especially true if outside mares are constantly added and taken from the pasture. This system is best if all the mares to be bred are put together and allowed to stabilize into a "band" before the stallion is introduced. No new mares should be added after this time. Hand mating is more common. In this system, the receptive mares are mated under close supervision. Mares are often placed in breeding hobbles to prevent them from kicking. Their tails are wrapped to help avoid injuries to the stallion, and the stallion is kept on halter and directed by a skilled person. Injuries can be minimized in this way. However, diseases can be transmitted during the act of mating and then passed on to other horses. An increasingly common method in use on progressive breeding farms is to collect the semen of the stallion with the use of an ***artificial vagina*** (Figure 21–21) and breed the receptive mares artificially. Although some breeds do not register the foals born to such matings, most do. This method has several advantages including disease prevention, virtual elimination of injuries, and an increase in the number of mares that can be bred to a given stallion each day.

The needs of the stallion manager to be able to collect, evaluate, extend, and ship semen have increased because several breed associations now allow the use of cooled shipped semen in breeding programs. This allows breeders to use genetics from across the country that they might not otherwise be able to access. Previously, most breed associations required the mare to be on the premises with the stallion when inseminated even if artificial insemination was used. The ability to use sires previously unavailable to many breeders is an exciting opportunity for horse breeders. Embryo transfer is also accepted by some breed associations.

## NUTRITION AND FEEDING OF HORSES

Poor feeding choices by horse owners are a major source of economic losses in the horse industry. They can also be a major source of emotional trauma for owners when the results of their feeding actions are illness, permanent disability, or even death for their animal. Obviously, people who have a recreation/companion horse, who pay good money for it, and who are perhaps emotionally attached to it, want to feed the animal

**Artificial vagina**
Device used to collect semen from a male.

**Tease Code**
1 – resistance
2 – indifferent
3 – interested
4 – winks
vulva,
urinates
5 – profuse
urination and
vulvular
activity

**Other Codes**
T – treated
C – culture
S – speculum
P – palpate
B – bred
Pr – pregnant
F – foaled
A – arrived
D – departed
U – ultrasound

TS – tease score
FS – follicle size
CX – cervix
FC – follicle
consistency

Mare _____ Color _____ Age _____ Farm number _____
In 20 _____ Book to _____ Mare owner _____
Results of last year's breeding _____

| | 1 2 3 4 5 6 7 8 9 10 11 12 13 14 15 16 17 18 19 20 21 22 23 24 25 26 27 28 29 30 31 |
|---|---|
| Dec. | |
| Jan. | |
| Feb. | |
| Mar. | |
| Apr. | |
| May | |
| June | |
| July | |

**Palpation**

| Date _____ Remarks _____ | Date _____ Remarks _____ | Date _____ Remarks _____ |
|---|---|---|
| TS: 1 2 3 4 5<br>FS: _____ mm<br>CX: 1 2 3<br>FC: T S O<br>Uterine tone _____ | TS: 1 2 3 4 5<br>FS: _____ mm<br>CX: 1 2 3<br>FC: T S O<br>Uterine tone _____ | TS: 1 2 3 4 5<br>FS: _____ mm<br>CX: 1 2 3<br>FC: T S O<br>Uterine tone _____ |
| Date _____ Remarks _____ | Date _____ Remarks _____ | Date _____ Remarks _____ |
| TS: 1 2 3 4 5<br>FS: _____ mm<br>CX: 1 2 3<br>FC: T S O<br>Uterine tone _____ | TS: 1 2 3 4 5<br>FS: _____ mm<br>CX: 1 2 3<br>FC: T S O<br>Uterine tone _____ | TS: 1 2 3 4 5<br>FS: _____ mm<br>CX: 1 2 3<br>FC: T S O<br>Uterine tone _____ |

**Teasing Code**
1 – Mare is visibly resistant to stallion
2 – Mare is indifferent to stallion
3 – Mare is slightly interested in stallion; may urinate, may wink vulva
4 – Mare is greatly interested in stallion; occasional urination, profuse vulva activity
5 – Mare is greatly interested in stallion; frequent urination, squatting, leans into stallion

**Follicle Size**
Commonly sized as <20, 30, 40, 50 millimeters or greater in diameter

**Follicular Consistency**
T – Turgid
S – Soft or breedable
O – Ovulated

**Cervix Size**
1 – ≤10 millimeters
2 – >10 but <30 millimeters
3 – ≥30 millimeters

**Uterine Tone**
Poor, fair, good, excellent, or pregnant

**FIGURE 21–20** Teasing and palpation record for mares. (*SOURCE:* Slusher et al., 1998. Used with permission.)

**FIGURE 21–21** An artificial vagina (a) is combined with such tools as a mounting dummy (b) to collect semen from stallions to be used in artificial insemination of mares.

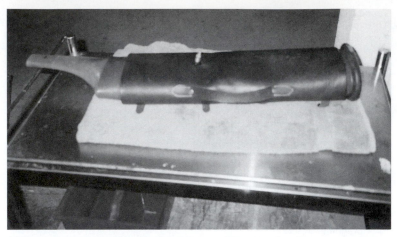

(a)

(b)

**Ration**

The feed allotment for an animal for a 24-hour period. Use of the term *ration* should be in conjunction with specific quantities of feeds being consumed.

well. People in the horse business have an equal, if different, reason to want to feed well: Properly fed horses make more money. It is unfortunate that both groups frequently fail to seek the advice of those who are trained to help them, such as extension agents and state specialists, and rather rely on advertising or the word of the person who has a horse in the next stall for their nutritional advice. As a result, horse owners spend millions of dollars every year on needless boxes and tubes of "magic dust" to put in horse *rations*. They pay too much for oats when several other grains have been proven to be just as good for horses. The horse suffers and is not as productive as it might be because its ration is unbalanced and/or deficient. A better and more economical approach is to learn the proper way to feed the horse. Doing so will no doubt help save for that new saddle. A word of caution: Learning what is in this chapter is not enough to make you a competent equine nutritionist. This is an introductory text. The material presented here is just designed to heighten awareness and get you started.

Nutrition and nutritional management of the horse revolve around the fact that the horse is a monogastric with a functional cecum. This allows it to use a significant amount of forage in its ration and do very well on it. As a nonruminant herbivore, the horse has characteristics of the simple monogastric and also some similarities to the ruminant in its ability to use feeds. The horse evolved using its speed as its major survival mechanism. Thus it developed the ***continuous eater strategy*** of eating frequently in small amounts and moving from place to place between grazing. Domestic horses on pasture still exhibit these same eating behaviors. Stabled or otherwise confined horses are generally handled in a way that makes them meal eaters instead of continual eaters. This creates problems because the stomach of the horse is small in comparison to the rest of the digestive tract, and poorly muscled. It is not designed to handle large amounts of feed at one time. Yet many horses are fed only once a day. This creates problems for many horses each year in the form of ***colic***, ***impacted intestine***, and other digestive problems. Chapters 5, 6, and 7 deal with the particulars of basic nutrition, digestive tract anatomy, and feeds. Refer to those chapters for information not covered here.

Different ***diets*** should be fed to horses in different classes. A mare nursing a foal has different nutrient requirements than when she is standing idle. The yearling has different requirements than the actively breeding stallion. *The Nutrient Requirements of Horses,* published by National Academy Press, should be used to determine the appropriate nutrients required for each particular horse or class of horses to be fed. Table 21–9 shows some examples of feed formulations for various classes of horses. These are example feeds and should not be used without considering the specific nutrient needs of the individual horse. By using the information specific to the horse(s) being fed, the horse will be healthier and the owner will be wealthier. Each of these feeds is formulated to be fed with forage to complete the ration for the animal.

**Continuous eater strategy**

Feeding strategy employed by many prey species as a mechanism of survival. They eat many small meals through the day and keep on the move throughout their range.

**Colic**

A broad term that means digestive disturbance.

**Impacted intestine**

Term used to describe constipation in the horse and some other species.

**Diet**

All the feeds consumed by animals, including water.

**TABLE 21–9    Example Feed Formulations for Horses[1]**

| Ingredient | All-Purpose Formulation[2] | Maintenance Formulation[3] |
|---|---|---|
|  | % of Total Formulation | |
| Corn, no. 2, yellow | 32.00 | 62.86 |
| Oats | 63.00 | — |
| Soybean meal (44%) | — | 14.30 |
| Molasses | 2.00 | — |
| Dehydrated alfalfa meal | — | 20.71 |
| Dicalcium phosphate | 1.00 | 0.71 |
| Limestone | 1.00 | 0.71 |
| Salt, trace mineral | 1.00 | 0.71 |
| **Nutrient Analysis** | | |
| Protein (%) | 12.00 | 17.00 |
| DE (Mcal/lb) | 1.3 | 1.5 |
| Calcium (%) | 0.7 | 0.8 |
| Phosphorus (%) | 0.5 | 0.4 |

[1]All formulations designed to be fed with bermuda grass hay to balance the daily ration.
[2]Designed to be fed as a textured ration.
[3]Designed to be fed as a pelleted ration.
*SOURCE:* Cooper, 2004. Used with permission.

**FIGURE 21–22** It is important to determine a horse's weight if it is to be fed accurately. Use of a scale is the best way to get an accurate weight, but heart girth tapes are fairly accurate.

It is important to determine a horse's body weight if it is to be fed accurately. Weighing the horse on a scale is ideal. If that is not feasible, then heart girth tapes can be purchased from most feed stores or ordered from equine and/or farm supply catalogs. These tapes give a good estimate of weight for most horses (Figure 21–22). A formula can also be used to estimate the horse's weight (Figure 21–23). A popular one is:

$$\text{Weight in pounds} = \frac{\text{Heart girth in inches}^2 \times \text{Body length in inches}}{330}$$

**Long hay**

Hay that has not been chopped or ground; the individual forage.

**Cubed hay**

Hay is forced through dies to produce an approximate 3-cm product of varying lengths.

The horse's digestive tract is such that it needs to have forage in its diet in long form. Generally, a horse should receive a minimum of 0.75–1% of its body weight daily in the form of roughage. Although *long hay* is probably the least expensive and most available to horse owners, research has shown that *cubed hays* can be fed without development of behavioral or digestive problems. Horses allowed to graze acceptable pastures consume enough forage to satisfy their requirement. However, stabled or otherwise confined horses may not receive enough forage (Figure 21–24). Many owners intentionally reduce the amount of forage to levels below 0.75–1% of body weight to keep their horses from having a large cecum, which is referred to as a "hay belly." The look of the horse is thus placed above its health. Failure to allow the horse adequate forage can lead to a wide variety of vices such as chewing wood, eating feces, eating bedding, cribbing, or chewing the manes and tails of their stable or pen mates. It can also lead to colic and other digestive upsets. Horses need forage! The grain portion of a ration for horses should be formulated to balance the forage the horse is being given. However, a cautionary note: Horses need hay of good quality. Feeding hay that is too mature to be easily digested, with weeds, insects, or foreign material, can also lead to digestive problems. It should also be free of mold and dust.

There is still something of an art to feeding horses. Even if you become skilled at using the nutrient requirement tables and at ration formulation, there is a good deal of

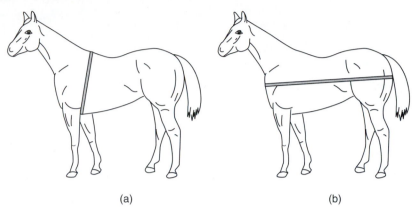

(a)                                           (b)

**FIGURE 21–23**   A formula can be used to estimate a horse's weight. Here is popular one:

$$\text{Weight in pounds} = \frac{\text{Heart girth in inches}^2 \times \text{Body length in inches}}{330}$$

(a)  Measure a horse's heart girth from the base of the withers down to a couple of inches behind the horse's front legs, under the belly, and up the opposite side to where you started.
(b)  Measure a horse's length from the point of the shoulder to the point of the hip.

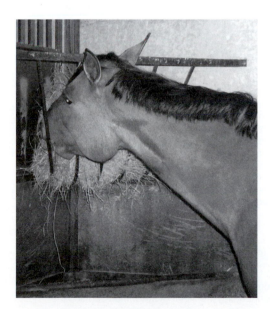

**FIGURE 21–24**   A horse should receive a minimum of 0.75–1% of its body weight daily in the form of roughage as pasture, hay, or cubed hays.

individual variation in horses. Anyone with much experience with horses has observed that some horses are "easy keepers" and others are "hard keepers." This means some horses are easier to keep in good condition than others are. Metabolisms, "normal" activity levels, and other factors vary from horse to horse. Thus feeding a horse to a particular *body condition* is another tool that horse nutritionists use (Figure 21–25). In general, horses should be fed so they have a moderate to fleshy body condition—neither too thin nor too fat. One of the inherent problems with this is that many horse owners like to see a horse in body condition that is fatter than is really healthy for the animal. By taking the time to learn how to body condition horses properly, the horse owner can save money and keep the horse healthier and performing better.

**Body condition**

The amount of fat on an animal's body.

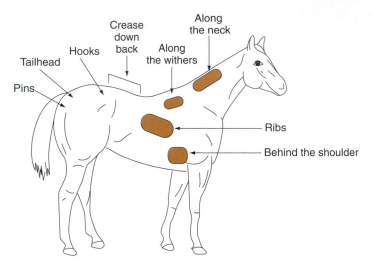

**Body Condition Scoring System**

Score

1    Poor.  Animal extremely emaciated.  Spinous processes (portion of the vertebra of the backbone which project upward), ribs, tailhead, and bony protrusions of the pelvic girdle (hooks and pins) projecting prominently.  Bone structure of withers, shoulders, and neck are easily noticeable.  No fatty tissues can be felt.

2    Very Thin.  Animal emaciated.  Slight fat covering over base of spinous processes, transverse processes (portion of vertebrae which project outward) of lumbar (loin area) vertebrae feel rounded.  Spinous processes, ribs, shoulders, and neck structures are faintly discernible.

3    Thin.  Fat built up about halfway on spinous processes, transverse processes cannot be felt. Slight fat cover over ribs.  Spinous processes and ribs are easily discernible.  Tailhead prominent, but individual vertebrae cannot be visually identified.  Hook bones (protrusion of pelvic girdle appearing in upper, forward part of the hip) appear rounded, but are easily discernible.  Pin bones (bony projections of pelvic girdle located toward rear, mid-section of the hip) not distinguishable.  Withers, shoulders, and neck accentuated.

4    Moderately Thin.  Negative crease along back (spinous processes of vertebrae protrude slightly above surrounding tissue).  Faint outline of ribs discernible.  Tailhead prominence depends on conformation, fat can be felt around it.  Hook bones are not discernible.  Withers, shoulders, and neck are not obviously thin.

5    Moderate.  Back level.  Ribs cannot be visually distinguished, but can be easily felt.  Fat around tailhead beginning to feel spongy.  Withers appear rounded over spinous processes.  Shoulders and neck blend smoothly into body.

6    Moderate to Fleshy.  May have slight crease down back.  Fat over ribs feels spongy.  Fat around tailhead feels soft.  Fat beginning to be deposited along the sides of the withers, behind the shoulders and along sides of neck.

7    Fleshy.  May have crease down back.  Individual ribs can be felt, but noticeable filling between ribs with fat.  Fat around tailhead is soft.  Fat deposited along withers, behind shoulders and along neck.

8    Fat.  Crease down back.  Difficult to feel ribs.  Fat around tailhead very soft.  Area along withers filled with fat.  Area behind shoulder filled in flush.  Noticeable thickening of neck.  Fat deposited along inner buttocks.

9    Extremely Fat.  Obvious crease down back.  Patchy fat appearing over ribs.  Bulging fat around tailhead, along withers, behind shoulders and along neck.  Fat along inner buttocks may rub together.  Flank is filled in flush with the rest of the body.

**FIGURE 21–25**    Location of fat deposits used in body condition scoring system. (*SOURCE:* Freeman, 1997. Used with permission.)

The concentrate portion of rations for horses can be mixed, ground, and then pelleted. *Pellets* have the advantage of ensuring that the horse receives all the proper nutrients in proper amounts because it cannot "select" its diet. Further advantages include a reduction in dust, being able to use less palatable feedstuffs, and ease of feeding. For operations with larger numbers of horses, the advantages will be greater than for those with just a horse or two (Figure 21–26). It is still recommended that horses receive 0.75–1% of their body weight as forage when pelleted rations are fed.

One of the big problems in horse nutrition is overfeeding. One of the causes of overfeeding is the major misperception that a coffee can holds the same amount of oats or corn as it does coffee. Not so! It holds 1.3 times as much oats and 1.8 times as much corn. Feeds have different densities, and processing further affects density. Horse feed—both hay and grain—should be weighed. Scales are cheap compared to the expense, trauma, and potential lifetime effects of colic or *laminitis*. This does not mean that every morsel given to the horse must be weighed. Scoops and cans can still be used. What it does mean is that the amount of the particular feed being used should be weighed periodically so the feeder can "calibrate" his or her eye and container accordingly (Figure 21–27).

Because the horse developed as a continuous eater, it is best served by feeding practices that increase the number of times it eats in a day. However, most horses are meal-fed. Because of management, labor, housing, and production needs, that practice will probably continue. If a horse is receiving grain in excess of 0.5% of its body weight, it is recommended that the grain be split into a minimum of two feedings per day. This reduces the amount of digestive upset the horse is likely to have. Once grain exceeds 1.0% of body weight, the grain should be fed in at least three equal portions. The feedings should be as evenly spaced apart as feasible. It is also important to feed very close to the same time each day, even on weekends. Any time a ration is changed, especially if the grain portion is increased, it should be done at an increasing rate of no more than a half lb per day until the new feeding level is reached. Even moving horses to a different pasture or turning them into pasture after a period of stall confinement should be done gradually by limiting the access to the pasture for several days before they are allowed to stay in it full time.

Many times a horse suffers digestive upset for reasons not directly related to the feed. Two prevalent reasons are parasites and water. Clean, palatable water should be

**Pellets**

Feeds that are generally ground and then compacted by forcing them through die openings.

**Laminitis**

An inflammation of the laminae of the hoof. It is commonly caused by overeating of grain but may also be related to eating lush pastures, road concussion, retained afterbirth in the mare, and other causes.

**FIGURE 21–26**  The concentrate portion of a horse's diet can be pelleted.

**FIGURE 21–27** A horse's feed should be weighed, or at least the containers being used to measure feed should be filled and weighed regularly so that the person doing the feeding can "calibrate" his or her eye to the container. This will help prevent the most common problem in horse nutrition: OVERFEEDING!

available to horses at all times. Waterers and buckets or troughs should be cleaned regularly. The only time that unlimited water is not a good idea is when a horse is hot from exercise. Unlimited water at this time can be dangerous. The horse should be cooled and allowed small quantities of water until normal body temperature is reached. Then the water supply should be made available again. Horses should be treated for internal parasites regularly. Several of the parasites to which horses are susceptible can cause digestive tract disorders and interfere with feed utilization by the animal.

For many horse owners, the safest way to feed is to purchase the grain portion of a horse's diet from a reputable company and feed according to the manufacturer's instructions. Forage should be provided at the recommended levels just discussed. The amount of grain can then be adjusted based on body condition of the horse. Horse owners should avoid the temptation to buy all of those cartons and bottles of "magic dust and tonic" on the shelves at the feed store. By feeding on time, and by keeping the horse exercised and dewormed, the benefits and joys of horse ownership can be enjoyed—at greatly reduced prices.

## TRENDS IN THE HORSE INDUSTRY AND FACTORS THAT WILL INFLUENCE THE INDUSTRY IN THE FUTURE

### EDUCATION AND RESEARCH

The typical owner of a recreational horse and the typical horse of a recreation-minded person are the most vulnerable beings in the horse industry. This is because of the general lack of knowledge of the typical owner. The uneducated horse owner is a menace to himself or herself, to horses everywhere, and to the entire horse industry. Their feed costs too much, their vet bills are too high, too many horses and people are injured, and, subsequently, overall satisfaction is lower than it should be. These owners often quit the horse in disgust and convince other would-be owners not to bother. Because they are also the most frequently occurring horses and people in this business, they are also the ones most in need of services. Yet they have proven to be very difficult to reach with good information. The valiant efforts of professionals in the business and in education don't seem to make a dent in the problem. Educating the new owner should be a top priority for everyone in the horse industry. Education is important because the "new" horse owner has much to learn to be a successful horse owner. "Old" horse owners are often woefully ignorant of any of the scientific information about horses. New owners tend to get most of their information from old owners. This creates a vicious cycle of ignorance, which needs to be broken for the good of all! One welcome trend is that many of the new horse owners now entering the industry seem hungry for information and readily seek educational opportunities. An unfortunate aspect is that not all the information they are receiving is credible.

There is a great need for more research and support for research on the horse (Figure 21–28). Currently, only a comparative pittance of the time and money spent on other species is directed to equine research. Scores of problems need answers that research could provide. Too many horses must be retired because of ***unsoundness***. The root of this problem is frequently genetic and/or management related. Diseases and parasites cost the industry millions per year. A huge percentage of this loss is unnecessary and could be prevented. The nutritional needs of horses are poorly understood compared to those of the food-producing species. All of these problems need to be much better addressed with research and information dissemination. It is hard to argue

**Unsoundness**

Any injury or defect that interferes with the ability of the animal to be used for its given purpose.

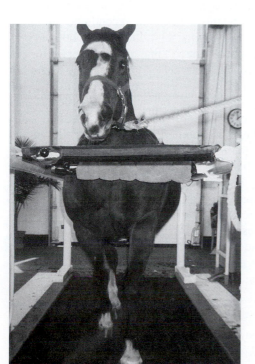

**FIGURE 21–28** Solid research work, like the exercise physiology research being conducted with this mare, is sorely lacking. The industry needs more research to help it prosper. (Photo courtesy of Dr. Steven R. Cooper, Department of Animal Science, Oklahoma State University. Used with permission.)

for integrating science and technology into the horse business when some of the science and technology is so poorly supported by research.

Two areas of research are receiving increased attention. Although not all the bugs have been worked out, artificial insemination techniques for the horse have been developed and are currently meeting with increased usage. As more and more breed associations allow the use of frozen semen, these techniques will get better. Also, the first draft of the horse genome map was completed in 2006. Once the genome is mapped, great advances can be made in horse genetics and health.

## COMPETITION

The recreational horse owner faces many choices for his or her leisure dollar and time. The horse industry must find ways to continue to attract the leisure dollar or risk shrinking in size. During the past decade and a half horse breed associations have expanded shows to include more divisions in which to compete. This is a positive development.

Examples of new ways that are being explored include a trend toward objectively timed or scored events. Such timed or scored events that seem to be growing in popularity include barrel racing, team roping, team penning, reining, dressage, and ranch horse competitions. In a move to attract the **baby boomers**, divisions for 50-and-over participants are already commonplace and growing in equine sports and other activities.

## NEW OWNERS

**Baby boomer**

Demographic term for the U.S. population born between 1946 and 1964. The U.S. Census estimates there are nearly 83 million boomers.

There is a growing trend among adults to compensate for the perceived missed opportunities of childhood. Many people, baby boomers in particular, were unable to have that much-dreamed-of horse or pony when they were children. Times are more affluent now and so are the baby boomers. They can now afford a horse or three if it suits them. On January 1, 2008, Kathleen Casey-Kirschling, recognized as the first baby boomer, also became the first one to start collecting Social Security. Born one second after midnight on January 1, 1946, the retired teacher leads the way for 83 million fellow boomers.

Others of us wish to recreate the part of our childhood that included horses for our children or grandchildren. Those young people are largely part of the "echo boom" or "Generation Y," names used to refer to the group of Americans born between 1977 and 1994. This echo boom generation is the second largest generation of Americans, second only to the baby boom generation. (In case you are counting on your fingers just now, I'll save you the trouble. Many of the echo boomers are the grandchildren and children of the baby boomers.) The baby boom generation is an affluent group that controls unprecedented and enormous collective economic resources. In the horse industry, this is being translated into baby boomer–fueled spending. They are either entering the horse industry as direct participants or funding horse-related opportunities for children or grandchildren. The demand for riding lessons for children is up, the market for "kids' horses" and well-trained horses for adults is up, and membership in horse-related youth groups as well as those for adults is also up (Figure 21–29). Also, adult first-time horse owners were recognized as a significant market segment in the 1990s. These trends show every sign of continuing.

## SAFETY CONCERNS

A concern for safety is working its way through the horse industry and is evident at competitive events and other horse-related activities. This is probably being fueled by concerns for personal safety by first-time horse owners of more advanced age and by the increased

**FIGURE 21–29** The generation of young Americans dubbed the "echo boom" generation is a growing part of the horse industry. Their entry into the industry has caused an increased demand for children's riding lessons, well-trained "kids' horses," and youth-oriented services from breed associations and extension services.

care and concern for safety that is a more general part of modern society. Here again the baby boom generation is contributing significantly. Remember, many of these people are introducing themselves and their grandchildren to the horse world. In addition, there is a heightened awareness of liability issues on the part of event facility management. Expect more and more safety gear and safety measures to be a part of the horse world.

## West Nile Virus

In 1999, the West Nile Virus was discovered for the first time in the Western Hemisphere. This disease is a threat to the human population, horses, dogs, and several species of birds, most notably the crow and its cousins. A vaccine and combination of vaccines exist and should be added to the complement of regularly administered vaccinations for horses.

## Social Issues

### Animal Rights/Animal Welfare

Animal rights/animal welfare concerns will continue to be a part of this industry. Because so much of the horse industry is sport and exhibition oriented, the industry is under more scrutiny than other animal industries and thus more vulnerable. It is also an easy target for activist groups because of the horse's place in the minds and/or hearts of people. The industry must learn to face legitimate concerns while fending off the unreasonable activist groups. Many breed organizations have adopted codes of conduct and addressed humane care issues. These efforts are to be applauded. However, nothing less than full care and attention to this issue is acceptable.

### Horse Slaughter

As of 2007, horse slaughter in the United State has effectively ceased. This was an emotional debate that will continue to rage. What remains to be seen is the effects this change will have on the horse industry.

### TECHNOLOGICAL INNOVATION

The horse industry is a poorly mechanized industry, partly because of the wishes of horse owners. Caring for the animal is important to horse owners. However, improvements such as simple, affordable automatic feeders could add to the horse's health and help the horse owner as well. It is hard to see how it would decrease the joy of horse ownership to have the horse fed at intervals, even in the owner's absence. Other labor-saving devices could help attract new owners.

### NATIONAL ANIMAL IDENTIFICATION SYSTEM

The USDA has recognized the Equine Species Working Group (ESWG), a broadly representative group from the horse industry, as the working group designated to review and evaluate the possible role of the National Animal Identification System (NAIS) within the equine industry. As with other animal industries, there are questions to be answered about NAIS before the horse industry as a whole will be willing to participate. The ESWG has a website, http://www.equinespeciesworkinggroup.com/home.html, designed to address these concerns and questions, clarify the purposes of NAIS and its potential benefits, explain how the horse industry is responding, and what NAIS means to horse owners and breeders.

## INDUSTRY ORGANIZATIONS

**American Connemara Pony Society**
P.O. Box 100
Middlebrook, VA 24459
Phone: (540)886-2239
Fax: (540)722-2277
http://www.acps.org

**American Cream Draft Horse Association**
193 Crossover Rd.
Bennington, VT 05201
Phone: (802)447-7612
Fax: (802)447-0711
http://www.acdha.org

**American Donkey and Mule Society**
P.O. Box 1210
Lewisville, TX 75067
Phone: (972)219-0781
Fax: (972)420-9980
E-mail: lovelongears@hotmail.com
http://www.lovelongears.com

**American Hackney Horse Society**
4059 Iron Works Parkway A-3
Lexington, KY 40511-8462
Phone: (859)255-8694
Fax: (859)255-0177
http://www.hackneysociety.com

**American Horse Council**
1616 H St. NW, 7th Floor
Washington, DC 20006

Phone: (202)296-4031
Fax: (202)296-1970
E-mail: ahc@horsecouncil.org
http://www.horsecouncil.org

**American Indian Horse Registry, Inc.**
9028 State Park Rd.
Lockhart, TX 78644
Phone: (512)398-6642
http://www.indianhorse.com

**American Miniature Horse Association, Inc.**
5601 South Interstate 35 W.
Alvarado, TX 76009
Phone: (817)783-5600
Fax: (817)783-6403
http://www.amha.org

**American Morgan Horse Association**
122 Bostwick Rd.
Shelburne, VT 05482-4417
Phone: (802)985-4944
Fax: (802)985-8897
E-mail: info@morganhorse.com
http://www.morganhorse.com/

**American Mule Association**
P.O. Box 1349
Yerington, NV 89447
Phone: (775)463-1922
E-mail: masmules@aol.com
http://www.americanmuleassociation.com

**American Mustang and Burro Association, Inc.**
P.O. Box 1013
Grass Valley, CA 95945-1013
E-mail: AMBAInc@bardalisa.com
http://www.bardalisa.com

**American Paint Horse Association**
P.O. Box 961023
Fort Worth, TX 76161-0023
Phone: (817)834-2742
Fax: (817)834-3152
http://www.apha.com/

**American Quarter Horse Association**
1600 Quarter Horse Dr.
Amarillo, TX 79104
Phone: (806)376-4811
Fax: (806)376-8304
http://www.aqha.com/

**American Quarter Pony Association**
Box 30
New Sharon, IA 50207
Phone: (641)675-3669
Fax: (641)675-3969
E-mail: jarrod@netins.net
http://www.aqpa.com

### American Saddlebred Horse Association, Inc.
4093 Iron Works Pike
Lexington, KY 40511
Phone: (859)259-2742
Fax: (859)259-1628
E-mail: saddlebred@asha.net
http://www.asha.net

### American Shetland Pony Club
81B Queenwood Rd.
Morton, IL 61550
Phone: (309)263–4044
Fax: (309)263-5113
E-mail: info@shetlandminiature.com
http://www.shetlandminiature.com/
Registers four breeds: The Classic American Shetland Pony, The Modern American Shetland Pony, The American Miniature Horse, and The American Show Pony.

### American Warmblood Registry
P.O. Box 197
Carter, MT 59420
Phone: (406)734-5499
Fax: (775)667-0516
E-mail: amerwarmblood@aol.com
http://www.americanwarmblood.com/

### Appaloosa Horse Club Inc.
2720 W. Pullman Rd.
Moscow, ID 83843
Phone: (208)882-5578
http://www.appaloosa.com/

### Arabian Horse Association
10805 E. Bethany Dr.
Aurora, CO 80014
Phone: (303)696-4500
Fax: (303)696-4599
http://www.arabianhorses.org

### Belgian Draft Horse Corporation of America
P.O. Box 335
Wabash, IN 46992
Phone/fax: (260)563-3205
http://www.belgiancorp.com/

### Cleveland Bay Horse Society of North America
P.O. Box 483
Goshen, NH 03752
Phone: (860)774-5433
http://www.clevelandbay.org

### Clydesdale Breeders of the United States
17378 Kelley Rd.
Pecatonica, IL 61063
Phone: (815)247-8780
Fax: (815)247-8337
http://clydesusa.com

**The Jockey Club**
40 East 52nd St.
New York, NY 10022
Phone: (212)371-5970
Fax: (212)371-6123
http://www.jockeyclub.com

**National Show Horse Registry**
10368 Bluegrass Parkway
Louisville, KY 40299
Phone: (502)266-5100
Fax: (502)266-5806
http://www.nshregistry.org

**Palomino Horse Breeders Association of America**
15253 E. Skelly Dr.
Tulsa, OK 74116-2637
Phone: (918)438-1234
Fax: (918)438-1232
http://www.palominohba.com/

**Paso Fino Horse Association, Inc.**
101 N. Collins St.
Plant City, FL 33566-3311
Phone: (813)719-7777
Fax: (813)719–7872
http://www.pfha.org

**Percheron Horse Association of America**
P.O. Box 141
Fredericktown, OH 43019
Phone: (740)694-3602
Fax: (740)694-3604
http://percheronhorse.org/

**Pinto Horse Association of America, Inc.**
7330 NW 23rd St.
Bethany, OK 73008
Phone: (405)491-0111
Fax: (405)787-0773
http://www.pinto.org/

**Pony of the Americas Club**
3828 S. Emerson Ave.
Indianapolis, IN 46203
Phone: (317)788-0107
Fax: (317)788-8974
http://www.poac.org

**Tennessee Walking Horse Breeders' and Exhibitors' Association**
250 North Ellington Parkway
Lewisburg, TN 37091
Phone: (931)359-1574
http://www.twhbea.com

**U.S. Trotting Association (Standardbred)**
750 Michigan Ave.
Columbus, OH 43215
Phone: (877)800-8782
http://www.ustrotting.com/

## SUMMARY AND CONCLUSION

In the United States, the horse began the 20th century as a partner with humans in the pursuit of food production and in everyday life. The majestic horse pulled plows, wagons, and threshing machines during the week, and it took the family to town on Saturday afternoon and to church on Sunday morning. Horses were an integral component of survival. All of this gave way to the petroleum-powered engine and the general onslaught of technology. The industrialization of the country took the population to the cities and left the farmer on a tractor with little need for the beasts that had so recently been his or her partners. By 1960, the horse was deemed to be so unimportant that the federal government quit counting them. With the growing prosperity of the nation, however, the horse made an interesting conversion around mid-century to become an animal of recreation rather than utility. A new hybrid industry developed around the horse, with interests and segments in agriculture, sports, recreation, and entertainment. People whose grandparents plowed with horses took to the horse as a tool to improve the quality of their urban, industrialized, high-tech life. The high-touch needs of the horse generated an industry very different from the other animal industries. It is an industry in which the values of people figure very prominently in the politics and actions of its participants.

---

### Facts about Horses

Terms Used in Describing Horses

| | |
|---|---|
| Head markings: | Star, stripe, snip, blaze, bald face, spot, race |
| Body markings: | Appaloosa, bay, black, brown, buckskin, chestnut, dun, gray, palomino, pinto, paint, roan |
| Leg markings: | Boot, sock, 1/2 stocking, stocking |

Mature Weight and Height

| | |
|---|---|
| Ponies: | 500–900 lbs; 11–14.2 hands |
| Saddle and light harness horses: | 800–1,400 lbs; 14.2–16 hands |
| Draft horses: | 1,700–2,200 lbs; 15–17 hands |

The weight will be dependent on breed, sex, nutritional state, and so on. The height of a horse is determined by standing it squarely on a level area and measuring the vertical distance from the highest point of the withers to the ground. The unit of measurement used in expressing height is the "hand"; each hand is 4 inches.

Breeding age (female): Horses reach puberty at 12–15 months, but should not be bred until after 2 years of age and up to about 15–20 years of age.

Stallion/mare ratio (# of hand-mating/year): 2-yr-old stallion/10–15; 3-yr.-old stallion/20–40; 4-yr.-old stallion/30–60; mature/80-100; over 18 yrs. old/20–40

Gestation: 336 days (a little over 11 months); varies from 310–370 days depending on age, size, and physical condition

Estrous cycle: 18–20 days

Duration of estrus: 5–7 days, breed on the 3rd day of estrus and on alternate days thereafter as long as the mare remains in heat

Age weaned: 4–6 months

Names of various sex classes: ***Foal***, ***filly***, ***colt***, mare, stallion, ***gelding***

---

**Foal**

A newborn horse of either sex. The term is sometimes used up to the time of weaning, after which *colt* and *filly* are more likely to be used.

**Filly**

A young female horse.

**Colt**

A young male horse.

**Gelding**

A castrated male horse.

It is impossible to present facts and figures for the horse that are comparable to those of the other livestock species because the federal government does not collect and summarize the same information on the horse as it does on other species. According to industry estimates, nearly 4 million people (horse owners, supportive family members, and volunteers) participate in the horse industry. The horse generates a direct economic effect of $29 billion on the U.S. economy.

## STUDY QUESTIONS

1. Why is it difficult to present the same kinds of statistics about the horse industry as there are for the other animal industries?

2. Describe the magnitude of the horse industry.

3. Explain what is meant when the horse industry is referred to as a "hybrid" industry.

4. Describe the structure of the horse industry. Compare and contrast it with the other animal industries discussed thus far in this text.

5. What makes the horse industry unique from the other animal industries?

6. Why is it so hard to say with certainty how many horses are in the United States?

7. Give a quick thumbnail sketch of the history of the horse in North America.

8. What are the major participant groups within the horse industry?

9. Discuss the various contributions the horse industry makes to the U.S. economy.

10. List some of the quantitative traits for horses and make a statement about how successful you would expect selection programs for those traits to be.

11. Horse genetics is complicated by the fact that so many different genes control it. Select your favorite horse color and write a genetic formula for it. If more than one is possible, just pick one.

12. Create a table like this for the breeds of horses listed in the text.

| Breed | Where Developed | Major Use(s) | Other Breeds Important to Development |
|-------|-----------------|--------------|---------------------------------------|

Complete the table for each breed.

Complete the table for each breed.

13. You have decided to breed your super-duper mare "Whiz-Kid." Outline how you will go about this from a reproductive management perspective.

14. What are the five most important things about feeding horses that you feel all horse owners should know? You must settle on only the five most important!

15. How can one estimate a horse's weight?

16. You are responsible for directing the Horse Owner Education Program for the XYZ Horse Breed Association. Explain what your educational program would be for each of the trends discussed in the chapter that are affecting or likely to affect the horse industry.

# REFERENCES

For the fourth edition, Sarah M. Teuschler-Stewart of Oklahoma State University reviewed the chapter and provided new material.

American Horse Council Foundation. 2005. *The economic impact of the U.S. horse industry on the United States.* Washington, DC: American Horse Council Foundation.

Barclay, H. B. 1980. *The role of the horse in man's culture.* London & New York: J. A. Allen.

Bowling, A. T. Aug. 1997. Coat color genetics. *The Quarter Horse Journal* 49(11).

Bowling, A. T. 1999. *Horse genetics.* Available from http://www.vgl.ucdavis.edu/~1vmillion/coats2.html.

Bowling, A. T., and A. Ruvinsky (eds.). 2000. *The genetics of the horse.* New York: CABI.

Cooper, S. R. 2004. Oklahoma State University, Stillwater. Personal communication.

Evans, J. W. 2006. *A guide to selection, care, and enjoyment of horses.* 3rd ed. New York: W. H. Freeman.

Freeman, D. W. 1997. *Body condition of horses.* Extension Bulletin No. F-3920. Stillwater: Cooperative Extension Service, Oklahoma State University.

Freeman, D. W. 1998a. *Ration formulation for horses.* Extension Bulletin No. 3997. Stillwater: Cooperative Extension Service, Oklahoma State University.

Freeman, D. W. 1998b. Cooperative Extension Service, Oklahoma State University, Stillwater. Personal communication.

Geddes, C. 1978. *The horse—The complete book of horses and horsemanship.* London: Octopus Books.

Griffin, J. M. and T. Gore. 1998. *Horse owner's veterinary handbook.* New York: Howell Book House.

Harland, J. R. 1976. The plants and animals that nourish man. *Scientific American* 3, 235.

Johnson, E. L. 1993. Basic equine genetics. In *Horse industry handbook.* Lexington, KY: American Youth Horse Council.

Jordan, R. M. 1998. *Horse nutrition and feeding.* Publication Number: FA-0480-GO. St. Paul: University of Minnesota Extension Service, University of Minnesota.

Kjersten, D., and J. M. Griffin. 1999. *Veterinary guide to horse breeding.* New York: Howell Book House.

Landers, T. A. 2002. *The career guide to the horse industry.* Albany, NY: Delmar Thomson Learning.

Loch, W. 1998. Agricultural publication G2780. Columbia, MO: Department of Animal Sciences, University of Missouri. http://muextension.missouri.edu/xplor/agguides/ansci/g02780.htm.

Loch, W., J. F. Lasley, and M. Bradley. 1993. *Genetics of coat color of horses.* Agricultural Publication G2791. Columbia: Department of Animal Sciences, University of Missouri.

McClure, J. J. 1993. Genetic abnormalities in horses. In *Horse industry handbook.* Lexington, KY: American Youth Horse Council.

Parker, R. 2007. *Equine science.* Albany, NY: Delmar.

Pilliner, S., and Z. Davies. 2004. *Equine science.* Ames: Iowa State Press.

Siegal, M., ed. 1996. *UC Davis book of horses: A complete medical reference guide for horses and foals.* New York: HarperCollins.

Slusher, S. H., C. Taylor-MacAllister, and D. W. Freeman. 1998. *Reproductive management of the mare.* Fact Sheet No. 3974. Stillwater: Oklahoma Cooperative Extension Service, Division of Agricultural Sciences and Natural Resources, Oklahoma State University.

Spoonberg, D. P., and A. T. Bowling. 1996. Champagne, a dominant color dilution of horses. *Genetics, Selection, and Evolution* 28: 457–462.

Thorson, J. S. 2001. 21st century horse trends. *Western Horseman,* January 2001, p. 50.

Turman, E. J. 1986. *Agricultural animals of the world.* Stillwater: Oklahoma State University.

Vogel, C. J. 1996. *An illustrated guide to veterinary care of the horse.* Ames: Iowa State University Press.

# Aquaculture

## LEARNING OBJECTIVES

After you have studied this chapter, you should be able to:

1. Define *aquaculture.*

2. Describe the purpose of aquaculture and its worldwide importance.

3. Describe the place of aquaculture in agriculture and explain its role in feeding the world.

4. Describe the growth rate of aquaculture as an industry segment.

5. Name the species that make up the bulk of U.S. aquaculture.

6. Discuss the various types of aquaculture systems.

7. Identify the general life cycles of aquatic species.

8. Compare the challenges of the geneticist who works with aquatic species with those faced by geneticists who work with the species previously studied.

9. Explain the nutritional benefits of aquatic products to humans.

10. Discuss trends affecting aquaculture.

## KEY TERMS

| | | |
|---|---|---|
| Aquaculture | Fingerlings | Postlarvae |
| Baitfish | Food-size | Spat |
| Blended fisheries | Fry | Spawn |
| Broodstock | Juvenile | Stocker fish |
| Capture fisheries | Larva | Stocking |
| Cold blooded | Polyploidy | |

# THE PLACE OF AQUACULTURE IN U.S. AGRICULTURE

**Aquaculture**

The farming of aquatic organisms including fish, mollusks, crustaceans, and plants in fresh water, brackish water, or salt water.

*Aquaculture* is the farming of aquatic organisms, including fish, ornamental fish, baitfish, mollusks, crustaceans, reptiles, and aquatic plants. This may be done in fresh water, brackish water, or salt water. Fish and other aquatic species have long provided a significant amount of food to the world's population. Much of this food is in the form of wild catch. However, because of the decline in the wild catch, interest in the cultivation and harvest of aquatic species has accelerated. Aquaculture also has recreational and aesthetic dimensions. Production of stock game fish and ornamental fish and plants fall into these categories.

Aquaculture is a rapidly growing segment of agriculture, although it is still small by comparison with some of the production giants (Figures 22–1 and Figure 22–2). It is predicted that this growth will continue and the industry will become a more important segment of agriculture. Fish and other aquatic species provide a substantial por-

**FIGURE 22–1**  Aquaculture farm cash receipts as a percentage of total U.S. farm cash receipts, 2005–2007. (*SOURCE:* USDA-NASS, 2007a and Fisheries of the United States, 2007.)

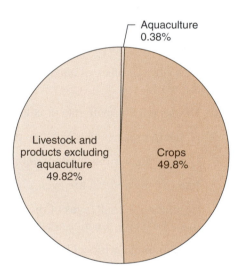

Aquaculture
0.38%

Livestock and products excluding aquaculture
49.82%

Crops
49.8%

**FIGURE 22–2**  Aquaculture yearly farm cash receipts as a percentage of total animal agriculture's cash receipts, 2005–2007. (*SOURCE:* USDA-NASS, 2007a and Fisheries of the United States, 2007.)

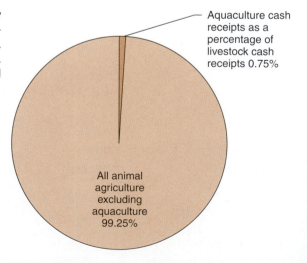

Aquaculture cash receipts as a percentage of livestock cash receipts 0.75%

All animal agriculture excluding aquaculture
99.25%

**FIGURE 22–3**    Fish and other aquatic species provide a significant amount of food for the world's population. Much is in the form of wild catch, such as this fisherman photographed off the coast of Corregidor, Philippines.

**FIGURE 22–4**    Aquaculture, which has been practiced in some parts of the world for centuries, is growing worldwide.

tion of the world's food (Figure 22–3). However, the harvest of seafood from the oceans and other waters appears to have hit a plateau. To meet the demand for products, the world is increasingly turning to aquaculture (Figure 22–4).

Aquaculture is practiced in a way that is quite analogous to other types of agriculture. The systems may be quite intensive, with high stocking densities or low-density systems integrated with other farm production as part of a diversified farm. The variety of enthusiasts ranges from those whose livelihood is taken from aquaculture to backyard hobbyists. Aquaculture is quite different from fishing for wild species, which is called ***capture fisheries.*** The term ***blended fisheries*** is used to describe the practice of using aquaculture techniques to enhance or supplement capture fisheries. Compared to activities surrounding land animals, think of fishing and hunting as analogous activities, aquaculture and farming as analogous, and blended fisheries as an enhanced form of wildlife management.

**Capture fisheries**

The harvesting of wild aquatic animal species.

**Blended fisheries**

A combination of aquaculture and capture fisheries. Aquaculture techniques are used to enhance or supplement capture fisheries.

**TABLE 22-1**    **U.S. Annual per Capita Consumption of Commercial Fish and Shellfish, Selected Years**

| Year | Per Capita Consumption (Pounds) | | | |
|------|------------------|--------|-------|-------|
| | FRESH AND FROZEN | CANNED | CURED | TOTAL |
| 1910 | 4.5 | 2.8 | 3.9 | 11.2 |
| 1920 | 6.3 | 3.2 | 2.3 | 11.8 |
| 1930 | 5.8 | 3.4 | 1.0 | 10.2 |
| 1940 | 5.7 | 4.6 | 0.7 | 11.0 |
| 1950 | 6.3 | 4.9 | 0.6 | 11.8 |
| 1960 | 5.7 | 4.0 | 0.6 | 10.3 |
| 1970 | 6.9 | 4.5 | 0.4 | 11.8 |
| 1980 | 7.9 | 4.3 | 0.3 | 12.5 |
| 1990 | 9.6 | 5.1 | 0.3 | 15.0 |
| 1995 | 10.0 | 4.7 | 0.3 | 15.0 |
| 2000 | 10.2 | 4.7 | 0.3 | 15.2 |
| 2005 | 11.6 | 4.3 | 0.3 | 16.2 |
| 2006 | 12.3 | 3.9 | 0.3 | 16.5 |

SOURCE: National Marine Fisheries Service, Fisheries Statistics and Economics Division, 2007.

## THE PURPOSE OF THE AQUACULTURE INDUSTRY

Aquaculture provides a means of expanding agriculture. Expansion may be through diversification by the addition of part-time employment opportunities and supplemental income to existing agricultural enterprises. It may also provide large-scale, full-time opportunities. Either way, the purpose of the aquaculture industry is to diversify the farming sector by providing a product that is in demand by several different sectors of the consuming public. Aquaculture species can be used to exploit land-surface area in a very efficient way because the space requirements of fish in relation to surface area are fairly small. In addition, fish are very efficient converters of feed to flesh and, as such, help provide high-quality food for the human population by using feedstuffs rather than foodstuffs, and doing so in a manner that is equal to or even more efficient than that done by poultry. Aquatic species are also quite capable of using wastes and waste by-products, thereby converting something with little or no use to a highly palatable and valuable food source.

Recent per capita consumption of seafood from all sources in the United States has been around 16 lbs (Table 22–1). Aquaculture's share of that consumption is growing. Table 22–2 shows the U.S. per capita consumption for the 10 most popular species.

## WORLDWIDE IMPORTANCE OF AQUACULTURE

The world's population consumes approximately 36 lbs of live weight equivalent of seafood per capita each year. That quantity has remained constant even though total human population continues to increase. Clearly, aquatic species are important in feeding the world. Worldwide, aquaculture production of fish, crustaceans, mollusks, and so on, is approximately 49 million metric tons (MT). It accounts for over a third of the seafood used for direct human consumption. Aquatic plants provide an additional 15 million MT of food.

| TABLE 22–2 | Approximate U.S. per Capita Consumption of Fish and Seafood | |
|---|---|---|
| Species | Pounds Per Capita | Rank |
| Shrimp | 4.40 | 1 |
| Tuna | 2.90 | 2 |
| Salmon | 2.03 | 3 |
| Pollock | 1.64 | 4 |
| Tilapia | 0.996 | 5 |
| Catfish | 0.969 | 6 |
| Crab | 0.664 | 7 |
| Cod | 0.505 | 8 |
| Clams | 0.440 | 9 |
| Scallops | 0.305 | 10 |

SOURCE: National Fisheries Institute, 2007.

One of the reasons for aquaculture's rapid climb as an agricultural industry is the increase in world population. However, of equal importance is the fact that the amount of wild fish harvested from the world's waters has been, at best, static since 1990 at approximately 93 million MT. This has been attributed to various reasons, including overfishing, environmental degradation, and social and economic influences. The social and economic uses include competition and conflicts for the fishery resource with recreation and commercial uses, conservation issues, and animal rights advocacy. If seafood supplies are to be maintained for the human population, then an alternative source is needed. On a worldwide basis, aquaculture production has been increasing at a rate of 7–9% per year since 1990. World leaders in aquaculture production include China (first by a wide margin), India, Japan, the Philippines, Thailand, Vietnam, and Indonesia.

On a worldwide basis, aquaculture is dominated by carp (considered a food crop) and salmon (Figure 22–5), shrimp, tilapia, and mollusks (cash crops). Tonnage produced by aquaculture exceeds wild catch on most of these. Carp will probably not be

**FIGURE 22–5** Salmon production from aquaculture in sea cages, like these off the coast of Ireland, is increasing rapidly. (Photographer Dale Sarver. Courtesy of Animals Animals/Earth Scenes.)

**TABLE 22-3**    **Major Aquaculture Species**

### AQUATIC PLANTS

| Microalgae | Macroalgae | Aquatic Plants |
|---|---|---|
| Spirulina | Irish moss | Water hyacinth |
| Tetraselmis | Giant kelp | |
| Ioschrysis | Nori | |

### INVERTEBRATES

**Annelids**

| Bloodworm | Lugworm | Sandworm | |
|---|---|---|---|

**Arthropods/Crustaceans**

| Brine shrimp | Copepods | Daphnia | Shrimp/prawn |
|---|---|---|---|
| Crab | Crayfish | Lobster | |

**Echinoderms**

Sea urchins

**Mollusks**

| Bivalves | Gastropods | Cephalopods |
|---|---|---|
| Clam | Abalone | Octopus |
| Mussel | Conch | Squid |
| Pen shell | | |
| Oyster | | |
| Scallop | | |

### VERTEBRATES

**Amphibians**

Frog

**Fishes**

| American shad | Flatfish | Pike | Sturgeon |
|---|---|---|---|
| Arctic char | Flounder | Pink salmon | Sunfish hybrids (bluegill, bream) |
| Atlantic salmon | Golden shiner (bait) | Plaice | Threadfin shad |
| Bowfin | Goldfish (ornamental) | Pompano (Florida) | Tilapia |
| Buffalo fish | Grass carp | Pumpkinseed | Tuna |
| Bull minnow (bait) | Grouper | Red drum (redfish) | Turbot |
| Channel catfish | Halibut (Atlantic) | Seatrout (weakfish) | Walking catfish |
| Cod (Atlantic) | King salmon | Smallmouth bass | Walleye |
| Coho salmon | Koi (ornamental) | Snapper | Yellow perch |
| Common carp | Largemouth bass | Snook | Yellowtail |
| Crappie | Milkfish | Sole | |
| Croaker | Mullet (striped) | Steelhead (rainbow) trout | |
| Dolphin fish | Muskellunge | Striped bass | |
| Eel | Paddelfish | | |

**Reptiles**

| Alligator | Crocodile | Turtle (incl. terrapin) | |
|---|---|---|---|

*SOURCE:* Adapted from Iverson and Hale, 1992, pp. 278–279.

exploited as a cash crop outside of Asia because its quality is not perceived to be as high. Even though it represents approximately 40% of the total poundage produced in aquaculture, most carp is grown in China and is expected to stay there. Table 22–3 shows the major groups of aquacultural species.

## HISTORICAL PERSPECTIVE

Aquaculture has been practiced in varying forms for centuries dating to the time of the Phoenicians. The Chinese also have a rich aquaculture heritage dating at least as far back as the first recorded references prior to 1100 B.C. Aquaculture began in the United States in the mid-1800s with the development of government-run fish hatcheries that produced fish for *stocking* ponds and lakes for recreation and resource management purposes. Similar to the ways that other agricultural industries have developed, the nation's colleges and universities initiated basic and applied research and education programs on aquaculture management, genetics, nutrition, reproduction, diseases, herbicide testing, and other areas. The commercial industry started shortly after World War II with the production of *baitfish.* Golden shiners, the fathead minnow, and goldfish were easy to raise and were all popular species as baitfish. During the 1950s and 1960s, species such as buffalo fish, European and Chinese carps, and several species of catfish such as the channel, blue, and white, were experimented with and culture practices were established. In addition, some sport fish species, including black bass, crappie, bluegill, and hybrid striped bass, were added to the growing list of aquaculture species. However, it was only in the last two decades of the 20th century that aquaculture became an important food supplier or an important economic sector in the United States.

### Stocking

The practice of placing fish from a hatchery or other source into a pond or lake to either establish a new species or augment an already existing species.

### Baitfish

Fish selected and produced to be used as bait to catch much larger fish.

## THE PLACE OF AQUACULTURE IN U.S. AGRICULTURE

In 1985, total aquaculture production in the United States was worth $347 million. By 1998, aquaculture production in the United States was worth $978 million. In 2005, it was an approximately $1.1 billion industry producing over twice as much product (Table 22–4). Aquaculture production and the value of that production (Figure 22–6) continue to increase in response to strong consumer demand, rapid technological advances, and limited supplies. Although all sectors are not increasing equally, the overall trend is upward. Further potential for increased aquaculture production in the United States is tremendous. Imported seafood accounts for approximately 45% of the U.S. seafood supply, much of it farm-raised shrimp, salmon, and tilapia. The aquaculture industry is increasing production to meet this demand. Although U.S. production is relatively small compared to that of other world producers (less than 1% of world production), the U.S. Department of Agriculture reports that aquaculture is the fastest growing segment of U.S. agriculture (Figure 22–7).

U.S. aquaculture includes the culture of catfish, trout, salmon, shrimp, mussels, clams, oysters, tilapia, hybrid striped bass, crayfish, ornamental fish, other species, and plants (Figure 22–8). The total industry is dispersed throughout the United States, although there is a definite regionality to the individual types of production. Operations range in size from the large and well established, employing hundreds of people, to hobby ventures.

**TABLE 22–4** Estimated U.S. Aquaculture Production

| SPECIES | 1985 THOUSAND POUNDS | 1985 THOUSAND DOLLARS | 1990 THOUSAND POUNDS | 1990 THOUSAND DOLLARS | 1995 THOUSAND POUNDS | 1995 THOUSAND DOLLARS | 2000 THOUSAND POUNDS | 2000 THOUSAND DOLLARS | 2005 THOUSAND POUNDS | 2005 THOUSAND DOLLARS |
|---|---|---|---|---|---|---|---|---|---|---|
| **Finfish** | | | | | | | | | | |
| Baitfish | 24,807 | 51,280 | 21,610 | 53,978 | 21,759 | 72,522 | 13,954 | 45,790 | — | 38,018 |
| Catfish | 191,616 | 138,922 | 360,435 | 273,210 | 446,886 | 351,222 | 593,603 | 445,919 | 607,933 | 429,245 |
| Salmon | 3,921 | 5,465 | 9,069 | 26,341 | 31,315 | 75,991 | 49,372 | 99,208 | 20,726 | 37,439 |
| Striped bass | NA | NA | 1,590 | 3,490 | 8,315 | 21,156 | 11,237 | 29,513 | 10,970 | 27,655 |
| Tilapia | 0 | 0 | 0 | 0 | 15,075 | 22,613 | 20,000 | 30,000 | 17,203 | 29,620 |
| Trout | 50,600 | 55,154 | 56,816 | 64,640 | 55,934 | 61,447 | 59,164 | 63,690 | 60,636 | 65,469 |
| **Shellfish** | | | | | | | | | | |
| Clams | 1,999 | 4,698 | 3,680 | 13,486 | 4,325 | 19,709 | 9,929 | 32,595 | 12,564 | 72,783 |
| Crawfish | 65,011 | 29,350 | 71,000 | 34,000 | 58,146 | 34,714 | 17,025 | 27,626 | 35,933 | 21,143 |
| Mussels | 1,210 | 642 | 607 | 1,173 | 410 | 1,221 | 424 | 525 | 962 | 4,990 |
| Oysters | 21,906 | 38,882 | 22,192 | 77,949 | 23,221 | 70,628 | 16,822 | 42,419 | 13,711 | 92,602 |
| Shrimp | 440 | 1,566 | 1,984 | 7,937 | 2,205 | 8,818 | 4,782 | 14,559 | 8,037 | 18,684 |
| **Miscellaneous** | 14,267 | 21,541 | 23,548 | 98,908 | 23,359 | 75,243 | 26,207 | 140,989 | — | 254,738 |
| Totals | 375,777 | 347,500 | 572,531 | 655,112 | 690,950 | 815,284 | 822,519 | 972,833 | 788,675 | 1,092,386 |

NA, not available.

*Note:* Table may not add due to rounding. Clams, oysters, and mussels are reported as meat weights (excludes shell); other identified species such as shrimp and finfishes are reported as whole (live) weights. Some clam and oyster aquaculture production are reported with U.S. commercial landings. Weights and values represent the final sales of products to processors and dealers. "Miscellaneous" includes ornamental/tropical fish, alligators, algae, aquatic plants, eels, scallops, crabs, and others. The high value and low production of "Miscellaneous" occurs because production value but not weight are reported for many species such as ornamental fishes.

*source:* Fisheries of the United States, 2007.

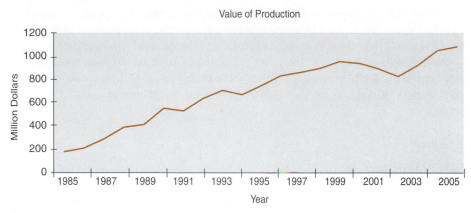

FIGURE 22–6    Value of the combined aquaculture sales of the United States. (*SOURCE:* Fisheries of the United States, 2007.)

FIGURE 22–7    Driven by strong consumer demand, aquaculture is the fastest growing segment of U.S. agriculture. (Photographer Doug Menuez. Courtesy of Getty Images, Inc./ PhotoDisc, Inc.)

FIGURE 22–8    Aquaculture also includes the production of seaweed. These cultured beds are off the coast of Okinawa on the Ryukyu Islands. (Photographer James Marshall. Courtesy of Corbis.)

Aquaculture is an agricultural activity whose importance is expected to increase in the United States. Factors influencing this growth include (1) an increasing demand for fish and seafood that exceeds supply; (2) the fact that the wild fish harvest has reached its limit and will probably decline; and (3) concerns over the foreign trade deficit in fishery products. Aquaculture can remedy these problems. U.S. aquaculturists cultivate approximately 30 species of fish and shellfish and a substantial number of aquatic plants.

## CATFISH

Table 22–4 confirms that the channel catfish segment is clearly the largest of this industry, generally accounting for approximately 40% of the income generated from U.S. aquaculture. Farmer sales of catfish to processors reached 520 million lbs in 1997, exceeding a half billion lbs for the first time (Figure 22–9). Table 22–5 shows the number of catfish operations, water surface, and total catfish sales for the United States. Catfish production is currently the largest aquaculture sector in the United States. It is an industry of the Southeast. The most important producing states are Mississippi, Alabama, and Arkansas, with Mississippi producing over half the total. The United States is the world's leading catfish producer, although production has declined in recent years (Figure 22–10).

## TROUT

Trout culture has the distinction of being the oldest form of U.S. aquaculture, dating to the 1800s (Figure 22–11). Trout farming began in the United States in a measurable way in the 1930s to provide fish for stocking and replenishing streams and lakes. USDA statistics for trout (Table 22–6) include sales of *food-size fish, stocker fish, fingerlings,* and eggs. Twenty states have food-size trout sales significant enough to be reported. Idaho is clearly the largest producer, with North Carolina, California, Pennsylvania, and Washington occupying a second tier of production. Idaho's dominance in this industry is related to its vast system of aquifers and springs.

---

**Food-size**

Fish that are grown commercially for food, usually ranging from three-fourths to 1 lb and over 12 in. in length.

---

**Stocker fish**

Fish usually 6 to 12 in. in length and less than three-fourths of a lb.

---

**Fingerlings**

A juvenile stage in fish. Fish usually from 1 to 6 in. long.

---

**FIGURE 22–9** Channel catfish production is the largest segment of U.S. aquaculture. These workers are harvesting catfish from a pond in the Mississippi delta, the center of U.S. catfish production. (Photo courtesy of Mississippi Department of Economic and Community Development/Division of Tourism Development.)

**TABLE 22–5  U.S. Catfish Production: Number of Operations, Water Surface Acres Used for Production, and Total Sales, by State**

| State | Number of Operations on January 1 (NUMBER) | | | | Water Surface Acres Used for Production, January 1–June 30 (ACRES) | | | | Total Sales (1,000 DOLLARS) | | | |
|---|---|---|---|---|---|---|---|---|---|---|---|---|
| | 1995 | 2000 | 2005 | 2007 | 1995 | 2000 | 2005 | 2007 | 1995 | 2000 | 2004 | 2006 |
| Alabama | 250 | 260 | 230 | 199 | 17,900 | 22,100 | 25,100 | 23,700 | 50,909 | 81,617 | 101,198 | 98,969 |
| Arkansas | 160 | 170 | 153 | 137 | 19,500 | 33,000 | 31,500 | 30,400 | 41,034 | 65,737 | 66,618 | 79,586 |
| California | 40 | 46 | 31 | 37 | 2,100 | 2,400 | 1,700 | 1,500 | 6,703 | 8,705 | 7,482 | 7,318 |
| Florida | 20 | 22 | 46 | 43 | 210 | 390 | 650 | 530 | 280 | 1,296 | 1,139 | 1,761 |
| Georgia | | 60 | 55 | 60 | | 990 | 1,090 | 1,100 | | 1,510 | 1,475 | 2,019 |
| Kentucky | 25 | 25 | 60 | 35 | 225 | 200 | 600 | 370 | 717 | 1,254 | 1,151 | 849 |
| Louisiana | 105 | 94 | 38 | 20 | 11,800 | 13,500 | 7,600 | 6,300 | 18,862 | 32,976 | 14,316 | 12,625 |
| Mississippi | 294 | 405 | 410 | 370 | 95,000 | 110,000 | 101,000 | 94,000 | 271,875 | 300,303 | 274,971 | 262,510 |
| Missouri | 75 | 40 | 24 | 21 | 2,100 | 1,860 | 1,320 | 1,100 | 2,924 | 3,117 | 1,358 | 2,060 |
| North Carolina | 47 | 44 | 49 | 44 | 1,100 | 1,300 | 2,000 | 2,000 | 1,961 | 2,743 | 7,021 | 7,213 |
| Texas | 125 | 49 | 62 | 57 | 2,200 | 580 | 1,030 | 1,700 | 875 | 1,031 | 3,446 | 5,910 |
| Total | 1,141 | 1,215 | 1,158 | 1,023 | 152,135 | 186,320 | 173,590 | 162,700 | 396,140 | 500,289 | 480,175 | 480,820 |

SOURCE: USDA-NASS, 2007a.

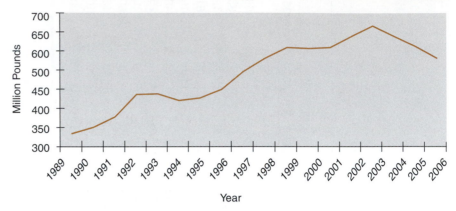

**FIGURE 22–10**    Total U.S. catfish processed annually. (*SOURCE:* USDA-NASS, 2007b.)

**FIGURE 22–11**    Trout culture is the oldest form of U.S. aquaculture.

## TILAPIA

Domestic production of tilapia is generally increasing in the United States (Table 22–4). Rapid expansion of the culture of this fish has been stalled by the ready availability of imported tilapia. Most imported tilapia is used in food service. It is a mild white-fleshed fish and priced so it offers an excellent way to add a seafood item to menus. Tilapia production is limited geographically by the biological needs of this fish to warm-water areas. They require water that stays above 10°C (50°F) to live. California is the production leader, with Florida and, increasingly, Arizona the major areas of production.

## CRAWFISH

Crawfish production is essentially restricted to Louisiana. There are over 140,000 acres of crawfish ponds in Louisiana. Production for recent years is shown in Table 22–4 and highly variable. Wild catch can add a third to a half again to the production (Figure 22–12).

**TABLE 22–6**  **U.S. Trout Production: Number of Operations, Value of Fish Sold and Distributed, by State**

| State | Number of Operations | | | Total Sales (THOUSANDS OF DOLLARS) | | | Value of Distributed Fish[1] (THOUSANDS OF DOLLARS) | | |
|---|---|---|---|---|---|---|---|---|---|
| | 2000 | 2005 | 2006 | 2000 | 2005 | 2006 | 2000 | 2005 | 2006 |
| Arkansas | 5 | 4 | 5 | — | | | —[2] | — | — |
| California | 28 | 33 | 35 | 5,033 | 6,077 | 5,573 | 7,676 | 8,127 | 10,845 |
| Colorado | 42 | 27 | 34 | 2,289 | 1,480 | 2,299 | 4,624 | 5,215 | 4,919 |
| Connecticut | 6 | 6 | 6 | 615 | 411 | 340 | — | — | — |
| Georgia | 11 | 14 | 15 | 737 | 844 | 630 | — | — | — |
| Idaho | 57 | 42 | 43 | 38,093 | 35,387 | 41,434 | 5,833 | 2,993 | 3,571 |
| Maine | 15 | 16 | 16 | 282 | 284 | 235 | — | — | — |
| Massachusetts | 15 | 15 | 16 | 468 | 424 | 424 | — | — | — |
| Michigan | 33 | 20 | 23 | 1,037 | 793 | 783 | — | — | — |
| Missouri | 14 | 15 | 14 | 1,968 | 2,649 | 2,345 | 2,279 | 2,236 | 2,526 |
| New York | 36 | 36 | 34 | 615 | 507 | 540 | — | — | — |
| N. Carolina | 59 | 49 | 47 | 5,247 | 6,590 | 7,232 | 317 | 729 | 1,458 |
| Oregon | 45 | 48 | 42 | 1,365 | 803 | 331 | 4,072 | 6,898 | 7,393 |
| Pennsylvania | 60 | 57 | 51 | 5,439 | 4,807 | 4,790 | 5,416 | 10,813 | 13,015 |
| Tennessee | 16 | 14 | 14 | 522 | 291 | 291 | — | — | — |
| Utah | 25 | 21 | 26 | 1,396 | 540 | 318 | 2,375 | — | — |
| Virginia | 29 | 20 | 24 | 1,644 | 1,256 | 1,475 | 1,346 | 2,071 | 2,103 |
| Washington | 65 | 69 | 65 | 3,033 | 4,124 | 4,007 | 5,915 | 6,633 | 6,698 |
| W. Virginia | 30 | 26 | 28 | 611 | 348 | 235 | 1,460 | 1,818 | 2,036 |
| Wisconsin | 77 | 70 | 66 | 1,732 | 1,573 | 1,573 | 2,367 | 2,120 | 2,933 |
| Total | 668 | 602 | 604 | 75,791 | 69,188 | 74,855 | 60,930 | 73,710 | 79,400 |

[1]Trout distributed for restoration, conservation, and recreational purposes, primarily by state and federal hatcheries.
[2]Values not listed to avoid disclosures of individual operations. The amounts are added to the total.
SOURCE: USDA-NASS, 2007c.

**FIGURE 22–12**    Crawfish production is restricted mostly to Louisiana. Production is highly variable from year to year. (Photographer Skip Nall. Courtesy of Getty Images, Inc./PhotoDisc, Inc.)

### SALMON

Salmon farming is predominantly practiced in Washington and Maine in ocean pens. Additional salmon are cultivated in hatcheries in various locations on the East and West coasts for release back to the wild. U.S. salmon production can be seen in Table 22–4.

### MOLLUSKS

Oysters, clams, and mussels are produced on the mid-Atlantic Coast, in the Gulf of Mexico, and in Washington state. Florida has recently expanded its clam production. Farmed mollusks account for about 40% of the world mollusk supply. They are also a significant percentage of total world aquaculture production. Concerns about alternative uses for shoreline, high labor demands, and pollution seem to be affecting the overall growth of this sector.

### ORNAMENTAL FISH

Not all of U.S. aquaculture is focused on food species. There is an active ornamental fish industry. Approximately 13% of U.S. households keep ornamental fish. The import/export market is always very dependent on the strength of the dollar against other currencies. Ornamental fish account for approximately 5% of U.S. aquaculture and amount to more than $50 million annually.

## TYPES OF AQUACULTURE SYSTEMS

Because environments, socioeconomic conditions, and species vary tremendously around the world, it is logical to expect a variety of aquaculture systems and practices. Aquaculture is also a dynamic industry with new segments being developed and tried in all parts of the world (Figure 22–13). However, the majority of aquaculture systems seem to fall into one of the following categories:

**FIGURE 22-13** The number of alligators raised in the United States is quite small but may increase in the future. (Photographer Zig Leszczynski. Courtesy of Animals Animals/Earth Scenes.)

- Extensively, semi-intensively, or intensively managed ponds, both fresh water and brackish water. These ponds may be earthen or concrete.
- Freshwater fish pens that may be large or small floating net enclosures, or cages in lakes or reservoirs.
- Intensively managed marine cages.
- Mussel rafts with mollusks on wooden stakes, hanging ropes, off-rafts, or floats in the intertidal zone.
- Seaweed strung on lines off the bottom of the ocean.
- Integrated agriculture/aquaculture that may include rice and fish, livestock or poultry and fish, vegetables and fish, or various combinations of these.
- Outdoor raceways, water recirculating tanks, or silos.

These systems are variably designed to take advantage of aquaculture's strengths in terms of productivity compared to terrestrial agriculture. Aquaculture is considered to have the following advantages in productivity compared to land systems: (1) First, a water environment provides a three-dimensional growing space rather than the two-dimensional agriculture field. This allows these systems to exploit the various niche environments available in the three-dimensional space. Different species can take advantage of different foods and different water spaces. (2) Further, aquatic species are *cold blooded* and therefore convert more food to growth than do other agricultural animals because they expend fewer nutrients to maintain body temperature. They also use both naturally occurring food sources and those provided by humans, which improves their gain-to-feed-provided ratio. (3) In addition, aquatic species have a higher flesh-to-bone ratio than do land species.

**Cold blooded**

Animals that cannot control their own body temperature.

## AQUACULTURE LIFE CYCLES

The breeding and management of the aquatic species is, of course, quite varied. However, the life cycle is broken into similar stages for each species and the production units are generally segmented based on the stages of the life cycle of the animals. The

**Broodstock**

The mature animals kept for reproductive purposes.

**Spawn**

To produce eggs, sperm, or young.

**Larva (plural, *larvae*)**

The first stage after hatching. An independent, mobile, developmental stage of development between hatching and juvenile.

**Fry**

Fish in the postlarval stage.

**Postlarvae**

Beyond the larval stage but not yet a juvenile.

**Spat**

For shellfish such as oysters, the stage when they settle down and become attached to some hard object.

**Juvenile**

Aquatic species between the postlarval stage and sexual maturity.

**Polyploidy**

Condition in which the number of chromosomes of an individual is some multiple of the haploid number. The animals are sterile and generally have increased growth rates.

hatchery is responsible for maintaining the **broodstock, spawning** them, hatching the fertilized eggs to produce **larvae,** and nurturing the larvae to the appropriate stage to move to the nursery. In the nursery, the young **fry** (fish), **postlarvae** (shrimp, prawns), or **spat** (shellfish) are raised to the **juvenile** stage. Juveniles attain a size at which they can be used to stock a production unit. The production unit is segmented into units named according to the stages in the life cycle of the animals. For some species, there are separate holding areas for broodstock. For some species, the broodstock is selected from the grow-out and moved to the hatchery.

## GENETICS

The application of genetic principles to aquaculture is no different than their application to any other species. The state of the science is that the techniques are perhaps not as widely known and practiced by aquaculturists as they are by land-based agriculturists. However, that is bound to change as aquaculture develops as an industry. Already, the types of practices used are becoming increasingly common and new practices are being introduced. Much is left to learn on this topic, and significant breakthroughs are expected.

Hybridization between different species in the same genus has yielded several hybrids with improved growth rates, better feed conversions, greater survival rates, greater disease resistance, and improved adult weights. The technique of **polyploidy** has been used to sterilize some species so they can be introduced into nonnative waters without running the risk they will reproduce. Transgenic animals have been developed and most probably will become a standard part of aquaculture as geneticists seek to change individual species in their resistance to diseases and parasites, growth rates and patterns, and ability to survive and produce in different environments. Research in transgenic fish has advanced more rapidly than has research for traditional farm animals. However, regulations from the FDA and/or USDA/APHIS could restrain the use of this and other biotechnologies that would assist the development of this industry.

One of the greatest challenges in the genetics of aquaculture species is the fact that several thousand species are available for the long term to potentially become a part of aquaculture systems. This contrasts with only a few dozen land animals that have been domesticated. The jobs of the "aqua-geneticist" are seemingly without end.

## NUTRITION OF AQUATIC SPECIES

The challenge of nutrition in aquaculture is the same as the challenge to other farming industries: Create balanced rations that will promote good production at reasonable costs. However, the nutrition of these species is complicated by the fact that many of them are carnivores. The U.S. catfish industry has successfully developed a cereal-based floating food that the naturally carnivorous catfish thrive on. For the other industries to develop and prosper, the development of equally successful feeding strategies and feeds seems essential. Luckily, generally good diets have been formulated for the major species. Most are available commercially, although a custom mix is probably more economical for large operations.

Many of the challenges of feeding the aquatic species are not related to the specific nutrient requirements, although those are certainly important. However, formulating feeds and correctly feeding the aquatic species includes giving consideration to:

**TABLE 22–7** **Example Diet Formulations for Selected Species of Fish**

| INGREDIENT | Channel Catfish Grow-Out | Channel Catfish Fingerling | Salmon | Trout |
|---|---|---|---|---|
| | PERCENTAGE OF INGREDIENT IN FORMULATION | | | |
| Fish meal | 4.0 | 12.0 | 52.0 | 30.0 |
| Soybean meal | 37.0 | 54.5 | — | 13.0 |
| Corn gluten meal | | | | 17.0 |
| Cottonseed meal | 15.0 | — | — | — |
| Poultry meal | — | — | 1.5 | — |
| Blood meal | — | — | 10.0 | — |
| Dried whey | — | — | 5.0 | 10.0 |
| Milk solubles | — | — | 3.0 | — |
| Corn, no. 2, yellow | 33.3 | 30.8 | — | — |
| Wheat middlings | 4.0 | — | 12.2 | 16.4 |
| Fat | 1.5 | 1.5 | 9.0 | 11.6 |
| Wheat germ meal | — | — | 5.0 | — |
| Dicalcium phosphate | 1.0 | 1.0 | — | — |
| Bone meal | 4.0 | — | — | — |
| Trace mineral and vitamin premix | 0.2 | 0.2 | 2.3 | 2 |

- Physical properties that must allow the species to be fed in water.
- Nutrient stability to avoid water damage.
- Particle stage appropriate to the species and life stage.
- Feeding technique that avoids waste because wasted feed affects water quality.
- The feed sources the species may take from its environment.
- The species' natural dietary habits.

Table 22–7 shows examples of diets that have been developed and successfully fed to some species.

## NUTRITIONAL BENEFITS OF FISH TO HUMANS

A 3-oz serving of farm-reared rainbow trout cooked with dry heat provides the following proportion of the recommended daily dietary allowance for a 25- to 50-year-old woman on a 2,200-calorie diet:

| | |
|---|---|
| Protein | 41% |
| Calories | 6.5% |
| Phosphorus | 28% |
| Iron | 2% |
| Zinc | 3.5% |
| Vitamin A | 30% |
| Thiamin | 9% |
| $B_{12}$ | 211% |
| Niacin | 50% |
| Folate | 11% |
| Riboflavin | 0.52% |

A 3-oz serving of farm-reared catfish cooked with dry heat provides the following proportion of the recommended daily dietary allowance for a 25- to 50-year-old woman on a 2,200-calorie diet:

| | |
|---|---|
| Protein | 32% |
| Calories | 5.9% |
| Phosphorus | 26% |
| Iron | 4.6% |
| Zinc | 7.4% |
| Vitamin A | 1.6% |
| Thiamin | 32.5% |
| $B_{12}$ | 119% |
| Niacin | 14% |
| Folate | 3.3% |
| Riboflavin | 0.5% |

Seafoods are nutritious foods that contain a variety of vitamins, minerals, and high-quality protein. The calories per serving are usually lower than those in most meats. Seafoods are considered to be excellent additions to an overall diet. The nutrient values for trout and catfish show each to be a nutrient-dense food. A complete nutrient analysis of trout and catfish is given in Appendix VIII.

## TRENDS AFFECTING AQUACULTURE

All indicators suggest a solid future for aquaculture in the United States. The techniques and the infrastructure are partially in place and the demand for the product, demonstrated successes, and people interested in developing this industry are all positive. However, there are challenges. This section discusses factors and trends affecting, or likely to affect, aquaculture and its development.

### SPECIES

The currently used species will continue to be used, and their use will even be expanded. However, other species will also become part of U.S. aquaculture. This expansion will be driven by the need to diversify, coupled with the different environments available in this country and the demand for different products. These factors will almost certainly encourage the development of appropriate culture techniques for additional species not currently grown. Species that consume vegetable matter will be of special interest. Many of the high-value fish now cultured are carnivorous and use fishmeal and trash fish as their food source. This feed source is probably not reliable for the long term. Species like carp and tilapia eat lower on the food chain.

### TECHNOLOGY

Technology will come to play a larger role in the aquaculture industry. Areas in which the adoption of available technologies and the development of new technologies will benefit are in monitoring and altering water quality; improved, more efficient harvesting techniques; inventory techniques; automated feeders; feed additives; and disease-prevention measures.

## STRUCTURAL CHANGES

There is a trend toward smaller numbers of larger farms in this industry, just as there is in all of animal agriculture. This trend is expected to continue. Economies of size favor larger operations. This will necessitate a more business-based management approach. As people in the industry become more interested in the bottom line, and not just in growing a better fish, the industry will mature.

The aquaculture industry will become even more market driven. Bursting on the scene as it has in the last two decades, the timing was good for it to observe the established livestock industries and their philosophies. It will mimic the poultry industry by adopting the philosophy that it is in the food business rather than in the fish business. To this end, value-added products with predictable consistency and quality will become a larger part of the industry as companies seek to capitalize on branded products that add value to their farm-raised product. This will become more evident as operations become larger and more vertically integrated.

## CONSTRAINTS TO EXPANSION

Expansion of the aquaculture industry will be a challenge in the United States. Aquaculture production sites need land and water, both of which are in decreasing supply.

Marine expansion is possible, but marine sites are being jealously guarded by environmental groups who fear unwanted ecological changes such as increased disease or genetic alterations in the indigenous species, pollution, and an altered aesthetic of the oceans. There is also competition for marine sites with recreational and industrial interests, as well as concerns over interference with navigation of the waters. Marine expansion is often opposed by the commercial fishing industry because of perceived competition. The actual numbers of marine sites with the requirements for good aquaculture are limited as well.

Expansion in the freshwater sector also faces challenges. Agriculture, industry, or municipalities are already using most of the freshwater resources in the United States. An exciting possibility for expansion of the aquaculture industry in the face of water shortages is a joint usage of water with row crop agriculture. Dubbed Joint Aquaculture/Agriculture Water Sharing (JAWS), the concept revolves around placing aquaculture facilities near the sources of agriculture irrigation water. The aquaculture enterprises use the water and then send it on to the row-crop partner to be used as irrigation water. This saves the aquaculture partner the need to clean the water while at the same time removing any environmental issue with the water use and providing the agriculture partner with water containing some nutrients, which are needed by the crops, thus saving on fertilizer costs. The partners share the cost of the water. Estimates suggest that adoption of this practice using just 4% of the available Western irrigation water would allow aquaculture to double in size.

Labor costs offer another impediment to aquaculture expansion. Relatively speaking, labor costs are high in the United States and many aquaculture production methods are labor intensive. Firms wanting to expand find this incentive to include other countries in their expansion plans. In addition to cheaper labor, the aquaculture industries of other countries can often offer more abundant resources, fewer environmental constraints, and few or no government regulations.

Feed costs are also relatively high for many of the currently cultured species and could get expensive enough to limit production in the future. As mentioned earlier, many of those species in current use are carnivores and their feed contains significant amounts of animal protein. With rising concerns over the practice of feeding animal-based feeds to food species, this practice is losing its consumer acceptance. In addition, other uses compete with aquaculture for availability of these feeds.

## CONSUMPTION PATTERNS

Increasingly, aquaculture will find ways to market products to a health-conscious population. A general perception persists that seafood is also healthy food. Increasing evidence is mounting that omega-3 fatty acids, which are found in substantial quantities in some fish, have nutritional benefits.

## DISEASE CONTROL

The aquaculture industry is bound to have to face some of the same problems as the intensive livestock industries, including disease. Already there seems to be a trend toward increased disease problems. Aquatic species are susceptible to several diseases that cause mass mortalities. Already, the world shrimp industry has had to deal with a variety of viral diseases that have affected the harvest in an important way. The major one is the Taura syndrome, a virus that deforms and kills pond-reared shrimp. Shrimp viral diseases have caused problems in California and caused some shrimp farms to close. Epizootic ulcerative syndrome (EUS) has caused massive losses in many countries in Asia. Norway now has a parasite in its salmon population that threatens its wild populations. The parasite was introduced by salmon brought from the Baltic. Many of the diseases that affect the aquaculture species currently have no cure. This problem is actually worse in aquaculture than in other livestock species. In poultry, the average mortality for broilers in the grow-out phase is only about 3%. With fish species, the loss is much higher. Any decrease in the mortality rate will also improve the economic efficiency of the farming operation.

## ENVIRONMENTAL CONCERNS

With the introduction of the new species will come the need to deal with the inevitable escapes of exotic species from aquaculture into indigenous waters. These are of concern because they have the potential to alter the new host environment, disrupt the indigenous animal and plant communities, contribute to genetic degradation of local fish, introduce parasites and diseases, and have socioeconomic effects.

Wastes are generated from all forms of aquaculture. Obviously, the more intensive the system, the greater the waste problem. These wastes include uneaten feed, excreta, chemicals, therapeutic agents, and dead fish. Issues of feed quality, methods of feeding, and general husbandry practices need to reflect concerns about waste management. Creative methods of cleansing and reusing the water from aquaculture need to be developed, the information disseminated, and the techniques incorporated. With greater concerns about the polluting effects of wastes on the environment, this industry certainly faces regulations at both the state and federal levels. In 2004, the EPA put in place new rules establishing wastewater controls for concentrated aquatic animal production facilities (see http://www.eps.gov/guide/aquaculture/fs-final.htm).

In every region of the world, the amount of clean water available per capita is declining. In North America, the amount of clean water has declined by 50% since 1950. Clean water is needed for aquaculture, and aquaculture pollutes the water it uses. Regulation of pollution and water availability issues will increase socioeconomic costs for this industry, just as in any other animal industry, and will affect its viability.

## PREDATOR LOSS

Predators of aquaculture species include mammals, turtles, and various fish-eating birds. The economic losses of predators to aquaculture amounts to tens of millions of dollars annually. Economic losses are caused directly by predators consuming and wounding aquatic species, as well as reduced production caused when the aquatic animals are forced to seek shelter and stop feeding. The stress of avoiding predators makes them more susceptible to disease. In addition, predators carry and transmit disease agents to aquatic animals and humans and damage roads and levees. The question is how to solve one problem without causing another. For example, how does one control predation by an endangered species? Increased understanding of predator behavior will help develop improved control strategies.

# CULTURE OF AQUATIC SPECIES

Because so many aquatic species are a part of aquaculture, it is not practical to handle the specifics of each in this chapter. However, information on a variety of topics ranging from culture to marketing to law is available. The County Cooperative Extension Service Office can usually help. They are commonly listed under "county government" in the phone directory. In addition, the U.S. Department of Agriculture has regional aquaculture centers. Their addresses are listed at the end of this chapter. As a means of exposing you to the type of information available, three representative publications have been selected. Edited versions appear in the following section. Literally thousands of other publications are available to the producer. Anyone considering an aquatic venture is encouraged to contact his or her extension service and the appropriate regional aquaculture center.

The Mississippi State University Extension Service provided the articles *Farm-Raised Catfish* and *Freshwater Prawns Pond Production and Grow-Out*. The fact sheet entitled *Reproduction of Angelfish* was provided courtesy of the Illinois–Indiana Sea Grant Program.

# FARM-RAISED CATFISH*

## PRODUCTION PROCESS

The production process begins with careful selection and mating of quality broodstock. Once the eggs are laid and fertilized, they are collected and placed in controlled hatchery tanks where they are closely monitored. The eggs hatch in 7 days at a controlled temperature of 78°F. When the eggs first hatch, the young fish are called *sac fry* because their yolk sacs are attached to their abdomens. The sac fry live off food stored in the yolk sacs. As the yolk sacs are depleted, the fish begin to swim and are ready to be put in a pond where they grow into fingerlings. When the fingerlings are about 4 to 6 in. long, they are transferred to catfish ponds at various rates per surface acre of water.

---

*By Dr. Martin W. Brunson, Extension Leader, Wildlife and Fisheries, and Dr. Robert Martin, Extension Economist, Department of Agricultural Economics. Information Sheet 1526, Extension Service of Mississippi State University. Ronald A. Brown, Director.

### FEEDING

Catfish are fed a high-protein, floating feed ration that is produced by several catfish feed mills located in the producing regions. This protein diet consists of soybean meal, corn, wheat, and fishmeal. Annual feed requirements are estimated to be about 5 tons of feed per surface acre of water. Catfish convert feed at an average yield of about 1 lb of fish for every 2.0 lbs of feed. Compared to other feed-converting animals, catfish produce one of the highest yields per pound of feed.

### WATER QUALITY

The oxygen content of the water must be checked on a regular basis because lack of oxygen can kill fish in a matter of minutes. When the oxygen content begins to reach a dangerously low level, such as on hot summer days, producers use paddle wheels or other devices to aerate catfish ponds. This technique introduces air into the water and replaces the oxygen that has been depleted. Other water chemistry parameters that affect fish health and production include ammonia and nitrite levels, alkalinity, and chloride levels.

### DISEASE MANAGEMENT

Intensive culture of channel catfish requires close attention to the stress situations that can predispose fish to disease. Several diseases affect channel catfish in production ponds. Some of the major diseases include ESC, channel catfish virus, winterkill, proliferative gill disease, columnaris, fungus, and assorted parasites.

### HARVEST AND MARKETING

By the time farm-raised catfish are 18 months old, they are usually ready for harvest. Averaging 1–1.5 lbs live weight, they are removed from the ponds by seines and placed in aerated tank trucks for live shipment to the processing plant (Figure 22–14). Some of the fish go directly to fish-out ponds or other fish markets, but the majority is processed. Catfish are kept alive right up to the time they are processed (Figure 22–15). Within just a matter of minutes they are processed and then placed on ice or frozen to temperatures of 40° below zero, using a quick-freeze method that allows the taste and

**FIGURE 22–14** Workers harvest catfish with seines. The fish are placed in aerated tank trucks for live shipment to processing plants. (Photographer Ken Hammond. Courtesy of USDA.)

quality to remain in the fish longer (Figure 22–16). Farm-raised catfish are taste-tested at the farm before the fish are harvested and again at the processing plant before the fish are unloaded. These quality control procedures ensure that the consumer receives a product of superior flavor.

Farm-raised catfish are marketed through three major channels: retail grocery store outlets, food service distributors, and catfish specialty restaurants. Each channel accounts for about one-third of total catfish sales. A self-funded promotion and marketing program called the Catfish Institute constantly works on behalf of farm-raised catfish and has greatly increased the stature and prominence of this food product.

## CONSUMER DEMAND

Farm-raised catfish has many nutritional advantages, including being low in fat, cholesterol, and calories, and high in protein. In addition, farm-raised catfish is a good

**FIGURE 22–15** Catfish fresh from the farms are ready for processing. To ensure freshness, catfish are kept alive until just before processing. (Photographer Ken Hammond. Courtesy of USDA.)

**FIGURE 22–16** Catfish processed and ready for flash freezing. The quick-freeze method used allows the taste and quality to remain in the fish longer. (Photographer Ken Hammond. Courtesy of USDA.)

source of many vitamins and minerals. Another advantage is that farm-raised catfish has less of the smelly fishy odor associated with cooking fish indoors. Farm-raised catfish has only large bones that are easy to locate and remove in whole fish. The catfish steak has only one bone section, which is easily removed after cooking. The filet strip and nugget have no bones. Farm-raised catfish is versatile and fits into many styles of preparation, including fried, baked, broiled, sautéed, smoked, and stuffed; and served as hors d'oeuvres, in sandwiches, gumbos, patés, soups, and quiches; in sushi, Mexican style, or barbecued. The products are available year-round at local retail outlets. Taste tests have shown that most people who have tasted catfish for the first time rate it high on fish preference charts.

### QUALITY ASSURANCE

Farm-raised catfish are not exposed to external environmental factors, so producers can more easily control the quality of their product. Even in cultured products, however, potential contamination by pesticides, drugs, or other environmental factors is becoming a major consumer concern. Catfish producers responded to increased consumer awareness and sensitivity with a Catfish Quality Assurance Program (CQA). This is an opportunity to demonstrate that catfish production practices are safe, carefully monitored, and that the safety and quality of farm-raised catfish can be assured. Catfish producers recognize that a producer–client relationship built on mutual trust and confidence is paramount to the survival of the industry.

Catfish producers are aware, however, that quality assurance goes far beyond the pond bank. Quality assurance begins, and hinges upon, activities at the production level. The quality of the product as it leaves the farm greatly influences the quality of the product reaching the consumer. Safety and quality cannot be added or increased at the processing or retail level; it must be guaranteed by the producers.

The strong and positive relationship that catfish enjoys with its consumers must be maintained, nurtured, and enhanced. CQA will serve to cement that relationship. Exposure of the public to the good management practices and quality assurance procedures involved in producing farm-raised catfish will heighten consumer awareness of how good, safe, and wholesome farm-raised catfish is. CQA demonstrates a strong commitment to quality assurance and will be an asset in maintaining consumer confidence and in confronting any false or negative perceptions about our product. The Catfish Quality Assurance Program can be reviewed online: Mississippi State University Extension Service, Mississippi State University, http://ext.msstate.edu/anr/aquaculture/aquapapers/catfish.html.

## FRESHWATER PRAWNS POND PRODUCTION AND GROW-OUT[*]

The final phase of freshwater prawn (shrimp) production is grow-out of juveniles to adults for market as a food product. Unless you have a hatchery/nursery, you must purchase juveniles for the pond grow-out phase. Commercial hatcheries in Texas, California, and Mexico produce postlarvae and juveniles.

---

[*]By Dr. Louis R. D'Abramo, Professor; Dr. Martin W. Brunson, Extension Leader/Fisheries Specialist; and Dr. William H. Daniels, former Research Assistant, all with the Department of Wildlife and Fisheries. Publication 2003. Extension Service of Mississippi State University. Ronald A. Brown, Director. Mississippi State University Extension Service. Mississippi State University, http://ext.msstate.edu/pubs/pub2003.htm.

## SITE SELECTION AND POND DESIGN

Ponds used for raising freshwater prawns should have many of the same basic features of ponds used for the culture of channel catfish. A good supply of freshwater is important, and the soil must have excellent water-retention qualities. Well water of acceptable quality is the preferred water source for raising freshwater prawns. Runoff from rivers, streams, and reservoirs can be used, but quality and quantity can be highly variable and subject to uncontrollable change. The quality of the water source should be evaluated before any site is selected. Locate ponds in areas that are not subject to periodic flooding. Before building ponds specifically for producing freshwater prawns, check the soil for the presence of pesticides. Prawns are sensitive to many of the pesticides used on row crops. Also, analyze the soil for the presence of residual pesticides. Do not use ponds that are subject to drift from agricultural sprays or to runoff water that might contain pesticides. The surface area of grow-out ponds ideally should range from 1 to 5 acres. Larger ponds have been successfully used; ideally, the pond should have a rectangular shape to facilitate distribution of feed across the entire surface area. The bottom of the pond should be completely smooth and free of any potential obstructions of seining. Ponds should have a minimum depth of 2 ft at the shallow end and a maximum depth of 3.5–5 ft at the deep end. The slope of the bottom should allow for rapid draining. You can obtain assistance in designing and laying out ponds by contacting a local office of the Natural Resources Conservation Service (formerly Soil Conservation Service).

Collect a soil sample from the pond bottom to determine whether lime is needed. Take soil samples from about six different places in each area of the pond, and mix them together to make a composite sample that is then air-dried. Put the sample in a soil sample box, available from your county extension agent, and send it to the Extension Soil Testing Laboratory, Box 9610, Mississippi State, MS 39762, and request a lime requirement test for a pond. There is a charge of $3 per sample for this service. If the pH of the soil is less than 6.5, you must add agricultural limestone to increase the pH to a minimum of 6.5, and preferably 6.8.

After filling the pond, fertilize it to provide an abundance of natural food organisms for the prawns and to shade out unwanted aquatic weeds. A liquid fertilizer, either a 10-34-0 or 13-38-0, gives the best results. Apply one-half to 1 gal of 10-34-0 or 13-38-0 liquid fertilizer per surface acre to the pond at least 1 to 2 weeks before stocking juvenile prawns. If a phytoplankton bloom has not developed within a week, make a second application of the liquid fertilizer. Do not apply directly into the water because it is denser than water and will sink to the bottom; liquid fertilizer should be diluted with water 10:1 before application. It can be sprayed from the bank or applied from a boat outfitted for chemical application.

At least 1 or 2 days before stocking the juvenile prawns, check the pond for aquatic insect adults and larvae that might eat the juvenile prawns. You can control the insects by using a 2:1 mixture of motor oil and diesel fuel at the rate of 1 to 2 gals per surface acre on a calm day. The oil film on the water kills the air-breathing insects and is more effective when applied on calm days.

If a water source other than well water is used, it is critically important to prevent fish, particularly members of the sunfish family (e.g., bass, bluegills, and green sunfish), from getting into the pond when it is filled. The effects of predation on freshwater prawns by these kinds of fish can be devastating. If there are fish in the pond, remove them before stocking prawns, using 1 quart of 5% liquid emulsifiable rotenone per acre-foot of water.

## STOCKING OF JUVENILES

Water in which postlarvae and juveniles are transported should be gradually replaced by the water in which they will be stocked. This acclimation procedure should not be attempted until the temperature difference between the transport and culture water is less than 6–10°F. The temperature of the pond water at stocking should be at least 68°F (20°C) to avoid stress because of low temperatures. Juvenile prawns appear to be more susceptible than adults to low water temperatures.

Juveniles, preferably derived from size-graded populations ranging in weight from 0.1 to 0.3 g, should be stocked at densities from 12,000 to 16,000 per acre. Lower stocking densities will yield larger prawns but lower total harvested poundage. The duration of the grow-out period depends on the water temperature of the ponds, and the time generally is 120 to 150 days in central Mississippi. Prawns could be grown year-round if you can find a water source that provides a sufficiently warm temperature for growth.

## FEEDING

Juvenile prawns stocked into grow-out ponds initially are able to obtain sufficient nutrition from natural pond organisms. At the recommended stocking densities, begin feeding when the average weight of the prawn is 5.0 g or greater. Commercially available sinking channel catfish feed (28–32% crude protein) is an effective feed at the recommended stocking densities. The feeding rate is based on the mean weight of the population. A feeding schedule has been developed by researchers at the Mississippi Agriculture and Forestry Experiment Station and is based on three factors:

1. A feed conversion ratio of 2.5:1.

2. One percent mortality in the population per week.

3. Mean individual weight determined from samples obtained every 3 weeks.

At the end of the grow-out season, survival may range from 60–85%, if you have practiced good water quality maintenance. Yields typically range from 600–1,200 lbs per acre. Weights of prawns range from 10 to 13 per lb.

## WATER QUALITY MANAGEMENT

Water quality is just as important in raising freshwater prawns as it is in raising catfish or any other species of aquatic animal. Dissolved oxygen (DO) is particularly important, and a good oxygen-monitoring program is necessary to achieve maximum yields. You should routinely check and monitor levels of dissolved oxygen in the bottom 1 ft of water, which the prawns occupy. Electronic oxygen meters are best for this purpose but are rather expensive and require careful maintenance to ensure good operating condition. The need for an electronic oxygen meter increases as the quantity of ponds to be managed increases. With only one or two small ponds, a chemical oxygen test kit is sufficient. Chemical oxygen test kits that perform 100 tests are commercially available from several manufacturers.

Use a sampler for collecting samples from an appropriate water depth for dissolved oxygen analysis. These sampling devices are commercially available or can be fashioned. It is important that the dissolved oxygen concentration in the bottom 1 ft of water does not fall below 3 parts per million (ppm). Dissolved oxygen concentrations of 3 ppm are stressful, and lower oxygen concentrations can be lethal. Chronically low levels of dissolved oxygen result in less-than-anticipated yields at the end of the growing season. Emergency aeration can be achieved by an aerator. The design and size of the aerator depend on the size and shape of the culture pond.

Oxygen depletions can be avoided. One method to predict low DO levels is to plot the level 1 hour after sunset and again approximately 2 hours later. Plot these two readings on a piece of graph paper and connect them with a straight line. Oxygen consumption during the late evening and early morning proceeds at a constant rate, caused by the respiration of the animals and plants in the water. By extending the line from these two points over time, you can quickly determine if the dawn DO concentration will decrease to a level that will stress or possibly kill the prawns. This method indicates whether emergency aeration is necessary and when to provide it.

Specific information on water quality requirements of freshwater prawns is limited. Although freshwater prawns have been successfully raised in soft water (5 to 7 ppm total hardness) in South Carolina, a softening of the shell was noticed. Hard water, 300-plus ppm, has been implicated in reduced growth and lime encrustations on freshwater prawns. Therefore, use of water with a hardness of 300-plus ppm is not recommended.

## NITROGEN COMPOUNDS

Nitrites at concentrations of 1.8 ppm have caused problems in hatcheries, but there is no definitive information as to the toxicity of nitrite to prawns in pond situations. High nitrate concentrations in ponds would not be expected given the anticipated biomass of prawns at harvest. High levels of un-ionized ammonia, above 0.1 ppm, in fish ponds can be detrimental. Concentrations of un-ionized ammonia as low as 0.26 ppm at a pH of 6.83 have been reported to kill 50% of the prawns in a population in 144 hours. Therefore, you must make every effort to prevent concentrations of 0.1 or higher ppm un-ionized ammonia.

## pH

A high pH can cause mortality through direct pH toxicity, and indirectly because a higher percentage of the total ammonia in the water exists in the toxic, un-ionized form. For more information on ammonia in fish ponds, request Extension Information Sheet 1333. Although freshwater prawns have been raised in ponds with a pH range of 6.0 to 10.5 with no apparent adverse effects, it is best to avoid a pH below 6.5 or above 9.5, if possible. High pH values usually occur in waters with total alkalinity of 50 or less ppm and when a dense algae bloom is present. Before stocking, liming ponds that are built in acidic soils can help minimize severe pH fluctuations.

Another way to avoid any anticipated problems of high pH is to reduce the quantity of algae in the pond by periodically flushing (removing) the top 12 in. of surface water. Alternatively, organic matter, such as corn grain or rice bran, can be distributed over the surface area of the pond. This procedure must be accompanied by careful monitoring of oxygen levels, which may dramatically decrease due to decay processes.

In some cases, dense phytoplankton growth may occur in production ponds. To control algae, do a bioassay before using any herbicide in a freshwater prawn pond. To do a bioassay, remove a few prawns, put them in several plastic buckets containing some of the pond water, and treat them to see if the concentration of herbicide you plan to use is safe. Be sure there is adequate aeration, and observe the response of the prawns for at least 24 hours afterward.

## DISEASES

So far, diseases do not appear to be a significant problem in the production of freshwater prawns, but as densities are increased to improve production, disease problems are bound to become more prevalent. One disease you may encounter is "blackspot," or "shell disease," which is caused by bacteria that break down the outer skeleton. Usually it follows physical damage and can be avoided by careful handling. At other times, algae or insect eggs may be present on the shell. This condition is not a disease, but rather an indication of slow growth, and is eliminated when the prawn molts.

## HARVESTING

At the end of the grow-out season, prawns may be seine or drain harvested. For seining, depth (or water volume) should be decreased by half before seining. Alternatively, ponds could be drained into an interior, large, rectangular borrow pit (ditch) where prawns are concentrated before seining. You can effectively drain harvest only if ponds have a smooth bottom and a slope that will ensure rapid and complete draining. During the complete drain-down harvest procedure, prawns generally are collected on the outside of the pond levee as they travel through the drain pipe into a collecting device. To avoid stress and possible mortality, provide sufficient aeration to the water in the collection device.

Selective harvest of large prawns during a period of 4 to 6 weeks before final harvest is recommended to increase total production in the pond. Selective harvesting usually is performed with a 1- to 2-in. bar-mesh seine, allowing those that pass through the seine to remain in the pond and to continue to grow, while the larger prawns are removed. Selective harvest may also be accomplished with properly designed traps. Prawns can be trapped using an array of traditionally designed crawfish traps.

## POLYCULTURE AND INTERCROPPING

Culture of freshwater prawns in combination with fingerling catfish has been successfully demonstrated under small-scale, experimental conditions, and appears possible under commercial conditions. Selective harvest can help to extend the duration of the availability of the fresh or live prawn product to the market. However, there is a lack of research to show whether selective harvesting or a complete bulk harvest is the most economical approach.

Before introduction of catfish fry, stock juvenile prawns at a rate of 3,000 to 5,000 per acre. Stock catfish fry at a density to ensure that they will pass through a 1-in. mesh seine used to harvest the prawns at the end of the growing season. Although polyculture of prawns and a mixed population of channel catfish has been successfully demonstrated, logistical problems arising from efficient separation of the two crops is inherent in this management practice. Moreover, when harvest of prawns is imminent due to cold water temperatures, catfish may not be a harvestable crop due to an "off flavor" characteristic. Polyculture of channel catfish and freshwater prawns may be best achieved through the cage culture of the fish.

Recently, a scheme for the intercropping of freshwater prawns and red swamp crawfish was developed and evaluated. Intercropping is the culture of two species that are stocked at different times of the year with little, if any, overlap of their growth and harvest seasons. Intercropping provides for a number of benefits that include:

1. Minimizing competition for resources.

2. Avoiding potential problems of species separation during or after harvest.

3. Spreading fixed costs of a production unit (pond) throughout the calendar year.

Adult mature crawfish are stocked at a rate of 3,600 per acre in late June or early July. Juvenile prawns are stocked at a density of 16,000 per acre in late May and harvested from August through early October. In late February, seine harvest of the crawfish begins and continues through late June before stocking of new adult crawfish. Prawns are small enough to pass through the mesh of the seine used to harvest crawfish during the May–June overlap period.

## PROCESSING AND MARKETING

Production levels and harvesting practices should match marketing strategies. Without this approach, financial loss owing to lack of adequate storage (holding) facilities or price change is inevitable. Marketing studies strongly suggest that a "heads off" product should be avoided and that a specific market niche for whole freshwater prawns needs to be identified and carefully developed.

To establish year-round distribution of this seasonal product, freezing, preferably individually quick frozen (IQF), would be an attractive form of processing. Block frozen is an alternative method of processing for long-term distribution. Recent research at the Mississippi Agriculture and Forestry Experiment Station suggests that adult freshwater prawns can be successfully live hauled for at least 24 hours, at a density of 0.5 lb per gallon, with little mortality and no observed effect on exterior quality of the product. Transport under these conditions requires good aeration. Distribution of prawns on "shelves" stacked vertically within the water column assists in avoiding mortality due to crowding and localized poor water quality. Use of holding water with a comparatively cool temperature (68–72°F) minimizes the incidence of water quality problems and injury by reducing the activity level of the prawns.

# REPRODUCTION OF ANGELFISH
## (PTEROPHYLLUM SCALARE)*

Since their introduction around 1911, angelfish have held a unique position in the fish-keeping world; angelfish have been called the "kings of the aquarium." They are extremely beautiful animals with highly varied finnage and color schemes. Angelfish are members of the Cichlidae family. The genus *Pterophyllum* comprises three species. The spectacular *Pterophyllum altum* (Pellegrin 1903), which can measure 13 in. from the tip of the dorsal to the tip of the caudal fin, is native to the upper Orinoco River basin in South America. The two remaining species, *P. scalare* (Lichtenstein 1823) and *P. dumerilii* (Castelnau 1855), are found throughout the Amazon basin and in the coastal rivers of the Guineas. Both *P. altum* and *P. dumerilii* are aquarist rarities. Even though *P. altum* is as attractive as *P. scalare*, both are seldom exported. The stringent water-quality requirements may partially expand the limited availability of these two species compared with the widely available congener *P. scalare*.

*Pterophyllum scalare* is without question the most popular and generally more available member of the entire family Cichlidae. Both the silver- and black-banded and a myriad of artificially selected color and finnage varieties are commercially produced. These Cichlids make a magnificent solo display, but there is no practical reason for

*By LaDon Swann, Aquaculture Extension Specialist, Illinois–Indiana Sea Grant Program, Purdue University, West Lafayette, IN.

excluding other fish from their aquarium. No aggressive tank mates or habitual fin-nippers belong in the company of any *Pterophyllum* species. Gouramis of the genera *Colisa* and *Trichogaster* are particularly well suited for this role. The only Cichlids that can be safely housed with angelfish are festivums, discus, keyhole acaras, and most of the South American and West African dwarf species.

Unfortunately, many specimens are purchased by neophyte aquarists whose ignorance of proper aquarium care dooms the overwhelming majority to a short and not-so-particularly pleasant life. The usual mistake entails introducing juvenile angelfish to a newly set-up aquarium. All of the laterally compressed Cichlids are extremely sensitive to nitrite and ammonia. They cannot cope with the fluctuations of these substances, which inevitably occur during the first few weeks of an aquarium's life.

The biological requirements and spawning techniques for *Pterophyllum scalare* are presented in this report.

## WATER QUALITY

As in any form of aquatic animal husbandry, excellent water quality should be maintained. Maintaining good water quality is as important to the ornamental fish producer using a spare room in the household as it is for the 1,000-acre catfish farm. In certain instances, it may be more difficult to provide ideal water quality requirements for a non-native fish species than for a species that evolved to fit the characteristics of its ecosystem. The water in which angelfish are naturally found is soft and slightly acidic. Angelfish will survive and grow in wide varieties of water hardness, but for good reproduction, the producer should attempt to provide the spawners with their preferred water. Water for broodstock reproduction should be 100 mg/l hardness and 6.8 to 7.2 pH. Controlling water temperature is essential to angelfish reproduction. Maintain angelfish at 24–26°C, and at 26–28°C for spawning. Day length for angelfish should be 8–12 hours. Nitrate levels should be maintained below 100 mg/l. Partial water changes are done weekly or biweekly by siphoning approximately 30% of the water from the bottom of the aquarium. If under gravel filters are used, the water is siphoned from under the gravel plate.

## NUTRITION

Angelfish are omnivores. Flaked foods are readily taken by angelfish of all sizes. Far more important to the well-being of angelfish is the proportion of roughage to protein in the diet. In nature, the food of most species comprises 50–85% roughage by weight, yet few food manufacturers take this into account in their formulations. Most angelfish can be kept in good condition on an exclusive diet of prepared foods and may even spawn freely on such a regime. However, all do better when regularly offered live and fresh food. This is particularly true when conditioning fish for breeding. It is not the superior nutritional value of such foods, as much as their superior palatability, that makes them so valuable.

Newly hatched brine shrimp (*Artemia salina*) nauplii are essential first food for newly hatched angelfish and are good for conditioning brooding angelfish. Brine shrimp and *Daphnia* exoskeletons rupture during freezing and the nutritional value after thawing decreases. Chrinonomid larvae, glassworms, and krill withstand freezing well and are preferred. Ground beef heart is also used as a staple angelfish food, but it degrades water quality more rapidly than other types of feed.

Some commercial operations use supplements of fresh vegetable food to maintain the full intensity of coloration as well as general well-being. This requirement is easily met, for a wide range of such foods is readily available. Romaine or other leaf lettuce varieties and spinach are particular favorites. Thinly sliced young zucchini or other marrow squashes are a superb food. These foods should be blanched by brief contact with boiling water, then cooled, before being offered to the fish. Live foods include shrimp, *Tubifex* worms, and mosquito larva. A variety of dried foods is also used. The higher the protein content, the better. The broodstock are fed to satiation twice per day.

As soon as the fry are free-swimming, they are fed exclusively newly hatched brine shrimp, three to four times a day. Within 15 minutes after each feeding, the bottom of each aquarium is siphoned clean and fresh water is added. This is done because live shrimp give off a tremendous amount of ammonia. The fry tank has a clean glass bottom and no gravel of any kind. A sponge filter raised approximately 6 mm off the bottom is used. The elevated filter prevents fry from being trapped underneath the filter.

## SPAWNING

The reproductive biology of Cichlids is extremely diverse and falls into two distinct categories—mouth brooders or substrate spawners. The mouth brooders incubate the fertilized eggs in the buccal cavity of the female. In a few species, the male will incubate the fertilized eggs. *Oreochromis niloticus* is an example of a maternal mouth brooder. Eggs of substrate spawners are incubated in a nest. The nest may either be formed on the river or lake bottom or, in the case of angelfish, the eggs are adhesive and are laid on plants or rocks. In nature, *Pterophyllum* sp. are monogamous, biparental, custodial substrate spawners. *Pterophyllum scalare* spawns freely under aquarium conditions. The altum angelfish has infrequently, but successfully, bred under aquarium conditions. No spawning by *P. dumerilii* has been reported in captivity.

The prospective angelfish breeder's chief problem is identifying males and females. Angelfish are not easily sexed. Large males typically have a more rounded cranial profile than do females. Apart from this less-than-convincing effort to produce a nuchal hump, they are somewhat larger than their consorts, and their ventral profile from the origin of the ventrals rearward slopes sharply downward. In contrast, the female's vertical profile is almost flat. These distinctions are virtually useless when dealing with young adults.

The extreme lateral compression of their bodies obscures the genital papillae sufficiently to render this otherwise infallible indicator of sex quite valueless. An accurate sign of imminent spawning is the appearance of the pair's genital papillae. The genital papilla of the female usually appears first and is more noticeable because it is larger and more blunt, while that of the male is more slender and pointed. These small protuberances that appear at the vent are used respectively for depositing the eggs and fertilizing them.

## Broodstock Selection

The easiest means of securing a pair is to raise a group of fry together and allow them to pair naturally. Professional breeders do not have time to wait for the fish to pair off on their own. They select approximately 20 to 30 fish as breeders and place them in a large aquarium, preferably 208 liters (55 gals) or larger. The water temperature should be approximately 27°C. Feed the fish as much live food as possible. Several slates measuring 30 cm × 10 cm are inclined along the walls of the aquarium. The fish pair off and attempt to breed at around 10 months, give or take a couple of weeks.

Courtship will begin if the fish are of mature age. Angelfish become very territorial during this process. Courtship works both ways, with the male selecting his mate or the female selecting hers. In either case, the pair selects a territory and protects it against all intruders. Once obvious courtship has started, the pair should be transferred to a separate tank depending on the spawning method chosen. The transfer will allow the pair to be alone and prevent aggressive behavior from tank mates.

## Parental Spawning

Parental spawning occurs when the eggs are laid and the parents provide parental care to the eggs and newly hatched fry until they are large enough to fend for themselves. This is an excellent method for the hobbyist who wants to observe the behavior of the parents. If one intends to allow the pair to rear their progeny undisturbed, a tank of at least 120-liter capacity is necessary to afford the fry sufficient living space. In nature, angelfish select a stout plant leaf as a spawning site. The aquarium strain of *P. scalare* will lay their eggs on any vertical surface that can be nipped clean. Usually 2 to 3 days before spawning, the pair selects and begins cleaning the spawning site, using their mouths to bite and scrub the surface of the leaf, slate, or whatever has been chosen.

After a few false passes at the site, the female passes over the site and deposits eggs, which adhere to the surface. The male makes alternate passes and releases spermatozoa, fertilizing the eggs. Continual movement of the angels over the eggs after the spawning serves the purpose of creating circulation through fanning movement of the pectoral fins.

Fish eggs usually are small (between 1.5 and 3 mm on the average) and round. Spawns numbering 500 eggs are not unusual. Egg size depends on the availability and quality of food fed to the spawners. Eggs are translucent when first laid. Infertile eggs turn white and are removed by the parents.

Eggs hatch in 36 to 48 hours. The pair chews the zygotes out of their eggshells 36 hours post-spawning. The larvae are initially shifted from one vertical resting place to another, but as they grow more active, their parents often move them to shallow pits in the substratum. The fry first attempt swimming 4 to 5 days later, but they usually require an additional day-and-a-half to 2 days to become fully proficient. At this stage they are called "swim-up fry."

Young pairs often eat their first few spawns, but given time, most settle satisfactorily into parenthood. Parental care can persist up to 8 weeks in captivity, but it is prudent to remove the fry from the breeding tank no later than the fourth week postspawning. By this time, most pairs show signs of wishing to respawn.

## Egg Removal Method

The majority of domestic angelfish are raised without parental care. The differences between parental spawning and the egg removal method occur after the eggs are fertilized. Once brood fish start to exhibit courtship behavior (either the male or the female begins cleaning slate), they are transferred to an 80-liter spawning tank. The spawning tank is aerated and has two sponge filters. This interruption will affect the pair for 2 or 3 days, after which they will resume the process for breeding. After fertilization, the slate with attached eggs is placed in a 12–20 liter aquarium containing enough methylene blue to give a dark-blue color. An air stone should be placed underneath the slate to provide circulation. After hatching, one-half of the aquarium water should be replaced each day, so that by the time the fry are free-swimming, the water is only slightly blue. Dead eggs should be removed each day to prevent the spread of fungus to live eggs.

When the fry are free swimming, they should be transferred to an aerated 60 liter long aquarium at 300 fry per aquarium. The aquarium should have a water depth of approximately 10 cm and should be filtered with a sponge filter. The shallow water depth facilitates the feeding of the fry. When the fry are approximately 15 mm in diameter, they should be transferred to a 120- to 200-liter aquarium with aeration and filtration. Fry should grow to a marketable size in 6 to 8 weeks.

Angelfish fry are not difficult to raise, provided every effort is made to keep metabolite concentrations as low as possible. If their finnage is to develop to its fullest degree, they must not be crowded during their first months of life. This is particularly true of the so-called veil strain. With heavy feeding and frequent partial water changes, the young grow quickly. Under exceptional circumstances, females begin spawning by the eighth month postspawning. In most instances, sexual maturity is attained 10 months to a year postspawning.

## DISEASES

Many disease outbreaks can be attributed to excessive parasitism complicated by secondary bacterial infections. When angelfish are purchased, they should be examined for external and internal parasites. Newly acquired fish should be strictly quarantined for at least 1 month before they are placed with established populations. This practice will substantially reduce the risk of introducing new pathogens to hatcheries or home aquariums.

Two of the most commonly encountered pathogens in angelfish are *Hexamita* and *Capillaria.* The prevalence of the enteric parasites can be reduced by periodically treating fish with metrinidazole and an anthelmintic. This is particularly important in commercial hatcheries. Treatment for other infectious agents, particularly bacterial diseases, should only be administered following identification of agents causing disease outbreaks. Sensitivity testing of bacteria is strongly encouraged to ensure proper use of antibiotics during disease outbreaks. Assistance is available from your aquaculture extension specialist and animal disease diagnostic laboratory. Currently, the diagnosis of viral disease is hampered by the lack of a cell culture system to isolate and thereby characterize viruses of angelfish. However, the structures can be observed in tissue and feces by electron microscopy, thereby permitting presumptive viral diagnosis.

## CONCLUSIONS

Producing angelfish is a relatively simple procedure if a few guidelines are followed:

1. Maintain good water quality. Angelfish prefer soft and slightly acidic water, a spawning temperature of 26–28°C, and 8–12 hours of daylength.

2. Provide high-quality feed to broodstock and newly hatched fry. The feed should consist of flakes and live foods.

3. Do not overstock tanks. Use only one brooding pair per spawning tank and do not stock more than 200 swim-up fry per 80-liter tank.

Spawning angelfish is a lot of fun for novice fish keepers and can be profitable for the more serious aquaculturists. Angelfish hatcheries can provide supplemental income to niche marketers or provide a primary income source for large-scale hatcheries that sell angelfish to wholesalers.

# SOURCES OF INFORMATION

**Aquaculture Information Center**
http://aquanic.org

**Center for Tropical & Subtropical Aquaculture**
http://www.ctsa.org/

**North-Central Regional Aquaculture Center**
http://www.ncrc.org/

**Northeastern Regional Aquaculture Center**
http://www.nrac.umd.edu/

**Southern Regional Aquaculture Center**
http://www.msstate.edu/dept/srac/

**Western Regional Aquaculture Center**
http://www.fish.washington.edu/wrac/

# STUDY QUESTIONS

1. Define *aquaculture*. What species are involved? What kinds of water? What are capture fisheries? Blended fisheries?
2. Describe the different types of aquaculturists.
3. What is the purpose of the aquaculture industry in the United States? Worldwide?
4. What is the per capita consumption of seafood in the United States?
5. How did aquaculture develop in the United States? What is the potential for future growth?
6. Describe the magnitude of each of the following species in relationship to the overall aquaculture industry: catfish, trout, tilapia, crawfish, salmon, mollusks, and ornamental fish.
7. What are the general types of aquaculture systems?
8. What are aquaculture's advantages over land-based systems?
9. What are the stages in the life cycle of most aquatic organisms? How are these segmented into production units?
10. Describe the state of the study of genetics of aquatic species compared to that of land-based livestock. How about nutrition?
11. What is the challenge of aquaculture where nutrition is concerned?
12. Describe the trends affecting aquaculture's growth as an industry.
13. What is the value of the state extension service to an aspiring aquaculturist? What are the ways to access the resources of the extension service?

# REFERENCES

Aquaculture Outlook. 1996–2004. Approved by the World Agricultural Outlook Board. A supplement to *Livestock, Dairy, and Poultry Monthly,* published twice a year by the Economic Research Service, U.S. Department of Agriculture, Washington, DC. LDP-AQS-6.

Avault, J. W. 1996. *Fundamentals of aquaculture.* Baton Rouge, LA: AVA.

Baird, D. J., M. C. M. Beveridge, L. A. Kelly, and J. F. Muir. 1996. *Aquaculture and water resource management.* Cambridge, MA: Blackwell Scientific.

Bardach, J. E. 1997. *Sustainable aquaculture.* New York: Wiley.

Beveridge, M. C. M. 1996. *Cage aquaculture.* Cambridge, MA: Fishing News Books, Blackwell Science.

Carlberg, J. M., and J. C. Van Olst. 2001. U.S. aquaculture: Current status and future directions. *Aquaculture Magazine* 27 (4):36–43.

Collins, C. 1998. *Warm water finfish aquaculture in the United States: Past, present and future.* Davis: University of California.

Costa-Pierce, B. A. 1998. *The role of aquaculture in farming systems.* Department of Environmental Analysis and Design, School of Social Ecology, University of California, Irvine, CA. Published by the Food and Agriculture Organization of the United Nations, Rome, Italy. http://darwin.bio.uci.edu/~sustain/FAO-may98version1.html.

FAO. 2006. *The state of world fisheries and aquaculture 2006.* Accessed online December 2007. http://www.fao.org/docrep/009/A0699e/A0699e00.htm.

Fisheries of the United States. 2007. Personal communication from the National Marine Fisheries Service, Office of Science and Technology, Silver Spring, MD. Accessed online November 2007. http://www.st.nmfs.gov/st1/fus/fus06/index.html.

Iverson, E. S., and K. K. Hale. 1992. *Aquaculture sourcebook: A guide to North American species.* New York: Van Nostrand Reinhold.

Mathias, J. A., A. T. Charles, and H. Bastong. 1994. *Integrated fish farming.* New York: CRC Press.

National Fisheries Institute. 2007. *Top 10 U.S. consumption of seafood by species.* Accessed online November 2007. http://www.aboutseafood.com/search_results.cfv.

National Marine Fisheries Service, Fisheries Statistics and Economics Division. 2007. Silver Springs, MD: NOAA. http://www.st.nmfs.gov/fus97/COMMERCIAL/ld-aquc.pdf.

New, M. B. 1997. Aquaculture and the capture fisheries—Balance of the scales. *World Aquaculture* (June 1997): 11–30.

Parker, R. 2002. *Aquaculture science.* 2nd ed. Albany, NY: Delmar.

Stickney, R. R. 1994. *Principles of aquaculture.* New York: Wiley.

USDA-NASS. 1998. *Census of aquaculture.* Washington, DC: National Agricultural Statistics Service, USDA.

USDA-NASS. 2001. *Catfish and trout production.* Washington, DC: National Agricultural Statistics Service, USDA.

USDA-NASS. 2007a. *Farm income data.* Accessed online November 2007. http://www.ers.usda.gov/Data/farmincome/finfidmu.htm.

USDA-NASS. 2007b. *Catfish production, February 2007.* Accessed online November 2007. http://usda.mannlib.cornell.edu/MannUsda/viewDocumentInfo.do?documentID=1016.

USDA-NASS. 2007c. *Catfish processing, January 2007.* Accessed online November 2007. http://usda.mannlib.cornell.edu/MannUsda/viewDocumentInfo.do;jsessionid=A6ECF3E6D32CB32243300C3FC02740F8?documentID=1016.

USDA-NASS. 2007d. *Trout production.* Accessed online November 2007. http://usda.mannlib.cornell.edu/usda/current/TrouProd/TrouProd-02-26-2007.pdf.

# Pet and Companion Animals

## LEARNING OBJECTIVES

After you have studied this chapter, you should be able to:

1. Describe the place of the pet species in the lives of the people of the United States.

2. Discuss the types of pets that find service in the United States.

3. Give a historical perspective to the keeping of pets.

4. Discuss the geographic differences in pet ownership in the United States.

5. Compare the roles of breeds and breeding programs in pet and livestock species.

6. Describe the rudiments of reproductive management in the dog and cat.

7. Explain the nutritional information found on a container or bag of pet food.

8. Cite the trends shaping the ownership of pets in the United States.

## KEY TERMS

AAFCO Dog or Cat
    Nutrient Profiles
AAFCO Feeding Trial
Association of American
    Feed Control Officials
    (AAFCO)
Bitch

Book
Companion animal
Digests
Human-developed cat
    breed
Natural breeds
Net quantity statement

Pet
Polymorphism
Queen
Spontaneous mutation

## THE PLACE OF PET AND COMPANION SPECIES IN THE UNITED STATES

Timely statistics on pet species are difficult to obtain. The American Veterinary Medical Association (AVMA) conducts surveys periodically. The common pet species in the United States are shown in Table 23–1, along with their estimated populations in selected

**TABLE 23 – 1** **Common Pet Species[1] and Their Populations in the United States**

| | Pet Populations (millions) Year | | | |
|---|---|---|---|---|
| TYPE OF PET | 1991 | 1996 | 2001 | 2007 |
| Dogs | 52.5 | 52.9 | 61.6 | 72.1 |
| Cats | 57.0 | 59.1 | 68.9 | 81.7 |
| Birds | 16.22 | 17.02 | 10.1 | 11.2 |
| Other birds (pigeons and poultry) | — | — | 2.894 | 4.97 |
| Fish | 23.997 | 55.554 | 49.25 | 75.8 |
| Ferrets | 0.275 | 0.791 | 0.99 | 1.06 |
| Rabbits | 4.574 | 4.940 | 4.81 | 6.2 |
| Hamsters | 1.316 | 1.876 | 0.881 | 1.2 |
| Guinea pigs | 0.838 | 1.091 | 0.629 | 1.0 |
| Gerbils | 0.619 | 0.764 | 0.319 | 0.4 |
| Other rodents | 0.875 | 1.053 | 0.786 | 0.9 |
| Turtles | 0.708 | 0.950 | 1.07 | 1.99 |
| Snakes | 0.735 | 0.900 | 0.661 | 0.59 |
| Lizards | 0.314 | 0.705 | 0.545 | 1.08 |
| Other reptiles | 0.281 | 0.924 | 0.598 | 0.199 |
| Livestock | 3.371 | 6.083 | 2.936 | 10.99 |
| All other pets | 0.638 | 1.225 | 2.013 | 3.66 |
| Total | 164.261 | 205.876 | 208.982 | 275.039 |

[1] Excluding horses.
SOURCE: AVMA, 1997, 2002, 2007.

years, according to the AVMA's survey results. A number of additional species, including llamas, several amphibian species, crabs, crickets, tarantulas, chinchillas (Figure 23–1), hedgehogs, and others, are also kept as pets. Horses are counted in the survey but are omitted here because they were covered in an earlier chapter. Many households have more than one type of pet. For instance, approximately 15% of the households in the United States have at least one cat and one dog. The AVMA study indicates that approximately 57.4% of U.S. households own at least one pet. Approximately 37.2% of U.S. households have a dog. Approximately 32.4% of U.S. households have at least one cat.

There are generalities that can be drawn about pet ownership. For example, families with children are the most likely of all demographic groups to have a pet; people older than age 65 who live alone are the least likely to have a pet. In general, the greater the household income, the more likely a pet will be found in the home. Homeowners are more likely to have a pet than those who rent their homes. Also, the larger the number of people in a household, the more likely it is to have a pet.

The pet industry generates billions of dollars of annual income from services such as training, boarding, and grooming; from recreation and entertainment industries such as racing and exhibitions; from the sale of pet care and accessory products; and from the small but growing specialty areas associated with therapy and service animals. Veterinary expenditures are another substantial source of economic activity associated with the pet industry. Combined veterinary expenditures for the pet species are approximately $25 billion yearly.

**FIGURE 23–1** Chinchillas are part of a group of exotic/specialty pets kept in a growing number of households. (Photo courtesy of Adele M. Kupchik. Used with permission.)

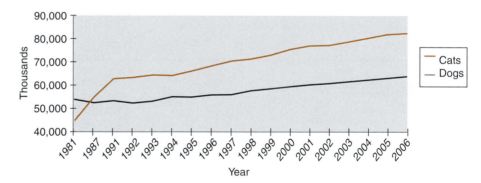

**FIGURE 23–2** Dog and cat population growth in the United States. (*SOURCE:* Pet Food Institute, 2007.)

It is obvious that pets have less to do with agriculture than they do with the larger picture of life in the United States. However, there are links to agriculture. The pet food market is linked to agriculture because agriculture provides most of the raw ingredients for pet food manufacture. The pet food industry is estimated at approximately $16 billion annually and is growing. Cat food sales are increasing more rapidly than dog food sales, which is reflective of the fact that cats are increasing in numbers more rapidly than dogs (Figure 23–2). The sales of U.S. manufactured pet foods to other countries exceeds $970 million and is increasing.

Because of the increasingly varied uses and needs that pet species provide, they have stimulated the growth of both new and existing careers and opportunities for people. Recognition of these opportunities and the fact that pets and companion animals are becoming a more important part of modern life have driven colleges and universities to change course offerings. Some have included material about pet species in existing courses. In addition, many are adding complete courses on pet and companion animal management. This is tacit recognition that animal science is no longer just about livestock species. Pets are an integral part of the daily lives of humans. With less than 2% of the American population involved in agriculture and close to 63% having an association with pets and companion animals, logic has prevailed.

# PURPOSE OF THE PET AND COMPANION ANIMALS INDUSTRY

The purpose of the pet and companion animal industry is to support the animals in service and companionship to people. Pet species provide many practical services to society. Guard dogs, police dogs, narcotics-sniffing dogs, search-and-rescue dogs, hunting dogs, herding dogs, service dogs, and therapy dogs are just a few of the important ways that dogs are in service to humans. Cats catch vermin just as they have from the dawn of their domestication. In addition to these uses, we have recently begun to understand and appreciate many other ways in which companion animals contribute to our lives. Once a source of derision, the human–companion animal bond is now recognized for its value as a contributing factor in the physical, mental, emotional, and social health of the owner. Dogs and cats provide the greatest service in this regard, but birds and small pets also do admirable service. The most common purpose for pet ownership is companionship (Figure 23–3). The value of pet species as companions sets the species apart from most domestic animals whose major purpose is to provide practical or economic benefit. The purpose of the industry can be more clearly seen by defining the pets themselves.

## PETS AND COMPANION ANIMALS DEFINED

In the classic definition, *pets* are differentiated from livestock in that they are kept for pleasure rather than for utility. The more modern term, *companion animal,* describes an animal whose owner has an intense emotional tie to the animal. People often have

**Pet**

An animal kept for pleasure rather than utility.

**Companion animal**

An animal to whom an owner has an intense emotional tie.

**FIGURE 23–3** The most common purpose of pet ownership is companionship.

**FIGURE 23–4** The keeping of tamed members of wild species as pets predates domestication and is still common, as shown with these African children and their pet Blue Duiker. (Photo courtesy of Hal Noss. Used with permission.)

relationships with animals that mirror relationships with humans, and describing these mirror relationships is how the term companion animal should be used.

Not all of the individuals of the companion/pet species qualify as either pets or companion animals. For instance, there are many barn cats across the country whose job in life is just to keep the rats and mice at bay. They are not kept for pleasure purposes, nor do they have a human companion. Neither term necessarily implies that the animals in question are domestic species (Figure 23–4). Although most pet/companion animals are undoubtedly of a domestic species, the keeping of tamed specimens of wild animals as pets predates domestication and is still a common practice around the world. As a means of clarifying the ambiguity of terminology, pets have been classified into six categories: ornamental pets, status symbols, playthings, hobby animals, helpers, and companions. These categories are not a complete solution to the nomenclature issue, but they go a long way in helping clarify the roles of pet/companions in modern society. Each category is discussed in the following sections.

## Ornamental Pets

Ornamental pets serve the same purpose that houseplants serve—they decorate and enhance the atmosphere. Ornamental pets are usually brightly colored birds or fish or some type of animal that adds aesthetic appeal to an environment. It is common to find an aquarium filled with brightly colored or otherwise interesting aquatic species in restaurants, professional offices, or homes (Figure 23–5). Decorators have been known to bring fabric swatches to pet stores to pick a bird that matches carpet and draperies. Outdoor environments are often graced by flashy species such as peacocks, pheasants, Sumatra chickens, swans, geese, and ducks. Rarely are these ornamental pets handled, named, or treated in any special way. They are not considered companion animals.

**FIGURE 23–5** Fish are a common ornamental pet in homes, businesses, and professional offices. (Photographer Renee Stockdale. Courtesy of Animals Animals/Earth Scenes. Used with permission.)

## Status Symbols

Strong evidence indicates that at least part of the domestication of the wolf was linked to the status its presence in camp gave the human occupants. A wolf as totem *and* companion would have conveyed a powerful message to rival clans or tribes. Sometimes we succumb to this same symbolism in modern life. This explains the motives of some people who keep poisonous snakes, piranhas, vicious dogs, big cats, bears, or wolves as pets. The animals are usually admired and well cared for as long as they satisfy the owner's expectations. In a more benign example, the symbolism of animals as totems for ancient people is not so different from that conveyed in modern society by what we generally refer to as "mascots." Status can also be conveyed by a pet kept for another primary reason. Pure-bred animals generally convey more status than mixed breed animals (Figure 23–6). Sometimes unusual, rare, and expensive animals are status symbols.

## Playthings

Pets as playthings may range from living toys given to children before they are old enough to appreciate the responsibilities to animals used in sporting events, such as hunting or riding. Children who are given a pet before they are capable of appreciating it are also not mature enough to understand the frailty of life. Thus many of these animals are treated poorly. Some of the people involved in sports that involve animals are only interested in the animal during the competitive season and lose interest and enthusiasm rather quickly at the close of the season. Often, these animals are poorly treated and may be discarded or destroyed by their owners when the animals lose their amusement value. Certainly this is not always the case. The fact that a dog will retrieve downed game hardly prevents it from being a treasured pet and valued companion as well, such as the Golden Retriever shown in Figure 23–7.

## Hobby Animals

The hobby animal category is intended specifically to identify those animals displayed by their owners in organized exhibitions. Frequently, the owners breed and exhibit their own

**FIGURE 23–6** The combination of size and striking markings of the harlequin Great Dane conveys more status to his owner than would a small, nondescript dog. (Photo courtesy of Kenneth J. Linsner. Used with permission.)

**FIGURE 23–7** This Golden Retriever, who fits into the "plaything" category of pets, is a field trial dog, a working sporting dog, and also a valued member of the family of one of the author's colleagues.

animals. The owners frequently belong to clubs and societies devoted to their animal's species. These clubs organize shows and events, which can be quite competitive (Figure 23–8). Many species are included in this category. Although dog and cat events are most common, there are clubs that sponsor events and shows for birds, rodents, and fish.

## Work, Helper, or Service Animals

Frequently, pet species perform vital services. Many of these are traditional services. For example, dogs have been trained to herd and track for millennia. Dogs still pull sleds and carts in some parts of the world (Box 23–1). In our modern society, they are trained as police dogs, search-and-rescue dogs, water rescue dogs, and drug dogs (Figure 23–9). Personal service dogs are an exciting and increasingly important part of modern society. In addition to seeing-eye dogs, there are service dogs that are trained to assist people with physical disabilities. Monkeys are also proving useful as service animals.

## Box 23–1   Sled Dog Racing

Sled dog racing is a sport that, much like horse racing, arose from the informal competition of dogs involved in work. Dogs have been a preferred means of hauling supplies in cold climates for centuries, and organized sled dog racing originated from hunters and trappers using their working dog teams for amusement as well as a vital part of their livelihood. In many cases, this arrangement persists in modern times, with a number of high-profile competitive mushers racing dogs that, between races, are used to work trap lines and haul supplies to and from remote areas.

There are numerous forms of sled dog racing, ranging from skijoring (using 1 to 3 dogs to pull a musher on skis) to sprint racing (from 6 to 18 dogs in a team, running 10- to 25-mile sprints, often on consecutive days for cumulative times), but the type of sled dog racing most widely recognized by the general public is the endurance and ultra-endurance racing typified by the annual Iditarod sled dog race. In these events, teams of 12 to 16 dogs race for hundreds of miles. Race officials establish checkpoints along the race course for judges and veterinarians to monitor the health, safety, and progress of all participants, and for mushers to cache supplies to avoid having to carry hundreds of pounds of food and gear throughout the race. The Iditarod, held every year on the first Saturday in March, attracts nearly 100 teams with a first-place purse of over $100,000 in cash and prizes. The race, which commemorates a historic lifesaving relay by dogsled of diphtheria antiserum to Nome in 1925, runs 1,100 miles from Anchorage to Nome. The winning team typically finishes in 9 to 10 days, with the last finisher arriving 5 to 6 days later. In an average race, a third of the teams fail to reach Nome and instead are flown out of a checkpoint after withdrawing from the race.

The relationship between a musher and his or her dogs is complex because of the nature of the sport. The dogs are highly trained athletes, and the musher's role is often very similar to the coach of any human sports team: part disciplinarian, part motivator, part decision maker and strategist. However, the bond between a musher and the dogs necessarily goes much deeper, for there must be mutual trust and devotion to have a successful team. Because of the vast distances covered during races, the teams must travel through raw wilderness in which there is a very real danger of injury and even death. A musher must earn and maintain the respect and trust of the dogs to complete a race safely, and many mushers have failed to complete a race when the dogs, having lost confidence and trust in their musher, elect to go on strike and refuse to leave a checkpoint. In some cases, mushers have suffered from frostbite and exposure when the dogs, well equipped to survive the extreme conditions of the wilderness, elect to go on strike between checkpoints, stranding the musher until such time that the dogs elect to continue on. Mushers place an extraordinarily high value on treating the dogs well, for it is no exaggeration that their lives may depend on their bond with their dogs.

Contributed by Michael S. Davis, DVM, PhD, Dipl ACVIM
*Department of Physiological Sciences*
*Center for Veterinary Health Services, Oklahoma State University*

Photo contributed by Dr. Kathy Williamson. Used with permission.

**FIGURE 23–8** Hobby pets are those displayed by their owners in *conformation* and performance events, such as this *Parson Russell Terrier* competing in an *agility* trial.

**FIGURE 23–9** Dogs are used to detect contraband of several kinds. Shown here is a member of the USDA/APHIS Beagle Brigade. Among other things, these dogs inspect luggage from international flights for food items that could bring disease into the United States. (Photo courtesy of USDA/APHIS.)

## Companion Animals

Regardless of where in the previously mentioned categories an individual animal may fall, it is clear that some animals are also included in the ranks of companion animals. The Parson Russell Terriers that own the author's family, for example, keep armadillos out of the yard, moles out of the flower beds, rabbits out of the vegetable garden, 'possums and 'coons out of the hen house, and raise a ruckus when a strange car enters the drive. They also fetch sticks, play tag, and elicit their share of "oohs and ahs" from admiring visitors. They are status-service-plaything pets at a minimum. However, their greatest value is that defined by their relationships with the young adults who have shared their lives with the dogs since childhood. They clearly view the Russells as companions, and thus they are (Figure 23–10). The Council for Science and Society states, "An animal employed for decoration, status-signaling, recreation, or hobby is being used primarily as an object—the animal equivalent of a work of art, a Rolls-Royce, a surfboard, or a collector's item. The companion animal, however, is typically perceived and treated as a subject—as a personality in its own right, irrespective of other considerations. With companion animals it is the relationship itself which is important to the owner." A companion pet may be treated as a member of the family, receiving presents at holidays, having its own chair, and so on (Figure 23–11).

Dogs, cats, small mammals, and birds are the most common species that become companion animals. Several reasons have been suggested for this. Perhaps most important is that each of these species is either easily restrained or does not require restraint. In addition, they can be easily house trained or don't require house training at all. They are large enough to be treated as an individual and still small enough to be nonthreatening (Figure 23–12).

**FIGURE 23–10**  This dog serves her family in a variety of roles. The service that transcends all practical and monetary considerations is as friend and companion to Joshua.

**FIGURE 23–11** Companion pets are treated as a subject with a personality, often receiving presents on holidays. (Photo courtesy of Kenneth J. Linsner. Used with permission.)

**FIGURE 23–12** Dogs, cats, small mammals, and birds are the species that most commonly transcend mere pethood to become companion pets. (Photo courtesy of Linda Guenther. Used with permission.)

## VALUE OF PETS

Parents often use pets as a tool in child rearing. Pets provide a mechanism for teaching children about basic biology as well as the larger lessons about life and death. Children learn responsibility by caring for the animal (Figure 23–13). In addition, the affection of a pet helps children cope with the stresses of life. With the high divorce rate of modern life, pets can help relieve the loneliness that comes from absent parents and siblings. Pets also provide children with willing play-acting partners. There is surely a special place in the hereafter for the dog or cat that allows itself to be dressed in doll clothes and pushed around in a baby stroller, or dressed in an army helmet and dragged across the yard in a tank (little red wagon). In such games, children develop their imaginations, emotional responses, and conflict resolution skills. Although it is difficult to prove cause and effect, studies reveal that adults who had pets as children are less likely to become criminals.

Medical studies have verified the benefits of pet ownership. The reasons appear to be physical, emotional, and social. The animals themselves provide companionship. However, multiple studies have demonstrated that pets provide a means for people to meet and interact, helping to alleviate loneliness and depression that is often a result of isolation. Pets help people to be more active because they need daily care. If the pet is a dog, this care often includes walks for the dog, which exercise the owner as well. Pet owners (both children and adults) have lower blood pressure than people who don't own pets. Studies on the aged have confirmed that pet owners have increased longevity and live more satisfying lives compared to their peers without pets. Elderly pet owners visit the doctor less often and use fewer medications. Pets give some elderly individuals the opportunity to nurture, to touch and be touched, and to feel a sense of safety and security. Interestingly, even birds are successful in making people feel safer. There is increasing interest in the ways that animals can be used to improve the physical and emotional health of the elderly. Studies suggest that animal visitation and full-time residence of pets with their owners in nursing homes and retirement communities should be more the norm than the exception.

The use of service animals is one of the most exciting of the developing interests in pet species, predominantly dogs. Nearly 20% of the population of the United States has

**FIGURE 23–13** Parents often provide pets to help teach their children responsibility.

some type of disability, with approximately 12% considered to have a serious disability. Over half of these people are disabled because of serious visual impairment or blindness, loss or lack of physical mobility, or hearing impairment or complete deafness. Service animals can help people with these handicaps. Dogs have been trained as seeing-eye dogs in the United States in a serious, formal way at least since the 1929 founding of The Seeing Eye Inc., of Morristown, New Jersey. Dogs for the Deaf of Central Point, Oregon, is a hearing-dog training and placement service that began in 1978. This group takes unwanted dogs from animal shelters and trains them to alert their partners to such sounds as telephones, doorbells, smoke alarms, and other important sounds. Canine Companions for Independence, with national headquarters in Santa Rosa, California, and regional offices across the United States, is perhaps the best known of several groups that train service, hearing, and social dogs. The dogs help individuals in wheelchairs by carrying packages, pulling wheelchairs, turning electric switches on and off, opening doors, and performing other chores to allow independent living for their human companions.

Animals are being increasingly used in mental and emotional therapy. Emotionally disturbed children and adults are often more willing to talk in the presence of an animal, often directing responses to questions by the therapist to the animal. This phenomenon is being expanded to help abused children, children with autism, persons with mental illness, dysfunctional families, and adult victims of violence. Rehabilitation centers use pets to help patients improve their strength, coordination, and mobility. Therapeutic riding programs have been started all across the country. With a growing body of research findings to support their value, these programs have developed into a legitimate health profession with special training and formal certification procedures.

In prisons, innovative programs have been started that help rehabilitate inmates, as well as providing benefits to broader society. Two programs associated with prisons have been established with the encouragement of Sister Pauline Quinn. The Wisconsin Correctional Liberty Dog Program is located at the Sanger B Powers Correctional Facility in Oneida, Wisconsin. One goal of the Liberty Dog Program is to meet the needs of people who have physical challenges by providing them with a service dog to help them live more independent lives. The other goal is to allow the prisoners the opportunity to serve their community. The Prison Pet Partnership Program, located at the Washington State Corrections Center for Women, helps inmates learn how to train, groom, and board dogs within the prison walls. Animals are placed with individuals and families dealing with disabilities. Other programs are found across the country. Inmates at 20 different Ohio prisons work with Pilot Dogs Inc., a Columbus-based organization that provides guide dogs for the blind. The inmates raise and socialize the dogs prior to their training as seeing-eye dogs. These and similar programs around the country benefit both the inmate and ultimate recipient of the animal.

## HISTORICAL PERSPECTIVE

Humanity's association with animals, and our uses for them, have long been dominated by their contributions to our needs for food and power, and by their association with religion. However, pet keeping is probably the use that led to the first domestications. Archaeological evidence from Paleolithic times suggests that people kept several different mammals as tame animals for short periods. An obvious explanation is that they were brought home as playthings for children (and probably adults, too). Tame wolves may have been kept as cave-mates as far back as 500,000 years ago. No doubt the young of other species of carnivores were also brought home and tamed. However, it

**FIGURE 23–14** *Canis lupis* has been established as the progenitor of the domestic dog. As unlikely as it seems, the Chihuahua and the gray wolf have the same blood coursing through their veins. (Photographer Renee Lynn. Courtesy of Photo Researchers, Inc.)

was the gray wolf, *Canis lupis,* that gave rise to the first domestic animal, the dog (Figure 23–14). The readily accepted reason for this is that humans and wolves share many of the same social characteristics. Human social structures at the time were similar to those of pack animals. Wolves easily assimilated into the human pack. Wolves and humans were competitors in the hunt for the same grazing species as a food source. Tame wolves may have been used by humans to help in the hunt. They would have, no doubt, retained their defensive nature and protected the camp. At a minimum, they could have warned of intruders. Some suggest that the motive for the earliest associations may have been the wolves' and that it was no more noble than to scavenge scraps from human encampments. Proximity led to tameness and then to domestication.

Regardless of who had what motive in the beginning of the association, at some point the nature of the relationship developed beyond any utilitarian motive. Evidence from a late Paleolithic burial cave dated to 10,000 B.C. suggests that the nature of the relationship between humans and the dog had evolved to include a bond. The human and the dog found in the tomb were arranged with the human's hand on the shoulder of the dog. The potential implications of the gesture (bonding, affection, devotion, and so on) are clear to anyone who has ever had a pet of any sort. Other archeological evidence suggests 12,000 B.C. as a date for the domestication of the dog. Recent DNA sequencing technology has suggested that the dog may well have been domesticated as long as 135,000 years ago. However, those methods are not accepted by all experts at this time and the research lacks corroboration. It is also likely that there were actually multiple domestication events in the road from *Canis lupis* to *Canis familiaris.*

Because modern dogs and wolves are different in so many traits, it is obvious that the tamed wolves were subjected to controlled breeding and subsequent domestication. Many of the physical traits that differentiate dogs from wolves were already in evidence by the time of the New Kingdom in Egypt. These traits included prick and pendant ears, solid and spotted coats, and curly and straight tails. Differentiation as to use was already apparent as well. Obviously the process has continued, as evidenced by the many breeds of dogs available today.

Domestic dogs were brought to the Western Hemisphere with the ancestors of the Native Americans who were living here when Europeans "discovered" the New World. Substantial archaeological evidence for domestic dogs in the Western Hemisphere dates as far back as 8000 B.C. Europeans brought very different dogs with them when they came to explore and settle the land.

It has long been believed that cats were not domesticated until after settled agriculture developed. The grain produced and stored by farmers provided a clear purpose for having them around human settlement—to help control vermin. Grain storage associated with settled agriculture attracts mice and rats. Wild cats were no doubt attracted to the vermin that any food store is likely to attract (Figure 23–15). The Egyptians had long been credited with domestication of the cat. Evidence indicates that the Egyptians had domestic cats as long as 6,000 years ago and it is known that cats were confined in temples and used for religious purposes 5,000 years ago. Priests adopted cats as objects for deification. The cult of the cat-headed goddess Bast lasted 2,000 years. New DNA sequencing evidence suggests that the cat was domesticated in approximately 7000 B.C. in the Near East from the Near Eastern Wildcat (*Felis silvestris lybica*) as agricultural villages developed in the Fertile Crescent (Driscoll et al., 2007). Archeological evidence from a gravesite in Cyprus along with the DNA findings now suggests the cat was domesticated by at least 7500 B.C. and possibly earlier. Some even suggest that cat domestication could have been as early as 10,000 B.C.

No doubt the unique cat ability of purring played a role in their domestication. It is not hard to imagine a parent Egyptian temple worker or priest bringing a kitten home for the amusement of his or her children. One can imagine such a scene, with the fa-

**FIGURE 23–15**   *Felis lybica,* ancestor of the domestic cat. (Photographer Nigel J. Dennis. Courtesy of Photo Researchers, Inc. Used with permission.)

ther telling the child to shut his or her eyes and placing the purring kitten up to the child's ear. The look of fascination and wonder that spread across the child's face was the clincher that kept the animal in the home longer than first planned. By 3,000 years ago, and probably earlier, cats had entered commerce. Ships and sailors crossing the seas and caravans crossing the desert spread cats throughout the known world.

After reaching Europe, cats were at first quite popular as rodent killers. Unfortunately, they became associated with Satan worship and its practitioners—witches and warlocks—during the Middle Ages. This led to the killing of large numbers of cats by religious zealots during the Dark Ages. With fewer cats, black rats proliferated and took with them the oriental rat flea that spread the Black Death across Europe. A third of the human population died. The Sabbath day for the Norse goddess Freya (cat goddess) became known as Friday. When the Christians barred her worship, Friday became known as the black Sabbath.

European ships brought cats to the Americas as pets and for their rodent-killing skills. Ships of the time were notorious for their rat and mouse populations. The pilgrims are known to have had at least one cat on board when they landed on Plymouth Rock. The cat fancy dates to 1871 with the first cat show held in England. This was followed by the first organized cat show in the United States in 1895. Today, the cat enthusiasts of the world are a strong and devoted group.

Whether viewed as rodent-slayer, goddess, harbinger of evil, or companion, the cat elicits unusually strong emotions from humans. Many contend that there are only two kinds of people in the world— cat-lovers and cat-haters. That may be an exaggeration (although only a mild one), but everybody knows which kind of person a cat seems to single out first for its affections.

The other species (except the horse and llama, which are companion species to many but have their own chapters in this book) have been relatively recent domestications. Many were domesticated for their value as laboratory species and have found homes in the hearts of companion/pet owners either for their novelty or because they offer price or space advantages. Hedgehogs, both the European and African varieties, were apparently only imported into the United States for zoos beginning in the 1980s and made the transition into pet status shortly thereafter (Figure 23–16). Hamsters were not even taken into captivity until 1930. They started appearing as pets sometime during the 1940s. Rats

**FIGURE 23–16** Hedgehogs have quickly found a place as novelty pets in the United States. (Photo courtesy of Adele M. Kupchik. Used with permission.)

**FIGURE 23–17** The guinea pig is considered a valuable food source in its native South America. A common laboratory species in the United States, they are being kept in increasing numbers as pets. (Photo courtesy of Phil M. Wanamaker. Used with permission.)

have been bred in captivity only for approximately 100 years, and they have been considered pets for a couple of decades at most. Guinea pigs have been kept as a food species in South America for centuries. However, the keeping of the familiar domestic variety as a pet is a decades-old phenomenon in this country (Figure 23–17). Parakeets, which are more appropriately called *budgerigars,* have been kept in captivity only since 1840 and are still quite capable of reverting to wild type if released into an environment warm enough to suit their needs. Many other birds kept as pets are probably not even classifiable domestic species because they are so difficult to breed in captivity.

## GEOGRAPHIC LOCATION

The AVMA surveys provide demographic information about pet ownership. Figure 23–18 shows the percentage of households that have pet animals within each region of the United States. It is important to note that all regions in the country have rates of ownership exceeding 50%.

Figure 23–19 shows the rate of dog ownership in the United States. Dogs are the most popular pet in terms of rate of ownership. When the AVMA survey was done in 1996, dogs were found in 31.6% of all households. The 2001 survey found the percentage had increased to 36.1%, very similar to the rate of ownership in 1991. However, in 2006, 37.2% of households owned a dog and the number of dogs per household had increased. This suggests that as the human population increases the dog population should also increase, which is the trend currently observed in both. About 60% of all dog-owning households own just one dog, and 42% also own a cat. Over half of dog owners consider their dog to be a member of the family.

Figure 23–20 shows the distribution of households that own cats. The total number of cats is greater than the number of dogs, but cats are found in fewer households, 31.6% in 2001, a number that increased from 27.3% in 1996. By 2006, 32.4% of U.S. households

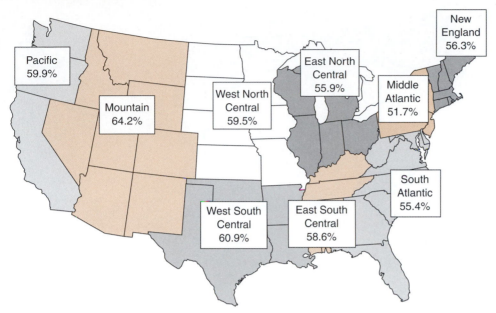

**FIGURE 23–18**  U.S. pet ownership—the percentage of households within each region that owned companion animals, 2007. (*SOURCE:* American Veterinary Medical Association.)

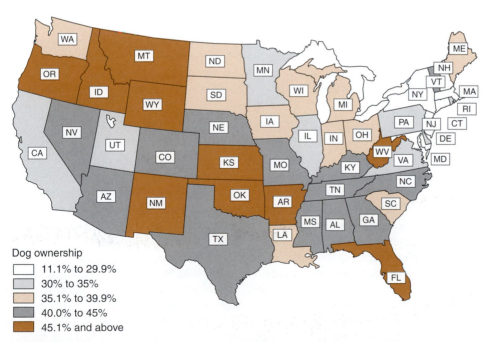

Dog ownership

☐ 11.1% to 29.9%
☐ 30% to 35%
☐ 35.1% to 39.9%
☐ 40.0% to 45%
☐ 45.1% and above

**FIGURE 23–19**  U.S. dog ownership—the percentage of households in each state that owned dogs in 2007. (*SOURCE:* American Veterinary Medical Association.)

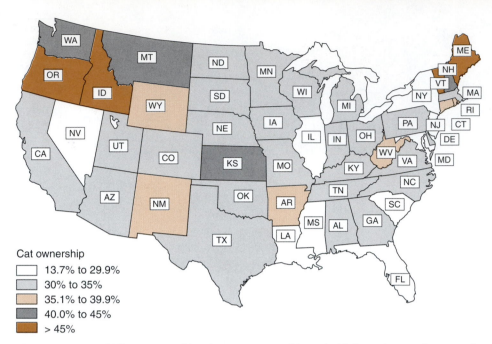

Cat ownership
☐ 13.7% to 29.9%
☐ 30% to 35%
☐ 35.1% to 39.9%
☐ 40.0% to 45%
☐ > 45%

**FIGURE 23–20** U.S. cat ownership—the percentage of households in each state that owned cats in 2007. (*SOURCE:* American Veterinary Medical Association.)

had a cat. Cats are found in greater numbers per cat-owning household than dogs are found in dog-owning households. Only 37.8% of dog-owning households have more than one dog; 51.8% of cat-owning households have more than one cat, and 47.2 % also own a dog. Almost half of cat owners consider their cat to be a member of the family.

Figure 23–21 shows the regional rates of ownership for the bird species in the United States. Birds are owned in a much smaller percentage of households in the United States.

## GENETICS AND BREEDING PROGRAMS

Genetics of the companion species follow the same laws of inheritance discussed in Chapters 8 and 9. However, the types of breeding programs discussed in Chapter 9 are nonexistent for the pet species because the goals of the breeding programs are entirely different. The pet species are selected for breeding based almost exclusively on phenotypes and pedigree selection. This approach to breeding has allowed many undesirable genetic diseases to become widespread problems in certain breeds of dogs and, to a lesser degree, in cats and other species as well. At best, the majority of breeders have only rudimentary skills in the scientific aspect of breeding. For those who do have expertise in genetics, it tends to be fairly specific for certain traits, often color. Certainly, a few knowledgeable breeders can be found for all of the species. Sadly, they are the exception rather than the rule. However, those who are knowledgeable tend to be highly competent. The cat fancy has a group of breeders who are excellent geneticists and have been very active in improving existing breeds and developing new ones. The increasing numbers of cat breeds are a testament to their work and dedication. Perhaps with the rapidly advancing state of DNA technology, the level of knowledge about overall genetics will increase for all pet species.

Advanced tools for improved selection and breeding of dogs will soon be available. The dog has a very large spectrum of ***polymorphism*** within the 400 or so breeds.

**Polymorphism**

The existence of two or more discontinuous, segregating phenotypes in a population.

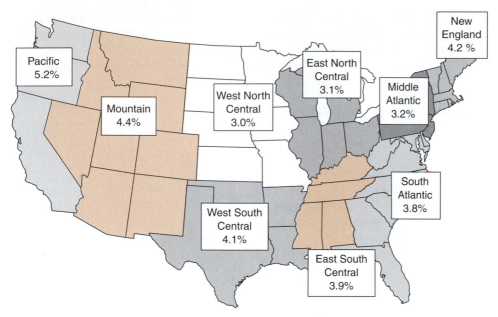

**FIGURE 23–21** U.S. bird ownership—the percentage of households in each region that owned birds in 2007. (*SOURCE:* American Veterinary Medical Association.)

This has made the dog a very attractive model for gene identification of phenotypic, behavioral, and pathological traits. The first draft of the dog genome sequence was made available in July 2004, and a complete sequence was published in 2005. The genome of the cat was published in 2007. Inexpensive technology is already available to positively identify parentage, but parental identification is just a small part of what is possible. In the very near future, it will be possible to compare genetic similarity between the individuals of a proposed mating. Dogs and cats can ultimately be screened for all genetic diseases. In dogs, the genes for herding instinct, protective instinct, scenting ability, running speed, and so on, will be identified and knowledgeable matings will be made. Even though the technology has not yet reached this stage, it is coming.

## BREEDS OF DOGS

It did not take humans long to begin shaping the genetics of the wolves that were to become domestic dogs. It is likely that some of the wolf disposition characteristics were altered fairly quickly. The progression of the other genetic changes is no more than a matter of speculation. It is known that by Roman times, wolf characteristics and behaviors had been refined into various categories of dogs, including herding, war, sight and scent hunters, terriers, and companions. Because dogs were domesticated earlier than any other species, it is logical that they were developed into numerous types and breeds. There are estimated to be between 400 and 450 dog breeds in the world today. The primary dog registry in the United States is the American Kennel Club (AKC). The AKC was established in 1884 as a nonprofit organization devoted to the advancement of purebred dogs. The AKC currently registers 160 breeds, with others in the stages of meeting eligibility requirements for recognition. The AKC classifies purebred dogs into seven categories: sporting dogs, hounds, working dogs, terriers, toy breeds, nonsporting dogs, and herding dogs. A miscellaneous group includes breeds that are not yet recognized but are considered purebred. This classification is a stepping-stone to full recognition in the AKC. Usually, six to eight breeds are in the miscellaneous class at any time. The current breeds of dogs classified by category are shown in Table 23–2. The 10 most popular breeds of purebred dogs are shown in Table 23–3. Photos of selected dog breeds are in the color plate section of this text.

### TABLE 23–2    American Kennel Club Dog Breeds by Category

#### GROUP I: SPORTING DOGS

| | | |
|---|---|---|
| American Water Spaniel | Field Spaniel | Nova Scotia Duck Tolling Retriever |
| Brittany | Flat-Coated Retriever | Pointer |
| Chesapeake Bay Retriever | German Shorthaired Pointer | Spinone Italiano |
| Clumber Spaniel | German Wirehaired Pointer | Sussex Spaniel |
| Cocker Spaniel | Golden Retriever | Vizsla |
| Curly-Coated Retriever | Gordon Setter | Weimaraner |
| English Cocker Spaniel | Irish Setter | Welsh Springer Spaniel |
| English Setter | Irish Water Spaniel | Wirehaired Pointing Griffon |
| English Springer Spaniel | Labrador Retriever | |

#### GROUP II: HOUNDS

| | | |
|---|---|---|
| Afghan Hound | Dachshund | Petit Basset Griffon Vendeen |
| American Foxhound | English Foxhound | Pharaoh Hound |
| Basenji | Greyhound | Plott |
| Basset Hound | Harrier | Rhodesian Ridgeback |
| Beagle | Ibizan Hound | Saluki |
| Black and Tan Coonhound | Irish Wolfhound | Scottish Deerhound |
| Bloodhound | Norwegian Elkhound | Whippet |
| Borzoi | Otterhound | |

#### GROUP III: WORKING DOGS

| | | |
|---|---|---|
| Akita | German Pinscher | Neapolitan Mastiff |
| Alaskan Malamute | Giant Schnauzer | Newfoundland |
| Anatolia Shepherd Dog | Great Dane | Portuguese Water Dog |
| Bernese Mountain Dog | Great Pyrenees | Rottweiler |
| Black Russian Terrier | Greater Swiss Mountain Dog | Saint Bernard |
| Boxer | Komondor | Samoyed |
| Bullmastiff | Kuvasz | Siberian Husky |
| Doberman Pinscher | Mastiff | Standard Schnauzer |
| | | Tibetan Mastiff |

#### GROUP IV: TERRIERS

| | | |
|---|---|---|
| Airedale Terrier | Irish Terrier | Scottish Terrier |
| American Staffordshire Terrier | Kerry Blue Terrier | Sealyham Terrier |
| Australian Terrier | Lakeland Terrier | Skye Terrier |
| Bedlington Terrier | Manchester Terrier | Smooth Fox Terrier |
| Border Terrier | Miniature Bull Terrier | Soft-Coated Wheaton Terrier |
| Bull Terrier | Miniature Schnauzer | Staffordshire Bull Terrier |
| Cairn Terrier | Norfolk Terrier | Welsh Terrier |
| Dandie Dinmont Terrier | Norwich Terrier | West Highland White Terrier |
| Fox Terrier (Smooth and Wire) | Parson Russell Terrier | Wire Fox Terrier |
| Glen of Imaal Terrier | | |

## TABLE 23-2 American Kennel Club Dog Breeds by Category—continued

| GROUP V: TOYS | | |
| --- | --- | --- |
| Affenpinscher | Italian Greyhound | Pomeranian |
| Brussels Griffon | Japanese Chin | Poodle (Toy) |
| Cavalier King Charles Spaniel | Maltese | Pug |
| Chihuahua | Manchester Terrier | Shih Tzu |
| Chinese Crested | Miniature Pinscher | Silky Terrier |
| English Toy Spaniel | Papillon | Toy Fox Terrier |
| Havanese | Pekingese | Yorkshire Terrier |

| GROUP VI: NONSPORTING DOGS | | |
| --- | --- | --- |
| American Eskimo Dog | Dalmation | Poodle (Standard and Miniature) |
| Bichon Frise | Finnish Spitz | Schipperke |
| Boston Terrier | French Bulldog | Shiba Inu |
| Bulldog | Keeshond | Tibetan Spaniel |
| Chinese Shar-pei | Lhasa Apso | Tibetan Terrier |
| Chow Chow | Löwchen | |

| GROUP VII: HERDING DOGS | | |
| --- | --- | --- |
| Australian Cattle Dog | Border Collie | German Shepherd Dog |
| Australian Shepherd | Bouvier des Flandres | Old English Sheepdog |
| Bearded Collie | Briard | Pembroke Welsh Corgi |
| Beauceron | Canaan Dog | Polish Lowland Sheepdog |
| Belgian Malinois | Cardigan Welsh Corgi | Puli |
| Belgian Sheepdog | Collie | Shetland Sheepdog |
| Belgian Tervuren | | Swedish Vallhund |

| MISCELLANEOUS CLASS | | |
| --- | --- | --- |
| Dogue de Bordeaux | Norwegian Buhund | Redbone Coonhound |
| Irish Red and White Setter | Pyrenean Shepherd | |

SOURCE: American Kennel Club, 2008.

## TABLE 23-3 Ten Most Popular Breeds of Dogs

| Rank | 2000 | 2005 | 2007 |
| --- | --- | --- | --- |
| 1 | Labrador Retriever | Labrador Retrievers | Labrador Retrievers |
| 2 | Golden Retriever | Golden Retriever | Yorkshire Terrier |
| 3 | German Shepherd | Yorkshire Terrier | German Shepherd |
| 4 | Dachshund | German Shepherd | Golden Retriever |
| 5 | Beagle | Beagle | Beagle |
| 6 | Poodle | Dachshund | Boxer |
| 7 | Yorkshire Terrier | Boxer | Dachshund |
| 8 | Chihuahua | Poodle | Poodle |
| 9 | Boxer | Shih Tzu | Shih Tzu |
| 10 | Shih Tzu | Miniature Schnauzer | Bulldog |

SOURCE: American Kennel Club, 2008.

# BREEDS OF CATS

**Natural breeds**

Cat breeds selected by human preference or natural conditions specific to a region.

**Human-developed cat breed**

Breeds that have been developed from existing breeds or crosses of existing breeds.

Cats have not been developed into as many different breeds as have dogs. This is because there have been fewer uses overall for cats. Cats have been kept predominantly for their vermin-controlling habits and as pets. Thus the differentiation into breeds with highly specialized functions has not occurred. There are approximately 50 breeds of cats. Table 23–4 lists cat breeds most commonly accepted by various breed associations. The 10 most popular cat breeds are listed in rank order in Table 23–5. Some breeds of cats were selected by human preference or regional diversity and have been in existence for hundreds of years. These breeds are referred to as the *natural breeds* and include Abyssinian, Birman, Burmese, Chartreux, Maine Coon, and Egyptian Mau. In some cases, modern cat fanciers have changed the natural breeds substantially through selection. The *human-developed breeds* are the live-stock equivalent of composite breeds. They were created by crossbreeding and sub-

**TABLE 23–4    Cat Breeds[1]**

| | | | |
|---|---|---|---|
| Abyssinian | Colorpoint Shorthair | Maine Coon | Selkirk Rex |
| American Bobtail | Cornish Rex | Manx | Siamese |
| American Curl | Devon Rex | Norwegian Forest Cat | Siberian |
| American Shorthair | Egyptian Mau | Ocicat | Singapura |
| American Wirehair | European Burmese | Oriental | Snowshoe |
| Balinese | Exotic | Persian | Somali |
| Birman | Havana Brown | Ragamuffin | Sphynx |
| Bombay | Japanese Bobtail | Ragdoll | Tonkinese |
| British Shorthair | Javanese | Russian Blue | Turkish Angora |
| Burmese | Korat | Scottish Fold | Turkish Van |
| Chartreux | La Perm | | |

[1] Names of breeds and standards for the breeds vary between associations.
*SOURCE:* Compiled from various breed associations and cat fancy publications.

**TABLE 23–5    Ten Most Popular Cat Breeds**

| Rank | 1996 | 2000 | 2005 | 2007 |
|---|---|---|---|---|
| 1 | Persian | Persian | Persian | Persian |
| 2 | Maine Coon | Maine Coon | Maine Coon | Maine Coon |
| 3 | Siamese | Siamese | Exotic | Exotic |
| 4 | Abyssinian | Exotic | Siamese | Siamese |
| 5 | Exotic | Abyssinian | Abyssinian | Ragdoll |
| 6 | Oriental | Oriental | Ragdoll | Abyssinian |
| 7 | Scottish Fold | Birman | Birman | Birman |
| 8 | American Shorthair | American Shorthair | American Shorthair | American Shorthair |
| 9 | Birman | Scottish Fold | Oriental | Oriental |
| 10 | Ocicat | Burmese | Sphynx | Sphynx |

*SOURCE:* The Cat Fancier's Association, Inc., Public Relations Dept.

sequent selection to fix type. Examples include crossing the Burmese and American Shorthair to develop the Bombay, and crossing the Siamese and Persian to create the Himalayan. *Spontaneous mutations* have also contributed to the development of new breeds that showcase the mutation. Examples include the American Curl (curled-back ears), American Bobtail (short-tailed), Cornish Rex (soft, short, wavy hair), Munchkin (short legs), and Scottish Fold (ears folded forward and down on the skull). There are several feline studbook organizations in the United States. The goals, rules of registration, recognized breeds, and breed standards for the different associations represent different perspectives on cat breeding and exhibition. Thus a cat that fits the standard of one association may be excluded or be inferior according to the standards of another.

**Spontaneous mutation**

A change in the DNA that creates new alleles.

## BREEDS OF OTHER PET SPECIES

Rabbits are separated into distinct breeds. This is because they were first domesticated as a food species and have been used around the globe for food and fiber for hundreds of years. The remainder of the pet species are not as breed oriented. For instance, guinea pig breeds do exist, but they are largely based on coat type and color variations. Likewise, color varieties exist for several of the other rodent species. The animals are not usually individually registered, and shows of the size and type held regularly for dogs and cats are fewer in number.

The bird fancy is divided along species lines. Interesting color variations have been developed in several species. Breeders of budgerigars and lovebirds have perhaps taken this color breeding to the greatest degree. Some splendid color varieties have been developed for both species and are exhibited at shows around the country. The budgerigar breeders have also developed the exhibition of their animals to the greatest degree. As the most popular of the caged bird species, their exhibition industry is perhaps expected to be the most developed. However, an active exhibition schedule is kept for several species of caged birds in the United States. In addition, interesting color variations can be found in most of the birds that make up the bird fancy.

## REPRODUCTIVE MANAGEMENT

The reproductive characteristics of the various pet species are too varied to cover adequately in one small section in one chapter. Table 23–6 explains some of the basic features of the reproductive cycles of selected mammalian species. Table 23–7 gives basic data on some avian species.

The basic structures in the reproductive tract of the *bitch* are shown in Figure 23–22. In most dogs, puberty begins at 6 to 9 months of age. The ovarian cycle of the bitch is monoestrous. The interval from cycle to cycle is influenced by breed differences and can be influenced by environmental factors as well. The time between cycles varies from 4 to 13 months, with an average of 7 months. The bitch begins to attract males before she is ready to mate. Exterior signs of heat in the bitch include behavioral changes like marking of territory, a swollen vulva, and a light bloody discharge from the vulva. Heat in the bitch is under the influence of luteinizing hormone (LH), progesterone, and estrogen

**Bitch**

A female dog.

**TABLE 23–6** **Features of the Reproductive Cycle of Selected Pet Species**

|  | Dog | Cat | Guinea Pig |
|---|---|---|---|
| Age at Puberty | 6–12 months | 6–15 months | 55–70 days |
| Cycle Type | Monoestrous, all year; mostly late winter and summer | Provoked ovulation, seasonally polyestrous, spring and early fall | Polyestrous |
| Cycle Length | 6–7 months | 15–21 days | 16 days |
| Duration of Heat | 4–14 days standing heat | 9–10 days in absence of male | 6–11 hours |
| Best Time for Breeding | 4–7 days after standing | Daily from day 2 of heat | 10 hours after start of heat |
| First Heat after Birth | 3–5 months | 4–6 weeks | 6–8 hours |
| Number of Young | 1–22 | 1–10 | 1–6 |
| Gestation Period | 58–70 days | 58–70 days | 59–72 days |

SOURCE: Adapted from USDA, 1984.

**TABLE 23–7** **Egg Production and Incubation for Some Bird Species**

| Type of Bird | Egg Production per Clutch | Incubation Time |
|---|---|---|
| Canaries | 2–7 | 14–15 |
| Cockatiels | 3–7 | 21–23 |
| Cockatoos | 2–4 | 28 |
| Conures | 4–6 | 24 |
| Lorikeets | 2–4 | 22–25 |
| Lovebirds | 3–8 | 23–24 |
| Macaws | 2–4 | 28 |
| Budgerigars | 4–8 | 18–20 |
| Finches | 2–4 | 14–19 |
| Parrots | 2–4 | 17–31 |

(Figure 23–23). Prior to the LH hormonal surge, the bitch generally refuses the dog. After the LH surge, she generally stands for mating. The ova are released from the follicles approximately 2 to 3 days after the LH surge. The ovulated eggs are not ready for fertilization until day 4 to day 7 following estrus. The greatest number of ovulations occur 24 to 72 hours after the LH peak. Breeding should ideally occur 4 to 7 days later. Plasma progesterone concentrations rise prior to ovulation in the bitch. Thus measurement of serum progesterone concentration in the bitch can be used to determine the day of ovulation and breeding. For anyone who has ever had an unwanted litter of puppies, this may seem to be too much information. However, now that the techniques and breed association rules are in place to allow artificial insemination, the timing of appropriate semen placement can be critical in mating success or failure. In addition, valuable stud dogs can be in high demand. A mating done on the wrong day can prevent the dog from being mated with another bitch that same day. For a popular stud, it is important to mate the bitches on the appropriate days to maximize the *book* on the dog. Some owners of valuable stud dogs insist on blood hormone monitoring on all bitches to be bred to their dogs.

**Book**
Refers to the number of females scheduled to be mated to a particular male.

|  | Hamster | Mouse | Rat | Gerbil |
|---|---|---|---|---|
| **Age at Puberty** | 5–8 weeks | 35 days | 37–73 days | 9–12 weeks |
| **Cycle Type** | Polyestrous | Polyestrous | Polyestrous, all year | Polyestrous |
| **Cycle Length** | 4 days | 4 days | 4–5 days | 4–6 days |
| **Duration of Heat** | 10–20 hours | 9–20 hours | 12–18 hours; usually begins about 7 P.M. | 12–15 hours |
| **Best Time for Breeding** | At start of heat, 8–10 P.M. | At start of heat | Near ovulation, which occurs close to midnight | Mid-heat |
| **First Heat after Birth** | 1–2 weeks after litter removed | 2–4 days after litter removed | Within 24 hours | 1–3 days |
| **Number of Young** | 1–12 | 1–12 | 2–20 | 2–15 |
| **Gestation Period** | 14 days | 17–21 days | 20–22 days | 24–26 days |

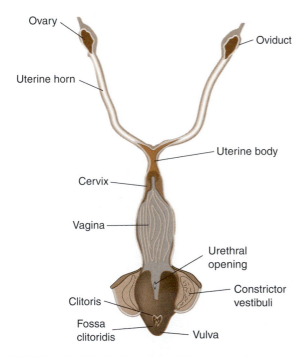

**FIGURE 23–22**  Basic structures in the reproductive tract of the bitch.

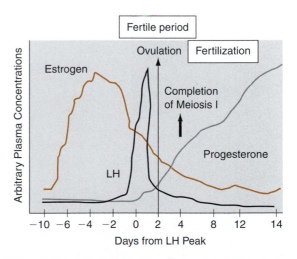

**FIGURE 23–23**  Hormonal patterns of the bitch during estrous cycle. (*SOURCE:* Adapted from Geisert, 1999. Used with permission.)

The basic structures of the female reproductive tract of the ***queen*** are shown in Figure 23–24. The average age of puberty in the queen is 10 months, with a range of 4 to 18 months. The queen is a seasonally polyestrous breeder. Photoperiod controls the period of cyclicity. Breeding season varies according to day length and is generally from March to September. Queens living indoors may cycle in the winter if lights in the house give them 12–14 hours of "daylight." Signs of heat in the queen include

**Queen**

A female cat.

**FIGURE 23–24** Anatomy of queen reproductive tract.

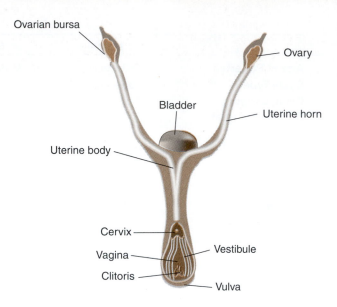

**FIGURE 23–25** Hormonal changes during the estrous cycle in the queen. (*SOURCE:* Adapted from Geisert, 1999. Used with permission.)

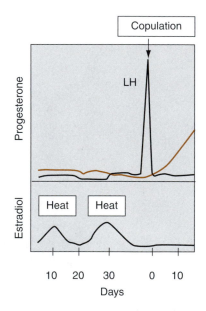

restlessness, vocalizations, and possibly marking of territory. The queen displays uniqueness in her reproductive function in that ovulation must be induced by copulation. Mating of the queen induces the LH surge required for ovulation (Figure 23–25). The average length of the estrous cycle is 14 to 21 days. Estrus lasts approximately 7 days, followed by a period of nonreceptivity if no mating occurs.

## NUTRITION OF THE PET SPECIES

In addition to dogs and cats, there are several dozen other pet species. These species vary from obligate carnivores to herbivores. Monogastrics, cecal fermenters, and ruminants are included. The bird species have avian tracts, but their diets are quite varied.

Thus a discussion of the nutrition of the individual species is beyond the scope of this text. Good nutrition for a pet is a responsibility of pet ownership. Learning what constitutes good nutrition for an animal can require a good deal of study and effort. The more uncommon and exotic the species being kept, the more daunting the task. See Chapters 5, 6, and 7 for basic discussions of nutrition and rations for pet species that are relevant to this discussion.

Most people who keep pets do so in limited numbers, which makes mixing rations on an individual animal basis impractical. In addition, most people do not have the knowledge to select ingredients and formulate their own rations. So most people do (and should) purchase a premixed feed for their pets. This makes the ability to choose a pet food based on the feed label information an important common denominator for all pet owners. Thus the discussion of nutrition in this chapter focuses on labels and how to interpret them. This discussion focuses on complete rations and does not include treats and specialty feeds. Figure 23–26 shows a representative pet food label.

Pet food labels are regulated, and the regulations are enforced by the FDA's Center for Veterinary Medicine (CVM). The FDA standards are the same as those that apply to livestock feeds and include proper identification of product, net quantity statement, manufacturer's address, and proper listing of ingredients. States also have the authority to establish and enforce state regulations. Many states have adopted model pet food regulations established by the *Association of American Feed Control Officials (AAFCO)* that are more specific. Aspects of feed labeling such as product name, guaranteed nutrient analysis, nutritional adequacy statement, feeding directions, and calorie statements are covered under the AAFCO standards. The requirements for labeling dog and cat foods are the most detailed and specific. This is because the needs of these species are known with enough confidence that the AAFCO can specify nutrient requirements.

## PRODUCT NAME

To keep pet food names honest to the ingredients they contain, names can only include certain words according to specific conditions named by four AAFCO rules. The "95%" rule states that for an ingredient to be used in the name of a pet food, 95% of the product must be the named ingredient, not counting the water added for processing and "condiments." If water is included, the named ingredient must still be at least 70% of the product. Ingredient lists must also be declared in the proper order of predominance by weight. Thus "Beef Dog Food" must contain at least 95% beef, and beef must be the first ingredient listed in the ingredient list. If the food is named "Lamb and Liver Dog Food," lamb and liver combined must equal 95% of the feed. The first ingredient in the name must be the one found in greatest quantity in the feed. This rule applies only to ingredients of animal origin. So grains and vegetables cannot be part of the 95% total. "Beef and Potatoes" must still be at least 95% beef to use "beef "in the name this way.

The "25%," or "dinner," rule works the same way as the "95%" rule except the minimum requirement is that the named ingredient must be at least 25% of the product, excluding the water added during processing. The name must include a qualifying descriptive term, such as dinner, platter, entree, nuggets, formula, or other similar term. If more than one ingredient is included in a "dinner" name, they must total 25% and be listed in the same order as found on the ingredient list. The first ingredient in the name must be the one found in greatest quantity in the feed. In addition, each ingredient named must be at least 3% of the total. Grains and vegetables cannot be part of the 25% total. "Beef and Potatoes" must still be at least 25% beef to use "beef" in the name this way. The "with" or "3%" rule allows a feed name to use the designation "with" if the

# SPARKLE 'N SHINE BRAND™ BEEF DOG FOOD

According to AAFCO model regulations this product must be 95% beef.

The "with" rule allows a statement such as this for ingredients found in a minimum of 3% in the feed.

Net Weight
5 LBS (2.26kg)

Net quantity statement tells the quantity of the product in the container.

INGREDIENTS. BEEF, POULTRY BY-PRODUCT MEAL, CORN, SOYBEAN MEAL, CORN GLUTEN MEAL, WHEAT FLOUR, CHEESE, FAT PRESERVED WITH BHA, FISH, CHICKEN, LIVER, PHOSPHORIC ACID, BREWERS DRIED YEAST, FUMARIC ACID, TRICALCIUM PHOSPHATE, SALT, SORBIC ACID (A PRESERVATIVE), CHOLINE CHLORIDE, CALCIUM PROPIONATE (A PRESERVATIVE), POTASSIUM CHLORIDE, TAURINE, DRIED WHEY, ZINC OXIDE, FERROUS SULFATE, NIACIN, VITAMIN SUPPLEMENTS (E, A, B-12, D-3), CALCIUM PANTOTHENATE, RIBOFLAVIN SUPPLEMENT, MANGANESE SULFATE, BIOTIN, THIAMINE MONONITRATE, FOLIC ACID, PYRIDOXINE HYDROCHLORIDE, COPPER SULFATE, MENADIONE SODIUM BISULFITE COMPLEX (SOURCE OF VITAMIN K ACTIVITY), CALCIUM IODATE.

Ingredients must be listed in the proper order of predominance by weight.

**GUARANTEED ANALYSIS:**

| | |
|---|---|
| CRUDE PROTEIN NOT LESS THAN | 24.0% |
| CRUDE FAT NOT LESS THAN | 8.5% |
| CRUDE FIBER NOT MORE THAN | 3.5% |
| MOISTURE NOT MORE THAN | 10.0% |
| CALCIUM | 0.1% |
| PHOSPHOROUS | 0.1% |

Minimums and maximums of these four must be on the guaranteed analysis label.

Other nutrients may be added to the guaranteed analysis label.

## — Daily Feeding Guidelines —

| Dog Weight | Up to 10 lbs | 11 to 20 lbs | 21 to 50 lbs | 51 to 90 lbs | Over 90 lbs |
|---|---|---|---|---|---|
| Amount to Feed Dry* | $3/4$ to $1^1/4$ cups | $1^1/4$ to $2^1/4$ cups | $2^1/4$ to $4^1/2$ cups | $4^1/2$ to 7 cups | 1 cup for each 14 lbs |

**AAFCO STATEMENT:**
Animal feeding tests using AAFCO procedures substantiate that "Sparkle 'N Shine Brand" Beef Dog Food provides complete and balanced nutrition for all life stages.

The nutritional adequacy statement must have the appropriate life stages and how the formulation was determined (either by AAFCO Nutrient Profiles or AAFCO feed trial).

Individual dog's requirements may vary depending upon breed, size, age, exercise, and environment. Pregnant and nursing or hard working dogs may require 2-3 times these amounts.

Feeding directions. Consider these a rough guide to feeding.

**FEEDING PUPPIES:**
For puppies 2 to 6 months of age multiply the amount to feed, from above, by two. For puppies 6 to 12 months of age multiply the amount to feed, from above, by 1.5. Puppies that are 12 months of age to adult should be fed the amount listed in the table above.
*Based on 8oz. measuring cup.

Manufactured by Josh N' Aubry Enterprises
P.O. Box 111,
Stillwater, OK 66666
Questions? Call 1-800-111-DOGS

"Manufactured by..." statement identifies party responsible for quality and safety and location of the party.

**FIGURE 23–26** Representative pet food label.

feed contains at least 3% of the ingredient that is named after the word "with." "Dog Food with Liver" must contain 3% liver (excluding water). "Lamb with Liver Dog Dinner" must contain 3% liver (excluding water) in addition to the minimum 25% lamb the term "dinner" indicates it contains. The feed could also be named "Lamb Dog Dinner" and have a separate entry such as "with liver" on the label that does not appear in the name. Likewise, "Doctors Brand" dog food could also carry a "with lobster" and would be required to have 3% lobster.

The "flavor" rule does not specify a minimum percentage. However, the product must have a detectable amount of the product as determined by a set of test methods using animals trained to prefer specific flavors. Pet foods often contain *digests*, which are materials treated with heat, enzymes, and/or acids to form concentrated natural flavors. Only a small amount of a "chicken digest" is needed to produce a " Chicken-Flavored Cat Food," even though no actual chicken is added. Stocks or broths are also occasionally added. Whey is often used to add a milk flavor. Often labels bear a claim of "no artificial flavors" implying something special about the feed. Actually, artificial flavors are rarely used in pet foods, so the inclusion of the wording is for perceived market advantage only. The major exception is artificial smoke or bacon flavors, which are added to some treats.

**Digests**

Produced by enzymatic degradation of animal tissues. Used to flavor pet foods.

In this discussion on the use of ingredients in product names, note that the ultimate purpose of dog and cat foods is to supply needed nutrients, not specific ingredients. Because the nutritional requirements can be met using a wide variety of ingredients, the presence or absence of a particular ingredient doesn't need to be a driving factor. However, if you choose to purchase a product on an ingredient basis, it is important to keep these rules in mind. In addition to scrutiny of the product name, also read the ingredient list to ensure that the preferred ingredient is present in a desirable amount.

## NET QUANTITY STATEMENT

The *net quantity statement* tells the quantity of product in the container. FDA regulations dictate the format, size, and placement of the net quantity statement on the container. The "manufactured by . . . " statement identifies the party responsible for the quality and safety of the product and its location. Regulations require that ingredients be listed in descending order of predominance by weight. Most ingredients on pet food labels have an AAFCO official definition.

**Net quantity statement**

FDA-required statement on a pet food package specifying the quantity of feed in the container.

A pet food label must state guarantees for its minimum percentages of crude protein and crude fat and the maximum percentages of crude fiber and moisture. Some labels have guarantees for other nutrients that manufacturers voluntarily include. These frequently include the maximum percentage of ash. Cat foods commonly have guarantees for taurine and magnesium. The minimum levels of calcium, phosphorus, sodium, and linoleic acid are often found on many dog food containers. The manufacturer may be willing to provide additional information on particular nutrients that are not guaranteed on the label. Nutrient guarantees are given on an " as-fed" basis.

According to AAFCO regulations, the maximum percentage of moisture content for a pet food is 78%, except for products labeled as a "stew," "in sauce," "in gravy," or similar terms. Exempted products have been found to be as high as 87.5% moisture. Moisture level can make a large difference in the amount of nutrients an animal receives from the feed. The amount of moisture in a canned food should be considered in any purchase, both in terms of the nutrients it contains and the price of the dry matter in the food.

## NUTRITIONAL ADEQUACY STATEMENT

The AAFCO nutritional adequacy statement is a very important part of a pet food label. A "complete and balanced," "perfect," "scientific," or "100% nutritious" pet food must have been substantiated for nutritional adequacy. This may be done by formulating the food to meet the *AAFCO Dog or Cat Food Nutrient Profiles,* or the product must be tested according to the *AAFCO Feeding Trial.* The nutritional adequacy statement must state the appropriate animal life stage(s) for which the product is suitable, such as maintenance, growth, reproduction, or all life stages. Products that are intended "for all life stages" must meet the requirements for growth and reproduction. Product labeling for a more specific use or life stage, such as "senior," or for a specific size or breed is done outside any rules governing these types of statements because the requirements have not been established in that much detail. Therefore, a geriatric diet is required only to meet the requirements for adult maintenance. Products developed outside these methods to establish nutritional adequacy must be labeled "This product is intended for intermittent or supplemental feeding only." Snack or treat foods are exempted from this rule because it is generally understood these foods aren't intended to be complete diets.

## FEEDING DIRECTIONS

Feeding directions are designed to give the purchaser guidance as to how much of a given food should be fed to an animal. Instructions usually indicate how many cups of feed per pound of body weight to offer the animal. Feeding directions simply offer guidance in feeding an animal because many factors interact to influence the amount an individual animal may need. Companies tend to overestimate the needs of the animal in making these feeding recommendations to be sure the animals are offered enough. The best strategy for an owner is to offer the recommended amount at first, and then use judgment to adjust the amount to fit the needs of the animal.

## CALORIE STATEMENT

AAFCO regulations allow calorie statements to be put on pet foods voluntarily. All products making calorie claims must include a calorie content statement on the label. Such statements must be expressed on a "kilocalories per kilogram" basis. In addition, they may appear on a per cup basis. The calorie statement is made on an " as-fed" basis, so corrections for moisture content must be made. A quick method to compare the caloric content values between a canned and a dry food is to multiply the value for the canned food by four and then compare.

For a dog food to be labeled "light," the calorie content of dry foods must be no more than 3,100 kilocalories/kilogram of metabolizable energy (ME). Canned dog foods must be no more than 900 kilocalories/kilogram ME. Light, dry cat foods must be no more than 3,250 kilocalories/kilogram ME. Canned cat foods must be no more than 950 kilocalories/kilogram ME. A dog or cat food labeled "less calories," "reduced calories," or similar words must include on the label the name of the product of comparison and the percentage of calories and feeding direction, which reflect a reduction in calories compared to feeding directions for the product of comparison. A comparison between products of different moisture content is considered misleading.

Low-fat dry dog foods must contain no more than 9% fat. Low-fat canned dog foods must contain no more than 4% fat. Low-fat dry cat foods must contain no more than 10% fat, and low-fat canned cat foods must have no more than 5% fat. A dog or

---

**AAFCO Dog or Cat Nutrient Profiles**
Nutritional standards on which nutritional adequacy statements are based.

**AAFCO Feeding Trial**
Standards under which a dog or cat food must be tested to qualify to use the AAFCO nutritional adequacy statement.

cat food labeled "less fat," "reduced fat," or similar terms must include on the label the name of the product of comparison and the percentage of fat reduction explicitly stated and a minimum crude fat guarantee in the Guaranteed Analysis immediately following the minimum crude fat guarantee in addition to the mandatory guaranteed analysis information. A comparison on the label between products in different categories of moisture content is considered misleading.

## OTHER LABEL CLAIMS

There is a growing trend for pet foods to be labeled as "premium," "super premium," "ultra premium," and "gourmet." None of these terms has any official regulatory standing or definition. In other words, they mean nothing in terms of a guarantee. They must be complete and balanced like any other feed, but there is no additional nutritional requirement.

AAFCO recommends the following guidelines for use of the term *natural* in the labeling of pet foods: "A feed or ingredient derived solely from plant, animal or mined sources, either in its unprocessed state or having been subject to physical processing, heat processing, or rendering, purification, extraction, hydrolysis, enzymolysis or fermentation, but not having been produced by or subject to a chemically synthetic process and not containing any additives or processing aids that are chemically synthetic except in amounts as may occur unavoidably in good manufacturing practices." The use of the term *natural* is only acceptable in reference to the product as a whole when all of the ingredients and components of ingredients meet the definition. However, exceptions for chemically synthesized vitamins, minerals, or other trace nutrients are acceptable if a disclaimer is on the bag, such as "Natural with added vitamins, minerals, and other trace nutrients." Also, a use such as "natural liver flavor" is not subject to the rule because this is not considered an implication that the whole product is natural.

Organic standards were put into effect for pet foods in March 2002 by the Department of Agriculture when it ruled that organic standards for pet foods are covered by the Organic Foods Production Act of 1990. However, comprehensive rules have not been agreed to by the USDA, the National Organic Standards Boards, and the pet food industry. Until comprehensive rules are put in place, the official USDA organic seal may not appear on pet foods. However, the terms "100% organic," "organic," and "made with organic ingredients" can be used.

AAFCO provides guidelines for labeling feeds that help control breath odor and reduce plaque and tartar buildup on teeth. The guidelines cover only purely mechanical (e.g., abrasive) mechanisms. Further, there can be no implication that the feed helps prevent or treat dental diseases, caries (cavities), or tooth loss. Products claiming to control breath odor may also do so if they contain chlorophyll or flavoring ingredients that are acceptable for use in animal feeds and not used in excess of amounts typical for flavoring foods.

Labels claiming that a pet feed contains ingredients that are "human grade," "human quality," "people foods," "ingredients you would eat," "food that you would feed your family," or similar claims, are considered false and misleading unless the entire product meets the USDA and FDA standards for foods edible by humans. Currently the terms "human grade" or "human quality" are not allowed by AAFCO because they are not defined. Any pet feed claiming to be human edible must be "manufactured, packaged, shipped and held under such conditions that conform to, and pass the standards set for, human edible products, and the manufacturer, shipper, distributor/wholesaler, and retailer have applicable current federal, state, and local permits, certificates, or licenses required for producing, shipping, handling, and selling products edible for people."

Additional guidelines exist for hairball claims for cat foods, for comparative claims ("Preferred by 9 out of 10 dogs over Brand X"), for fat claims, for feed ingredients, feed additives, and drugs. It is also important to note that drug claims are allowed only on prescription diets.

## TRENDS IN THE PET INDUSTRY

### HUMANIZING THE PET

The trend toward humanizing pets is growing, and no trend is quite so important as this one in defining both pet owners' views of their animal friends and their spending habits. People wish to reward their pet in human terms (Figure 23–27). This translates into demand for pet orthodontics, hotels, designer clothes, designer birdcages, rhinestone tiaras, computerized identification tags, self-cleaning litter boxes, and services from doggy spas to companies that regularly clean dog owners' yards. An impressive $41 billion is expected to be spent on pets in 2008. Much of these expenditures seem to be recession proof. At the time of publication of this text, it was expected that Americans would spend as much as $52 billion on their pets in 2009 and $60 billion in 2010 (Figure 23–28).

### PET POPULATION

The total U.S. pet population is expected to grow and continue to be dominated by dogs and cats (Figures 23–29, 23–30, and 23–31). The factors driving the overall increased pet population include:

- The baby boom generation is in the most common pet-owning years (35 to 65 years old)
- Increased resources for pet ownership because of smaller family size

**FIGURE 23–27**  Hawkeye owns empty nesters Hank and Pam. He is shown here on "his" float. The ball in the background belongs to his dog cousin Sketch. Pam works at home, and Hawkeye spends most of his day in her company. When Hawkeye gets bored, he goes to the pool to hang out. Not much of a fan of swimming, he prefers to float his time away (Sketch is the real athlete of the family). He has a regular group of dog friends that visit frequently with their people for poolside gatherings. Hawkeye welcomes his cousin Sketch for the birthday bash he shares with Pam and her daughter-in-law, Brooke (one of Sketch's people) every September. When Hank and Pam are out of town, a sitter comes to the house to stay with Hawkeye. Hawkeye is typical of humanized pets: They enjoy and are provided many of the same things as their humans. (Photo courtesy of Pamela Damron Knight. Used with permission.)

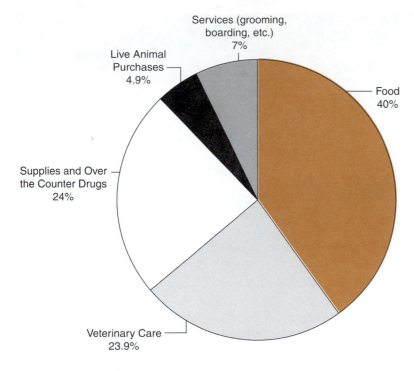

Services (grooming, boarding, etc.)
7%

Live Animal
Purchases
4.9%

Food
40%

Supplies and Over
the Counter Drugs
24%

Veterinary Care
23.9%

**FIGURE 23–28** Approximate percentage of spending on pets in the major categories.

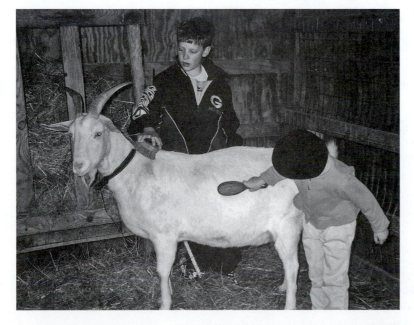

**FIGURE 23–29** Keeping livestock as pets is consistent with hobby farming.

**FIGURE 23–30**  Keeping unusual pets, such as the Red-bellied African frog and toads, is popular. (Photo courtesy of Adele M. Kupchik. Used with permission.)

**FIGURE 23–31**  Reptiles of many types, such as the iguana, are kept as pets. (Photo courtesy of Adele M. Kupchik. Used with permission.)

- Security and companionship of a pet against factors of modern life such as divorce, job anxiety, and geographic dispersal of families
- Surrogate bonding with the natural world through pets
- Increased ecological awareness, which is having spillover effects on a desire for pet ownership
- The dog's traditional role as protector (Figure 23–32)

**FIGURE 23–32**    Pet animal ownership is increasing for several reasons. Included are the emotional security and companionship they provide to their owners from the factors of modern life as well as the dog's traditional role as protector. (Photo of Mandy and Taylor courtesy of Kathryn Beckett Traicoff. Used with permission.)

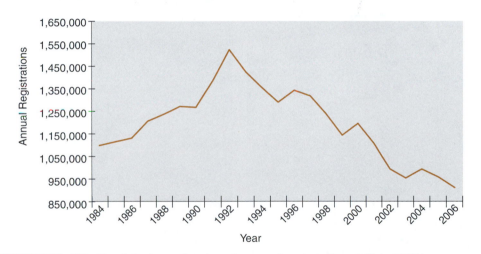

**FIGURE 23–33**    Trends in dog registrations. (*SOURCE:* American Kennel Club, 2008.)

## REGISTERED ANIMALS

The number of annual registrations for dogs and cats is declining (Figures 23–33 and 23–34). The figures are based on data from the American Kennel Club and The Cat Fancier's Association, Inc., and thus do not represent all registrations. However, each organization represents the largest registry for the respective species and, as such, is a good indicator of total registrations. The reasons for the decrease in purebred registrations are probably multiple. However, one possible answer is that many cities are placing restrictions on breeding dogs and cats. Another is that people as a whole may be less interested in a purebred animal (Figure 23–35). Hybrid dogs, in particular, are becoming popular. Such crosses include Labrador Retriever and Poodle (Labradoodle), Pug and Beagle (Puggle), Bichon Frise and Papillon (Papichon), and dozens more.

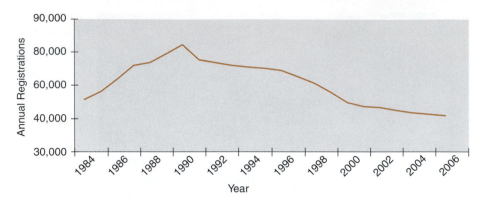

**FIGURE 23–34**  Trends in purebred cat registrations. (*SOURCE:* The Cat Fancier's Association, Inc. and privately collected data for 2005–2007.)

**FIGURE 23–35**  Reasons for the decline in purebred dog and cat registrations are unclear. (a) For some, owning a recognizable purebred animal like this elegant Borzoi is important. (b) For others, whether an animal is purebred is of secondary importance to simply having an animal to love. (Photos of Taylor with Alex and Thomas with Stripes courtesy Kathryn Beckett Traicoff. Used with permission.)

## PET FOODS

Trends in the U.S. pet food industry are truly worldwide trends because U.S. manufacturers are the leaders in the worldwide market. With a foreign market potential greater than the domestic market, manufacturers have a good deal at stake. The production of specialized products such as pet snack foods, low-calorie feeds, medicated diets for such ailments as diabetes, breed- and/or size-specific feeds, and nutrient-packed rations for athletic dogs is a major focus of the industry. Life-stage concept feed formulations and marketing are now universal throughout the pet food industry, and further offerings should be forthcoming in the near future. Diets marketed for specific types of performance are also expected to proliferate.

Owners are concerned about more than nutrition; they wish to mirror their own preferences for their pets. This has spurred the growth of organic, vegetarian, kosher, and locally sourced pet foods. The massive pet food recall of melamine-contaminated pet food in 2007 caused the same level of concern among pet owners as a baby food recall might. Consumers are demanding that ingredients put in pet foods be human-consumption grade. Manufacturing methods are being changed to reflect consumer demands such as feeds with visible vegetables. Even condiments for pet foods are becoming available on the market. The organic and "natural" pet food markets are especially strong. Specialty products generally cost more. With demand strong, companies have profit opportunities in the pet food sector and an incentive to continue developing more products.

## NUTRITION

The U.S. pet food market is expanding at a rapid rate. One of the major reasons for this expansion is the introduction of more specialized pet products. Up to 400 new products are introduced each year, with many directed at particular market segments. This trend is escalating. Presumably, the nutritional value of the foods will improve as companies do the research necessary to create the new feeds and comply with label regulations at the same time. An expert panel is currently at work on pet and exotic bird nutrition. The committee's purpose is the development of nutrient profiles for these species. Once accomplished (this will take years), the number of feeds available for bird species and the labeling of those feeds will improve.

An updated *Nutrient Requirements of Dogs and Cats* (The National Academies Press) was released in 2006. This publication provides nutrient recommendations based on physical activity and stage in life. It looks at how nutrients are metabolized in the bodies of dogs and cats, indications of nutrient deficiency, and diseases related to poor nutrition. This volume is a valuable resource for formulating diets, setting research agendas, developing regulations for pet food labeling, and assisting knowledgeable pet owners.

It is estimated that one in four dogs and cats in the United States is obese. With obesity comes a whole set of ailments that pets of normal weight rarely suffer. Nutrition is clearly linked to health. This pet obesity epidemic is sorely in need of attention.

## VETERINARY EXPENDITURES

The total costs of veterinary care are rising 10% annually. Americans now spend approximately $10 billion annually (2007) on veterinary services for pets. As more medical technology becomes available to pet owners, and as the baby boomer owners increasingly take advantage of those services, these costs are expected to continue to increase. People are now demanding care for their pets that parallels what they want for themselves. In addition, there is a greater tendency for pet owners to seek regular services for pocket pets and other animals that historically received little if any veterinary care. Added to this are the increasing array of pharmaceuticals for pets for problems such as obesity and separation anxiety. Pet hospitals now routinely provide wellness plans. Pet owners are demanding and are willing to pay for human-quality care for their pets. One can even find pet liposuction, rhinoplasty, eye lifts, and other cosmetic procedures for pets.

## MINI-TRENDS

A variety of trends are affecting how much money is being spent on pets and on what. Several are worthy of awareness.

**Upgrading necessities.** Pet owners are buying many of the same things they always have but spending more for items of higher perceived quality or status. For instance, a leather collar with charms or the must-have Signature Pet Carrier from Coach. Then there are the down-filled or memory foam dog beds, the stone-washed jean jacket for dogs, the Asian-inspired antique brown iron pet feeder, dog and cat matching towel and bathrobe sets, china food bowls, and so much more!

**Pet services.** Much of the growth in pet services revolves around providing services for pets that mirror services for people. Many kennels are now near hotel quality in the accomodations they provide for pets. Doggie day care with pick-up and return services are common and expanding. Pet health insurance has entered the mainstream for pet owners. Funeral services are starting to provide services for pets.

**Bequeathing wealth to pets.** In most states it is legal to establish a trust fund to be used to care for pets if the owner dies first. Legal service providers are increasing to provide for this and other needs of pet owners.

**Big companies enter the market or expand.** The big pet specialty retailers (Petco and Petsmart) continue to grow and add services. By adding services they hope to become the consumer's sole shopping place for both products and services. Other major retail outlets like Wal-Mart and Target are expanding the products they offer. A clue to which companies are investing heavily in their pet divisions can be seen in their advertising. Notice the retailers who use pets in their advertising.

**Travel with pets.** A certain segment of the population takes their pets on trips, short and long. This is causing hotels to ramp up their services for pets traveling with their owners. Travel with pets is so important to some people that they consider the ease with which their pet can enter and exit a vehicle when buying transportation.

# INDUSTRY ORGANIZATIONS

The following list of organizations and other resources is provided to assist you in finding additional information.

**American Cat Fancier's Association**
P.O. Box 1949
Nixa, MO 65714
Phone: (417)725-1530
Fax: (417)725-1533
http://www.acfacat.com

**American Federation of Aviculture, Inc.**
P.O. Box 7312
Kansas City, MO 64116
Phone: (816)421-2473
Fax: (816)421-3214
http://www.afabirds.org

**American Kennel Club**
5580 Centerview Dr.
Raleigh, NC 27606
Phone: (919)233-9767
Fax: (919)854-0161
http://www.akc.org

**American Veterinary Medical Association**
1931 North Meacham Rd., Suite 100
Schaumburg, IL 60173-4360
Phone: (847)925–8070
Fax: (847)925–1329
http://www.avma.org

**The Cat Fancier's Association, Inc.**
P.O. Box 1005
Manasquan, NJ08736-0805
Phone: Central Office (732)528–9797
Fax: (732)528-7391
http://www.cfainc.org

**Delta Society**
875 124th Ave. NE, Suite 101
Bellevue, WA 98005
Phone: (425)679–5500
Fax: (425)679–5539
E-mail: info@deltasociety.org
http://www.deltasociety.org
An organization dedicated to promoting animals that help people improve their
health, independence, and quality of life.

**The International Cat Association, Inc.**
P.O. Box 2684
Harlingen, TX 78551
Phone: (956)428-8046
Fax: (956)428-8047
http://www.tica.org

*Service Dogs*

**Canine Companions for Independence**
National Headquarters
P.O. Box 446
Santa Rosa, CA 95402–0446
Phone: (707)577-1700; (707)577-1756 (TDD)
http://www.caninecompanions.org/

CCI trains four types of dogs:

*Service dog:* A service dog placement is made with an individual with a physical
disability who can work independently with the canine companion. The service
dog performs practical tasks to assist the individual.

*Hearing dog:* A hearing dog placement is made with an individual who is hard of
hearing or deaf and can independently work with the canine companion. A hearing
dog is trained to alert the individual to various sounds, including the telephone
ring, the alarm of a clock, and a smoke alarm.

*Assisted service (social) dog:* An assisted service dog placement is made with a person with a disability who requires the assistance of another person to work with the dog. The dog is taught to perform interactive and practical tasks.

*Facility dog:* A facility dog placement is made with a trained professional who utilizes the dog through pet-facilitated therapy and interactions.

### Friends for Folks

P.O. Box 260
Lexington, OK 73051–0260
Phone: (405)527-5676, ext. 630
The Friends for Folks Program provides senior citizens with a well-trained and behaved companion trained by prison inmates. The dogs are rescued from a dog shelter. Long-term offenders are used to train the dogs for rehabilitation purposes as a means of improving prison life.

### International Association of Assistance Dog Partners

38691 Filly Dr.
Sterling Heights, MI 48310
Phone: (513)245-2199
http://www.iaadp.org/
IAADP is a nonprofit organization formed in 1993 as an independent, cross-disability consumer organization that could represent all assistance dog partners and advance consumer interests in the assistance dog field. "IAADP's mission is to (1) provide assistance dog partners with a voice in the assistance dog field; (2) enable those partnered with guide dogs, hearing dogs, and service dogs to work together on issues of mutual concern; and (3) to foster the disabled person/assistance dog partnership."

### National Association for Search and Rescue

P.O. Box 232020
Centerville, VA 20120-2020
Phone: (888)893-7788
http://www.nasar.org

### NEADS Dogs for Deaf and Disabled Americans

P.O. Box 213, West Boylston, MA 01583
500 Colony Rd., Gardner, MA 01440
Phone: (978)422-9064
Fax: (978)422-3255
www.neads.org

### The Winn Feline Foundation

1805 Atlantic Ave.
P.O. Box 1005
Manasquan, NJ 08736
Phone: (732)528-9797
http://www.winnfelinehealth.org

### *General Resource*

### American Dog Trainers Network

www.The DogSite.org
Provides an excellent website with information on dozens of dog-related topics. The network offers dog-training articles, safety tips, a recommended book list, seminars, and free referrals to a variety of resources. "Our primary goal is to promote humane education, responsible pet care, and positive motivational dog training."

## SUMMARY AND CONCLUSION

Pet keeping probably led to the first domestication of animals. Pets and companion animals contribute depth and enrichment to the lives of people in all life stages and circumstances. We enjoy, appreciate, and benefit from pets for many reasons. It is impossible to assign a monetary value to such social, societal, and personal benefits. However, there are large economic segments to the pet industry. The AVMA indicates that more than 58 million U.S. households have at least one pet.

The purpose of the pet and companion animal industry is to support the animals in service and companionship to people. The purpose of pets is to serve in one or more roles that include ornamental pets, status symbols, playthings, hobby animals, helpers, and companions. All regions of the United States have rates of pet ownership exceeding 50%.

There are estimated to be between 400 and 450 breeds of dogs in the world today. Only about a fourth of those have gained enough popularity to be included in the primary dog registry of the United States, the American Kennel Club (AKC). There are only approximately 50 breeds of cats. Cats have not been developed into as many different breeds as dogs have because cats are put to fewer uses. However, cat breeders are actively developing new breeds. The bird fancy is divided along species lines. The remainder of the pet species are not as breed oriented as are the dog and cat.

The FDA regulates labels on pet foods. States also have the authority to establish and enforce state regulations. Many states have adopted model pet food regulations established by the Association of American Feed Control Officials (AAFCO). The regulations are more specific. Aspects of feed labeling such as product name, guaranteed nutrient analysis, nutritional adequacy statement, feeding directions, and calorie statements are covered under the AAFCO standards. The requirements for labeling dog and cat foods are the most detailed and specific. This is because the needs of these species are known with enough confidence that the AAFCO can specify nutrient requirements.

The total U.S. pet population is expected to grow modestly for at least the next decade. However, the percentage of registered animals is declining. A major focus of the pet food industry is the production of specialized products such as pet snack foods, low-calorie feeds, medicated diets for animals with such ailments as diabetes, and nutrient-packed rations for athletic dogs. The industry is expanding to include more sales of dog and cat foods by veterinarians. Up to 400 new pet products are introduced each year, with many directed at particular market segments. The total cost of veterinary care is rising, and this trend is expected to continue.

## STUDY QUESTIONS

1. Describe the rate of ownership for the various pet species in U.S. households.
2. Describe some of the generalities that can be drawn about pet owner demographics.
3. What is the monetary value of the particular industry segments associated with the pet species that were discussed in the chapter?
4. Compare the percentage of U.S. households involved in the livestock industry to the percentage that have a pet or companion animal.
5. Define and differentiate among all the categories of pet animals. Include their functions.
6. In outline form, give all "value" uses for pets in modern society.
7. Give a brief history of the dog as a domestic species.

8. Give a brief history of the cat as a domestic animal.

9. Describe the distribution of pet ownership across the geographic regions of the United States.

10. Compare and contrast the types of breeding and genetics resources available for pet species to those for livestock species.

11. Describe the role of breeds in the pet species. Compare the role of breeds in the dog and cat fancies to that in the reptile fancy or the bird fancy.

12. Describe in general terms the process of mating a bitch.

13. Compare the basic structure of the reproductive tract of the dog to that of the cat. What are the similarities and differences?

14. Explain the information found on the label of a dog or cat food.

15. Describe the major trends affecting pets and the pet industry.

## REFERENCES

AAFCO. 2007. *Official publication 2007*. College Station, TX: American Feed Control Officials, Inc.

American Kennel Club. 2008. Personal Communication, Michelle Baker, Director, Customer Service.

American Pet Products Manufacturers Association, Inc. 2007. *2005–2006 APPMA national pet owners survey*. Greenwich, CT: American Pet Products Manufacturers Association.

AVMA. 1997. *U.S. pet ownership & demographics sourcebook*. Schaumburg, IL: Center for Information Management, American Veterinary Medical Association.

AVMA. 2002. *U.S. pet ownership & demographics sourcebook*. Schaumburg, IL: Center for Information Management, American Veterinary Medical Association.

AVMA. 2007. *U.S. pet ownership & demographics sourcebook*. Schaumburg, IL: Center for Information Management, American Veterinary Medicine Association.

Bennett, L. 2007. *Pet food trends for 2008*. Accessed online January 2008 through Small Business Trends. http://www.smallbiztrends.com/category/2008-trends.

Campbell, J. R., M. D. Kenealy, and K. L. Campbell. 2003. *Animal sciences: The biology, care, and production of domestic animals*. 4th ed. New York: McGraw-Hill.

Case, L. P. 2003. *The cat: Its behavior, nutrition, and health*. Oxford, UK: Blackwell.

Case, L. P. 2005. *The dog, its behavior, nutrition, and health*. Ames: Iowa State University Press.

Council for Science and Society. 1988. *Companion animals in society*. Oxford, UK: Oxford University Press.

Gage, L. J., and R. S. Duerr. 2007. *Hand-rearing birds*. Oxford, UK: Blackwell.

Geisert, R. 1999. *Learning reproduction in farm animals*. Stillwater: Oklahoma State University.

Irlbeck, N. A. 1996. *Nutrition and care of companion animals*. Dubuque, IA: Kendall/Hunt.

Klug, W. S., and M. R. Cummings. 2000. *Concepts of genetics*. Upper Saddle River, NJ: Prentice Hall.

Morris, D. 1999. *Cat breeds of the world*. New York: Viking.

Pet Food Institute. 2007. *Pet incidence trend report*. Pet Food Industry Reference Desk. http://www.petfoodinstitute.org.

Schwartz, M. 1997. *A history of dogs in the early Americas*. New Haven, CT: Yale University Press.

Tabor, R. 1995. *Understanding cats: Their history, nature and behavior*. Pleasantville, NY: The Reader's Digest Association.

USDA. 1984. *Yearbook of agriculture: Animal health, livestock and pets*. Washington, DC: USDA.

# Lamoids

## LEARNING OBJECTIVES

After you have studied this chapter, you should be able to:

1. Describe the relationships among the six genera of the family Camelidae and list the major contributions each has made to society.

2. Describe the history of the llama in the United States.

3. Identify the physical differences among the genera and the unique characteristics they possess that make them a desirable and/or valuable animal.

4. Describe the important health concerns of llamas and explain methods of maintaining good llama herd health.

5. Explain lamoid behavior in relation to age, stage of reproduction, and social status.

6. Compare and contrast the physiology of the lamoid digestive system with that of livestock species in terms of efficiency.

7. Explain the basics of lamoid feeding.

8. Discuss some of the major challenges facing the llama industry.

## KEY TERMS

| | | |
|---|---|---|
| Alpaca | Erythrocytes | Maiden |
| Bloat | Flehmen response | Multiparous |
| Cria | Hemoglobin | Orgle |
| Dystocia | Llama | Sternal recumbancy |

## SCIENTIFIC CLASSIFICATION OF CAMELIDS

There are six members of the camelid family. The *camel* genus (*Camelus*) has two members: the one-humped dromedary camel (Figure 24–1) and the two-humped Bactrian camel (Figure 24–2). The llama genus (*Lama*) includes the guanaco (Figure 24–3), the

**FIGURE 24–1** *Camelus dromedarius,* the dromedary camel. (Photo courtesy of Dr. Thomas Thedford.)

**FIGURE 24–2** *Camelus bactrianus,* the Bactrian camel. (Photo by Four by Five. Courtesy of SuperStock, Inc.)

llama (Figure 24–4), and the alpaca (Figure 24–5). The *Vicugna* genus includes only the vicuna (Figure 24–6). Members of the camel family are commonly referred to as *camelids,* and members of the *Lama* and *Vicugna* genera are called *lamoids.* It is also fairly common to refer to camels as Old World camelids and lamoids as New World camelids.

| | |
|---|---|
| Phylum: | Chordata |
| Subphylum: | Vertebrata |
| Class: | Mammalia |
| Order: | Artiodactyla |
| Suborder: | Tylopoda (means "padded foot") |
| Family: | Camelidae |
| Genus: | *Camelus* (Old World camelids); *Lama* (South American camelids); *Vicugna* (South American camelid) |
| Species: | *dromedarius* (dromedary camel), *bactrianus* (Bactrian camel); *glama* (llama), *pacos* (alpaca), *guanicoe* (guanaco); *vicugna* (vicuna) |

**FIGURE 24–3** *Lama guanicoe.* A young guanaco. (Photographer Ernesto Rios. Courtesy of PhotoDisc, Inc.)

**FIGURE 24–4** *Lama glama.* A pack train of llamas in Ecuador. (Photographer Janet M. Cummings. Courtesy of Embassy of Peru.)

**FIGURE 24–5** *Lama pacos,* the alpaca. (Photographer Bruna Stude. Courtesy of Omni-Photo Communications, Inc.)

# THE PLACE OF LAMOIDS IN THE UNITED STATES

The only camelids found to any extent in the United States are the llama and alpaca; they are thus the most economically important. The New World camelids originated in the high elevations of the Andes Mountains in South America. They consist of four species in two genera: *Lama glama,* the llama; *Lama pacos,* the alpaca; *Lama guanicoe,* the guanaco; and *Vicugna vicugna,* the vicuna. All four species have 37 pairs of chromosomes and can interbreed and produce fertile hybrids. The most commonly produced hybrids are between the llama and alpaca (known as "huarizo")

**FIGURE 24–6**  *Vicugna vicugna,* the vicuna, on its native range in Peru. (Photographer Ted Levin. Courtesy of Animals Animals/Earth Scenes.)

and between the alpaca and vicuna (paco-vicuna). The guanaco and vicuna are not domesticated, although some are tame. The vicuna produces wool with the finest fiber of any animal, grading 120's on the Bradford scale. Higher numbers on the Bradford scale indicate finer wool fiber, which is more valuable for spinning and weaving into very expensive woolen material. Because of the demand for this very luxurious wool, the vicuna was hunted almost to extinction, because the animal had to be killed to obtain the fleece. To preserve the species, strict government regulations forbidding the killing of vicunas have been imposed in their native South American countries.

Llamas are very plentiful in the Andes Mountains from southern Peru to northwestern Argentina. As the largest of the lamoids, llamas have been kept primarily as a pack animal since their domestication. They are also valuable as a source of food, wool, hides, tallow for candles, dried dung for fuel, and as a provider of offerings to the gods (especially the white llamas). The alpaca, which is smaller than the llama, is often described as looking like a large goat with a camel's head and neck. The alpaca is kept and bred for its wool, which in the Suri breed can grow long enough to touch the ground.

Because of their relatively small numbers in the United States, llamas and alpacas still have a specialty status in the animal industries.

## THE PURPOSE OF THE LLAMA AND ALPACA INDUSTRIES

Since the early 1970s, there has been interest in llamas and alpacas in the United States, where they are not considered a food species. However, they are used for work (packing and driving) (Figure 24–7); as guard animals (primarily for sheep and goats) (Figure 24–8); as pets and therapy animals; for fiber production; and for exhibition in shows, parades, and fairs (Figure 24–9). The llama is popular as a project animal for 4-H, Scouts, FFA, and other youth activities. Their most common use is as a pet, with

**FIGURE 24–7** Llamas are a useful pack animal. They carry heavy loads relative to their body weight, and cause minimal damage to fragile trails because of their padded feet. (Photo courtesy of Dr. Thomas Thedford. Used with permission.)

**FIGURE 24–8** Llamas serve a useful function as guard animals for sheep and goats. (Photo courtesy of Dr. Rebecca L. Damron. Used with permission.)

**FIGURE 24–9** A hobby actively pursued by many in the United States is exhibiting llamas in shows. (Photo courtesy of Dr. Thomas Thedford. Used with permission.)

wilderness packing as the second most common use. Llamas can carry very heavy loads, as much as a third to a half of their own weight. In the United States, llamas are often used as pack animals in wilderness areas because of their low cost and stamina on the trail. Their feet are less damaging to trails than those of horses, mules, and burros. There are several dozen commercial packers in business, and the USDA's Forest Service also uses llamas for packing. Undoubtedly, their popularity comes from a combination of need, novelty, and general appeal. The fact that they are easy to feed (grain, grass, and browse), easy to train, and generally easy to care for also contributes to their overall popularity.

## HISTORY OF THE LLAMA IN THE UNITED STATES

The family Camelidae was domesticated about 4,000 to 6,000 years ago. Llamas were exported to the United States from South America in the late 19th century predominantly as zoo animals. Other importations, like that of William Randolph Hearst in the early 1900s, were for exotic species displays in game parks or on vast estates. Few alpacas were imported because Peruvian legislation enacted in 1843 prohibited the export of live alpacas. Imports of llamas were restricted when the United States banned importation of all hoofed stock from South America in 1930 to prevent the importation of foot-and-mouth disease. At about the same time, all of the Andean countries united in an attempt to prevent the exploitation of llamas and alpacas by other countries. The only legal exportations that occurred from then until the 1980s, when the ban was lifted, were from a small herd in Canada. Thus North American llamas are descended predominantly from importations made from South America prior to 1930, some animals from Canada, a few illegal importations from Mexico, plus a few animals imported from other countries at various times. Importation from Chile began again in 1984 after Chile was recognized as being free of foot-and-mouth disease. Foot-and-mouth disease makes importations into the United States from South America difficult and expensive. The USDA's Animal and Plant Health Inspection Service (APHIS) is in charge of regulating such importation. A long, costly quarantine is necessary, and most imports have been of breeding stock. Unfortunately, it seems that the small number of quality imported animals has had minimal effects on expanding the gene pool of the U.S. population. The first substantial importation of alpacas occurred in 1984.

The llama and alpaca are different from traditional meat animal species in the United States in terms of who owns them and the general attitude of the owners. The closest comparison is with the horse. Llama and alpaca owners are often not previously experienced with large animals, a situation similar to horse owners. Also, like horses, the llama and alpaca are often kept as a companion rather than as an investment.

## GEOGRAPHIC DISTRIBUTION

There are estimated to be around 160,000 llamas and 100,000 alpacas in the United States based on registration statistics. The West Coast is the most important area of the country for llama and alpaca ownership, with Oregon, Washington, and California each having greater than 10,000 llamas. Texas, Colorado, and Ohio also have substantial numbers of llamas. Alpacas are distributed similarly. Ohio is the most important non–West Coast state for alpacas.

# PHYSICAL DESCRIPTION

## CAMELIDAE

Members of the Camelidae family are very different from one another, but they also share certain characteristics. They all have long necks, small heads, and no horns or antlers. Each has a prehensile upper lip similar to that of a rabbit. They are herbivores with compartmentalized forestomachs like the ruminant, but their stomach has only three chambers, rather than the four of other ruminants. Camelidae can make good use of plant material and they graze many different species of plants. They are generally believed to be more efficient than most ruminants on poor-quality feed. The red blood cells (*erythrocytes*) of camelids are elliptical, whereas in all other mammals they are circular. These elliptical erythrocytes can swell to 150% of normal size without rupturing. This allows a dehydrated animal to be rehydrated immediately. Lamoid blood contains more red blood cells per unit volume of blood than does the blood of other mammals. In addition, the *hemoglobin* of the blood reacts faster with oxygen. This gives the animals the ability to exert themselves strenuously at high altitudes and makes them well adapted to their environment. Their feet have two toes, with a very flat hoof or nail at the end of each. A thick, callused pad forms the sole of the foot.

**Erythrocytes**
Red blood cells.

**Hemoglobin**
Oxygen-carrying pigment found in erythrocytes.

## LAMOIDS

The llama and alpaca are very similar. Although the llama is larger than the alpaca, there is a marked range of size. Adult llamas weigh 240–500 lbs, with little weight difference between males and females. Llamas are 40–50 in. at the shoulder, and 65–72 in. at the poll. Males should reach mature size at 3 years of age and females at 2. The alpaca ranges from 100–185 lbs and is about 36 in. at the shoulder.

Colors vary widely in the lamoids. They may be white, black, several shades of brown, red, and mixtures of these colors. They may be roan, solid, frequently spotted, or marked in a variety of other patterns. Alpaca wool comes in 22 colors from black to tan to white.

On the llama, the fiber is found on the neck, back, and sides. Hair covers the head, belly, and legs. The fiber is 3–8 in. long and gives the llama great protection against cold, wet weather. The hair-covered areas help the llama to dissipate heat in warm climates. The fiber is oil free (no lanolin as in sheep) and lightweight. It has good spinning quality, with the fleece of the llama considered coarse and inferior to that of the alpaca. The fiber has a fine undercoat and longer, tough outer guard hair. The guard hair helps shed rain while the undercoat holds air and insulates the animal against cold. In cold weather, lamoids rest by facing the wind with their legs tucked under them. The hair-covered belly and legs are protected. In hot weather, they sit in a manner that allows air to flow under the body. The woolless areas then dissipate heat.

The alpaca is raised primarily for its wool. Alpaca wool has an extremely fine fiber that grades better than the finest wool of any sheep. The fiber of the alpaca is superior to that of the llama. It is very fine, lightweight, has good insulation value, and is dense and soft. It is a warm fiber that doesn't scratch the wearer. It is used in making items like parkas, sleeping bags, and fine coat linings, among many other things. An adult alpaca produces about 4 lbs of fiber per year and can be sheared yearly. However, in cooler climates, it is desirable to shear every other year so that the fiber is longer. In hot climates, yearly shearing is necessary to avoid heat stress. Alpacas have been selected to be gentle and submissive to facilitate shearing. These characteristics make them easy to handle as well. Because of fiber-producing ability, it is probable that the alpaca was domesticated from the vicuna.

Adult male alpacas and llamas are referred to as males. Breeding males are sometimes called studs. Females are not called any other name, and babies are either called babies or **crias,** which is Spanish for baby. "Gelding" refers to neutered males. Parturition is commonly called birthing.

**Cria**

Baby llama.

## GENETICS AND BREEDS

Organized breeding programs such as those that exist for livestock species are not a part of llama and alpaca breeding and genetics. As with the companion animal species, breeding decisions are based primarily on subjective criteria. Knowledgeable breeders do exist and have made progress in improving these animals. Since its organization in 1988, the Alpaca Registry has required blood typing of any animal before it can be registered, and as of November 1, 1998, it began requiring DNA testing. Also on November 1, 1998, the Llama Registry began requiring DNA testing, instead of simple blood typing, on sires used for outside breeding and all that have more than 10 crias. In this way, both registries protect the genetics of their species.

Because the alpaca was developed for its fiber and is the primary fiber producer of the Andes regions, it is logical that breed differences should be primarily in wool type. There are two breeds of alpaca—the Suri, which produces a long, wavy fiber, and the Huacaya, which produces a more desirable, shorter, crimped fiber resembling the wool of Corriedale sheep. The predominant Peruvian breed, the Huacaya, is the breed that has been most exported to other countries. Two llama breeds are recognized in Peru. The chaku is the woolly breed. The breed with less wool is called *ccara* (*q'ara* in some literature). In the United States, two llama types are recognized, but no breeds are recognized. One type has been bred for work and is taller, heavier framed, and has shorter wool. The other is smaller, broader, and has a much heavier fleece. This type is preferred in the show ring. It is probable that as the U.S. industry expands, breeds will be developed.

## HEALTH CARE

Llamas and alpacas are considered very hardy and easy to care for, are generally resistant to many diseases, and have few maintenance problems other than deworming and an occasional foot trimming. Their ruggedness is no doubt related to the harsh environment where they evolved. However, they are not immune to diseases and, in fact, are susceptible to a wide variety of diseases and parasites, with more reported each year. A good herd health program is important, and will become more so as lamoids increasingly come in contact with other domestic animals and with others of their own kind. An effective herd health program is influenced greatly by such factors as the goals of the owner (economics and purpose for the animals), the number of animals in a group, the other species with which they have contact, and geographic location.

It is often difficult to identify an ill llama or alpaca. They are very stoic and don't show many signs of illness. Owners should develop the habit of routinely and carefully observing feed intake and grazing patterns and take the temperature of animals who are not acting "normally" to determine illness.

When the sum of temperature in degrees Fahrenheit and humidity equal or exceed 180, lamoids can suffer from heat stress. As a means of prevention, heavily wooled animals should be sheared and provided with the means to cool themselves. Shade shelters, ponds for wading, sprinkler systems, fans, or an abundance of shade trees can all be effective, depending on other environmental conditions.

Enterotoxemia (overeating disease) has been frequently observed in llamas, especially the young. Apparent preventive success has been reported by immunization of the female with *Clostridium perfringens* types C and D, followed by immunization of the cria at 4 to 6 weeks of age. Tetanus can be prevented by using the *Clostridium* vaccine that contains tetanus toxoid. Llamas should also be vaccinated for rabies because they are very curious and apt to get bitten by rabid animals. Tuberculosis, anthrax, malignant edema, and Johne's disease are also possible diseases for the llama. Internal parasites to be concerned about are coccidia, liver flukes, meningeal worms, tapeworms, lungworms, nasal bots, and gastrointestinal nematodes. External parasites of potential concern are mange, ticks, mites, and lice. Both internal and external parasites can be treated with medicines and pesticides that are approved for cattle, sheep, and goats. Other reported diseases are leptospirosis, equine rhinopneumonitis, eperythrozoonosis, and toxoplasmosis.

## REPRODUCTION

There are few differences of any importance in the reproductive practices of any of the South American camelids other than some behavioral differences. Female llamas and alpacas are usually ready for their first mating at 15 to 18 months of age, but weight is an important variable in determining first mating. Reproductive ability ends at 15–18 years of age, with a normal life span of 20–25 years. In rare cases, individuals can live to be 30 years or older. Males reach full sexual maturity at approximately 3 years of age and can be mated at 2½ years. However, younger males may be fertile and should be separated from females.

Lamoids are unusual in that they are induced ovulators and they ovulate after mating. As induced ovulators, lamoids do not have a heat cycle and will mate anytime they are not pregnant. Their wild ancestors are seasonal breeders, but this characteristic has apparently been bred out of the domestic species. Females should not be rebred until 14 to 21 days after the birth of a cria. Waiting longer than 14 to 21 days to provide additional rest between breedings may actually reduce the chances of a subsequent pregnancy. Very cold and very hot and/or humid seasons should probably be avoided as birthing seasons unless sophisticated facilities with heating and/or cooling are available.

Llamas mate with the female in ***sternal recumbency*** (sitting on her legs with her belly on the ground) (Figure 24–10). Because copulation lasts on the average about 18 minutes for the llama and 20 minutes for the alpaca (range of 5 to 55 minutes), the male llama will seek to assure himself he is in safe surroundings before he will copulate. If moved to a new paddock or breeding enclosure, he will inspect the surroundings, including the dung pile, and may exhibit the ***flehmen response*** to the scents there.

Once he feels safe from rivals, he will approach an open female and begin to ***orgle***. This is a term for the distinctive glutteral noise, which has been described as sounding like a cross between a gurgle and a snort. The orgling continues for the duration of the mating. An interested female clucks to the male, "fans" her tail back and forth, and begins to run. The male chases her. The female either stops running and sits or is forced down on the ground by the male. The male then mounts and penetrates.

**Sternal recumbancy**

Mating position in the llama and alpaca. The female lies on the ground with her legs under her and her belly on the ground.

**Flehmen response**

Sexual behavior of the male of several species where the male curls his upper lip and inhales.

**Orgle**

The distinctive noise made by a male llama before and during mating.

**FIGURE 24–10** Llamas and alpacas mate in sternal recumbancy. (Photographer Susan Jones. Courtesy of Animals Animals/Earth Scenes.)

During the chase, the prepuce of the male, which normally faces to the rear for urination, is pulled forward. After the male has settled on the female and positioned himself properly, he extends his penis. The penis penetrates the corkscrew-shaped opening of the cervix and the male makes rhythmic thrusts into the uterine horns, alternating between them, and deposits semen there. The male raises his tail during ejaculation. There are reports of males copulating up to 18 times per day. The female remains still, calm, and quiet during the mating. Some lie down in a lateral position. Ovulation occurs 26 to 45 hours after copulation in the female and is stimulated through hormonal action initiated by the action of the penis during copulation, the leg clasp of the male, and probably by a neural response stimulated by the male's orgling. Ninety percent of the receptive females ovulate after one copulation.

A female who has ovulated and might be pregnant (corpus luteum begins development) refuses further matings. If the male insists, the female "spits him off" or actually chases him away. Some breeders leave the mating pair together for a few days, and others hand-breed on successive days. These practices can help increase the conception rate because the female may be between ripened eggs (there is a 12-day cycle of follicular growth, degeneration, and regeneration), and also because some females need repeat matings to stimulate ovulation. Pregnancy can be confirmed by a very experienced ultrasound technician as early as 12 to 15 days postmating. A more practical time for such a procedure is 21 to 28 days postmating. Rectal palpation, in the hands of a very experienced and careful palpator, can also be used to determine pregnancy. Blood tests have also been developed to determine pregnancy. The most common blood test is the progesterone test. Blood should not be drawn from the female until 21 days after the last breeding to avoid a false positive from the corpus luteum (CL) of ovulation rather than the CL of pregnancy. Gestation length for llamas is normally 350 days, plus or minus 2 weeks. The average gestation is 335 days, plus or minus 2 weeks, for alpacas.

Most llama females give birth standing up between the hours of 9 A.M. and 3 P.M. (Figure 24–11). This is no doubt a trait selected in response to the cold environment in which these animals developed. The wild vicuna of South America gives birth almost

**FIGURE 24–11** Llamas give birth standing up, generally between 9 A.M. and 3 P.M. A new birth arouses the curiosity of the herd. (Photographer Susan Jones. Courtesy of Animals Animals/Earth Scenes.)

exclusively in the morning. This behavioral adaptation helps them avoid the afternoon storms in the Andes and gives the babies a better chance of survival. A baby born early in the day is able to dry, stand, and nurse before nightfall. Young born during the colder hours of the night or the wet of the afternoon are much less likely to survive. A female approaching parturition usually appears restless and loses her appetite. She separates herself from the herd, may urinate frequently, "hums," and lies down and rises again repeatedly. The whole process, including placenta expulsion, takes 2 to 3 hours in a female who has had young before (*multiparous*), and somewhat longer in the *maiden* female. *Dystocia* is rare, as is retained placenta. Although females are generally good, attentive mothers, they do not lick their young or eat the placenta. A thin cutaneous membrane that covers the cria at birth, but does not restrict breathing or movement, dries up and falls from the cria's coat soon after birth. The female nudges, nuzzles, and hums to the cria to encourage it and bond with it.

### REARING THE CRIA

Because of their hardy nature, little attention is usually required by the newborn cria. However, general husbandry practices such as dipping the navel cord in 7% tincture of iodine to prevent navel ill can be beneficial. Given the fact that these animals can be very valuable, other husbandry practices may be desirable.

The female llama has four teats and raises her baby on milk. Crias, like young ruminants, must have colostrum soon after birth for antibody protection. They should receive colostrum to equal 5% of their body weight within the first 6 hours after birth and 10% within the first 12 hours. If there is a problem and colostrum must be fed artificially to the cria, it should be given 4 to 8 oz at a time at 2-hour intervals. Generally, the cria is up and nursing within 90 minutes of birth. If it has not nursed naturally by 6 hours of age, the dam should be milked and the cria fed with a bottle or a stomach tube (Figure 24–12). Goat or cow colostrum can be given if the mother's colostrum is not available. Collecting excess colostrum and freezing it for future use is also a good strategy. However, excess colostrum is available from llamas and alpacas only in the case

**Multiparous**

Refers to a female that has had previous pregnancies and offspring.

**Maiden**

In this context, a female who has never given birth.

**Dystocia**

Difficulty in birthing.

**FIGURE 24–12** Generally cria are up and nursing within 90 minutes of birth. If a baby llama or alpaca fails to nurse, it must be given colostrum within 6 hours of birth by bottle or stomach tube. (Photographer Bill Tarpenning. Courtesy of USDA.)

of a stillborn. Frozen colostrum lasts many months and can be collected from healthy, vaccinated cattle, goats, or sheep and stored for later use. The alternative is to transfuse needy babies with plasma or blood from a healthy adult llama or alpaca at a rate of 8–10% of their body weight, or to feed blood plasma in place of, or as a supplement to, colostrum.

Early growth can be monitored by weighing crias as soon as they are dry and then daily for 2 weeks. They should gain between 0.5 and 1 lb of body weight per day for the first 2 weeks and should double their birth weight in the first month of life. Normal weight at birth is 18–40 lbs. Crias are weaned at 5–6 months and should weigh 100–150 lbs. Baby alpacas weigh 11–20 lbs at birth and baby llamas weigh 8–40 lbs. at birth.

Suggested immunizations for lamoids include:

- *Clostridium perfringens* C/D tetanus toxoid annually for all animals and a booster to females 1 month before birthing
- *Clostridium perfringens* C/D antitoxin and tetanus toxoid at birth for the cria with boosters at 30 and 60 days of age and then annually with the herd
- Futher immunization options that may be warranted and should be considered are twice-yearly leptospirosis vaccination of brood females, rabies, malignant edema, and equine herpes virus-1 killed vaccine.

Crias have a functional fermentation by 2 months of age. At this age, their stomach proportions are like those of adults. They begin eating solid foods within the first 1 to 2 weeks of life and should be encouraged to do so. This early eating habit can be exploited for the cria of poor-milking mothers by providing a creep area with alfalfa and some grain. Crias who are growing well should probably be excluded from the creep area.

## NUTRITION AND FEED USE

Lamoids are herbivores with a complex stomach, but they have two compartments prior to the glandular stomach as compared to "standard" ruminants, which have three forestomachs (Figure 24–13). Their digestion is very similar to that of ruminants. Regurgitation and remastication are essential features of their digestion. Anaerobic fermentation is accomplished by the presence of microorganisms in the first two compartments of the digestive system. Roughage is converted to products the animal can use to its own benefit. In general, the food mass found in the forestomachs of the lamoid is drier and not stratified into the same layers as that found in the true ruminant. *Bloat* is rare, and this is probably the reason.

By nature, llamas graze and browse. The alpaca is a grazer only. The incisor teeth are firmly fixed in the mandible like those of sheep and goats. These teeth are pressed against an upper dental pad, which assists them in shearing plant material. Their mouthparts enable them to graze close to the ground, which could allow them to harm plant communities by grazing too closely. However, their grazing habit is to move about over the grazing area and thus not damage the available forage. The habit of moving around as they graze also helps prevent them from consuming enough toxic plants to cause them harm.

**Bloat**

Gas collecting in the fermentative portion of the digestive tract. Can be life threatening.

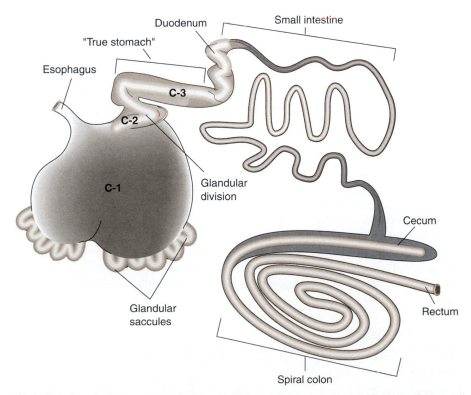

**FIGURE 24–13** Digestive tract of the lamoid. C-1 and C-2 are used primarily as the fermentation vat. The distal portion of C-3 is glandular and can be thought of as the "true stomach," analogous to the abomasum of a ruminant.

**FIGURE 24–14**    Lamoids have a cleft lip that allows them to manipulate each side of the lip independently. (Photographer Susan Jones. Courtesy of Animals Animals/Earth Scenes.)

The prehensile upper lip of the lamoids has a labial cleft, which allows each side of the lip to be manipulated independently of the other. This allows lamoids to check their feed carefully and prevents them from consuming rocks and other potentially harmful objects (Figure 24–14). Unlike most domestic farm species commonly found in the Western world, the lamoid tongue does not readily protrude from the mouth. This is why they do not groom their young or themselves with their tongue. This characteristic also hampers their ability to lick a salt block. They usually chew it instead.

The camelids are more efficient at feed utilization than the ruminants, especially in their ability to digest and use protein and energy. This explains why llamas and alpacas are sometimes too easily fattened. Llamas are able to gain weight and prosper on native South American grasses that average 7% crude protein. The digestible energy requirement for maintenance in a llama is 75 kcal per $kg^{w.75}$. For sheep, it is 119 kcal per $kg^{w.75}$—a 37% advantage to the llama. Volatile fatty acids are the principal end products of carbohydrate fermentation in the lamoids.

There are several possible reasons for the camelid's efficiency of feed utilization, including:

■ Unique anatomy
■ Rapid forestomach motility, with more frequent rumination cycles
■ Increased feed retention
■ Increased urea hydrolysis
■ Rapid liquid absorption and passage rate from the forestomachs
■ More rapid and complete VFA absorption

- Low basal metabolism
- Low maintenance requirements for energy and protein

The practical implication of the efficiency of feed utilization is that camelids can be kept on low-quality pasture and range and stocked more densely than other species. In fact, they seem to have greater efficiency when consuming poor-quality forages than when consuming good-quality feeds. A llama needs 10–20% as much feed as a horse and can manage on 5% of the grazing area required by a horse. The stocking rate for alpacas is 5 to 8 animals per acre, and for llamas, 3.5 to 5.5 animals per acre. Variations depend on animal size, function, and quality of the pasture.

The large intestine and cecum are not designed for fermentation. Instead, the large intestine is spiral shaped (Figure 24–13), and decreases in size by two thirds as it spirals. Fecal pellets are formed in the large intestine.

## LAMOID FEEDING

The nutrition of the lamoids has been poorly researched and much is unknown. This is further complicated by the fact that much of the available literature concerns South American conditions and plant species. This leaves producers in other parts of the world with a tremendous lack of information. Because the basic nutrient requirements for these animals are not known, data from other species must be used to estimate the requirements. Recommended levels of the most important nutrients for llamas are shown in Table 24–1. The complex digestive system of the llamas, which makes them much easier to feed than monogastrics and modestly easier to feed than true ruminants, gives the llama owner a tremendous advantage. Llamas may also have some

**TABLE 24–1     Recommended Levels of Most Important Nutrients for Llamas**

| Nutrient | Common Sources | Recommended Level in Diet |
|---|---|---|
| Crude protein | Hays, grains, and supplements | 10–14% |
| TDN | Grains | 55–65% |
| Fiber | Forages/browse | 25% |
| Calcium | Forages, especially alfalfa and supplements | 0.6–0.75% |
| Phosphorus | Grains or supplements | 0.3–0.5% |
| Magnesium | Hays, especially alfalfa and supplements | 0.2–0.4% |
| Sulfur | Protein sources | 0.2–0.25% |
| Copper | Forages and grains from nondeficient areas | 4–10 ppm |
| Selenium | Forages in nondeficient areas, marine-origin supplements | Varies with vitamin E in diet; 0.3–0.4 ppm |
| Zinc | Forages/browse, grains, and supplements | 25–50 ppm |
| Vitamin A | Green forages | 3,000–5,000 units/# |
| Vitamin E | Green forages | Varies with selenium level in diet (20 ppm) |

SOURCE: Johnson, 1989a, p. 52.

instinct about balancing their overall ration that is generally lacking in other domestic animals. Knowledgeable owners suggest that if concentrates and grains are overfed, llamas will seek very low-quality, empty browse to dilute the nutrient density of the concentrates. In practice, many llama breeders in the United States and Canada have been successfully breeding and keeping llamas since the 1930s, and a wealth of information has been gathered and disseminated to owners. Although lacking the stature of scientific information, it nevertheless gives us valuable knowledge on how to feed the llama. In addition, some excellent, basic research about llama nutrition has been published recently.

If we presume that lamoids can eat 1.8% of their body weight (BW) per day in forage, and do some simple calculations on commonly available feeds, it becomes evident that many readily available feedstuffs will meet the nutrient needs of lamoids if they are provided free choice. Some feeds, if eaten at 1.8% of body weight, could easily be eaten to excess and cause obesity. Table 24–2 shows how small quantities of some common feeds meet the maintenance needs of a 330-lb (150-kg) llama. It would be necessary to limit the amounts of the higher-quality feeds listed in the table to prevent llamas from getting too fat. A good rule of thumb is to start with 1% of body weight as dry forage and increase, if necessary, to maintain body condition up to 1.5% BW. As with all species, it is necessary to increase nutrient quantities for production functions such as growth, lactation, gestation, and work; for environmental conditions; and so on (Figure 24–15). Grain is probably needed only in rare situations by a minority of llamas. Grain feeding may be warranted when pasture conditions are poor, if animals are being worked heavily, for the crias of poor milk-producing dams, for rapidly growing animals, for heavy-milking females, and for animals in very cold environments.

Water needs can be met for lamoids by providing clean water free choice to the animals. They adjust their intake to meet their needs, just as other animals do. Care must be taken if automatic waterers are used that require the animals to push a lever to receive their water because it may take lamoids some time to adjust. Trail llamas frequently refuse to drink during the day, especially from running water, if they are accustomed to standing water. This is not a cause for alarm—water should simply be provided at the end of the day. It is also likely that when grazing lush pastures, camelids will get most of their water from the feed, just as other herbivores do. In addition, the natural water economy of llamas allows them to take advantage of the water from dew and frost.

Vitamin and mineral needs of lamoids are very poorly known, and it is difficult to make recommendations. Lamoid rations that contain mineral levels that fill the needs of sheep are considered adequate. Because llamas and alpacas do not use their tongues to

**TABLE 24–2**    **Feed Necessary to Meet Daily Maintenance Energy Requirements of a 330-lb (150-kg) Llama**

| Feedstuff | Dry Matter (kg) | % Body Weight |
|---|---|---|
| Alfalfa hay | 1.57 | 1.04 |
| Corn silage | 0.77 | 0.52 |
| Grass hay | 1.71 | 1.14 |
| Oat hay | 1.77 | 1.18 |
| Oat straw | 1.92 | 1.28 |
| Rye grass pasture, fertilized | 1.21 | 0.81 |
| Grass pasture, high quality | 1.25 | 0.83 |
| Grass pasture, average | 1.49 | 1.00 |

SOURCE: Compiled from Johnson, 1994b, p. 197, and Sell, 1993, p. 3.

**FIGURE 24–15** High-quality feeds should be used sparingly with lamoids to prevent them from getting too fat. A rule of thumb is to start feeding 1% of body weight as dry forage and increase as needed to maintain body condition. Pictured here is a llama being allowed access to a round bale of low-quality prairie hay.

lick, many never learn to use a trace mineral block. It may be necessary to include minerals in a small amount of grain-based feed to get the animals to consume the minerals they need. Loose mineral mixes in weatherproof feeders are better than mineral blocks. Lamoids grazing or eating hay in temperate zones seem to do well enough on the forage without supplemental minerals, although problems have been observed in tropical zones. Toxicity should be carefully avoided by excluding excessive levels of minerals in the diet. Toxicity can easily occur, and overzealous and/or novice lamoid owners need to be very careful not to overfeed. New lamoid owners should visit with reputable lamoid breeders in their area to find out how to handle mineral nutrition. Sheep breeders, extension agents, and local veterinarians should also be consulted. In lieu of other recommendations, a mixture of 50 lbs each of steamed bonemeal, dry powdered molasses, and trace mineralized salt, along with 10 lbs of ZinPro 100, offers a suitable all-around mineral mix for llamas.

## General Feeding Recommendations

Although it is difficult to make broad generalizations about feeding lamoids, some observations can be made to guide the novice. Good pasture, hay, or silage should meet most nutrient needs of lamoids kept in North America. Reasonable pasture, or almost any hay of 8–10% protein, is acceptable for most stages of a llama's life. Recent research has indicated that even weanlings probably need no more than 10% protein feed. Alpaca owners frequently feed protein at 14–16% of the ration, reasoning that this level is necessary to produce a good fleece, although there is probably no need to do so. However, sulfur needs should be carefully met. In stress situations, such as late gestation and lactation and for the weanling, alfalfa hay can be used as a protein supplement. The addition of a mineral supplement finishes the needs.

Overfeeding is consistently more of a problem than any other nutritional ill and results in a too-fat animal. This can be difficult to detect because of the wool. Owners should weigh animals yearly to assess feeding regimes and should learn to assess the condition of their animals (Figure 24–16). A simple condition score index should be used. Many species of animals are condition scored on a scale of 1 to 10, with 1 being thin and 10 representing the ideal. Just such a scale has been established for llamas.

**FIGURE 24–16** It is easy to allow lamoids to overeat and become too fat because their wool makes it hard visually to evaluate their body condition.

Lamoids should be adapted slowly to new feed, especially if it includes concentrates, to avoid acidosis, which can be a problem just as it is with ruminants. If the animals learn to eat trace mineral salt mixes formulated for sheep found in the same geographic location, their mineral needs will probably be met. Body condition should be monitored, and obesity avoided because it interferes with reproduction and causes heat stress. Llamas are good at clearing brush because of their browsing habit. However, several authors warn that they can become *browse starved*. If they are allowed sudden, unlimited access, they may overeat and become ill.

## BEHAVIOR

Llamas are social creatures and do best when kept with others of their species. They are often described as shy, independent, gentle, and curious (Figure 24–17). They are calm by nature and are considered easy to handle and train (Figure 24–18). However, they are naturally territorial, a trait no doubt inherited from their wild guanaco ancestors whose herd survival depended on balancing the number of animals to the carrying capacity of a territory.

Llamas use ear and tail position, various body postures, and both a humming sound and a shrill alarm call to communicate (Figure 24–19). Invasion of territory is signaled by a combination of several of these communication mechanisms. Generally, females are able to resolve disputes without physical contact. Males often fight using butts, kicks, ramming of the body, and a good deal of noise. The fight is won when one combatant is put on the ground in the breeding position, which signals submissiveness.

Males have fighting teeth—two uppers and one lower on each side, which should be removed at 3 years of age and again later if they grow out. Although llamas rarely bite humans, precaution is warranted. Llamas are prone to biting each other.

Spitting and body charging are normal herd dominance behavior patterns in males. Spitting is used by a female to signal her lack of interest to an unwanted

**FIGURE 24–17** Llamas are generally curious creatures. This gentle young animal went from person to person, apparently smelling everyone's breath.

**FIGURE 24–18** Llamas are generally easy to train and handle. The young llama here seems intent on determining what his young handler wants of him. This was only the second time the animal had been haltered.

suitor. It is also used to discourage a threat from another llama and to set the pecking order for meals. If directed at a human, spitting probably signals that the llama has been excessively or poorly handled by a human. This behavior is most evident in bottle-fed males, who tend to develop unacceptable behavior as adults, referred to as "berserk male syndrome." These animals are dangerous and should be destroyed.

Restraint and handling techniques are described in detail by Fowler (1989b) and McGee (1994). Those interested should consult these references.

Lamoids are communal dung-pile users. This behavior can be considered an advantage if the dung pile is in a good place and a disadvantage if one is started in a barn or near a water supply (Figure 24–20).

**FIGURE 24-19**    Llamas use vocalizations, body postures, and both ear and tail position to communicate. In this photo, the female is signaling alarm and potential aggression with her ear and tail positions and general body posture.

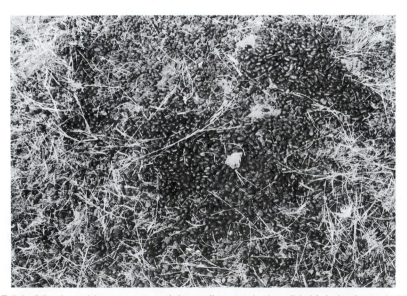

**FIGURE 24-20**    Lamoids are communal dung-pile users. In the wild, this behavior probably serves to mark the territory of a family group and perhaps help control diseases.

# TRENDS AND CHALLENGES TO THE LLAMA INDUSTRY

## RESEARCH NEEDS

Much research needs to be done in many areas if owners are to increase their ability to care for lamoids. Finding ways and means of supporting this research is a challenge to the entire llama and alpaca industry. Identification of genetic markers to assist breeders in identifying and dealing with genetic defects would be very beneficial. Other information is needed on nutrition and health—especially approved pharmaceuticals.

## EDUCATIONAL NEEDS

Owner education is important in all species, and the lamoids are no exception. Many llama and alpaca owners lack experience with large animals. The unique attributes of these animals make it especially important that owners be educated. Unfortunately, the standard livestock education routes—colleges and universities, state specialists, extension services, numerous authoritative books and publications, experienced and knowledgeable feed salespeople—are sorely lacking for the lamoid enthusiast. It is a challenge for every owner, and even for the industry itself, to develop a sound continuing education program. Also, llama owners should be educated about resources to help them with pasture management, poisonous plants, pasture rotation, economic fencing, and so on. The local extension agent has a wealth of information for the llama owner, even though he or she may know little about llamas.

## HEALTH CARE

Every llama owner should find a good veterinarian who either knows something about lamoids or is willing to learn. One can't afford to wait until an animal is critically ill to find medical help.

## STANDARD OF PERFECTION

A logical and commonly agreed on standard of perfection for the llama and alpaca is essential to provide guidance to both novice and experienced owners. Because of the great differences in what owners want in their animals, this may necessitate the creation of multiple standards—one for working animals and one for companion/exhibition animals. This could lead to formal type or breed designations, which would add diversity to the industry and potentially benefit all involved.

# LLAMA ORGANIZATIONS

The following list of organizations is provided to assist those who wish to learn more about the lamoids.

**Alpaca Llama Show Association**
607 California Ave.
Pittsburgh, PA 15202
Phone: (412)761-0211
Fax: (412)761-0212
alsa@nauticom.net

**Alpaca Owners and Breeders Association**
5000 Linbar Dr., Suite 297
Nashville, TN 37211
Phone: (615)834-4195
Fax: (615)834-4196
Email: info@aobamail.com
http://www.alpacainfo.com/

**International Llama Registry**
P.O. Box 8
Kalispell, MT 59903
Phone: (406)755-3438
Fax: (406)755-3439
E-mail: ilr@alpacaregistry.com
http://www.lamaregistry.com

**Rocky Mountain Llama and Alpaca Association**
11818 West 52nd Ave.
Wheat Ridge, CO 80033
Phone: (303)422-4681
http://www.rmla.com

**The Alpaca Registry, Inc.**
4665 Innovation Dr., Suite 160
Lincoln, NE 68521
Phone: (402)437-8484
Fax: (402)437-8488
E-mail:ari@alpaceregistry.net
www:alpacaregistry.net

## SUMMARY AND CONCLUSION

Although lamoids are certainly considered a nontraditional agricultural animal in the United States, they are nevertheless becoming more and more common. This suggests that mainstream agriculturalists, veterinarians, and hobbyists could benefit from having more knowledge about them. Whether one is a backwoods enthusiast, a hobby farmer, a

### Facts about Llamas

| | |
|---|---|
| Birth weight: | 18–40 pounds |
| Mature weight: | 240–500 pounds |
| Weaning age: | 5–8 months |
| Breeding age: | 15–18 months (female); 2½–3 years (male) |
| Normal season of birth: | Year-round, spring and fall seasons preferred |
| Gestation: | 350 days with $+/-$ 14 days normal |
| Estruous cycle: | Induced ovulation rather than heat cycle |
| Birthing interval (months): | 12 desirable |
| Names of various sex classes: | Male or stud for intact males and gelding for castrated males. Female. Baby, or cria. |
| Types of digestive system: | Herbivores with two fermentation compartments prior to the true stomach. Pseudo-ruminant. |

companion animal owner, or someone who makes a living with these animals, it is imperative to understand their physiology, their habits, and the industry that surrounds them.

## STUDY QUESTIONS

1. List the genera that fall under the family Camelidae. List the species within each genus.

2. What are the two most economically important camelids in the United States?

3. Which camelid produces the highest quality wool? How does the Bradford scale work?

4. Where did llamas and alpacas originate?

5. What are the estimated populations of llamas and alpacas in the United States?

6. List the most common uses of llamas.

7. What type of digestive system do the Camelidae have? How do they compare to the more popular domestic species that have the same type of digestive system?

8. At what age do llamas reach sexual maturity? Describe the size of a mature llama.

9. How does the fiber produced from llama wool differ from that of domestic sheep?

10. Other than "baby," what are newborn llamas called?

11. List five measures that can be taken to prevent heavily wooled animals from suffering from heat stress.

12. List six possible diseases of the llama. What are its most common internal and external parasites?

13. What is a more important determinant than age for when a female should be bred?

14. How long are the average gestation lengths of llamas and alpacas?

15. During what time of the day does parturition most often occur? What is the reason for this?

16. As a percentage of body weight, how much colostrum should a newborn cria consume within the first 12 hours of life?

17. What are some common methods of providing colostrum to the newborn if it has problems receiving it from the mother?

18. At what age do crias have a functional rumen?

19. Why is bloat a rare occurrence given the type of food mass found in the forestomachs?

20. What level of forage crude protein is usually sufficient to meet maintenance and growth requirements for a mature llama?

21. List the possible reasons that contribute to the increase in efficiency of lamoid feed utilization as compared to other ruminants.

22. In terms of consumption, mature llamas on average will eat what percent of their body weight per day? To what percentage should this level be limited to prevent overconditioning?

23. What are some examples of production functions that can aid in determining the level of nutrition?

24. List the two most common dominance behavior patterns in males.

25. Describe the "berserk male syndrome" and its probable cause.

## REFERENCES

Belknap, E. B. 1994. Medical problems of llamas. In *The veterinary clinics of North America, food animal practice,* L. W. Johnson, ed., Vol. 10, No. 2. Philadelphia: W. B. Saunders.

Blalock, C. 1993. Feeding and nutrition I. Llamas. *Herd Sire Edition,* June 1993, 116–123.

Bravo, P. W. 1994. Reproductive endocrinology of llamas and alpacas. In *The veterinary clinics of North America, food animal practice,* L. W. Johnson, ed., Vol. 10, No. 2. Philadelphia: W. B. Saunders.

Bravo, P. W., and L. W. Johnson. 1994. Reproductive physiology of the male camelid. In *The veterinary clinics of North America, food animal practice,* L. W. Johnson, ed., Vol. 10, No. 2. Philadelphia: W. B. Saunders.

*Compton's interactive encyclopedia.* 1995. Compton's New Media, Inc.

Cotton, J. 1996. Personal communication.

Ebel, S. 1989. The llama industry in the United States. In *The veterinary clinics of North America, food animal practice,* L. W. Johnson, ed., Vol. 5, No. 1. Philadelphia: W. B. Saunders.

Fowler, M. E. 1978. *Restraint and handling of wild and domestic animals.* Ames: Iowa State University Press.

Fowler, M. E. 1989a. *Medicine and surgery of South American camelids.* 3rd ed. Ames: Iowa State University Press.

Fowler, M. E. 1989b. Physical examination, restraint and handling. In *The veterinary clinics of North America,* L. W. Johnson, ed., Vol. 5, No. 1. Philadelphia: W. B. Saunders.

Fowler, M. E. 1992. Feeding llamas and alpacas. In *Current veterinary therapy XI (small animal),* R. W. Kirk and J. D. Bonagwa, eds. Philadelphia: W. B. Saunders.

Fowler, M. E. 1994. Hyperthermia in llamas and alpacas. In *The veterinary clinics of North America, food animal practice,* L. W. Johnson, ed., Vol. 10, No. 2. Philadelphia: W. B. Saunders.

Franklin, W. L. 1982a. Biology, ecology and relationship to man of South American camelids. In *Mammalian biology in South America,* M. A. Mares and H. H. Genoways, eds. Spec. Publ. 6, 457–489. Linesville, PA: Pymatuning Laboratory of Ecology, University of Pittsburgh.

Franklin, W. L. 1982b. Llama language. *Llama World* 1(2): 6–11.

Hintz, H. F., H. F. Schruyner, and M. Halbert. 1973. A note on the comparison of digestion by new world camelids, sheep and ponies. *Animal Production* 16: 303–305.

Johnson, L. W. 1987. Llama restraining chutes. *Llamas* 11(5): 30–32.

Johnson, L. W. 1989a. Llama reproduction. In *The veterinary clinics of North America, food animal practice,* L. W. Johnson, ed., Vol. 5, No. 1. Philadelphia: W. B. Saunders.

Johnson, L. W. 1989b. Nutrition. In *The veterinary clinics of North America, food animal practice,* L. W. Johnson, ed., Vol. 5, No. 1. Philadelphia: W. B. Saunders.

Johnson, L. W. 1994a. Llama herd health. In *The veterinary clinics of North America, food animal practice,* L. W. Johnson, ed., Vol. 10, No. 2. Philadelphia: W. B. Saunders.

Johnson, L. W. 1994b. Llama nutrition. In *The veterinary clinics of North America, food animal practice,* L. W. Johnson, ed., Vol. 10, No. 2. Philadelphia: W. B. Saunders.

Mason, I. L. 1969. *A world dictionary of livestock breeds, types, and varieties.* Bucks, UK: Commonwealth Agricultural Bureaux, Farnham Royal.

McGee, M. 1994. Llama handling and training. In *The veterinary clinics of North America, food animal practice,* L. W. Johnson, ed., Vol. 10, No. 5. Philadelphia: W. B. Saunders.

Paul-Murphy, J. 1989. Obstetrics, neonatal care, and congenital conditions. In *The veterinary clinics of North America, food animal practice,* L. W.

Johnson, ed., Vol. 5, No. 1. Philadelphia: W. B. Saunders.

San Martin, F., and F. C. Bryant. 1989. Nutrition of domesticated South American llamas and alpacas. *Small Ruminant Research* 2: 191.

Sell, R. 1993. Llama. In *Alternative agriculture service,* D. Aakre, series ed., No. 12. NDSU Extension Services, North Dakota State University of Agriculture and Applied Science, and U.S. Department of Agriculture.

Thedford, T. R. 1992. *Some general information for the potential llama owner*. OSU Extension Facts F-9122. Stillwater: Oklahoma Cooperative Extension Service, Division of Agricultural Sciences and Natural Resources, Oklahoma State University.

# Rabbits

## LEARNING OBJECTIVES

After you have studied this chapter, you should be able to:

1. Describe the place of rabbits in U.S. society and agriculture.

2. Explain the many uses for rabbits in the United States.

3. Give a brief history of the rabbit as a domestic animal.

4. Describe the rabbit industry segments.

5. Explain the value of breeds in the rabbit industry.

6. Discuss the basics of managing rabbits for reproductive efficiency.

7. Identify the major methods of feeding rabbits.

8. Explain the nutritional benefits of rabbits to humans.

## KEY TERMS

| | | |
|---|---|---|
| Angora | Doe | Kits |
| Animal Welfare Act | Fancy | Leporarium |
| Buck | Feed conversion ratio | Nest box |
| Cecotrophs | Fryer | Rabbitry |
| Cecotrophy | Kindling | |

## SCIENTIFIC CLASSIFICATION OF THE RABBIT

| | |
|---|---|
| Phylum: | Chordata |
| Subphylum: | Vertebrata |
| Class: | Mammalia |
| Order: | Lagomorpha |
| Family: | Leporidae |

Genus:　　　　*Oryctolagus* (rabbits), *Lepos* (hares), *Ochotona* (pikas), *Sylvilagus* (cottontails)

Species:　　　*cuniculus forma domestica* (domestic rabbit), *cuniculus* (wild rabbit)

## THE PLACE AND PURPOSE OF RABBITS IN THE UNITED STATES

The domestic rabbit is a descendant of the European wild rabbit. Compared to other animal industries in the United States, the rabbit industry is small. The industry associated with rabbits is difficult to measure because it is quite diverse and contains many sectors. The rabbit industry has meat-producing, pet, and hobbyist/fancy segments. In terms of the number of people involved, the pet segment is the largest. According to the American Veterinary Medical Association, 6.17 million rabbits were kept as pets in the United States in 2006, with an estimated 1.6% of all households having a domestic rabbit as a pet (Figure 25–1). The *fancy* (owners who exhibit) is hard to measure. Of the approximately 30,000 people who belong to the American Rabbit Breeders Association, the majority raise rabbits as a hobby or to exhibit at shows. Over 1,500 shows are sanctioned annually by the American Rabbit Breeders Association in the United States. It is common to have several hundred participants at a show.

Many people keep a few rabbits in a backyard operation to produce meat for the family. In addition, a viable commercial meat-producing segment of people is attempting to make a livelihood from raising meat rabbits. Although the commercial meat-producing segment probably has the fewest people of all segments, they own the largest number of rabbits per owner. Large commercial operations keep hundreds or thousands of rabbits for meat production. Combined, it is estimated that approximately 200,000 producers market 6 to 8 million rabbits for meat purposes annually in the United States.

**Fancy**

The industry segment associated with exhibitions and shows.

**FIGURE 25–1**　In terms of the numbers of people involved, the pet rabbit segment of this industry is the largest. (Photographer Todd Yarrington. Courtesy of Merrill Education.)

However, according to the Census of Agriculture, there are only approximately 4,300 rabbit farms producing $1,000 or more in product per year. These market less than 1 million meat rabbits per year (Figure 25–2).

Schools and universities use rabbits as teaching tools. Rabbits were used extensively in the early studies of coat color and Mendelian inheritance. They are still useful as models for teaching genetic principles. Rabbits are also raised for their skins (Figure 25–3), for their wool, and as 4-H and FFA projects. At one time, there was a viable market for the fiber produced from the Angora strains of rabbit. However, this market is currently restricted to a few hand spinners.

**FIGURE 25–2**   Rabbit producers who raise meat rabbits are smaller in number than the number of owners who raise rabbits for other uses, but they raise more rabbits per producer. (Photo courtesy of Corbis Digital Stock.)

**FIGURE 25–3**   The rabbit produces an excellent fur for the making of fine apparel. (Photo courtesy of Corbis Digital Stock.)

The rabbit is an important laboratory species. The rabbit has been a very useful animal in immunology studies, diagnostic research, eye research, pyrogen testing, fetal drug-induced teratology, parasite research, and for scores of other conditions. The rabbit has also been used extensively in the testing of human use products. The rabbit is a useful research model because it provides good repeatability of animal model studies; it is large enough for single samples; there are many stocks/strains available; it is easily managed; it provides high-quality immunologic products; and it is easy to control its reproduction (Figure 25–4).

There are also several alternative uses for rabbits in the United States. These alternatives are often tapped by the meat-producing segment as markets for its surplus meat-type rabbits. These uses include:

- "Feeder rabbits" for carnivorous pet reptiles and the larger pet snakes. The increase in the number of more exotic pets has created a need for a steady food source for them. Rabbits are more economical than rats as a food source for these pets.

- Feed for endangered carnivorous species. The Fish and Wildlife Departments of several states manage reservations for the breeding of such species as eagles, condors, alligators, and wolves. Their goal is to return the animals or their offspring to the wild. Domestic rabbits are often used as the food species of choice. In addition, most states have centers that rehabilitate injured birds of prey. Because these birds are returned to the wild if at all possible, the domestic rabbit provides a food similar to the food that the animal can find in the wild. The rabbit is also generally accepted by the rehabilitating bird because of its similarities to wild species.

- Breeding stock for new producers. New producers need a source of quality breeding stock. Well-established breeders can get higher prices for quality breeding stock than if they sell the same rabbits as fryers.

- Military training. Some military bases require large numbers of rabbits for use in training soldiers in survival techniques. Part of the soldiers' training includes finding and killing their own food. Although native, wild species would probably be preferred, such species are quickly depleted in the training camps because of the

**FIGURE 25–4** Rabbits are an important laboratory species. Schools and universities also make use of the rabbit as a teaching tool. (Photographer Mark Richards. Courtesy of PhotoEdit.)

large number of soldiers being trained. Thus domestic rabbits are purchased and released.

- Sport/recreation. Greyhound racing dogs are trained to race using live rabbits. So in areas where Greyhounds are trained, a supply of rabbits is needed by trainers.
- Craft uses. For the producer with a "craft" side, tanning pelts for craft items and making items such as rabbit foot key chains can offer an additional income.

Rabbits are an important source of meat worldwide with annual production of approximately 1 million MT of meat. China is the world's largest producer with 39% of all production followed by Italy, Spain, France, Egypt, and the Czech Republic. The United States slaughters only approximately 1% of the world's total. Worldwide, rabbit numbers are increasing approximately 3.8% annually.

## HISTORICAL PERSPECTIVE

**Leporarium**

Specially designed enclosure for keeping rabbits that was popular in Roman times. Both rabbits and hares were kept as a ready food supply.

Fossil remains suggest that both hares and rabbits developed in the Western Hemisphere and then migrated to Asia and Europe. North and South America still have the greatest number of species compared to the rest of the continents. The exact dates of domestication are unknown. Most sources credit Spain as being the place of origin of the domestic rabbit. Some sources suggest that rabbits were domesticated as early as 600 B.C. The fact that Phoenicians traded the rabbit to all regions of the world as it was then known to be is evidence to support this early date of domestication. Romans also kept hares and then rabbits brought from Spain in the Roman *leporarium*, from which they were easily harvested for kitchen use, and thus spread the rabbit to many lands. However, neither of these facts necessarily means that rabbits were domesticated. Rather, the domestic breeds of rabbits we know today have their roots in the French monasteries where, from the early Middle Ages, rabbits were kept in hutches and raised as a food source. These species were *Oryctolagus cuniculus*, the progenitor of the modern domestic rabbit. Several breeds were known by the 16th century. In the early 17th century, "types" were described. Some present-day breeds appear little changed from those developed in the 18th and 19th centuries.

Rabbits were probably brought to the United States before the early 1900s; however, evidence to support this estimate is scant. Rabbit breeding has never had the widespread appeal in the United States that it had, and continues to have, in Europe. However, there is a thriving, if small, rabbit industry in the United States.

## GEOGRAPHIC LOCATION OF THE RABBIT INDUSTRY IN THE UNITED STATES

Some meat-rabbit production exists in every state. However, the number of meat rabbits produced commercially is declining. According to the 2002 Census of Agriculture, Pennsylvania, Oregon, Arkansas, and California were the only states producing 75,000 or more per year. This is a small industry. Pet rabbits are widely dispersed across the country with the human population, with the more heavily populated states having the most pet rabbits. The rabbit fancy has a healthy contingent in every state in the union.

# THE STRUCTURE OF THE RABBIT INDUSTRY

## THE RABBIT FANCY

The ease of rearing, small space requirements, low initial investment required, and great variety of rabbits available in different sizes, coat types, and colors made the rabbit a natural animal around which a fancy could develop. Today an enthusiastic group of rabbit breeders and owners produce rabbits for exhibition. The animals are exhibited according to a set of rules and regulations at a large number of shows around the country. The rabbit fancy is largely centered on the activities of the American Rabbit Breeders Association (ARBA) and the activities of specific breed organizations. Youth activities are supported, and many 4-H and FFA groups encourage rabbit clubs.

## COMMERCIAL MEAT PRODUCTION

Many producers have part-time operations consisting of 20 to 100 does. For an operation to be a full-time enterprise, it should probably have at least 600 does. Obviously, there are many operations that are both larger and smaller. Does produce from 25 to 50 live offspring a year, which yield 125–250 lbs of meat. Restaurants, wholesalers, custom meat stores, and individual buyers are the predominant markets for rabbit meat in the United States. The meat-producing sector is highly dependent on the availability of slaughtering facilities, packaging requirements, transportation costs, and potential buyers. Typical slaughter weight for rabbits is 5 lbs, which they reach at about 10 weeks of age. Rabbits at this slaughter stage are typically referred to as *fryers.*

**Fryer**

A rabbit weighing approximately 5 lbs and no older than 12 weeks of age. Some processors stipulate no older than 10 or 11 weeks of age.

Rabbit meat sold in commercial establishments (food stores, restaurants, and so on) must be processed in accordance with local or state health codes. Requirements vary from state to state. Producers can check with the extension service and/or state meat inspection agency to determine the policies for their area. The USDA has set standards for the grading of rabbit carcasses. The program is voluntary and the costs of the grading program are paid by the user of the service. The grading program, the *Regulations Governing the Voluntary Grading of Poultry Products and Rabbit Products,* and the *United States Classes, Standards, and Grades for Rabbits* establish a basis for quality and price relationship and enable more orderly marketing. Effective December 4, 1995, the *United States Classes, Standards, and Grades for Rabbits* was removed from the Code of Federal Regulations (7 CFR Part 70) and is now maintained by the Agricultural Marketing Service, U.S. Department of Agriculture, as AMS 70.300 et seq. These standards can be accessed at http://www.ams.usda.gov/poultry/standards/AMSRABST.html.

## LABORATORY SPECIMEN PRODUCTION

Laboratory rabbit production is aimed at research laboratories, hospitals, and universities. Although the market is profitable for some, the requirements for entering the market make it difficult for individuals to become established. The specifications of the purchaser are generally stringent and often have little tolerance. In addition, the market is somewhat mercurial.

Producers for this market must be licensed under the provisions of the Animal Welfare Act (AWA). The AWA specifies that dealers selling animals to biomedical research must be licensed with APHIS, which conducts inspections to ensure compliance with the act. Violations of the provisions of the act can lead to license suspensions and/or civil penalties. The AWA requires that regulated individuals and businesses provide animals with care and treatment according to standards established by APHIS. The standards include requirements for record keeping, housing, sanitation, food, water, transportation, and veterinary care. The law regulates the care of animals that are sold as pets at the wholesale level, transported in commerce, used for biomedical research, and used for exhibition purposes.

### BREEDING STOCK PRODUCTION

Almost anyone with any size and kind of operation, a good reputation, and a knack and willingness for promotion can offer breeding stock to other breeders or new producers. The keys to success in the long term depend on how well the stock offered to the public performs in the *rabbitry* of the purchaser.

**Rabbitry**

A place where domestic rabbits are kept.

### ANGORA PRODUCTION

Angora rabbits produce 8 to 10 in., or 12 to 16 oz, of wool per year. Most owners who raise the Angora for its wool sell their product directly to individuals or organizations who buy for mills. This market is extremely small. Some producers spin their wool and market the yarn to the general public through craft organizations and publications or on the Internet. The fiber produced by the Angora rabbit breeds is a high-quality wool.

### EQUIPMENT AND SUPPLIES

Rabbit equipment and supplies are specialized enough to provide a market segment. Some manufacturers specialize in cages and accessories. Frequently, these businesses have regular distribution routes, provide equipment to wholesalers, and even have mail-order segments.

## BREEDS AND GENETICS

Many rabbit breeds are produced in the United States. Although the vast majority do not have registration certificates, it is possible to register rabbits with the ARBA. The procedure involves an inspection of the rabbit to be registered. Some of the major rabbit breeds are listed in Table 25–1. Pictures of selected rabbit breeds are shown in the color plate section of this text. The ARBA maintains a website at http://www.arba.net/ where additional pictures can be viewed. Rabbits are generally classified by weight or hair/fur type. The weight categories are small (2–4 lbs), medium (9–12 lbs, although several breeds are smaller than this range), and large (14–16 lbs, although some breeds are larger). The hair/fur classifications are normal, rex, satin, and wool. The medium-weight New Zealand White is considered the best meat-producing breed. Californians are considered the second best meat-producing breed. Many different breeds are used in laboratories. All breeds can likely be found as pets somewhere; however, the smaller varieties and those with the most interesting coat colors are favored. The Angora breed is the only breed used for wool production. The American Rabbit Breeders Association publishes a standard of perfection that gives a standard for the perfect rabbit in each breed. The standards are updated from time to time to reflect progress in the breeds.

The meat-type breeds of rabbits are Champagne d'Argent, Californian, Cinnamon, American Chinchilla, Creme d'Argent, French Lop, Hotot, New Zealand, Palomino, Rex, American Sable, Satin, Silver Fox, and Silver Martens. The most commonly used is the New Zealand White, which is considered the best meat-type breed because of

**TABLE 25-1**   **Selected Breeds of Rabbits**

| Breed | Size | Hair/Fur | Use | Mature Weight (lbs) |
|---|---|---|---|---|
| English Angora | Medium | Wool | Fur, fancy, pet, meat | 9–12 |
| American Chinchilla | Medium | Normal | Fur, pet, fancy | 9–12 |
| Californian | Medium | Normal | Meat, pet, fancy, laboratory | 8–11 |
| Champagne d'Argent | Medium | Normal | Meat, pet, fancy | 9–12 |
| Checkered Giants | Large | Normal | Pet, meat, fancy, fur | 11+ |
| Dutch | Small | Normal | Laboratory, pet | 3–6 |
| English Spot | Medium | Normal | Meat, laboratory | 9–13 |
| Flemish Giants | Large | Normal | Pet, fancy | 13+ |
| Himalayan | Small | Normal | Laboratory, pet, fancy | 2–6 |
| New Zealand | Medium | Normal | Meat, pet, fancy, laboratory | 9–12 |
| Polish | Small | Normal | Laboratory, pet, fancy | 3–4 |
| Rex | Medium | Rex | Fur, pet, fancy | 8–11 |
| Silver Martens | Medium | Normal | Fur, pet, fancy | 6–10 |
| Satin | Medium | Satin | Fur, pet, fancy | 9–10 |

both husbandry and processor preference. The Californian is second in popularity. These two breeds have the best size, growth rates, feed conversion ratios, dress-out weights, and meat-to-bone ratios. In addition, their albino characteristics give them a cleaner, lighter dressing carcass than colored rabbits have. The pelts from these two breeds are easier to remove in processing. The production goal for meat-producing operations is to produce 4- to 5-lb fryers at 8 weeks of age. Fryers are expected to gain an average of 0.5 lb per week with a ***feed conversion ratio*** of 4:1.

A sound breeding program for a commercial operation should include selection for the traits that most affect profitability. These include litters per doe per year, number of ***kits*** per litter, number of kits weaned, herd longevity, weaning weights, feed conversion, and dressing percentage.

The rabbit fancy and the pet segment both rely on the purebred animal. Thus breeding of these animals revolves around breed improvement and selection for both eye appeal and disposition. Sometimes crossbred animals are seen in the pet industry as well. Crossbreeding is a useful strategy for producing meat rabbits because of the heterosis it provides. The most popular cross is Californian bucks on New Zealand White does. The fryers from this cross, called "smuts," are highly regarded by processors.

## REPRODUCTIVE MANAGEMENT

The medium-weight breeds are mature and can be put into the breeding colony at 6 to 7 months of age. Given the same environmental conditions and breeding, ***bucks*** mature approximately 1 month later than does.

Outward signs of heat are not always evident in ***does,*** nor do does have a well-defined heat such as that found in the other livestock species. Their cycles revolve

**Feed conversion ratio**
Feed needed to produce 1 lb of gain.

**Kits**
Baby rabbits.

**Buck**
Intact male rabbit.

**Doe**
Female rabbit.

around the development and degeneration of follicles on the ovary, which occurs on a 16- to 18-day cycle. For all but 4 days of the cycle, a doe mates and is fertile. The most successful mating schemes breed on a schedule. Does that are not receptive on a given day are scheduled for rebreeding a few days later. For mating, the doe is put in the buck's cage. The buck should never be put in the doe's cage because she will fight to protect her territory. If the doe is ready, mating occurs immediately. Directly after mating, the doe should be returned to her cage to prevent fighting and possible injury. The stress of a fight can also cause the doe's temperature to rise, which is not good for the intended conception. The doe is an induced ovulator and ovulates approximately 10 hours after mating.

It is generally recommended that one buck service about 10 does. Recommendations vary on how frequently each buck should be used for mating. Certainly, using a buck two to three times a week is acceptable. Bucks may be used as frequently as once a day. Although some producers use bucks more frequently, at least for short periods, small litters are likely to result. Artificial insemination is an option in rabbits. Rabbit semen can be extended and frozen and remain viable. Although labor intensive, artificial insemination in rabbits offers the same potential for genetic progress as it does in other livestock species. Another advantage of AI is that bucks from anywhere in the world can be mated with does found anywhere else in the world.

Gestation is 31 to 32 days. On day 28 of gestation, a *nest box* (Figure 25–5) filled half full of nesting material should be placed in the doe's hutch. A doe normally pulls some of her own fur to line a nest for the kits. An average litter is 8–10 kits. *Kindling* (birthing) normally occurs in the early morning and requires approximately 30 minutes. Within 24 and no longer than 48 hours after kindling, the nest box should be examined and any dead kits removed. Litters can be equalized between does if the kits are approximately the same age. A desirable number is seven to eight kits for each doe. Excessively large litters should either be fostered or culled to no more than eight kits. Does rarely have more than eight nipples for feeding the young. Owners in commercial operations should record the number of live and dead kits. The information is useful in making culling and selection decisions.

The nest box is removed 15 to 21 days after birth or after all of the kits have left it. Some producers prefer to leave the nest boxes; however, the young generally soil the box. Removing it allows for better sanitation and offers an opportunity to clean the box and ready it for the next litter. The kits are weaned at 28 to 30 days of age in commercial operations. At this time, they are eating well. Producers in the fancy frequently prefer to wean at 6 to 8 weeks of age. In a commercial operation, it is imperative to wean early or the operation will not be profitable. Most profitable commercial operations rebreed on a more accelerated schedule, either 14 or 21 days after kindling. A 14-day rebreeding schedule allows more litters per year. Does who lose a litter at kindling can be rebred the same day, but it is generally recommended to wait 3 days. Those who lose a litter a few days after kindling can be rebred immediately.

Fall and winter breeding slumps occur frequently in rabbitrys. Use of artificial light to extend the light hours to at least 14, and up to 16 hours, is beneficial. Recommendations suggest a 40-watt bulb at 10-foot intervals set on a timer for consistency as adequate. Bucks are frequently the problem because they are very sensitive to heat. A buck exposed to 85° temperatures for 5 or more days can become sterile for as long as 3 months. Thus summer temperatures in much of the country can contribute to breeding problems in the fall and winter.

The reproductive tracts of the buck and doe are shown in Figures 25–6 and 25–7, respectively.

**Nest box**

A box provided for the doe in which to give birth and rear the young for their first few weeks.

**Kindling**

Parturition in rabbits.

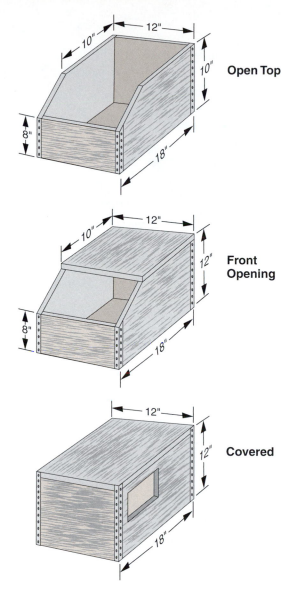

**Open Top**

**Front Opening**

**Covered**

**FIGURE 25–5** Nest boxes should provide enough room for the doe and litter but be small enough to keep the litter close together. Nest boxes can be constructed of nontreated wood, wire mesh, or sheet metal. Wood boxes should have metal reinforced edges. For winter use, the box should be enclosed with a small opening for the doe's entrance. Summer nest boxes can be open.

# NUTRITION

The principles of nutrition are discussed in Chapters 5, 6 and 7. Refer to those chapters for more detailed information than is presented here.

The rabbit is a cecal fermenter that practices *cecotrophy.* The cecotrophic animal consumes the contents of the cecum, usually directly from the rectum. Once a day, usually at night, the contents of the large intestine are emptied of fecal material, which are hard fecal pellets. The animal allows the pellets to drop to the ground or through the cage floor. The contents of the cecum are then emptied into the lower large intestine as *cecotrophs.* These cecotrophs are higher in water, electrolytes, and protein and lower in fiber. The cecotrophs are redigested in the upper portion of the digestive tract and then become the next hard fecal pellets in the cycle. As much as 20% of the protein and

## Cecotrophy

The act of consuming the material from the cecum in the form of cecotrophs.

## Cecotrophs

Large soft pellets of cecally fermented feed, which move through the colon in a distinctly different manner than feces and are reconsumed for upper digestive tract digestion.

**FIGURE 25–6**    Reproductive tract of the buck rabbit.

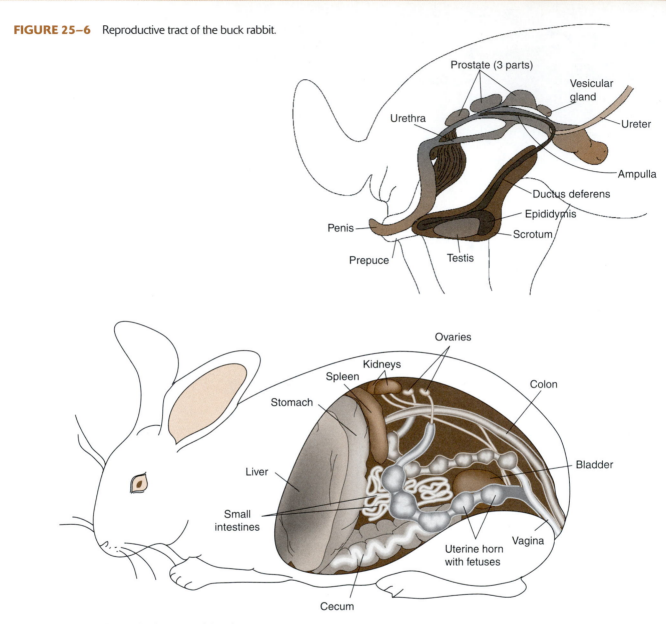

**FIGURE 25–7**    Reproductive tract of the doe.

10% of the energy needs of the animal may be met by the microbial protein and the energy of the cecotrophs. Animals deprived of the practice of cecotrophy frequently develop nutritional deficiencies.

For practical purposes, there are two ways to feed rabbits. The first and most convenient method is to feed a commercially prepared and balanced complete diet. These feeds are almost exclusively pelleted rations. They are easy to feed in automatic feeders that can be fitted on the outside of the cage for ease in servicing. For large numbers of rabbits, this type of feed is the overwhelming choice. The alternative method is to feed forage, usually hay and grain. Because rabbits are herbivores, they may also be fed fresh grass and legume forage, extra garden produce such as turnips and other root

**TABLE 25-2    Sample Diets for Rabbits**

| INGREDIENT | Maintenance Diet | Growth Diet | Growth Diet | Lactation Diet | Gestation Diet |
|---|---|---|---|---|---|
| | PERCENTAGE OF INGREDIENT IN FORMULATION | | | | |
| Alfalfa hay | 52.4 | 45.0 | 60.0 | 36.4 | 50.0 |
| Barley | 18.1 | 34.0 | 15.0 | — | — |
| Corn | — | 2.7 | 22.0 | 32.5 | — |
| Oats | 9.6 | — | — | — | 45.5 |
| Soybean meal | — | — | 3.0 | 16.3 | 4.0 |
| Sunflower meal | 3.5 | 11.1 | — | — | — |
| Wheat bran | 13.0 | 2.1 | — | 10.2 | — |
| Cane molasses | 1.5 | 3.0 | — | 1.6 | — |
| Dicalcium phosphate | 1.1 | 1.3 | — | 2.2 | — |
| NaCl | 0.5 | 0.5 | — | 0.5 | 0.5 |
| Vitamin, mineral, and amino acid supplement | 0.3 | 0.3 | — | 0.3 | — |

SOURCE: Compiled from Adams and Berry, 1987, p. 10; Kellems and Church, 1998, p. 461; and Overstreet, 1993, p. 96.

crops, and limited amounts of garden greens. Owners of small numbers of rabbits may find this an attractive and low-cost alternative. Cages can even be built that are portable and can be moved across fresh pasture, allowing rabbits to harvest their own forage.

If forage and grain rations are fed, they require individual balancing for the rabbits' needs. Commercial pelleted feeds can be purchased in one of three forms: all grain, all forage, and complete rations. Commercially available complete, pelleted rations are prebalanced to meet the rabbit's complete nutritional requirements. Pregnant and nursing does should be fed free choice. Bucks and nonpregnant, nonlactating does should be fed 6–8 oz of commercial pellets daily. It is recommended that Angora rabbits be fed complete, commercial diets exclusively because their wool gets contaminated with dust and chaff (debris) from hay, which lowers its quality. Feeding commercial pellets does not preclude the addition of small amounts of greens to the diet. Thus pet owners can use the commercial diets for their animal and still feed greens as a treat. Examples of some diets for rabbits are shown in Table 25–2.

Rabbit owners commonly underestimate the amount of water that rabbits require. Commercial operations generally rely on automatic watering systems that offer a continuous clean water supply. A doe and her litter may need as much as 1 gal of water a day in warm weather. Standard nipple water bottles hold only a fraction of that amount. Thus several bottles will need to be available on the cage to meet the animals' needs.

## HEALTH PROGRAM

The principles of rabbit health maintenance are much the same as those for other species. Chapter 14 provides a general guide to animal health. Refer to that chapter for more complete information than is presented here.

For pet rabbits, a good program of care by a veterinarian versed in rabbit diseases and health maintenance is the safest course of action. Some of the recommendations for health maintenance in commercial operations are not practical for the pet animal and animals that participate in the show circuit. Thus diseases may be more difficult to control for the pet and fancy animal.

Providing sufficient amounts of feed and water, maintaining stringent levels of cleanliness, providing good ventilation, practicing close observation, and providing protection from sun and precipitation go a long way in protecting the health of the rabbitry. Rabbits are susceptible to several diseases. *Pasturella multocida* is an organism that causes a respiratory disease that results in decreased productivity and a high mortality rate in does. When starting a commercial operation, owners should purchase *Pasturella*-free animals as foundation stock.

The principles of biosecurity can be applied to a rabbitry as a means of preventing disease. At a minimum, neither human nor animal visitors should be permitted inside the rabbitry. Visitors can introduce disease and cause stress to the rabbits. Sick animals should be isolated immediately after observation. If the sick animal is in a group cage within a larger housing arrangement, the rest of the group should be isolated as well. The cage that housed the sick animal should be disinfected. Once the diseased animal leaves the isolation, that cage should be disinfected as well. In operations of all sizes, nest boxes should be cleaned and sanitized after each use, and the accumulated hair should be burned from the cages. New breeding stock should be isolated from the general herd for a period of observation. Then, a few animals from the herd should be introduced to the new animals before the new animals are introduced to the general herd. If signs of disease develop in the animals from the established herd, then a knowledgeable decision can be made about adding the new animals to the general herd. It may be prudent to dispose of both the new and exposed rabbits rather than risking exposure in the general herd.

Good records are the cornerstone of good health maintenance. Periodic review of the records with an objective eye directed at cause and effect can help spot problems and suggest long-term solutions for prevention of disease. Ear tattoos with unique numbers provide a good means of permanent identification so that individual health records can be maintained. Individual identification cards with health and reproductive information should also be kept on each cage or hutch for ease of recording. Several computer programs have been developed to assist rabbit producers with record-keeping chores. If the rabbitry contains as many as 20 does, one of these programs is probably justifiable.

## NUTRITIONAL VALUE OF RABBIT MEAT TO HUMANS

Rabbit meat is high-quality meat that is often considered a delicacy. Rabbit meat is considered white meat. It is fine grained, delicately flavored, and adaptable to many styles of cooking—from stew pot, to the barbecue, to the heights of fine Italian and French cookery. (One of the author's favorites is a Vietnamese recipe.) It is considered high in protein and low in fat, cholesterol, sodium, and calories. A 3-oz serving of cooked rabbit meat provides the following proportion of the recommended daily dietary allowance for a 25- to 50-year-old woman on a 2,200-calorie diet:

| | |
|---|---|
| Protein | 56% |
| Phosphorus | 28% |
| Iron | 13% |
| Zinc | 16% |
| Riboflavin | 14% |
| Thiamin | 6% |

| B$_{12}$ | 352% |
| Niacin | 48% |
| Calories | 8% |
| Fat | 9% |

Rabbit meat is very high in nutritive value with high nutrient density. Rabbit meat contains all essential amino acids needed by humans, many needed minerals, and all required vitamins except vitamin C. Although not a traditional food in many households, rabbit meat is healthful food that can be part of the diet of virtually all people. A complete nutrient analysis is given in Appendix IX.

# TRENDS AND FACTORS IN RABBIT PRODUCTION

## ADVANTAGES OF COMMERCIAL RABBIT PRODUCTION

Commercial rabbit production offers the advantage of a supplemental income that can be earned in the owner's spare time. It has a small land requirement (a huge rabbitry can be placed on less than 1 acre of land). The labor is sustained but less physically demanding than that required in many other agricultural enterprises. By comparison, the investment in facilities is less than that in most other agricultural enterprises. In addition, another source of income can be gained from producing and selling earthworms in the rabbit manure.

## DISADVANTAGES OF COMMERCIAL RABBIT PRODUCTION

Although the investment is less than that for most agricultural enterprises, there is nevertheless an initial investment that amounts to $70 to $120 per doe unit, depending on such factors as availability of a building and the need for cages, equipment, and breeding stock. If a building is available, initial costs can generally be halved.

As an agricultural enterprise, the rabbit industry must be considered high risk. The industry structure is simply not adequately developed to provide consistent and stable markets and support to the segment. The ready availability of multiple markets for livestock is one of the strongest features of the U.S. livestock industry. A similar market structure does not exist for the rabbit industry. Individual producers must take responsibility for developing and maintaining markets. The prospects for developing those markets seem slim. In the 5 years between the 1992 and 1997 agricultural census reports, the number of rabbits sold commercially in the United States declined by 40%. Between 1997 and the 2002 census, another 25% decrease occurred. In addition, ample supplies of meat can be acquired on the international market and that does not bode well for a domestic industry.

## INCOME

Regional differences exist; however, most estimates place net annual income at about $18 to $24 per doe. A 1,000-doe operation, considered a full-time operation for two people, would then net a maximum $24,000 income. Income estimates for some parts of the country are less than half that number. Clearly this is not a get-rich proposition. Rather, in much of the country, rabbit production simply does not support a reasonable full-time income for a producer.

## MANAGEMENT

Raising rabbits has a reputation of being easy. This is probably because of the fabled reproductive rates of the animal. However, rabbits are not overly easy to raise. They are like any other living creature—they require the application of good management and husbandry skills. New producers are invariably plagued with such a variety of problems that maximum production is seldom achieved for the first few years. Raising rabbits is also confining and labor intensive. During the early years of an operation, the cost of hired help can quickly deplete any profit. Attention to detail, which includes meticulous record keeping, is essential to success. Not all people are suited to these rigors.

## ORGANIZATIONS AND SOURCES OF INFORMATION

**American Rabbit Breeders Association, Inc.**
P.O. Box 5667
Bloomington, IL 61702
Phone: (309)664-7500
Fax: (309)664-0941
E-mail: info@arba.net
http://www.arba.net

This organization is devoted to the rabbit fancy and to commercial production.

**Maryland Small Ruminant Page**
www.sheepandgoat.com/lvstk.html

**Mississippi State University Extension Service – Commercial Rabbit Production**
http://www.msstate.edu/dept/poultry/rabbits.htm

**Professional Rabbit Meat Association**
http://www.prma.org/

**Small Farms Program**
135c Plant Science Building, Cornell University
Ithaca, NY 14853
www.smallfarms.cornell.edu

## SUMMARY AND CONCLUSION

Modern domestic rabbits are descended from the European wild rabbit. The rabbit industry in the United States is a small, but quite diverse, industry. The predominant uses for rabbits in the United States are as pets, for laboratory use, and as a hobby. Many people who keep rabbits as a hobby are fanciers who participate in some part of the exhibition industry. Rabbit breeding has been evident in the United States since roughly the beginning of the 20th century but has never been practiced here to the degree that it has in Europe.

The rabbit industry is composed of the fancy, commercial meat production, laboratory specimen production, breeding stock production, fiber production, and equipment and supplies segments. Many people participate in more than one segment of the industry. More than 100 breeds of rabbits can be found worldwide, but less than half that number are found in any numbers in the United States. New breeds are being developed. Rabbit breeds are generally divided into weight categories and fur/hair categories.

Reproductive management in rabbits is not as simple as the reputation of the rabbit implies. Commercial units must work at effective reproductive management programs to remain cost effective. The rabbit practices cecotrophy, in which cecotrophs are consumed from the anus and redigested. Rabbits deprived of cecotrophs generally develop nutritional deficiencies. Most rabbits are fed complete commercial diets. However, information on grain and forage diets is readily available.

The principles of rabbit health maintenance are much the same as those for other species. Providing sufficient amounts of feed and water, maintaining stringent levels of cleanliness, providing good ventilation, practicing close observation, and providing protection from weather will do much to protect the health of rabbits.

Rabbit meat has high nutrient density, contains all essential amino acids, many minerals, and all required vitamins except vitamin C.

Commercial rabbit production offers the advantage of a supplemental income that can be earned in the owner's spare time, but must be considered high risk. The industry structure is not adequately developed to provide consistent and stable markets. With a net annual income of $18 to $24 per doe, rabbit production simply will not support a reasonable full-time income for a producer in many parts of the country. Raising rabbits has a reputation of being easy, but rabbits are like any other animal—they require the application of good management and husbandry skills. Attention to detail, good record keeping, and hard work are required. However, a single or small number of rabbits can provide an enjoyable pet or hobby.

## Facts about Rabbits

| | |
|---|---|
| Birth weight: | Varies with breed and sex, 1.75–2.3 oz |
| Mature weight: | Varies with breed, sex, and condition. The weight categories are small (2–4 lbs), medium (9–12 lbs, although several are smaller than this range), and large (14–16 lbs, with some being larger). |
| Hair/fur classifications: | Normal, rex, satin, and wool |
| Slaughter weight: | 4–5 lbs |
| Weaning age: | 4–8 weeks depending on goals of the breeding program |
| Breeding age: | 6–7 months, bucks within a breed mature approximately one month later than does |
| Normal season of birth: | Year round; may require artificial lighting for year-round breeding |
| Gestation: | 31–32 days |
| Estrous cycle and duration of estrus: | Induced ovulation, breeds all but 3–4 days in a 16–18 day period |
| Kindling interval: | As little as 31–32 days; can rebreed on day of kindling; practically, 45–52 days desirable |
| Normal mating success: | 70–90% of matings should result in a litter; seven to eight kits are the preferred litter size |
| Names of various sex classes: | Buck, doe, kits, fryers, capons |
| Weight at weaning: | At 4 weeks—1.25 lbs, 4 lbs at 8 weeks |
| Type of digestive system: | Monogastric, cecal fermentation, practices cecotrophy |

## STUDY QUESTIONS

1. Describe the place of the rabbit in the United States.

2. How do the uses of the rabbit compare with the uses of the other species discussed in this text?

3. What are the primary and alternative uses for rabbits in the United States?

4. Give a brief accounting of the history of the domestic rabbit.

5. Describe the individual segments of the rabbit industry.

6. What are the classification categories used to distinguish among rabbit breeds?

7. In outline form, describe the basics of reproductive management in the rabbit.

8. What are the general steps necessary to get a doe's cage ready for her to kindle?

9. Describe nest box management after the kits are born.

10. Pretend you are a county agent who has been asked to give a talk to a group of grade school students who are starting a rabbit project in 4-H. What would you tell them about nutrition? What would you tell their parents?

11. What would you tell a group of potential rabbit producers about the general need for a biosecurity program if they expect to have a serious rabbit breeding program?

12. In each of the chapters on food-producing species covered in this text, a section has discussed the nutritive value of the food product to humans. Create a table comparing and contrasting the information on rabbit meat to at least two other food products.

## REFERENCES

Adams, A. W., and J. G. Berry. 1987. *Raising rabbits.* Circular E-8222. Stillwater: Cooperative Extension Service, Division of Agriculture, Oklahoma State University.

Clutton–Brock, J. 1999. *A natural history of domestic animals.* Cambridge, UK: Cambridge University Press.

FAO, 2008. *FAOSTAT statistics database.* Agricultural Production and Production Indices Data. http://apps.fao.org/.

Gomez, E. A., M. Baselga, O. Rafel, and J. Ramon. 1998. Comparison of carcass characteristics in five strains of meat rabbit selected on different traits. *Livestock Production Science* 55:53–54.

Gonzalez-Mariscal, G., J. I. McNitt, and S. D. Lukefahr. 2007. Maternal care of rabbits in the lab and on the farm: Endocrine regulation of behavior and productivity. *Hormones and Behavior* 52:86–91.

Kellems, R. O., and D. C. Church. 1998. *Livestock feeds and feeding.* 4th ed. Upper Saddle River, NJ: Prentice Hall.

Lebas, F., P. Coudert, H. de Rochambeau, and R. G. Thebault. 1997. *The rabbit: Husbandry, health and production.* 2nd ed. FAO Animal Production and Health Series. No. 21. Food and Agriculture Organization of the United Nations.

Lebas, F., P. Coudert, R. Rouvier, and H. de Rochambeau. 1986. *The rabbit: Husbandry, health and production.* FAO Animal Production and Health Series, No. 21. Rome: Food and Agriculture Organization of the United Nations.

Lukefahr, S. D. 2002. Opportunities for rabbit research and human development in the Western Hemisphere: A rabbit revolution? *World Rabbit Science.* (France) 10(3):111–115.

Lukefahr, S. D., P. R. Cheeke, J. I. McNitt, and N. M. Patton. 2004. Limitations of intensive meat rabbit production in North America. *Canadian Journal of Animal Science*. 84:349–360.

McCrosky, R. 2000. *Raising rabbits in the Pacific Northwest.* BC Canada: Canadian Center for Rabbit Production Development. Accessed online July 2004. http://pan-am.uniserv.com.

McNitt, J. I., N. M. Patton, S. D. Lukefar, and P. R. Cheeke. 2000. *Rabbit production.* Upper Saddle River, NJ: Prentice Hall.

Morrow, M., G. L. Greaser, G. M. Perry, J. K. Harper, and C. C. Engle. 1998. *Agricultural alternatives: Rabbit production.* Small and part-time farming project at Penn State with support from the U.S. Department of Agriculture-Extension Service. Document Number: 28503258.

*Official guidebook to raising better rabbits and cavies.* Bloomington, IL: American Rabbit Breeders Association, Inc.

Overstreet, N. 1993. *Rabbit production handbook.* College Station, TX: Instructional Materials Service, Texas A&M University.

Payne, W. J. A., and R. T. Wilson. 1999. *An introduction to animal husbandry in the tropics.* 5th ed. London: Blackwell Science.

Sherman, D. M. 2002. *Tending animals in the global village.* Philadelphia: Lippincott Williams & Wilkins.

Smith, T. W. 1997. *Starting a rabbit enterprise.* Poultry Science Home Page, College of Agriculture & Life Sciences, Mississippi State University, Mississippi State, MS. http://www.msstate.edu/dept/poultry/rabenter.htm. E-mail: tsmith@poultry.msstate.edu.

*Standard of perfection.* 1996. Bloomington, IL: American Rabbit Breeders Association, Inc.

Taboada, E., J. Mendez, G. G. Mateos, and J. C. De Blas. 1994. The response of highly productive rabbits to dietary lysine content. *Livestock Production Science* 40: 329–337.

Thompson, H. V., and C. M. King. 1994. *The European rabbit: The history and biology of a successful colonizer.* Oxford, UK: Oxford University Press.

Wright, M. A. 2005. *Raising meat rabbits.* Cornell University: Cornell Small Farms Program & Department of Animal Science Livestock Fact Sheets.

# PART FOUR

# Animals and Society

# Careers and Career Preparation in the Animal Sciences

## LEARNING OBJECTIVES

After you have studied this chapter, you should be able to:

**1.** Describe the general job market in agriculture and animal science.

**2.** Identity the general areas of curriculum in animal science and the careers associated with each area.

**3.** Develop a strategy for directing your education toward a satisfying career.

## KEY TERMS

| | | |
|---|---|---|
| Agribusiness | Husbandry | Scientific literacy |
| Careers | | |

## INTRODUCTION

Agriculture in general, and animal science specifically, offer a wide variety of challenging and rewarding career choices. The demand for college graduates with agriculturally related college educations has been strong and is expected to stay that way for the foreseeable future. One of the reasons for the opportunities is that many jobs in great demand in agriculture would not have been recognized as "agricultural" in the recent past. The diversification of agriculture and its changing structure have brought many more opportunities. As in any industry, the demand for specific occupations waxes and wanes. For instance, fewer production specialists are needed than are marketing, merchandising, and sales representatives. However, production specialists will still be needed, as will qualified people in many other areas. This is good news for students with a focus on agriculture, food, and natural resources. Agriculture was hard-pressed to find enough good people to fill its positions during most of the 1990s, and the same trend is anticipated through the first decade of the new century and beyond. For job seekers, this has meant and continues to mean more and better opportunities. The agricultural industries need the requisite human resources to grow, advance, and flourish. Many see the lack of college graduates as the most limiting factor in agriculture's growth. A government report summed it up this way: "The strategic importance

**TABLE 26-1    Employment Opportunities for College Graduates in the U.S. Food, Agricultural, and Natural Resources System, 2005–2010**

| Major Occupation Clusters in Food, Agriculture, and Natural Resources | Percentage of Total Employment Opportunities for Graduates | Annual Employment Opportunities | Agricultural and Natural Resources Graduates | Allied Fields Graduates |
|---|---|---|---|---|
| Management and business | 46% | 24,125 | 12,198 | 7,990 |
| Scientific and engineering | 25% | 12,916 | 7,423 | 5,298 |

of our food, agricultural, and natural resource system will grow . . . and . . . require even stronger leaders, more creative scientists, greater international business understanding, and increased sensitivity for consumers and the environment. So, it's largely human resources which will chart the course of the U.S. food, agricultural, and natural resources" (Coulter et al., 1990). These words are just as true today as when written.

The average yearly employment opportunities for agricultural and foods graduates for the entire decade of the 1990s and early in the new century exceeded the number of

| Total Annual Qualified Graduates | Total Qualified Graduates Shortage/ Surplus | Types of Opportunities | Strong Employment Opportunities Expected |
|---|---|---|---|
| 20,188 | 16.3% shortage | Accountant, account executive, advertising manager, appraiser, auditor, banker, business manager, commodity broker, consumer information manager, consultant, contract manager, credit analyst, customer service manager, economist, export sales manager, financial analyst, financial manager, food broker, food service manager, forest products merchandiser, golf course superintendent, grain merchandiser, government program manager, grants manager, human resources manager, insurance agency manager, insurance agent, insurance risk manager, landscape contractor, landscape manager, market analyst, marketing manager, policy analyst, purchasing manager, real estate broker, research and development manager, retail manager, risk manager, sales representative, technical service representative, wholesale manager | Technical sales representatives, food brokers, accountants and financial managers, forest products salespersons, market analysts, fruit and vegetable marketing representatives, sales managers, landscape managers, small animal health care product distributors, and international business specialists. |
| 12,721 | 1.5% shortage | Agricultural engineer, animal physiologist, animal scientist, biochemist, cell biologist, entomologist, environmental scientist, fisheries scientist, food engineer, food scientist, forest scientist, geneticist, landscape architect, microbiologist, molecular biologist, nanotechnologist, natural resources scientist, nutritionist, plant breeder, quality assurance specialist, rangeland scientist, research technician, resource economist, soil scientist, statistician, toxicologist, veterinarian, waste management specialist, water quality specialist, weed scientist | Precision agriculture, functional genomics and bioinformatics, forest science, plant and animal breeding, biomaterials engineering, food quality assurance, nanotechnology, animal health and well-being, nutraceuticals development, and environmental science. |

*(continued)*

graduates. Table 26–1 also shows the projected employment opportunities for graduates. It is easy to see that significant opportunities exist in all the employment clusters. Table 26–2 gives projected employment opportunities and selected representative occupations for 2016 and compares them to actual 2006 employment numbers for specific occupations within the employment clusters shown. The projections are overwhelmingly positive. Many opportunities are available in agriculture.

**TABLE 26–1**    **Employment Opportunities for College Graduates in the U.S. Food, Agricultural, and Natural Resources System, 2005–2010—continued**

| Major Occupation Clusters in Food, Agriculture, and Natural Resources | Percentage of Total Employment Opportunities for Graduates | Annual Employment Opportunities | Agricultural and Natural Resources Graduates | Allied Fields Graduates |
|---|---|---|---|---|
| Education, communication, and government services | 13% | 6,967 | 6,323 | 2,915 |
| Agricultural and forestry production | 16% | 8,022 | 6,381 | 755 |
| Total/overall | 100% | 52,030 | 32,325 | 16,958 |

*SOURCE:* Goecker et al., 2005.

| Total Annual Qualified Graduates | Total Qualified Graduates Shortage/ Surplus | Types of Opportunities | Strong Employment Opportunities Expected |
| --- | --- | --- | --- |
| 9,238 | 32.61% surplus | Agricultural science and business teacher, agricultural science reporter, animal inspector, college teacher, computer software designer, computer systems analyst, conference manager, conservation officer, cooperative extension educator, dietician, editor, educational specialist, environmental impact analyst, farm and ranch advisor, farm service agency manager, food inspector, forest service administrator, high school teacher, illustrator, information specialist, information systems analyst, journalist, land-use planner, naturalist, outdoor recreation specialist, nutrition counselor, park manager, peace corps representative, personnel development specialist, plant inspector, public relations specialist, radio/television broadcaster, regulatory agent, training manager, youth program director | Plant and animal inspection, public health administration, biotechnology impact assessment, foods and nutrition services, outdoor recreation, food system security, consumer information technologies, environmental management, high school agricultural science and business teaching, and land-use planning occupations. |
| 7,136 | 11% surplus | Animal breeder, aquaculturist, equine operator, farm manager, farmer, feedlot manager, forest resources manager, fruit and vegetable grower, greenhouse manager, nursery operator, nursery products grower, rancher, seed producer, tree farmer, turf producer, viticulturist, wildlife manager | Producers of fruits and vegetables, growers of specialty crops that provide raw materials for medical and energy products, managers of specialized livestock operations, forest managers, growers of landscape plants and trees, managers of aquaculture operations, turf producers, equine operators, organic farmers, and providers of outdoor recreation |
| 49,283 | 5.28% shortage | | |

**TABLE 26−2** Occupations, Employment, and Projected Employment in Agricultural Areas

| Occupation | 2006 EMPLOYMENT | | PROJECTED 2016 EMPLOYMENT | | CHANGE, 2006−2016 | |
|---|---|---|---|---|---|---|
| | Number | Percentage Distribution | Number | Percentage Distribution | Number | Percentage |
| Total, all occupations | 150,620,175 | 100.00 | 166,220,300 | 100.00 | 15,600,125 | 10.4 |
| Agricultural and food science technicians | 25,804 | 0.02 | 27,516 | 0.02 | 1,712 | 6.6 |
| Agricultural engineers | 3,133 | 0.00 | 3,401 | 0.00 | 268 | 8.6 |
| Agricultural equipment operators | 59,451 | 0.04 | 56,482 | 0.03 | −2,969 | −5.0 |
| Agricultural inspectors | 16,176 | 0.01 | 15,998 | 0.01 | −178 | −1.1 |
| Agricultural workers, all other | 20,350 | 0.01 | 20,360 | 0.01 | 10 | 0.0 |
| Animal breeders | 10,505 | 0.01 | 10,967 | 0.01 | 462 | 4.4 |
| Animal control workers | 15,202 | 0.01 | 17,097 | 0.01 | 1,895 | 12.5 |
| Animal scientists | 5,371 | 0.00 | 5,899 | 0.00 | 529 | 9.8 |
| Animal trainers | 42,924 | 0.03 | 52,682 | 0.03 | 9,758 | 22.7 |
| Appraisers and assessors of real estate | 101,125 | 0.07 | 118,197 | 0.07 | 17,072 | 16.9 |
| Biochemists and biophysicists | 20,131 | 0.01 | 23,326 | 0.01 | 3,196 | 15.9 |
| Biological scientists, all other | 29,067 | 0.02 | 30,138 | 0.02 | 1,071 | 3.7 |
| Biological technicians | 78,690 | 0.05 | 91,288 | 0.05 | 12,598 | 16.0 |
| Biomedical engineers | 14,379 | 0.01 | 17,415 | 0.01 | 3,036 | 21.1 |
| Butchers and meat cutters | 131,352 | 0.09 | 133,852 | 0.08 | 2,501 | 1.9 |
| Chefs and head cooks | 114,784 | 0.08 | 123,521 | 0.07 | 8,737 | 7.6 |
| Combined food preparation and serving workers, including fast food | 2,502,891 | 1.66 | 2,954,811 | 1.78 | 451,919 | 18.1 |
| Conservation scientists | 19,777 | 0.01 | 20,830 | 0.01 | 1,053 | 5.3 |
| Dietitians and nutritionists | 57,126 | 0.04 | 62,038 | 0.04 | 4,913 | 8.6 |
| Economists | 14,792 | 0.01 | 15,900 | 0.01 | 1,107 | 7.5 |
| Environmental engineering technicians | 21,126 | 0.01 | 26,362 | 0.02 | 5,236 | 24.8 |
| Environmental engineers | 54,341 | 0.04 | 68,161 | 0.04 | 13,819 | 25.4 |
| Environmental scientists and specialists, including health | 83,267 | 0.06 | 104,142 | 0.06 | 20,874 | 25.1 |

| Occupation | 2006 EMPLOYMENT | | PROJECTED 2016 EMPLOYMENT | | CHANGE, 2006–2016 | |
|---|---|---|---|---|---|---|
| | Number | Percentage Distribution | Number | Percentage Distribution | Number | Percentage |
| Farm and home management advisors | 14,969 | 0.01 | 15,735 | 0.01 | 766 | 5.1 |
| Farm equipment mechanics | 30,672 | 0.02 | 31,087 | 0.02 | 415 | 1.4 |
| Farm, ranch, and other agricultural managers | 258,156 | 0.17 | 261,032 | 0.16 | 2,876 | 1.1 |
| Farmers and ranchers | 1,058,444 | 0.70 | 968,838 | 0.58 | −89,606 | −8.5 |
| Farmworkers and laborers, crop, nursery, and greenhouse | 603,083 | 0.40 | 582,746 | 0.35 | −20,337 | −3.4 |
| Farmworkers, farm and ranch animals | 106,843 | 0.07 | 109,772 | 0.07 | 2,929 | 2.7 |
| Fish and game wardens | 8,030 | 0.01 | 8,017 | 0.00 | −13 | −0.2 |
| Fishers and related fishing workers | 38,372 | 0.03 | 32,179 | 0.02 | −6,193 | −16.1 |
| Foresters | 13,188 | 0.01 | 13,868 | 0.01 | 679 | 5.2 |
| Landscape architects | 27,839 | 0.02 | 32,402 | 0.02 | 4,563 | 16.4 |
| Landscaping and groundskeeping workers | 1,220,054 | 0.81 | 1,441,326 | 0.87 | 221,272 | 18.1 |
| Meat, poultry, and fish cutters and trimmers | 144,228 | 0.10 | 159,947 | 0.10 | 15,719 | 10.9 |
| Microbiologists | 17,357 | 0.01 | 19,306 | 0.01 | 1,949 | 11.2 |
| Nonfarm animal caretakers | 156,624 | 0.10 | 185,414 | 0.11 | 28,789 | 18.4 |
| Slaughterers and meat packers | 122,074 | 0.08 | 137,614 | 0.08 | 15,539 | 12.7 |
| Soil and plant scientists | 15,790 | 0.01 | 17,110 | 0.01 | 1,320 | 8.4 |
| Tree trimmers and pruners | 40,560 | 0.03 | 45,071 | 0.03 | 4,510 | 11.1 |
| Veterinarians | 62,196 | 0.04 | 83,956 | 0.05 | 21,760 | 35.0 |
| Veterinary assistants and laboratory animal caretakers | 74,534 | 0.05 | 86,228 | 0.05 | 11,694 | 15.7 |
| Veterinary technologists and technicians | 71,178 | 0.05 | 100,373 | 0.06 | 29,195 | 41.0 |
| Vocational education teachers, middle school | 15,805 | 0.01 | 15,003 | 0.01 | −802 | −5.1 |
| Vocational education teachers, secondary school | 95,534 | 0.06 | 91,133 | 0.05 | −4,401 | −4.6 |
| Zoologists and wildlife biologists | 20,091 | 0.01 | 21,830 | 0.01 | 1,739 | 8.7 |

*SOURCE:* The information in this table was compiled using the National Industry-Occupation Employment Matrix. It can be found at http://data.bls.gov/oep/nioem/empiohm.jsp. The Bureau of Labor Statistics home page is http://stats.bs.gov.bls/home.htm.

# ANIMAL SCIENCE STUDIES AND CAREERS

Students vary in their approach to animal science and in what they want out of their education as it prepares them for a career. Some students, perhaps even the majority, are primarily interested in a specific animal species. For example, they may want to be in the horse industry and are willing to consider many different career paths as long as the careers involve horses. Others are more interested in a career track and less interested in a specific species. They may wish to be a production specialist but would be just as content running a swine farrowing unit, a beef feedlot, or a dairy. Others are more interested in a specific biologic function and want to be involved in the industry as a nutritionist or reproductive specialist or a management consultant, for example. There is plenty of room for all of these orientations. To accommodate the necessary learning that must take place to prepare students for their chosen path, colleges and universities design study options and curricula with all of the career choices in mind. As you might expect, there are several approaches to curricula design. However, if you look at them objectively, you will generally see that most college and university curricula in the animal sciences are designed around similar categories. These curriculum categories usually include a set of core animal science courses combined with all of the other course opportunities that institutions of higher learning provide to create a meaningful college learning opportunity. Brief descriptions of career options follow.

## PRODUCTION

The primary goals of those in production are to improve the quality of agricultural products and the efficiency of farms. We generally think of these occupations as those most directly involved with the raising of animals or those who work in a service position closely tied to production. People in this occupational cluster direct the activities of the world's largest and most productive agricultural sector. They produce the food and fiber for the consumers of this country and enough surplus to generate a large and positive trade balance in agricultural products and goods to the other countries of the world (Figure 26–1).

College curricula designed to prepare students for careers in production generally stress more ***husbandry*** practices than do the other approaches, but they also include supporting course work to help a student prepare for specific production positions. For instance, a future ranch manager most certainly needs to know good agronomic practice as it relates to producing good forage. A future poultry producer would find value in a course on environmental stress. All students need basic knowledge in business and accounting. Areas of employment for the production specialist include aquaculturist, farmer, rancher, farm or ranch manager, feedlot manager, and unit managers in intensive production systems, and careers in marketing, feed manufacturing, international opportunities, and many more. Related areas include breed association personnel, animal breeders, animal scientists, dairy scientists, poultry scientists, county agricultural agents, state extension specialists, feed and farm management advisors, and stud managers. A large number of good opportunities exist and will continue to exist in production positions and on farms and ranches throughout the country. However, the general trend to larger and fewer production units is causing a decrease in the number of available positions. This will increase competition for these positions. Another avenue for the production-minded student who also wishes to work with people is to combine an animal science production degree with a double major in education. Students who choose this path can teach in public schools, in technical programs, or in one of the

---

**Husbandry**

The combined animal care and management practices.

**FIGURE 26–1**  For many students, the goal of their education is to own or manage a farm or ranch. (Photo courtesy of Dr. J. Robert Kropp, Department of Animal Science, Oklahoma State University. Used with permission.)

many trade and technical schools found all over the United States. Master's degrees can further enhance a career in these areas.

## SCIENCE AND MEDICINE

Agriculture is scientifically based and scientifically dependent. Virtually all who work in agriculture are well served by having minimal *scientific literacy.* However, some specialties require advanced scientific competency and training. Science curricula are designed for those who wish to pursue further education after the BS degree in graduate and professional schools and also for those who may wish to fill the need for technicians and industry science representatives (Figure 26–2). Perhaps because of the rigor of science-based curricula, perhaps because so many other opportunities exist, or perhaps because additional years in school are usually needed, this general area of agriculture and animal science is barely producing enough graduates. This translates into good opportunities for students who choose this area. The foods areas offer an especially large number of opportunities for those with advanced degrees owing to the changing eating habits of our population and the pressing needs of making the food supply safer. Good demand is also seen for other animal science specialties including molecular geneticists, veterinarians, toxicologists, pathologists, physiologists, nutritionists, microbiologists, animal behaviorists, and waste management specialists to become professors, industry and university scientists, technical consultants, research technicians, and so on (Figure 26–3). All of these "biotechnology" areas are in need of quality people.

The work of animal and food scientists helps to improve agricultural productivity and efficiency by increasing quantity and quality of products, reducing labor, controlling diseases and parasites, protecting natural resources, and converting raw products into safe, convenient, appealing, and healthy food products for consumers. Those in the science area use biology, chemistry, physics, mathematics, and other sciences to meet challenges in animal science, with foods, and in the companion animal market. Obviously, curricula designed to prepare students in these areas are laced with chemistry,

**Scientific literacy**

The minimum knowledge necessary to stay abreast of further scientific development and innovation.

**FIGURE 26–2**    Science curricula are for those who wish to pursue advanced degrees and pursue research with universities and industry and for those who wish to be technicians and industry science representatives. Pictured here, animal scientists examine a cross-sectional magnetic resonance image (MRI) from the abdominal area of a pig. MRI technology is one of many tools used in modern animal research. (Photographer Scott Bauer. Courtesy of Agricultural Research Service, USDA.)

**FIGURE 26–3**    The area of biomedical research needs qualified researchers and technicians. (Photographer Hisham F. Ibrahim. Courtesy of Getty Images, Inc./PhotoDisc, Inc. Used with permission.)

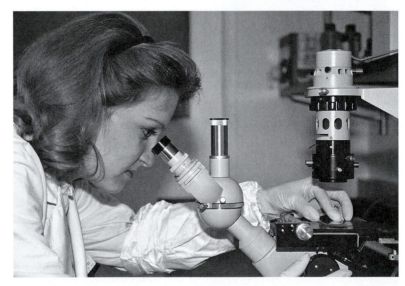

physics, biology, microbiology, biochemistry, and more. Animal and food scientists work in basic or applied research and development, administer research and development programs, manage marketing or production operations in companies, and serve as consultants to businesses, producers, and government agencies. Opportunities exist in each of the species groups (dairy, beef, companion animal, and so on), and in discipline groups (nutrition, reproduction, food microbiology, production, and so on), and in the full array of opportunities from conception to consumption.

One very popular career opportunity that animal science can lead to is that of veterinarian. The majority of veterinarians are in private practices where they pursue the clinical work necessary for the prevention, diagnosis, and treatment of diseases, disorders, and injuries in animals. About half of these veterinarians are predominantly or exclusively small-animal practitioners. Most of the remaining veterinarians are in mixed-animal practices or exclusive large-animal practices. Other veterinarians care

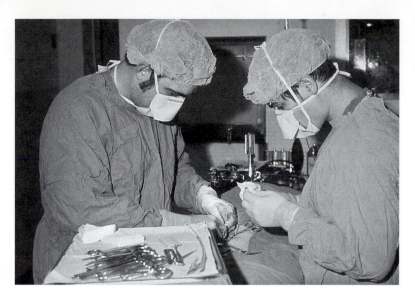

**FIGURE 26–4**  Veterinary medicine offers the opportunity to enter a profession in which medical knowledge and working with animals and people can be combined into a rewarding career.

for zoo or aquarium animals, or for laboratory animals. Additional careers for veterinarians include clinical researchers on human and animal diseases, drug therapies, antibiotics, or new surgical techniques. The U.S. government employs veterinarians as livestock inspectors, and as meat, poultry, or egg product inspectors. In addition to the inspection service, U.S. government positions include disease-control workers, epidemiologists, and commissioned officers in the U.S. Public Health Service, U.S. Army, or U.S. Air Force. Some veterinarians choose to specialize beyond the DVM degree and may do so by completing internships, residencies, and sitting for board certification examinations. Specialties include a focus on specific types of animals; specialization in a clinical area, such as pathology, surgery, radiology, or laboratory animal medicine; and board certification in internal medicine, oncology, radiology, surgery, dermatology, anesthesiology, neurology, cardiology, ophthalmology, or exotic small-animal medicine (Figure 26–4).

## FOODS

The food industry represents one of the most significant areas of current and potential employment in animal science and all of agriculture. Food scientists and technologists have a variety of career opportunities in the food-processing industry, with universities, and in state and federal government (Figure 26–5). For those trained in the science and/or the business of food production, career track options are available in all areas, including food safety, new product development, market testing, nutrition, processing, packaging, distribution, quality assurance, plant management, technical service, sales, and regulation compliance, and as agricultural commodity graders, consumer safety inspectors and officers, environmental health inspectors, food inspectors, and consumer educators. Career tracks are plentiful for those with BS degrees as well as advanced degrees (Figure 26–6). Concerns over food safety and changing demographics relating to how we eat will keep demand high for this area. The high starting salaries for food science graduates have begun to pull more and more students into the area, but a shortage of graduates is still projected.

Universities and colleges vary somewhat in where foods curricula are found. In some institutions, they are housed in animal science departments. In others, they may

**FIGURE 26–5**  The food industry represents one of the most significant areas of current and potential employment in all of agriculture. (Photo courtesy of U.S. Department of Agriculture. Used with permission.)

**FIGURE 26–6**  Because of the science and engineering know-how needed to produce a safe, convenient, and economical food supply, career tracks for those with advanced degrees are plentiful in the foods industry. In this photo, a food technologist is using computer images of steak samples to predict beef carcass composition. (Photographer Keith Weller. Courtesy of Agricultural Research Services, USDA.)

be found in separate food departments or as cross-disciplinary studies. Food technology centers have been built at several universities, acknowledging both the importance and the complexity of the foods area. Mechanization, microbiology, product development, personnel management, and many other areas are appropriate areas of study for food science students. In addition to meat and milk products, there are also a wide variety of opportunities in the foods area that deal with plant-based products and careers that deal with both.

## AGRIBUSINESS

**Agribusiness**

The segment of agriculture that deals in sales and service to production agriculturists and consumers.

*Agribusiness* careers have proliferated over the last two decades as all of agriculture has come to increasingly take on the structure of the big business it is. As production consolidated, international trade expanded, and the products of the agricultural industries proliferated, the need for managers, financial specialists, marketing specialists, and people with general business knowledge has opened a wide variety of opportunities for animal science students. These careers require a knowledge of the animal industries, but they also require business acumen, skilled communication, and people skills. Animal science curricula have been designed to incorporate a wide variety of business, human and resource management, marketing, computer, communications, accounting, and financial management courses. In many places, an agricultural economics, marketing, or management minor has been incorporated into the animal science major. Graduates of such programs have traditionally taken promotion, livestock marketing, and industry positions. In addition, they are now accepting positions and excelling as brokers, purveyors, pharmaceutical salespeople, market analysts, advertising consultants, technical service representatives, computer software developers, appraisers, bankers, business managers, service managers, sales representatives, and fair/exposition managers. In retail, there are opportunities in advertising and public relations, and with livestock organizations, cooperative marketing and service organizations, and trade journals. This area represents an opportunity for many more graduates because colleges and universities are chronically short of qualified graduates for these positions, with little prospect of catching up anytime soon. These are intensely people-related jobs and, for many people, they offer the best of both worlds for career opportunities.

## COMMUNICATION AND EDUCATION

A combination of strong communication skills and technical knowledge in animal science, foods, or other agricultural disciplines prepares students for careers as communication and education specialists. Opportunities for computer information specialists, advertising representatives, public relations specialists, secondary school agriculture and business teachers, and internal communications specialists are expected to be plentiful. Other opportunities will be available as writers, editors, newscasters, and extension agents. Computer skills, especially those necessary to provide education, goods, and services via the Internet will be in demand.

Many colleges and universities have recognized that traditional programs have not kept up with the needs of people who will fill these roles in life. As a result, a variety of cross-disciplinary and double major programs have been developed to assist people in acquiring the skills necessary to compete in this job market. It is common now to find animal science–agricultural communications double majors and animal science–agricultural education double majors at many universities (Figure 26–7).

## ANIMAL CARETAKERS

Universities, commercial laboratories, pharmaceutical companies, animal hospitals and clinics, kennels, animal shelters, stables, and zoos and aquariums employ animal caretakers (Figure 26–8). Responsibilities, job titles, and long-term opportunities vary tremendously. Many of these jobs require no formal training but rely on on-the-job training. However, advancement can depend on both experience and/or formal training. Research labs and zoos, especially those involved in conservation of endangered species, often require caretakers to have a bachelor's degree. Because competition for higher level positions in zoos is so keen, the only entry level into this career track is as a caretaker. Advancement can be painfully slow because positions are limited and few openings occur for the better positions. Those who aspire to curator positions would be advised to receive a postbaccalaureate degree. With experience and formal training, caretakers in animal shelters may become shelter directors. Laboratory animal caretakers can become animal research facility managers and research assistants.

**FIGURE 26–7**  The animal industries offer many opportunities, among them the opportunity to work with both animals and young people in vo-ag and 4-H programs. (Photo courtesy of Dr. Rebecca L. Damron. Used with permission.)

**FIGURE 26–8**  The opportunity to work with zoo and exotic animals is a goal of many students. (Photo courtesy of Corbis. Used with permission.)

Many veterinarians employ caretakers to assist in caring for the animals in their charge. Veterinary technician programs can be found at approximately 65 accredited veterinary technology programs around the country. Laboratory animal caretakers work in research facilities with scientists, physicians, veterinarians, and laboratory technicians and assist in conducting vital research work. On the whole, this area is not full of opportunities for the college graduate, but opportunities do exist.

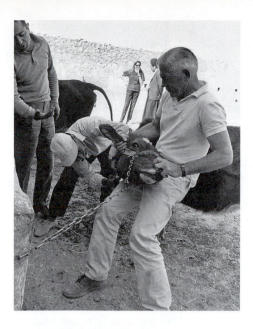

**FIGURE 26–9** More and more international opportunities are available. Here, Dr. Frank Olney (USA), epizootiologist of the Near East Animal Health Institute, and Dr. Siadet, his Iranian counterpart, take a blood sample. (Photographer V. U. Contino. Courtesy of FAO, Food and Agriculture Organization of the United Nations.)

**FIGURE 26–10** The Peace Corps and other humanitarian organizations offer opportunities for people of all ages to contribute to the greater good of humanity as well as gain valuable perspectives on the world. Students with agricultural degrees are always in demand for these assignments, but other skills are also needed. Pictured here is a Peace Corps volunteer in Nicaragua serving in urban youth development. He taught English and organized a soccer team. (Photo courtesy of Peace Corps.)

## INTERNATIONAL OPPORTUNITIES

As the world has become increasingly open and accessible to trade, cultural exchange, and humanitarian intervention, more and more opportunities are created for those who wish to work in the international arena (Figure 26–9). Federal government programs in the Peace Corps, Vista, and the foreign Agricultural Service of the USDA (USAID) offer both long- and short-term opportunities. The Food and Agricultural Organization, World Health Organization, and World Bank also offer opportunities (Figure 26–10). Domestic and multinational firms offer the full range of business and trade positions. With the future economic growth of the food and agricultural sector largely dependent on being competitive in the international arena, the individual with language training in addition to basic discipline education will be increasingly able to command premium salaries and have access to the best opportunities. A wealth of international exchanges, study tours, and study abroad opportunities are available at colleges and universities around the United States. These are valuable experiences, whether you wish to make a long-term commitment to international work or not. Even for those who do not wish to live outside the United States, ample and ever-increasing opportunities exist for involvement in the international arena and many offer fast-track advancement opportunities within a company.

## TRENDS AFFECTING EMPLOYMENT*

The following major trends are expected to influence employment in food, agricultural, and natural resources.

**Consumer demands.**  Consumers are increasingly demanding more specific products and services. Much of what they demand is associated with particular lifestyle choices. Those choices are in turn being affected by such demographic factors as domestic and world population growth, changing ethnicity of the U.S. population, the graying of America (over 80 million baby boomers started retiring on January 1, 2008), and strong consumer preferences influenced by social beliefs and health concerns. As these factors affect consumption, they will also require specific expertise in production, marketing, and distribution to meet consumer demands.

**Changing industry structures.**  Major factors influencing the direction of the changes in the business structure include globalization and consolidation. The U.S. food system is big business with well-developed business structures in place to produce, collect, process, and distribute its products. However, the system constantly changes as it works to meet consumer demands, market opportunities/shifts, challenges in energy availability and cost, and the needs of food distribution and supply chains that literally extend around the globe. To meet the demands of the changing workplace, graduates must have an excellent base set of skills in business, leadership, and international understanding and be willing to use those skills creatively to bring about innovation and productive structural evolution in the agribusiness sector.

**Science and technology.**  The current rate of advancement in science and technology is having profound effects on employment. Science and its technological applications are joined inexorably to the solutions for biosecurity, population growth, health issues, depleting natural resources, and climate change. New and powerful biotechnologies and nanotechnologies are being developed. This creates the need for both graduates with basic science skills to create the new tools of science and technology and those with the ability to use the creations and innovations in practical and useful ways.

**Public policy.**  We are in an unprecedented era of public policy regarding the food we eat and the system that provides it, especially regarding food safety, diet and health, and environmental protection. Consumers are demanding and governments are providing policy. This is creating increasing demand for graduates to manage/oversee regulatory programs and provide public services to all sectors of the food and natural resources system in education, natural resource management, food assistance, and recreation.

*The material in this section was based in part on Goecker et al., 2005, p. 12.

## A JOB, A CAREER, A LIFETIME: WHAT ARE YOU DECIDING WHEN YOU CHOOSE A CAREER AND A MAJOR?

Certainly one of the most important and difficult decisions we ever make in life is our choice of a career. There are usually competing motives at work trying to sway our choice, making it a difficult decision. There may be pressures from family or community to contend with. We usually have certain expectations for ourselves in terms of the amount of money we want to make or the perceived prestige associated with certain careers. We may have strong preferences about the working conditions we prefer or the part of the country we are willing to live in. Unless one has been very fortunate and been able to travel and have a variety of life experiences, we often find ourselves making this very important decision from a position of inexperience. It often seems that the most important choices in life are left to rank amateurs! All of this can be very confusing and disconcerting, leaving us in a position of indecision.

This is one of those life decisions that calls for a plan. Having said that, this is perhaps the only time in this book that the author will step out of the third person and offer some advice. Here is how I think this decision should be approached. First of all, choose an occupation you can enjoy. Ample evidence exists that one must be happy in an occupation to truly excel at it. Doing something just for the money gets old after awhile because money is an external motivator. To continue doing something for a lifetime and not get burned out on it, we have to gain satisfaction from what we do. That provides internal motivation—the only kind that lasts. Consider that we spend from two thirds to three fourths of the most alert and productive hours of our lives in the workplace. It seems a shame to spend so much of life's quality time on something we're doing just for the money. Today's agriculture offers a wealth of opportunity and choice. Everyone can find challenging, enjoyable, and productive life's work (Figure 26–11). Even if the area that most interests you seems to have an oversupply of interested and qualified graduates, opportunities are available. The question becomes simply, "How good do I have to

**FIGURE 26–11**  Regardless of career choice, it is important to be happy in your life's work. So far, this young lady seems to be enjoying herself immensely. (Photo courtesy of Dr. Rebecca L. Damron. Used with permission.)

be to be placeable?" Also, demand varies from region to region. An area that has too many graduates on a national basis may have too few in specific geographic areas. Do your homework! Don't dismiss a potentially rewarding career just because finding employment might be challenging in general terms. You are an individual, not a statistic!

Once you've settled on an area of interest, a specific career, or even a career cluster, there is still work to do to ensure that your decision is a good one. Perhaps the most important part of the process is to adopt the attitude that this is a tentative choice. Even though we probably feel more secure with a firm decision, this is not the time to limit your options. Use all of the resources at your disposal to seek additional information about the career and its opportunities. Explore other options. As a safety net, develop a "Plan B" that includes other career areas that also interest you. Seek summer, part-time, and/or internship positions in your career area. Reevaluate your choice periodically. Plan your course of study so it supports your career choice and take your courses in the sequence that best supports preparation for the chosen career. Above all, be realistic and keep your perspective. If, for instance, you find that you just don't like physical, biological, and chemical science courses, then perhaps veterinary medicine isn't a good choice after all, and you need to explore some other career and focus of studies. Always be ready to consider a different path.

Don't neglect the areas that transcend a specific focus of study. Employers may be attracted to certain people because of their knowledge in specific areas. However, they never lose sight of the fact that they are hiring a person, not a degree plan. The classes a student takes in college are certainly important, but they are only part of an education. It is equally important to develop the talents of leadership and problem solving. All walks of life are well served by the ability to communicate and the knowledge of how to work with people. Developing a personal philosophy and a work ethic are also important. Join and participate in clubs and organizations. Participating in extracurricular activities gives us a laboratory to learn how to deal with people. In developing people skills, there is no substitute for practice!

It also helps to be aware that an education is more than job training. Although it is certainly true that an education should provide the basis for securing a position and earning a living, it should also provide a broad enough base to allow a person to grow both professionally and personally. An education should also add quality to life in ways not associated with a career. It should further challenge the student in ways to help him or her develop basic life skills that transcend any given occupation. An education should also be broad enough in design to allow a given individual to be adaptable and flexible throughout life. This enables us to take advantage of career and life decisions that offer the most of what we want from life. We should also develop an appreciation that one of the most important lessons of a formal education is that an education never ends. A formal education is no more than the basis for continued learning throughout life. The world is changing rapidly. We must be willing to learn and change with it.

And now for a little role clarification. Education works best if students make best use of the resources available to them. Some of your best resources can be found in your professors and advisers. As professionals in the field of education, it is our charge to help facilitate your learning process and your transition into the work world. As advisers and faculty members, we can tell you what is important. However, we can't make you do any of it. We can tell you that good course choice is important, but we cannot make you learn the material in those courses. We can tell you that practical experience is important, but we can't force you to take summer, part-time, or internship positions. We can tell you the value of good communication skills, but we cannot compel you to place yourself in positions where you have to function in front of a group or sharpen your writing skills. We can tell you the value of a good work ethic and good work skills. We can tell you that being on time, being honest, and caring about what you do are important qualities to have in this world, but we cannot make you do or be anything you don't choose. All of these things must come from within. All we can do is hope that you will take advantage of all opportunities to prepare for a productive and enjoyable career in some area that interests you and for which you have an aptitude. Your success depends on you. This is an area of life in which we must all accept the responsibility.

## SUMMARY AND CONCLUSION

Career opportunities are available in agriculture and its component parts. It is common consensus that a workforce that is too small has been an impediment to agriculture for the entire decade of the 1990s. Projections of future opportunities hold the promise of continued opportunities. In animal sciences, students may choose a production, science and medicine, or foods or agribusiness orientation. Further education will be required for some careers. Ultimately, a career is a personal choice filled with personal and professional responsibility. All students can prepare for useful, satisfying careers by working closely with professors and advisers at their school.

## STUDY QUESTIONS

1. What is the current, overall job outlook for agricultural graduates?
2. What is the projected yearly ratio of graduates to jobs?
3. The employment figures projected to the year 2016 show a decrease in the numbers of farmers and ranchers that will be needed. However, the number of jobs is still quite high in this category. Where will the absolute numbers of farmers and ranchers rank in 2008 compared to all others listed in Table 26–2?
4. In a brief paragraph, present a rationale for having an educational path that leads to a specific career.
5. Even though many production specialists will still be needed in the future, there is a downward trend in the total number that will be needed. Why?
6. Describe the variety of opportunities open to those who have an interest in science and medicine.
7. Food science careers have proliferated in the last few years and are expected to continue doing so. Look around you at the way people eat and think of the food-related news items you've seen in the past few months. Based on these observations, speculate as to why foods careers are so available and variable.
8. Compare the support courses in an agribusiness curriculum with those in a science curriculum.
9. The opportunities for those with BS degrees in the general area of animal caretaking is limited. What might cause those prospects to improve?
10. For those who wish to have an international career, what are the most important skills to learn and experiences to have?

11. In the last section of this chapter, the author presented a personal view of the career decision-making and educational process. However, you should recognize that you have the right to disagree with any or all of it. After all, advice is just that. After reading that section, develop a brief "point–counterpoint" essay in response to the section as a means of comparing your own thoughts to those presented.

## REFERENCES

Bureau of Labor Statistics. 2008. National Industry–Occupational Employment Matrix. This is a searchable database available at http://stats.bls.gov.

Coulter, K. J., A. D. Goecker, and M. Stanton. 1990. *Employment opportunities for college graduates in the food and agricultural sciences.* Higher Education Programs, Cooperative State Research Service, U.S. Department of Agriculture, Washington, DC. Printed and distributed by Food and Agricultural Careers for Tomorrow, Purdue University, West Lafayette, IN.

Goecker, A. D., K. J. Coulter, and M. Stanton. 1995. *Employment opportunities for college graduates in the food and agricultural sciences.* Prepared and published by the School of Agriculture, Purdue University, West Lafayette, IN. Supported by Science and Education Resources Development, Cooperative State Research, Education, and Extension Service, U.S. Department of Agriculture, Washington, DC.

Goecker, A. D., C. M. Whatley, and J. L. Gilmore. 2000. *Employment opportunities for college graduates in the food and agricultural sciences.* Prepared and published by the School of Agriculture, Purdue University, West Lafayette, IN. Supported by Science and Education Resources Development, Cooperative State Research, Education, and Extension Service, U.S. Department of Agriculture, Washington, DC.

Goecker, A. D., J. L. Gilmore, E. Smith, and P. G. Smith. 2005. *Employment opportunities for college graduates in the U.S. food, agricultural, and natural resources system 2005–2010.* Produced through a cooperative agreement between USDA's Cooperative Research, Education, and Extension Service and Purdue University. Accessed online January 2008. http://www.csrees.usda.gov/newsroom/news/2005news/USDA_05_Report2.pdf.

*Occupational outlook handbook 2008–2009 edition.* 2008. Washington, DC: Bureau of Labor Statistics, U.S. Government. Available online at http://www.bls.gov/oco/home.htm.

# 27

# Animals as Consumers of Grain: Asset or Liability?

## LEARNING OBJECTIVES

After you have studied this chapter, you should be able to:

1. Discuss the flaws of the "grain to people" argument as a solution to world hunger.

2. Describe the advantages of feeding grain to animals.

3. Explain how using animals in agriculture provides for good land use.

4. Discuss how consumer preferences affect animal product production.

## KEY TERMS

| | | |
|---|---|---|
| By-products | Feedstuff-to-foodstuff | Pharmaceuticals |
| Consumer preference | Food grains | Power |
| Efficiency | Land use | Soil fertility |
| Feed grains | Manure | |

## INTRODUCTION

As humans, we have relatively simple digestive tracts. This makes us incapable of digesting and using a fibrous diet. We require easily digested, concentrated foods. Both plant and animal foods are available that humans can use. Use of both types of foods makes selecting a nutritionally adequate diet much more likely. If offered a choice, most humans include both plant and animal products in their diet. How much animal product is consumed is almost always related to the economic status of the people in question and their level of agricultural development. Because plant-derived foods are generally more available and less expensive, and hungry people are always poor people, most of the world's hungry and malnourished people derive the majority of their food from grain sources. Cereal grain is the most important single component of the world's food supply, accounting for nearly half of the total food produced. Plant foods combined account for about 83% of all the calories consumed by humans worldwide. The situation is very different in the developed world, where animals eat much more grain than does the human population. For example, 80% of the U.S. corn crop is used as animal feed. In some developing countries, less than 3% of the total grain is used for animals.

Many people feel that because the world population is increasing, we are becoming less and less able to afford agricultural animals. Animal foods have been especially criticized because feeding plant products to animals is often viewed as an inefficient use of potential human food. As the population increases and the challenges of food distribution create more hunger and starvation, some members of our society are increasingly suggesting that grain should no longer be fed to livestock so people can use it directly. This view has been ardently championed by some in society, and others have been persuaded to agree. The argument is further extended to say that we should do away with all agricultural animals and divert all of the resources currently used by them directly to the human population. Proponents of this view further assert that the system of agriculture in place around the world is causing people to go hungry and that it is morally indefensible to continue animal agriculture. Experience teaches us that such sweeping assertions rarely hold up under appropriate scrutiny. However, the argument is put forth that it is more efficient for humans to eat plant products directly rather than to allow animals to convert them to human food. On the surface, this argument appears to be logical because, at best, animals produce only 1 lb or less of human food for each 3 lbs of plants eaten. To this, add the prestige of a celebrity spokesperson to champion the cause, and many never think to challenge the argument and look for its flaws. Yet, there are flaws in this line of thinking. As Kinsman (1982) said, "Without the use of animals as food, it would be impossible to support the world's population either presently or in the future. In fact, it would be irresponsible, irrational, wasteful, unconscionable and even immoral not to utilize this natural source of high-quality nutrition for human purposes."

A major flaw in the "inefficiency of conversion argument" is that this inefficiency only applies to those plants and plant products that the human can utilize. The fact is that over two thirds of the feed fed to animals consists of substances that are either undesirable or completely unsuited for human food. Thus, by their ability to convert inedible plant materials to human food, animals not only do not compete with humans for food, but rather aid greatly in improving both the quantity and the quality of the diets of human societies. This is a very important function that animals have provided for millennia.

There are other flaws in the argument as well. Most notable is the fact that such things as differences in productive capacities of land are rarely even acknowledged, much less factored into the arguments. Nor is the tremendous diversity in climate, culture, and/or economic constraints found in various parts of the world considered. Another serious flaw in the argument is that all types of animal production systems are considered to have the same efficiencies. Nothing could be further from the truth. All of the reasons for feeding grain are not tied to producing meat or milk.

If animal grain consumption is curtailed, this would have tremendous implications for the world agricultural order. For all of the world's developed countries, removing grain as animal feed would severely restrict animal agriculture. Poultry and swine production would be especially hard hit. Developing countries would be forced to change their approach to agriculture as well. Before taking that step, we should consider the benefits of animals and see if these benefits justify their continued use. We also need to examine the arguments put forth against animal use in a more extensive manner and see if a defensible argument can be presented for continued animal use. That is the responsible course of action. It is not enough just to say "you're wrong." A working dialogue demands that counterassertions be spelled out so they can be considered, and people with opposing views be given the opportunity to consider and agree or rebut. That's what this chapter is about.

Chapter 27/Animals as Consumers of Grain: Asset or Liability?

703

# THE "NO ANIMAL" ARGUMENT

The major argument for the direct human consumption of grain with no use by animals is that more people can be fed if we eliminate the grain feeding of animals and consume that grain directly. Animal agriculturists may not like this, but taken as an isolated fact, it is true. When animals consume feeds, they use much of it for their own body maintenance before they produce any useful product. This amount is thus "wasted" in terms of human consumption. More hunger would be alleviated if humans simply ate the grain. At first glance, the information in Table 27–1 appears to demonstrate this fact clearly on a dry-matter, calorie, or protein conversion basis. However, further exploration of this and other information must occur before a rational conclusion can be drawn. Let us now consider the value to be had from the use of grain in animal production.

# THE ARGUMENT FAVORING ANIMAL GRAIN CONSUMPTION

## JUDICIOUS USE OF GRAIN TO ANIMALS

The United States and other developed countries feed a considerable amount of grain to animals, even to ruminants that have the ability to be productive entirely on roughage materials that humans cannot digest. This is done because it is currently profitable to do so. However, only a sixth of the energy needed to produce livestock products comes from cereal grains. When grains are used, the overall productivity of the agricultural system improves because the efficiency of use of forages, crop residues, and by-products is increased. In addition, ruminants and other agricultural animals serve an important function as a buffer for fluctuating grain supplies (Figure 27–1). On a year-to-year basis, animal usage of grains is flexible. In this way, grain farmers have a reliable secondary market for their product, thus providing stability and flexibility to grain farming. The demand created by livestock goes up and down in response to supply. Thus grain farmers stay in business. Without this reliable second market, less grain would be raised to keep prices high. Consumers would suffer. Feeding surplus grain keeps the price that farmers receive for their grain at a level that allows them to make a profit. However, if a shortage of grain, for whatever reason, causes the price of grain to rise sufficiently to make it unprofitable to feed to ruminants, the practice would cease. If this happens, ruminant animals would still be needed for their primary function of converting inedible plant materials to edible human food (Figure 27–2).

**TABLE 27–1    Feed-to-Food Efficiency Ratings[1]**

|  | Beef | Lamb | Turkey | Layer | Hog | Broiler | Fish | Dairy |
|---|---|---|---|---|---|---|---|---|
| Feed efficiency[2] | 9.0 | 8.0 | 5.2 | 4.6 | 4.0 | 2.1 | 1.6 | 1.1 |
| Energy efficiency[3] | 34.2 | 44.1 | 18.9 | 12.1 | 18.8 | 12.4 | 6.9 | 5.8 |
| Protein efficiency[4] | 10.6 | 16.5 | 3.2 | 3.9 | 4.1 | 1.9 | 2.1 | 2.7 |

[1]*SOURCE:* Compiled from Ensminger, 1991, pp. 20–22, various figures.
[2]Pounds of feed required to produce 1 lb of product.
[3]Kilocalories of energy required to produce 1 kilocalorie of product.
[4]Pounds of feed protein required to produce 1 lb of product protein.

**FIGURE 27–1**   Feeding grain to livestock, such as these finishing steers, provides an important function in serving as a buffer for fluctuating grain supplies.

**FIGURE 27–2**   One of the great contributions of livestock to humans is converting crop residues, like the cornstalks in this photo, to usable products. (Photo courtesy of Dr. Jerry Fitch, Department of Animal Science, Oklahoma State University. Used with permission.)

Monogastric animals like swine and poultry can be competitors with humans for food if they are produced by the intensive confinement systems widely used in the developed countries. In fact, the highest proportion of feed grains and other concentrates, such as oilseed meals, fed to livestock in the United States are fed to swine and poultry. As is the case for ruminants, this is done because it is profitable to do so. Likewise, the feeding of grain to these species will be stopped if the demand for grain for human consumption forces the price too high. In this case, unlike the situation with ruminants, the numbers of swine and poultry will probably be reduced somewhat in the developed countries. However, they will not disappear as an agricultural animal because they have a unique ability to convert waste products to human food and to scavenge for their own food (Figure 27–3). This is one of the most important functions of swine in agriculture today. Although practiced less in the United States today than in the past, much of the

**FIGURE 27–3** In developing countries, pigs and poultry are often allowed to scavenge and are then slaughtered for food. Larger animals are reserved for work.

world relies extensively on the animals scavenging for themselves and feeding food scraps to chickens and pigs.

How does meat production fit into agricultural development? The lower the level of agriculture, the less likely it is that the primary reason for raising large farm animals is to produce meat. Instead, pigs, chickens, and small ruminants such as sheep and goats are kept as meat animals, and cattle are raised for other purposes (primarily). A major reason for this is related to the lack of proper meat-storage facilities in hot climate areas of the world. A single family can consume a chicken, small pig, goat, or sheep carcass before it spoils, but not a cow carcass. The goat is often referred to as the poor person's refrigerator because of this. This is the reason that meat dishes in poor areas are often very heavily spiced. The spices mask off-flavors of meat. Curry and peppers are often used this way. Cattle are used for meat but only in a salvage situation. Old, crippled, or sick animals that are near death may be slaughtered or they are eaten after they die naturally. A cow or bull may be killed for a village feast in which it can all be consumed in a short time. Another reason for not eating cattle is that the cattle are more important for some other purpose such as draft use or milk production.

## FOOD VERSUS FEED

Rice, wheat, and other grains commonly grown for and consumed by people are called *food grains.* Farm animals are not fed any significant amount of food grains in the U.S. and other developed countries. Animals are fed *feed grains*. Although people could eat these grains in quantity, we prefer not to. Besides, there is a huge surplus of these grains. Should there be a need by the human population for that grain, it is readily available. In addition, animals use a tremendous amount of by-product feeds. Animals consume such things as milling by-products, cotton seed hulls, distillery wastes, beet and citrus pulp, and fruit and vegetable processing wastes. There is no demand for these items for human use. By feeding them to animals, the huge problem of their disposal is handled and the cost of the primary product for the human market is kept lower. Animals turn a liability into an asset.

**Food grains**

Those grains for which humans provide the primary consuming market.

**Feed grains**

Those grains that are produced primarily as livestock feed.

## NUTRITION

Animal products provide needed nutrients to the human diet. High-quality and easily digestible vitamins, minerals, and energy are provided. However, protein is the most important contribution animals make to the quality of the human diet. Animal protein is of higher quality than plant protein. The quality that animal products add to diets is perhaps the most important contribution they make to food. This is because animal products have every amino acid needed by the human body. Especially important are lysine, tryptophan, and methionine, which are generally deficient in plant-based foods. By providing well-balanced amino acids in the protein, animals help humans balance their diet. This factor alone negates much of the advantage of consuming the protein as grain. Ruminant animals, with their bacterial fermentation, can even use nonprotein nitrogen sources, such as urea, and the nitrogen in poultry litter to produce high-quality protein for people.

## PRODUCTIVE ABILITY OF LAND

Much of the world's land cannot be tilled and farmed, but it does grow a wide variety of forage and browse that animals, especially ruminants, can use (Figure 27–4). Table 27–2 presents some statistics that are ignored by those who suggest we can no longer afford to use animal-produced foods. Only about 37% of the land area of the world is classified as agricultural. Thus roughly two thirds of the land area of the world is not suited for any sort of agricultural use because it is covered by cities, mountains, deserts, swamps, snow, and so on. Of the 37% devoted to agriculture, only 30% (or about 11% of the total world land area) can be cultivated to produce plant products that the human can digest. The remainder of the world's agricultural land is not capable of producing food crops but is covered by grass, shrubs, or other plants that ruminant animals can digest (Figure 27–5). Where this resource is concerned, the inefficiency of animals is not a major concern because they represent the only way these plants can be converted to human food. The relationship between the ruminant and humans is not competitive because humans cannot use the feeds that the ruminant uses. Most of the

**FIGURE 27–4** Much of the world's land is too dry to be cultivated but does grow forage and browse that ruminants can use. (Photo courtesy of Dr. J. Robert Kropp, Department of Animal Science, Oklahoma State University. Used with permission.)

Chapter 27/Animals as Consumers of Grain: Asset or Liability?

707

**TABLE 27–2**    **Characteristics of Agricultural Land in Various Geographic Regions**

| Geographic Region | Total Land Area (1,000 HA[1]) | % of Total Land That Is Agricultural | % of Agricultural Land That Is Cultivated Land | % of Agricultural Land That Is Permanent Pastures |
|---|---|---|---|---|
| World | 13,048,300 | 37 | 30 | 70 |
| Developed countries | 5,462,356 | 35 | 36 | 64 |
| Developing countries | 7,585,948 | 39 | 27 | 73 |
| Africa | 2,963,468 | 36 | 18 | 82 |
| Asia | 2,678,234 | 47 | 37 | 63 |
| Europe | 472,578 | 45 | 63 | 37 |
| Oceania | 849,135 | 57 | 12 | 88 |
| North America | 1,838,009 | 27 | 47 | 53 |
| South America | 1,752,925 | 35 | 19 | 81 |
| USA | 915,912 | 47 | 44 | 57 |

[1]HA=hectare=2.47 acres.

*SOURCE:* FAO, 1997.

**FIGURE 27–5** Some of the world's fertile land cannot be cultivated because of the terrain, altitude, or other problems, but it is fertile and produces forage for animal production.

world's ruminants do not compete with humans for land or feed. As the human population of the world increases, it is likely we will be forced to depend more and more on animals, especially ruminants, to meet our increased demands for food.

There are several things that stand out in Table 27–2. The developed countries of the world (the richer countries) are not that way because most of their land is agricultural land. The developing (poorer) countries have about the same percentage of their total land area that is classified as agricultural. It is true, however, that the two most highly developed continents (Europe and North America) have more than the average amount of cultivated land. Most of the world's domestic livestock are herbivorous, and plants provide the majority of their feed. Humans have domesticated those species that make the best use of the land we have. In the United States, grazing lands amount to 57% of our nonforest agricultural land. In addition, our crop lands produce a tremendous volume of very low-quality roughages, such as cornstalks, straw, and other crop residues. Pastures and other roughages supply 94% of the total feed of sheep and goats, 85% of the feed of beef cattle, 59% of the feed of dairy cattle, and 62% of the feed of all livestock.

## CONSUMER PREFERENCES

Agriculture is driven by the needs and wants of the consumer. In the United States, we enjoy the highest quality food at the lowest cost. Grain consumption by animals is an important reason for this, and thus consumers benefit from this practice. To eliminate grain feeding would radically curtail the supply of meat because animal numbers would have to be reduced drastically. The remaining animals would take a longer time to reach market and would yield less per animal. The product would be less acceptable to the consumer and the price would be higher. Because of great dependence on seasonally produced forages, animal products would be more seasonal, revolving more around the grazing season. Large supplies would be available in the fall with scarcities at other times. Milk would be similarly affected.

## OTHER PRODUCTS

Another factor that cannot be overlooked is that animals provide many products other than the generally acknowledged work, fiber, meat, and milk. This is true for the people of the developed as well as developing world. Slaughter by-products are used in making everything from candles to cosmetics. Without these by-products, the health and lifestyle of many people would be altered. In addition, well over 100 pharmaceutical products used by people are by-products of animal slaughter. A list of these products for the pig alone is shown in Table 27–3.

## ANIMALS AND SOIL FERTILITY

Animals help maintain soil fertility for crop production. Animals provide manure for the fields, a fact that is very important in the developing countries. In addition to nutrients, manure adds organic matter, which improves soil texture. This is very important because the crude tools available in poorer societies are much more effective in a manure-enhanced soil. The positive effects of using animals and their products to improve soil fertility are central to the success of many sustainable systems. Organic farmers also find many good reasons to use animal-generated fertilizer.

## POWER

Animals provide needed power in most of the developing nations of the world. Cattle, water buffalos, donkeys, and horses are the major species used for power. In this capacity, they contribute to the human food supply from plant sources. Although they may be offered only small amounts of grain (if at all), that grain may be crucial to sustaining the animal until the next planting season. The product here is the work necessary to help provide the very grain in question, as well as a surplus for humans (Figure 27–6).

**TABLE 27-3** **Pharmaceutical By-Products Provided by Hogs**

Pharmaceuticals rank second only to meat in the important contributions hogs make to society. New uses are being continually added to the list of life-supporting and life-saving products. All told, hogs are a source of nearly 40 drugs and pharmaceuticals.

**Adrenal Glands**
Corticosteroids
Cortisone
Epinephrine
Norepinephrine

**Blood**
Blood fibrin
Fetal pig plasma
Plasmin

**Brain**
Cholesterol
Hypothalamus

**Gallbladder**
Chenodeoxychlolic acid

**Heart**
Heart valves[1]

**Intestines**
Enterogastrone
Heparin
Secretin

**Liver**
Desiccated liver

**Ovaries**
Estrogens
Progesterone
Relaxin

**Pancreas Gland**[2]
Insulin
Glucagon
Lipase
Pancreatin
Trypsin
Chymotrypsin

**Skin**
Porcine burn dressings[3]

**Spleen**
Splenin fluid

**Stomach**
Pepsin
Mucin
Intrinsic factor

**Thyroid Gland**
Thyroxin
Calcitonin
Thyroglobulin

**Pineal Gland**
Melatonin

**Pituitary Gland**
ACTH-Adrenocorticotropic hormone
ADH-Antidiuretic hormone
Oxytocin
Prolactin
TSH-Thyroid-stimulating hormone

[1]Hog heart valves, specially preserved and treated, are surgically implanted in humans to replace heart valves weakened by disease or injury. Since the first operation in 1971, tens of thousands of hog heart valves have been successfully implanted in human recipients of all ages.

[2]Hog pancreas glands are an important source of insulin hormone used to treat diabetics. Hog insulin is especially important because its chemical structure most nearly resembles that of humans.

[3]Specially selected and treated hog skin, because of its similarity to human skin, is used in treating massive burns in humans, injuries that have removed large areas of skin, and in healing persistent skin ulcers.

*SOURCE:* Adapted from National Pork Producers Council, 1997.

**FIGURE 27-6** Animals provide a much needed source of power in developing countries.

## SUMMARY AND CONCLUSION

Feeding grain to animals has become a source of concern for some in our society. Some people argue that because feedstuff-to-foodstuff conversions are less than 100%, it is not a good use of feed resources to use them in producing meat, milk, and other products. In this chapter, we saw that the assumptions about efficiency on which these assertions are based are faulty. In fact, using grains for animals can be a very efficient use because doing so helps us maximize feed resources of low quality. Returns of human edible outputs/human nonedible inputs show that using grain can be very efficient and is thus a strategy for helping to feed the world's population rather than a hindrance. Animals will continue to help feed the world because, without them, it would be impossible to do so.

## STUDY QUESTIONS

1. What is the single most important component of the world's food supply? What percentage of food produced does this account for?
2. Plant foods account for what percentage of all calories consumed by humans?
3. According to the argument favoring direct human consumption of grain, can more people be fed with direct grain consumption, or with animal consumption of grains followed by humans using the products?
4. It is said that animals provide elasticity and stability to grain farming and stimulate grain production. Explain this statement.
5. If worldwide grain shortages force the cost of grains too high, swine and poultry numbers might suffer. Would these animals cease to be agricultural animals? Justify your answer.
6. What is the difference between a feed grain and a food grain? How much of the corn produced in the United States is used for human consumption? Animals utilize by-products of other agriculture as feeds. List five examples.
7. Explain why animal-derived protein is of such high quality.
8. Discuss the differences between the digestibility of plant and animal proteins.
9. Another argument favoring animal use is that much of the land in the world is not cultivable but does produce forages. How does that justify animal use?
10. If grain were not fed to animals, what would happen to animal product availability?
11. What percentage of agricultural land in the United States is grazing lands?
12. Another advantage of ruminant animals is that they use low-quality roughages. Why is this such an important argument for animal agriculture?

Chapter 27/Animals as Consumers of Grain: Asset or Liability?

711

# REFERENCES

Baldwin, R. L., K. C. Donovan, and J. L. Beckett. 1992. An update on returns of human edible input in animal agriculture. Proceedings of the California Animal Nutrition Conference, 13–14 May, Fresno, CA.

Beckett, J. L., and J. W. Oltjen. 1993. Estimation of the water requirement for beef production in the United States. *Journal of Animal Science* 71: 818–826.

Bywater, A. C., and R. L. Baldwin. 1980. In *Animals, feed, food and people: Alternative strategies in food animal production*, R. L. Baldwin, ed. Boulder, CO: Westview Press.

Ensminger, M. E. 1991. *Animal science: Food and animals—A global perspective*. Danville, IL: Interstate.

FAO. 2008. *FAOSTAT statistics database*. http://faostat.fao.org/site/291/default.aspx.

Kinsman, D. M. 1982. Would it be beneficial to eliminate the use of animals as food? Paper presented at Symposium on Animals and Ethics: Should We Use Animals for Food and Experimentation? 9 October, Yale University, New Haven, CT.

National Pork Producers Council. 1997. http://www.nppc.org/pork-products-med.html.

Smith, N. E. 1980. In *Animals, feed, food and people: Opportunities for forage, waste and by-product conversion to human food by ruminants*, R. L. Baldwin, ed. Boulder, CO: Westview Press.

USDA-ERS. 2008. *Land use, value, and management: Major uses of land*. Accessed online January 2008. http://www.ers.usda.gov/Briefing/LandUse/majorlandusechapter.htm.

Wheeler, R. D., G. L. Cramer, K. B. Young, and E. Ospina. 1981. *The world livestock product, feedstuff and food grain system*. Morrilton, AK: Winrock International.

# Food Safety and Consumer Concerns

## LEARNING OBJECTIVES

After you have studied this chapter, you should be able to:

1. Describe the complexities of the food safety issues facing the food industry.

2. Discuss the basics of the history of food safety during the 20th century.

3. Describe the magnitude of the problem of food safety to consumers.

4. Identify the most important of the foodborne pathogens.

5. Differentiate between the roles of various government agencies in providing for a safe food supply.

6. Describe the current and changing roles of FSIS in food safety.

7. Describe HACCP and state its purpose and principles.

8. Explain the level of safety associated with bovine somatotropin, growth-promoting hormones, and antibiotics in animal production.

9. Describe the value of food irradiation.

## KEY TERMS

BSE

BST

Delaney clause

Electronic pasteurization (e-beams)

Emerging pathogens

Epidemiology

Fight BAC!

Foodborne illness

GRAS list

Guillain-Barré syndrome

HACCP

Hemolytic-uremic syndrome (HUS)

Irradiated foods

Pathogen

Residue avoidance

Residue monitoring program

Septicemia

Withdrawal time

Zero tolerance

# INTRODUCTION

It must be said at the outset that the American food supply is among the safest in the world, if not the safest. It must also be acknowledged that, despite our safe food supply, millions of Americans are made ill each year by the food they eat. According to the Food and Consumer Economics Division of the USDA, microbial pathogens in foods cause from 6.5 to 33 million illnesses and 9,000 deaths in the United States each year, often among the very young and the elderly. Estimates indicate that the cost of human illness for just six specific foodborne pathogens range from $2.9 billion to $6.7 billion annually. The six most important foodborne **pathogens** are *Campylobacter jejuni, Clostridium perfringens, Escherichia coli O157:H7, Listeria monocytogenes*, Salmonella (nontyphoid), and *Staphylococcus aureus*. The highest costs can be traced to salmonellosis. Foodborne illness can result from consuming contaminated fruits and vegetables, water, or animal-based foods. Threats range from disease-causing microorganisms to contamination with toxins of chemical and microbiological origin.

**Pathogen**
A bacterium or virus that causes disease.

# HISTORY OF FOOD SAFETY AS A PUBLIC ISSUE

Food safety has not always been such a high-profile public issue, although it is by no means a new issue. Concerns about the processing of food began at the turn of the 20th century. *The Jungle,* a novel by Upton Sinclair, caused great public concern by describing deplorable conditions in the meat-processing industry (Figure 28–1). Public pressure consequently resulted in the passage of the Pure Food and Drug Act of 1906 and the Meat Inspection Act of 1906, beginning the government regulation of food processing. In 1958, the Federal Food Additives Amendment made food-additive testing the responsibility of the food product's manufacturer and established that any new food additive must be demonstrated as safe before it can be added to food. A part of the 1958 Food Additive Amendment was the **Delaney clause,** which prohibited the addition of any substance shown to cause cancer in any animal at any dose. Almost from its adoption, the clause was under attack as being unrealistic. Extremely high doses of some

**Delaney clause**
Prohibited the addition of any substance to human food shown to cause cancer in any animal at any dose.

**FIGURE 28–1** *The Jungle*, a novel by Upton Sinclair, described deplorable conditions in the meat-processing industry. Public outrage sparked by the novel led directly to the passage of the Pure Food and Drug Act of 1906 and the Meat Inspection Act of 1906. (Photo courtesy of Culver Pictures, Inc. Used with permission.)

substances may cause cancer but are perfectly safe at much lower doses. To remedy this problem, Congress passed the Food Quality Protection Act (FQPA) in August 1996. It repealed the Delaney clause, replacing *zero tolerance* with a more science-based standard of "reasonable certainty of no harm" for raw and processed food tolerances. The EPA was given 10 years to implement the law.

There was also another important outcome of the 1958 Food Additives Amendment. In effect, it created a grandfather clause for a group of compounds that were already in use when the legislation passed and were considered "generally recognized as safe (GRAS)." Compounds on this ***GRAS list*** were given special status and continued to be used. If they were removed from the list, it was because they had been proven to be a hazard. This list of compounds has been under constant review ever since its creation. By applying the strict provisions of the Delaney clause, several compounds have been removed from the GRAS list. The new provisions will result in some of those being returned to the GRAS list.

These measures have been highly effective and generally satisfied the concerns of the consuming public. For several decades, when food safety was discussed, the concerns focused on proper home-canning techniques to avoid botulism toxicity and the proper cooking of pork to avoid trichinosis. If there was an outbreak of food-related illness, it was usually associated with something like bad potato salad at a church social or Fourth of July picnic. Being very local, these outbreaks rarely created any media interest beyond the local paper (Figure 28–2). Perhaps the issues that most served to refocus the national attention on food safety were two incidents relating to fruits. The first was an incident with apples and the growth-regulator alar, which received sensationalized publicity because of the involvement of a well-known Hollywood actress, Meryl Streep, as spokesperson for the Natural Resources Defense Council and coverage in two *60 Minutes* programs. The second was an incident involving contaminated grapes.

The 1990s saw several episodes of food-related diseases. The most widely recognized and publicized incidents involved *Escherichia coli* (*E. coli*) O157:H7 in meat in 1993 and 1997, and in apple juice in 1996. Several incidents involved *Cryptosporidium* in drinking water. *Salmonella* caused outbreaks traced to cantaloupe in 1991, in ice cream in 1994, on alfalfa sprouts in 1995, and in homemade mayonnaise (traced to

**Zero tolerance**

Common term used to indicate restrictions imposed by the Delaney clause.

**GRAS list**

A list of common food additives given special safe status under the 1958 food additive amendment because of their previous records as safe food additives.

**FIGURE 28–2** Prior to the 1990s, most outbreaks of food-related illnesses were usually associated with small groups of people at a picnic or a potluck social gathering. (Photo courtesy of Getty Images, Inc./PhotoDisc, Inc. Used with permission.)

eggs) in 1996. In 1996, *Cyclospora* on raspberries from Guatemala caused an outbreak of that parasite; in 1997, oysters infected several hundred people with Norwalk virus; and in 1997, Hepatitis A–contaminated strawberries from Mexico caused an outbreak. However, the galvanizing event that brought public attention to the issue of food safety was probably in 1993 when *E. coli* O157:H7 contamination occurred in hamburgers from a well-known fast-food restaurant. The hamburger was unknowingly contaminated with the bacteria. Had it been cooked to the proper temperature, tragedy would have been averted. However, it was not properly cooked before it was served. The tragic deaths of children and the illnesses of several other individuals served to make *E. coli* a household word. From the time of that tragedy, food safety has been a national issue.

## IMPORTANCE OF FOOD SAFETY TO CONSUMERS

There is no doubt that the food safety issue is important to consumers. In a land of plenty, our consumers want the plenty to also be safe. The predominant food safety concerns of consumers are chemical residues such as pesticides and herbicides, food additives, antibiotics and hormones used on animals, ***irradiated foods,*** foodborne pathogens, and naturally occurring toxicants. Some are real threats, and some are exaggerated consumer perceptions of threats that are minimal. In actuality, the most significant threat to the average consumer is foodborne pathogens. Table 28–1 shows the sources, symptoms, onset, and duration of illnesses associated with common food-borne, disease-causing organisms.

Consumers generally do not understand that it is impossible to choose a diet free from all risk. This has always been true and will always be true. In addition, the potential for new risks comes with each change in food, agricultural, or processing technology, with each new trade agreement; and with each shift in eating habits. Such changes are responsible for some of the risks our food supply currently faces and have made today's food supply different than that from any time in history. We import over 46 million tons of food, much of it from developing countries (Figure 28–3).

Another issue is that the food supply is highly centralized. This is both a "curse and a blessing." It is a blessing because it makes monitoring easier and a curse because

**Irradiated foods**

Foods treated with ionizing pasteurization, which kills insects, bacteria, and parasites.

**FIGURE 28–3** There are many obstacles to providing a safer food supply including control over the millions of tons of food we import, much of it from developing nations. (Photographer Ken Hammond. Courtesy of USDA. Used with permission.)

**TABLE 28–1**    **Sources, Symptoms, Onset, and Duration of Illness Associated with Common Foodborne, Disease-Causing Organisms**

| Microbe | Sources | Symptoms | Onset | Duration |
|---|---|---|---|---|
| **Bacteria Causing Foodborne Infections** | | | | |
| Salmonella | Raw meat, especially poultry; eggs | Nausea, abdominal pain, diarrhea, headache, fever | 6–48 hrs | 1–4 days |
| Campylobacter jejuni | Raw milk, raw chicken | Fever, headache, diarrhea, abdominal pain | 2–5 days | 1–2 wks |
| Listeria monocytogenes | Raw mild and cheese, raw meat | Fever, headache, chills, nausea, vomiting | >12 hrs | 6 wks |
| Vibrio parahaemolyti-cus | Contaminated seafood | Crampy abdominal pain, weakness, watery diarrhea, fever, chills | 4–96 hrs | 1–7 days |
| *Staphylococcus aureus | Human contamination, undercooked meat | Nausea, vomiting, diarrhea, abdominal pain, prostration | 1–6 hrs | 3 days to 2 wks |
| Clostridium perfringens | Temperature abuse of prepared foods, meats, meat products, gravies | Abdominal cramps diarrhea, putrifactive diarrhea, sometimes nausea and vomiting | 8–22 hrs | 6–24 hrs |
| Escherichia coli enteric type | Human contamination, undercooked meat | Abdominal pain, bloody diarrhea, kidney failure | 5–48 hrs | 3 days to 2 wks |
| *Clostridium botulinum | Canned foods, deep casseroles, honey | Vomiting, abdominal pain, dizziness, respiratory failure, paralysis | 12–36 hrs | 10 days |
| Shigella | Contaminated water or foods, especially salads like chicken, tuna, shrimp, and potato salad | Diarrhea, abdominal pain | 12–50 hrs | 5–6 days |
| **Viruses** | | | | |
| Norwalk virus | Contaminated seafood | Diarrhea, nausea, vomiting | 1–2 days | 4–5 days |
| Hepatitis A virus | Human contamination | Jaundice, fatigue, liver damage | Up to 50 days | Months to years |
| **Parasites** | | | | |
| Giardia lamblia | Contaminated water and uncooked contaminated foods | Diarrhea, abdominal pain, gas, anorexia, nausea, vomiting | 5–25 days | Until treated |
| Trichinella spiralis | Pork, game meat | Muscle weakness, flu symptoms | Weeks | Months |
| Anisakis | Raw fish | Severe abdominals pain | 12 hrs | Until removed |

*Bacteria causing foodborne intoxications
SOURCE: Adapted from *Nutrition: Science and Application,* 2nd Edition, by Lori A. Smolin and Mary B. Grosvenor, copyright © 1997 by Saunders College Publishing, reproduced by permission of the publisher.

when there is a problem, it has greater potential to be a big problem. Hamburger from one plant can be processed and distributed to a dozen states in scores of food outlets per state in a manner of hours. Contaminated food can be processed and distributed in many different forms and can cause illnesses before it can be isolated (Figure 28–4). Changes in the demographics of our population also change the number of people susceptible to risk from foodborne microorganisms. The very young, the elderly, and those with compromised immune systems are at greatest risk. All of these factors make foodborne illness a different issue today than it ever was in the past.

**FIGURE 28–4**  The food-processing system is centralized in a way that brings lower prices and other consumer benefits but creates problems as well. Meat processed in a given batch can be packaged and distributed to many states and distributed in many consumer outlets in a matter of hours. Should contamination exist in a batch, thousands of people could potentially be affected.

Survey after survey has shown the level of concern consumers have for the problem. There is also a good deal of confusion on the part of the consumer as to what is a threat, how large the threats may be, and what to do to protect themselves. Much of the focus of consumer demands for solutions have centered on the government. This concern has sparked several actions at local, state, and national levels. It is outside the scope of this chapter to give a complete accounting of the actions taken and planned. Significant changes in the food system are being addressed and will continue to unfold. If properly implemented, many of the changes will lead to a safer food supply. Many already have. The much-publicized recalls of hamburger that occurred in the late 1990s are evidence of a system that is doing a better job of detecting contamination. One of the interesting ironies associated with the recalls is that segments of the consuming public and some consumer advocate groups have reacted negatively to the recalls. They have cited the recalls as evidence of an unsanitary food supply. Logically, the recalls should be comforting because the product did not get into the consumers' hands. The system worked to detect the problem before anyone became ill as a result. Unfortunately, this has not been the general response. Of course, recalls are not desirable. In 1997, Hudson Foods had to recall 25 million lbs of ground beef that was potentially contaminated with *E. coli* O157:H7. The publicity from the event was quite widespread. One of Hudson's major customers announced that it would no longer purchase from the company. A short time later, the company was sold.

One of the difficult parts of the issue from a consumer-industry perspective is that many of the problems with food and its safety are the responsibility of the consumer or the food handlers. Yet there seems to be an abdication of that responsibility. Whether that is from a lack of knowledge on the part of the consumer, or a general feeling that it's simply someone else's job, is unclear. The following remarks by Michael T. Osterholm, then state epidemiologist and chief of the Minnesota Department of Health's Acute Disease Epidemiology Section, were made in the wake of the massive 1997 recall of ground beef by Hudson Foods:

> Most people desperately want to believe that someone else will look out for them. We'd like to believe that just as we can drive over a bridge and not have to get out of the car and check it for safety, we can be equally confident that the government has declared our food safe. But it's not that easy. Despite our improvements . . . the problem is still present. This is not meant

as a criticism either of those who raise or produce beef or of the public-health community. It is simply very difficult to eliminate the organism. . . . *Newsweek*, Sept. 1, 1997, p. 33.

The livestock producer is responsible for only a very small percentage of the problem. Food safety experts estimate that 77% of all illnesses can be prevented if the food handler and preparer take proper measures. Most foodborne illnesses are caused by foods prepared in the home.

## PREVENTING FOODBORNE ILLNESSES

To prevent foodborne illnesses, food preparers should be aware of a few simple, but highly effective, measures:

- Cook foods thoroughly. Cooking kills most bacteria, parasites, and viruses that are found on foods. Meat, dairy, and egg products should be cooked before being eaten. Ground beef, veal, lamb, and pork should be cooked to a temperature of 160°F. Ground turkey and chicken should reach 165°F. Reheated foods should be heated to at least 165°F (Figure 28–5).

- Prevent cross-contamination. The juices and drippings of raw meat products should not come into contact with other foods, especially those that will receive no further cooking. Be careful of refrigeration practices. Wash hands, cutting boards, spills on countertops, and other items with warm, soapy water to prevent meat juices from coming into contact with other foods.

- Refrigerate foods, including leftovers, properly. Bacteria have a hard time growing in properly refrigerated conditions. Refrigerator temperature should be set at 40°F or lower. Don't overpack the refrigerator. Keep the refrigerator clean (Figure 28–6).

- Select only the freshest meat, poultry, fish, and other food products. Do not purchase dented, bulging, or rusted cans of food. Observe the "purchase by" and "use by" dates on products.

**FIGURE 28–5**   Cooking to the proper temperature is extremely important in preventing foodborne illness. Meat thermometers help ensure that safe internal temperatures are reached during cooking.

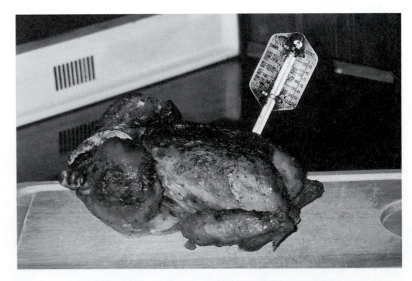

**FIGURE 28–6** For food safety, don't overpack refrigerators. Segregate products, especially those that will not be further cooked. Keep the refrigerator clean and make sure the temperature is no warmer than 40°F. Can you spot the problems in this refrigerator?

■ Freeze fresh meat, poultry, and fish that will not be used within a couple of days. Proper freezer temperature is 0°F. Thaw foods in the microwave, in the refrigerator, or under cold, drinkable running water.

## SOME IMPORTANT MICROBIAL PATHOGENS ASSOCIATED WITH FOODBORNE ILLNESS

The surveillance and identification of new emerging foodborne diseases has always been a challenge for public health officials. Several agencies are all responsible for different aspects of food safety monitoring and testing. Historically, the communication between these agencies has been seriously lacking and as a result, made the tracking and identification of foodborne diseases very difficult. In an attempt to improve the surveillance of foodborne diseases in the United States, the CDC, in cooperation with the USDA and FDA, established in 1995 the Foodborne Disease Active Surveillance Network (Food-Net, http://www.cdc.gov/foodnet). The FoodNet surveillance started with five sites around the United States (West Coast—California; East Coast—Connecticut; South—Georgia; North—Minnesota; Pacific Northwest—Oregon). Officials felt that monitoring a site of population at different regions across the United States would prevent a data overload to the system and provide a better picture of the types of foodborne diseases occurring in the United States. Since 1995, the selected monitoring sites have expanded to 10 states with about 14% of the U.S. population being monitored. FoodNet is helping public officials better understand the *epidemiology* of foodborne disease in the United States. The following information was excerpted and edited from FDA (1997) and is found in Appendix B of that report. It is presented here to provide information about the specific pathogens and to demonstrate that the federal agencies involved do indeed have an understanding of the magnitude of the problem facing them.

**Epidemiology**

Medical service that involves the study of the incident distribution of diseases in large populations and conditions influencing the spread and severity of diseases.

## BACTERIA

### Salmonella

**Septicemia**

Invasion of the bloodstream by virulent microorganisms from a focus of infection.

*Salmonella* species cause diarrhea and **septicemia,** which can be fatal in particularly susceptible persons such as the immunocompromised, the very young, and the elderly. Animals used for food production are common carriers of *Salmonella,* which can subsequently contaminate foods such as meat, dairy products, and eggs. Foods often implicated in outbreaks include raw meats, poultry, eggs, milk and dairy products, fish, shrimp, frog legs, yeast, coconut, sauces and salad dressings, cake mixes, cream-filled desserts, dried gelatin, peanut butter, cocoa, and chocolate. An estimated 2 to 4 million infections occur each year in the United States, most of them as individual cases apparently unrelated to outbreaks. Between 128,000 and 640,000 of those infections are associated with *Salmonella enteritidis* (Figure 28–7) in eggs. During the 1990s, more than 500 outbreaks were attributed to *S. enteritidis,* with more than 70 deaths. In 1994, an estimated 224,000 people became ill from consuming ice cream in one outbreak alone.

### Campylobacter

**Guillain-Barré syndrome**

An inflammatory disease of the peripheral nerves characterized by weakness and often paralysis of the arms, legs, breathing muscles, and face.

The bacterium *Campylobacter* (Figure 28–8) is the most frequently identified cause of acute infectious diarrhea in developed countries and the most commonly isolated bacterial intestinal pathogen in the United States. It has been estimated that between 2 and 4 million cases of campylobacteriosis occur each year with an associated 120–360 deaths. *Campylobacter jejuni* and *Campylobacter coli* (two closely related species) are commonly foodborne, and they are the infectious agents most frequently described in association with **Guillain-Barré syndrome,** as frequently as 1 in 1,000 cases. Approximately 50% of infections are associated with eating inadequately cooked or recontaminated chicken meat or from handling chickens. It is the leading cause of sporadic (nonclustered cases) diarrheal disease in the United States. Unpasteurized milk and untreated water have also caused outbreaks.

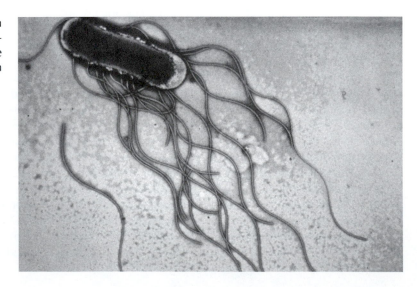

**FIGURE 28–7** *Salmonella enteritidis* is often the cause of foodborne illness. It is a difficult organism from which to protect consumers because animals used for food production are common carriers of the disease. (Courtesy of USDA).

## Shiga-Like Toxin-Producing *Escherichia coli*

Several strains of the bacterium *E. coli* cause a variety of diseases in humans and animals. *E. coli* O157:H7 (Figure 28–9) is a type associated with a particularly severe form of human disease. *E. coli* O157:H7 causes hemorrhagic colitis, which begins with watery diarrhea and severe abdominal pain and rapidly progresses to passage of bloody stools. It has been associated with **hemolytic-uremic syndrome (HUS),** a life-threatening complication of hemorrhagic colitis characterized by acute kidney failure that is particularly serious in young children. *E. coli* O157:H7 is found in cattle, but there may be other reservoirs; the dynamics of *E. coli* O157:H7 in food-producing animals are not well understood. Approximately 73,000 cases of foodborne illness can be attributed to *E. coli* O157:H7 each year, with as many as 60 deaths resulting. *E. coli* O157:H7 outbreaks have been associated with ground beef, raw milk, and minimally processed vegetables and

**Hemolytic-uremic syndrome (HUS)**

A rare condition that mostly affects children under the age of 10; characterized by damage to the lining of blood vessel walls, destruction of red blood cells, and kidney failure.

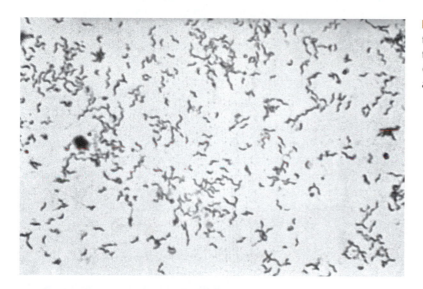

**FIGURE 28–8** *Campylobacter* is considered the most frequently identified cause of acute infectious diarrhea in developed countries. *C. jejuni* and *C. coli* are the infectious agents most frequently associated with Guillain-Barré syndrome.

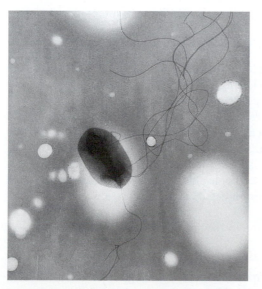

**FIGURE 28–9** Only recognized as a pathogen in 1982, *Eschericia coli* O157:H7 is a gram-negative bacteria that has been associated with HUS, which is a life-threatening condition. Pictured here is a transmission electron micrograph of *E. coli* O157:H7 showing flagella. (Micrograph by Elizabeth H. White. Courtesy of Centers for Disease Control and Prevention.)

fresh fruit juices. A much-publicized outbreak occurred in the spring of 2007 associated with contaminated fresh-cut spinach. Although investigators are still not completely clear about the source of the contamination, it is believed that vectors (likely feral pigs) transferred the *E. coli* from adjacent cattle ranches.

### Vibrio

*Vibrio* species are gram-negative bacteria most commonly associated with seafood dishes. *Vibrio parahemolyticus* is the species most commonly reported as a cause of foodborne disease; it generally causes watery diarrhea and abdominal pain lasting 1–7 days, and commonly follows consumption of improperly handled cold seafood salads. *V. vulnificus* is one of the more serious foodborne pathogens, with a case-fatality rate for invasive disease that exceeds 50%. Most cases of foodborne *V. vulnificus* infections occur in persons with underlying illness, particularly liver disorders, who eat raw mollusks and shellfish. Since the late 1980s, the FDA, the CDC, and the Gulf Coast states have intensified efforts to collect information on *Vibrio* infections, and on the microorganisms' ecology, to improve our ability to prevent foodborne infections.

### PARASITIC PROTOZOA

### Toxoplasma Gondii

*T. gondii* is a parasitic protozoan responsible for some 1.4 million cases of toxoplasmosis and 310 deaths annually. Otherwise healthy adults who become infected usually have no symptoms but might get diarrhea. Pregnant women who become infected can pass the disease to their fetuses. In infants infected before birth, fatality is common. Should the infant survive, the effects of infection are typically severe (i.e., mental retardation). The disease can be life threatening in persons with weakened immune systems and often is fatal to people with HIV/AIDS. *T. gondii* has been found in virtually all food animals. The two primary ways that humans become infected are consumption of raw or undercooked meat containing *T. gondii* or contact with cats that shed cysts in their feces during acute infection. A single cat can shed millions of oocysts after eating rodents, birds, or other animals infected with *T. gondii*. Under some conditions, the consumption of unwashed fruits and vegetables can contribute to infections.

### Cryptosporidium Parvum

*C. parvum* is a parasitic protozoan. The most common consequence of infection in healthy people is profuse watery diarrhea lasting up to several weeks. Children are particularly susceptible. Cryptosporidiosis can be life threatening among people with weakened immune systems. The largest recorded outbreak of cryptosporidiosis was a waterborne outbreak in Milwaukee, Wisconsin, in 1993, affecting more than 400,000 people. More recently, a waterborne outbreak in Las Vegas resulted in at least 20 deaths. The first large outbreak of cryptosporidiosis from a contaminated food occurred in 1993. That outbreak was attributed to fresh-pressed apple cider. *Cryptosporidium* also is found in animal manure. Farm workers have been infected from this source.

### VIRUSES

### Norwalk Virus

Norwalk viruses are important causes of sporadic and epidemic gastrointestinal disease that involve overwhelming, dehydrating diarrhea. An estimated 181,000 cases occur annually with no known associated deaths. In January 1995, a multistate outbreak of viral gastroenteritis related to Norwalk virus was associated with the consumption of oysters. A 1993 Louisiana outbreak of Norwalk virus gastroenteritis involved 70 ill people and was associated with the consumption of raw oysters. In 1992, another outbreak resulted in 250 cases. Outbreaks of Norwalk virus intestinal disease have been linked to contaminated water and ice, salads, frosting, shellfish, and person-to-person contact, although the most common food source is shellfish. Several such outbreaks are believed to have been caused by oysters contaminated by sewage dumped overboard by oyster harvesters and recreational boaters.

### Hepatitis A

Hepatitis A (HAV) is a virus that infects the liver and causes hepatitis A, an illness with an abrupt onset that can include fever, malaise, nausea, abdominal discomfort, dark urine, and jaundice after a prolonged incubation period (e.g., more than 2 months). In children less than 6 years old, most (70%) infections are asymptomatic, but in older children and adults, infection is usually symptomatic, with jaundice occurring in more than 70% of patients. Signs and symptoms of hepatitis A usually last more than 2 months, and there are no chronic consequences. About 130,000 infections with HAV and 100 deaths occur each year in the United States. The primary mode of transmission for HAV is person-to-person by the fecal–oral route. Recognized foodborne hepatitis A outbreaks account for only 2–5% of hepatitis A cases reported in the United States each year, most of which are caused by an infected food handler. Outbreaks owing to foods contaminated before preparation, although uncommon, are associated with widely distributed products such as shellfish, lettuce, frozen raspberries, and frozen strawberries. Hepatitis A can be prevented by good personal hygiene and safe food-handling practices. It can also be prevented before exposure by hepatitis A vaccine, and after exposure by immune globulin, if given within 14 days of exposure.

## GOVERNMENTAL AGENCIES AND FOOD SAFETY

Six different federal agencies bear the responsibility for the government's role in food safety. These include two agencies under the Department of Health and Human Services (HHS)—the Food and Drug Administration (FDA) and the Centers for Disease Control and Prevention (CDC); three agencies under the Department of Agriculture (USDA)—the Food Safety and Inspection Service (FSIS), the Agricultural Research Service (ARS), and the Cooperative State Research, Education, and Extension Service (CSREES); and the Environmental Protection Agency (EPA). This system has long been criticized as being unnecessarily cumbersome and wasteful as well as inefficient in the way it conducts the business of providing a safer food supply. There are changes in the making to address some of these problems. However, it is easy to criticize something with 20–20 hindsight. Remember that this system was built over several decades and across several distinctly different sets of economic and political landscapes. Does it need reworking? Most agree it does, including the agencies themselves. However, most would also agree that, in spite of its flaws, the system has created a very high level of safety in our food supply.

The following sections explore the roles of some of these agencies in assuring a safe food supply with emphasis on animal products. Meat exported into the United

States must come from countries with residue avoidance efforts equivalent to or more rigorous than the U.S. program. As a safeguard, imported meat is reinspected and tested for residues when it enters the country.

## THE ROLE OF THE FDA

The concerns that consumers express over food additives and the hormones and antibiotics used in animal production are largely without foundation. The government requires a rigorous approval procedure before these products can be offered for sale. This system assures the safety of the products used. The FDA is in charge of the approval process. Animal health companies are required to show that any new product they wish to offer for sale is both safe and effective. In addition, the company must provide a reliable method for detecting the drug in slaughtered animals. A new drug is approved for use only after the company has done these things. In addition to its product supervision/approval role, the FDA also limits the amounts of drug residues that can be found in animal tissue. The tolerance level for a given drug is intentionally set at 100 to 1,000 times less than a potentially harmful amount to provide a margin of safety for the consumer. Thus withdrawal times are calculated for additives and antibiotics that ensure the compound is cleared from the animal's body before it goes to slaughter. It then becomes the job of the USDA to monitor tissue samples from slaughtered animals for the compound. The agency currently monitors for over 100 compounds.

The specific FDA requirements for approval of agricultural chemicals and drugs are as follows:

1. The product must be effective at the proposed dosage level for the proposed use.
2. The drug must not create a residue in the edible tissue of the animal or bird that is at a level judged to be harmful to the consumer.
3. The drug or agricultural compound must serve a useful purpose in the production of a feed crop, animal, or bird.
4. An analytical detection method must be available that is capable of detecting the substance at or below the tolerance level.
5. The drug must be used in a manner that will not contaminate the environment or food supply.

The FDA also provides the guidelines to states for regulating the safety of milk, dairy foods, and foods served in restaurants. However, FDA delegates the responsibilities of inspecting restaurants, groceries, and other food-related operations to the states.

## THE ROLE OF FSIS

The USDA, through the FSIS, has a comprehensive program in place to protect consumers from drug and other chemical residues. The FSIS inspection of carcasses at slaughter plants, as one means of protecting consumers from contaminated meat and poultry, is central to the success of the program. The sampling is done in different phases, each with slightly different goals. The comprehensive inspection is designed to ensure the safety of the consumer from drug residues and other substances.

## Residue Monitoring Program

Carcasses are monitored to provide information on the occurrence of residues. Compounds tested include drugs, pesticides, and environmental and agricultural chemicals. In the monitoring program, carcasses are sampled at random in a way that produces a statistically valid sample. This requires the annual sampling of approximately 22,000 animals. Multiple samples are taken from some animals (muscle, liver, kidney, and fat). The goal of the sample procedure is to detect a 1% incidence of illegal residues and have 95% statistical confidence in the result. Results from the monitoring program suggest an industry largely in compliance with withdrawal times and appropriate use of drugs and additives. The number of illegal drug residues detected annually has been less than 1%.

If the residue data suggest an emerging problem, a special surveillance program may be implemented. One such program was initiated in 1977 because the number of sulfonamide residue violations in swine was unacceptably high. A special program of increased sampling that was specific for sulfonamides was implemented. In addition to protecting the consumer in the short run, such programs tend to focus the attention of the industry on the problem. This greatly heightened awareness helps bring the problem under control. When the sulfa drug problem occurred, those involved in industry publications, extension agents, university personnel, news reporters, and others were all working "overtime" on this issue.

The monitoring program is also used as a means of avoiding problems. Individual producers are notified when a sample produces a level of a compound that is found to be within 80–100% of the allowable residue tolerance level. Follow-up testing of animals from that producer helps the producer maintain residue levels below allowable levels.

When illegal substances are discovered, the FDA launches follow-up investigations to determine why the residues are present and who was responsible. Depending on the results of the investigation, the FDA and FSIS may initiate the surveillance program.

## Surveillance Program

The surveillance program is initiated if illegal residues are found or if there are other indications of a potential for illegal residues. The surveillance program may also be used to sample sick or diseased animals. Any time an illegal drug residue is found, FDA can initiate an investigation and FSIS will prevent future shipments from the producer until tissue samples are free of illegal residues. Then, and only then, will the producer's next shipments be approved.

## Residue Avoidance Program

This portion of the overall effort is educational and is designed to help producers avoid problems. It is designed to help producers understand how to prevent illegal levels of residues from occurring. The program seeks to prevent problems from birth through slaughter and processing by educating producers on the proper use of drugs and other chemicals. Producer groups and the extension service have been instrumental in working with FSIS in this portion of the overall effort.

## In-Plant/On-Farm Testing

Several tests are employed to determine if illegal residues are present. The Swab Test on Premises (STOP) is used by public health veterinarians or designated consumer supply inspectors to determine if antibiotics are present. It is fast and can detect problems with a carcass before the carcass leaves the slaughter facility. A recently developed test, Fast Antimicrobial Screen Test (FAST), reduces test time from 8 hours to 6 hours. Field tests have also been developed to screen for potential antibiotic residues. These tests are not USDA/FSIS approved, but they do allow the producer to test animals on the farm. The Live Animal Swab Test (LAST) tests the urine of live animals for illegal antibiotic levels. Live animals can thus be tested before they are marketed. The Sulfa-on-Site (SOS) test screens swine serum, urine, or feed for sulfamethazine. Other tests are used in specific circumstances, and more are under development.

## RESPONSIBILITIES OF THE FEDERAL AGENCIES

The FDA, the USDA, and the EPA have specified responsibilities in the residue monitoring program and responsibilities for working with states to correct detected problems. FDA has the responsibility to investigate illegal residues in animals. In some instances, the responsibility may be delegated to the state agency if there is an agreement with the state. If pesticide residues are the problem, either from direct applications or environmental contamination, EPA joins FDA and FSIS in taking corrective measures because the EPA is responsible for the proper use of pesticides. FSIS has the primary responsibility for the carcass. FDA has the authority for enforcing the proper use of animal drugs and medicated feeds. Both FDA and FSIS can seize or condemn contaminated food. In addition, the FDA has the same power with regard to animal feed. Depending on the nature of the violations, those responsible for illegal residues may face criminal prosecution, resulting in fines or imprisonment.

## CHANGES IN FSIS

FSIS also has had the responsibility for checking carcasses for diseases that may affect the wholesomeness of food. This practice originated at the turn of the 20th century when diseased animals posed a real threat to the welfare of the consuming population. This is rarely the case anymore. By and large, the animals slaughtered for human consumption are young and healthy. In recognition of this fact and in response to the need to direct more effort to foodborne pathogens, the FSIS is in the process of changing the way it works to fulfill this portion of its mission. The agency describes this change as moving from plant-based inspections to a farm-to-table consumer safety system.

The greatest risks to the food supply currently come from microbial pathogens. The type of inspection system that has been in place cannot detect those problems. Recommendations springing from studies conducted by the National Academy of Sciences (NAS), the U.S. General Accounting Office (GAO), and FSIS itself have established the need for fundamental change in the FSIS inspection program. In response to this need, the FSIS is moving to reduce its reliance on the direct inspection of carcasses and shifting to prevention-oriented inspection systems. This change will shift resources away from inspection to a broader approach intended to minimize hazards throughout the farm-to-table food chain and thus reduce foodborne illness.

This change in focus and responsibility is tied to the Pathogen Reduction and Hazard Analysis and Critical Control Point (HACCP) Systems Final Rule, published in July 1996. The Pathogen Reduction and HACCP rule was designed to establish a system that ensures appropriate and feasible measures are taken at each step in the food-production process to prevent or reduce and, ideally, eliminate foodborne disease hazards. FSIS set as its initial goal a 25% reduction in foodborne illnesses attributed to meat and poultry and put a comprehensive program in place to achieve this goal. The program as outlined, published, and widely distributed by the FSIS includes:

- Requiring establishments to develop and implement written Sanitation Standard Operating Procedures (SSOPs) to prevent contamination.

- Requiring that all meat and poultry establishments develop and implement the HACCP system of preventive controls designed to improve the safety of their products.

- Establishing food safety performance standards and microbiological testing requirements.

- Placing clear responsibility for safe products on the establishments producing those products, backed up by rigorous FSIS monitoring and verification, with regulatory action where warranted. FSIS enforcement authorities are used to the fullest extent to address operators who violate food safety regulations and put consumers at risk.

- Enhancing FSIS's present food safety activities beyond slaughter and processing plants. This will include increased oversight of the activities and systems that affect food safety after products leave the plant, including transportation, distribution and retail, and restaurant or food service sale of meat and poultry products. FSIS will work collaboratively with other federal, state, and local agencies to help ensure this coverage.

- Encouraging research, education, and voluntary adoption of preventive strategies on the farm.

- Reshaping its workforce and the way it deploys that workforce. The agency is redeploying its resources and improving the skills and qualifications of its workforce to meet its goal of reducing foodborne illness and providing appropriate regulatory oversight within its statutory authorities along the farm-to-table continuum.

- Deploying the workforce in a manner that will enable it to protect the integrity of the mark of inspection on a product as the product moves from the controlled environment of the plant through distribution, transportation, and retail to the consumer.

These historic changes include changes to the in-plant inspection procedures. Under HACCP rules, plants assume full responsibility and are held accountable for the safety of their products. The focus of FSIS in-plant inspection changes is to verify compliance with the standards and rules. Inspection schedules are being changed. They are becoming less prescribed and instead are being designed to enhance the flexibility and effectiveness of the inspectors and their abilities to make day-to-day decisions about what food safety and consumer protection activities should be carried out in individual plants. Another part of the planned changes includes the shifting of some consumer protection activities that have nothing to do with food safety issues to other agencies or units within FSIS. This will allow inspectors to focus more on specific food safety tasks. Several other changes are being discussed and tested. All plants were required to be in compliance with HACCP as of the year 2000.

FSIS intends to improve its service in a number of ways once HACCP rules and procedures are sufficiently tested. Several of these improvements involve changes in regulatory activities beyond the processing plant. FSIS intends to monitor retail food stores, restaurants, commercial kitchens, hotels, and other institutions. FSIS will work to establish standards for postprocessing transportation, storage, and distribution systems. Educational programs are planned. The objective is to create a seamless national food safety system.

## HAZARD ANALYSIS AND CRITICAL CONTROL POINTS (HACCP)

**HACCP**

Process control system designed to identify and prevent health hazards in food.

The following explanation of HACCP is excerpted from documents published by FSIS to explain the HACCP final rule.

Hazard Analysis and Critical Control Points *(HACCP)* is a process control system designed to identify and prevent microbial and other hazards in food production. It includes steps designed to prevent problems before they occur and to correct deviations as soon as they are detected. Such preventive control systems with documentation and verification are widely recognized by scientific authorities and international organizations as the most effective approach available for producing safe food. HACCP is endorsed by such scientific and food safety authorities as the National Academy of Sciences and the National Advisory Committee on Microbiological Criteria for Foods (NACMCF), and by such international organizations as the Codex Alimentarius Commission and the International Commission on Microbiological Specifications for Foods.

Under the Pathogen Reduction and HACCP systems regulations, USDA is requiring that all meat and poultry plants design and implement HACCP systems. Plants will be required to develop HACCP plans to monitor and control production operations. HACCP was implemented first in the largest meat and poultry plants, with 75% of slaughter production under HACCP-based process control systems on January 26, 1998, with the remainder allowed to be phased in by January 25, 2000.

### THE SEVEN HACCP PRINCIPLES

HACCP systems must be based on the seven principles articulated by the NACMCF. The seven principles are (1) hazard analysis, (2) critical control point identification, (3) establishment of critical limits, (4) monitoring procedures, (5) corrective actions, (6) recordkeeping, and (7) verification procedures.

**Principle 1:** Conduct a hazard analysis. Plants determine the food safety hazards and identify the preventive measures the plant can apply to control these hazards.

**Principle 2:** Identify critical control points. A critical control point (CCP) is a point, step, or procedure in a food process at which control can be applied and, as a result, a food safety hazard can be prevented, reduced to an acceptable level, or eliminated. A food safety hazard is any biological, chemical, or physical property that may cause a food to be unsafe for human consumption.

**Principle 3:** Establish critical limits for each critical control point. A critical limit is the maximum or minimum value to which a physical, biological, or chemical hazard must be controlled at a critical control point to prevent, eliminate, or reduce risk to an acceptable level.

**Principle 4:** Establish critical control point monitoring requirements. Monitoring activities are necessary to ensure that the process is under control at each critical control point. FSIS requires that each monitoring procedure and its frequency be listed in the HACCP plan.

**Principle 5:** Establish corrective actions. These are actions to be taken when monitoring indicates a deviation from an established critical limit. The final rule requires a plant's HACCP plan to identify the corrective actions to be taken if a critical limit is not met. Corrective actions are intended to ensure that no product injurious to health or otherwise adulterated as a result of the deviation enters commerce.

**Principle 6:** Establish recordkeeping procedures. The HACCP regulation requires that all plants maintain certain documents, including hazard analysis and a written HACCP plan, and records documenting the monitoring of critical control points, critical limits, verification activities, and the handling of processing deviations.

**Principle 7:** Establish verification procedures. Validation ensures that the plans do what they were designed to do; that is, they are successful in ensuring the production of safe product. Verification ensures the HACCP plan is adequate; that is, it is working as intended. (Verification procedures may include such activities as review of HACCP plans, CCP records, critical limits, and microbial sampling and analysis.) FSIS requires that the HACCP plan include verification tasks to be performed by plant personnel. Verification tasks are also performed by FSIS inspectors. For example, both FSIS and industry conduct microbial testing as one of several verification activities.

## ADDITIONAL CHANGES AT FSIS AND OTHER FOOD SAFETY INITIATIVES

Additional changes at FSIS are being heralded as a new era for food safety. One of the initiatives at FSIS is to emphasize that everyone has a responsibility for food safety. An important part of these initiatives is emphasizing that the consumer is the final critical link in preventing foodborne illness. To accomplish this, consumer education programs emphasizing "safety from farm to table" have been developed in cooperation with public and private groups to educate consumers on safe food handling. Information is provided by the USDA Meat and Poultry Hotline, through direct consumer inquiries, e-mail, and the FSIS website. The cooperation of the media, extension and public health offices, and other public and private educators has also been sought to distribute food safety information. Various programs have been developed for different audiences and are delivered in different ways. Programs are based on scientifically substantiated information. One major information campaign includes the safe handling information label: *food handling reminders.* Other food safety educational initiatives have made additional information available. Many of these initiatives are in cooperation with the FDA, the CDC, and the EPA. One example is the national education campaign called Fight BAC!, which is sponsored and coordinated by the Partnership for Food Safety Education whose membership includes federal agencies, industry organizations, and consumer groups. The Fight BAC! campaign focuses consumer attention on four critical food safety messages:

- Clean: Wash hands and surfaces often.
- Separate: Don't cross-contaminate.
- Cook: Cook to proper temperatures.
- Chill: Refrigerate promptly.

## THE BIOTERRORISM ACT OF 2002

The events of September 11, 2001, reinforced the need to enhance the security of the United States. Congress responded by passing the Public Health Security and Bioterrorism Preparedness and Response Act of 2002 (the Bioterrorism Act), which President Bush signed into law on June 12, 2002. The Bioterrorism Act is divided into five titles:

- Introduction
- Title I – National Preparedness for Bioterrorism and Other Public Health Emergencies
- Title II – Enhancing Controls on Dangerous Biological Agents and Toxins
- Title III – Protecting Safety and Security of Food and Drug Supply
- Title IV – Drinking Water Security and Safety
- Title V – Additional Provisions

FDA is responsible for carrying out certain provisions of the Bioterrorism Act, particularly Title III, Subtitle A (Protection of Food Supply) and Subtitle B (Protection of Drug Supply). In September 2004, the USDA, in partnership with the FDA and the Department of Homeland Security (DHS), signed a cooperative agreement with the National Association of State Departments of Agriculture (NASDA) to further develop integrated federal-state response plans for food and agricultural emergencies. USDA's Food Safety and Inspection Service (FSIS), FDA, and DHS's Information Analysis and Infrastructure Protection are funding the development of an integrated approach to prepare for and respond to emergencies affecting national agriculture and food infrastructure. The state departments of agriculture gain technical expertise from FSIS, FDA, and DHS officials. Best practices and guidelines for federal and state food regulatory officials will be developed to address lessons learned from case studies and threat assessments. A result of this coordinated effort was the development by NASDA of a Food Emergency Response Plan (FERP) template for states to coordinate their activities with the National Response Plan (NRP).

## ENSURING SAFETY OF THE MILK SUPPLY

The federal Food, Drug, and Cosmetic Act is the legislation that covers the safety procedures for milk and milk products as well as other foods shipped from state to state. FDA is responsible for enforcing the law. The Grade A Pasteurized Milk Ordinance (PMO) is the standard used in the Cooperative State Public Health Service (PHS)/FDA Program for certification of Interstate Milk Shippers (IMS). The PMO was recommended by PHS and FDA. State agencies assume the responsibility for routine inspection and sampling of milk. Each tank of milk from each producer must be tested before it is picked up for transport by the milk handler. In addition, FDA spot-checks hundreds of milk-processing plants annually. According to Anderson et al. (1991), over 99% of all samples test negative for contamination of any kind. If a milk sample is found to be contaminated, the sale of the milk is prevented. The producer is prevented from selling any subsequently produced milk until the milk tests clear of the contamination found in the earlier shipment.

As a part of the PMO, all dairy farms must be inspected at least twice a year. Some states require more frequent inspections. Part of the inspection includes examinations of animal drugs on the premises. Unapproved and/or improperly labeled drugs cannot be used or stored in the milk house, milking barn, stable, or parlor. This helps to ensure proper drug use and residue avoidance. This system has an extremely good record for ensuring the quality of dairy products. Very few incidents of foodborne diseases have ever been linked to dairy products. Most of the incidents that have been connected to dairy products were caused by mishandling during further processing or consumption of raw milk or milk products.

## OTHER ISSUES OF CONCERN TO CONSUMERS

Science has brought a variety of tools to the modern livestock producer. Several of these are of concern to consumers because they are perceived to affect food quality and safety. These products are used because they produce meat and milk more efficiently or they keep animals healthier, benefitting the producer and the consumer. A brief discussion of each product follows, along with information to explain why each is considered safe. It is important to remember that each of these products has FDA approval.

## BOVINE SOMATOTROPIN

Bovine somatotropin (BST) is a protein hormone produced naturally by cattle from the pituitary gland. When BST is administered to dairy cows, their milk production generally increases. Some consumers have been concerned that milk from cows treated with BST will cause negative health effects when consumed by humans. Milk from BST-treated cows is safe for humans to consume and has no ill effects on human health (Figure 28–10). BST was approved by the FDA and has undergone rigorous follow-up study. There are three major reasons why BST use in cows is safe. First, BST is a species-specific protein that is active only in dairy cows. It is a protein that is structurally different from human somatotropin. The receptors for human somatotropin are sensitive to the structure (shape) of the protein. BST does not "fit" into the human structure. Second, over 90% of the BST found in milk is destroyed during the pasteurization process. Third, any BST found in milk is digested in the human digestive system as any other protein would be. (Remember, naturally occurring BST has always been found in milk.) Slightly elevated levels of insulin-like growth factor (IGF-I) can be found in milk from cows treated with BST. This is not a cause for concern for three reasons. First, these levels are not above normal ranges for cows. Second, human milk has higher levels of IGF-I and it has not caused any problems. Third, IGF-I is not biologically active when ingested by humans and is digested by the human digestive system just as BST is. Multitudes of studies have demonstrated that BST has no effect on humans. In what should be the last word on the topic, the Institute of Food Science and Technology, through its Public Affairs and Technical and Legislative Committees, authorized a position statement on June 11, 1998. In the summary, the institute states that objective scientific assessment of the use of bovine somatotropin (BST) to improve milk yield in cows indicates that it carries no harmful effects to humans, to the treated animals, or to the environment. The report goes on to state that the resulting milk and meat are not significantly different from milk and meat from untreated cows,

**FIGURE 28–10** Milk from BST-treated cows is processed along with all other milk for human consumption because BST-produced milk poses no threat to the consumer. (Photo courtesy of U.S. Department of Agriculture. Used with permission.)

in composition or quality; and in consequence there is no scientific or ethical basis for requiring distinctive labeling of milk or meat from BST-treated cows.

## HORMONES

Hormonal growth promotants (implants) are a tremendous tool for animal production, especially for beef production (Figure 28–11). Benefits include faster growth rate, improved feed conversion, increased amount of lean tissue gain, and decreased fat deposition. On an industry-wide basis, hormone implants improve beef production by an estimated 750,000 lbs and save 3 million tons of feed. They are a very cost-effective tool for beef production. Consumers' concerns about growth promotants are directed at whether or not the hormonal compounds can be found in the meat of implanted animals. The level of hormones that could be consumed from implanted cattle is less than 0.006% and 0.02% per 100 g consumed of estrogen and progesterone found in prepubescent boys who carry the lowest level of both hormones (CCABIC, 2006). In fact, many other foods have higher levels of hormones. According to the Council for Agricultural Science and Technology, "In a meal of mashed potatoes, whole wheat bread, green salad, green peas, and ground round steak from estrogen-treated cattle, the food that would contain by far the least estrogenic potency is the ground round steak."

## ANTIBIOTICS

Antibiotics are used in animal production to treat animals suffering from infections. In addition, subtherapeutic antibiotic levels are used to maintain the health of pigs, veal calves, and poultry. This second use has been declining in recent years. Consumers are concerned with the potential for development of superbugs that are resistant to antibiotics that might threaten human health. It is important that consumers be aware that antibiotic use is tightly regulated. The length of time an antibiotic must be taken away

**FIGURE 28–11**   Hormonal implants of several types are used in animal production. Consumers express concern over these products. All such implants must meet rigorous approval procedures before they are licensed for use. (Photo courtesy of W. Humphrey, Mississippi State University. Used with permission.)

from the animal before it goes to slaughter, or its milk can be used for feed, is called its *withdrawal time.* These withdrawal times are closely observed. Antibiotics are among the compounds that the FSIS Monitoring Program takes special interest in. Milk, meat, and poultry all have extremely good records of compliance with proper antibiotic use.

## FOOD IRRADIATION

Irradiation, or ionizing pasteurization or *electronic pasteurization (e-beams)* or cold pasteurization as it is also called, subjects the product to radiation from radioactive or machine sources. Three types of rays are used: gamma rays, electron beams, and X-rays. Gamma rays use a source of radioactive material (Cobalt 60 or Cesium 137). They emit high-energy protons that do not make the irradiated product "radioactive." This is an old technology that has been used for years to sterilize medical, dental, and household products. Electron beams, or e-beams, are produced by an electron gun. No radioactive material is used. This technology has been used in the medical field for at least 15 years. X-ray technology is the newest technology. Only commercial units have been built since 1996, and like e-beam, it does not use radioactive material (CDC, 2007). The radiation kills insects, pathogenic bacteria, and parasites (Figure 28–12). Irradiation does not make food radioactive. Its many uses include preservation of food by destroying organisms that cause spoilage and decomposition, thereby extending the shelf life of foods; sterilization so that foods may be stored without refrigeration; controlling sprouting, ripening, and insect damage in potatoes, tropical and citrus fruits, grains, spices, and seasonings; and controlling foodborne illness by destroying pathogens that cause foodborne illness. This latter use has tremendous implications for controlling *Escherichia coli* O157:H7 and *Salmonella* species. Irradiation, although a potentially useful tool for helping reduce the risks of foodborne disease, is a complement to, not a replacement for, proper food-handling practices by producers, processors, and consumers.

The history of food irradiation dates to 1895 when the first paper was published on the idea of irradiating food. Over the course of the 20th century, the process was proven

**Withdrawal time**

The length of time an antibiotic must be taken away from an animal before the animal can be legally slaughtered.

**Electronic pasteurization (e-beams)**

Ionizing radiation from a focused beam of energy created by the acceleration of electrons using magnetic and electric fields.

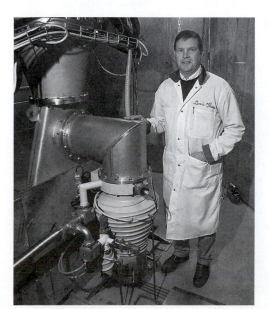

**FIGURE 28–12** Dennis Olson, director of irradiation research at Iowa State University, stands next to the linear accelerator in a lab on November 3, 1997, in Ames, Iowa. Research at Iowa State and elsewhere shows that *E. coli, Campylobacter, Salmonella,* and other dangerous pathogens are destroyed when they are zapped with radiation. (Photographer Charlie Neibergall. Courtesy of AP/Wide World Photos. Used with permission.)

effective and approved for many foods such as spices, fruits, vegetables, and grains in both the United States and several other countries. The first approval of its use on animal products in the United States came about when the FDA, in July 1985, and FSIS, in January 1986, issued rules to allow pork to be irradiated to control *Trichinella*. Subsequent rules have declared irradiation safe for poultry, raw meat, and fresh eggs. Additional products are in the process of being approved, and more will be submitted for approval.

## MAD COW DISEASE

Bovine spongiform encephalopathy (BSE) is a disease in cattle. It is a degenerative, central nervous system disease that causes cattle to become nervous, lose coordination, lose weight, and have difficulty walking. Cattle die 2 weeks to 6 months after these symptoms appear. The incubation period is 5–8 years. In March 1996, it was announced that a new variant of Creutzfeldt-Jakob disease in humans might be linked to BSE. Evidence accumulated since that time strongly suggests this is the case.

It is believed that the disease is spread in cattle through the feeding of mammalian-derived protein by-product feeds. On June 5, 1997, the Food and Drug Administration Center for Veterinary Medicine published a rule banning mammalian-derived protein by-product feeds for all ruminants to protect the cattle population of the United States. Since 1989, the United States has restricted importation of ruminants, ruminant products, and ruminant by-products from BSE-positive countries.

Despite all of these precautions, on December 23, 2003, the FDA was alerted to the first U.S. case of bovine spongiform in the state of Washington. A "downer" cow that suffered partial paralysis as a result of birthing difficulties had been slaughtered and, as required, a sample was sent to a USDA laboratory in Ames, Iowa, for BSE testing. Test results were positive. Unfortunately, because the results came back about 2 weeks after slaughter, the carcass had already been processed. Edible meat had been converted to hamburger and steak and edible by-products were ground and rendered to make animal feed and produce fat for soap and other products. Once notified of the positive test, the FDA took immediate action and notified the public. During the next month, over 30 officials from the FDA and various state agencies would accomplish the diligent task of tracing down all the infectious material. Within 96 hours all potentially inedible by-products had been found. By the end of the month, over 2,000 tons of meat and by-product had been destroyed.

No other BSE-infected cows were found. Quick response and the cooperation and dedication of many professionals helped to minimize damage and protect the consumer. As a result of this incident, FSIS quickly worked to implement further protections against BSE. A surveillance program to test high-risk cattle was implemented. In addition, effective December 30, 2003, carcasses from cattle intended for human food that are sampled and submitted to APHIS for BSE testing are held from further processing until results are reported. In addition, the definition of meat was clarified by FSIS not to include brain, trigeminal ganglia, spinal cord tissue, or dorsal root ganglia (all of which are central nervous system–type tissues).

Finally, slaughter and processing establishments are required to develop procedures that demonstrate specified risk materials (SRMs) are removed and not present in meat. SRMs include the brain, skull, eyes, trigeminal ganglia, spinal cord, and vertebral column and dorsal root ganglia of cattle 30 months of age and older. SRMs also include the tonsils and distal ileum of all cattle (the removal of distal ileum requires removal of the entire small intestine). SRMs have been banned because science indicates this is where infectious material accumulates.

## GENETICALLY ENGINEERED PRODUCTS

Consumers abroad, and to a lesser degree in the United States, have expressed concern about consuming genetically engineered food products. Most of this controversy has centered on plant products. However, it is inevitable that those concerns will also be directed at animals fed those plant products and then ultimately at transgenic animals themselves.

Livestock have been fed products from plants whose genetics have been altered by recombinant DNA technology (biotech crops) since those crops were first introduced in 1996. Two important types of biotech crops are crops tolerant to pesticides and crops protected from insect pests. Expected in the future are more biotech crops, many with enhanced levels of nutrients or other beneficial substances in the plant.

For those biotech crops currently available, both the levels of nutrients and anti-nutrients are the same as conventional crops. When fed to livestock, livestock digest and absorb the nutrients from biotech crops just as they do from conventional ones. Livestock grow and produce when fed biotech crops just as they do when fed conventional ones. The meat, milk, and eggs produced from animals fed biotech crops are the same as from conventional feeds. The proteins found in the biotech plants are broken down into smaller components during digestion and thus do not become part of the animal products. Thus meat, milk, and eggs from animals fed biotech feeds are no different than those from animals fed conventional feeds, and they are safe for human consumption.

The use of animals with genetic alteration and of cloned animals as food is a consumer issue in the United States. At press time, the FDA is preparing to allow the first biotech meats and meat products onto the market. The issues have been debated for a number of years. It is certain that once the products are allowed on the market, the debate will continue rather than subside, at least for a time.

## SOURCES OF INFORMATION

More information on the **Partnership for Food Safety Education** and the **Fight BAC!** campaign is available through the Internet at http://www.fightbac.org/. Because of interest in BAC and food safety materials, the partnership now offers a number of items for sale through the BAC Store.

The **National Food Safety Information Network** was part of the Food Safety Initiative of the Clinton administration. FSIS and FDA are working together to coordinate food safety information through the network. The network will coordinate and link work currently being done by both agencies' hotlines and enhance services provided by the USDA/FDA Foodborne Illness Education Information Center, which develops and maintains a database of education materials. Food safety information from FSIS is available on its website at http://www.fsis.usda.gov. Interagency food safety information is available at http://www.foodsafety.gov.

**EdNet** is an electronic network allowing food safety educators around the country to keep abreast of new federal food safety education projects. It is a direct e-mail communication from the federal government. To subscribe, send the following message to *listserv@foodsafety.gov:* Subscribe EdNet-L firstname lastname (substitute your name for firstname and lastname).

The **Food Safety Educator** newsletter, produced by FSIS, provides a forum to communicate new initiatives and research information. The newsletter can be accessed through the FSIS website at http://www.fsis.usda.gov/OA/educator/educator.htm. To subscribe, send or fax your name and mailing address to USDA/FSIS, Food Safety Education Staff, Room 1180-South Building, Washington, DC 20250; fax (202)720-9063. Or e-mail your name and mailing address to *fsis.outreach@usda.gov.*

**USDA Meat and Poultry Hotline**. (202)720-3333; TTY: 1(800)256-7072; 1(888) 674-6854.

**General food safety information**. For further information, contact FSIS Food Safety Education and Communications Staff, Public Outreach and Communications; phone: (202)720–9352; fax (202)720-9063; http://www.usda.gov/.

**HACCP** information. Contact: FSIS Food Safety Education and Communications Staff, Public Outreach and Communications; phone: (202)720-9352; fax (202)720-9063.

## SUMMARY AND CONCLUSION

**Emerging pathogens**

Pathogens that have mutated to become more virulent or have only recently been recognized as a safety issue.

The food supply in the United States is arguably the safest in the world. However, in spite of the safety of the U.S. food supply, people still become ill from their food. Most illnesses are traced to one of "seven bad bugs." Actually, around 14 organisms cause problems with enough frequency to be of concern. Several of these are referred to as ***"emerging" pathogens*** because the problem has just begun to be realized or because some mutation in the microorganism itself has made it more of a problem than it previously was. Understandably, food safety has become an important issue with consumers. However, consumers do not always do a good job of evaluating what is and what is not a real risk. Nor are consumers especially good at taking responsibility for the safety of their own food supply.

Even though the overwhelming majority of foodborne illnesses could be prevented if better choices were made by the food preparer, the tendency of the average consumer is to look to the government to provide food safety. This task is not totally possible in a practical world. However, government agencies do have a role, and they are trying to expand that role. Various agencies in the federal government are responsible for segments of the food safety network. This is a system that has evolved since the passage of the Pure Food and Drug Act of 1906 and the Meat Inspection Act of 1906. Currently, these agencies are undergoing a transformation in the way they do business. The Food Safety and Inspection Service is dramatically changing its practices. This has been made possible by the adoption of the Pathogen Reduction Hazard Analysis and Critical Control Point Systems Final Rule, published in July 1996. Other changes are unfolding and many are still in the planning stage.

## STUDY QUESTIONS

1. Describe the real versus perceived threats that the food supply poses to the average consumer. What are some events of the 1990s that focused the attention of the consumer on food safety?

2. What is the GRAS list? The Delaney clause? The Food Quality Protection Act of 1996? How do they relate to food safety?

3. What are the "bugs" that cause the majority of foodborne disease?

4. What role does the consumer have in protecting and preparing his or her own safe food supply?

5. Describe the FDA's role in ensuring a safer food supply.

6. What is the role of FSIS in ensuring a safer food supply? What are the purposes of the FSIS's monitoring program? Surveillance program? Residue avoidance program? In-plant/ on-farm testing?

7. FSIS is currently implementing and evaluating for implementation a myriad of changes in how it does business. Describe what the goals of the changes seem to be.

8. What is the Partnership for Food Safety Education and its Fight BAC! campaign? Where can you find more information about these programs?

9. Describe HACCP, its purpose, the seven HACCP principles, and when and where it was implemented.

10. Describe the safety/danger of using bovine somatotropin on cattle as it relates to the human food chain.

11. Is the use of growth-promoting hormones on food-producing animals a safe practice? Why or why not?

12. What are the benefits and risks of food irradiation? Would you eat irradiated food? Why or why not? Do you think you have already consumed any irradiated food? Why do you think consumers have been reluctant to consume foods they knew to be irradiated? If your job was to develop an advertising campaign to convince consumers to eat irradiated products, what would your slogan be?

13. What changes in animal products occur when producing animals are fed genetically engineered plant products?

14. What is FoodNet and what is its purpose?

## REFERENCES

For the second edition, Dr. Christina DeWitt, Assistant Professor of Animal Science, Oklahoma State University, reviewed the chapter. In addition, Dr. DeWitt contributed new material to the chapter. For the third and fourth editions, Dr. DeWitt assumed coauthorship of the chapter. The author gratefully acknowledges her contributions.

Anderson, P. T., B. A. Crooker, and M. M. Pullen. 1991. *Animal products: Contributors to a safe food supply.* University of Minnesota, Educational Development System, Minnesota Extension Service.

Buzby, J. C., T. Roberts, C. T. J. Lin, and J. M. MacDonald. 1996. *Bacterial food-borne disease: Medical costs and productivity losses.* Food & Consumer Economics Division, Economics Research Service, U.S. Department of Agriculture. Agricultural Economics Report No. 741. 1301 New York Ave NW, Washington D.C. 20005-4788.

CCABIC. 2006. *Understanding hormone use in beef.* FAQ sheet prepared by the Canadian Cattlemen's Association & Beef Information Centre. Accessed online July 2, 2007. www.cattle.ca/factsheets/hormal.pdf.

CDC. 2007. *Food irradiation.* Center for Disease Control & Practice. Report of Health & Human Services. Accessed online July 2, 2007. www.cdc.gov/rcidod/dbmd/diseaseinfo/foodirratiation.htm#whatis.

Crooker, B. A., D. E. Otterby, J. G. Linn, B. J. Conlin, H. Chester-Jones, L. B. Hansen, W. P. Hansen, D. G. Johnson, B. D. Marx, J. K. Reneau, M. D. Stern, J. F. Anderson, B. E. Seguin, J. D. Olson, R. J. Farnsworth, and W. G. Olson. 1994. *Dairy research & bovine somatotropin.* University of Minnesota Extension FO-06337. Accessed online July 2, 2007. www.extension.umn.edu/distribution/livestocksystems/DI6337.html.

FDA. May, 1997. *Report to the president. Food safety from farm to table: A national food safety initiative.* Food and Drug Administration, U.S. Department of Agriculture, U.S. Environmental Protection Agency, Centers for Disease Control and Prevention.

FSIS. 1998a. *Key facts: HACCP final rule.* Revised January 1998. Washington, DC: Food Safety and Inspection Service. http://www.fsis.usda.gov/index.htm.

FSIS. 1998b. Various reports. Washington, DC: Food Safety and Inspection Service, U.S. Department of Agriculture.

Mead, P. S., L. Slutsker, V. Dietz, L. F. McCaig, J. S. Bresee, C. Shapiro, P. M. Griffin, and R. V. Tauxe. 1999. Food related illness and death in the United States. *Emerging Infectious Diseases* 5(5): 1–38.

Smolin, L. A., and M. B. Grosvenor. 1997. *Nutrition science and applications.* 2nd ed. Orlando, FL: Saunders.

USDA. 2004. *Bad bug book. Foodborne pathogenic microorganisms and natural toxins handbook.* Available online at http://vm.cfsan.fda.gov/~mow/intro.html.

# Animal Welfare and Animal Rights

## LEARNING OBJECTIVES

After you have studied this chapter, you should be able to:

1. Describe the basis for the general concern, but the lack of a consensus opinion, relating to animal welfare and animal rights.

2. Compare and contrast animal welfare issues and animal rights issues. If possible, reconcile whether animal rights and animal welfare are different or the same issue to you.

3. Cite the major pieces of legislation that have been passed in the United States regarding animal welfare and animal rights.

4. Describe the major philosophical differences among various groups that have an interest in this subject.

5. Outline a view of the issues likely to be debated for legislative action.

## KEY TERMS

| | | |
|---|---|---|
| Abolitionists | Downer cow | Speciesism |
| Animal rights | Reformists | 3 Rs |
| Animal welfare | Sentient | Vegan |
| Anthropomorphism | | |

## INTRODUCTION

Since the 1980s, no other animal-related social issue has generated as much emotion, rhetoric, and ill will as the discussion surrounding animal welfare/animal rights. Some of the uproar is simply a function of how a democratic society resolves disputes in the postmodern age. When a major social issue emerges and beliefs and policies are challenged, anger and conflict arise. Media hype, distorted information dissemination, emotional responses, and chaotic debate usually follow. Such is true of this debate. Yet there is more to this issue. People on both sides are affected in a very visceral way. Belief systems are challenged, people feel threatened, and reason is often hard to find.

Additional factors add to the confusion on this issue. Perhaps the most important is that this is not just one issue, but many, often lumped together in people's minds. The animal rights/animal welfare issue affects *sentient* animals used for every conceivable purpose, including those used for food, research, companionship, and recreational activities. Thus many levels of society are affected (Figure 29–1). In addition, this issue surfaced at a time when agriculture was experiencing an economic crisis of such magnitude that it changed the structure of agriculture profoundly, and with it the lives of agriculturists. The biomedical research community was also facing tough economic times, yet rapidly increasing the world's knowledge and ability to help humanity with cures and preventatives for a myriad of diseases. The research community was outraged at the notions of those who would say, "A rat is a dog is a pig is a boy" as justification for ending lifesaving research. When hunters, trappers, and product testers also came under attack, a wide spectrum of the population discovered this issue affected them.

The idea of giving an animal rights is rooted collectively in several philosophical ideas and theologies, some of which come to us from ancient times. Thus the notion of having animal rights tied up in the human value system is certainly not a new one. Writings from many religions have successfully incorporated animal use, rights, and

**Sentient**

Creatures that experience pain and pleasure.

**FIGURE 29–1** Animal rights/animal welfare. The animal rights/animal welfare issue includes concerns over animals used for every conceivable purpose.

admonitions about care into human mythology. Some assert animal equality on several levels, and the very existence of these writings serves to point out that humans have long considered the relationship to animals as something to ponder. Why then has this issue gained so much attention at this point in history?

The answer is complex. Several factors have contributed. In many ways, attitudes on animal use have changed because of startling advances in technology, and the efficiency those advances have brought. Technological innovation has led to profound changes in society's structure, organization, and goals. Agriculture's efficiency has freed millions of people to pursue other avenues of livelihood. This has left few people with any contact with agriculture and no reason to share the modern agrarian ethic. If they have any concept of the agrarian ethic at all, it is either that of the antique or relic family farm of 50 years ago, or it is one developed as a complete outsider of the modern reality of agriculture. Being so far removed from agriculture, the general population does not have the same contact with animals that previous generations had (Figure 29–2). Their most common contact is with companion animals that yield no economic benefit to their owners and are often treated as a member of the family. As Rollin (1995) points out, "a primarily utilitarian view of animals has been superseded by a more personal and comradely view." This, as well as popular culture icons such as those in Disney movies, has led many people to a highly **anthropomorphic** view of animals. With this view, it is hard for them to be objective and critical of information they receive. Technological innovations in human health are also the products of highly specialized production systems. Yet most people do not really understand that they are living longer because of research done with animals because they have no contact with the research or the researchers that make the discoveries. Nor do they tie medicinal production or product safety to animals. In sum, they do not really understand how much of their way of life depends on both agricultural and research animals. Even if they do hold such general understanding, the philosophies of the animal rights movement make the case that the animal rights philosophy does not allow humans the luxury of these pragmatic reasons for using animals. Those who argue for animal rights hold the view that humanity's quest for efficiency, productivity, knowledge, medical progress, and product safety is responsible for most of animal suffering and that these ends do not justify the means.

**Anthropomorphism**

Attributing human thoughts, emotions, and characteristics to animals, gods, objects, and so on.

**FIGURE 29–2** Agriculture's efficiency has freed millions from the need to grow their own food. With no ties to agriculture, most people have no reason to share the agrarian ethic. (Photographer C. W. McKeen. Courtesy of The Image Works. Used with permission.)

At the core of the focus on farm animal rights is the fact that farming is very different today than it was a half century ago. That has led to a great difference in the way animals are treated today in "production systems" compared to the more traditional role and treatment of animals in yesteryear's small family farm. Today's farm is often called a "factory farm" and the implication is intended to condemn. As Rollin (1995) points out, "The key feature . . . of traditional agriculture was good husbandry." In this traditional view, the animal's interests and the farmer's interests were so closely intertwined that few questioned that animal care was as good as it could be. Thus very few would ever suggest that there should be laws dictating husbandry practices. However, modern, highly productive, industrial approaches to agriculture have little resemblance to the quaint family farm of the first part of the 20th century. Barnyard chicken flocks have been moved into battery cages (Figure 29–3). Baby calves that once ran by their mothers have been put into veal crates (Figure 29–4). Fenced pastures have been replaced by concrete and cable enclosures. Although many animals can be housed

**FIGURE 29–3** Modern confinement systems of animal production, such as battery cages for laying hens, bear little resemblance to production practices of the past. Many people are uncomfortable with what they view as "factory farming."

**FIGURE 29–4** The modern practice of raising veal calves in crates has proven particularly irksome to animal rightists and some animal welfarists.

efficiently and food can be grown more economically in these modern systems, the general population does not have the comfort level with modern animal agriculture that preceding generations had with traditional agriculture.

Add to all of this the fact that activism is an established phenomenon of the post-modern age. The Vietnam War, women's issues, and the environmental movement were training grounds for a very competent breed of activist (Figure 29–5). People are also better educated on the whole than at any other point in history, and they are richer with more discretionary time. Added to this is the influence of several very effective writer/philosophers. The writings of Jeremy Bentham in 1789, Albert Schweitzer in 1965, Peter Singer in 1975, and Tom Regan in 1983 have all influenced modern thinking. Bentham is credited by some as the originator of the animal rights movement. Singer's influence cannot be underestimated. The publication of Ruth Harrison's *Animal Machines—The New Factory Farming Industry* in 1964 brought agriculture and its animal production methods under real public scrutiny. The establishment and emergence of People for the Ethical Treatment of Animals (PETA) must be viewed as a pivotal event in the animal rights activism observed in this country today. During the decade of the 1980s, PETA and other restructured animal welfare/animal cruelty groups were able to organize, collecting millions in donations, and influence many Americans to become more animal rights minded. They successfully took their idea through the emerging stages of social consciousness and have turned it into a contemporary social issue. That is no small feat, and their skill and tenacity in this accomplishment was really extraordinary.

The new animal ethic accepted and openly advocated by some, while stringently opposed by others, presents challenges to individuals who work with animals in any capacity. Who prevails, on what issues, and how society comes to view the overall issues will determine the standards society adopts for all in the future. People educated on the issues and willing to enter the public debate should determine those standards. That debate is far from over. However, both providing and receiving education on emotional and controversial issues are usually difficult. When we are angry and embroiled in conflict, opinion is easily polarized. This makes it hard to get information one can trust as being objective, and hard to truly listen to it when it is presented. Even so, responsible citizenship demands of us the ability to differentiate between opinion and

**FIGURE 29–5** Activism is an established part of postmodern life. A very competent breed of activist has been generated through such issues as the Vietnam War, the women's movement, and the environmental movement. (Photo courtesy of AP/Wide World Photos. Used with permission.)

fact, emotional responses from objective ones, and rhetoric from information. With so much at stake, it is necessary for us all, on both sides of any issue, to come to understand the varying viewpoints. It is also morally incumbent that we find a position that is ethically defensible to support in the debate. Regardless of which position we adopt, it is further important that we be willing to inform others of our position and defend it.

## ANIMAL RIGHTS VERSUS ANIMAL WELFARE

**Animal welfare**

The treatment and well-being of animals while they provide for human needs; humane use.

Although lines between the two tend to blur, animal rights philosophy and animal welfare philosophy are generally argued as separate issues. *Animal welfare* concerns began with the first domesticators of animals and have continued to the present with progressively responsible animal husbandry practices. The animal welfarist is concerned with an animal's treatment and well-being while the animal provides for human needs. The basis of the welfarist position is the idea that using animals obligates people to tend to basic needs considered to be good husbandry. These needs include feed, water, protection, shelter, health care, alleviation of pain and suffering, and other similar needs. Most would call these the necessary elements of humane care. In agriculture, providing for animal welfare determines whether or not animal production systems make money. Attending to welfare is an issue of biology (Figure 29–6). Providing for the welfare of an animal does not necessarily require giving it rights. Agriculturists fear that ascribing rights to animals will affect costs. As long as ascribing rights to animals does not change the way the production systems function, then there are no costs associated with the rights. However, if ascribing rights to the animals causes changes to be made purely for the sake of animal rights, then the rights have economic consequences.

**Animal rights**

Philosophy, sociology, and public policy as they deal with the standing of animals in relation to human society.

*Animal rights* deals with philosophy, sociology, and public policy as they apply to the standing of animals in relation to human society. The most extreme animal rightists assign rights to animals that most human societies reserve for people alone, thus removing the moral barriers between people and animals. Their view is that animals have a right not to be used by humans for any purpose. To those taking the extreme

**FIGURE 29–6** The animal welfarist is concerned with the animal's treatment and well-being while it is being used to provide for human needs.

**FIGURE 29–7** Animal rights deals with philosophy, sociology, and public policy as they apply to animals in relation to humans. The most extreme animal rights position ascribes rights to animals that most reserve for humans alone. (Photo courtesy of Paul Conklin. Used with permission.)

animal rights position, the idea of humane care is an oxymoron (Figure 29–7). Of course, not all animal rights supporters take the extreme position. Many are more moderate and practical in their approach and work to abolish what they consider unacceptable situations of animal suffering. Many animal rightists support animal use if they can believe that both animal and human benefit. An example is keeping animals as pets. By contrast, the extreme animal rights position views pet keeping as an inappropriate exploitation that should cease (Figure 29–8). To some, biomedical research with animals is acceptable, but calf-roping events at rodeos are not. To those who hold this position, the ultimate value of the use is what decides whether the use is justified (Figure 29–9). The anti-use animal rights philosophy leaves no room for compromise and is not open to change. It is based on a philosophy of no use. Many who subscribe to this view are *vegans,* people who use no animal products of any kind. By contrast, treatment of animals under a pro-use animal rights philosophy can change as the expectations of the greater society change and as new scientific understandings of animals develop. Some rights are already preserved in law, such as humane slaughter, and are not considered extreme by any segment of society. The new concern probably stems from the change in focus of animal rights advocates. Prior to the 1980s, almost all animal rights concerns focused on cruelty issues and generally exempted animals used for economic and recreational benefit. This distinction kept animal agriculture, research, racing, rodeo, hunting, and other similar uses from being considered in the cruelty laws. Animal rights advocates now include all of these animal uses in their agenda for change. Laws and policies are beginning to reflect the change in thought.

With this difference in philosophy, we naturally see different kinds of rights advocates—usually either reformists or abolitionists. *Reformists* may believe in the views of philosophers of the movement but generally are willing to work within the system to achieve their goals. *Abolitionists* focus on promoting total abandonment of any animal use. Some abolitionists have resorted to violence, vandalism, and theft as methods to destroy the veal and fur industries, and to stop product testing, animal research,

**Vegan**

Someone who eschews the use of any animal product, including the nonfood products.

**Reformists**

Animal rights proponents who focus on changing methods of animal use.

**Abolitionists**

Animal rights proponents who advocate the total abandonment of any animal use.

**FIGURE 29–8**  Among the most extreme animal rights views is the idea that pet keeping is an inappropriate exploitation of animals that should cease. (Photo courtesy of Whitney Linsner. Used with permission.)

**FIGURE 29–9**  To some animal rightists, biomedical research with animals is acceptable, but rodeo events, such as calf roping, are unacceptable exploitation. (Photographer Fotopic. Courtesy of Omni-Photo Communications, Inc. Used with permission.)

**Speciesism**

Placing the interests of one species above that of another species.

and hunting. Regardless of the tactics, their goal is to stop animal use. Most members of these abolitionist groups use the term *speciesism,* or the placing of the interests of one's own species above the interests of another, to describe what they view as mankind's unconscionable arrogance against other species.

Bernard Rollin (1995) argues that the new animal ethic is not defined by the abolitionist view and that those opposed to animal rights make a mistake by characterizing the mainstream movement in abolitionist terms. His argument with specific regard to agriculture follows:

> Rather, it is an attempt to constrain *how* they can be used, so as to limit their pain and suffering. In this regard, as a 1993 *Beef Today* article points out, the thrust for protection of animal natures is not at all radical; it is very conservative, *asking for the same sort of husbandry that characterized the overwhelming majority of animal use during all of human history, save the last fifty or so years.* It is not opposed to animal use; it is opposed to animal use that goes against the animals' natures and tries to force square pegs into round holes, leading to friction and suffering. If animals are to be used for food and labor, they should, as they traditionally did, live lives that respect their natures. If animals are to be used to probe nature and cure disease for human benefit, they should not suffer in the process. Thus this new ethic is conservative, not radical, harking back to the animal use that necessitated and thus entailed respect for the animals' natures. It is based on the insight that what we do to animals matters to them, just as what we do to humans *matters* to them, and that consequently we should respect that mattering in our treatment and use of animals as we do in our treatment and use of humans. And since respect for animal nature is no longer automatic as it was in traditional agriculture, society is demanding that it be encoded in law.

Not all people agree that animal welfare and animal rights are separate issues. Some have begun to argue that continuing to separate the ideology of animal welfare from animal rights is no longer very practical. If it is true that the vast majority of society believes in both animal use and animal rights, then the argument has merit. Rollin (1995) suggests a new way of confronting the issue: "[I]f it is the case that the notion of animals rights is a pivotal part of the emerging mainstream social ethic for the treatment of animals, there is little value in maintaining the dualism of animal welfare versus animal rights. It is rather that the traditional notion of animal welfare is being socially augmented and explicated by the notion that animals have certain rights." However, it will be difficult for some in society to accept this view. Traditional agriculturists are highly unlikely to agree anytime soon to the idea that there is no difference between animal welfare thinking and animal rights thinking. Nor are animal rights supporters likely to be in accord with this idea. In *Rain without Thunder,* Gary Francione (1996) states, "The welfarists seek the *regulation* of animal exploitation; the rightists seek the abolition. The need to distinguish animal rights from animal welfare is clear not only because of the theoretical inconsistencies between the two positions but also because the most ardent defenders of institutionalized animal exploitation themselves endorse animal welfare."

Animal agriculturists usually take the position that they are practicing animal welfarists because a well-cared-for animal performs better and thus is more profitable. However, we should also be aware that social issues do not generally arise without provocation. Animal rightists have raised concerns about some common practices in the animal industries that reasonable people in the livestock industry have been fairly quick to support. The entire *"downer cow"* handling that animal rightists exposed at some slaughter plants provoked a cry of outrage from many levels of agriculture.

Clearly, it is difficult to discuss this issue without becoming confused about who is a rightist and who is a welfarist, and what each wants. As a means of distinguishing between animal welfare and animal rights, the author favors using the term *animal rights* in its political sense. In this sense, animal rights describes ideas and actions that have potential political action as a possibility or goal. It pertains to such goals regardless of the species. Animal rights, then, becomes any change or attempted change in the legal status of animals that challenges the status quo. *Animal welfare,* in the author's view, refers to well-being and care in its biological sense.

**Downer cow**
A nonambulatory cow.

# PHILOSOPHY, HISTORY, AND LEGISLATION

Modern animal rights philosophy has been influenced by many thinkers and their ideas. However, the philosophies of Peter Singer (*Animal Liberation,* 1990) and Tom Regan (*The Case for Animal Rights,* 1983) have unquestionably had tremendous influence. It is outside the scope of this text to put forth the particulars of their philosophies. Gary Francione (1996) does a good job of distilling these two theories to their basics in the first chapter of *Rain without Thunder* for those of you who wish to explore them further. Of course, a thorough understanding requires a reading of the books (and considerable reflection on the ideas). The point that is important for this discussion is that there have been philosophies put forth for people to believe in. The implications for the animal movement have been enormous.

Pro-animal use has a philosophical base also. In most of the Western world, including the United States, pro-animal use philosophy is based on the Judeo-Christian philosophy. The justification for animal use is essentially tied to the Old Testament concept of humans' dominion over animals. Interestingly, few people who oppose animal rights or abolitionist views ever evoke their philosophical reasons for doing so. Instead, they fall back on pragmatic arguments like "research saves lives" and "animal products help feed the world." Animal rights arguments, however, are usually based on adherence to a philosophical stance that is in opposition to the use of animals for any use. Peter Singer himself acknowledges the relationship between Christianity and the Western attitude toward animals. He goes on to say, "A more enlightened view of our relations with animals emerges only gradually, as thinkers begin to take positions that are relatively independent of the Church" (Singer, 1990).

In recent years, some moral philosophers have taken a stance against enhancing the moral standing of animals. In the preface to *The Animals Issue,* Peter Carruthers (1992) states that although "almost all the books and articles recently published on this issue have argued in favor of the moral standing of animals" this is simply because "most of those who take the opposite view have chosen to remain silent." His book sets forth a nonreligious, moral basis for animal use and against "forbidding hunting, factory farming, or laboratory testing of animals." He concludes his book with the flat assertion "that those who are committed to any aspect of the animal rights movement are thoroughly misguided."

In 1873, the first federal law was enacted to prevent animal cruelty in the United States. Table 29–1 lists the most important pieces of legislation enacted since that time.

# ANIMAL WELFARE/ANIMAL RIGHTS GROUPS

Because opinions, beliefs, and views vary widely from individual to individual, it is logical that the same would be so of the organizations that develop around this complex issue. Groups exist on a continuum, from the most extreme militant groups on one end to pro-animal use groups on the other. Depending on how the various groups are categorized, there are probably 10,000 or so of them, all told. They are decidedly mixed in their goals and styles, and range from animal producer groups to humane societies to terrorist organizations. They may be local, national, or international. They may be pro-use or anti-use. They may be reformist, abolitionist, or protectors of the status quo. They may be political action groups or militant activists. They may count among their ranks agriculturists, militant animal rights activists, animal scientists, philosophers, and others from service industries,

**TABLE 29–1  Some Important Laws Governing Humane Animal Use**

The **Federal Humane Slaughter Act of 1958,** as amended in 1978, regulates slaughter practices. It requires federally inspected meat plants to meet humane slaughter conditions including rendering animals unconscious before slaughtering.

The **Horse Protection Act of 1970** was enacted in order to prevent soring of Tennessee Walking Horses but covers all breeds of horses.

The **Marine Mammal Protection Act of 1972** regulates the killing, capturing, and harassing of marine mammals.

The **Animal Welfare Act** (PL-89-544) with amendments in 1970 (PL-91-579), 1976 (PL-94-279), 1985 (PL-99-198), and 1990 (PL-101-624). This federal legislation and its amendments protects pets from theft; defines minimum holding times for pets in pounds sold to dealers; deals with basic animal management and veterinary care; and regulates research facilities, animal dealers, animal exhibitors, intermediate handlers of animals, including air and truck lines; and prohibits certain forms of animal fighting.

The **Food Security Act of 1985** (PL-99-198) was the first omnibus farm legislation to include animal welfare issues. As a result of this legislation the Secretary of Agriculture is required to set standards for humane care, treatment, and transportation of animals by dealers, research facilities, and exhibitors. Included were minimum requirements for handling, housing, feeding, watering, sanitation, ventilation, shelter, veterinary care, separation of species, exercise for dogs, and the provision of an environment considered adequate for the psychological well-being of nonhuman primates. Further, research facilities were required to have a committee to monitor animal care and practices, and to provide training for anyone handling or using research animals. An information service on employee training and animal experimentation to reduce animal pain and stress was established at the National Agricultural Library.

The **Health Professions Educational Assistance Amendments of 1985** provided for federal fund availability so that schools of veterinary medicine could develop curricula for humane care of laboratory species, for distress limiting methodology, and for the use of alternatives to animals in research and testing.

The **Health Research Extension Act of 1985** (PL-99-158) put forth standards for research animals, which had been a matter of Public Health Service policy since 1971. Animal care committees, research plans for the reduction of or alternatives for animals in research, as well as the reduction of animal pain and discomfort were all addressed in this legislation. It was an amendment to the **Public Health Service Act** [U.S.C. 42 §§ 289d(b) & (c)], which regulates projects funded by the Public Health Service or one of its agencies.

The **Public Health Service Policy on Humane Care and Use of Laboratory Animals** (OPRR-NIH) was first established in 1971 and was revised several times, most recently in September of 1986 to implement the Health Research Extension Act of 1985. The policy required institutions to use the *Guide for the Care and Use of Laboratory Animals* (NIH Publication 86-23, revised 1985) as the basis for developing the institution's animal use program.

The **Animal Facilities Protection Act of 1992** made destruction of animal research or production facilities a federal crime if damage exceeds $10,000. This legislation was prompted by incidents of vandalism, theft, and threats to research workers.

**Good Laboratory Practice Standards** were established under 3 different pieces of legislation: Toxic Substances Control Act (TSCA) (15 U.S.C. § 2603 *et seq.*) (40 C.F.R. 792.1 *et seq.*); Federal Insecticide, Fungicide, and Rodenticide Act (FIFRA) (7 U.S.C. § 136 *et seq.*) (40 C.F.R. § 160.1 *et seq.*); and Federal Food, Drug, and Cosmetic Act (FFDCA) (21 U.S.C. § 321 *et seq.*) (21 C. F. R. § 58.1). Included are regulations for the care and housing of test animals in addition to regulations for other areas of laboratory management. Data to be submitted to the Food and Drug Administration and/or the Environmental Protection Agency must be collected under these rules.

**State and Local Laws** address issues such as pound animals, cruelty, regulation of research facilities, and educational use of animals.

*SOURCE:* Compiled from Bennett et al., 1994, pp. 3–10; Katz, 1999; and Guither and Swanson, 1999.

**TABLE 29–2**    **Animal Rights and Animal Welfare Groups**

**Groups targeting animal use**

Farm Animal Concerns Trust
Animal Welfare Institute
Humane Society of the United States
Humane Farming Association
Animal Rights International
Animal Legal Defense Fund
Farm Animal Reform Movement
People for the Ethical Treatment of Animals
American Vegan Society
Animal Liberation Front

**Groups defending animal agriculture and biomedical research**

National Livestock and Meat Board
American Farm Bureau Federation
Animal Industry Foundation
Farm Animal Welfare Coalition
Americans for Medical Progress Educational Foundation
Foundation for Biomedical Research

**Groups promoting voluntary guidelines for animal use and care**

American Medical Association
American Association for the Accreditation of Laboratory Care
American Psychological Association
American Veal Association
American Veterinary Medical Association
Fur Farm Animal Welfare Coalition
Livestock Conservation Institute
National Cattlemen's Association
Pork Producer's Council
Southeastern Poultry & Egg Association
Numerous other scientific organizations

environmental groups, wildlife conservation groups, and nature groups. Almost all people involved have a polarized viewpoint, or they probably would not have joined any group at all. Table 29–2 offers a sampling of groups whose agendas favor either the expansion of animal rights or the opposition to the expansion of animal rights. Some focus on specific issues, while others direct their attention to a wide range of issues.

## ANIMAL RIGHTS ISSUES PRESENT AND FUTURE

Animal rights issues involve all species of animals and could conceivably develop in an infinite number of directions. Thus predicting the issues of the future can be difficult. Nevertheless, many indicators suggest the probable issues of the future. The most likely are discussed in this section. The directions that public policy could take include protecting the status quo, leaving the issue to individual states, controlling public lands and not private lands, and setting policy for some issues at the national level while leaving others to the states.

## DEFINING AND MEASURING ANIMAL WELFARE

Currently, agriculturists choose management practices based on economic principles (i.e., they choose the production practices that will yield the most profit within their constraints). If government is to regulate care of agricultural animals by passing agricultural animal rights legislation, then an acceptable definition of humane care must be developed within the context of agricultural production systems. The methods for handling, transporting, and confining animals will need legal definition. Presumably, the development of such definitions would be a part of legislation aimed at restricting and directing animal management practices within production systems.

One problem with determining practices that promote the best welfare for livestock is in defining and measuring physiological welfare. Often, the production of useful product is measured. The supposition is that if the animal is gaining well or producing milk at a high level, its needs are being met. However, modern animal production systems are not always designed to measure individual production at all and, if so, only at the end of the production phase. Another problem is that good production does not prevent animals from being subjected to distress. Other indicators such as physiological measurements (blood parameters, and so on), animal behavior, preference tests, or other measures, including simple observation, may help measure an animal's state of care and distress. Measuring the psychological well-being of animals is an even more challenging task. Ultimately, a combination of measures will probably be used to measure the welfare provided in various management systems. Currently, the research necessary to make such determinations is not receiving the funding and the effort it sorely needs. Even though precise systems for measuring animal welfare need further development, those who work with animals have been developing guidelines for animal use rapidly and working to educate and implement standard practices (Table 29–3).

Regulation may involve establishment of boards and commissions that would define humane care and management, and then interpret, oversee, and arbitrate the legislative mandates. Alternatively, laws could be written through the public hearing process and enforced through regular law enforcement mechanisms. Still another alternative might include a system of labeling for food products that describes the production practices under which they were manufactured, allowing consumers to make informed spending choices.

This would no doubt increase the costs associated with food production and the cost of the products, and reduce the returns to the agricultural sector. Reduced supplies and increased consumer prices could result. The addition of new policies would also lead to regulatory bureaucracies, added taxes, and other forms of public support. However, some analysts argue, quite persuasively, that if all producers are required to make the same changes in production techniques, thereby raising costs to all and giving advantage to none, that consumers would then absorb the costs. They further argue that those costs would be so little per consumer as to be inconsequential to all but the very poorest (see Webster, 2001). If these arguments are correct, then producers would be compensated for providing improvements in the welfare of their livestock, consumers would absorb the costs, and production would continue.

It must be pointed out that certain avenues for defining animal welfare do not require broad consensus or even majority opinion and methods of enforcing adherence to the definition, even if not all agree with the rubric assigned. Food companies have been facing mounting pressure to develop animal welfare plans, which their suppliers of animal products are then expected to follow. This gives the food companies a way to assure their consumers that the food animals are treated in a fashion

**TABLE 29–3    Some Important Guidelines for Care and Use of Animals**

**American Association for the Accreditation of Laboratory Animal Care (AAALAC).** A voluntary accreditation body that requires compliance with the *Guide for the Care and Use of Laboratory Animals* and the *Guide for the Care and Use of Agricultural Animals in Agricultural Research and Teaching.* Peer evaluation is used to ensure the proper care and use of research animals, as well as to protect people from dangers associated with conducting research with animals, and minimizing variables that can negatively affect the quality of research.

***Guide for the Care and Use of Laboratory Animals*** (NIH Publication 86-23, revised 1965, 1968, 1972, 1978, 1985, and 1996). The *Guide* sets forth recommendations on policy, veterinary care, husbandry practices, and requirements for research facilities and the animals used for research in them. It further emphasizes that the responsibility for animal care is with the institution. Published by the National Institutes of Health, the *Guide* is the resource explaining the requirements enforced under the Public Health Service Policy on Humane Care and Use of Laboratory Animals.

***Guide for the Care and Use of Agricultural Animals in Agricultural Research and Teaching.*** This guide was first published in 1988. It is voluntary but has received wide acceptance, support, and use by those who use agricultural animals in research and teaching. It includes guidelines for institutional policies, general husbandry guidelines, health care, physical plant, beef cattle, dairy cattle, horses, poultry, sheep, goats, swine, and veal calves.

**Policy on Personnel Ethics in Youth Livestock Activities.** Oklahoma Cooperative Extension Service. This is an excellent example of a policy designed to retain and promote what is good about youth livestock programs. Many states have adopted policies on youth livestock programs relating to exhibitions. Such policies spell out unethical and illegal practices and the penalties for infractions.

**Report of the AVMA Panel on Euthanasia, 1993.** Presents acceptable methods of euthanasia. Public Health Service Policy requires methods of euthanasia to be consistent with this report and all regulatory agencies recognize it as the standard for selecting and evaluating methods of euthanasia.

**Guidelines for Ethical Conduct in the Care and Use of Animals.** Developed by the American Psychological Association's Committee on Animal Research and Ethics.

*SOURCE:* Compiled from Bennett et al., 1994, pp. 3–10; Katz, 1999; and Guither and Swanson, 1999.

they are willing to defend to their consumers. Several high-profile companies have taken this step. There seems little doubt that others will do the same. One is forced to consider the possibility that the marketplace may be the ultimate arbitrator and that laws will be unnecessary.

Webster (2001) offers the following view of welfare, "The welfare of a sentient animal is good if it can sustain fitness and avoid suffering: i.e. stay fit and happy." An avenue for supporting that definition can be found in the "Five Freedoms" (Table 29–4) put forth by the Farm Animal Welfare Council, an independent advisory body established by the British government in 1979. Ultimately, a consensus will be reached through one or more avenues.

## Transgenic Animals

The debate over transgenic animals is likely to center on four questions:

1. Should their development be allowed?
2. Should public funds be used for their development?

**TABLE 29–4** **The Five Freedoms of the Farm Animal Welfare Council**

The welfare of an animal includes its physical and mental state and we consider that good animal welfare implies both fitness and a sense of well-being. Any animal kept by man must at least be protected from unnecessary suffering.

We believe that an animal's welfare, whether on farm, in transit, at market or at a place of slaughter should be considered in terms of **five freedoms.** These freedoms define ideal states rather than standards for acceptable welfare. They form a logical and comprehensive framework for analysis of welfare within any system together with the steps and compromises necessary to safeguard and improve welfare within the proper constraints of an effective livestock industry.

1. Freedom from hunger and thirst—by ready access to fresh water and a diet to maintain full health and vigour.
2. Freedom from discomfort—by providing an appropriate environment including shelter and a comfortable resting area.
3. Freedom from pain, injury or disease—by prevention or rapid diagnosis and treatment.
4. Freedom to express normal behavior—by providing sufficient space, proper facilities, and company of the animal's own kind.
5. Freedom from fear and distress—by ensuring conditions and treatment that avoid mental suffering.

**Stockmanship—The Key to Welfare**

Stockmanship, plus the training and supervision necessary to achieve required standards, are key factors in the handling and care of livestock. A management system may be acceptable in principle but without competent, diligent stockmanship the welfare of animals cannot be adequately safeguarded. We lay great stress on the need for better awareness of welfare needs, for better training and supervision.

SOURCE: Farm Animal Welfare Council. http://www.fawc.org.uk/index.htm.

3. Should their introduction into the environment be allowed and/or controlled?
4. Should patents be allowed?

## Cloning

When the world found out about Dolly, the cloned sheep, the inevitable debate ensued over the ethics of cloning animals and humans (Figure 29–10). Cloning is likely to continue as a hotly contested issue in the future. In a policy paper issued in January 2008, the FDA declared cloned animals safe to eat. Many have couched the opposition argument in animal rights terms. Legislative agendas have already developed around this issue and will continue to do so.

## Control of Predatory Species

Predators (Figure 29–11) and wild grazers affect the ability of production agriculturists to produce inexpensive food because they increase the costs of producing crops and livestock. Sheep have long been the target of coyotes. Deer and other related species can do tremendous damage to crops. In addition, several predator species have adapted to human

**FIGURE 29–10** Dolly, the cloned sheep, meets the press on February 25, 1997. Dolly's presence sparked a debate on the ethics of creating clones. Many have called this issue an animal rights issue. (Photographer Reuters/Jeff Mitchell. Used with permission.)

**FIGURE 29–11** Predators and agriculturists have long had an uneasy relationship because some predators kill domestic livestock. Yet there is much pressure to return predators, such as the wolf, to their natural range. (Photo courtesy of USDA.)

encroachment on their space by harvesting our pets as their food supply. Herbivores freely roam some suburbs, damaging expensive landscapes. At issue is whether or not to control the problem species in order to protect the domestic plants and animals. Policy alternatives range from banning any control, to allowing almost any form of control. Alternatives in between those options include restricting some practices to minimize damage to nontargeted species, developing better control methods, and allowing the damage but instituting a method of reimbursement to those who suffer economic losses.

## Hunting and Trapping

For the short term, the debate over this issue will probably center on whether or not to allow hunting on public property, whether those who interfere with hunters on public

**FIGURE 29–12** Animal rights groups tend to object to hunting wildlife species, whereas many other groups support hunting for many reasons. (Photographer Ken Hammond. Courtesy of USDA.)

property should be restricted, and whether or not leg-hold traps should be allowed. Many groups have interests in this issue. Animal rights groups generally object to all hunting and trapping (Figure 29–12). Conservationists favor restrictions on hunting and trapping to ensure species health and survival, including both endangered and nonendangered species. The sale of hunting rights on private property gives agriculturists an economic stake. Property rights debates ensue when restrictions on private property are considered. Revenues from hunter's fees and licenses are used in conservation programs aimed at preserving wildlife species.

## Research

Legislation that regulates and controls management of animals used in research already exists. The next step would be to determine if the government would exercise value judgments and determine the exact uses for animals in research, or to determine if they are to be used at all. This could mean prohibiting all animal research, or restricting some types, such as agricultural animal or cosmetics research, but allowing other types, such as biomedical research, to continue. This could negatively affect development of new drugs, cures for diseases, and the general health and well-being of the human population, and could increase the costs of health care. It is estimated that advances from animal research have added 15 to 25 years to the average human life span. Medical advances that will add even more years certainly remain to be discovered.

The number of animals used in research in the United States has been decreasing for several years, partly because of the costs of the regulations that have been imposed. However, a significant amount of reduction in animal use is also due to the increased use of in vitro cultures, mathematical models, and the substitution of lower organisms (Figure 29–13). Researchers have accomplished much of this through the concept of replacement, reduction, and refinement (the *3 Rs*). The success that researchers achieve in carrying out this concept may ultimately decide what laws get passed.

**3 Rs**

Replacement (substitute something else for higher animals), reduction (reduce the number of animals needed), and refinement (decrease in inhumane procedures).

**FIGURE 29–13**  Advanced laboratory techniques are giving researchers tools to reduce the number of animals needed in biomedical research. Cell cultures can sometimes be used as alternatives for live animals in research. Here, animal physiologist Caird Rexroad inspects bovine embryonic cells. (Photographer Keith Weller. Courtesy of USDA.)

## SOURCES OF INFORMATION

The Animal Welfare Information Center (AWIC) is an information service established at the National Agricultural Library in Beltsville, Maryland. It was established as a result of the 1986 amendments to the Animal Welfare Act. Its mandate includes providing information on subjects relating to animal welfare. It is the single most important resource on this topic in the United States. Address: Animal Welfare Information Center, National Agricultural Library, 10301 Baltimore Ave, 4th Floor, Beltsville, MD 20705-2351. Phone: (301)504-6212; fax (301)504-7125; awic//www.nal.usda.gov/awic/.

## SUGGESTED READINGS

Caruthers, P. 1992. *The animals issue: Moral theory in practice.* Cambridge, UK: Cambridge University Press.

Cohen, C. 1986. The case for the use of animals in biomedical research. *New England Journal of Medicine* 315:865.

Duncan, I., and J. Petherick. 1991. The implication of cognitive process for animal welfare. *Journal of Animal Science* 69:5017.

Finsen, L., and S. Finsen. 1994. *The animal rights movement in America.* New York: Twayne Macmillan.

Harrison, R. 1994. *Animal machines.* London: Vincent Stuart.

Jasper, J. M., and D. Nelkin. 1992. *The animal rights crusade: The growth of a moral protest.* New York: The Free Press.

Leavitt, S. E. 1978. *Animals and their legal rights.* Washington, DC: Animal Welfare Institute.

Marquardt, K., H. M. Levine, and M. LaRochelle. 1993. *Animal scam: The beastly abuse of human rights.* Washington, DC: Regnery Gateway.

Mench, J., and A. Van Tienhoven. 1986. Farm animal welfare. *American Scientist* 74:598.

Nicoll, C. 1991. A physiologist's view on the animal rights/liberation movement. *The Physiologist* 34:303.

Pardos, H., A. West, and H. Pincus. 1991. Physicians and the animal-rights movement. *New England Journal of Medicine* 324:1640.

Regan, T., and G. Francione. 1992. A movement's means create its ends. *The Animals' Agenda,* Jan/Feb:40.

Regan, T., and P. Singer, 1976. *Animal rights and human obligations.* Englewood Cliffs, NJ: Prentice-Hall.

Singer, P. 1990. *Animal liberation.* New York: Random House.

## SUMMARY AND CONCLUSION

All things considered, there is little wonder that widely divergent attitudes, philosophies, and ethics that challenge more traditional values regarding animal use have developed. Thus, although animal rights is not merely some modern notion, at no time in history has the idea of rights for animals been accepted by so many and thus debated by so many. Nor have the numbers of people involved directly with noncompanion animals ever been so few. Consequently, at no other time have the conditions ever been so ripe for the issue actually to gain a following and grow.

Changing animal welfare concerns and emerging animal rights considerations have broad potential benefits and costs to humans and our society on many levels. These include philosophical, social, legal, economic, biological, emotional, and political perspectives. The interested parties encompass a range of people from agriculturists and consumers, to the most extreme animal rights abolitionist advocate. The different viewpoints pit such issues as the hard, practical consideration that animal agriculture is a major part of the economy of this country against the more esoteric considerations associated with the belief systems of those who have equated animal rights to human rights. Should the most extreme approach to animal rights be accepted by society, then all uses of animals for food, clothing, leisure, or research purposes would cease. Few believe this could ever occur.

The level of concern for animal rights issues among the general population of the United States indicates that more concern for these issues will grow in the future. Proponents of animal rights issues will join with environmental, diet, health, and food safety groups to pursue an agenda to support their goals. It is probable that they will influence future policies. Many argue that it is only a matter of time until farm animal welfare is legislated in the United States. Agriculturists, agribusiness entities, consumers, and all the rest, along with their organizations, must join the discussion if they are to be part of the policy-making process. In fact, as Getz and Baker (1990) suggest, the process may ultimately improve animal agriculture because challenges to the status quo often lead to positive change. Many argue that it has already done so for biomedical research. Those who graduate with degrees in agriculture, biomedicine, and related fields are likely to be thought leaders of the future on this issue. It is necessary to work intelligently, follow this issue diligently, distribute good information effectively, and respond to criticisms from outside groups. The entire future of animal use in all its forms depends on informed debate.

## REFERENCES

Appleby, M. C., and B. O. Hughes. 1997. *Animal welfare*. CAB International. Cambridge, UK: University Press.

Azjen, I., and M. Fishbein. 1980. *Understanding attitudes and predicting social behavior*. Englewood Cliffs, NJ.: Prentice Hall.

Bennett, B. T., M. J. Brown, and J. C. Scofield. 1994. *Essentials for animal research: A primer for research personnel*. Washington, DC: Animal Welfare Information Center, United States Department of Agriculture, National Agriculture Library.

Bennett, R. 1998. Measuring public support for animal welfare legislation: A case study of cage egg production. *Animal Welfare* 7:1–10.

Bowd, A. D., and A. C. Bowd. 1989. Attitudes toward the treatment of animals: A study of Christian groups in Australia. *Anthrozoös* 3:20–24.

Carruthers, P. 1992. *The animals issue: Moral theory in practice*. Cambridge, UK: Cambridge University Press.

Davis, S. L. 2000. What is the morally relevant difference between the mouse and the pig? *Proceedings of the 2nd Congress of the European Society for Agricultural and Food Ethics,* 24–26 August. Copenhagen, Denmark: Royal Veterinary and Agricultural University, pp. 107–109.

Davis, S. L. 2001. The least harm principle suggests that humans should eat beef, lamb, dairy, not a vegan diet. *Proceedings of the 3rd Congress of the European Society for Agricultural and Food Ethics,* 3–5 October. Florence, Italy. Milan, Italy: A and Q, University, pp. 449–450.

Davis, S. L., and P. R. Cheeke. 1998. Do domestic animals have minds and the ability to think? A provisional sample of opinions on the question. *Journal of Animal Science* 76:2072–2079.

Dennis, J. U. 1997. Morally relevant differences between animals and human beings justifying the use of animals in biomedical research. *Journal of the American Veterinary Medical Association* 210:612–618.

Farm Animal Welfare Council. 1993. *Second report on priorities for research and development in farm animal welfare*. Ministry of Agriculture, Fisheries and Food.

Farm Animal Welfare Council. 2004. *The five freedoms*. Accessed January 25, 2008. http://www.fawc.org.uk/freedoms.htm.

Francione, G. L. 1996. *Rain without thunder*. Philadelphia: Temple University Press.

Fraser, D. 1999. Animal ethics and animal welfare science: Bridging the two cultures. *Applied Animal Behavior Science* 65:171–189.

Fraser, D. 2001. The "new" perception of animal agriculture: Legless cows, featherless chickens, and a need for genuine analysis. *Journal of Animal Science* 79:634–641.

Friend, T. H. 1990. Teaching animal welfare in the land grant universities. *Journal of Animal Science* 68:3462.

Getz, W. R., and F. H. Baker. 1990. Educational methodology in dealing with animal rights and welfare in public service. *Journal of Animal Science* 68:3468.

Graf, B., and M. Senn. 1999. Behavioural and physiological responses of calves to dehorning by heat cauterization with or without local anaesthesia. *Applied Animal Behaviour Science* 62(2–3):153–171.

Guither, H. D., and J. Swanson. 1999. *Animal rights and animal welfare*. University of Illinois and Kansas State University. http://unlvm.unl.edu/aniright.htm.

Heleski, C. R., A. J. Zanella, and E. A. Pajor. 2003. Animal welfare judging teams: A way to interface welfare science with traditional animal science curricula? *Applied Animal Behaviour Science* 81:279–289.

Hemsworth, P. H., and G. J. Coleman. 1998. *Human-livestock interactions: The stockperson and the productivity and welfare of intensively farmed animals.* Oxon, UK: CAB International.

Hemsworth, P. H., G. J. Coleman, J. L. Barnett, S. Borg, and S. Dowling. 2002. The effects of cognitive behavioral intervention on the attitude and behavior of stockpersons and the behavior and productivity of commercial dairy cows. *Journal of Animal Science* 80:68–78.

Hodges, J., and I. K. Han (eds.). 2000. *Livestock, ethics and quality of life.* Wallingford, UK: CABI.

Katz, L. S. 1999. *Animal rights vs. animal welfare.* Publication No. FS753. New Brunswick, NJ: Rutgers Cooperative Extension Service Agricultural Extension Service.

Kellert, S. R. 1980. American attitudes toward and knowledge of animals: An update. *International Journal for the Study of Animal Problems* 1:87–119.

Kunkel, H. O. 2000. *Human issues in animal agriculture.* College Station: Texas A&M Press.

McInerney, J. P. 1998. The economics of welfare. In *Ethics, welfare, law and market forces: The veterinary interface,* A. R. Michell and R. Ewbank, eds. Wheathampstead, Herts: UFAW.

Molony, V., J. E. Kent, and I. J. McKendrick. 2002. Validation of a method for assessment of an acute pain in lambs. *Applied Animal Behaviour Science* 76:215–238.

Paul, E. S. 2000. Empathy with animals and with humans: Are they linked? *Anthrozoös* 13:194–202.

Paul, E. S., and A. L. Podberscek. 2000. Veterinary education and students' attitudes towards animal welfare. *Veterinary Record* 146:269–272.

Pifer, L., K. Shimizu, and R. Pifer. 1994. Public attitudes toward animal research: some international comparisons. *Society and Animals* 2:95–113.

Preece, R., and D. Fraser. 2000. The status of animals in Biblical and Christian thought: A study in colliding values. *Society and Animals* 8:245–263.

*Public health care policy on humane care and use of laboratory animals.* 1986. Bethesda, MD: Department of Health and Human Services.

Regan, T. 1983. *The case for animal rights.* Berkeley: University of California Press.

Report of the AVMA Panel on Euthanasia. 1993. *Journal of the American Veterinary Medical Association* 202(2):229–240.

Rollin, B. E. 1992. *Animal rights and human morality.* 2nd ed. Buffalo, NY: Prometheus Books.

Rollin, B. E. 1995. *Farm animal welfare: Social, bioethical and research issues.* Ames: Iowa State University Press.

Rowan, A. 1989. The development of the animal protection movement. *The Journal of NIH Research* 1(Nov.–Dec.):97–100.

Russell, W. M. S., and R. L. Burch. 1959. *The principles of humane experimental technique.* London: Methuen.

Singer, P. 1990. *Animal liberation.* New York: Random House.

Taylor, A. A., and D. M. Weary. 2000. Vocal responses of piglets to castration: identifying procedural sources of pain. *Applied Animal Behaviour Science* 70:17–26.

Thompson, P. B. 1998. *Agricultural ethics: Research, teaching, and public policy.* Ames: Iowa State University Press.

Thompson, P. B. 1999. From a philosopher's perspective. How should animal scientists meet the challenge of contentious issues? *Journal of Animal Science* 77:372–377.

Webster, A. J. F. 2001. Farm animal welfare: The five freedoms and the free market. *The Veterinary Journal* 161:229–237.

Wells, D. L., and P. G. Hepper. 1997. Pet ownership and adults' views on the use of animals. *Society and Animals* 5:45–63.

# Animals in Sustainable Agriculture

## LEARNING OBJECTIVES

After you have studied this chapter, you should be able to:

1. Define and describe sustainable agriculture.

2. Describe sustainable practices.

3. Explain the "systems philosophy" of sustainable agriculture.

4. Elaborate on the place of animals in sustainable systems.

5. Identify a monoculture system and contrast it to a diversified system.

6. Identify areas of concern for making livestock systems more sustainable.

7. Explain the "lifestyle element" of sustainable agriculture.

## KEY TERMS

Agroforestry
Crop rotation
Diversification
Ecology
Environmentalism

Integrated pest
   management
Lifestyle
Monoculture
Organic agriculture

Recombinant DNA
   technology
Sustainable agriculture

## INTRODUCTION

**Sustainable agriculture**

Agriculture that meets the needs of the present generation without jeopardizing the ability of further generations to meet their own needs and without limiting their choices.

There is an active movement in the United States and other industrialized countries to make modern agriculture more sustainable in nature, hence the term *sustainable agriculture.* At the core of the movement is the belief that modern agriculture, with its technologically driven, petrochemical-dependent specialization that is designed to maximize production, has brought costs that in some cases may outweigh the benefits. The costs, often referred to as external costs, most frequently mentioned are to natural resources such as:

1. Soil damage from erosion, topsoil depletion, and contamination

2. Loss of soil health from reduced microbial populations, loss of soil tilth, increased soil compaction

3. Habitat loss and loss of species

4. Groundwater depletion, pollution from lagoons, and nitrogen in water wells

5. Damage from animal waste, usually associated with water quality, especially municipal surface water supplies

Social costs assessed to modern agriculture include:

1. The decline of rural areas, which has come about with the loss of the family farm

2. The tendency of corporate farming to transport profits out of communities and provide mostly minimum-wage jobs

3. Community structure changes as packing plants employ migrants

4. Loss of communities and local businesses

5. Reduced human satisfaction with and within agriculture, associated with quality of life for farmers and their neighbors as well as discomfort of greater society with agriculture's means of producing food

Sustainable agriculture is often characterized as simply "organic farming" (Box 30–1) by people who are unfamiliar with its complexities (Figure 30–1). But sustainable agriculture is vastly more complicated than that term implies. Yet defining sustainable agriculture is perhaps the most difficult chore of this chapter. There seems to be no single definition to which all agree, perhaps because the concept is still actively evolving. A definition your author likes is this: Agriculture that meets the needs of the present generation without jeopardizing the ability of further generations to meet their own needs and without limiting their choices.

Sustainable agriculture is as much concept as anything else, and it blends the social with the scientific in a way that many of us are either unwilling to think about or,

**FIGURE 30–1** This organically produced produce may or may not have been produced in a sustainable system. The terms *organic gardening* and *sustainable agriculture* are not synonymous. (Photographer Owen Franken. Courtesy of Corbis. Used with permission.)

## Box 30–1    USDA's National Organic Program

The Organic Foods Production Act (OFPA) of 1990 required the U.S. Department of Agriculture (USDA) to develop national standards for organically produced agricultural products to assure consumers that agricultural products marketed as organic meet consistent, uniform standards. The OFPA and the National Organic Program (NOP) regulations require that agricultural products labeled as organic originate from farms or handling operations certified by a state or private entity accredited by USDA. The NOP is a marketing program housed within the USDA Agricultural Marketing Service.

The NOP developed national organic standards and established an organic certification program based on recommendations of the 15-member National Organic Standards Board (NOSB). The NOSB, appointed by the secretary of agriculture, is comprised of representatives from the following categories: farmer/grower, handler/processor, retailer, consumer/public interest, environmentalist, scientist, and certifying agent. The NOP regulations are flexible enough to accommodate the wide range of operations and products grown and raised in every region of the United States.

The regulations prohibit the use of genetic engineering, ionizing radiation, and sewage sludge in organic production and handling. Generally, all natural (nonsynthetic) substances are allowed in organic production and all synthetic substances are prohibited. The National List of Allowed Synthetic and Prohibited Non-Synthetic Substances, a section in the regulations, contains the specific exceptions to the rule. Production and handling standards address organic crop production, wild crop harvesting, organic livestock management, and processing and handling of organic agricultural products. Organic crops are raised without using most conventional pesticides, petroleum-based fertilizers, or sewage sludge-based fertilizers. Animals raised on an organic operation must be fed organic feed and given access to the outdoors. They are given no antibiotics or growth hormones.

For consumers, the NOP standards and certification process define organic food. For farmers, the standards provide clear guidelines on how to take advantage of the exploding demand for organic products. For the organic food industry, the NOP serves as an important marketing tool for domestic and export markets.

Organic farming is one of the fastest growing segments of U.S. agriculture. The value of retail sales exceeded $15 billion in 2006, with consumer demand increasing by approximately 20% yearly.

Adapted from http://www.ams.usda.gov/NOP/FactSheets/Backgrounder.html.

---

because it is difficult to assign values to external costs, choose to reject. All wrapped up in the concept are elements of environmentalism, economics, ecology, and social concerns. Furthermore, the complexities of the topic will prevent this chapter from dealing with this timely topic in a comprehensive way. However, it will lay out the fundamentals. You are encouraged to look at the lists of articles and books at the end of the chapter and to take advantage of other sources of information to pursue the topic further.

## WHAT IS SUSTAINABLE AGRICULTURE?

A good place to start in understanding sustainable agriculture is with a document from the Alternative Farming Systems Information Center titled "Sustainable Agriculture: Definitions and Terms" (Gold, 1994). In that paper, Gold points out that "In its congressionally mandated annual reports, the Joint Council on Food and Agricultural Sciences has consistently listed 'attain sustainable production systems and ensure their compatibility with environmental and social values' as first in a set of long-term objectives for food and agricultural sciences." She goes on to say, "The goal of sustainability requires addressing philosophical, economic, and sociological issues, as well as environmental and scientific questions." This last sentence demonstrates the complexity of understanding the underpinnings of the sustainable agriculture movement by pointing out its many elements. It is also instructive to note that the first item in her list of what "requires addressing" is *issues* from the social sciences, and the last item is biological *questions*. It seems that the essence of the sustainable agricultural movement can be found here. That essence is the struggle to bring a set of social issues to modern agriculture with the goal of changing some of the current and most common biological approaches to agriculture so that they pass muster as being ecologically sound, economically viable, and socially responsible. These three items are often referred to as the "triple bottom line" of sustainability. The three "lines" represent society, the economy, and the environment. (The phrase was coined by John Elkington in his 1998 book *Cannibals with Forks, the Triple Bottom Line of 21st Century*.)

It is important for the long-term implementation strategy of sustainable systems that producers not be backed into absolutist corners where it appears that adopting sustainable practices is an all-or-none proposition. For example, sustainable agriculture, in its pure definition, refers to any farming system that can be productive and can last forever. The movement often seems to take the view that much of modern agricultural practice is nonsustainable, and presumably has the goal of converting current agriculturists away from modern practices that it deems as nonsustainable to those that will last forever. In reality, all farming practices can be viewed on a continuum from not sustainable to very sustainable. In addition, there is plenty of ground for disagreement on what is and is not sustainable. A multitude of rational arguments can be made defending many modern practices as meeting Ikerd's 1990 standard that sustainable agricultural systems "must be resource conserving, socially supportive, commercially competitive, and environmentally sound." It is the absolutism of movements that often turns otherwise rational people away. The best interests of all sides get lost when purist "all-or-none" thinking and philosophy dominate.

The definition for a sustainable agriculture found in the 1990 Farm Bill is "an integrated system of plant and animal production practices having a site specific application that will, over the long term: (a) satisfy human food and fiber needs; (b) enhance environmental quality and the natural resource base upon which the agricultural economy depends; (c) make the most efficient use of nonrenewable resources and on-farm resources and integrate, where appropriate, natural biological cycles and controls; (d) sustain the economic viability of farm operations; and (e) enhance the quality of life for farmers and society as a whole." In the view of the author, there is much in this definition to like and support. This view leaves us the option of incorporating farming practices that are more sustainable without having a perfectly sustainable system. This frame of reference paves the way for a mind-set that allows all to objectively review and consider sustainability as something we might have a stake in and move toward incrementally. Keeping an open mind is the key to learning.

One of the central tenets of some who promote sustainable systems is that cropland used for animal feed is less efficient in producing calories and protein for human use than if the land were to be used to produce food crops. Several examples that dispute this conclusion can be found in the literature. The following is from Beckett and Oltjen (1996). "[I]n California, alfalfa yields average 15 t/ha annually whereas wheat averages 5.4 t/ha. . . . Alfalfa is about 20% protein; wheat is about 12% protein. Hence a hectare yields 3,010 kg of protein from alfalfa, or 647 kg of protein from wheat (21.5% that of alfalfa). Dairy cows convert the protein in alfalfa to milk protein at about 25% efficiency, giving 753 kg of food protein per hectare, compared to 647 for wheat.

In addition, milk and milk protein is of much higher quality and biological value." A study of the various estimates with different species in different systems actually yields several examples of systems of animal production that produce more protein and/or energy than those same resources would have produced from the common, humanly edible grains found in those same areas. From these examples we learn that animals yield the most sustainable use of certain resources. Those who work to bring about a more sustainable overall agriculture would do well to keep this in mind.

This illustrates a drawback to achieving sustainability and something that should also be pointed out to the reader. Not everyone who supports and/or practices sustainable agriculture agrees on exactly what is sustainable, on the best methods to reach sustainability, what practices are acceptable, which are best, and in what direction the movement should go. This should not be interpreted negatively. Thinking people don't necessarily reach consensus overnight. This lack of consensus is rather a sign of life in a movement that many say has reached a critical social threshold and is here to stay. Mature social movements tend to get boring. This one still has lots of "interest factor" left in it, with the probability of a few surprises yet to come.

### PRACTICES THAT ARE PART OF SUSTAINABLE FARMING

Ideally, you have not been drawing the conclusion that sustainable agriculture means going back to milking cows by hand, choppin' cotton, sloppin' hogs, and sheathing cornstalks. Nothing could be further from fact. Sustainable farming practices can be simple but don't have to be simplistic. They can actually be quite sophisticated (Figure 30–2). The following list of some commonly used practices, and some not-so-commonly used practices, provides but a few examples of sustainable practices that can be implemented as a part of modern agricultural systems:

1. Rotating crops to reduce weeds, disease, and insect pests. This practice also provides alternative sources of soil nitrogen, reduces soil erosion, and reduces risk of water contamination by agricultural chemicals.

2. Using intensive grazing and rotation systems for pasture and range management to improve carrying capacity and build the land rather than take from it.

**FIGURE 30–2** This farmstead incorporates many of the practices used in sustainable agriculture industry: crop rotation, pasture rotation, the use of animal and green manures to augment fertilizer, and the use of trees as windbreaks to protect the soil. (Photo courtesy of Stanley Trimble, Natural Resources Conservation Service/USDA. Used with permission.)

3. Using integrated pest management, which uses advanced entomology and agronomic practices of pest scouting, resistant cultivators, biological controls, and other methods to reduce pests and lessen pesticide use.

4. Increasing mechanical/biological weed control, more soil and water conservation practices, and using animal and green manures to replace and augment chemical fertilizers.

5. Using swine to scavenge behind ruminants in intensive dairy or beef operations.

6. Using ruminants to manage small-grain crops through the winter prior to grain harvest.

7. Using ruminants to eat food-processing wastes such as citrus pulp, beet pulp, and cannery wastes that would otherwise present huge disposal problems.

8. Using mixed ruminant species to graze stockpiled forages and "wild" growth after the plant-growing season in dry areas. The animals return the nutrients to the soil and reduce the danger of and need for burning.

9. Using ducks and geese to eat grass between rows, and flocks of chickens to create an insect buffer zone around plantings for market gardening operations.

10. Using riparian buffer strips to protect water quality and to provide habitat for wildlife.

## THE SYSTEMS PHILOSOPHY

To understand and practice sustainable agriculture, it is essential to have an appreciation for the systems philosophy. Feenstra et al. (1998) explain it this way: "The system is envisioned in its broadest sense, from the individual farm, to the local ecosystem, and to communities affected by this farming system both locally and globally. An emphasis on the system allows a larger and more thorough view of the consequences of farming practices on both human communities and the environment. A systems approach gives us the tools to explore the interconnections between farming and other aspects of our environment. A systems approach also implies interdisciplinary efforts in research and education. This requires not only the input of researchers from various disciplines, but also farmers, farmworkers, consumers, policymakers, and others."

Ample evidence supports the idea that sustainable systems can work (Figure 30–3). Sustainable agricultural systems using both animal and plant crops have been in place in some parts of the world for thousands of years. In these systems, animal and plant interactions combine to cycle carbon, nitrogen, minerals, and energy to create a sustainable system. The animals eat surplus plant material and in turn return waste products that the plants use to the plant system. Body wastes from animals contain virtually all of the nutrients plants need to grow. In these agricultural systems, humans have been able to take those things we need with little noticeable damage to the environment. The only part of the "resource conserving, socially supportive, commercially competitive, and environmentally sound" test those historically successful sustainable systems do not meet for modern times is commercial viability, and that is because the concept is vastly different in the subsistence systems where this agriculture has existed for so long. There is a problem in making such systems commercially viable. However, it is a misconception to say that none of the sustainable or near-sustainable farming systems are economically viable. Admittedly, most of the viable systems tend to be smaller, but in many areas of Canada and North Dakota in the United States there are large-scale farms (some even organic) that are doing very well. A healthy market has developed for

**FIGURE 30–3**  Sustainable systems, such as the continuous rice culture found in the humid tropics, have been in place for thousands of years.

consumers who will pay a premium for produce that they perceive as fresh. These consumers also value knowing who grew their food and how it was grown. The number of farmers' markets around the United States has markedly increased in the last few years and is still doing so. Because of such successes, some in sustainable agriculture pose an interesting question: "What if we had invested more heavily in agroecological farming systems research? Where would we be?" Instead, we have directed the bulk of our research efforts to the industrial system.

In the not-so-distant past, U.S. agriculture came much closer to meeting the goals of sustainability. This has been changing fairly rapidly since the early 1950s when agriculture began moving away from the small family farm to much more specialized systems. As Feenstra et al. (1998) point out, it was at this time that "productivity soared due to new technologies, mechanization, increased chemical use, specialization and government policies that favored maximizing production. These changes allowed fewer farmers with reduced labor demands to produce the majority of the food and fiber in the U.S." In this sense, then, the current increased emphasis on sustainable systems can be viewed as "back to the future" in terms of both knowledge and practice.

## ANIMALS IN SUSTAINABLE AGRICULTURAL SYSTEMS

In the big picture, the uses of animals in sustainable agriculture mirror the arguments put forth for animal use in various places earlier in this text. Those uses include adding value to crops, serving as a buffer against fluctuating grain prices and supplies, providing a storage of food against catastrophe, creating year-round employment, increasing the number of products from the production unit, storing of capital, providing capital for purchase of inputs, using lands not fit for cropping, and so on. In fact, animals are indispensable in maintaining highly functioning sustainable systems where there are dense populations (Figure 30–4).

The inclusion of both crops and animals has been at the heart of many of the world's historically successful sustainable systems. These systems were/are sustainable be-

**FIGURE 30–4** Animals are an indispensable part of an overall sustainable system. In some cases, animal use is the only means of creating value from renewable vegetation.

cause they rely on the interrelationships between diverse plant, animal, and human segments of the system to nurture each other. These systems were the pre-1950s, family-farm model in this country. The small farms that had adequate cropland were sustainable. In turn, the entire network of small rural communities that once flourished in this country were sustained by these farms. As Baker et al. (1990) point out, in these family farms "human and community systems can be integrated with the agroecosystem into a larger system." Although this type of farm still exists in this country, specialized systems with many fewer enterprises are much more prevalent in all of agriculture, both plant and animal. Some farmers grow thousands of acres of corn to the exclusion of all else. Others have tens of thousands of chickens, thousands of dairy cows or pigs, and nothing else. As Feenstra et al. (1998) observed, "The integration now most commonly takes place at a higher level—between farmers, through intermediaries, rather than within the farm itself." In the process, some of the elements of sustainability have been lost (Figure 30–5). Bringing a more sustainable approach to the current system certainly presents challenges. Yet, there are recommendations for doing so. Not all fit the rubric of being perfectly sustainable. However, being more sustainable is a goal that most could achieve if they chose to.

One of the basic tenets of sustainable systems is that they are "site specific." Because sites vary, it then follows that sustainable production practices must include different techniques. Even so, the general principles listed in Table 30–1 may guide producers.

Diversity has its advantages. Diversified operations have more sources of income than do modern, specialized operations, so they are not as susceptible to the unexpectedly low prices that occur from time to time in virtually all products of crop and livestock farms. It is unusual for many farm products to have bad prices at the same time. For example, the price of pigs and the price of wheat aren't usually low at the same time. Likewise, it is unusual for multiple crop failures to occur simultaneously, or for egg production to be subpar at the same time that the corn crop fails. Diversified operations have an added hedge against catastrophe in this way. In addition, producers are generally able to market their multiple products at different times. Wheat can be sold in the spring or early summer. Excess hay can be sold whenever cash is needed. Corn, milo, and other grains are generally harvested and sold in the fall. Eggs from a laying flock or milk from a herd of dairy cows provide year-round income to help with cash flow. Multiple enterprises also tend to make best use of labor. For example, calves that nurse their mothers and run on pastures in summer with little labor needed can be put in feedlots and fattened for market for the winter, when more labor is needed.

**FIGURE 30–5**   Concentrating livestock also concentrates wastes. Recycling wastes in a revenue-neutral fashion is a pressing need to make animal agriculture more sustainable.

**TABLE 30–1**    **Principles of Sustainability**

1. Select species and varieties that are well suited to the site and to conditions on the farm.
2. Diversify crops (including livestock) and cultural practices to enhance the biological and economic stability of the farm.
3. Manage soil and water to conserve, protect, and enhance their quality.
4. Manage livestock waste and other agricultural system by-products.
5. Manage pests ecologically and with minimal effects on the environment.
6. Protect wildlife habitat and encourage biodiversity.
7. Maximize solar energy use and place less reliance on petroleum-based chemicals and fuels.
8. Use inputs efficiently and humanely.
9. Restore farm economic profitability.
10. Consider the farmer's goals and lifestyle choices.
11. Ensure that sustainable farm communities are built and that those communities are resilient.
12. Ensure social equity in today's actions and ensure intergenerational equity for our actions.

*SOURCE:* Adaped from Feenstra et al., 1998, and Horne and McDermott, 2001.

Obviously, an excellent opportunity to diversify is to include both animal and plant enterprises on the same farm. Further diversity is brought about by including multiple species of livestock and multiple crops. Crop and livestock farms that did just this very thing were once the predominant type of farm across the farming belt of the United States. These mixed operations have distinct advantages in best using the natural resources of the farm and reducing the need for costly inputs. Land with slopes that would suffer from erosion if cropped can be pastured or used for hay and the level land can be tilled for growing crops (Figure 30–6). In addition, livestock can glean crop residue from harvested fields and benefit from them. This residue may be in the form of stalks and stover

**FIGURE 30–6**   Soil erosion is one of the most persistent threats to the long-term productivity of the nation's land. Erodable lands are often best in pasture and used for grazing.

**FIGURE 30–7**   Using grazing animals properly increases soil quality, reduces erosion, reduces water runoff, and thus benefits water table levels and quality.

or grain left by harvesting machines. Crop failures can be harvested by animals to provide value from what would otherwise be ruin, and add the nutrients back to the soil at the same time. Animals allow for a short-term productive use for land that lies fallow so it can rejuvenate. They can graze a cover crop for part of the year and put nutrients back as manure. Further, if grazed properly, they serve as a natural way to control weeds. An additional use of grazing animals is to use them for grass and weed control under high-growing crops. Sheep can graze under corn and geese under cotton. The land benefits because soils can be improved and better crops can be raised. The farmer benefits by selling product from fallow land or additional product from cropland while getting a rejuvenated soil. The "system" works to maximum benefit for total productivity (Figure 30–7).

In a single-site, mixed, diverse system the animal enterprises must be chosen with all the maxims of sustainability in mind. The differing goals of a sustainable unit, with its optimum production goals, dictate different choices of species, breeds, feeds, and system inputs than a specialized animal unit might. Careful choices must be made. Wastes and by-products almost certainly form a significant portion of the feed for the animal unit(s). Lower production levels are tolerated for increased longevity and disease and parasite resistance. Maximum production in each livestock unit is not the goal. The goal is that the unit produces as much as possible while contributing to the sustainability of the whole system. However, the greater the biological and physical efficiency of animals and plants, the greater the potential to develop sustainable animal-plant systems.

One of the most exciting approaches to mixed systems is **agroforestry.** In these systems, animal feed and other crops are produced in conjunction with forests, orchards, and woodlands. Baker et al. (1990) classified such systems as "(1) producing forage for livestock in land harvested primarily for lumber, (2) producing livestock-forage under orchard tree crops such as walnuts, (3) managing forests to increase off-take of wildlife used for food, (4) reducing livestock-grazing damage to forests while maintaining animal off-take, and (5) planting 'fodder trees' for animal feed to slow deforestation in the humid tropics and desertification in the arid tropics."

### MONOCULTURE

Large operations that focus on producing only one type of animal in intensive units seem to bear the brunt of much objection from those who promote sustainable agriculture. Although not all operations offend to the same degree, the general feeling is that there is no place for large **monoculture** operations in a sustainable agriculture. Part of the objection seems to be on the heavy dependency on grain usage. Such animal systems, generally referred to as confinement feeding, concentrated animal feeding operations, and factory farms, are not considered efficient converters of energy and protein to human food. This is a case in which generalizations have gotten in the way of examining the facts. Simply stated, the efficiency argument is not valid in all cases. Much depends on the particular system in question. Several articles and research studies have demonstrated that animal systems can return edible energy and protein for people at close to or exceeding 100% efficiency. (See Baldwin et al., 1992; Beckett and Oltjen, 1996; Bywater and Baldwin, 1980; and Wheeler et al., 1981.) However, not all who subscribe to the sustainable paradigm disagree with feeding grain to any kind of animal. Many sustainable agriculturists believe in maximizing forage and the use of limited grain. Normally, sustainable agriculturists are believers in systems such as grass-fed and grass-finished beef, free-range turkey, and pasture/self-harvest pig production. Further, these systems are free of many of the ethical concerns related to animal welfare (e.g., the use of crates for sows, layers, and/or veal).

Many sustainable agriculturists consider factory farms, especially vertically integrated corporations, as exploiters of animals, people, and communities. Although often located in depressed communities, workers for these operations are frequently brought in from outside the community. Much, or even most, of the profits of the operations are exported out of the community. The environmental impact is considered significant, and the external costs on community services are high. These vertically integrated companies are also viewed as exploitative of contract farmers. As an added negative, large vertically integrated operations, with their emphasis on consistency in product and standardized production conditions, are contributing to the loss of biodiversity in livestock (Figure 30–8).

**Agroforestry**
Includes land-use systems and practices using woody perennials on the same unit as livestock and/or crops.

**Monoculture**
Producing only one crop or livestock species.

**FIGURE 30–8** Breeds like the Pineywoods cattle (pictured) are rapidly disappearing. Such breeds are often adapted to harsh conditions and are tolerant of parasites and infectious diseases. Large vertically integrated operations, with their emphasis on consistency in product and standardized production conditions, use high-production breeds that require high inputs, skilled management, and controlled environments. Worldwide, multiple livestock breeds become irretrievably lost each month.

Even for the systems that do not return energy or protein at greater than 100% efficiency, it is probably unrealistic to expect that U.S. agriculture will return to diversified, mixed systems and abandon the current methods on a large scale, at least in the short term. The reasons are many, but chief among them are that current profit margins in agriculture are small, current economic and social systems do not promote them, and the world's population is growing and needs feeding. Even if the public policy, economic institutions, social values, and world population issues are all addressed satisfactorily, there are still production restraints to the mixed systems. From a biological perspective, the problem with small, mixed systems is that they lose their efficiency. Generally, they are not able to produce their multiple crops and livestock with the same efficiency as larger, specialized operations can. This is true of both agronomic enterprises and animal enterprises. They lose the ability to incorporate the latest technologies, equipment, and techniques into their operations. Farmers and managers of mixed systems are required to stay abreast of rapidly changing information about a wide variety of crops, which is a difficult feat. Thus a more realistic view holds that, for some producers, systems must be viewed from the much larger picture of area, regional, and overall national sustainability rather than as a return to single, diversified, mixed-production units. There are practices that even monoculture production can put into place that are more sustainable than other practices. Although this type of approach will most assuredly not satisfy the purists, it is a positive and realistic approach with a good deal of potential. Areas of concern and needs for creating more sustainable total agriculture using animals include the following:

1. Concentrated livestock systems also concentrate wastes. Recycling these wastes into animal feed, fertilizer, or some other use that provides biological benefit and revenue neutrality is a pressing need (Figure 30–9).

2. Using grazing animals properly as part of agronomic systems increases soil quality, reduces erosion, reduces water runoff, and thus benefits water table levels and water quality. Current agronomic, monoculture practice could benefit from the integration of grazing animals with crops (Figure 30–10). Use of cover crops with row crops in minimum- or no-tillage systems is gaining in popularity. Using

**FIGURE 30–9**   To minimize the negative environmental effects of confinement animal feeding operations and create more sustainable agricultural systems, it is clear that the wastes from the animals must be handled correctly. Here, a district conservationist discusses the benefits of composting chicken manure. (Photographer Bob Nichols. Courtesy of USDA.)

**FIGURE 30–10**   Areas adjacent to monoculture practice can benefit from simple practices. (a) Wetlands and streamside vegetation serve as a buffer to filter excess nutrients from water running off agricultural land. (b) Switchgrass or other grasses planted in a buffer area not only protect nearby streams but also provide habitat for ground-nesting birds and forage for livestock. (Photographer Scott Bauer. Courtesy of Agricultural Research Service.)

(a)                                             (b)

animals as tools to manage the cover crop and/or crop residues is a logical addition to these systems. Using animals to help recycle soil nutrients in this way was once common practice. Often residues are now burned as a means of removal. Contractual relationships between livestock owners and crop producers for grazing services seem a logical solution.

**3.** Intensive grazing of forage resources coupled with substantial rest between grazing cycles has been shown to increase total short-term productivity of the forage by harvesting it at its optimal nutritive value, as well as long-term productivity by improving soil and other environmental features of the grazing system. These practices can be carried out on higher-quality land to produce milk or on lower producing range or marginal croplands for beef, sheep, or goats. Intensive grazing systems have spread all across the country in a variety of environments. Further education and promotion and subsequent practice of these techniques would improve the sustainability of grazing lands as well as their productivity.

4. Recombinant DNA technology and other biotechnological applications hold great promise for sustainable practices. There seems to be a thread that runs through some otherwise reasonable sustainable agriculture writings that has an "all-natural" overtone that flies in the face of reason. The goal is sustainable agriculture systems, not "all-natural" ones. This mind-set is dangerous to the overall development of sustainable systems when it causes the exclusion of some of the incredible advancements that this area of inquiry has brought and will continue to bring. Crops that need no insecticides or herbicides, cereal grains that can fix nitrogen like legumes, tender beef finished on grass, animals born with a lifetime immunity to disease—all of these and countless more opportunities await the promise of biotechnology. As these tools become available, their use may prove to be the best means of achieving sustainability.

A factor that must be again mentioned is that monoculture reduces the biodiversity of a farm. A crop failure is more difficult to overcome on a farm that has only one crop than it is on a farm that has multiple crops and animal products to market.

Another factor that deserves to be visited again relates to the inefficiencies of many sustainable systems compared to large monoculture systems. How would they compare today if there had been a greater investment in research in agroecological farming systems? Most agricultural research in the United States in the last half century has been directed at making industrial-type agriculture more efficient. The sustainable agriculturist view is that food security is best achieved when there are many participants. This is at every level of production, marketing, processing, and retailing. Sustainable agriculturists further argue that the use of genetic engineering to produce food must consider ethical, spiritual, and societal issues. The conduct of industry may ultimately be the determining factor in whether genetically modified organisms (GMOs) will become more acceptable.

A final point relates to the applications of biotechnology to sustainable systems. Often, GMOs are patented by the companies that developed them. This puts the power of distribution and costs in the hands of a few decision makers. Thus the argument that can be put forth is that the issue is not about the GMOs themselves but about who controls them and, by extension, who controls agriculture. Because those same companies also often have tremendous control over traditional breeding programs for plants and animals, one has to be concerned that traditional animal and plant breeding programs will be neglected in favor of producing more GMOs. There is still a need for both.

## SOURCES OF INFORMATION

Many elements of understanding are needed to comprehend sustainable agriculture completely and how animals work to enhance sustainability. It is outside the scope of this text to present a complete picture of this complex issue. Read the articles listed in the Reference section and seek information from the following sources to get a larger understanding.

**Alternative Farming Systems Information Center**
National Agricultural Library, Room 304
10301 Baltimore Ave.
Beltsville, MD 20705-2351
Phone: (301)504-6559
Fax: (301)504-6409
E-mail: afsic@nal.usda.gov
http://afsic.nal.usda.gov/nal_display/index.php?info_center=2&tax_level=1&tax_subject=285
This site provides a wealth of information and links to other sites.

**Kerr Center for Sustainable Agriculture, Inc.**
P.O. Box 588
Poteau, OK 74953
Phone: (918)647-9123
Fax: (918)647-8712
E-mail: mailbox@kerrcenter.com
http://www.kerrcenter.com

**National Sustainable Agriculture Information Service**
P.O. Box 3657
Fayetteville, AR 72702
Phone: (800)346-9140 (English) 7A.M. to 7 P.M Central Time
(800)411-3222 (español) 8 A.M. to 5 P.M. Pacific Time

**Sustainable Agriculture Library**
World Wide Web Virtual Library
http://www.floridaplants.com/sustainable.htm

**UC Sustainable Agriculture Research and Education Program**
University of California
One Shields Ave.
Davis, CA 95616
Phone: (530)752-7556
Fax: (530)754-8550
E-mail: sarep@ucdavis.edu
http://www.sarep.ucdavis.edu./

## SUMMARY AND CONCLUSION

Sustainable agriculture is typical of a new set of concerns and considerations that agriculture is seeking to come to grips with. It is not just about efficient production of food and fiber. There are also ethical, social, philosophical, and lifestyle choices. In fact, these issues are at its core (Box 30–2). Consider this passage from Feenstra et al. (1998): "Management decisions should reflect not only environmental and broad social considerations, but also individual goals and lifestyle choices. For example, adoption of some technologies or practices that promise profitability may also require such intensive management that one's lifestyle actually deteriorates. Management decisions that promote sustainability, (necessarily) nourish the environment, the community and the individual." There is a strong element of personal satisfaction with one's lifestyle and one's own place within the greater scheme of things that forms a common thread within the sustainable agriculture movement. Baker et al. (1990) pointed out how animals fit into this scheme: "Livestock . . . increase the family's pride in their farm and their satisfaction with farm life. Many people find it easy to identify with an-

## Box 30–2   The Lifestyle Element of Sustainable Agriculture

The following set of references represents the entire citation list cited by Ikerd (1997) in his publication "Understanding and managing the multi-dimensions of sustainable agriculture." Notice the titles. More specifically, notice the words that are not in the titles. Do you see the words *farm* or *agriculture*, for instance? Clearly, something more is involved. Although Ikerd is a prolific writer on the topic and is sharing his own perceptions in this article, which affects the citation list, nevertheless, the citation list is clearly one that covers a spectrum of thought outside what most of us might expect in an article about agriculture.

Barker, J. 1993. *Paradigms: The business of discovering the future*. New York: HarperBusiness, a Division of HarperCollins Publishing.

Capra, F. 1982. *The turning point: Science, society, and the rising culture*. New York: Simon and Schuster.

Covey, S. 1989. *Seven habits of highly effective people*. New York: Simon and Schuster.

Drucker, P. 1989. *The new realities*. New York: Harper and Row.

Drucker, P. 1994. *Post-capitalist society*. New York: HarperBusiness, a Division of HarperCollins Publishing.

Hock, D. W. 1995. The chaordic organization: Out of control and into order. *World Business Academy Perspectives* 9(1). San Francisco: Berrett-Koehler Publishers.

Naisbitt, J., and P. Aburdene. 1990. *Megatrends 2000*. New York: Avon Books, The Hearst Corporation.

Peters, T. 1994. *The pursuit of WOW!* New York: Vintage Books, Random House, Inc.

Reich, R. B. 1992. *The work of nations*. New York: Vintage Books, Random House, Inc.

Savory, A. 1988. *Holistic resource management*. Covelo, CA: Island Press.

Senge, P. M. 1990. *The fifth discipline*. New York: Doubleday Publishing Co.

Toffler, A. 1990. *Power shifts*. New York: Bantam Books.

imals and enjoy being producers and caretakers of livestock. The relationship with animals often brings farm operators in closer association with the biology of farming and with nature." Indeed one of the elements of the definition of sustainable agriculture given in the 1990 Farm Bill is that it will "enhance the quality of life for farmers and society as a whole."

The lifestyle element of sustainable agriculture is something well known to those whose roots are in agriculture. Family farmers of yesteryear in the United States had a satisfaction with what they did and how they did it. Their lifestyle sustained them. Although it is easy to say they lived in simpler times, fairness leads us to concede that they had their share of adversity and issues to deal with. The "good old days" have a tendency to look much better than they actually were to those who lived the life. Every generation defines the elements it holds as important. Every generation defines its own lifestyle and thus defines for itself what will enhance that lifestyle. Those who wish to practice sustainable or more sustainable agriculture in this and future generations will do the same thing. It must be a conscious decision. However, the practice of sustainable agriculture affects many more people than just those who produce the food. Thus decisions must be made by many different stakeholders, including farmers, farm workers, and farm advisers, and also researchers, suppliers, union members, processors, retailers, consumers, and policymakers. In short, all members of society must make these decisions. Achieving sustainability will mean changing laws and public policies, redirecting economic goals, rethinking government's approach to agriculture, and changing the values of millions of people. It is naive to think this will be an easy job. To paraphrase Feenstra et al. (1998), making the transition to sustainable agriculture is a process requiring a series of small, realistic steps. Family economics and personal goals influence how fast or how far participants can go in the transition, but each small decision can make a difference. The key to moving forward is the will to take the next step. Visit Box 30–3 to discover 72 ways to make agriculture more sustainable.

---

### Box 30–3  72 Ways to Make Agriculture Sustainable

**Conserve and Create Healthy Soil**

- Stop soil erosion by terracing, strip cropping, and repairing gullies
- Add organic matter to soil (with "green manure," cover crops, compost, manures, crop residues, and organic fertilizers)
- Practice conservation tillage
- Plant wind breaks
- Rotate cash crops with hay, pasture, or cover crops

**Conserve Water and Protect Its Quality**

- Stop soil erosion in field and pasture
- Reduce use of chemicals
- Establish conservation buffer areas
- Grow crops adapted to rainfall received
- Use efficient irrigation methods

**Manage Organic Wastes So They Don't Pollute**

- Test soil and apply manures and litters only when needed
- Compost dead birds and litters
- Store litter piles out of the rain and snow
- Raise pastured or free-range poultry
- Raise hogs in hoop houses or free range

**Manage Farm Chemicals and Trash So They Don't Pollute**

- Look for alternatives to chemicals
- Use the least amount necessary
- Buy the least toxic chemicals
- Recycle
- Dispose according to label instructions

**Manage Weeds with Minimal Environmental Impact**

- Mechanical Approaches
  - Mowing
  - Flaming
  - Flooding
  - Tillage
  - Controlled burning
- Cultural Approaches
  - Crop rotation
  - Smother crops
  - Cover crops
  - Allelopathic plants
  - Close spacing of plants
- Biological Approaches
  - Multispecies grazing
  - Rotational grazing
- Chemical Approaches
  - Integrated pest management
  - Use of narrow spectrum, least-toxic herbicides
  - Properly calibrated sprayers
  - Application methods that minimize amount used, drift, and farmer contact

**Manage Insects and Diseases with Minimal Environmental Impact**

- Introduce or enhance existing populations of natural predators, pathogens, sterile insects, and other biological control agents
- Use traps
- Maintain wild areas or areas planted with species attractive to beneficial insects
- Selective insecticides or botanical insecticides that are less toxic
- Trap crops
- Rotate crops (avoid monoculture)
- Practice intercropping, strip cropping
- Maintain healthy soil (prevents soil-based diseases)
- Keep plants from becoming stressed

**Select Plants and Animals Adapted to the Environment**

- Grow crops and crop varieties well suited to the specific climate
- Match crops to the soil

## Box 30–3    72 Ways to Make Agriculture Sustainable—continued

- Experiment with older, open pollinated varieties that do well without chemical inputs
- Raise hardy breeds of livestock adapted to climate
- Raise livestock that gain well on grass and native forages

**Encourage Biodiversity of Domesticated Animals, Crops, Wildlife, Native Plants, and Microbic and Aquatic Life**

- Diversify crops and livestock raised
- Leave habitat (field margins, unmowed strips, pond and stream borders, and so on) for wildlife
- Maintain the health of streams and ponds
- Provide wildlife corridors
- Rotate row crops with hay crops

**Conserve Energy Resources**

- Reduce number of tillage operations
- Cut use of chemical fertilizers
- Develop production methods that reduce horsepower needs

- Recycle used oil
- Use solar-powered fences and machines
- Use renewable, farm-produced fuels: ethanol, methanol, fuel oils from oil seed crops, methane from manures and crop wastes

**Increase Profitability and Reduce Risk**

- Diversify crops and livestock
- Substitute management for off-farm resources
- Maximize the use of on-farm resources
- Work with, not against, natural cycles
- Keep machinery, equipment, and building costs down
- Add value to crops and livestock
- Try direct marketing (subscription farming, farmers' market, farm stores, mail order)

*SOURCE:* Adapted from Horne and McDermott, 2000. Used with permission.

## STUDY QUESTIONS

1. Write a short paragraph defining organic farming. Then do the same for sustainable agriculture. What is the primary difference in your definitions?

2. What are the "issues" and "questions" mentioned by Gold (1994) relative to sustainable agriculture?

3. Make a table like the one shown here:

| Farming Practices | Resource Conserving | Socially Supportive | Commercially Competitive | Environmentally Sound |
|---|---|---|---|---|
| 1 | | | | |
| 2 | | | | |
| 3 | | | | |
| 4 | | | | |
| 5 | | | | |
| 6 | | | | |
| 7 | | | | |
| 8 | | | | |
| 9 | | | | |
| 10 | | | | |

Under farming practices, list 10 common farming practices. Analyze each practice to determine how it fits under the remaining four headings, and answer simply *yes* or *no*.

4. Make a list of resources for which animals are the most sustainable use.

5. Describe some sustainable practices that are part of, or could be part of, modern agricultural systems.

6. Describe the systems philosophy necessary for an understanding of a sustainable agriculture.

7. What are the "big-picture reasons" for having animals as part of sustainable systems?

8. Describe what site specificity means with regard to sustainable agriculture.

9. Describe ways to integrate both animals and plants into a diversified farm site.

10. Why does efficiency play such a role in whether or not specific plants and animals are useful in a sustainable farm?

11. What is agroforestry?

12. Why is it likely that U.S. agriculture will continue to contain monoculture of plants and animals as part of an overall sustainable agriculture, at least in the near future?

13. How does the lifestyle element factor into sustainable agriculture philosophy and practice?

14. How many ways can you name to make agriculture more sustainable?

## REFERENCES

For the second, third, and fourth editions, Dr. James Horne, president of the Kerr Center for Sustainable Agriculture, Inc., Poteau, Oklahoma, reviewed and contributed new material to the chapter. The author gratefully acknowledges these contributions.

Baker, F. H., F. E. Busby, N. S. Raun, and J. A. Yazman. 1990. The relationship and roles of animals in sustainable agriculture and on sustainable farms. *The Professional Animal Scientist* 6(3):35.

Baldwin, R. L., K. C. Donovan, and J. L. Beckett. 1992. An update on returns on human edible input in animal agriculture. *Proceedings California Animal Nutrition Conference*, 97–111.

Beckett, J. L., and J. W. Oltjen. 1996. Role of ruminant livestock in sustainable systems. *Journal of Animal Science* 74:1406–1409.

Bywater, A. C., and R. L. Baldwin. 1980. Alternative strategies in food animal production. In *Animals, feed, food and people*, R. L. Baldwin, ed. Boulder, CO: Westview Press.

Dagget, D. 2005. *Gardeners of Eden: Rediscovering our importance to nature*. Reno: University of Nevada Press.

Dagget, D. 2000. *Beyond the rangeland conflict: Toward a west that works*. Reno: University of Nevada Press.

Farm Bill. 1990. Food, Agriculture, Conservation, and Trade Act of 1990, Public Law 101-624, Title XVI, Subtitle A, Section 1603. Washington, DC: Government Printing Office.

Feenstra, G., C. Ingels, D. Campbell, D. Chaney, M. R. George, and E. Bradford. 1998. Education program: What is sustainable agriculture? UC

Sustainable Agriculture Research and Education Program, University of California, Davis, CA. http://www.sarep.ucdavis.edu./concept.htm.

Fraser, D. 2001. The "new" perception of animal agriculture: Legless cows, featherless chickens, and a need for genuine analysis. *Journal of Animal Science* 79:634–641.

Gliessman, S. R. 1998. *Agroecology: Ecological processes in sustainable agriculture*. Boca Raton, FL: CRC Press.

Gold, M. V. 1994. *Sustainable agriculture: Definitions and terms*. National Agricultural Library Cataloging Record: (Special reference briefs; 94–05) 1. Sustainable agriculture—Terminology. I. Title. aS21.D27S64. No.94–05. Available online at http://www.nal.usda.gov/afsic/AFSIC_pubs/srb94-05.htm.

Hawken, P., A. Lovins, and L. H. Lovins. 2004. *Natural capitalism: Creating the next industrial revolution*. New York: Little, Brown.

Horne, J., and M. McDermott. 2000. *72 ways to make agriculture sustainable*. Kerr Center Fact Sheet. Poteau, OK: Kerr Center for Sustainable Agriculture.

Horne, J., and M. McDermott. 2001. *The next green revolution: Essential steps to a healthy, sustainable agriculture*. Binghamton, NY: Haworth Press.

Ikerd, J. E. 1990. Sustainability's promise. *Journal of Soil and Water Conservation* 45(1):4.

Ikerd, J. E. 1997. Understanding and managing the multi-dimensions of sustainable agriculture. Paper presented at the Southern Region Sustainable Agriculture Professional Development Program Workshop, SARE Regional Training Consortium, January 15, 1997, Gainesville, FL. Available online at http://www.ssu.missouri.edu/faculty/jikerd/papers/NC-MULTD.htm.

O'Connell, P. F. 1992. Sustainable agriculture: A valid alternative. *Outlook on Agriculture* 21(1):6.

Owens, F. N., H. M. Arleovich, B. A. Gardner, and C. F. Hanson. 1998. Sustainability of the beef industry in the 21st century. In *Proceedings from the World Conference on Animal Production*, Seoul, South Korea.

Peart, R. M., and W. D. Shoup. 2004. *Agricultural systems management: Optimizing efficiency and performance*. New York: Marcel Dekker.

Raeburn, P. 1996. *The last harvest—The genetic gamble that threatens to destroy American agriculture*. Lincoln: University of Nebraska Press.

Rifkin, J. 1992. *Beyond beef: The rise and fall of the cattle culture*. New York: Dutton.

Roling, N. G., and M. A. E. Wagemakers. 1998. *Facilitating sustainable agriculture*. Cambridge: Cambridge University Press.

Sayre, N. F. 2001. *The new ranch handbook: A guide to restoring western rangelands*. Santa Fe, NM: The Quivira Coalition Books.

Wheeler, R. D., G. L. Kramer, K. B. Young, and E. Ospina. 1981. *The world livestock product, feedstuff, and food grain system*. Morrilton, AK: Winrock International.

Wilson, L. L. 1998. Sustainability as applied to farm animal systems. *The Professional Animal Scientist* 13:55–60.

# Appendixes

**APPENDIX I**  **Nutritive Value of Beef (Composite of Trimmed Retail Cuts, Separable Lean and Fat, Trimmed To ¼ in. Fat, All Grades, Cooked)**

| Nutrient | Units | Value per 100 Grams of Edible Portion |
|---|---|---|
| **Proximates** | | |
| Water | g | 51.43 |
| Energy | kcal | 305 |
| Energy | kj | 1276 |
| Protein | g | 25.94 |
| Total lipid (fat) | g | 21.54 |
| Carbohydrate, by difference | g | 0 |
| Fiber, total dietary | g | 0 |
| Ash | g | 1.06 |
| **Minerals** | | |
| Calcium (Ca) | mg | 10 |
| Iron (Fe) | mg | 2.62 |
| Magnesium (Mg) | mg | 22 |
| Phosphorus (P) | mg | 203 |
| Potassium (K) | mg | 313 |
| Sodium (Na) | mg | 62 |
| Zinc (Zn) | mg | 5.85 |
| Copper (Cu) | mg | 0.11 |
| Manganese (Mn) | mg | 0.015 |
| **Vitamins** | | |
| Vitamin C (ascorbic acid) | mg | 0 |
| Thiamin | mg | 0.08 |
| Riboflavin | mg | 0.21 |
| Niacin | mg | 3.64 |
| Pantothenic acid | mg | 0.35 |
| Vitamin $B_6$ | mg | 0.33 |
| Folate | mcg | 7 |
| Vitamin $B_{12}$ | mcg | 2.44 |
| Vitamin A (IU) | IU | 0 |
| Vitamin A (RE) | mcg RE | 0 |
| Vitamin E | mg ATE | 0.2 |
| **Lipids** | | |
| Fatty acids (saturated) | g | 8.54 |
| 4:00 | g | 0 |
| 6:00 | g | 0 |
| 8:00 | g | 0 |

*(continued)*

**APPENDIX I** **Nutritive Value of Beef (Composite of Trimmed Retail Cuts, Separable Lean and Fat, Trimmed To ¼ in. Fat, All Grades, Cooked)—continued**

| Nutrient | Units | Value per 100 Grams of Edible Portion |
|---|---|---|
| **Lipids (continued)** | | |
| 10:00 | g | 0.06 |
| 12:00 | g | 0.05 |
| 14:00 | g | 0.68 |
| 16:00 | g | 5.24 |
| 18:00 | g | 2.5 |
| Fatty acids (monounsaturated) | g | 9.22 |
| 16:01 | g | 0.89 |
| 18:01 | g | 8.29 |
| 20:01 | g | 0.04 |
| 22:01 | g | 0 |
| Fatty acids (polyunsaturated) | g | 0.78 |
| 18:02 | g | 0.53 |
| 18:03 | g | 0.22 |
| 18:04 | g | 0 |
| 20:04 | g | 0.03 |
| 20:05 | g | 0 |
| 22:05 | g | 0 |
| 22:06 | g | 0 |
| Cholesterol | mg | 88 |
| **Amino acids** | | |
| Tryptophan | g | 0.29 |
| Threonine | g | 1.133 |
| Isoleucine | g | 1.166 |
| Leucine | g | 2.05 |
| Lysine | g | 2.158 |
| Methionine | g | 0.664 |
| Cystine | g | 0.29 |
| Phenylalanine | g | 1.013 |
| Tyrosine | g | 0.871 |
| Valine | g | 1.262 |
| Arginine | g | 1.639 |
| Histidine | g | 0.888 |
| Alanine | g | 1.564 |
| Aspartic acid | g | 2.37 |
| Glutamic acid | g | 3.897 |
| Glycine | g | 1.415 |
| Proline | g | 1.145 |
| Serine | g | 0.992 |

*SOURCE:* USDA Nutrient Database for Standard Reference, Release 12, March 1998.

## APPENDIX II   Nutritive Value of Milk

| Nutrient | Units | Milk, Fluid, 3.25% Milkfat NDB No: 01077 | Milk, Nonfat, Fluid, with Added Vitamin A (fat-free or skim) NDB No: 01085 |
|---|---|---|---|
| | | 1 cup = 244.000 g | |
| **Proximates** | | | |
| Water | g | 214.696 | 222.46 |
| Energy | kcal | 149.916 | 85.534 |
| Energy | kj | 627.08 | 357.7 |
| Protein | g | 8.028 | 8.354 |
| Total lipid (fat) | g | 8.15 | 0.441 |
| Carbohydrate, by difference | g | 11.37 | 11.883 |
| Fiber, total dietary | g | 0 | 0 |
| Ash | g | 1.757 | 1.862 |
| **Minerals** | | | |
| Calcium (Ca) | mg | 291.336 | 302.33 |
| Iron (Fe) | mg | 0.122 | 0.098 |
| Magnesium (Mg) | mg | 32.794 | 27.832 |
| Phosphorus (P) | mg | 227.896 | 247.205 |
| Potassium (K) | mg | 369.66 | 405.72 |
| Sodium (Na) | mg | 119.56 | 126.175 |
| Zinc (Zn) | mg | 0.927 | 0.98 |
| Copper (Cu) | mg | 0.024 | 0.027 |
| Manganese (Mn) | mg | 0.01 | 0.005 |
| Selenium (Se) | mcg | 4.88 | 5.145 |
| **Vitamins** | | | |
| Vitamin C (ascorbic acid) | mg | 2.294 | 2.401 |
| Thiamin | mg | 0.093 | 0.088 |
| Riboflavin | mg | 0.395 | 0.343 |
| Niacin | mg | 0.205 | 0.216 |
| Pantothenic acid | mg | 0.766 | 0.806 |
| Vitamin $B_6$ | mg | 0.102 | 0.098 |
| Folate | mcg | 12.2 | 12.74 |
| Vitamin $B_{12}$ | mcg | 0.871 | 0.926 |
| Vitamin A, IU | IU | 307.44 | 499.8 |
| Vitamin A, RE | mcg RE | 75.64 | 149.45 |
| Vitamin E | mg ATE | 0.244 | 0.098 |
| **Lipids** | | | |
| Fatty acids (saturated) | g | 5.073 | 0.287 |
| 4:00 | g | 0.264 | 0.022 |
| 6:00 | g | 0.156 | 0.002 |
| 8:00 | g | 0.09 | 0.005 |
| 10:00 | g | 0.205 | 0.01 |
| 12:00 | g | 0.229 | 0.007 |
| 14:00 | g | 0.82 | 0.042 |
| 16:00 | g | 2.145 | 0.13 |
| 18:00 | g | 0.988 | 0.047 |

*(continued)*

## APPENDIX II    Nutritive Value of Milk—continued

| Nutrient | Units | Milk, Fluid, 3.25% Milkfat NDB No: 01077 | Milk, Nonfat, Fluid, with Added Vitamin A (fat-free or skim) NDB No: 01085 |
|---|---|---|---|
| | | 1 Cup = 244.000 g | |
| **Lipids (continued)** | | | |
| Fatty acids (monounsaturated) | g | 2.355 | 0.115 |
| 16:01 | g | 0.183 | 0.017 |
| 18:01 | g | 2.05 | 0.093 |
| 20:01 | g | 0 | 0 |
| 22:01 | g | 0 | 0 |
| Fatty acids (polyunsaturated) | g | 0.303 | 0.017 |
| 18:02 | g | 0.183 | 0.012 |
| 18:03 | g | 0.12 | 0.005 |
| 18:04 | g | 0 | 0 |
| 20:04 | g | 0 | 0 |
| 20:05 | g | 0 | 0 |
| 22:05 | g | 0 | 0 |
| 22:06 | g | 0 | 0 |
| Cholesterol | mg | 33.184 | 4.41 |
| Phytosterols | mg | 0 | |
| **Amino acids** | | | |
| Tryptophan | g | 0.112 | 0.118 |
| Threonine | g | 0.364 | 0.377 |
| Isoleucine | g | 0.486 | 0.505 |
| Leucine | g | 0.786 | 0.818 |
| Lysine | g | 0.637 | 0.662 |
| Methionine | g | 0.203 | 0.211 |
| Cystine | g | 0.073 | 0.078 |
| Phenylalanine | g | 0.388 | 0.404 |
| Tyrosine | g | 0.388 | 0.404 |
| Valine | g | 0.537 | 0.559 |
| Arginine | g | 0.29 | 0.301 |
| Histidine | g | 0.217 | 0.225 |
| Alanine | g | 0.276 | 0.289 |
| Aspartic acid | g | 0.61 | 0.635 |
| Glutamic acid | g | 1.681 | 1.749 |
| Glycine | g | 0.171 | 0.176 |
| Proline | g | 0.778 | 0.809 |
| Serine | g | 0.437 | 0.453 |

SOURCE: USDA Nutrient Database for Standard Reference, Release 12, March 1998.

**APPENDIX III**  **Nutritive Value of Poultry Products**

| Nutrient | Units | Chicken[1] Value per 100 Grams of Edible Portion | Duck[2] Value per 100 Grams of Edible Portion | Goose[3] Value per 100 Grams of Edible Portion | Eggs[4] Value per 100 Grams of Edible Portion (2 eggs) |
|---|---|---|---|---|---|
| **Proximates** | | | | | |
| Water | g | 59.45 | 51.84 | 51.95 | 74.62 |
| Energy | kcal | 239 | 337 | 305 | 155 |
| Energy | kj | 1,000 | 1,410 | 1,276 | 649 |
| Protein | g | 27.3 | 18.99 | 25.16 | 12.58 |
| Total lipid (fat) | g | 13.6 | 28.35 | 21.92 | 10.61 |
| Carbohydrate, by difference | g | 0 | 0 | 0 | 1.12 |
| Fiber, total dietary | g | 0 | 0 | 0 | 0 |
| Ash | g | 0.92 | 0.82 | 0.97 | 1.08 |
| **Minerals** | | | | | |
| Calcium (Ca) | mg | 15 | 11 | 13 | 50 |
| Iron (Fe) | mg | 1.26 | 2.7 | 2.83 | 1.19 |
| Magnesium (Mg) | mg | 23 | 16 | 22 | 10 |
| Phosphorus (P) | mg | 182 | 156 | 270 | 172 |
| Potassium (K) | mg | 223 | 204 | 329 | 126 |
| Sodium (Na) | mg | 82 | 59 | 70 | 124 |
| Zinc (Zn) | mg | 1.94 | 1.86 | 2.62 | 1.05 |
| Copper (Cu) | mg | 0.066 | 0.227 | 0.264 | 0.013 |
| Manganese (Mn) | mg | 0.02 | 0.019 | 0.023 | 0.026 |
| Selenium (Se) | mcg | 23.9 | | | 30.8 |
| **Vitamins** | | | | | |
| Vitamin C (ascorbic acid) | mg | 0 | 0 | 0 | 0 |
| Thiamin | mg | 0.063 | 0.174 | 0.077 | 0.066 |
| Riboflavin | mg | 0.168 | 0.269 | 0.323 | 0.513 |
| Niacin | mg | 8.487 | 4.825 | 4.168 | 0.064 |
| Pantothenic acid | mg | 1.03 | 1.098 | 1.53 | 1.398 |
| Vitamin $B_6$ | mg | 0.4 | 0.18 | 0.37 | 0.121 |
| Folate | mcg | 5 | 6 | 2 | 44 |
| Vitamin $B_{12}$ | mcg | 0.3 | 0.3 | 0.41 | 1.11 |
| Vitamin A (IU) | IU | 161 | 210 | 70 | 560 |
| Vitamin A (RE) | mcg RE | 47 | 63 | 21 | 168 |
| Vitamin E | mg ATE | 0.265 | 0.7 | 1.74 | 1.05 |
| **Lipids** | | | | | |
| Fatty acids (saturated) | g | 3.79 | 9.67 | 6.87 | 3.267 |
| 4:00 | g | 0 | 0 | 0 | 0 |
| 6:00 | g | 0 | 0 | 0 | 0 |
| 8:00 | g | 0 | 0 | 0 | 0.003 |
| 10:00 | g | 0 | 0 | 0 | 0.003 |
| 12:00 | g | 0.02 | 0.04 | 0.04 | 0.003 |

*(continued)*

## APPENDIX III    Nutritive Value of Poultry Products—continued

| Nutrient | Units | Chicken[1] Value per 100 Grams of Edible Portion | Duck[2] Value per 100 Grams of Edible Portion | Goose[3] Value per 100 Grams of Edible Portion | Eggs[4] Value per 100 Grams of Edible Portion (2 eggs) |
|---|---|---|---|---|---|
| **Lipids (continued)** | | | | | |
| 14:00 | g | 0.11 | 0.17 | 0.11 | 0.035 |
| 16:00 | g | 2.78 | 6.8 | 4.53 | 2.349 |
| 18:00 | g | 0.77 | 2.43 | 1.81 | 0.828 |
| Fatty acids (monounsaturated) | g | 5.34 | 12.9 | 10.25 | 4.077 |
| 16:01 | g | 0.73 | 1.11 | 0.68 | 0.31 |
| 18:01 | g | 4.4 | 11.52 | 9.52 | 3.725 |
| 20:01 | g | 0.13 | 0.26 | 0.02 | 0.03 |
| 22:01 | g | 0 | 0 | 0 | 0.003 |
| Fatty acids (polyunsaturated) | g | 2.97 | 3.65 | 2.52 | 1.414 |
| 18:02 | g | 2.57 | 3.36 | 2.24 | 1.188 |
| 18:03 | g | 0.11 | 0.29 | 0.18 | 0.035 |
| 18:04 | g | 0 | 0 | 0 | 0 |
| 20:04 | g | 0.11 | 0 | 0 | 0.149 |
| 20:05 | g | 0.01 | 0 | 0 | 0.005 |
| 22:05 | g | 0.02 | 0 | 0 | 0 |
| 22:06 | g | 0.04 | 0 | 0 | 0.038 |
| Cholesterol | mg | 88 | 84 | 91 | 424 |
| **Amino acids** | | | | | |
| Tryptophan | g | 0.305 | 0.232 | 0.332 | 0.153 |
| Threonine | g | 1.128 | 0.773 | 1.123 | 0.604 |
| Isoleucine | g | 1.362 | 0.872 | 1.183 | 0.686 |
| Leucine | g | 1.986 | 1.465 | 2.109 | 1.075 |
| Lysine | g | 2.223 | 1.486 | 1.988 | 0.904 |
| Methionine | g | 0.726 | 0.475 | 0.608 | 0.392 |
| Cystine | g | 0.364 | 0.299 | 0.39 | 0.292 |
| Phenylalanine | g | 1.061 | 0.752 | 1.055 | 0.668 |
| Tyrosine | g | 0.879 | 0.64 | 0.805 | 0.513 |
| Valine | g | 1.325 | 0.938 | 1.232 | 0.767 |
| Arginine | g | 1.711 | 1.284 | 1.566 | 0.755 |
| Histidine | g | 0.802 | 0.462 | 0.7 | 0.298 |
| Aspartic acid | g | 2.434 | 1.814 | 2.262 | 1.264 |
| Glutamic acid | g | 3.991 | 2.798 | 3.739 | 1.644 |
| Glycine | g | 1.764 | 1.624 | 1.594 | 0.423 |
| Proline | g | 1.322 | 1.172 | 1.216 | 0.501 |
| Serine | g | 0.963 | 0.804 | 1.002 | 0.936 |

[1]Chicken, broilers or fryers, meat and skin, cooked, roasted; NDB No: 05009.
[2]Duck, domesticated, meat and skin, cooked, roasted; NDB No: 05140.
[3]Goose, domesticated, meat and skin, cooked, roasted; NDB No: 05147.
[4]Hens' eggs, whole, cooked, hard-boiled, 2 large eggs; NDB No: 01129.

*SOURCE:* USDA Nutrient Database for Standard Reference, Release 12, March 1998.

**APPENDIX IV**    **Nutritive Value of Pork (Composite of Trimmed Retail Cuts Leg, Loin, and Shoulder, Separable Lean Only, Cooked)**

**NDB No: 10093**

| Nutrient | Units | Value per 100 Grams of Edible Portion |
|---|---|---|
| **Proximates** | | |
| Water | g | 60.310 |
| Energy | kcal | 212.000 |
| Energy | kj | 887.000 |
| Protein | g | 29.270 |
| Total lipid (fat) | g | 9.660 |
| Carbohydrate, by difference | g | 0.000 |
| Fiber, total dietary | g | 0.000 |
| Ash | g | 1.180 |
| **Minerals** | | |
| Calcium (Ca) | mg | 21.000 |
| Iron (Fe) | mg | 1.100 |
| Magnesium (Mg) | mg | 26.000 |
| Phosphorus (P) | mg | 237.000 |
| Potassium (K) | mg | 375.000 |
| Sodium (Na) | mg | 59.000 |
| Zinc (Zn) | mg | 2.970 |
| Copper (Cu) | mg | 0.061 |
| Manganese (Mn) | mg | 0.018 |
| Selenium (Se) | mcg | 45.000 |
| **Vitamins** | | |
| Vitamin C (ascorbic acid) | mg | 0.300 |
| Thiamin | mg | 0.846 |
| Riboflavin | mg | 0.345 |
| Niacin | mg | 5.172 |
| Pantothenic acid | mg | 0.684 |
| Vitamin $B_6$ | mg | 0.434 |
| Folate | mcg | 6.000 |
| Vitamin $B_{12}$ | mcg | 0.750 |
| Vitamin A (IU) | IU | 7.000 |
| Vitamin A (RE) | mcg | 2.000 |
| Vitamin E | mg | 0.260 |
| **Lipids** | | |
| Fatty acids (saturated) | g | 3.410 |
| 4:0 | g | 0.000 |
| 6:0 | g | 0.000 |
| 8:0 | g | 0.000 |
| 10:0 | g | 0.010 |
| 12:0 | g | 0.010 |
| 14:0 | g | 0.120 |
| 16:0 | g | 2.110 |
| 18:0 | g | 1.090 |

*(continued)*

**APPENDIX IV**    **Nutritive Value of Pork (Composite of Trimmed Retail Cuts Leg, Loin, and Shoulder, Separable Lean Only, Cooked)—continued**

**NDB No: 10093**

| Nutrient | Units | Value per 100 Grams of Edible Portion |
|---|---|---|
| **Lipids (continued)** | | |
| Fatty acids (monounsaturated) | g | 4.350 |
| 20:1 | g | 0.090 |
| 22:1 | g | 0.000 |
| Fatty acids (polyunsaturated) | g | 0.750 |
| 18:2 | g | 0.650 |
| 18:3 | g | 0.020 |
| 18:4 | g | 0.000 |
| 20:4 | g | 0.050 |
| 20:5 | g | 0.000 |
| 22:5 | g | 0.000 |
| 22:6 | g | 0.000 |
| Cholesterol | mg | 86.000 |
| **Amino acids** | | |
| Tryptophan | g | 0.372 |
| Threonine | g | 1.337 |
| Isoleucine | g | 1.371 |
| Leucine | g | 2.348 |
| Lysine | g | 2.632 |
| Methionine | g | 0.775 |
| Cystine | g | 0.373 |
| Phenylalanine | g | 1.168 |
| Tyrosine | g | 1.020 |
| Valine | g | 1.588 |
| Arginine | g | 1.819 |
| Histidine | g | 1.169 |
| Alanine | g | 1.705 |
| Aspartic acid | g | 2.715 |
| Glutamic acid | g | 4.582 |
| Glycine | g | 1.390 |
| Proline | g | 1.176 |
| Serine | g | 1.209 |

*SOURCE: USDA Nutrient Database for Standard Reference, Release 12, March 1998.*

## APPENDIX V

**Nutritive Value of Lamb (Domestic, Composite of Trimmed Retail Cuts, Separable Lean and Fat, Trimmed to ¼ in. of Fat, Choice, Cooked)**

**NDB No: 17002**

| Nutrient | Units | Value per 100 Grams of Edible Portion |
|---|---|---|
| **Proximates** | | |
| Water | g | 53.72 |
| Energy | kcal | 294 |
| Energy | kj | 1230 |
| Protein | g | 24.52 |
| Total lipid (fat) | g | 20.94 |
| Carbohydrate, by difference | g | 0 |
| Fiber, total dietary | g | 0 |
| Ash | g | 1.04 |
| **Minerals** | | |
| Calcium (Ca) | mg | 17 |
| Iron (Fe) | mg | 1.88 |
| Magnesium (Mg) | mg | 23 |
| Phosphorus (P) | mg | 188 |
| Potassium (K) | mg | 310 |
| Sodium (Na) | mg | 72 |
| Zinc (Zn) | mg | 4.46 |
| Copper (Cu) | mg | 0.119 |
| Manganese (Mn) | mg | 0.022 |
| **Vitamins** | | |
| Vitamin C (ascorbic acid) | mg | 0 |
| Thiamin | mg | 0.1 |
| Riboflavin | mg | 0.25 |
| Niacin | mg | 6.66 |
| Pantothenic acid | mg | 0.66 |
| Vitamin $B_6$ | mg | 0.13 |
| Folate | mcg | 18 |
| Vitamin $B_{12}$ | mcg | 2.55 |
| Vitamin A, IU | IU | 0 |
| Vitamin A, RE | mcg RE | 0 |
| Vitamin E | mg ATE | 0.14 |
| **Lipids** | | |
| Fatty acids (saturated) | g | 8.83 |
| 4:00 | g | 0 |
| 6:00 | g | 0 |
| 8:00 | g | 0 |
| 10:00 | g | 0.05 |
| 12:00 | g | 0.09 |
| 14:00 | g | 0.82 |
| 16:00 | g | 4.48 |
| 18:00 | g | 2.84 |
| Fatty acids (monounsaturated) | g | 8.82 |

*(continued)*

**APPENDIX V**   **Nutritive Value of Lamb (Domestic, Composite of Trimmed Retail Cuts, Separable Lean and Fat, Trimmed to ¼ in. of Fat, Choice, Cooked)—continued**

**NDB No: 17002**

| Nutrient | Units | Value per 100 Grams of Edible Portion |
|---|---|---|
| 16:01 | g | 0.61 |
| 18:01 | g | 8 |
| 20:01 | g | 0 |
| 22:01 | g | 0 |
| Fatty acids (polyunsaturated) | g | 1.51 |
| 18:02 | g | 1.14 |
| 18:03 | g | 0.3 |
| 18:04 | g | 0 |
| 20:04 | g | 0.07 |
| 20:05 | g | 0 |
| 20:06 | g | 0 |
| 22:05 | g | 0 |
| 22:06 | g | 0 |
| Cholesterol | mg | 97 |
| **Amino acids** | | |
| Tryptophan | g | 0.287 |
| Threonine | g | 1.05 |
| Isoleucine | g | 1.183 |
| Leucine | g | 1.908 |
| Lysine | g | 2.166 |
| Methionine | g | 0.629 |
| Cystine | g | 0.293 |
| Phenylalanine | g | 0.998 |
| Tyrosine | g | 0.824 |
| Valine | g | 1.323 |
| Arginine | g | 1.457 |
| Histidine | g | 0.777 |
| Aspartic acid | g | 2.159 |
| Glutamic acid | g | 3.559 |
| Glycine | g | 1.198 |
| Proline | g | 1.029 |
| Serine | g | 0.912 |

SOURCE: *USDA Nutrient Database for Standard Reference*, Release 12, March 1998.

**APPENDIX VI** — **Nutritive Value of Game Meat and Goat (Cooked, Roasted)**

NDB No: 17169

| Nutrient | Units | Value per 100 Grams of Edible Portion |
|---|---|---|
| **Proximates** | | |
| Water | g | 68.21 |
| Energy | kcal | 143 |
| Energy | kj | 598 |
| Protein | g | 27.1 |
| Total lipid (fat) | g | 3.03 |
| Carbohydrate, by difference | g | 0 |
| Fiber, total dietary | g | 0 |
| Ash | g | 1.46 |
| **Minerals** | | |
| Calcium (Ca) | mg | 17 |
| Iron (Fe) | mg | 3.73 |
| Magnesium (Mg) | mg | 0 |
| Phosphorus (P) | mg | 201 |
| Potassium (K) | mg | 405 |
| Sodium (Na) | mg | 86 |
| Zinc (Zn) | mg | 5.27 |
| Copper (Cu) | mg | 0.303 |
| Manganese (Mn) | mg | 0.042 |
| **Vitamins** | | |
| Vitamin C (ascorbic acid) | mg | 0 |
| Thiamin | mg | 0.09 |
| Riboflavin | mg | 0.61 |
| Niacin | mg | 3.95 |
| Vitamin $B_6$ | mg | 0 |
| Folate | mcg | 5 |
| Vitamin $B_{12}$ | mcg | 1.19 |
| Vitamin A, IU | IU | 0 |
| Vitamin A, RE | mcg RE | 0 |
| Vitamin E | mg ATE | 0.05 |
| **Lipids** | | |
| Fatty acids (saturated) | g | 0.93 |
| 4:00 | g | 0 |
| 6:00 | g | 0 |
| 8:00 | g | 0 |
| 10:00 | g | 0 |
| 12:00 | g | 0 |
| 14:00 | g | 0.04 |
| 16:00 | g | 0.43 |
| 18:00 | g | 0.43 |
| Fatty acids (monounsaturated) | g | 1.36 |
| 16:01 | g | 0.06 |
| 18:01 | g | 1.24 |

*(continued)*

## APPENDIX VI    Nutritive Value of Game Meat and Goat
(Cooked, Roasted)—continued

**NDB No: 17169**

| Nutrient | Units | Value per 100 Grams of Edible Portion |
|---|---|---|
| 20:01 | g | 0 |
| 22:01 | g | 0 |
| Fatty acids (polyunsaturated) | g | 0.23 |
| 18:02 | g | 0.13 |
| 18:03 | g | 0.02 |
| 18:04 | g | 0 |
| 20:04 | g | 0.08 |
| 20:05 | g | 0 |
| 22:05 | g | 0 |
| 22:06 | g | 0 |
| Cholesterol | mg | 75 |
| **Amino acids** | | |
| Tryptophan | g | 0.403 |
| Threonine | g | 1.29 |
| Isoleucine | g | 1.371 |
| Leucine | g | 2.258 |
| Lysine | g | 2.016 |
| Methionine | g | 0.726 |
| Cystine | g | 0.323 |
| Phenylalanine | g | 0.941 |
| Tyrosine | g | 0.833 |
| Valine | g | 1.452 |
| Arginine | g | 1.989 |
| Histidine | g | 0.565 |

*SOURCE: USDA Nutrient Database for Standard Reference,* Release 12, March 1998.

## APPENDIX VII    Nutritive Value of Goat Milk
### NDB No: 01108

| Nutrient | Units | 1 Cup<br>244.000 g |
|---|---|---|
| **Proximates** | | |
| Water | g | 212.353 |
| Energy | kcal | 167.904 |
| Energy | kj | 702.72 |
| Protein | g | 8.686 |
| Total lipid (fat) | g | 10.102 |
| Carbohydrate, by difference | g | 10.858 |
| Fiber, total dietary | g | 0 |
| Ash | g | 2.001 |
| **Minerals** | | |
| Calcium (Ca) | mg | 325.74 |
| Iron (Fe) | mg | 0.122 |
| Magnesium (Mg) | mg | 34.087 |
| Phosphorus (P) | mg | 270.108 |
| Potassium (K) | mg | 498.736 |
| Sodium (Na) | mg | 121.512 |
| Zinc (Zn) | mg | 0.732 |
| Copper (Cu) | mg | 0.112 |
| Manganese (Mn) | mg | 0.044 |
| Selenium (Se) | mcg | 3.416 |
| **Vitamins** | | |
| Vitamin C (ascorbic acid) | mg | 3.148 |
| Thiamin | mg | 0.117 |
| Riboflavin | mg | 1.337 |
| Niacin | mg | 0.676 |
| Pantothenic acid | mg | 0.756 |
| Vitamin $B_6$ | mg | 0.112 |
| Folate | mcg | 1.464 |
| Vitamin $B_{12}$ | mcg | 0.159 |
| Vitamin A, IU | IU | 451.4 |
| Vitamin A, RE | mcg RE | 136.64 |
| Vitamin E | mg ATE | 0.22 |
| **Lipids** | | |
| Fatty acids (saturated) | g | 6.507 |
| 4:00 | g | 0.312 |
| 6:00 | g | 0.229 |
| 8:00 | g | 0.234 |
| 10:00 | g | 0.634 |
| 12:00 | g | 0.303 |
| 14:00 | g | 0.793 |
| 16:00 | g | 2.223 |
| 18:00 | g | 1.076 |
| Fatty acids (monounsaturated) | g | 2.706 |
| 16:01 | g | 0.2 |

(continued)

**APPENDIX VII** **Nutritive Value of Goat Milk—continued**
**NDB No: 01108**

| Nutrient | Units | 1 Cup<br>244.000 g |
|---|---|---|
| 18:01 | g | 2.384 |
| 20:01 | g | 0 |
| 22:01 | g | 0 |
| Fatty acids (polyunsaturated) | g | 0.364 |
| 18:02 | g | 0.266 |
| 18:03 | g | 0.098 |
| 18:04 | g | 0 |
| 20:04 | g | 0 |
| 20:05 | g | 0 |
| 22:05 | g | 0 |
| 22:06 | g | 0 |
| Cholesterol | mg | 27.816 |
| **Amino acids** | | |
| Tryptophan | g | 0.107 |
| Threonine | g | 0.398 |
| Isoleucine | g | 0.505 |
| Leucine | g | 0.766 |
| Lysine | g | 0.708 |
| Methionine | g | 0.195 |
| Cystine | g | 0.112 |
| Phenylalanine | g | 0.387 |
| Tyrosine | g | 0.437 |
| Valine | g | 0.586 |
| Arginine | g | 0.29 |
| Histidine | g | 0.217 |
| Alanine | g | 0.288 |
| Aspartic acid | g | 0.512 |
| Glutamic acid | g | 1.527 |
| Glycine | g | 0.122 |
| Proline | g | 0.898 |
| Serine | g | 0.442 |

SOURCE: *USDA Nutrient Database for Standard Reference,* Release 12, March 1998.

## APPENDIX VIII — Nutritive Value of Various Fish (Finfish, Trout, Rainbow, Farmed, Cooked, Dry Heat. Finfish, Catfish, Farmed, Cooked, Dry Heat)

| Nutrient | Units | Trout NDB No: 15241 Value per 100 Grams of Edible Portion | Catfish NDB No: 15235 Value per 100 Grams of Edible Portion |
|---|---|---|---|
| **Proximates** | | | |
| Water | g | 67.53 | 71.58 |
| Energy | kcal | 169 | 152 |
| Energy | kj | 707 | 636 |
| Protein | g | 24.27 | 18.72 |
| Total lipid (fat) | g | 7.2 | 8.02 |
| Carbohydrate, by difference | g | 0 | 0 |
| Fiber, total dietary | g | 0 | 0 |
| Ash | g | 1.6 | 1.17 |
| **Minerals** | | | |
| Calcium (Ca) | mg | 86 | 9 |
| Iron (Fe) | mg | 0.33 | 0.82 |
| Magnesium (Mg) | mg | 32 | 26 |
| Phosphorus (P) | mg | 266 | 245 |
| Potassium (K) | mg | 441 | 321 |
| Sodium (Na) | mg | 42 | 80 |
| Zinc (Zn) | mg | 0.49 | 1.05 |
| Copper (Cu) | mg | 0.061 | 0.122 |
| Manganese (Mn) | mg | 0.02 | 0.02 |
| Selenium (Se) | mcg | 15 | 14.5 |
| **Vitamins** | | | |
| Vitamin C (ascorbic acid) | mg | 3.3 | 0.8 |
| Thiamin | mg | 0.237 | 0.42 |
| Riboflavin | mg | 0.08 | 0.073 |
| Niacin | mg | 8.79 | 2.513 |
| Pantothenic acid | mg | 1.31 | 0.617 |
| Vitamin $B_6$ | mg | 0.396 | 0.163 |
| Folate | mcg | 24 | 7 |
| Vitamin $B_{12}$ | mcg | 4.97 | 2.8 |
| Vitamin A, IU | IU | 287 | 50 |
| Vitamin A, RE | mcg RE | 86 | 15 |
| **Lipids** | | | |
| Fatty acids (saturated) | g | 2.105 | 1.789 |
| 14:00 | g | 0.26 | 0.112 |
| 16:00 | g | 1.321 | 1.309 |
| 18:00 | g | .0377 | 0.319 |
| Fatty acids (monounsaturated) | g | 2.096 | 4.155 |
| 16:01 | g | 0.319 | 0.292 |
| 18:01 | g | 1.427 | 3.738 |
| 20:01 | g | 0.326 | 0.106 |
| Fatty acids (polyunsaturated) | g | 2.33 | 1.392 |

*(continued)*

**APPENDIX VIII**    **Nutritive Value of Various Fish (Finfish, Trout, Rainbow, Farmed, Cooked, Dry Heat. Finfish, Catfish, Farmed, Cooked, Dry Heat)—continued**

| Nutrient | Units | Trout NDB No: 15241 Value per 100 Grams of Edible Portion | Catfish NDB No: 15235 Value per 100 Grams of Edible Portion |
|---|---|---|---|
| 18:02 | g | 0.949 | 1.029 |
| 18:03 | g | 0.082 | 0.082 |
| 18:04 | g | 0.068 | 0.016 |
| 20:04 | g | 0.036 | 0.041 |
| 20:05 | g | 0.334 | 0.049 |
| 22:06 | g | 0.82 | 0.128 |
| Cholesterol | mg | 68 | 64 |
| **Amino acids** | | | |
| Tryptophan | g | 0.272 | 0.21 |
| Threonine | g | 1.064 | 0.821 |
| Isoleucine | g | 1.118 | 0.862 |
| Leucine | g | 1.972 | 1.521 |
| Lysine | g | 2.229 | 1.719 |
| Methionine | g | 0.718 | 0.554 |
| Cystine | g | 0.26 | 0.201 |
| Phenylalanine | g | 0.947 | 0.731 |
| Tyrosine | g | 0.819 | 0.632 |
| Valine | g | 1.25 | 0.964 |
| Arginine | g | 1.452 | 1.12 |
| Histidine | g | 0.714 | 0.551 |
| Alanine | g | 1.468 | 1.132 |
| Aspartic acid | g | 2.485 | 1.917 |
| Glutamic acid | g | 3.623 | 2.794 |
| Glycine | g | 1.165 | 0.898 |
| Proline | g | 0.858 | 0.662 |
| Serine | g | 0.99 | 0.764 |

SOURCE: *USDA Nutrient Database for Standard Reference*, Release 12, March 1998.

**APPENDIX IX**  **Nutritive Value of Game Meat and Rabbit (Domesticated, Composite of Cuts, Cooked, Roasted)**

**NDB No: 17178**

| Nutrient | Units | Value per 100 Grams of Edible Portion |
|---|---|---|
| **Proximates** | | |
| Water | g | 60.61 |
| Energy | kcal | 197 |
| Energy | kj | 824 |
| Protein | g | 29.06 |
| Total lipid (fat) | g | 8.05 |
| Carbohydrate, by difference | g | 0 |
| Fiber, total dietary | g | 0 |
| Ash | g | 1.04 |
| **Minerals** | | |
| Calcium (Ca) | mg | 19 |
| Iron (Fe) | mg | 2.27 |
| Magnesium (Mg) | mg | 21 |
| Phosphorus (P) | mg | 263 |
| Potassium (K) | mg | 383 |
| Sodium (Na) | mg | 47 |
| Zinc (Zn) | mg | 2.27 |
| Copper (Cu) | mg | 0.189 |
| Manganese (Mn) | mg | 0.032 |
| **Vitamins** | | |
| Vitamin C (ascorbic acid) | mg | 0 |
| Thiamin | mg | 0.09 |
| Riboflavin | mg | 0.21 |
| Niacin | mg | 8.43 |
| Pantothenic acid | mg | 0.93 |
| Vitamin $B_6$ | mg | 0.47 |
| Folate | mcg | 11 |
| Vitamin $B_{12}$ | mcg | 8.3 |
| Vitamin A (IU) | IU | 0 |
| Vitamin A (RE) | mcg RE | 0 |
| **Lipids** | | |
| Fatty acids (saturated) | g | 2.4 |
| 14:00 | g | 0.21 |
| 16:00 | g | 1.81 |
| 18:00 | g | 0.38 |
| Fatty acids (monounsaturated) | g | 2.17 |
| 16:01 | g | 0.26 |
| 18:01 | g | 1.86 |
| Fatty acids (polyunsaturated) | g | 1.56 |
| 18:02 | g | 1.24 |
| 18:03 | g | 0.32 |
| Cholesterol | mg | 82 |

(continued)

**APPENDIX IX**    **Nutritive Value of Game Meat and Rabbit (Domesticated, Composite of Cuts, Cooked, Roasted)—continued**

**NDB No: 17178**

| Nutrient | Units | Value per 100 Grams of Edible Portion |
|---|---|---|
| **Amino acids** | | |
| Tryptophan | g | 0.384 |
| Threonine | g | 1.3 |
| Isoleucine | g | 1.379 |
| Leucine | g | 2.264 |
| Lysine | g | 2.544 |
| Methionine | g | 0.727 |
| Cystine | g | 0.365 |
| Phenylalanine | g | 1.193 |
| Tyrosine | g | 1.035 |
| Valine | g | 1.477 |
| Arginine | g | 1.795 |
| Histidine | g | 0.815 |
| Alanine | g | 1.753 |
| Aspartic acid | g | 2.839 |
| Glutamic acid | g | 4.662 |
| Glycine | g | 1.578 |
| Proline | g | 1.42 |
| Serine | g | 1.288 |

SOURCE: *USDA Nutrient Database for Standard Reference,* Release 12, March 1998.

# Glossary

**AAFCO**   Association of American Feed Control Officials. An organization that provides a mechanism for developing and implementing uniform and equitable laws, regulations, standards, and enforcement policies to regulate the manufacture, distribution, and sale of animal feeds.

**AAFCO Dog or Cat Nutrient Profiles**   Nutritional standards on which nutritional adequacy statements are based.

**AAFCO Feeding Trial**   Standards under which a dog or cat food must be tested to qualify for use in the AAFCO nutritional adequacy statement.

**Abomasum**   The true glandular stomach in the ruminant.

**Accuracy**   The measure of reliability associated with an EPD. If little or no information is available, accuracies may range as low as 0.01.

**Active immunity**   Immunity to a disease developed by exposure to the disease or by receiving a vaccine for the disease.

**Acute diseases**   Diseases that are sudden or severe in onset and effect on the animal.

**Ad libidum**   Having feed available at all times.

**Adaptation**   The sum of the changes an animal makes in response to environmental stimuli.

**Additive gene action**   When the total phenotypic effect is the sum of the individual effects of the alleles.

**Aggressive behavior**   Threatening or harmful behavior toward others of the same or different species.

**Agribusiness**   The segment of agriculture that deals in sales and service to production agriculturists and consumers.

**Agriculture**   The combination of science and art used to cultivate and grow crops and livestock and process the products. A person who practices agriculture is an agriculturist or agriculturalist. Agricultural is the adjective that refers to things used in, relating to, or associated with agriculture.

**Agroforestry**   Land-use systems and practices using woody perennials on the same land as livestock and/or crops.

**AI stud**   Company that markets semen from high-quality males.

**Alleles**   One of two or more alternative forms of a gene occupying corresponding sites (loci) on homologous chromosomes.

**All-in, all-out animal management**   Adding animals to a facility, such as a farrowing house, at the same time and then removing them at the same time.

**Alveoli**   Spherical-shaped structures making up the primary component of the mammary gland. Consists of a lumen, which is surrounded by secretory cells that produce the milk, and myoepithelial cells, which contract to squeeze the milk out of the lumen into the mammary gland ducts.

**Anaerobic**   Conditions that lack molecular oxygen.

**Anestrus**   Period of time when a female is not having estrous cycles.

**Aneuploidy**   A condition where an organism has a chromosome number that is not an exact multiple of the monoploid (m) number.

**Animal breeding**   The use of biometry and genetics to improve farm animal production.

**Animal health**   The study and practice of maintaining animals as near to a constant state of health as is possible and feasible.

**Animal rights movement**   A political and social movement that concerns itself with philosophy, sociology, and public policy as each deals with the standing of animals in relation to human society.

**Animal science**   The combination of disciplines that together comprise the study of domestic animals. Animal science has traditionally been associated only with livestock species. However, the expanded interests of the population have caused an expansion of animal science to include additional domestic species, chiefly those in the pet and companion animal classification.

**Animal welfare**   Concern for an animal's treatment and well-being while it is being used to provide for human needs. Humane use of animals.

**Angora**   A specialized fiber-producing breed of goat.

**Anorexia**   Inappetence or unwillingness to eat.

**Anthropomorphism**   Attributing human thoughts, emotions, and characteristics to animals.

**Antibodies**    Proteins produced by the body that attack infectious agents and neutralize them.

**Applied ethology**    The study of domestic animal behavior.

**Applied or production nutritionist**    A practical production-oriented nutritionist. A basic nutritionist might discover that compound X improves rate of growth. An applied nutritionist would work on practical questions such as cost effectiveness, method of delivery, carcass effects, and so on.

**Aquaculture**    The farming of aquatic organisms including fish, mollusks, crustaceans, and plants in fresh water, brackish water, or salt water.

**Artificial insemination**    The procedure for placing semen in the reproductive tract of a female animal through means other than the natural mating act in the hopes of causing a pregnancy. Commonly used in animals.

**Artificial selection**    The practice of choosing the animals in a population that will be allowed to reproduce.

**Artificial vagina (AV)**    Device used to collect semen from a male. Following erection, the penis is directed into the artificial vagina and the male ejaculates, depositing the ejaculate in a reservoir of the AV.

**Ash**    The incombustible residue remaining after complete combustion at 500–600°C of a sample, such as feed, animal tissue, or excreta to remove the organic matter. Considered to be the mineral matter of a feed.

**Atresia**    The degeneration of follicles that do not make it to the mature, or Graafian, stage. The majority of follicles in the ovary undergo atresia.

**Auditory**    Related to hearing.

**Autosomes**    All chromosomes other than the sex chromosomes.

**Aversive event**    A negative experience that may be painful, frightening, or nauseating. It may even involve the senses, such as a foul taste or odor or a sound like that made by a can full of rocks when tossed to the floor near a misbehaving puppy.

**Avian**    Pertaining to poultry and/or fowl.

**Babcock Cream Test**    Test for determining the fat content in milk.

**Baby boomer**    Demographic term for the U.S. population born between 1946 and 1964. The U.S. Census estimates there are nearly 83 million boomers.

**Backfat**    The subcutaneous fat on an animal's back. For hogs, backfat is highly correlated to total body fat and is often measured and used as a means of selecting lean brood stock.

**Baitfish**    Fish selected and produced to be used as bait for catching larger fish.

**Balance trial**    A type of metabolism trial designed to determine the retention of a specific nutrient in the body.

**Bantam**    Fowl that are miniatures of full-sized breeds. Some are distinct breeds. Usually a fourth to a fifth the weight of standard birds. Considered ornamental.

**Barrow**    A castrated male hog.

**Barter**    Trading services or commodities for one another.

**Bases**    One of the four chemical units on the DNA molecule that form combinations that code for protein manufacture. The four bases are adenine (A), cytosine (C), guanine (G), and thymine (T).

**Basic breeder**    In poultry production, a unit that produces parent stock used for multiplication of poultry, either by outcrossing, inbreeding, or other methods. Also referred to as *primary breeder.*

**Basic nutritionist**    A person interested in elucidating basic metabolism and nutrient action and interaction. Basic science in any discipline is necessary if practical implications are to be discovered. Think of basic science as the acquisition of knowledge for the sake of knowledge.

**Beef cycle**    Historic fluctuations in beef cattle numbers, which occur over roughly 10-year periods.

**Behavioral ecology**    The study of the relationships between a species' behavior and its environment.

**Biofuel**    Fuel derived from biological materials such as crops and animal waste.

**Biometry**    The application of statistics to topics in biology.

**Biopsy**    Surgical removal and examination (usually microscopic examination) of a tissue from a living body to reach a diagnosis.

**Biosecurity**    Procedures designed to minimize disease transmission from outside and inside a production unit.

**Biotechnology**    A collective set of tools and applications of living organisms (or parts of organisms) to make or modify products, improve plants or animals, or develop microorganisms for specific uses. Much of the recent innovation associated with biotechnology has been brought about by the tools of recombinant DNA technology.

**Bitch**    A female dog.

**Blastocyst**    More differentiated embryo consisting of an inner cell mass, blastocoele, and trophoblast.

**Blended fisheries**    A combination of aquaculture and capture fisheries. Aquaculture techniques are used to enhance or supplement capture fisheries.

**Bloat**    Abnormal quantities of gas collecting in the fermentative portion of the digestive tract. Can be life threatening.

**Blood spots**    Small spots of blood inside an egg, probably caused by blood vessels breaking in the ovary or oviduct during egg formation.

**Boar**   An intact male hog. Boars are kept only for breeding purposes. Boar meat has a characteristic unpleasant taste.

**Body condition**   The amount of fat on an animal's body.

**Bolus**   A rounded mass that is ready to swallow. In the ruminant, a bolus may stay intact and be regurgitated to be remasticated during rumination.

**Bomb calorimeter**   A device into which a substance can be placed and ignited under a pressurized atmosphere of oxygen. An insulated water jacket surrounds the "bomb" portion where the ignition takes place. By measuring the change in the temperature of the water, the amount of energy in the substance can be determined. Energy values for solids, liquids, or gases can be determined.

**Book**   Refers to the number of females scheduled to be mated to a particular male.

**Bovine somatotropin (BST)**   Growth hormone in cattle. Produced and secreted by the anterior pituitary gland, BST acts on various target tissues throughout the body. One of the primary mechanisms by which BST works is through a second hormone, insulin-like growth factor-1 (IGF-1). In cattle, BST increases milk production dramatically, presumably by increasing nutrient availability to the secretory cells, or by increasing the production of milk components directly.

**Breed**   Animals with common ancestry that have distinguishable, fixed characteristics. When mated with others of the breed, they produce offspring with the same characteristics.

**Breeding complementarity**   When the characteristics of different breeds complement each other in cross-breeding systems.

**Breeding soundness exam**   Examination to determine the physical capacity of an individual to breed.

**Breeding value**   The worth of an individual as a parent.

**Breeds Revolution**   Period of great expansion in numbers of breeds of beef cattle in the United States.

**British breeds**   Hereford, Angus, and Shorthorn; breeds of cattle that originated in England.

**Broiler**   A chicken of either sex produced and used for meat purposes. Generally slaughtered at 6 weeks of age or younger. The term *fryer* is often used interchangeably.

**Broiler duckling, also fryer duckling**   A young duck usually under 8 weeks of age of either sex. It must have a soft bill and weigh from 3 to 6 ½ lbs.

**Brood**   A group of baby chickens or other poultry or birds. As a verb, it can also refer to the growing of baby chicks.

**Brooder**   A device with controlled heat and light used to warm chicks from the day of hatch to approximately 5 weeks of age. The heat is usually contained in a large reflector or hover under which the birds congregate.

**Broodiness**   When a hen stops laying eggs and prepares to sit on the eggs to incubate them. Once the eggs hatch, the hen will care for them. Such hens are referred to as being *broody*.

**Brooding**   The act of raising young poultry under environmentally controlled conditions during the first few weeks of life.

**Broodstock**   Animals kept for reproductive purposes.

**Browse**   The tender twigs and leaves from brush and trees.

**Buck**   Term used for both an intact male rabbit and an intact male goat.

**Calorie**   A measurement of food energy. A kilocalorie is the amount of heat required to raise the temperature of 1 gram of water 1°Celsius, from 14.5 to 15.5°C. The calories we discuss in human diets are actually kilocalories (kcal).

**Candling**   Inspection of the inside of an intact egg with a light to detect defects. Incubating eggs can also be tested for dead germs and infertile eggs.

**Capital requirements**   In this context, a combination of the capital required to integrate sufficiently to target a large enough market segment.

**Capture fisheries**   The term used to designate the harvesting of wild aquatic animal species.

**Carbohydrates**   Chemically defined as polyhydroxy aldehydes or ketones, or substances that can be hydrolyzed to them.

**Carnivore**   Animals that subsist on meat.

**Casein**   The major protein of milk.

**Case-ready product**   A business system in which centrally prepared meats and meat items are used to increase consumer satisfaction and related profits.

**Cash crop**   A crop grown specifically with the intent of marketing its product.

**Cecotrophs**   Large soft pellets of cecally fermented feed, which move through the colon in a distinctly different manner than feces and are reconsumed for upper digestive tract digestion.

**Cecotrophy**   A process in which hard fecal material is expelled from the large intestine. The cecum contracts and expels its contents, which are subsequently covered with mucus. These mucus-covered, soft fecal pellets are propelled through the large intestine by peristaltic action and consumed by the animal directly from the anus.

**Cellulase**   An enzyme that specifically attacks and digests cellulose. Mammals do not make cellulases but some microorganisms do.

**Cellulose**   A carbohydrate composed of thousands of glucose molecules that forms the support structure of plants.

**Centrally planned economy**   An economy under government control. Prices, labor, and other economic inputs are controlled and are not allowed to fluctuate in accordance with supply and demand.

**Centromere**   The region of a chromosome where spindle fibers attach.

**Chick**   A young chicken or gamebird of either sex from 1 day old to about 5 to 6 weeks of age.

**Chromosome**   The DNA-containing structures in cells. Composed of segments called genes.

**Chronic diseases**   Continuing over a long period or having a gradual effect.

**Chyme**   The mixture of food, saliva, and gastric secretions as it is ready to leave the stomach and move into the duodenum.

**Civilization**   In modern context, refers to what we consider a fairly high level of cultural and technological development. An important contribution of the cultural revolution that farming brought to the process of civilization was giving people the time and lifestyle that led to the development of written language, record keeping, and artistic endeavors.

**Class**   In poultry, a group of breeds originating in the same geographic area. Names are taken from the region where the breeds originated.

**Classical conditioning**   The type of conditioning the Russian physiologist Ivan Pavlov demonstrated, hence the name Pavlovian conditioning, in which a reflex-like response can be stimulated by a neutral stimulus. In Pavlov's studies, food (an unconditioned stimulus) produced a reflex-like response in dogs, namely salivation. He coupled the ringing of a bell (neutral stimulus) with the food. He was then able to use only the bell to produce the salivation.

**Clinical infection**   An infectious disease in which clinical signs of the disease are expressed. Clinical signs allow detection and identification of the disease.

**Clinical sign**   Observable difference in an animal's normal function or state of health that indicates the presence of a bodily disorder or disease. Examples are fever, weight loss, reduced performance, decreased appetite, depression, edema, or any other signs of disease.

**Cloning**   The process of producing a genetic copy of a gene, DNA segment, animal, or embryo. If cloning a gene or DNA segment, the segment is inserted into a bacteria or virus that is then cultured to make the copies. For organisms, animals, or embryos, the nucleus from a cell of the animal or embryo is transplanted into an egg with the nucleus (and thus its original genetic material) removed. The new egg is then allowed to develop into an animal.

**Closed herd**   A herd into which no new animals are introduced.

**Cock**   A rooster aged 1 year or older.

**Cockerel**   A rooster less than 1 year old.

**Codominance**   Both alleles are expressed in the phenotype when present in the heterozygous state.

**Cold blooded**   Animals that cannot control their own body temperature. Poikilotherm.

**Colic**   A broad term that means digestive disturbance. The major symptom of colic in horses is pain. Many things can cause colic, with parasites and infections the most likely causes. However, rapid changes in diet, overeating, moldy feed, lack of water, and eating of shavings or sawdust bedding are also causes. Considered the most prevalent, costly, and dangerous internal diseases of horses.

**Colostrum**   Specialized milk produced in the early days following parturition to provide extra nutrients and immune function to the young. Colostrum is higher in vitamins, minerals, and protein compared with normal milk, and generally carries antibodies to aid the young in developing an immune response against potential antigens.

**Colt**   A young male horse.

**Companion animal**   An animal to whom an owner has an intense emotional tie.

**Comparative method of study**   A systematic method of comparing the behavior of two or more species as a way of discovering the mechanism of behavior.

**Comparative psychology**   The study of the mechanisms controlling behavior, learning, sensation, perception, and behavior genetics in animals and making extrapolations to humans or other animals.

**Complete diet**   Diet formulated to meet all the nutritional needs of an animal.

**Composite breed**   A breed developed from two or more previously established breeds.

**Compost**   Decayed organic matter used for fertilizing and conditioning land.

**Conception**   When the sperm fertilizes the ovum.

**Condensed milk**   Milk with water removed and sugar added.

**Conditioning**   The learned response of an animal to a stimulus.

**Conformation events**   Term that usually refers to a competition in which an animal's conformation is judged.

**Contagious disease**   A disease capable of being transmitted from animal to animal.

**Contemporary group**   A group in which animals of a given sex and age, having similar treatment, are given an equal opportunity to perform.

**Continuous eater strategy**   Feeding strategy employed by many prey species as a mechanism of survival. They eat many small meals throughout the day and keep on the move throughout their range. Thus they are harder to stalk by predators, generally are more alert to danger, and can take instant flight. This leads to digestive tract modifications. For example, the horse has a small stomach in relation to body size because it doesn't need storage space for huge meals. Also, it does not have a gallbladder because it has food coming to the small intestine fairly continuously and has no need for large amounts of bile at any one time.

**Contract grower**    Producers who contract with an organization to produce a product for a price determined through a contractual arrangement.

**Contract integration**    When one firm from one industry phase contracts with a firm at an adjacent phase for products and/or services.

**Coprophagy**    Eating feces.

**Corpus luteum (CL)**    Structure resulting from the conversion of the follicle (after disruption) into a structure that is responsible for the production of progesterone for the support of pregnancy. If pregnancy is not detected, the corpus luteum undergoes lysis, allowing the female to initiate another estrous cycle (plural, *corpora lutea*).

**Cortisol**    A hormone produced by the adrenal cortex. It is elevated during stress and has been used as a gauge for the degree of stress an animal is under.

**Creep**    An area where young nursing animals can have access to starter feeds. Creep feeds are generally high-quality feeds made available to young animals.

**Cria**    Baby llama.

**Cribbing**    An undesirable behavior in horses in which they bite and/or hold on to objects such as posts, fences, feed troughs, stall doors, and so on. Thought to be brought on by boredom because of stabling or confinement in a small area.

**Critical period**    Similar to *sensitive periods*, but with a more definite beginning and end.

**Crossbreeding**    Mating animals of diverse genetic backgrounds (breeds) within a species.

**Cross-fostering**    Moving young from their dam and placing them with another female for rearing.

**Crude fiber**    In proximate analysis, the insoluble carbohydrates remaining in a feed after boiling in acid and alkali.

**Crude protein**    An estimate of protein content obtained by multiplying the nitrogen content of a substance by a factor, usually 6.25. Both true protein and nonprotein nitrogen is included in the calculation, hence the use of the designation *crude*.

**Cubed hay**    Hay that is forced through dies to produce an approximate 3-cm product of varying lengths.

**Culture**    In this context, the set of occupational activities, economic structures, beliefs/values, social forms, and material traits that define human actions and activities.

**Cutting event**    A competition in which a horse cuts one cow from the herd and holds the cow away from the herd. The horse must work the cow without assistance from the rider for 2 ½ minutes.

**Dairy product science**    The science of providing milk and milk products as food.

**Darwin, Charles**    (1809–1882) English naturalist who published *On the Origin of Species by Natural Selection* in 1859, *The Variation of Animals and Plants Under Domestication* in 1868, and *The Descent of Man* in 1871. Among other things, he proposed the theory of evolution by natural selection.

**Dead germs**    Embryos that have died.

**Defecation**    The act of expelling fecal matter from the large intestine via the rectum or cloaca.

**Defect**    Unacceptable deviation from perfection. Most defects are inherited.

**Deglutition**    The act of swallowing. Passing material from the mouth through the esophagus to the stomach or first fermentation compartment.

**Delaney clause**    Act that prohibited the addition of any substance to human food shown to cause cancer in any animal at any dose. It was a portion of the 1958 Food Additive Amendment.

**Delmarva**    Geographic region comprised of Delaware, Maryland, and Virginia.

**Denature**    In protein chemistry, to disrupt the structure of a native protein, causing it to lose its ability to perform its function. Denatured proteins are generally easier to digest because the enzymes can better attack the chemical bonds.

**Desertification**    The degradation or destruction of the biological potential of land, leading to desert-like conditions.

**Diagnosis**    The process of determining the nature and severity of a disease; art of distinguishing one disease from another.

**Diagnostician**    An expert on diagnosing disease.

**Dichromat**    Ability to perceive only two colors.

**Diet**    All of the feeds consumed by an animal, including water. One can describe the diet for a full year. The diet of the animal includes mixed pasture in the summer, alfalfa hay in the winter, and trace-mineralized salt year-round.

**Differentiated product**    A value-added product with a brand name.

**Digestibility**    A measure of the degree to which a feedstuff can be chemically simplified and absorbed by the digestive system of the body.

**Digestion** The physical, chemical, and enzymatic means the body uses to render a feedstuff ready for absorption.

**Digestion trial** An experimental tool used to determine the digestibility of a specific feedstuff, nutrient, or ration.

**Digests** Produced by enzymatic degradation of animal tissues. Used to flavor pet foods.

**Diluter gene** A type of modifier gene that changes a base color to a lighter color.

**Diploid** Having two sets of chromosomes as opposed to the one set found in gametes.

**Direct causes of disease** Exposure to, or contact with, pathogens or other substances that cause a decrease in animal health.

**Dirties** Eggs with dirt, fecal material, or other material on the shell.

**Disease** State of being other than that of complete health. Disturbance of normal function of the body or its parts.

**DNA (deoxyribonucleic acid)** Chemically, a complex molecule composed of nucleotides joined together with phosphate sugars. Chromosomes are large molecules of DNA. Two strands are joined together in the shape of a double helix. The sequence of the nucleotides in a segment of the chromosome called a gene determines the characteristics of the organism.

**DNA polymerase** The enzyme that forms the sugar-phosphate bonds between adjacent nucleotides in a chain so that replication can occur.

**DNA replication** The cellular process of making a copy of a DNA molecule.

**Doe** A female goat; a female rabbit.

**Dolly** A normal Finn Dorset sheep who also happened to be the first clone of an adult mammal. Cloning an adult animal was thought to be impossible before Dolly was developed. She was created by a research team led by Dr. Ian Wilmut at the Roslin Institute in Scotland. Since Dolly was born, adult mice and cattle have been cloned.

**Domestic animals** Those species that have been brought under human control and have adapted to life with humans. Individuals in some species can be tamed, but have not necessarily been domesticated, such as species of exotic birds, elephants, and big cats. In general, domestic species must be able to adapt to a wide range of physical environments, must adapt to the changes that humans make in their immediate environment (artificial environment), and must respond to selection for some specialized need of humans.

**Dominance** In behavior, an animal's place in the social ranking. The most dominant animal in the group exerts the major influence over other animals. A term often used to describe this is *pecking order.*

**Dominant** When one member of an allele pair is expressed to the exclusion of the other.

**Dorsal** The back on an animal.

**Downer cow** Nonambulatory cow.

**Draft** To move loads by drawing or pulling. A draft animal is one that is used to draw or pull loads.

**Draft animal** An animal whose major purpose is to perform work that involves hauling or pulling. An oxen or horse pulling a plow or wagon is a draft animal.

**Dressage** A competition in which horses are required to perform highly advanced maneuvers in a specified pattern. This competition may be held alone or may be part of a three-day event.

**Dry matter** Everything in a feed other than water.

**Drylot** A confined area generally equipped with feed troughs, automatic watering devices, shelter, and working facilities where animals are fed and managed.

**Duodenum** The first segment of the small intestine. Many important digestive secretions enter the small intestine here.

**Dystocia** Difficulty in birthing.

**Economy of size** A relatively simple concept revolving around the maximization of the use of equipment, labor, and other costly items. Example: An owner of two poultry houses will need a tractor and spreader to clean the houses and spread the litter. However, if he or she builds two more houses, one tractor will probably suffice. The cost of owning the tractor can now be spread over the production of four houses rather than two.

**Electric prods** Small handheld devices designed to give a small electrical shock. Used to keep animals moving in a chute or other handling situation. A poor substitute for good facilities and knowledge of animal handling technique.

**Electronic pasteurization (e-beams)** Ionizing radiation from a focused beam of energy created by the acceleration of electrons using magnetic and electric fields.

**Embryo cloning** The splitting of one embryo to produce many offspring.

**Embryo transfer** Collecting the embryos from a female and transferring them to a surrogate for gestation.

**Emerging pathogen** Pathogen that has mutated to become more virulent and/or has only recently been recognized as a safety issue.

**Ensiling** The process of producing silage from forage. High-moisture forage is stored under anaerobic conditions and allowed to ferment. The acids produced by the fermentation preserve the feed.

**Enzymes** Proteins capable of catalyzing reactions associated with a specific substrate.

**EPD (expected progeny difference)**    A value equal to half the breeding value for an animal.

**Epididymis**    Duct connecting the testis with the ductus deferens. Responsible for sperm storage, transport, and maturation. It consists of a head, body, and tail.

**Episodic**    The pulsatile manner in which the gonadotropic hormones are secreted by the anterior pituitary gland. Controlled by the pulse-generating center of the brain.

**Epistasis**    Interaction among genes at different loci. The expression of genes at one locus depends on alleles present at one or more other loci.

**Equine**    Pertaining to horses.

**Eructation**    Belching. Removing gas from the rumen via the esophagus.

**Erythrocytes**    Red blood cells.

**Essential amino acids**    Those amino acids required by the body that must be consumed in the diet.

**Essential fatty acids**    Fatty acids required in the diet.

**Estrous cycle**    The time from one period of sexual receptivity in the female (estrus or heat) to the next.

**Estrus**    The period when a female is receptive to mating. Synonymous with *heat.*

**Ether extract**    In proximate analysis, the portion of a sample that is removed by extraction with a fat solvent such as ethyl ether.

**Ethogram**    A catalog or inventory of all of the behaviors an animal exhibits in its natural environment. Originally this was the study of only wild animals, but domestic species are also studied in their surroundings. These studies are used to compare species.

**Ethologist**    An individual who studies the discipline of ethology.

**Ethology**    The study of animals in their natural surroundings. The focus is on instinctive or innate behavior. When practiced on domestic species, especially livestock, it is often referred to as *applied ethology.*

**Etiology**    The factor that causes a disease or the study of the factors that cause disease.

**Eutrophication**    Promotion of excess growth of one organism to the disadvantage of other organisms in the ecosystem.

**Ewe**    A female sheep.

**Exotic fowl**    Nonindigenous, nonlivestock species often kept for ornamental reason. Examples include peacocks and various ornamental pheasants such as Melanistic Mutant Pheasant and Lady Amherst Pheasant.

**Expected progency difference (EPD)**    A prediction of the performance of an individual's progeny compared to all contemporaries for the progeny.

**Expression**    In genetics, the manifestation of a characteristic that is specified by a gene.

**Extensive agriculture**    Agriculture systems practiced in a manner that spreads human time and attention across vast acreages and/or large numbers of animals.

**Extensive rearing systems**    Usually associated with range conditions in which hardy animals such as beef cattle receive little individual attention. They may be handled only once or twice per year.

$F_1$    Two-breed cross animals.

**Fancy**    In the rabbit and pet industries, the segment associated with exhibitions and shows.

**Farmer**    Anyone who practices agriculture by managing and cultivating livestock and/or crops.

**Farrow**    In swine, the term used to indicate giving birth.

**Fats**    One of a class of biomolecules called lipids. Lipids are substances that are insoluble in water but are soluble in organic solvents. Chemically, fats are triacylglycerides, which are composed of the alcohol glycerol, with three fatty acids attached.

**Feed**    Foods used to feed animals.

**Feed conversion ratio**    Amount of feed needed to produce 1 lb of gain or other product such as milk or eggs. A 3:1 ratio means that three units of feed are needed to produce one unit of gain. The smaller the ratio, the better.

**Feed efficiency**    See *feed conversion ratio.*

**Feed grains**    Grains that are produced primarily as livestock feed.

**Feeder pig**    Generally thought of as a pig weighing between 30 and 90 lbs. There is some regional difference in this range.

**Feeding trial**    A comparatively simple experimental tool in which animals are fed to determine their performance on specific feeds or substances added to feeds.

**Feedstuff**    Any substance that is used as animal feed. Corn, salt, prairie hay, and silage are examples.

**Filly**    A young female horse.

**Fingerlings**    A juvenile stage in fish. Fish usually from 1 to 6 inches long.

**Finishing**    (phase) The final feeding stage when animals are readied for market. Finishing was once synonymous with the term *fattening.* However, the emphasis on lean meat production has made the goals of this period broader.

**First-limiting amino acid**    The first amino acid whose lack of availability in the diet restricts the performance of an animal.

**Flehmen response**    Sexual behavior of the male of several species in which the male curls his upper lip and inhales.

**Flight zone**    The distance which an animal is caused to flee from an intruder.

**Flighty**    The tendency of an animal to take sudden flight when alarmed. Also called **mobile alarm.** May be used to describe a breed or strain (i.e., Salers are more flighty than Herefords). It can also be used to describe individual animals (i.e., Bossy is more flighty than Big Bess). A type of temperament.

**Flock**   A group of sheep.

**Flocking instinct**   A type of shelter-seeking behavior that has been selected for in sheep. At the least hint of danger, they move close together and move as a group. This can allow one person and a dog to shepherd a few thousand sheep.

**Flushing**   Feeding extra feed to stimulate estrus and ovulation rates.

**Foal**   A newborn horse of either sex. The term is sometimes used up to the time of weaning, after which *colt* and *filly* are more likely to be used.

**Foaling**   To foal. Parturition in the horse.

**Follicle-stimulating hormone**   Gonadotropic hormone responsible for growth, development, and maintenance of follicles in females, and the production of sperm in males. Produced and secreted by the anterior pituitary gland in a pulsatile manner in response to GnRH.

**Food**   Material containing essential nutrients. Humans are generally considered to eat food, while animals eat feed, even if they are eating the same thing.

**Food and Agricultural Organization of the United Nations (FAO)**   The largest autonomous agency within the United Nations system. FAO was founded in 1945 with a mandate to raise levels of nutrition and standards of living, to improve agricultural productivity, and to better the condition of rural populations. FAO works to alleviate poverty and hunger by promoting agricultural development. It offers development assistance; collects, analyzes, and disseminates information; and consults with governments on policy and planning issues. Perhaps most important, it provides an international forum for debate on food and agricultural issues.

**Food grains**   Food grains are those grains for which humans provide the primary consuming market.

**Food-size fish**   Fish that are grown commercially for food, usually ranging from 3/4 to 1 lb and over 12 inches in length.

**Forage**   Fiber-containing feeds like grass or hay. Can be grazed or harvested for feeding. Contain at least 18% fiber but are have high digestible energy (>70%).

**Forestomachs**   The name given to the three digestive compartments of the ruminant tract that are placed anatomically before the true stomach.

**Founder**   In biotechnology, a transgenic animal that is subsequently used to establish a transgenic line of animals.

**Fowl**   Any bird, but generally refers to the larger birds. In this context, it refers to poultry species only.

**Freemartin**   Condition in cattle in which the female-born twin to a bull is infertile because of improper development of the female anatomy.

**Fresh-packaged product**   Traditional fresh product sold with minimal processing.

**Fry**   Fish in the postlarval stage.

**Fryer rabbit**   A rabbit of approximately 5 lbs and no older than 12 weeks of age. Some processors stipulate an age no older than 10 or 11 weeks.

**Full-time equivalent**   In referring to employment, a 40-hour equivalent position.

**Gait**   Forward movement of a horse. The three natural gaits for most horses are the walk, trot, and canter or gallop. Some breeds have additional or different gaits such as the pace, foxtrot, running walk, rack, and others, and they are considered five-gaited.

**Game birds**   Fowl for which there is an established hunting season. Also refers to fighting chickens.

**Gametes**   Mature sperm in the male and the egg or ova in the female. The reproductive cells.

**Gametogenesis**   The formation of gametes.

**Gander**   A male goose.

**Gelding**   A castrated male horse.

**Gene**   A short segment of a chromosome. Genes direct the synthesis of proteins or perform regulatory functions.

**Gene enhancement**   The process of changing the genetic potential of an individual for a trait such as height or eye color.

**Gene frequency**   The proportion of loci in a population that contain a particular allele.

**Gene map**   The locations of specific genes along a chromosome marked with probes.

**Gene probe**   A short segment of DNA used to identify and map an area of a specific chromosome or entire genome.

**Gene therapy**   The process of replacing a missing or incorrectly functioning gene with a correct one to treat a disease. Also called *gene replacement therapy*.

**General combining ability**   A term that describes a strain that contributes positively to the genetic makeup of offspring that result from mating it with several different strains.

**Generation interval**   In a herd, the average age of the parents when their offspring are born. Replacing breeding animals with their progeny can shorten generation intervals. Genetic progress is slowed when generation intervals are long, and hastened when the generation interval is shortened, presuming a sound selection program is being practiced.

**Genetic code**   The set of rules by which information encoded in genetic material (DNA or RNA sequences) is translated into proteins (amino acid sequences) by living cells.

**Genetic correlation**   The situation in which the same or many of the same genes control two traits.

**Genetic drift**   A change in gene frequency of a small breeding population owing to chance.

**Genetic engineering**   The term most frequently used to describe the technologies for moving genes from one animal or one species to another.

**Genetic markers**   Biochemical labels used to identify specific alleles on a chromosome.

**Genetics**   The science of heredity and the variation of inherited characteristics.

**Genome**   The complete genetic material of an organism.

**Genotype**   The genetic makeup of an organism.

**Genotypic frequency**   The frequency with which a particular genotype occurs in a population.

**Germ-line therapy**   The process of changing reproductive cells so that an individual that carries a genetic defect need not pass it on to his or her offspring.

**Gestation**   The period when the female is pregnant.

**Gilt**   Any female pig that has not yet given birth. Sometimes producers continue to use the term after the first litter is born and call a first-litter sow a first-litter gilt.

**Gomer bull**   Bull rendered incapable of mating naturally that is used to detect cows in heat.

**Gonads**   Sex organs; testis in male, ovary in female.

**Gosling**   An immature goose weighing about 8 lbs.

**Grading up**   In animal species, the process of improving animals for some productive function by consecutive matings with animals considered to be genetically superior.

**Grain-fed beef**   Meat from cattle that have undergone a significant grain feeding.

**GRAS list**   A list of common food additives given special, safe status under the 1958 Food Additive Amendment because of their previous records as safe food additives.

**Green Revolution**   Dramatic improvements in grain production in developing countries during the 1960s to the 1980s because of technological innovation and application.

**Gross national product**   The total of all goods and services produced by a nation in a given period, usually 1 year.

**Growth**   The process of adding tissues similar to those already present in the body to increase the size of an organism toward the goal of maturity when growth stops.

**Guillain-Barré syndrome**   An inflammatory disease of the peripheral nerves characterized by weakness and often paralysis of the arms, legs, breathing muscles, and face.

**Habituation learning**   A type of operant conditioning. It refers to an animal's ability to eventually ignore something that occurs often. A cat will ignore a collar after it has been on awhile. A horse will eventually ignore a train that passes by the stable or the pasture.

**Hand**   One hand equals 4 inches. Horses are measured for height at the withers in hands.

**Handling**   In this context, any manipulation necessary to care for or evaluate animals. It may or may not include physically touching them. Moving animals from pasture to pasture is handling of one type. Vaccinating, tagging, deworming, and so on, is more invasive handling.

**Haploid**   A cell with half the usual number of chromosomes. Sex cells are haploid.

**Hatching**   The process of a chick leaving the egg and emerging into the world. Birth for fowl.

**Hazard Analysis and Critical Control Points (HACCP)**   A process control system designed to identify and prevent microbial and other hazards in food production.

**Hemoglobin**   Oxygen-carrying pigment found in erythrocytes.

**Hemolytic-Uremic Syndrome (HUS)**   A rare condition that mostly affects children under the age of 10; characterized by damage to the lining of blood vessel walls, destruction of red blood cells, and kidney failure.

**Hen**   A mature female chicken or turkey.

**Herbivore**   Animals that eat a diet of only plant material.

**Herd or flock health management program**   A comprehensive and herd-specific program of health management practices.

**Herd health**   A comprehensive and herd-specific program of health management practices.

**Heredity**   The transmission of genetic characteristics from parent to offspring.

**Heritability**   A measure of the proportion of phenotypic variation that is caused by additive gene effects. The proportion of differences between individuals that is genetic.

**Heritability estimate**   A measure of the genetic progress that can be made from generation to generation in a given trait.

**Heterosis**   The superiority of an outbred individual relative to the average performance of the parent populations included in the cross.

**Heterozygous**   When two genes in a pair are not the same.

**Homologous chromosomes**   Chromosomes with the same size and shape, occurring in pairs, and affecting the same traits.

**Homozygous**   When two genes of a pair are the same.

**Horse slaughter**   Aged, unsound, and surplus horses are sold to slaughter in the United States. The animals are slaughtered and most of the meat is shipped overseas to countries that view the consumption of horse meat as a normal part of the diet. This includes several European countries and Japan.

**Horsepower**   Term that originated as a measure of the pulling power exerted by a horse. Technically equal to the rate of moving 33,000 lbs a distance of 1 foot in 1 minute.

**Human-developed cat breed**   Breeds that have been developed from existing breeds or crosses of existing breeds.

**Hunter under saddle**   At a horse show, an English division class in which movement and mannerisms are judged. Intended to depict a horse on the hunt chasing the hounds.

**Hunter-gatherer**   Before agriculture, all people were hunter-gatherers. Hunter-gatherer peoples support their needs by hunting game, fishing, and gathering edible and medicinal plants.

**Husbandry**   The combined animal care and management practices.

**Hybridoma**   A cell line that is used to produce specific antibodies. They are created by fusing an antibody-producing lymphocyte with a cancer-causing cell.

**Hypothalamus**   Area of the brain responsible for many homeostatic functions. In reproduction, it is the region responsible for the production of GnRH, which is released into a specialized vasculature to the anterior pituitary gland. The hypothalamus is also responsible for producing oxytocin.

**Ileum**   The last, short portion of the small intestine.

**Impacted intestine**   Constipation in the horse and some other species.

**Imprint learning**   Learning that has restrictive conditions and times when it can occur.

**In vitro**   In a test tube or other environment outside the body.

**Inbred line**   An established line of chickens created by intensive inbreeding. They are usually mated to other inbred lines to produce commercial varieties.

**Inbreeding**   Mating system in which mated individuals have one common ancestor appearing several times at least three to four generations back in the pedigree.

**Inbreeding depression**   A loss or reduction in vigor, viability, fertility, or production that usually accompanies inbreeding. The reverse of hybrid vigor.

**Incomplete dominance**   When neither allele is dominant to the other. Both influence the trait.

**Incrossbred hybrid**   Chickens developed by crossing inbred lines within the same breed. This technique is generally used to produce laying hens.

**Incubation**   The process of sitting on eggs by a hen to warm them with body heat so that the eggs develop into young. Can also be done artificially in an incubator.

**Incubator**   A machine that provides the environmental conditions that encourage embryonic development.

**Individual cage**   Small pens, perhaps 8 to 12 in. wide and 18 in. long, designed to hold one or more hens. Commercially, cages are arranged in rows with automatic watering devices and easy access to feed, and they are generally arranged so the eggs roll to an automated system of collection.

**Infectious diseases**   Diseases caused by living organisms, which invade and multiply in or on the body and result in damage to the body.

**Infertile**   An egg that is unfertilized or in which the embryo dies.

**Inheritance**   The transfer of gene-containing chromosomes from parent to offspring.

**Intensive agriculture**   Any agriculture system in which human attention and focus is directed to a small plot of land or to each animal.

**Intensive systems**   Animals produced in intensive systems are kept in more concentrated conditions, fed higher quality feeds, and have much more labor, technology, and overall attention directed to them.

**Intergeneric microorganisms**   Microorganisms created to contain genetic material from organisms in more than one taxonomic genus. Another term for *transgenic.*

**Involution**   An organ's return to a normal state or normal size. Often used in describing the uterus after calving, or the mammary gland after completion of the lactation cycle.

**Irradiated foods**   Foods treated with ionizing pasteurization that kills insects, bacteria, and parasites.

**Jejunum**   The second and longest portion of the small intestine. The bulk of digestion and absorption of nutrients occurs in this segment.

**Juvenile**   Aquatic species between the postlarval stage and sexual maturity.

**Kidding**   Parturition in goats.

**Kindling**   Parturition in rabbits.

**Kits**   The young of some small mammals, including rabbits.

**Kosher**   Meat that is ritually fit for use as sanctioned by Jewish religious law.

**Lactation**   The process of producing milk.

**Lactation curve**   A plot of milk production over the life of the lactation period. Different species tend to have characteristic lactation curves.

**Lactose**   Disaccharide comprised of one glucose unit attached to a lactose sugar. Made only by the mammary gland. Broken down by the enzyme lactase.

**Lactose intolerance**    Condition in humans where the lack of the enzyme lactase leads to stomach distress following milk consumption.

**Lamb**    A sheep under 1 year of age. Also, the meat from a sheep under 1 year of age.

**Lambing**    Parturition in sheep.

**Laminitis**    Commonly referred to as *founder,* an inflammation of the laminae of the hoof. It is commonly caused by overeating of grain but may also be related to eating lush pastures, to road concussion, to retained afterbirth in the mare, and to other causes.

**Larva**    The first stage after hatching. An independent, mobile, developmental stage between hatching and juvenile. The larva stage bears little resemblance to the juvenile or adult stage (plural, *larvae*).

**Layer**    A hen in the physiological state of producing eggs regularly.

**Laying**    Technically, the expulsion of an egg (i.e., "That hen is *laying* an egg"). However, commonly used to refer to hens in egg production (i.e., "Barn 12 is a *laying* hen barn").

**Least-cost ration**    A ration formulated to meet the animal's nutritional needs at the lowest cost from the feeds available.

**Leporarium**    Specially designed enclosure popular in Roman times for keeping rabbits and hares.

**Lesion**    Abnormal changes in body organs because of injury or disease.

**Libido**    Sexual drive.

**Lignin**    Polymers of phenolic acids found in plants as part of the structural components of the plant. Lignin is indigestible and can cause a decrease in the digestibility of the other fibers in the plant. Lignin generally increases as the plant ages.

**Linebreeding**    A system of mild inbreeding.

**Lipoproteins**    Compounds in the bloodstream that carry the lipids (fats) in the blood. Because lipids are not soluble in water, lipoproteins provide a special mechanism by which the lipids can carry fats from the intestines and liver to the peripheral parts of the body where they can be used as energy, or as building materials for cells. Made up of protein, fatty acids, cholesterol, and cholesterol esters, the lipoproteins exist as low-density lipoproteins (LDLs), very low-density lipoproteins (VLDLs), high-density lipoproteins (HDLs), and chylomicrons.

**Literacy**    The ability to read and write.

**Litter**    Material such as wood shavings, straw, or sawdust used to bed the floor of a poultry house.

**Locus**    The specific location of a gene on a chromosome.

**Long hay**    Hay that has not been chopped or ground; the individual forage.

**Lordosis**    Posture assumed by females in estrus in response to pressure applied to the back.

**Luteinizing hormone**    Gonadotropic hormone primarily responsible for providing the signal to disrupt the mature follicle in females, and the production of testosterone by the Leydig cells of the testes in the male. Produced and secreted by the anterior pituitary gland in a pulsatile manner in response to GnRH.

**Luteolysis**    Breakdown or degeneration of the corpus luteum. Occurs at the end of the luteal phase of the estrous cycle if pregnancy is not detected.

**Maiden**    In this context, a female who has never given birth.

**Maintenance**    The nutritional needs of the animal exclusive of those required for a productive function such as growth, work, milk production, and so on.

**Mare**    A mature female horse.

**Marker-assisted selection**    Selection for specific alleles using markers such as linked DNA sequences.

**Market economy**    Economies in which prices are freely determined by the laws of supply and demand.

**Market gardening**    Specialized production of fruits, vegetables, or vine crops for sale.

**Mash**    Finely ground and uniformly mixed feeds. Animals cannot separate feed ingredients; thus each bite provides all the nutrients in the diet.

**Mastication**    The process of chewing.

**Mastitis**    Inflammation of the mammary gland, most often caused by bacterial infection. Symptoms include redness, fever, and sloughing of white blood cells into the milk. Causes damage to the secretory cells and can be fatal if left unattended.

**Maternal effect**    Any environmental influence that the dam contributes to the phenotype of her offspring. The contribution of the dam is environmental with respect to the calf (mothering ability, milk production environment, and maternal instinct). The genetics of the dam allow her to create this environment for her calf. Maternal effects are important during the nursing period but have diminishing effects postweaning.

**Mature duck or old duck**    Ducks of either sex with toughened flesh and hardened bill. Their meat is used in processed products.

**Mature goose or old goose**    A spent breeder whose meat is used in processed products.

**Meat**    The flesh of animals used for food.

**Meat science**    The science of handling, distributing, and marketing meat and meat products.

**Meiosis**    The process that forms sex cells. Cells formed through meiosis have half the number of chromosomes of the parent cells.

**Mendel, Gregor**    (1822–1884) Austrian monk whose pioneer work with garden peas laid the foundation for the science of genetics.

**Messenger RNA (mRNA)**   Nucleic acid that carries instructions to a ribosome for the synthesis of a particular protein.

**Metabolism trial**   An advanced form of digestion trial that measures the body's use of nutrients.

**Metric ton**   Approximately 1.1 U.S. tons. Equal to 1 million grams, or 1,000 kilograms.

**Micropropagation**   Individual cells of a plant used to generate another plant. Also called *tissue culture*.

**Micturate**   Urination.

**Migration**   The process of bringing new breeding stock into a population.

**Minerals**   In nutrition, the specific set of inorganic elements thus far established as necessary for life in one or more animal species.

**Mitosis**   The process of somatic cell division.

**Mob**   A group or herd of goats.

**Modifier gene**   Gene that influences the expression of another gene or genes.

**Mohair**   The fiber produced by the Angora goat.

**Molting**   The shedding of feathers by chickens. Usually egg laying is reduced or stops during molting.

**Monoclonal antibodies**   Antibodies produced by the daughter cells of a hybridoma that recognize, and thereby identify, only one antigen. Used to develop diagnostic tests and in research.

**Monoestrus**   Exhibiting only one estrous cycle; for example, the bitch is seasonally monoestrus.

**Monogastric**   Having only one stomach. A term used to differentiate between animals who have a rumen and those who do not. Some nutritionists prefer the term *nonruminant* rather than monogastric because even ruminants have only one true stomach. The additional "stomachs," or forestomachs, have very different functions than the true stomach has.

**Monosomy**   The absence of one chromosome from an otherwise diploid cell.

**Morula**   Early stage embryo, after cell division multiplies cell numbers in the zygote.

**Multiparous**   A female that has had previous pregnancies and offspring.

**Multiple alleles**   Genes with three or more alleles.

**Mustang**   The term used to describe the feral horse of the American West. At one time, these were largely the descendants of the Spanish horses brought to this country beginning in 1600 to work the Spanish missions. Today, they are the descendants of many nondescript horses whose lineage and characteristics hardly compare with the original.

**Mutations**   Changes in the chemical composition of a gene.

**Myoepithelial cells**   Specialized muscle cells that surround the secretory cells of the alveoli and cause them to contract when stimulated by oxytocin.

**Natural breeds**   Cat breeds selected by human preference or natural conditions specific to a region.

**Natural selection**   Selection based on factors that favor individuals better suited to living and reproducing in a given environment.

**Necropsy**   The examination of a body after death.

**Nest box**   A box provided for a rabbit doe in which to give birth and rear the young for their first few weeks. Also, a box provided for avian species to lay their eggs.

**Net merit dollars**   Index that includes feed costs, mastitis, milk quality costs, and productive life.

**Net quantity statement**   FDA required statement on a pet food package specifying the quantity of feed in the container.

**Nitrogen-free extract**   In proximate analysis, a measure of readily available carbohydrates calculated by subtracting all measured proximate components from 100.

**Nose tongs**   Small clamp-like restraining device put in an animal's nose.

**Novel gene**   A previously nonexistent gene created in a laboratory.

**Novel organism**   An organism with DNA from an outside source.

**Novelty**   Anything new or sudden in an animal's environment.

**Nucleotide**   The building blocks of nucleic acids. Each nucleotide is composed of sugar, phosphate, and one of four nitrogen bases.

**Nursery pig**   Term often used to indicate an early-weaned pig of light weight that is housed in special, environmentally controlled nursery facilities. These pigs may be only 10–15 lbs when weaned and 14–28 days of age.

**Nursing pig**   A pig still nursing the sow.

**Nutraceuticals**   Products perceived to have both nutrient and pharmaceutical properties. Examples: Vitamins that are also antioxidants or omega-3 fatty acids that may protect against heart disease as well as supply fats in the diet.

**Nutrient**   A chemical substance that provides nourishment to the body. Essential nutrients are those necessary for normal maintenance, growth, and functioning.

**Nutrient density**   A measurement of the essential nutrients found in a food compared to the caloric content of the food.

**Nutrient-dense food**   A food that has a variety of nutrients in significant amounts. Meat has protein, vitamins, minerals, and other nutrients in good supply. On a per calorie basis, the consumer gets many nutrients.

**Nutrition**   The study of the body's need and mechanism of acquiring, digesting, transporting, and metabolizing nutrients.

**Olfactory**   Relating to the sense of smell.

**Omnivore**   Animals that eat both plant- and animal-based foods.

**Oocyte**   The gamete from the female.

**Operant or instrumental conditioning**   Learning that is primarily influenced by its effects. A positive or negative reward that either precedes or follows an act can induce an animal to repeat the act. Many types of learning fall under this category. B. F. Skinner did extensive early work in this area.

**Optimal growth**   When optimizing growth, such factors as cost of the ration, environment, labor, and other nonfeed inputs are considered. Sometimes optimal and maximal growth rates are the same. However, in a large number of practical production situations, they are different. For instance, a diet that maximizes growth may be too expensive to be economically feasible. A diet that produces 90% as much growth may do so in a more cost-effective way than the more expensive diet that maximizes growth. Thus, optimal growth may be 90% of maximal growth. Likewise, a growth promotant may help maximize growth. However, if the cost of the growth promotant is too high, the added gain will not pay for it.

**Orgle**   The distinctive noise made by a male llama before and during mating.

**Osmotic pressure**   Pressure exerted related to differences in osmotic gradient. Osmotic gradient is developed because of the tendency of water to follow certain compounds across a membrane or filter. For example, high salt concentrations on one side of a cell membrane cause water to travel from low salt concentrations to high salt concentrations to maintain equal osmotic pressure.

**Outbreeding**   The process of mating less closely related individuals when compared to the average of the population.

**Outcrossing**   The practice of mating unrelated breeds or strains.

**Ovulation**   Release of the ovum or egg from the ovary.

**Ownership integration**   An integrated operation that is under one ownership.

**Oxytocin**   A small hormone produced in the hypothalamus and secreted by the posterior pituitary gland that stimulates contraction of muscle cells in the uterus to aid in parturition and stimulates contraction of the myoepithelial cells surrounding the alveoli to force milk out of the lumen into the mammary gland ducts.

**Pacifier cow**   A cow that has been trained to accept moving, restraint, and other types of management. They are usually very tame and nonflighty. When mixed with less-conditioned cows, they facilitate handling of individuals or groups and have a calming influence on them. Especially useful for work in corrals and squeeze chutes.

**Palatability**   The acceptability of a feed or ration to livestock. Influenced by taste, texture, form of the feed (whole or processed), dustiness, and several other factors, depending on the species.

**Palpation**   Physically touching and examining with one's hands.

**Papillae**   Small finger-like projections that greatly increase the surface area of the small intestine.

**Parasite**   An organism that lives at the expense of a host organism. Generally must live on or in the host; a form of symbiosis.

**Parent average**   The average predicted transmitting ability of the parents of a dairy cow.

**Particle gun**   A "gun" that shoots DNA into the cells of an organism.

**Parturient paresis**   Metabolic disorder generally occurring within 72 hours of calving. Caused by low blood serum calcium level.

**Parturition**   Process of giving birth.

**Passive immunity**   Immunity conferred to an animal through preformed antibodies that it receives from an outside source. Antibodies are harvested from the mother's bloodstream by the mammary gland to put into colostrum as a means of conferring passive immunity to newborn mammals.

**Pasteurization**   Controlled heating to destroy microorganisms.

**Pathogen**   Any living disease-producing agent.

**Pathogenicity**   The capability of an organism to produce disease.

**Pathology**   The branch of medicine that deals with the essential nature of disease and the structural or functional changes that cause or are caused by disease. A pathologist is an expert in disease and disease processes.

**Peer review**   A process in which those with expertise and experience similar to the expertise and experience of an author of an article (his or her peers) review and critique the article for scientific merit before it is published. These reviews are usually done anonymously through an editor to help ensure candor and thoroughness of the review. Reviewers may suggest changes and recommend whether or not to publish the paper.

**Pelleting**   A method of processing feed in which the feed is first ground and then forced through a die to give it a shape. Pellets are generally well accepted by all classes of livestock.

**Pellets**   Feeds that are generally ground and then compacted by forcing them through die openings.

**Per capita**   Per unit of population.

**Percentile**   Ranking for dairy cows based on predicted transmitting ability dollars.

**Performance**   The sum of the inherited abilities and environmental effects. Measured as gain, milk production, and so on.

**Performance testing**   Evaluating an individual in terms of performance such as weight gain or milk production.

**Peristalsis**   The progressive, squeezing movements produced by the contraction of muscle fibers in the wall of the digestive tract. The movements proceed in a wave down the length of the tract or some portion of it. The primary purpose of peristaltic action is to move material down the tract. However, it can also be used to mix the contents of the tract.

**Pesticides**   Any agent or poison used to destroy pests, including fungicides, insecticides, herbicides, and rodenticides.

**Pet**   An animal kept for pleasure rather than utility.

**Pharming**   The production of pharmaceuticals from livestock or crops.

**Phenotypic frequency**   The proportion of individuals in a population that express a particular phenotype.

**Phenotypic value**   A measure of individual performance for a specific unit.

**Phobia**   Excessive and unwarranted fear.

**Photoperiod**   The length of time when light is present.

**Physiology**   The study of the physical and chemical processes of an animal or any of the body systems or cells of the animal.

**Pica**   A craving for and willingness to eat unnatural feedstuffs. Often caused by nutritional deficiencies.

**Pig meat**   The meat from a hog. Synonymous with *pork*.

**Pituitary gland**   Gland sitting directly below the hypothalamus. It is divided into the posterior pituitary gland and the anterior pituitary gland. The posterior pituitary releases oxytocin. The anterior pituitary produces and secretes several hormones, including luteinizing hormone, follicle-stimulating hormone, prolactin, and growth hormone.

**Placenta**   The organ that surrounds the fetus and unites it to the female while it develops in the uterus. Organ responsible for the exchange of oxygen, nutrients, and waste between the fetus and the mother. Derives from the trophoblast cells of the embryo.

**Plumage**   The total body feathering of poultry.

**Points**   The legs, mane, and tail of a horse. In other species, the head may be substituted for the mane.

**Polyestrus**   Exhibiting more than one estrous cycle.

**Polymerization**   The process of building high molecular weight molecules by repeatedly chemically bonding the same compound to itself.

**Polymorphism**   The existence of two or more discontinuous, segregating phenotypes in a population.

**Polyploidy**   Condition in which the number of chromosomes an individual has is some multiple of the haploid number.

**Population genetics**   The study of how genes and genotypic frequencies change, and thus change genetic merit in a population.

**Porcine stress syndrome**   Genetic defect in which pigs are heavily muscled but have poor carcass quality and may die when subjected to stress.

**Pork**   The meat from a hog. In many parts of the world, the term *pig meat* is preferred.

**Pork Quality Assurance Program**   A voluntary, multilevel, management education program for producers that was introduced by the National Pork Producers Council in 1989 as a tool to enhance the quality of pork sold to the world's consumers.

**Pork somatotropin**   The porcine version of somatotropin. Produced by the pituitary gland.

**Posilac**   Monsanto Company's approved, commercially available BST.

**Possible change**   The measure of the potential error associated with EPD values.

**Postlarvae**   In aquatic species, animals that are beyond the larval stage but are not yet a juvenile. They may resemble juveniles, but lack certain developmental characteristics of the juvenile.

**Postpartum**   After parturition.

**Postpartum interval**   Period of time from parturition to first estrus in the female.

**Poult**   Baby turkeys. Once sex can be determined, they are called young toms (males) or young hens (females).

**Poultice**   A soft, moist mass held between layers of cloth, usually warm, which is applied to some area of the body for therapeutic reasons.

**Poultry**   Domestic birds raised for eggs and meat. However, this designation has some flexibility. For example, most peafowl are kept for their ornamental plumage even though they produce eggs and can be eaten. Who would think of eating a swan? Included are the economically important species of chickens, turkeys, ducks, and geese, and those kept as hobbies, for sport, and for ornamental and aesthetic reasons. Includes ostriches, peafowl, pigeons, swans, guinea fowl, pheasants, and an assortment of other game birds.

**Predicted transmitting ability**   Half the breeding value.

**Predicted transmitting ability dollars**   An economic index that includes milk and component traits combined and weighted for producer value.

**Predisposing causes of disease**    Any condition or state of health that confers a tendency and/or susceptibility to disease.

**Pregnancy disease**    Also referred to as *pregnancy toxemia*. A form of ketosis in females that occurs in late pregnancy because the female cannot eat enough of the feed she is provided or is not provided enough feed. Usually occurs in cases of multiple fetuses. Common in sheep and goats.

**Prehension**    The act of seizing and grasping.

**Prepotent**    An animal that transmits its characteristics to its offspring in a consistent fashion.

**Principle of independent assortment**    Mendel's second law. It says that in the formation of gametes, separation of a pair of alleles is independent of the separation of other pairs.

**Principle of segregation**    Mendel's first law, often called the law of segregation. The law states that when gametes are formed, the genes at a given locus separate so that each is incorporated into different gametes.

**Probe**    See *gene probe*.

**Production**    The general term that describes the output of usable products and services by animals.

**Progesterone**    Female sex steroid produced by the corpus luteum or the placenta. Has "progestational" effects, including suppression of gonadotropic secretion, decrease in uterine motility, and production of the cervical seal.

**Prostaglandin**    A group of fatty acid hormones, one of which is prostaglandin F2$\alpha$, which breaks down the corpus luteum, allowing the female to return to estrus.

**Protein**    Compounds composed of combinations of alpha-amino acids.

**Protein quality**    A measure of the presence and digestibility of the essential amino acids in a feedstuff.

**Proventriculus**    The glandular stomach in fowl.

**Puberty**    Transitional state through which animals progress from an immature reproductive and hormonal state to a mature state. Varies with species, but timing primarily depends on age and weight.

**Pullet**    A young female chicken.

**Purines and pyrimidines**    Organic ring structures made up of more than one kind of atom (heterocyclic compounds). Purines and pyrimidines contain nitrogen in addition to carbon.

**Qualitative traits**    Traits such as coat color for which phenotypes can be classified into groups.

**Quality grade**    Scale that indicates quality and value of the carcass such as *prime, choice,* and so on.

**Quantitative traits**    Those traits that are numerically measured and are usually controlled by many genes,

each having a small effect. Examples are milk and egg production.

**Queen**    A female cat.

**Rabbitry**    A place where domestic rabbits are kept.

**Rate of gain**    Pounds of gain per day over a specific period.

**Ration**    The specific feed allotment given to an animal in a 24-hour period. For example, the animal's ration is 10 lbs of hay, 3 lbs of oats, 1.5 lbs of soybean meal, and trace mineral salt. The use of the term *ration* should be in conjunction with specific quantities of feed.

**Ratite**    Ostrich, emu, cassowary, and rhea.

**Recessive**    The member of an allele pair that is expressed only when the dominant allele is absent from the animal's genome.

**Recipients**    Females used to carry the embryos of a donor animal throughout gestation.

**Recombinant DNA**    DNA molecules that have had new genetic material inserted into them. A product and tool of genetic engineering.

**Recreational horses**    Horses of the light breeds kept for riding, driving, or nonprofessional racing and show.

**Reformists**    Animal rights proponents who focus on changing methods of animal use.

**Reining event**    A competition in which horse and rider perform a specified pattern of advanced maneuvers. These include long sliding stops, spins (up to four 360° turnarounds at high speeds), rollbacks (180° turns after a sliding stop), and circles of different sizes.

**Reliability**    A measure of accuracy in diary records.

**Renewable resources**    Those resources that can be replaced or produced by natural ecological cycles or management systems.

**Reproduction**    The combined set of actions and biological functions of a living being directed at producing offspring.

**Resistance**    The natural ability of an animal to remain unaffected by pathogens, toxins, irritants, or poisons.

**Resorption**    The loss of bone tissue through normal physiological means or through a pathological process.

**Ribonucleic acid (RNA)**    Long chains of phosphate, ribose sugar, and several bases.

**Ribosomal RNA (rRNA)**    Ribosomal RNA and protein are the components of ribosomes. Functions in the manufacture of proteins coded by the DNA.

**Ribosomes**    A component of cells that contains protein and tRNA. Ribosomes synthesize proteins.

**Roaster chicken**    A young meat chicken, generally 12 to 16 weeks old, weighing 4–6 lbs.

**Roaster duckling**    A young duck of either sex usually under 16 weeks of age. The bill must not have hardened completely.

**Rooster** A mature, male chicken. Also referred to as a *cock.*

**Roughage** A bulky feedstuff with low weight per unit volume. Contains at least 18% fiber but can range up to 50%. Less digestible than forages.

**Rumen** The largest of the ruminant forestomachs. Contains microorganisms that degrade complex carbohydrates and produce volatile fatty acids, amino acids, and vitamins to the host animal.

**Ruminal bloat** More correctly called *ruminal tympany.* An overdistention of the rumen and reticulum with the gasses of fermentation. Commonly referred to as *bloat.*

**Ruminal tympany** See *ruminal bloat.*

**Ruminant** Hooved animals that have a rumen and chew their cud.

**Rumination** The process in ruminants where a cud or bolus of rumen contents is regurgitated, remasticated, and reswallowed for further digestion.

**Saddle seat pleasure** An English event in which the horse's movement and mannerisms are judged.

**Salivation** The elaboration of the mixed secretion (saliva) produced primarily in three bilateral pairs of glands in the mouth known as salivary glands.

**Saturated fats** Fatty acids that are completely saturated with hydrogens, and thus do not contain any hydrogen bonds.

**Savanna** Tall-grass vegetation belts in the hot areas of the world. The same zone found in temperate zones is referred to as *prairie.*

**Scientific literacy** The minimum knowledge necessary to stay abreast of further scientific developments and innovation.

**Sclera** The tough white outer coat of the eyeball.

**Seasonal market** A time of the year when there is increased demand for a product.

**Seasonally polyestrus** When an animal has repeated estrous cycles but only in response to some environmental factor associated with the seasons, such as the photoperiod.

**Secondary enterprise** In diversified farming operations, the enterprise secondary to the one providing the bulk of the income.

**Secondary sex characteristics** Characteristics that differentiate the sexes from each other; occur most profoundly during and after puberty because of changes in hormonal concentrations occurring during puberty. Examples in males include such things as humps on the necks of bulls, beards on men, increased musculature in the males of most species, changes in the sound of vocalization (voice change in boys that happens at puberty). For females, includes the many characteristics lumped together that we refer to as femininity: added body fat, which creates curves where angles were once

visible, mammary development, smoother hair coats, and so on. Behavioral characteristics such as "marking" territory or aggression in males are also part of the complex.

**Secretory cells** The functional units of the alveoli that absorb nutrients, make the milk components, and transport the milk into the lumen of the alveoli.

**Seedstock** Broodstock intended for future production.

**Selection** The process of allowing some animals to be parents more than others.

**Selection differential** The phenotypic advantage of those chosen to be parents. The difference in the mean of those chosen to be parents and the mean of the population.

**Self-sufficient** Providing for one's own needs.

**Semen** Fluid from the male that contains sperm from the testes and secretions from several other reproductive organs.

**Sensitive periods** Times in an animal's life when certain types of learning are more easily accomplished. For example, imprinting behavior between the mother and offspring has a sensitive period near birth. Closely related, but more specific, is the term *critical period.* The critical period for imprinting in species X may be the first hour after birth. The critical period for puppy socialization is 3–12 weeks.

**Sentient** Animals that experience pain and pleasure.

**Setting** Placing eggs to incubate. A setting hen is a broody hen that is incubating eggs.

**Sex-influenced inheritance** The same genotype is expressed differently depending on the sex of the animal.

**Sex-limited inheritance** Traits expressed in one sex or the other, such as milk production in females. Both sexes carry genes for the trait.

**Sex-linked cross** Sex-linked chicks can be sexed at birth by their color. Males are one color and females are another color or color pattern.

**Sex-linked inheritance** Traits inherited on the X or Y chromosome and therefore inherited only when that respective chromosome is passed on.

**Sheep** An animal of the genus *Ovis* that is over 1 year of age.

**Shelf life** The period of freshness and wholesomeness of food items. In poultry, the length of time fresh-dressed, iced, packed poultry can be held without freezing.

**Shelter-seeking behavior** Overall, behaviors that an animal exhibits to escape from weather, insects, or danger. It may include crowding together for warmth and protection, standing close to cedar trees for their insect-repelling properties, or standing head-to-tail as horses often do to swish away flies from each other's faces.

**Show pigs**   Pigs bred for exhibition, usually by 4-H and FFA students.

**Silo**   Structure in which silage is made and stored.

**Singleton**   An offspring born singly.

**Single-trait selection**   Selection for only one trait or characteristic.

**Sire summary**   Genetic information published on sires available within a breed.

**Slash-and-burn agriculture**   The practice of clearing a plot of land from the forest by cutting the trees and shrubs and then burning them. The ash from the burning fertilizes the soil. The practice has gained an undeserved bad reputation in modern times because it is poorly understood. Practiced as it has been for centuries on small rotating plots in remote, tropical areas, it has the long-term effect of preserving the forest and being a very sustainable system.

**Small grains**   Such grains as oats, wheat, and barley.

**Social structure**   The organization of a group. The patterns of the relationships of one to another.

**Sociobiology**   The study of the biological basis of social behavior. Especially of interest is behavior that helps pass the gene pool on to the next generation.

**Somatic cells**   All cells in the body other than gametes.

**Sow**   Female pig that has given birth.

**Spat**   For shellfish such as oysters, the developmental stage when they settle down and become attached to some hard object.

**Spawn**   To produce eggs, sperm, or young.

**Specialty market**   A term that suggests that a product is generally aimed at a specific segment of the overall market.

**Speciesism**   Placing the interests of one's own species above those of another species. This term was coined by Richard D. Ryder, a British psychologist.

**Specific combining ability**   When a strain only contributes positively to a cross when mated with certain, specific other lines.

**Sperm**   The gamete from the male.

**Spontaneous mutation**   A change in the DNA that creates new alleles.

**Squeeze chute**   A restraining device used to handle livestock. Consists of a head gate to catch the animal's head and side panels that can be adjusted to restrict the animal's movement. Used to contain the animal for management procedures like deworming and artificial insemination.

**Stadium jumping**   A competition in which horses jump a course of fences (jumps) in a specified order. In the first round of competition, the event is scored on a mathematical basis. Horses that receive a perfect score compete in a timed jumpoff to determine the winner.

**STAGES**   Swine Testing and Genetic Evaluation System. A series of computer programs that analyzes performance data of purebred swine and their crossbred offspring.

**Stallion**   A mature, male horse that is not castrated. Castrated male horses are *geldings*.

**Standard of Perfection**   A standard published in book form by the American Poultry Association. It lists the recognized breeds and varieties of poultry and their characteristics.

**Steppes**   Short-grass vegetation zones. Steppe vegetation accounts for most of the land area of the world devoted to range livestock production.

**Stereotyped behaviors**   A nonfunctional, repetitive, and intentional behavior. Stall walking, weaving, and pawing are examples in horses. Often the behaviors are rhythmic. Stalled horses often develop them in response to boredom, lack of exercise, and infrequent pasture time. Weavers appear to walk in place. Stall walkers or pacers walk or trot in a fixed pattern. This may be a line or sometimes a circle. Some incorporate a head toss at specific points. Caged primates, kenneled dogs, and caged, large carnivores also sometimes develop similar behaviors for similar reasons.

**Sternal recumbancy**   Mating position in the llama and alpaca. The female lies on the ground with her legs under her and her belly on the ground.

**Stock horse**   Any horse of the light breeds trained and used for working livestock, mainly cattle.

**Stocker calf**   Weaned calf being grown prior to placement in a feedlot for finishing.

**Stocker fish**   Fish usually 6–12 inches in length that weigh less than 3/4 lb.

**Stocking**   The practice of placing fish from a hatchery or other source into a pond or lake to either establish a new species or augment an already existing species.

**Strain**   Families or breeding populations within a breed. They have been more rigorously selected for some trait or set of traits than the average of the breed.

**Strain cross**   Mating different strains of the same breed and variety. Generally, the strains have been inbred to some degree and selected for different strengths to get increased production in the offspring.

**Strategic alliances**   Partnerships between various independent segments of an industry to maximize cooperation, value, and return on investment.

**Stress**   A physical, emotional, or chemical factor causing body or mental strain or tension. Coping with stress is a natural part of life. However, excessive amounts of stress can cause the body to suffer distress in which the body can undergo extensive, detrimental changes. Excessive stress is a contributing factor to disease.

**Stud**   A unit of male animals kept for breeding. Also a term for a stallion.

**Stud book**    The set of records a breed association keeps on the animals registered with it.

**Stud fee**    The fee that a mare owner pays to a stud owner for the rights to breed the mare to the stallion and to register the resulting foal.

**Subclinical**    No readily observable clinical signs associated with disease exist.

**Submissive behavior**    Behaviors that a less-dominant animal exhibits toward a more-dominant animal to prevent being subjected to aggression.

**Subtherapeutic**    Quantities too low to treat disease.

**Survival of the fittest**    Natural selection. A process whereby those best equipped for the conditions in which they are found survive and pass their genetic material on to the next generation.

**Sustainable agriculture**    Agriculture that meets the needs of the present generation without jeopardizing the ability of further generations to meet their own needs and without limiting their choices.

**Symbiotic relationship (symbiosis)**    A relationship in which dissimilar organisms live together or in close association. If the relationship is beneficial to both, it is referred to as *mutualism*. Rumen microorganisms and ruminants share a mutualistic form of symbiosis.

**Taboo**    A prohibition imposed by social custom against some action or object. Frequently found as part of religious codes and laws.

**Teasing**    Placing a stallion and a mare in proximity to each other and observing the mare's actions. Mares that are coming into heat display specific behaviors. Signs of estrus in mares include winking of the vulva, urination, squatting, and seeking the stallion. Mares that are not in heat will either not respond or will respond in a negatively aggressive way to the stallion.

**Temperament**    Characteristic behavior or mode of response. Long-term temperament for an individual generally stabilizes after the animal reaches sexual maturity.

**Testcross**    Mating with a fully recessive tester animal to determine if an individual is homozygous or heterozygous.

**Testosterone**    Male steroid sex hormone. Derived from cholesterol in the testes.

**3 Rs**    Acronym for *replacement* (substitute something else for higher animals), *reduction* (reduce the number of animals needed), and *refinement* (decrease in inhumane procedures). Research philosophy that helps researchers gain good information from experiments using animals while reducing the pain and suffering of animals.

**Three-day eventing**    An event of the Olympic Games since 1912. It is comprised of three parts—stadium jumping, dressage, and cross-country.

**Tissue culture**    See *micropropagation*.

**Tom**    A male turkey. Also called a *gobbler*.

**Toxin**    One of several poisonous compounds produced by some microorganisms, plants, and animals.

**Trail event**    A western event in which the horse is scored on its ability to pick cleanly through a set course that mimics outdoor trail riding.

**Transcription**    In protein manufacture, the process of building RNA that is complementary to DNA.

**Transfer RNA (tRNA)**    Molecules of RNA coded by DNA to bond with a specific amino acid. tRNA "collects" the amino acids from the cytoplasm that the ribosomes use to manufacture proteins.

**Transformation**    Transferring DNA into a new cell.

**Transgenic**    An animal or plant that has had DNA from an external source inserted into its genetic code.

**Transhumance**    The practice of moving animals seasonally from a permanent base to more abundant feed and water and then returning to the permanent base when the season changes.

**Translation**    In protein manufacture, the process of building an amino acid sequence according to the code specified by mRNA.

**Trapnest**    A nest that traps a hen while she is on the nest so that her production and egg quality can be recorded.

**Trimester**    A third of a pregnancy. The last trimester in a ewe is approximately the last 50 days.

**Trisomy**    The presence of a third chromosome in an otherwise diploid cell.

**Type production index**    Index used by the Holstein Association to rank sires on their ability to transmit a balance of traits.

**Ultrasonic scan measures**    Measurements of body tissues taken with ultrasound waves.

**Ultrasonography**    Using ultrasound waves to visualize the deep tissues of the body.

**Unsaturated fats**    Fatty acids that are not saturated with hydrogens but instead have double bonds. The number of double bonds changes the physical, biochemical, and metabolic characteristics of the fats, as compared with saturated fats.

**Unsoundness**    Any injury or defect that interferes with the ability of an animal to be used for its given purpose.

**Value-added product**    A product processed in some way that has enhanced its value.

**Variety**    A subdivision of a breed distinguished by color, pattern, comb, or some other physical characteristic.

**Vector**   Animal, usually an arthropod, that transfers an infectious agent from one host to another.

**Vegan**   A view that eschews the use of any animal product, including nonfood products.

**Vertical coordination**   The process of organizing, synchronizing, or orchestrating the flow of products from producers to consumers and the reverse flow of information from consumers to producers.

**Virulence**   Degree of pathogenicity.

**Vitamin**   Term that is used to group together a dissimilar set of organic substances required in very small quantities by the body. The term comes from a reference in 1912 by Polish scientist Casimir Funk to a "vital amine" he thought he had discovered in rice hulls. The compound he discovered was thiamin, the first vitamin to be discovered. In many ways, modern nutrition was born with that discovery.

**Wean**   To remove offspring from the dam and prevent them from nursing.

**Weathering**   Loss in nutritive value through exposure to the elements.

**Western pleasure**   A western event in which the manners and movement of the horse are judged. A good western pleasure horse is quiet, responsive, and gives a very smooth ride.

**Western riding**   A western event in which the horse is scored on lead changes through one of three potential patterns, as well as its manners.

**Withdrawal time**   The length of time an antibiotic must not be administered or fed to an animal before the animal can be legally slaughtered.

**Wool**   The fiber that grows instead of hair on the bodies of sheep.

**Wool chewing**   A problem of sheep. Often referred to as *wool pulling*. Animals nibble away at their fleece until they make bald spots. There may be nutritional causes.

Animals on high-concentrate, low-fiber diets seem especially prone. Both confinement and crowding may be at the root of the problem.

**Wool sucking**   One of two major types of prolonged sucking syndrome. It is most frequently observed in cats. In both forms, cats continue to suck on objects and perhaps knead with the forepaws long after weaning. Objects may be other cats, people, dogs, themselves, and so on. Wool sucking is specifically linked to Siamese and Siamese-cross animals. The animal sucks and/or chews on wool or otherwise fluffy objects, including clothing, furniture, and bed clothes.

**Work**   Physical exertion as a production function.

**Working cow horse**   A western event broken into two scored performances with those scores combined for a grand score to determine the winner. The two categories include "dry work," which is the reining portion, and "cow work," in which the horse shows its ability to control the cow.

**Xenotransplantation**   Transplanting animal organs into humans. Scientists hope to someday have animals grow donor organs for humans that will function in the human body without being rejected. These organs could help alleviate the chronic shortage of donor organs.

**Young goose**   A young goose weighing from 12–14 lbs.

**Zero tolerance**   Common term used to indicate restrictions imposed by the Delaney clause.

**Zoonotic**   The ability to be passed from animals to humans under natural conditions.

**Zygote**   Cell resulting from the fusion of the sperm and oocyte.

# Index